Venomous and Poisonous Marine Animals:

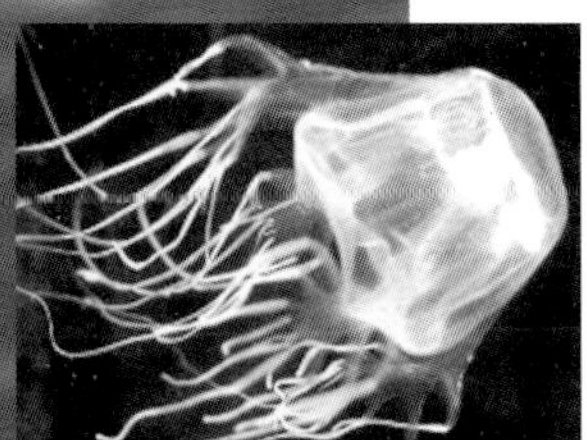

A Medical and Biological Handbook

Venomous and Poisonous Marine Animals: A Medical and Biological Handbook

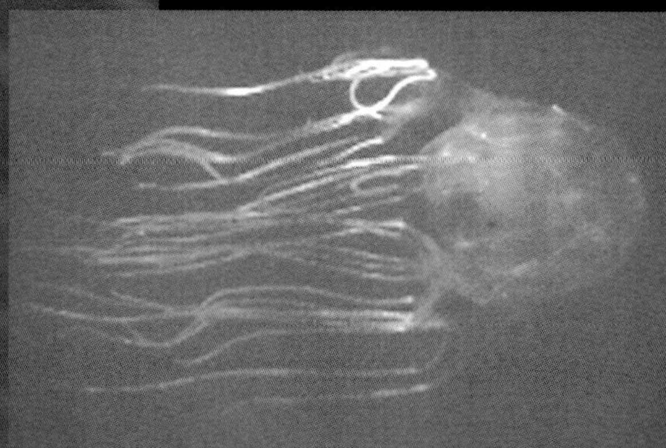

Medical Editors:

John A. Williamson
Peter J. Fenner
Joseph W. Burnett

Biology Editor:

Jacqueline F. Rifkin

Published by
UNIVERSITY OF NEW SOUTH WALES PRESS
Sydney 2052 Australia
Telephone 61 2 9398 8900
Fax: 61 2 9398 3408

Surf Life Saving Queensland Inc.
PO Box 2136 Fortitude Valley Qld 4006
Australia
Telephone 61 7 3852 1496
Fax: 61 7 3252 4511

First printed in 1996

National Library of Australia
Cataloguing-in-Publication entry.

Venomous and Poisonous Marine Animals: Medical and Biological Handbook

Bibliography.
Includes index.
ISBN 0 86840 279 6.

1. Dangerous marine. 2. Jellyfishes.
I. Williamson, John A. II. Fenner, Peter J.
III. Burnett, Joseph W. IV. Rifkin, Jacqueline F.

Designer: DiZign Pty Ltd
Managing Editor: Nada Madjar
Printer: South China Printing, Hong Kong
Production Manager: Di Quick
Publisher: John Elliot

FOREWORD

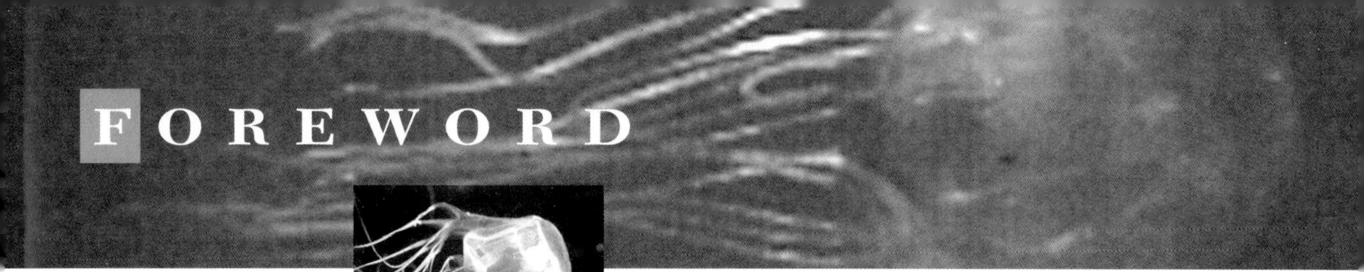

Struan K. Sutherland, MD, DSc, FRACP, FRCPA

Foundation Director,
The Australian Venom Research Unit, and
Associate Professor of Pharmacology and Toxicology,

University of Melbourne,

Melbourne Vic. 3052, Australia

The series of Marine Stinger Books put out by Surf Life Saving Queensland has grown steadily in status since the first edition appeared in 1974. That was a slender but beautifully illustrated volume of less than 50 pages, which, for the first time in Australia, combined important details of venomous marine wildlife with carefully thought-out plans for first aid and management of injuries they could inflict. Prevention of these injuries was also addressed. The hard slog of gathering this material was done by Associate Professor John Williamson.

This present book is greatly revised over the third edition (1985) of the Marine Stinger Book, propelling it from a small but highly valuable guide to marine injuries, to the definitive text on the subject. The data presented are up to date, and the advice is both highly practical and thoroughly tested. The clarity of the writing is excellent, and readers from many walks of life will enjoy the informative style, pleasured here and there with a touch of humour.

Scanning the contents list gives us an idea of the comprehensive nature of this work. All important venomous marine creatures are covered and marine biological expertise has been added. The chapter on the difficult problem of marine wound infections is a useful and welcome inclusion.

Associate Professor John Williamson must be mentioned again, since he is the driving force behind this book. His enthusiasm only just exceeds that of Dr Peter Fenner, Professor Joseph Burnett, biologist Dr Jacquie Rifkin and the others who banded together to form the International Consortium to investigate jellyfish stings. The usefulness of this organisation, and the wonderful network of co-operating physicians, biologists, biochemists and others that has been built up, is clearly seen in this book.

All Australians are indebted to the international contributors to this book, as should be patients and their doctors in many countries around the world. It is the culmination of many years of research, clinical observation and critical assessments of techniques that will undoubtedly save lives. Finally, if the authors' warnings are heeded, many marine injuries will be avoided, and many beautiful marine animals and their precious environments will also be protected from unnecessary harm.

In Case of Emergency

This is not the Contents list, but an alphabetical list of the first aid treatment for specific species, followed by their medical treatments. Under their own sub-heading, jellyfish are listed alphabetically by both their genus names and by their 'common' names. This list is not designed to replace the general index.

First aid for specific marine envenomations

SPECIFIC MEDICAL TREATMENTS

DIVERS — for medical advice in a marine envenomation emergency, phone Divers Emergency Service (DES) Australia on:
1800 088 200 (from within Australia)
or
61 8 8373 5312 (from outside Australia)

CONTENTS

SECTION III

ECHINODERMS, MOLLUSCS AND SPONGES

SECTION IV

TOXIC (POISONOUS AND VENOMOUS) MARINE VERTEBRATES

List of Colour Plates

CONTRIBUTORS

DR CHRISTOPHER ACOTT,
Sea snakes
FANZCA, DipDHM
Senior Staff Specialist
Department of Anaesthesia
and Intensive Care
Royal Adelaide Hospital
North Terrace
Adelaide SA 5000
Phone: 61 8 8222 5116
Fax: 61 8 8232 4207

DR JUNAID MAHMOOD ALAM,
Venomous marine animals of Pakistan
MSc (Pak.), MACTM (Aust.), PhD
Senior Research
Associate and Coordinator
c/o Professor Dr Syed
Mahmood Alam,
Department of Pathology
College of Physicians and
Surgeons — Pakistan
7th Central Street, Phase II,
Defence Housing Authority
Karachi 75500
Pakistan
Phone: 92 21 589 2801
Fax: 92 21 588 7513

DR PHIL ALDERSLADE,
Cubozoan field observations and aspects of mortality
MSc, Senior Curator
Northern Territory
Museum of Arts and
Sciences
GPO Box 4646
Darwin NT 0801
Australia
Phone: 61 8 8999 8252
Fax: 61 8 8999 8289

DR YASUTETSU ARAKI, PhD
Photographic material
Okinawa Prefectural
Institute of Health and
Environment
2085 Ozato, Ozato Village
Okinawa 901-12
Japan
Phone: 81 98 945 0781
Fax: 81 98 945 9366

THE LATE DR LES ASHDOWN,
Marine wound infections
PhD, FASM, FACTM
Consultant Microbiologist
Department of Pathology
Townsville General Hospital,
and Anton Breinl Centre for
Tropical Health and Medicine
James Cook University of
North Queensland
Townsville Qld 4811
Australia

PROFESSOR JOSEPH W. BURNETT, Editor
AB(Yale), MD(Harvard) Chairman,
Department of Dermatology, and
Professor of Oncology University of
Maryland School of Medicine
Associate Professor of Pharmaceutical
Science University of Maryland
Pharmacy School Professional
address:
405 West Redwood Street, 6th Floor
Baltimore MD 21201 USA
Phone: 1 410 328 5766
Fax: 1 410 328 0098
Email: jburnett@umabnet.um.ab.edu

ASSOCIATE PROFESSOR VIC CALLANAN FANZCA, FICANZCA,
FACTM, DipDHM
Marine wound, infections
Director of Anaesthesia
and Intensive Care, Townsville
General Hospital, and Clinical
Associate Professor of Anaesthetics
University of Queensland
Townsville Qld 4810 Australia
Phone: 61 77 81 9211
Fax: 61 77 81 9616

ASSOCIATE PROFESSOR PRATAP CHAND, MRCP, DM
Marine stings in the Gulf of Oman
Associate Professor in Neurology
College of Medicine
Sultan Qaboos University
PO Box 32485
Al-Khod 123 Muscat
Sultanate of Oman
Phone 968 51 5102
Fax: 968 51 3950

DR PAUL F. S. CORNELIUS, *Sanderia malayensis*, field observations; proof reader
MSc, PhD, DSc,
Curator of Coelenterates
The British Museum of Natural History, Cromwell Road
London SW7 58D
United Kingdom
Phone: 44 171 938 9453
Fax: 44 171 938 9158
Email: p.cornelius@nhm.ac.uk

ASSOCIATE PROFESSOR BART CURRIE, Chirodropid mortality, morbidity, epidemiology
FRACP, FACTM
Head, Clinical Unit
Menzies School of Health Research
PO Box 41096
Casuarina NT 0811 Australia
Phone: 61 8 8922 8196
Fax: 61 8 8927 5187

MS BERTHA M. CUTRESS, Field data and Photographic Material
Associate Investigator, on behalf of the late Professor Charles ('Chuck') E. Cutress, PhD
Department of Marine Sciences
College of Arts and Sciences
University of Puerto Rico, Mayagüez Campus
PO Box 500 Mayagüez
PR 00681 5000 Puerto Rico
Phone: 1 809 832 4040

DR DAVID DAVIES, Proofreader
FANZCA, DipDHM
Suite 6, Killowen House
St Anne's Hospital Ellesmere Road
Mt Lawley WA 6050
Australia
Phone: 61 9 370 1711
Fax: 61 9 370 4541

MR DAVID EXTON, Sea urchins, Sea cucumbers
BSc(Qld), DipEd(Qld), BDSc(Qld), BEdSt(Qld), MEdAdmin(Qld), MACE
Human Resource Management Branch Queensland Health
147–163 Charlotte Street Brisbane Qld 4000 Australia
Phone: 61 7 3234 1346
Fax: 61 7 3234 0062

DR PETER FENNER, Editor
MB BS, DRCOG, FRCGP, FACTM
PO Box 3080
North Mackay Qld 4740 Australia
Phone: 61 79 57 7800
Fax: 61 79 57 7824
Email: peter.fenner@130.aone.net.au

DR JOSEP-MARIA GILI, PhD
Photographic material, Stingers from the Mediterranean Sea
Consajo Superior de Investigaciones Cientificas
Instituto de Ciencias Del Mar
Passeo Nacional s/n
08039 Barcelona Spain
Phone: 34 93 310 6416
Fax: 34 93 319 9842

PROFESSOR ROGERIO A. F. GONZAGA, MD
Marine stings in Portugal
Serviço de Cirurgia I
Hospital do Porto
04200 Porto
Portugal
Fax: 351 2 59 8119

MS KAREN GOWLETT-HOLMES, Photographic material, Field observations
BSc(Biol)
Collection Manager
Marine Invertebrates
South Australian Museum
North Terrace Adelaide
Australia 5000
Ph: 61 8 8207 7500
Fax: 61 8 8207 7222

RONALD GRENDON
Artistic illustrations
Grendon Design Services
PO Box 5692 Mackay Mail Centre, QLD 4741
Australia
Fax: 61 79 52 4715

BOB (AND DINAH) HALSTEAD
Photographic and case report contributions
'Telita' Cruises
PO Box 141 Earlville Qld 4870
Australia
Phone: 61 70 354 5401
Fax: 61 70 354 7436

PROFESSOR DR THOMAS HEEGER, PhD
Field observations, jellyfish photographs, proofreading
Marine Biology Section
University of San Carlos
Cebu City
Philippines 6000
Phone: 63 32 8 5021
Fax: 63 32 46 0351
Email: <theeger@mangga.usc.edu.ph7

JAN HYNES, DipA, BFineA
Biological drawings
Brisbane Qld
Australia

DR GEOFFREY KING, MD
Medical aspects of ciguatera poisoning
FRACGP
Medical Superintendent,
Cairns Base Royal Flying Doctor Service of Australia
(Queensland Section)
PO Box 187 Edge Hill Qld 4870
Phone: 61 70 53 1952
Fax: 61 70 32 1776

DR MASAHIRO KOHAMA, MD
Japanese marine medical data
Hyperbaric Medicine Unit, R.A.H.
Visiting Research Fellow, Adelaide University, North Terrace,
Adelaide S.A. 5000 Australia
Phone: 61 8 8222 5116
Fax: 61 8 8232 4207

DR FRANCO KOKELJ, MD
Jellyfish stings in the Mediterranean Sea
Senior Consultant
Dermatologist Istituto di Clinica Dermatologica
c/o Ospedale di Cattinara
Universita' Degli Studi di Trieste
Strada di Fiumi 34149 Trieste Italy
Phone: (w) 39 40 91 0415
Fax: 39 40 91 0690

MS REBEKAH LEWIS
Biological drawings
21 Torrens Street Karalee
Qld 4306 Australia
Phone: 61 7 3294 6200

DR RICHARD J. LEWIS, PhD
Ciguatera
Principal Scientist
Queensland Agricultural
Biotechnology Centre Queensland
Department of Primary Industry
Level 4, Gehrmann Laboratories
University of Queensland
St Lucia Qld 4072 Australia
Phone: 61 7 3365 1924
Fax: 61 7 3365 1990

DR HERMES W. MIANZAN, PhD
***Olindias sambaquiensis* stings in the South Atlantic**
Laboratory de Zooplancton
Instituto Nacional de Investigación y Desarrollo,
Pesquero (INDEP)
PO Box 175
7600 Mar del Plata Argentina
Phone: 54 23 86 2404
Fax: 54 23 86 1292 or 86 1831
Email: hermes@inidep.gov.ar

MR SIMON MOORE,
Marine animals in the United Kingdom
MIScT, FLS
Conservator of Natural Sciences
Hampshire Museum
Chilcomb House Chilcomb Lane
Winchester Hampshire SO23 8RD
United Kingdom
Phone: 44 1962 84 6337
Fax: 44 1962 86 9836

DR PETER J. MORAN,
Crown-of-thorns sea star
BSc(Hons), PhD
Principal Research Scientist
Program Leader, Coral Reef Ecosystems, Member, Crown-of-Thorns Sea Star Research Committee and Member of the Monitoring Advisory Committee, and Associate Director Australian Institute of Marine Science
PMB No 3, MC
Townsville Qld 4810 Australia
Phone: 61 77 53 4211
Fax: 61 77 53 4386
Email: postmaster@aims.gov.au

MR MAURICE (MAURIE) MULCAHY
Marine stinger history
Honorary Life Member,
Surf Life Saving Queensland,
and Surf Life Saving Australia
17 Payne Street
Ayr Qld 4807
Australia
Phone: 61 77 83 1759

ANDREW CAMPBELL

DR GAEL E. PHILLIPS,
Artistic and technical assistance
MB BS, FRCPA
Acting Director of Cytopathology
Department of Pathology
PO Box 124
Royal Brisbane Hospital
Brisbane Qld 4029
Australia
Phone: 61 7 3253 7124 or 3253 7126
Fax: 61 7 3253 7478

DR FERNANDO C. RAMÍREZ,
***Olindias sambaquiensis* in the South Atlantic**
PhD
Laboratory de Zooplancton
Instituto Nacional de Investigación
y Desarrollo,
Pesquero (INDEP)
PO Box 175
7600 Mar del Plata
Argentina
Phone: 54 23 86 2404
Fax: 54 23 86 1292 or 86 1831

DR JACQUIE F. RIFKIN,
Editor
PhD
Consultant in Marine Biology
Lot 105, Riverside Avenue
Karalee Qld 4306 Australia
Phone: 61 7 3294 7815
Fax: 61 7 3365 1655
Email: w.stablum@mailbox.uq.oz.au

PROFESSOR YASUHIRO TOMIHARA, PhD, and colleagues
Stingers from Okinawa, Photographic material
Okinawa Prefectural
University of Arts
4 Tonokura 1-chome,
Shuri, Naha Okinawa 903 Japan
Phone: 81 98 831 5027
Fax: 81 98 831 5033

PROFESSOR R. VICTOR, PhD
Marine stingers in the Gulf of Oman
Head, Department of Biology
College of Science
Sultan Qaboos University
P.O. Box 36, Al-Khod 123 Muscat
Sultanate of Oman
Phone: 968 51 4080
Fax: 968 51 4080
Email: sce2293@squ.edu

MR BRETT WILLIAMSON
Dedication to Jack McMaster, Publication Supervision, Administration
General Manager
Surf Life Saving Queensland Inc.
P.O. Box 2136
Fortitude Valley Qld 4006
Australia
Phone: 61 7 3852 1496
Fax: 61 7 3252 4511

ASSOCIATE PROFESSOR JOHN WILLIAMSON,
Editor
SBStJ, BSc, MB BS(Qld),
DA(Melbourne), FANZCA,
DipDHM, FACTM
Clinical Associate Professor of
Anaesthesia, Faculty of Medicine,
University of Adelaide; and
Director of Hyperbaric Medicine
Department of Anaesthesia
and Intensive Care Royal Adelaide
Hospital North Terrace
Adelaide 5000 Australia
Phone: 61 8 8224 5116
Fax: 61 8 8232 4207

DR YURI M. YAKOVLEV,
Gonionemus vertens
PhD Senior Scientist,
and Dr Victor E. Vaskovsky, PhD
(Doctor of Science)
Institute of Marine Biology
Far East Branch
Russian Academy of Sciences
690041 Vládivostok Russia
Fax: 7 4232 31 0900
Email: faribm@visenet.iasnet.com

DR MINGLIANG ZHANG, MD
***Stomolophus nomurai* stings in China**
Department of Cell and Animal
Biology, Institute of Life Sciences
The Hebrew University of Jerusalem
Jerusalem 91904 Israel
Phone: 972 2 658 5890
Fax: 972 2 61 7918
Email: ming@vms.huji.ac.il

JOHN WEST MCMASTER, AM, DFC
1920–1993
SURF LIFE SAVER, 1934–93
ADMINISTRATOR AND GENERAL MANAGER, SURF LIFE SAVING QUEENSLAND INC., 1971–1991

The full spectrum of the life and achievements of 'Mr Life Saving', 'Jack' McMaster, has been documented with distinction by Ralph Devlin, QC, in the Sixty-Third Annual Report (1992–93), of Surf Life Saving Queensland Inc. Ralph, himself a prominent lifesaver and barrister, said in his eulogy at Jack's funeral, 'If a modern association is a monument to his memory then the love and affection, especially from Queensland lifesavers, which Jack enjoyed, would rate as the greatest monument of all. The contribution he made to this great movement is perhaps his greatest legacy to the people of Queensland.'

That contribution came over 58 years of unselfish service to surf life saving at club, branch, state, national and international levels, where he had been a competitor and served as captain, instructor, president, superintendent, delegate and team manager.

'Big Jack', as he was affectionately called by his colleagues, or 'Mr Mac' by his staff, gave almost six decades of dedicated service to build one of the greatest community services in Australia. Seventy-three when he died suddenly after a heart attack, Jack McMaster was a former President and Superintendent of the Queensland Surf Life Saving Association, and served for 20 years as State Administrator or General Manager of what became known as Surf Life Saving Queensland. He became an expert in every section of surf life saving activity and competition, was one of sport's most able administrators, for many years one of Australia's leading carnival officials, and practically rewrote and re-organised Association manuals and carnival competitions.

Jack, the second son of Hugh and Lily McMaster, was born at Marrickville, Sydney, on 2 September 1920. His grandfather was John Edward West, one of the longest serving members of Australia's Federal parliament and Labor member for East Sydney until his death. Jack was educated in Brisbane at Ascot and New Farm Primary Schools and Brisbane Grammar School. Starting work as an apprentice carpenter in 1934, he rose to become managing director of the company, Stuart Bros Pty Ltd, and remained with them for 35 years. Two of Queensland's building landmarks constructed by the company were the Channel Nine television studios at Mt Coot-tha and the Southport Bridge.

He married Jean Veronica Cruickshank in Brisbane in 1940 and was fiercely proud of their two sons, two daughters, 16 grandchildren and a great grandchild. For the first five years of the McMasters' marriage they had to endure the separation brought by Jack's service in the Royal Australian Air Force (RAAF) during the Second World War. He flew mainly in New Guinea with 22 Squadron, the only Boston Bomber squadron in the RAAF, and saw a lot of action in the Salamanda Bay and Lae areas. He rose to the rank of Flight Lieutenant, gaining the Distinguished Flying Cross. Tragedy struck for Jack in the early part of the war when his brother, who was also an RAAF pilot, was killed in the Middle East.

A top all-round sportsman, Jack chose surf life saving as his first love. He gained his Bronze Medallion in 1935-36 as a member of Burleigh Heads–Mowbray Park Club and continued through a long list of achievements at club, state and national level, which included membership of Queensland representative winning teams. In 1957 he gained his Examiner's Certificate.

Always regarded as a hands-on operator who tended to shun publicity and kudos, he came to the movement practically in its infancy. He served and steered his great organisation through some turbulent periods, often involved in some hard-headed battles between the organisation of his home state, Queensland, and the Sydney-based Australian hierarchy.

Jack helped to develop modern resuscitation techniques, revolutionised surf carnivals throughout Australia and was one of the people responsible for introducing to Australia the ironman and taplin relay competitions.

At the South Coast of Queensland Branch he was president for six seasons from 1960-66 (a 'surf season' straddles the New Year period), Queensland state president for four years and a state delegate on the national body for five years. Ultimately he was awarded life membership of club, branch, state and Australian levels of the association.

In his early days he was a strong swimmer and competitor in state championships winning teams in Rescue and Resuscitation competition (R and R) at both junior and senior levels, and was a standard-bearer in march-past competition. He was a state Carnival official for 30 years and an Australian official for 20 years. He was sectional referee for 16 years, boat referee for 13 years and also a board and ski referee and competition judge of water, R and R, march-past, boards, skis and boats. He was the instructor for two state R and R championships and two state march-past titles. In 1956 (the year of the Olympics) he was manager of the Queensland team to Torquay. As a boat referee at national level he was regarded as 'the Big Fella' with a firm but fair hold on all proceedings. He was first appointed referee of the boat events at the Australian championships at Collaroy in 1964.

According to Jack, boat and craft sections faced many difficulties. 'You had instances of kerosene tins being used as boat buoys or buoys that went under as boats rounded them, thus causing disqualifications. You had judges being knocked over in boat finishes and an overall untidy setup.' Jack went to work in cleaning up the mess and linked with the boat and craft people to see how they could help. He called the first meeting of boat rowers to gather their ideas about the running of their events. This led to a boat rowers' panel, with which officials could confer; he then made a submission to update the manual concerning boats.

He was on representative Australian teams to Hawaii in 1939, the mainland United States in 1965, manager of the state team of examiners to the Philippines in 1975 and of a similar team to the USA in 1981. From the 1965 trip came the idea to introduce the ironman and taplin relay into Australian competition, which happened from 1966.

Jack also revised and rewrote the boat racing rules and was involved in the development and research of carnival gear, designed and introduced towing sleds for boats (first used at Burleigh Heads, Queensland, in 1973) and chaired the first official National Boat Rowers Convention in Sydney in 1967–68. He also served on the organising committee for the Australian championships at Mooloolaba in 1959, Coolangatta in 1966, Burleigh Heads in 1973, Maroochydore in 1980 and Kurrawa in 1984.

In 1971, after already giving 35 years of outstanding voluntary service to Queensland and Australian surf life saving, he devoted his working life entirely to the service, becoming Administrator to the Queensland State Centre, in Brisbane. Even when he finally retired in 1991 as General Manager of the association, he still couldn't stay away from the state office and offered his advice wherever he could. Jack had a burning ambition in his later years to convince governments and local authorities of the need for greater surf safety, particularly with the rapid surge of tourism. He became an advocate of aerial patrols combining with ground operations, to adequately safeguard the widespread areas of Queensland beaches. He also believed in specialised youth training.

Jack had a very close relationship with a group of colleagues who became known as the 'Dirty Dozen'. During the last 15 years of his life he had a special camaraderie with this group who were the band of workers for major surf championships in Queensland.

Jack enjoyed a lifetime love of rugby football, gaining his baptism watching his father play rugby league for the famous South Sydney 'Rabbitohs'. After moving to Queensland he became a top rugby union player. He played more than 100 matches, was captain of GPS Rugby Union Football Club in Brisbane for four years and a coach for four years, and twice represented the city of Brisbane. He later was awarded life membership of the GPS Club.

For many years Jack was a member of Rotary International and the Masonic Lodge. He played a prominent role in the establishment of the Rotary Foundation, which has provided major assistance to many community activities, including Surf Life Saving Queensland.

He played a prominent role in the Major Sporting Bodies Association of Queensland and over the years he was not only a pacesetter in his field but earned the respect of professional administrators from all the major sports in Queensland. He certainly lifted the organisation's profile throughout his career and never ceased to sell his message about the need for greater surf safety.

Jack McMaster left a great legacy to his state and to the nation. The regard in which he was held is illustrated by the fact that after his death a Jack McMaster Foundation was established to perpetuate his memory and to recognise his contribution, for the benefit of all his fellow Australians.

Brett Williamson
General Manager,
Surf Life Saving Queensland Inc.

Here the editors of this book record the pioneering and pivotal role played by Jack McMaster in the involvement of Queensland and Australian Surf Life Saving organisations with Australian marine envenomation studies and research.

The stinger problem in northern Australia had reached such dread significance by the 1950s and early 1960s, with mortalities and serious morbidity from stingings, that the Surf Life Saving organisation's actual existence in north Queensland became threatened. Jack's characteristic farseeing perception of the need for Surf Life Saving to involve itself with the north Queensland stinger problem led to a chain of events of fundamental importance for the protection of the public from *Chironex fleckeri* envenomation. It also led to a resurgence of a 'raison d'être' for Surf Life Saving in the north, which has never waned since. Some of these events are recorded chronologically in Chapter 1.

As testimony to his ability to see far ahead, and to his skill in human relations, he succeeded in recruiting and welding together a bunch of rugged individualists, to form a team of expert, on-the-beach, marine biological and medical investigators, to tackle the many unknowns in the subject. This feat must have been akin to 'herding cats', but he did it, and the gains that this team has achieved are fitting memorials to the man.

Following the impact on surf life saving of the brilliant field research and taxonomic studies of Dr Hugo Flecker, Dr Jack Barnes and Dr Ron Southcott — soon to be followed by fundamental marine biological work of Dr Bob Hartwick and Dr Bob Endean — in the early 1970s Earle Smith, Maurie Mulcahy and John Williamson hatched the idea of a Surf Life Saving 'Stinger Book'. Jack McMaster immediately threw his administrative wisdom and talents behind this project and, with his uncanny knack to anticipate, warned us that we were making, albeit happily, 'a rod for our own backs' (including his!), in that future editions would become necessary. Time has proved him absolutely correct. He did not stop there but was instrumental in creating the inaugural Marine Envenomation Committee of World Life Saving in 1980, to the lasting credit of Surf Life Saving Australia. It is fitting that this, the first international book in the series that began with *The Marine Stinger Book*, should belong to Jack McMaster. His spirit is between the lines; his wise, calming and far-seeing influence helped to shape its character; and he, as much as anyone, projected its influence out into the rest of the globe.

Jack, this one is for you.

John Williamson
Peter Fenner
Joseph Burnett
Jacquie Rifkin

PREFACE

This book has been a long time coming. Incredibly, it is a decade since its forerunner, The Marine Stinger Book Third Edition appeared. This delay has been partly a consequence of the enormous expansion of knowledge about marine envenomation during the decade. The subject is no longer the province of the 'eccentric specialist'; a wider audience is becoming aware of the frequency with which humans and their livelihood are affected by encounters with poisonous and venomous marine creatures. Not that these are any fault of the animals — we hope that one of this book's firm messages is that no venemous marine creature has malice of forethought towards humans in the water, (though some fish will mount 'unprovoked' attacks [see Chapter 16 Venomous fishes], and the frenzied attacks of sharks, barracudas and the stealthy attacks of the crocodile are well known. These 'attacks' result from instinctive self-preservation. Humans, as usual, are the cause of the problem, as they increasingly and at times thoughtlessly invade the territory of marine creatures.

In keeping with *The Marine Stinger Books*, all contributors and the editors have striven for accuracy, objectivity and the documentation of only verifiable events. Where uncertainty remains despite rigorous investigation, the event is identified as such, or else not included. We acknowledge that this may result in the omission of some genuine events from our data. Nevertheless, the research for this text has shown that the incidence of morbidity and especially of human mortality attributed to some marine animals has been inaccurate in the past. Some 'mortality' claims evaporated when traced to their sources. We further observe that some earlier data from different world regions, valuable though they are, often reflected the presence of an enthusiastic resident investigator at the time, and not necessarily the true relative incidence of the problem when viewed in international perspective. Thus some of the earlier epidemiological data has been skewed.

Purists have cast aspersions upon numerator research, which is naturally the bulk of morbidity (and some mortality) data in marine envenomation. In our opinion denominator research remains an 'ivory tower' concept in many fields and as such may lead to people actually doing nothing. More importantly, numerator research is a validated tool when pursued with discipline.* We also note that the relatives and friends of a victim of a human fatality or serious envenomation are singularly unimpressed by the epidemiological or statistical revelation that such an event is 'rare'.

As in earlier books, this text makes minimal reference to the activities of non-venomous (and already overemphasised) sharks and crocodiles. In contrast there is an emphasis on cnidarians (now the preferred term to 'coelenterates', which once included comb jellies [phylum Ctenophora]). Particularly emphasised are those élite of the cnidarian world, the chirodropid cubozoans, with their phenomenal envenomation capabilities. We remind the reader that jellyfish neither bite nor attack human beings, and we also attempt to explain the misuse and the appropriate use of those two confusing terms, 'box jellyfish', and 'sea wasp'.

Many marine animal venoms have been shown *in vitro* to have haemolytic properties. We suggest that, allowing for any general toxinological light that observation may shed, haemolytic properties appear to date to have translated into little of direct human clinical relevance. Further studies will test this view.

Although this is a text on marine animals that may injure humans, in this edition we do refer to certain significant estuarine animals (e.g. the extraordinary Amazonian catfish, 'candirú' {page 374} and the Australian bullrout *Notesthes robusta* {page 385}). Also there is a special chapter devoted to the equally extraordinary, predominantly freshwater, and venomous Australian platypus.

Venomous and Poisonous Marine Animals hopes to attract the first-aider and the medical officer wanting a user-friendly quick-reference

* Symposium — The Australian Incident Monitoring Study. 1993. Anaesth Intens Care 21:495–695.

guide to the management of a particular marine envenomation, as well as the reader with a more cosmopolitan appetite. Consequently we provide not only a special 'In Case of Emergency' list and a detailed index but also extensive cross-referencing within the text (identified by the brackets { } when distant); and we have tried to make each topic 'stand alone', so that it may be read in isolation. As a consequence there is some repetition of information, particularly certain identifying features, and first aid and medical management protocols, but for this we make no apology.

The text of this book also naturally tells a little about the animals, their beauty and the elegance of their structure and function. Many people believe that the taxonomy of the earth's marine creatures is largely a settled scientific matter. This is far from the truth, even in the confined field of this book, as the following pages show. Indeed, taxonomic names have changed and required updating during the actual period of preparation of this text. Much uncertainty still exists, and extensive epidemiological and taxonomic research are still needed. Taxonomists, including one of the editors (JFR), are currently grappling with the formidable uncertainties surrounding chirodropid taxonomy — uncertainties with which this text also struggles, and which bear upon first aid and medical considerations. New information is presented in Chapter 6 by JFR, concerning particularly the taxonomy of the 'non-Australian' Indo-Pacific chirodropids. The Indo-Pacific region really includes the northern Australian continental coastline, but mindful of Barnes' (1965) work, chirodropid distribution studies are simplified by regarding the Australian and the Indo-West Pacific regions as distinct, at least for the purposes of this text.

Excluding cubozoans, we have adopted mainly the taxonomic approach of Kramp,* as well as those of various researchers in worthy institutions both in Australia and overseas. Our efforts may not be considered sufficiently comprehensive (especially to the professional biologist), but in a text with a primary first-aid and medical focus, a serious attempt has been made to communicate basic taxonomic concepts to the 'beginner' — a term that applies to most medical persons, including three of the editors! Thus we include a Glossary at the beginning of the book.

In different parts of the world there are differences in spelling of both generic and species names (e.g. *Rhizostoma/Rhyzostoma*, ...*tuberculata/tubercolata*, ...*hysoscella/hyoscella*). Under the guidance of its biologist editor, and for the sake of uniformity, this text uses the English-language spelling, which is currently accepted among biologists (e.g. *Rhizostoma*, ...*tuberculata*, ...*hysoscella*).

The overall material is grouped into seven major sections, while the 'Front Matter' deals with preliminaries including the important (to the editors and to Surf Life Saving Queensland) dedication to Jack McMaster and acknowledgments of the work of several pioneer researchers. Section I includes generic information on the subject. For example, Chapter 1, on the history of marine envenomation, focuses particularly (but not exclusively) on Australian contributions of the past 40 years. Sections II and III commence the consideration of specific animals and cover toxic marine invertebrates, and include original contributions from Argentina, the Caribbean region, the United States of America, eastern Russia, Japan, China, Malaysia, Indonesia, Pakistan, Sri Lanka, the Gulf of Oman, South Africa, the United Kingdom, Portugal, Spain, Italy and the Adriatic/Mediterranean regions, as well as from Papua New Guinea and the Australian, New Zealand and Pacific areas. Section IV deals with specific toxic marine vertebrates, again with many valuable international contributions. Section V contains a unique chapter on marine envenomation in the Gulf of Oman, and two other smaller but similarly original contributions from Japan and Portugal. Section VI (a favourite of the editors) gathers together the disparate topics of marine antivenoms, scuba diving perspectives, research aspects, the enticing 'mystery syndromes' and a somewhat 'maverick' chapter on Australian platypus envenomation. Section VII contains the large bibliography, the index, and a special 'Stop Press' chapter which, by the generosity of the University of New South Wales Press, enables the editors to include very recently

* KRAMP PL. 1961. Synopsis of the medusae of the world. J Marine Biol Assoc UK 40:1-469.

published and (in their views) significant information on the subject.

It quickly became apparent to the editors embarking on this more comprehensive book that, with regard to the layout of the material, a conflict presented itself between the 'conventional' biological approach and practical medical needs. The two approaches were often incompatible. While clearly separating invertebrates from vertebrates, and identifying the phyla in which various animals are classified, we have chosen to discuss the animals using groupings that favour the placing together of similar injuries, toxinologies, or first aid and medical management strategies. We also attempt to place the more commonly encountered and/or more serious injuries earlier in the text. This makes paramedical and medical sense, and after all *Venomous and Poisonous Marine Animals* is primarily 'hands-on' first aid and medical in its purpose. In keeping with its practical orientation, specialist medical and biochemical terminology is used minimally within the text; but when it is, it is explained in the Glossary.

However, retrieval and testing of natural products from marine organisms has led to the development of new compounds with potential for use as drugs against the effects of diseases such as cancer, AIDS, and Alzheimer's disease. The biochemical actions of some such compounds is described at times in specialist jargon, and it is expected that persons specifically interested in those areas would search for further details in the references provided.

Within chapters the choice of order presented similar difficulties. The cubozoan jellyfish are the most important envenomating cnidarians in the sea, but cubozoan envenomation and its management are currently the more complex. In the interests of first aid comprehension we have chosen, within the cnidarian chapters (Section II), to deal with the medically simpler cnidarian injuries (e.g. coral abrasions and stings) first, and the cubozoan stings last. Despite this, the simple headings within chapters of the earlier books have been retained: 'Distribution', 'Appearance', 'The Sting' or 'Envenomation', and 'Management' or 'Treatment' ('First Aid' and 'Medical').

The editors of a book combining medicine and biology have long been aware of the unfortunate reaction engendered among some investigators, when their cousin-colleagues from other disciplines venture to write on topics considered to belong to someone else's scientific or medical 'territory'. In our opinion, cross-fertilisation among scientific and medical disciplines is highly desirable, even crucial to progress; advances are built upon the results of many causes and the work of many minds.

Although this book does contain some new and evolving concepts relating to chirodropid taxonomy from Dr Rifkin, we hasten to emphasise it has no pretensions as a marine biology text and should not be regarded as such. During the writing this biologist editor struggled to keep the other three editores on the biological rails, and to maintain gender non-discrimination in the text; where the text deviates from the conventions used by biologists, the three medical editors accept responsibility for the consequent textural shortcomings, as well as for deficiencies of the medical content. All four editors would however be grateful to learn from any reader who detects errors of fact.

This new book has taken some other 'brave' new directions. The first significant departure from previously is in the international approach of the book. At the invitation of the editors, besides colleagues who provided original material, there are 36 additional contributors to the contents of this edition, from all corners of the globe. Many of them are world authorities in their respective fields; all of them are wonderful people who have provided their skill, knowledge and time for no remuneration whatever, and for the good of life saving and their fellow humans. Much of their material is original work and has never before been published in any language, and all of it is that most precious of data, 'first-hand'. All the contributors are listed with their contact details in the Front Matter. As one would expect from such a galaxy of international talent, there are a number of innovations (not to mention controversies) recorded in the pages that follow, which the authors and editors hope may provoke thought and further investigation.

Secondly, this book has four editors: an Australian, an English-born Australian, an American and a Canadian permanent resident in Australia. A task of this magnitude required a breadth of approach and experience; the editors between them have brought at least a measure of experience in zoology, marine biology, cell biology, field research, taxonomy, general medical practice, surf life saving, resuscitation, dermatology, toxinology, immunology, laboratory and microscopic procedures, clinical trials, diving medicine, pharmacology, anaesthesia and analgesia, wound healing, and editorial activities. With good will, a team approach, and a love of the sea and its creatures, we have found this task largely a joy.

Third, mindful of current non-uniformity of referencing in the world literature, we have adopted a slightly unusual referencing style on the basis of what we believe is a relative reader-friendliness; we also wanted to highlight the names of authors who are referenced. Within the body of the text, authors followed by the year of publication are identified in upper case, bracketed; two or more publications placed together are in chronological order; more than two authors are shown as the name of the first author (in upper case), followed by '*et al*'. This pattern (but with all authors' names and initials) is repeated in the large, strictly alphabetically listed (then chronologically, for two or more references with identical authors) bibliography, which appears at the conclusion to the book. The exceptions to this location for references are the additional free-standing lists of references that relate to and appear immediately following the sections on 'Historical Background' {pages 55–57}, 'Fatalities' {pages 74–77}, 'Morbidity' {pages 78–79} (some references inevitably are repeated in these different lists) and within the 'Stop Press' chapter (page 442). Great effort has been expended in seeking accuracy and completeness of the references. Some of the older references are, we hope, traceable from the information provided. Where it may assist in the more difficult searches, the issue number (if known) appears in parenthesis immediately following the volume number. We record here the usefulness of the 1985 publication, *Bibliography of Venomous and Poisonous Marine Animals and their Toxins*, by Findlay E. Russell, Henry Gonzalez, Sue B. Dobson and John A. Coats.

The editors are also concerned with ethical issues and deplore the needless suffering inflicted on inadequately analgesed laboratory animals in the name of 'research'. We hope this will no longer be tolerated by a late but emerging world human conscience. Thankfully many countries, including Australia, have implemented stringent bioethical guidelines to control this sort of research. We also argue strongly for an increased awareness of the important ecological role of all marine animals in the earth's seas (especially those threatened because they are venomous to humans) and the need to protect them and their environment from further destruction. The editors wish this book to reinforce the now-urgent plea for global environmental conservation,* a responsibility of every human being.

So what of future editions? A publication such as this brings with it the pleasure of new friendships and harmonious international co-operation between people of goodwill and integrity, devoid of political, military or commercial distortions. In the face of continuing disruptive forces on our troubled planet, such cooperation and harmony are sorely needed. We suspect another edition, the fifth since we began, may happen. Besides, the contributors to this book, in their own fields of interest and regions of the world, have tasted that heady elixir, 'Truth' — once experienced, allowing no rest! As the Australian author Xavier Herbert remarked:

'There is only one thing I enjoy more than a good laugh, and that's the truth.'

This pervading spirit is at the heart and soul of Surf Life Saving Australia.

John Williamson, Adelaide,
South Australia
Peter Fenner, Mackay, Queensland,
Australia
Joseph Burnett, Baltimore, Maryland,
USA
Jacquie Rifkin, Brisbane, Queensland,
Australia

* Journal of Wilderness Medicine. 1992. 3(4):331–337 and 348–350.

Acknowledgments by the Editors

We stand on the shoulders of many great international pioneers, who were in fact early conservationists. Four twentieth-century works have influenced us throughout: P.L. Kramp's classic text, 'Synopsis of the medusae of the world', Sir John Cleland and Dr Ron Southcott's brilliant text, *Injuries to Man from Marine Invertebrates in the Australian Region*, Dr Bruce Halstead's monumental *Venomous and Poisonous Marine Animals of the World*, and J.S. Nelson's definitive *Fishes of the World*. We make special reference to Dr Southcott below. The influence continues of the late Dr Jack Barnes, a phenomenon of the field, whose life and contributions have been recorded in many other publications. The writings of Dr Findlay Russell, Dr Gary Calton (see below) and Australia's Associate Professor Struan Sutherland, Dr Julian White, and Dr Carl Edmonds have been frequently consulted. Dr Carl Edmonds has also generously provided some photographic material and information from his personal storehouse of knowledge in the field. We wish to pay particular tribute to the life and original works of the late Dr Z. Maretić of Yugoslavia.

The stimulus, guidance and example of Professor Struan Sutherland, Australia's internationally eminent animal toxinologist is gratefully recorded by the editors. Professor Sutherland has kindly written the Foreword to this book.

This is the place to record the brilliant (and until recently, barely acknowledged) initiating field and laboratory researches of Dr Hugo Flecker and Dr Ron Southcott (see below), Dr Harold Baxter, Dr Alec Marr, Dr William Lane and Dr Saul Wiener, and Dr Shirley Freeman, Dr Hugh Crone and Dr A.R Coulter. Many basic premises on which present thinking in this field of study rests, sprang from their works.

Other 'present-day pioneers' on whose work we have drawn include our close colleagues Dr Bob Hartwick PhD and Associate Professor Vic Callanan (the latter a contributor to this book), Professor John Pearn (an outstanding chronicler and catalyst), the innovative and perceptive Dr Bob Endean PhD, Dr Ann Cameron PhD, Dr Des Gurry, Loisette Marsh, Shirley Slack-Smith, Professor Thomas Heeger, Jeanette Covacevich, Dr Jim Tibballs, Dr Ian Audley, Dr Col Limpus PhD, Dr Harold Heatwole PhD and Professor Malcolm Fisher, all of Australia; in Malaysia, Dr Lim Yu Ho of Penang, Dr Nor Azila PhD and Dr Iekhsan Othman PhD of Kuala Lumpur; in the USA, Mr David G. Cargo MS of the University of Maryland, Professor Paul Auerbach MD formerly of Stanford University Hospital, and Professor Ken Kizer MD of the University of California, Davis; in Canada, Dr Dale Calder PhD of the Royal Ontario Museum in Toronto; in the United Kingdom, Simon Moore of Hampshire Museum, Winchester, and the internationally recognised Dr Paul Cornelius PhD, DSc of the Natural History Museum, London.

As part of our stimulating collaboration with Dr Robert Hartwick over many years, we acknowledge the on-going support of his entire Department of Biological Sciences (in particular its inaugural Chairman, Professor Cyril Burdon-Jones), and the Sir George Fisher Centre for Tropical Marine Studies (in particular its former head, Dr J.T. (Joe) Baker), all at the James Cook University of North Queensland. We also gratefully acknowledge the non-stop assistance and encouragement of the celebrated Australian Institute of Marine Science (AIMS) at Cape Ferguson, north Queensland, and many of its distinguished scientists, with special thanks to former Directors Dr John Bunt and Dr Joe Baker.

This certainly is the place to record the names of some of the Queensland surf life savers who made this book and its forerunners, and much of the field work herein documented, a reality. Maurie Mulcahy (see below), Don Johnson, Brian Coomber (a superb contributor, sadly recently deceased), Peter Craddick, Reg Butler,

Flo Payne and Glen Jones of North Barrier Branch, Ron Pollard (see below) and 'Tubby' Mathews, among many others, of North Queensland Branch, and Graham McKenzie of South Barrier Branch, were ever stalwarts. At the then Queensland State Centre of SLSA (now Surf Life Saving Queensland Inc.) in Brisbane, Earle Smith and Jack McMaster (see the Dedication) translated the concept of *The Marine Stinger Book* into a reality, while the current Queensland Surf Life Saving team at State Centre are referred to below.

There emerged a remarkable team of in-water 'catchers and handlers' of chirodropids in association with the early field researches into *Chironex fleckeri* in north Queensland, whose work was pivotal to progress. They developed the original techniques now used to capture intact specimens safely, and have had no peers since. These were Ron Pollard of Cairns (Jack Barnes' indispensable lieutenant over his marine research lifetime), Maurie Mulcahy of Ayr and Townsville (whose many contributions appear throughout this book), and Charles Fitzpatrick of Townsville whose perfectly captured and preserved specimens (on behalf of Dr Ron Southcott) stand today in the South Australian Museum, Adelaide, as testimony to his great skill.

The *Medical Journal of Australia*, with its succession of editors, has published almost the entire saga of Australian marine envenomation in all its facets. Some of these papers have become pivotal in the subject, and the journal now rightfully occupies a distinguished international place in the field. The editors of this book wish to record their appreciation of the journal's preeminent role over 40 years. In the same context the editors gratefully acknowledge the support of the editorial staff of another distinguished international journal in the field, *Toxicon*.

Our thanks also go to our colleagues in Australia and the USA — surf life savers, medical officers and divemasters — who have faithfully reported stings and other marine injuries to us over the years, and continue to do so. They helped to provide the raw data essential to progress.

ACKNOWLEDGMENT OF THE WORK OF DR R.V. SOUTHCOTT, OF ADELAIDE

The editors wish to acknowledge the brilliant and seminal role played by Dr Ron Southcott, MD, DSc, DTM, in the field of world marine taxonomy clinical documentation and research, over a period of 50 years. Initially a graduate in medicine from the University of Adelaide, he has become an international authority in several fields of medical biology and one of Australia's greatest (and still somewhat unsung) 'sons of science'. His research and documentation all bear the characteristic imprint of his meticulous attention to detail and accuracy, absolute integrity and great scholarship. He is the author of scores of published works — papers, book chapters and monologues, and some texts. Many of these are perennial classics in their field and are referred to repeatedly throughout the world, including in this book. His text, coauthored with the late Sir John Cleland, *Injuries to Man from Marine Invertebrates in the Australian Region*, published in 1965 by the National Health and Medical Research Council, stands as a magnificent testimony to Dr Southcott. This work is regarded by the editors as one of the most original and accurate documentations of clinical biomedical science in the English language. Dr Southcott has made invaluable contributions to cnidarian biology, especially the cubozoans. He has enquired into venomous and poisonous plants and animals all his life and has just completed major works on the medical biology of mites and of plants harmful to humans. The editors express their deep appreciation to Dr Southcott for the additional personal guidance and assistance rendered to them, unstintingly and with customary precision, during the compilation of this text.

Dr Ron Southcott, MD, DSc, DTM

ACKNOWLEDGMENT OF THE WORK OF DR G. J. CALTON, OF BALTIMORE

Dr Gary Calton was born in July 1943 and educated at Eastern New Mexico University where he received Bachelor's and Master's degrees in chemistry in 1965 and 1968 respectively. He was awarded a PhD degree from Texas A & M University in 1971, then completed a postdoctoral fellowship in the Division of Dermatology at the University of Maryland, School of Medicine. On successfully passing this mile post he ascended to the rank of Associate Professor of Medicine before he founded some biotechnical companies and went into business for himself. His first company, Purification Engineering Inc., was purchased by Rhone Poulenc of France. In addition to these activities he served as a consultant for other science/business groups. His interest in marine venoms stems from the fact that he liked to scuba and snorkel while in Texas and has continued to publish prolifically and be interested in the field of marine venoms.

Dr Gary Calton, PhD

For more than a decade he collaborated with one of the editors (JWB) in a fruitful research alliance, which produced 54 peer-reviewed papers and 61 other publications in the field of marine toxinology and immunology. He brought his intellect to bear upon what was then a somewhat neglected area in science and, with his colleague JWB and others, has been instrumental in establishing marine envenomation within the ranks of disciplined medical research. The editors thank him for his major contributions, many of which have become essential reading and are featured in the following pages.

SOME PERSONAL ACKNOWLEDGMENTS

In every task of this nature, colleagues in the background have a burden to bear. Lip-service can be paid to such contributions, but the editors have no such intention here. Had we not received this support, always freely and pleasantly given, this book could not have started, let alone been completed.

JW's secretary, Debbie Lombardi, bore the brunt of his incessant requests over a five-year period, for assistance, processing, photocopying, phoning, faxing and posting, not to mention giving answers to endless questions, cups of coffee, and counselling. By her lovely attitude, endless patience, high intelligence and quite fabulous efficiency, she really warrants the fifth editorship of this text. Only his medical colleagues and staff of the Hyperbaric Medicine Unit at the Royal Adelaide Hospital know the irritations to which JW subjected them during the same period. For their forbearance and robust sense of humour, he thanks them. The patience of the hospital's ever-helpful mailroom staff and the premature ageing of the fax machine are noted! JW also acknowledges the 'eleventh-hour' knowledgable contributions of Dr Paul Cornelius of London, accompanied by some therapeutic good humour. A special thankyou goes to Dr Bob Webb, whose computer skills saved not only time but also a 'lost' major segment of the manuscript at a critical stage in its preparation. JW would also like to acknowledge the humour, zest and interest provided by his children, Kylie, Ken, Kirsten and Robina, who all refused to join him in his editorial groans! Finally there are few departments that, in the face of a large clinical workload, contain the tolerance and flexibility to encourage undertakings of this magnitude; such is the Royal Adelaide Hospital's Department of Anaesthesia and Intensive Care under the leadership of Professor Bill Runciman.

PJF would like to dedicate his work in this book to his mother, who passed away just prior to its publication. Always interested in its subject, and supportive, she remained so understanding of the time demanded by its preparation, which PJF could otherwise have spent with his family. He is likewise deeply grateful to his wife for her similar patience and encouragement (see below). He also expresses gratitude to his partners and staff at the Ambrose Medical Group in Mackay, and to those involved in his own family medical practice, for their endurance of his recent peripatetic existence, for the tolerance of the doctor trying to arrange his roster, for the use of the Group's phone and fax facilities (which has extended over many years) and for their moral support throughout. Also to his son Paul, without whose computer skills the task would have taken interminably longer, and his daughter Clare, who lost much Harley Davidson riding time while her father worked on the manuscript. PJF's thanks also go to Ronald Grendon of Mackay for the many fine illustrations he drew.

JWB wishes to record his deep appreciation of the support of his colleagues and staff in his University of Maryland Department during the preparation period. In particular, thanks to his long-time research colleagues Gary Calton, Helene Rubenstein, Dave Cargo and Dale Calder, Laure Aurelian, Karen Long-Rowe, Carrington Cobbs, I. Neeman, I. Othman, Bob Endean, J. Warnick, Ann Cameron, A. S. Kreger, D. Bloom, K. Hartman, E. Adams, J.T. Tilden, D. Lal and the three Burnett sons.

JFR is deeply grateful to her friend and mentor Dr Ron Southcott, for his ever-enthusiastic participation in discussions concerning living things. To her research colleague, Dr Robert Endean, she attributes her basic enthusiasm for the study of toxic animals. She gratefully acknowledges her husband, Bill Stablum, for the collection of toxic animals; for his persistence with the library computer system to obtain the huge amount of reference material required for the project; also for providing several photographs for use in the text. She thanks Professor Tom Heeger of the University of San Carlos in the Philippines for helpful discussions concerning aspects of jellyfish behaviour, anatomy and nematocysts; for donation of excellent slide material on jellyfish, and for checking. JFR is especially grateful to Associate Professor Jack Greenwood of the University of Queensland and Dr George Matsumoto of Flinders University, South Australia for kindly checking parts of the manuscript at short notice.

For his guidance with the modern taxonomy of cnidarians she thanks Dr Paul Cornelius of the British Museum of Natural History. She also thanks Dr John Hooper, Curator of Lower Invertebrates at the Queensland Museum for provision of reference material and photographs of toxic sponges. She thanks Dr Richard Willan, Curator of Molluscs at the Museums and Art Galleries of the Northern Territory for advice on taxonomic matters and for checking parts of the manuscript. For the taxonomy of venomous fish, she thanks Mr Roland McKay, Curator of Fishes, Queensland Museum, Brisbane; and for advice on fish from an angler's point of view she thanks Mr David Lems of the Queensland Valley Times. Her thanks also go to Dr Phil Alderslade, Curator of Cnidarians at the Museums and Art Galleries of the Northern Territory for generously providing specimens and information; to Dr Wolfgang Zeidler, Curator of Lower Invertebrates at the South Australian Museum for valuable advice; to Karen Gowlett-Holmes, collection manager at the South Australian Museum for provision of specimens and first-hand information on jellyfish behaviour; and to Mr Martin Jones of the Great Barrier Reef Marine Park Authority (Great Barrier Reef Aquarium) in Townsville for advice and assistance.

She also acknowledges the Electron Microscope Centre at the University of Queensland whose facilities she utilised for the production of the scanning electron micrographs included in this publication.

All the editors, but especially JFR, thank the multi-talented Dr Gael Phillips of the Department of Pathology, Royal Brisbane Hospital, for her patience and advice and her gift of crucial texts. A very special thankyou goes to the distinguished Australian artist Jan Hynes, who not only came lightly to our rescue when the artistic burdens began to overwhelm us but ultimately undertook to draw or redraw a majority of the illustrations with consummate skill. Dr Matt Sheehy, formerly of the University of Queensland Department of Zoology redrew many of the complex drawings for the nematocyst keys provided in Chapter 6. Ms Rebekah Lewis generously, and with great loyalty, sacrificed an entire school holiday period for several artistic tasks. Last, JFR's mother Lucy kindly provided some artistic advice.

The work of some brilliant and celebrated underwater and marine biology photographers is displayed in this text. The editiors acknowledge in particular and with gratitude, Neville Coleman, Karen Gowlett-Holmes, Dr Robert Hartwick, Professor Thomas Heeger, Ben and Eva Cropp, Dr Ronald Larson, Dr Yasuhiro Tomihara, David Doubilet and Bill Stablum.

However, perhaps most significant of all was the love and encouragement we received from our spouses, Noeleen, Maggie, Kitsie and Bill (the 'ghost editors'), who were subjected to the inevitable delays and disruptions to normal family life and relaxation time over several years, as well as to the grumbles of their editor-spouses, traumatised by self-imposed deadlines and by the frustrations that accompany a task of this nature. Without these people, nothing!

It is a pleasure to thank our contract publishers, the University of New South Wales Press, the hard working editors Beverley Barnes and Karen Trutwein, Production Manager Diane Quick, and the other highly skilled staff, for their patience (well tried) and attention to detail with what was a difficult manuscript. We acknowledge especially the outstanding professional publishing guidance of the Publisher at the Press and Managing Editor Nada Madjar. Errors of fact, omission or commission, however, remain ours alone.

We express our abiding gratitude and admiration to each of the contributors to this volume. We thank them for their information, gathered by their own research and studies, and yet given so freely and unselfishly to almost total strangers, for the benefit of a voluntary organisation responsible for the safety of the general public. We thank them also for their patience and persistence in the face of, at times, ragged communication and editorial demands, and for their adherence to excellence. Individually they represent international leadership in their respective fields of study, and together they must form one of the most powerful teams of marine medical and biological scientists to be found.

Finally we again pay homage here to the actual publishers of all three editions of *The Marine Stinger Book* and the present book, Surf Life Saving Queensland, in furthering these studies. They continue at the forefront of international surf life saving, and we remember the key pioneer roles of Jack McMaster, Maurie Mulcahy, Ron Pollard, Earle Smith and Graham McKenzie. The key role in this text's appearance, played by the General Manager of Surf Life Saving Queensland, Brett Williamson, is especially and gratefully acknowledged. The current core team of the General Manager Brett Williamson, the Manager of Administration and Resources Mrs Penny Weir, the State President Ron Rankin, and Chief Superintendent Don Burchill, were ever-encouraging, provided practical support and (in keeping with the best traditions of Surf Life Saving) exhibited courage, tenacity and good cheer in the face of repeated editorial delays and demands during the preparation of this book. The editors only hope the following aphorism will apply:

People forget how fast you did something;
they only remember how well you did it.

Standard Abbreviations and Units of Measurement

Å	angstrom; 1Å = 10^{-8} cm.
atm.	atmosphere.
ATA	atmosphere absolute.
ca.	abbreviation for the latin 'circa', meaning 'approximately' or 'about'.
cf.	abbreviation for the latin 'confer', meaning 'compare with'.
cm	centimetres (2.54cm = 1 inch).
Da	dalton. An arbitrary unit of mass, equal to one-twelfth the mass of carbon 12, or 1.657 x 10^{-24} gram (1000 daltons = 1 kilodalton).
ICJS	International Consortium for Jellyfish Stings.
kPa	kilopascals; a unit of pressure equal to just less than one-hundreth of an atmosphere; approximately equal to 7.4 mmHg.
kg	kilogram; 1 kg = 1000 grams, or 2.2 lb.
LD_{50}	the dose of a drug (e.g. toxin) which causes death (i.e. lethal dose) in 50 per cent of the experimental living cells, organs or animals tested.
lb	pounds weight; 1 lb = 16 ounces.
m	metres; 1 m = 100 cm, 1000 mm, or 39.37 inches.
mins	minutes.
ml or cc	millilitres or cubic centimetres (units of volume).
mm	millimetres; one-thousandth of a metre (25.4 mm = 1 inch).
mmHg	millimetres of mercury pressure (760 mmHg = 1ATA).
MPa	megapascals; (1 megapascal = 1000 kilopascals).
mw	molecular weight.
p. and pp.	page (p) or pages (pp).
pH	an inverse measure of acidity (i.e. hydrogen ion concentration of [usually] a liquid). Neutrality = pH 7.0 (as in pure water). Human blood in health has a pH = 7.35 to 7.44.
pl.	plural (i.e. more than one).
µm	micrometre; a microscopic unit of length equal to one-thousandth of a millimetre (10^{-3} mm).
pO_2	partial pressure of (in this case) oxygen.
$<$	less than.
$>$	more than.
$\approx$	approximately equal to.
$Sp0_2$	haemaglobin saturation with oxygen as measured by pulse oximetry.

Abbreviations for the States of Australia

ACT	Australian Capital Territory
NSW	New South Wales
NT	Northern Territory
Qld	Queensland
SA	South Australia
Tas	Tasmania
Vic	Victoria
WA	Western Australia

GLOSSARY

The editors acknowledge, among other publications, the following as sources of information: 25th edition of *Steadman's Medical Dictionary* (International Edition 1990), Williams & Wilkins, Baltimore.
The World Book Dictionary (2 volumes) (Clarence L. Barnhart, Editor in Chief), Doubleday & Company, Inc. Chicago, 1972.

Aboral A region opposite the mouth.
Acontium (pl. acontia) a long filament equipped with stinging cells (nematocytes) arising from a septum of some sea anemones
Adradial Situated near or beside a radially arranged structure.
Aerobic Dependant upon or consuming oxygen.
Aetiology Cause of a medical condition.
Allergy An abnormal or harmful body reaction resulting from the interaction between an antigen and an antibody.
Anaerobic Functioning without oxygen.
Analgesia Reduction or relief of pain.
Anaphylaxis A major, usually rapidly developing, often life-threatening allergic reaction.
Antibody A body molecule that is produced in response to the body's exposure to an antigen (a venom is usually a mixture of antigens) and which, as part of the body's immune defence against harmful foreign substances (such as bacterial antigens), attacks the antigen.
Antigen A foreign substance (e.g. a venom) introduced into the body, triggering an immune response and the production of antibodies. Antigens are commonly proteins.
Antivenom A mixture of antibodies produced in the blood stream of animals, after they have been exposed to non-harmful injected doses of a venom harmful to humans; the antibody mixture is then collected from the animal's blood, purified, concentrated and prepared for injection into humans seriously affected by that particular venom. Each antivenom is specific for the venom against which it is prepared, and will not normally neutralise other venoms.
Aposematic Describes natural colours or markings that serve to warn or alarm enemies.
Arterial tourniquet A very tight ligature applied around a body structure (e.g an arm or a leg) to exert a pressure sufficient to obstruct the arterial blood supply to the part of the body that lies on the far side (ie. furthest from the heart.)
Asporogenous Happening without the development of spores.
Asystole Absence of visible contraction of the heart, and consequent circulation of the blood, resulting rapidly in death. This may occur after untreated envenomation.
Axil The upper angle between angulated or branching parts.
β-cell A lymphocyte that makes circulating antibody.
Cardio-pulmonary resuscitation (CPR) (see Resuscitation) A combination of artificial ventilation of the lungs (i.e. expired air resuscitation) to provide oxygen to the blood, and external chest compression to propel the oxygenated blood to the victim's brain (i.e. external cardiac compression).
Cardiac arrest Absence of a palpable pulse, and thus of circulation of blood around the body, by the heart's contraction. The cause may be asystole, ventricular fibrillation or pulseless ventricular tachycardia.
Catecholamines Hormones and neurotransmitters released by the body under any stressful reaction, or after envenomation (e.g. Irukandji), that affect the circulatory system, often increasing heart rate and blood pressure.
Centripetal canals Blind canals in the body of a jellyfish, extending from the ring canal towards the centre of the umbrella or bell (body).
Chondrophora A suborder of free-floating cnidarians of the class Hydrozoa. Some authorities consider these to be polymorphic colonies, while others treat them as inverted polyps related to tubularian polyps.
Cilium (pl. cilia) A threadlike appendage of locomotion, projecting from the tip of a cell lining a cavity: hence 'ciliary action', in which a group of cilia, beating in unison, aids in passage of substances across the apical ends of cells.
Circular canal See Ring canal.
Class A taxonomic grouping of related, similar orders. It is the taxonomic category below 'phylum' but above 'order'.
Cnida (pl. cnidae) The general term used for all 3 categories of capsular organoid manufactured by cnidarians, whether they bear spines or filaments or are devoid of either. Cnidae include the 'nematocysts', the 'spirocysts' and the 'ptychocysts'.
Cnidaria Members of the phylum Cnidaria, which are mainly marine, radially symmetrical, usually bearing tentacles and containing stinging cells called cnidocytes produced by the animal itself (cf. nudibranchs). They possess a body wall composed of epidermis and gastrodermis, with mesogloea (connective tissue) in between; one body cavity opens via a mouth. They occur in polyp and/or medusa stages.
Cnidocil The flagellum-like structure in the cnidocyte complex situated on the apical surface of nematocytes of scyphozoans, hydrozoans and cubozoans. It

is part of the discharge triggering mechanism (hair trigger) for nematocysts.

Cnidocytes Mature cnidarian cells that contain mature nematocysts. ('Cnidoblasts' contain developing nematocysts.)

Cnidom The collective name for the different cnidae possessed by a single species of cnidarian; the nematocyst complement.

Coelenterata The term 'Coelenterata' is used less commonly than 'Cnidaria'. Coelenterata originally contained Spongiaria, Cnidaria and Ctenophora. The term 'coelenterates' is still occasionally used to include the cnidarians and ctenophores. However ctenophores are now allocated to their own phylum Ctenophora. Thus the phylum Cnidaria does not include the ctenophores and the two terms 'coelenterate' and 'cnidarian' and therefore not considered interchangeable.

Compression/immobilisation bandage A broad, firm elastic bandage applied over the bite or sting site on a human body (usually a limb), applied as tightly as for a sprained ankle (but not cutting off the arterial blood flow, i.e. it is not a tourniquet). It aims to trap the injected venom in the tissues at the site of the bite or sting.

Cormidium: A group or cluster of zooids.

Crinotoxin: A poison produced by specialised glands in the skin of a marine organism. Crinotoxic secretions are not associated with a venom apparatus and may be released into the water.

Cubomedusae A term that includes all box jellyfish species.

Cubozoa The taxonomic class of box-shaped jellyfish consisting both of chirodropids and carybdeids.

Dactylozooid A tentacle in a siphonophore colony; also referred to as a palpon or feeler. These stinging tentacles, which are separate from the feeding polyps, are also called fishing tentacles in *Physalia* species. The tentacles have numerous 'batteries' or 'buttons' full of nematocysts, in order to capture prey.

Demersal Living on or near the bottom.

Dermoptic A reaction or response to different colours (i.e. wavelengths) of the light spectrum.

Diathesis (pl. diastheses) The constitutional or inborn state predisposing someone to one or more diseases or abnormal conditions.

Dichotomous key A systematic list of taxonomic characteristics, used to identify members of a group of organisms. The adjective 'dichotomous' signifies that the identifier (the reader) is faced with two choices at each step.

Dioecious Male and female organs in separate individuals of the same species.

Distal Description of part of the body that is further from the heart than another part.

Dysphagia Difficulty with swallowing.

Echinoid test Loosely fitted calcareous plates arranged to form the usually globular or spherical skeleton enclosing the soft parts of sea urchins.

Ecthyma A pyogenic skin infection characterised by adherent, bloody, crusted lesions which penetrate into the dermis and may rarely scar.

Ectoderm The outer (embryonic) tissue layer of an organism, containing future nerve, gland and nematocyst cells.

Embolism A blockage of blood vessels either by blood clot, fat or gas.

Emollient An agent (e.g. lotion, cream or ointment) that softens the skin or soothes irritation in the skin or mucous membranes.

Endoderm The innermost (embryonic) tissue layer of an organism.

Endogenous Produced within the body.

Endotoxaemia Presence of endotoxins in the blood.

Endotoxin A (usually bacterial) toxin not freely liberated into the surrounding medium (e.g. human tissues).

Enteric Pertaining to the intestine.

Envenomation The process of introduction of venom into the tissues by some puncturing or traumatising apparatus, e.g. teeth, spines, miniature harpoons or drills.

Epitope The portion of a complex molecule that determines the antigenic property of the molecule.

Erythema Redness of the skin resulting from an inflammatory reaction.

Erythrocyte Red blood cell.

Estuarine Pertaining to the estuary of a river.

Expired air resuscitation (EAR) (see Resuscitation) The process or technique of blowing expired air from the rescuer into the lungs of the victim who is not breathing sufficiently to sustain life.

External cardiac compression (ECC) (see Resuscitation) The process or technique of compressing the anterior chest wall with usually the hands, in order to propel the blood from the heart to the brain.

Extrinsic Originating outside the part where found, or upon which it acts (cf. 'intrinsic').

Exumbrella Outside surface of a jellyfish bell.

Faecal Pertaining to faeces (discharge from the anal end of the bowel).

Family A taxonomic grouping of related, similar genera: taxonomic category below 'order' but above 'genus'.

Fibrous basket A component of a nematocyte consisting of a collar of interwoven fibrils. The longitudinally-running fibrils are called supportive rods and emanate from the small microvilli; other fibrils run in a cross-wise direction.

Flagellum (pl. flagella) A long cellular appendage specialised for locomotion and usually formed from a core of nine outer doublet microtubules and two inner single microtubules, ensheathed in an extension of cell membrane.

Fulminant Very rapidly progressing.

Ganglion (pl. ganglia) A cluster of nerve cell bodies, which may produce a lump or swelling normally in association with a nerve.

Gastric cirri (filaments) Small finger-like appendages projecting into the stomach, bearing nematocysts and gland cells.

Gastrodermis The cell layer that lines the gastrovascular cavity.

Gastrovascular cavity The digestive-excretory 'system' of the cnidarian, with its outpouchings and canals. The canals form a primitive vascular 'system'.

Genus (pl. genera) A taxonomic grouping of closely related species: taxonomic category below 'family' but above 'species'.

Glandular triangle A term originally used to designate a cross-section of the ventrolateral-glandular grooves of the stingray's sting, but which was later expanded to include cross-sections of the glandular grooves of other fishes regardless of the anatomical position of the groove (HALSTEAD 1988).

Gonad Aggregation of male and/or female reproductive cells. (A hermaphrodite has gonads.)

Gonophore The medusa individual of a siphonophore which bears eggs or sperm.

Granulomatous Describes a non-specific nodular inflammatory lesion.

Halophilic Requiring a high concentration of salt for growth.

Hermaphrodite A person, animal or plant having the reproductive organs of both sexes (see 'Monoecious').

Humoral Relating to the body fluids and their physiological and biochemical activities.

Hyperpnoea and **Hypopnoea** Excessive (hyper-) or insufficient (hypo-) breathing (-pnoea).

Hypersensitivity Extreme sensitivity to any protein, over and above its normal effect. It usually occurs in certain sensitive people after more than one exposure to the offending protein.

Hypertension High blood pressure — usually above 150/95 mm Hg.

Hypotension Low blood pressure — usually with the diastolic (the lower level) below 60 mm Hg, and sufficient to cause symptoms such as dizziness or collapse.

Hypoxia Low oxygen saturation (levels) in a gas mixture of the body tissues.

Iatrogenic Caused by, or as a result of, medical management.

Ichthyocrinotoxin A toxic fish poison produced as a glandular secretion (usually in the skin of the fish), which may be released into the surrounding water, poisonous to other fishes, but which is not associated with a venom apparatus.

IgG (immunoglobulin G) One of a class of immunoglobulin antibodies found in the serum in response to challenge by an antigen, and capable of subsequent identification by serological testing.

IgE (immunoglobulin E) One of a class of immunoglobulin antibodies found in the serum in response to challenge by an antigen. Its level rises immediately in response to challenge by an antigen and can be measured in the serum. However, it may quickly return to a non-measurable level — a sign of 'acute' or recent envenomation or illness. High levels of this antibody are thought to be associated with anaphylaxis.

Immunoglobulins A class of structurally related proteins that function in the body as antibodies (e.g. blood serum).

Interradius (pl. interradii) Second-order radius. Interradii make up four radial axes, which each bisect the right angle between the perradii.

Intrinsic: Inherent; belonging entirely to a part or function (cf. 'extrinsic').

In vitro In an artificial environment.

In vivo In a living body.

Ischaemia Inadequate local blood perfusion, usually due to impaired local arterial blood flow.

Kingdom A taxonomic grouping of related, similar phyla; the highest level category in biological classification (e.g. the plant kingdom).

Lability Instability, especially chemical or molecular.

Lappet Lobe-like extensions of the umbrella margin.

Lymphadenopathy Swelling of the body lymph glands which is sometimes painful, especially after envenomation. Swollen lymph glands may occur almost anywhere in the body, but are more easily felt in the neck, under the arms (axillae) and in the groins.

Mannitol: A hexahydric alcohol derived from the chemical reduction of fructose. It is used intravenously as an osmotic diuretic.

Manubrium The tube between the stomach and the mouth; also called the oesophagus.

Medusa (pl. medusae) Free-swimming jellyfish stage of life cycle of jellyfish (except for one order, which remains attached to the substrate). In Greek mythology Medusa was one of the three Gorgons or monsters who had snakes for hair.

Mesogloea The jelly of a cnidarian, found between the epidermis and gastrodermis, forming the 'supporting' substance.

Microtubules Hollow rods of tubular protein found in the cytoplasm of all eukaryotic cells and found in cilia and flagella and as part of the cytoskeleton.

Microvillus (pl. microvilli) Elongate protrusions at the apical end of the nematocyte.

Mollusc A member of the phylum Mollusca, a large group of animals having no backbone and a soft body. It includes land-based slugs and snails, many species of edible shellfish, and species lacking a protective shell (e.g. blue-ringed octopus and nudibranchs).

Monoecious Having both male and female organs in the same organism (see Hermaphrodite).

Motile Having the power of spontaneous movement.

Motor neurones Nerve cells that transmit signals from the central nervous system out to body organs (e.g. muscles, glands).

Muscle pump The descriptive term for the action of contracting limb muscles propelling the blood in the low-pressure limb veins back to the central blood volume within the chest.

Myofibrils Fibrils arranged in longitudinal bundles in muscle fibres, composed of thick and thin filaments.

Nematocyst The stinging capsules of cnidarians, which contain an inverted tubule and venom. Nematocysts are a secretory product of nematocytes.

Nematocyte Stinging cell containing a nematocyst.

Neritic That part of the ocean between the low-tide mark and a depth of about 180 metres (585 feet), i.e. the shallower part of the oceans.

Neurone A nerve cell, capable of conducting signals.

Ocellus (pl. ocelli) Photosensitive organ ('eye') consisting of pigment. Ocelli are usually found at the margin of a jellyfish bell.

Operculum (pl. opercula) The lid found at the apical end of some cnidae (phylum Cnidaria); the posterior bone in the protective cover over the gills (Phylum Chordata, Class Osteichthyes).

Oral The term used to refer to the region or surface of a medusa or polyp with the mouth opening.

Order The taxonomic term for a group, above genus, but below class.

Osculum (pl. oscula) A mouth or mouth-like opening.

Osmotic The tendency (pressure, force) of two fluids of different concentrations, separated by a semi-permeable membrane (e.g. blood vessel walls) to equalise their solute composition.

Ostium (pl. ostia) Opening in stomach wall in cnidarians.

Pacemaker A specialised cluster of somatic cells, which set the rate of contraction.

Parenteral Administration of a substance by some means other than through the gastro intestinal tract (i.e. by mouth), e.g intramuscular, intravenous, subcutaneous, per-tracheal, conjunctival, percutaneous or intramedullary.

Paresis Weakness or paralysis of (usually voluntary) muscles.

Paresthesiae Tingling and burning in the skin frequently described as 'pins and needles'. It is caused by irritation of cutaneous nerves by a variety of causes including trauma and envenomation.

Pari passu At an equal rate.

Pathogenic Causing disease or injury.

Pathognomonic Characteristic or indicative of a particular disease or diagnosis.

Peau d'orange Literally 'the skin of an orange'; describes the punctate pitted appearance that human skin can assume in certain types of oedema (swelling).

Pedalium (pl. pedalia) Blade-, paddle- or hand-like projection from bell-edge, with tentacles, as in the Cubozoa and Coronatae.

Pelagic Of the open sea.

Perisarc The external horny or chitinous covering of certain hydrozoan animals.

Perradius (pl. perradii) Primary or first-order radii of four axes corresponding to the position of the oral lobes. They are situated at angles of 45° between two interradii.

Phacella A group of gastric cirri containing cells with digestive enzymes and nematocysts which help to subdue prey.

Phylum (pl. phyla) A taxonomic grouping of related, similar classes, below 'kingdom' and above 'class'.

Piliation Possession by some bacteria of functional fine filamentous appendages known as pili. They are somewhat analogous to flagella, but pili are short, entirely of protein, more numerous and may be specialised (e.g. for conjugation).

Placebo An inert substance, object or procedure with no inherent action, but which may confer benefit to the patient by its suggestive effect. Placebo effect is classically demonstrated in the relief of pain (analgesia).

Polymorphism Occurrence in more than one form or morphological type.

Poison A toxin that gains entry to the body via the gastrointestinal tract, the respiratory tract, or via absorption through intact body layers (e.g. ciguatoxin).

Polyp The attached stage in cnidarian life cycles.

Polypeptides The collective chemical term for linked smaller groups of amino-acids, with molecular weights generally lower than actual proteins.

Pouch Pocket-shaped extensions of stomach cavity of a jellyfish, connected to each other by septal ostia.

Preoperculum (pl. preopercula) (Class Osteichthyes) The bone between the cheek and the operculum forming the front part of the gill cover.

Primary and secondary bands The widest and second widest transverse 'bars' or rings in each band series on a cubozoan tentacle. They appear in series along the tentacle length.

Prognosis The predicted outcome of an illness or injury.

Protean Changeable in form.

Pruritus Itching.

Pseudostratified The false appearance of multiple layers (e.g. of an epithelial surface) because of overlapping of folds and convolutions of a single-layered surface.

Pyrexia Raised body (core) temperature; fever.

Radial canal A canal that runs between the central stomach and the periphery of a medusa.

Radial fibres Broad fibres belonging either to the nematocyte or to neighbouring accessory cells in some cnidarians. The radial fibres are firmly anchored in the mesogleal (connective tissue) layer.

Radial symmetry Structure like that of a pie or barrel, with many equal parts radiating outward like the spokes of a wheel.

Resuscitation The process of preserving or restoring life by sustaining an oxygen supply to the brain.

Rhopalium (pl. rhopalia) Projections housing statocysts and ocelli which are found in scyphozoan and cubozoan jellyfish. In scyphozoans they are located between scallops of the bell margin. In cubozoans they are usually found within sensory niches.

Ring canal (circular canal) The continuous circular canal that runs around the periphery of a medusa. It is connected with the central stomach via the radial canals.

Scuba Self-contained underwater breathing apparatus.

Scyphistoma A polypoid stage in development of scyphozoan jellyfish just after the larva becomes attached to the substrate.

Semelparous Giving birth singly.

Septicaemia Systemic illness caused by the multiplication and spread of microorganisms and their toxins via the circulating blood.

Septum (pl. septa) A dividing wall between two cavities.

Serum sickness A complex systemic reaction that may become evident any time up to 14 days after antivenom use. Symptoms are fever, generalised lymphadenopathy and an urticarial rash. Severe cases of serum sickness may have to be treated with oral steroids. The incidence of serum sickness is often related to the amount of antivenom used.

Shock A physiological state in which there is potential or actual tissue damage because of inadequate tissue perfusion by the blood.

Somatic Relating to the body in general.

Species The lowest taxonomic grouping of closely-related varieties — below a Genus.

Spirocyst Capsular organelle in which the undischarged capsule is thin-walled and contains a tubule that is coiled like a spring in the undischarged state and which bears fine hollow filaments emanating from the discharged tubule.

Sporotrichoid Pertaining to fungal infection (e.g. sporotrichosis from *Sporotrichum schenkii*) in animals and humans, which spreads along lymph vessels.

Statocyst An organ of balance, a receptor organ consisting of a pit or vesicle provided with small concretions (statoliths), which give position 'information' to cnidarians.

Statolith Small granule, made of gypsum or calcium phosphate and magnesium carbonate, which rests on the tufts of cilia (hair cells) in the statocysts; it responds to gravity.

Sting The painful injection of a venom into the body through the intact skin or mucous membranes.

Stinging button A grouping or 'battery' of nematocysts on a jellyfish tentacle, particularly *Physalia* species.

Stomach The digestive and excretory cavity, without considering the outpouchings, canals, mouth and manubrium of the gastrovascular system.

Stomocnide Large group of nematocysts once thought to possess openings at the tips of fully discharged tubules.

Strobila The sessile polypoid form of jellyfish in which the larva forms a series of medusae (ephyrae). The medusae form from saucer-like buddings, a process called 'strobilisation'.

Subumbrella The undersurface or trailing surface of a jellyfish.

Suppuration The formation of pus, usually indicating infection.

Symbiotic The description of an intimate association between species, usually to the mutual advantage of each species.

Systemic Relating to the entire organism, as distinct from any of its individual parts.

T Cell A lymphocyte derived from a thymocyte, of immunological importance, and responsible for cell-mediated immunity.

Taxon (pl. taxa) A group of organisms sharing common characteristics in varying degrees of distinction that constitute one of the categories in taxonomic classification, such as phylum, order, family, genus or species.

Taxonomy The science of the systematic classification of living things.

Tentacle Long flexible process containing the nematocysts (stinging cells); used for food getting.

Test See 'Echinoid test'.

Tetanus/Tetany Violent spasms, muscle contraction ('lock-jaw') caused by a toxin from abacillus (microorganism) which penetrates the body though a puncture wound, and usually leading to death. The organism occurs in water and may occur after envenomation (e.g. in stingray spine puncture wounds), as well as the more commonly-known soil contamination (e.g. in war wounds). It may be prevented by vaccination with tetanus toxoid.

Tetramerous Divided into four symmetrical parts or tetrameres.

Tourniquet (see arterial tourniquet) A tight ligature applied around a part of the body (e.g. an arm or a leg) which cuts off the arterial blood supply to that part.

Toxin A molecule that is harmful to the (human) body tissues. (Although this text is oriented to clinical requirements, toxins naturally affect all living things.) See venom, poisons.

Trauma Injury from mechanical force.

Trirhopaloids The preferred name used in this book for certain nematocysts found in the tentacles of *Chironex fleckeri*. Some act as 'grappling hooks' in the stinging process. This term replaces the original diagnosis 'heterotrichous microbasic euryteles' which is inaccurate.

Velar canals Extensions of the gastrovascular system found in the velarium, in cubomedusae.

Velarium The horizontal flap-like extension of the edge of the bell in the Cubozoa; it contains velar canals.

Velum The circular shelf-like extension that partially blocks the mouth-opening of the bell of a hydromedusan. The velum does not contain velar canals.

Venom A toxin or a mixture of toxins that normally gain access to the body by injection through the intact skin (e.g. by a fish spine).

Vesicle A serous or clear fluid-containing blister on the skin.

Zooid Each of the distinct individuals which make up a colonial or compound animal organism.

SECTION

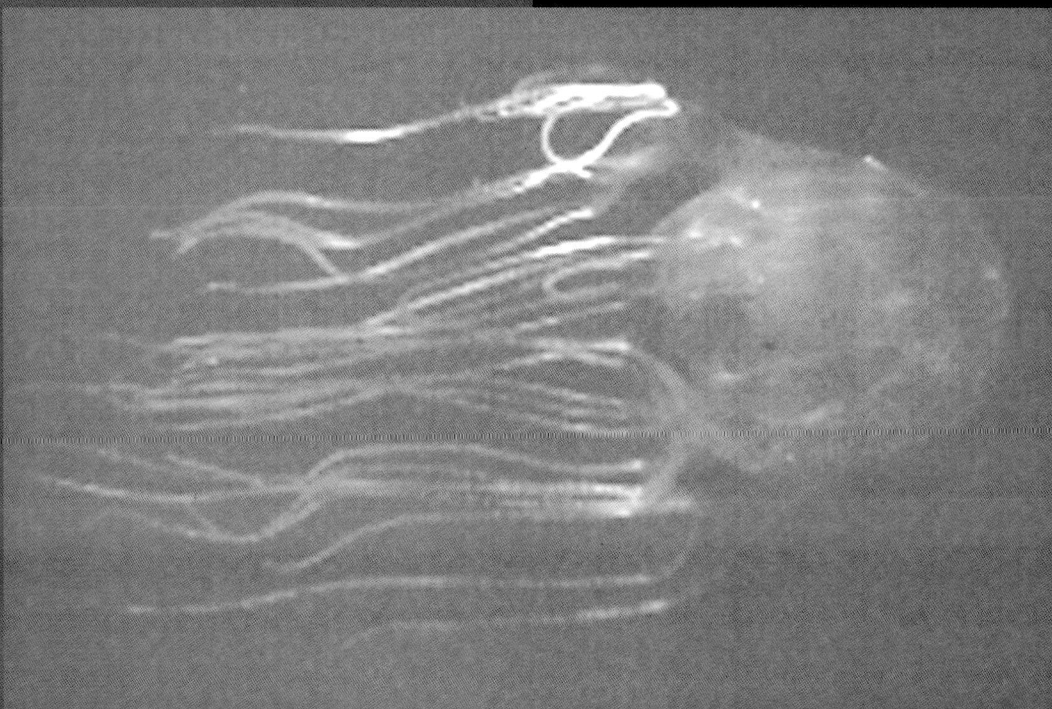

ONE

GENERAL INTRODUCTION

CHAPTER 1

HISTORICAL BACKGROUND TO MARINE ENVENOMATION

PETER FENNER,

MAURIE MULCAHY AND

JOHN WILLIAMSON

INTERNATIONAL SYNOPSIS

The presence of poisonous, biting and stinging marine creatures was acknowledged by humans early in recorded history. HALSTEAD (1988) cites GAILLARD (1923) in describing an inscription of the poisonous puffer *Tetraodon stellatus* from the tomb of the Egyptian pharaoh Ti (Fifth Dynasty 2700 BC).

Aristotle (384–322 BC) documented jellyfish (STORER & USINGER 1957), stingray (GUDGER 1943) and scorpionfish *Scorpaena poreus* stinging (COUTIÈRE 1899). A Greek surgeon, Dioscorides (50–100 AD), serving in the Roman army, wrote about bristle worms, sea snakes and a stingray. The great Roman naturalist Pliny (29–79 AD) described stingray and weeverfish injuries (HALSTEAD 1988). The subject of drugs from fish (READ 1939) is contained among the vast medical literature of ancient China, in the work of Ch'en T'sang-chi (Tang Dynasty 618–907 AD).

The earliest study efforts in marine toxinology were intermingled with magic and sorcery (Plate 1.1) (RICCI 1982), a characteristic human response to inadequate understanding. This has its modern counterpart in the still-present tendency for anecdotal exaggeration and superstition which has bedevilled the subject (WILLIAMSON 1992). However, among such tales lie germs of fact whose dis covery remains a constant challenge.

For a fuller appreciation of the international evolution of marine envenomation studies, the reader is referred to the writings of Grevin (1538–1570),

Linnaeus (1707–1778), Autenrieth (1799–1874), Poey y Alay (1799–1891) on ciguatera, Fayrer (1824–1907) on sea snakes, Savtschenko (1846–1885) on poisonous and venomous marine fishes, Tahara (1855–1935) on marine toxin chemistry, Phisalix (1861–1946), Pavlovskiy (1884–1965) and Muir Evans (1866–1947) on toxinology of venoms, Myer (1884–1974) and Sommer (1899–1950) on shellfish poisoning, Gudger (1905–1951) on medical ichthyology, Mortensen (1868–1952) on sea urchins, Herre (1868–1962) and Coutière (1869–1952) on marine biotoxicology, and Whitley (1903–1976) on venomous and poisonous Australian fish. Many important scientific milestones attributable to marine toxinological studies feature in the lives, times and works of such pioneers. For example, it was Portier and Richet who in 1902 induced and described anaphylaxis for the first time (PORTIER & RICHET 1902a, 1902b) while experimenting with marine cnidarian proteins.

This book is concerned primarily with international events since the Second World War and particularly since 1970. The classical works of CLELAND & SOUTHCOTT (1965), RUSSELL (1975), MARETIĆ (1975) and HALSTEAD (1988) are recommended, supplemented by the papers and monographs of LIGHT (1914a, 1914b), FLECKER (1945, 1952, 1956), SOUTHCOTT (1956, 1959, 1967), BARNES (1960, 1966), SUTHERLAND (1983, 1994), MARSH & SLACK-SMITH (1986), AUERBACH (1987a, 1987b), EDMONDS (1989), and FENNER (1990b), and the writings of HARTWICK and colleagues (1980), RIFKIN & ENDEAN (1983, 1988), GONZAGA (1985) and BURNETT, CALTON and colleagues (1977, 1987, 1988). International progress from here to the present time is the story contained in the following pages, which also can be supplemented by reference to recent texts of HALSTEAD *et al* (1990), EDMONDS *et al* (1992), AUERBACH (1995) and MEIER & WHITE (1995).

At present there are some 31 countries participating in the affairs of the International Consortium for Jellyfish Stings {see Chapter 23, page 428}. Because of the extraordinary concentration of venomous and poisonous marine (and terrestrial) animal life in the Australasian region, this book highlights the significant Australian contributions to the subject.

AUSTRALIAN PERSPECTIVE

Surf Life Saving Queensland (the Queensland State headquarters), and Surf Life Saving Australia (SLSA — the national co-ordinating body of Australian Surf Life Saving, which also houses the autonomous New South Wales State headquarters), have played an influential role in the story of marine envenomation in Australia. In 1936, six years after the formation of the Queensland State Centre of Surf Life Saving, clubs from Mackay to Cairns in north Queensland established the North Barrier Branch. In 1949 this area was divided into two branches: clubs from Cairns to Mission Beach near Tully forming the North Queensland Branch, and those south to Mackay remaining in the North Barrier Branch.

When the earliest clubs were formed in north Queensland (Ayr in 1926 and Picnic Bay in 1927), all surf life savers were very much aware of the jellyfish problem, as they were regularly called upon to treat sting victims during their normal patrols. At this time the treatment was very rudimentary, consisting mainly of rubbing the area of the sting with wet sand to remove adherent tentacles. This idea was based on the old Aboriginal treatment (although Aborigines actually used hot, dry sand).

Later treatment was extended to include a 'sting lotion', introduced to try to stop the burning and irritation of the skin. Many older surf life savers still remember the large medicine bottle dispensed by the local chemist and kept in the club's First Aid Room (Table 1.1).

TABLE 1.1
CONTENTS OF A STING LOTION DISPENSED TO EARLY SURF LIFE SAVING CLUBS

Picric acid	35 grains
Camphor	90 grains
Methylated spirits	3 fluid ounces
Distilled water	4 fluid ounces

The picric acid and camphor were separately dissolved in the methylated spirits, before the distilled water was added. This mixture was allowed to stand for three hours and then was filtered and stored for use. It was used (with unknown results) by surf life savers on any stings they treated.

FATAL STINGS

There have been at least 63 properly documented fatal stings from the north Australian box jellyfish *Chironex fleckeri* since reliable records were first kept (CLELAND & SOUTHCOTT 1965) {see Chapter 3, page 63}, but others certainly occurred, as more past fatalities are being identified with further historical and epidemiological research (CURRIE 1991, 1992; INTERNATIONAL CONSORTIUM FOR JELLYFISH STINGS 1993) {see also Chapter 9, page 236}.

The first documented Australian death was in December 1884 at Ross Creek, Townsville (Table 3.4A, page 70). The cause of death in these early years was often classified simply as 'drowning' (CLELAND & SOUTHCOTT 1965).* Many of these fatal stings were in coastal areas covered by the North Barrier and North Queensland branches of Surf Life Saving Queensland, Inc., but it is worth noting that there has never been a fatal sting on a *patrolled* beach, even to the present day (SURF LIFE SAVING QUEENSLAND 1993). With the fatal sting of a five-year-old boy in January 1987 (LUMLEY *et al* 1988), records of deaths now extend from Gladstone in central Queensland, all along the north tropical Australian coastline and south to Broome in Western Australia — approximately 60 per cent of the continent's coastline (CLELAND & SOUTHCOTT 1965).

According to records dating back to the time when many North Barrier surf life saving clubs were being formed, fatal stings occurred at the following locations: Proserpine in 1923; Googarra Beach (now more commonly known as Coogarra) near Tully in 1934; Bramston Beach near Babinda in 1937; Googarra and Proserpine in 1939; North Mission Beach near Tully in 1949; Kissing Point Baths, Townsville, in 1951; Saltwater Creek, Townsville, in 1953; Cardwell in 1955 (CLELAND & SOUTHCOTT 1965).

At the time, confusion existed because victims who died quickly were left with very few visible skin marks on arrival at the local hospital {see Plates 7.2 and 9.70B} whereas victims who survived often had severe skin damage {see Figure 1.2 page 45, and Plates 9.91 to 9.94}. This led to the belief that certain victims were 'not in good health', dying of 'shock' or 'heart failure' rather than from the massive endotoxaemia of the sting. Many victims were children (WILLIAMSON 1983; VIMPANI *et al* 1988; CURRIE 1992), which tended to support the theory that victims could have been in 'weak' or 'delicate health'. This theory was itself weakened in 1956 when a fit six-foot man died in Townsville within 10 minutes of a large sting mainly confined to below his knees (CLELAND & SOUTHCOTT 1965 p. 64)

Another widely held opinion (mainly by people who did *not* live in north Queensland) was that the stings were all caused by the Portuguese man-o'-war or 'bluebottle' (*Physalia* sp.) {see Chapter 8, page 192}, which was thought to be more venomous in summer, causing death in this season only, despite being present most months of the year.†

It became appreciated that death could occur within minutes ('sudden' [BARNES 1960]) from cardiac arrest (FREEMAN 1974; ENDEAN & SIZEMORE 1988; LUMLEY *et al* 1988). It was also argued that in less fulminant but fatal envenomations, perhaps death could be contributed to by respiratory arrest (WILLIAMSON *et al* 1984a), i.e. hypoxic hypoxia. It is now understood that with rapid death the victim does not have sufficient time to develop the severe inflammatory skin reaction {Plates 7.2 and 9.70B}. Surviving victims of stings that caused severe skin damage may carry the scars for many years (WILLIAMSON *et al* 1980) {see Plate 5.20C}. Most stings occur on the lower limbs (with the hands secondarily involved {Plate 9.70B} because most encounters occur in shallow water as the victim wades in

* Such an erroneous diagnosis of 'drowning' was recorded against many past autopsies on Australians who died while immersed in the sea (e.g. scuba divers) {see Chapter 22, page 426}, whatever the actual primary cause of their demise.

† This argument had an interesting parallel, in that bushmen and Aborigines believed that the fruit of the north Queensland finger cherry bush would cause blindness only when eaten at certain times of the year CLELAND & SOUTHCOTT 1965).

{Plate 9.75}, and where children with their supervising mothers tend to frolic. Women and children, with their small body size and little 'natural' protection from body hair (Plates 1.2 and 1.3), are particularly vulnerable to serious jellyfish envenomation (WILLIAMSON 1983) {see Chapter 5, page 98 and Chapter 7, page 174}.

STING IDENTIFICATION

DR HUGO FLECKER

MB BS(SYDNEY), LRCP, MRCS, FRCS(EDIN.) (1884–1957)

Dr Hugo Flecker (Figure 1.1), a pioneer radiologist and radiotherapist from Melbourne, moved to Cairns in 1932 at the height of the Great Depression (FENNER 1990b; PEARN 1994). He was a keen naturalist and immediately began to study the plants and animals of north Queensland and their medicinal properties. Within a few years he presented papers to journals on the finger cherry tree, scorpions, and a fatal cone shell sting on Hayman Island in 1935 (FLECKER 1936).

As a result of his efforts a Registry of Animals and Plants Injurious to Man was set up in 1935. From this time onwards Dr Flecker collected case histories and conducted an extensive correspondence with many medical practitioners on the jellyfish problem in north Queensland. Many of these early papers were published in the *Medical Journal of Australia* and made available to the northern surf life saving clubs. This material generated considerable interest but provided little information on the type of jellyfish causing the problem or the first aid treatment for a victim.

FIGURE 1.1
Dr Hugo Flecker, a great Australian pioneer of marine envenomation research in North Queensland. (Photograph courtesy of Professor John Pearn)

The deaths continued, as did the speculation on the cause, until 1955 when a five-year-old boy was fatally stung at Cardwell, south of Tully. At Dr Flecker's instructions, Inspector Anthony of Cairns organised local police to net the beach in the area of the boy's death. Several large cubomedusae were netted and then forwarded to Dr Ronald Southcott in Adelaide for investigation.

The specimens sent to Southcott were subsequently identified by him as a new species, which he named *Chironex fleckeri* (Plate 1.4) in recognition of Flecker's work (SOUTHCOTT 1956). Thus by the late 1950s the 'Portuguese man-o'-war' had been exonerated, and the terms 'sea wasps', or later 'box jellies' were introduced to describe this new, large cubomedusan ('box-shaped') jellyfish {Plate 9.1}, which was believed to be the cause of the problem. In the early 1980s the name 'sea wasp' was not encouraged, and the Commonwealth Serum Laboratories' Box Jellyfish Antivenom {see Plate 9.87} ceased to be called 'Sea Wasp Antivenom', after visitors from southern Australia were seen wading in *Chironex*-infested waters using umbrellas and hats to keep away the 'wasps'.

DR RON SOUTHCOTT

MD, DSC, DTM&H

Dr Ron Southcott, who is acknowledged earlier in this book (FENNER 1990b), had been stationed in north Queensland during the Second World War and was well aware of the jellyfish problem in 1943, although his article on the subject was not published in *Military Medicine* until 1959 (SOUTHCOTT 1959). In it he described two types of severe stings:

'Type A' stings: The patient had severe aches all over the body and was 'very sick', but had no visible wheals (FLECKER 1945). These symptoms were later identified as Flecker's 'Irukandji syndrome' (*Carukia barnesi* [BARNES 1964]). {See below, and Chapter 9, page 246.}

'Type B' stings: The patient had severe local pain and wheals; although some victims were 'quite sick', no directly observed deaths were recorded at that time. These stings were later shown to be caused by *Chironex fleckeri*, the northern Australian box jellyfish (SOUTHCOTT 1956).

Flecker himself was also aware of both these jellyfish sting groups and wrote several papers on symptoms of stings, including the Type B group; he kept very accurate records. Flecker was also closely interested in Type A stinging and was responsible for the first known description of the syndrome in the medical literature (FLECKER 1945); he published a classic paper in the *Medical Journal of Australia*, naming the bizarre variety of systemic symptoms the 'Irukandji syndrome' (FLECKER 1952). He derived this name from the 'Irukandji', a tribe of Aboriginal people living in the area around Palm Cove (the present site of the Cairns Surf Life Saving Club), where many stings occurred.

Flecker spent many years trying to identify the marine organism causing this Irukandji syndrome but died in 1957 before he succeeded. His death concluded a wonderful early impetus to the study of marine envenomation. Sadly the Register of Stings that he so carefully kept disappeared with his death; its rediscovery would be of great value to marine and medical science.

Two human stingings and a major international medical advance that occurred at the time demonstrated to the now-alarmed public that extensive *Chironex fleckeri* stings were not invariably fatal. In 1956 a ten-year-old girl was severely stung in Mackay but was saved by expired air resuscitation (EAR) (QUEENSLAND SLSA REPORTS 1956). In the early 1960s, in response to the pioneer work of Elam, Safar and colleagues (SAFAR *et al* 1958), cardiopulmonary resuscitation (CPR) techniques — i.e. expired air resuscitation (EAR)* plus external cardiac compression (ECC) — were adopted and taught nationwide in Australia by the Surf Life Saving Association of Australia. The second stinging occurred on New Years Day in 1968 when a 13-year-old boy was stung with 53 feet (16 metres) of tentacle at Bramston Beach, Babinda. He was saved after prompt and extensive CPR (MAGUIRE 1968). This rescue and the subsequent careful documentation of the medical events represents both a triumph for all those rescuers and medical staff involved, and one of the most valuable early contributions to this subject.

* It is interesting to note that in the first known death due to a jellyfish sting in 1884, 'artificial respiration' had been attempted (CLELAND & SOUTHCOTT 1965).

Dr Southcott continues his contributions to this subject (and several other toxinological subjects) with great distinction, right up to the present day; this has included some invaluable collaboration with the late Dr Jack Barnes, of Cairns.

DR JACK BARNES

OBE, MB BS (1922–1985)

Another Cairns medical doctor of the highest calibre, general practitioner Dr Jack Barnes (Plate 1.5), was approached to carry on Flecker's work. Thus began a brilliant chapter in the research into cubozoan jellyfish, and a long and fruitful association between Dr Barnes and Queensland State Centre of Surf Life Saving Australia (SLSA).

In 1964 the northern Queensland Surf Life Saving Clubs, worried by their inability to help sting victims, held a special meeting at Ingham. This was immediately prior to the Queensland state surf championships, conducted at Forrest Beach, near Ingham on the Australia Day weekend 25–26 January 1964. At the meeting the North Queensland SLSA Branch Superintendent, Don Campbell, spoke at length about the jellyfish problem. Others present included the President of the North Queensland Branch, Maurie Mulcahy, Queensland State Superintendent Jack Dearlove and Deputy Superintendent Jack McMaster {see the Dedication}.

After this meeting Maurie Mulcahy convened a special conference in Cairns in July 1964 with Dr Jack Barnes, North Queensland Branch officials and club members, to develop a plan of action. This included the netting of shallow water adjacent to SLSA-patrolled beaches at Ellis Beach (Cairns) and Etty Bay (Innisfail), with recording of seawater temperatures, specific gravity, time and state of the tidal flow, and wind strength and direction. Thirty years later the clubs in North Queensland and North Barrier branches still regularly net their beaches in this way to detect the presence of *Chironex fleckeri*.

At this time the North Queensland Branch of the SLSA undertook the printing of 1000 posters at seven shillings and sixpence each (Figure 1.2), an outlay of £750, which exhausted

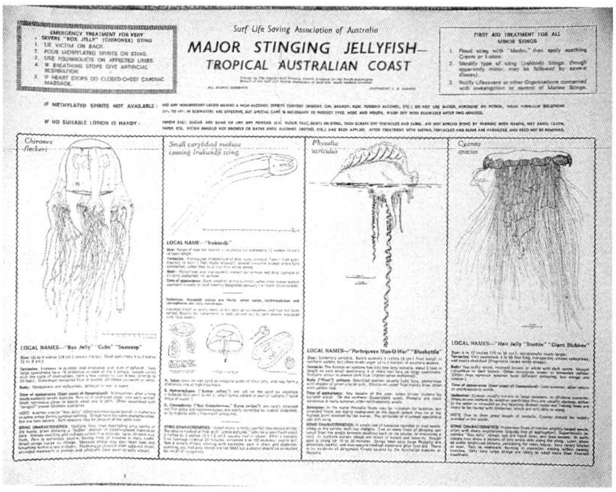

FIGURE 1.2
The 1964 poster prepared by Dr Jack Barnes and printed by the North Queensland Branch of Surf Life Saving Australia.

the Branch funds. These posters were prepared by Jack Barnes and identified the common jellyfish using line diagrams, and the then-recommended treatment. These posters proved to be another landmark, highlighting as they did the efforts of the SLSA, and the expertise of Jack Barnes. As a direct result, and somewhat fortunately at the time, local government authorities in the areas of the north Queensland clubs made substantial donations towards the cost of the posters.

The information from the SLSA clubs was forwarded monthly to Jack. Many jellyfish were preserved in formalin, stored in large plastic bags by the life savers, and sent to Jack for his research.

Jack had suggested in 1963 that the old method of rubbing wet sand onto adherent *Chironex fleckeri* tentacles after a sting made the envenomation worse and in 1964 suggested that 'household methylated spirits' (a mixture of 96% ethanol plus 4% water and denaturants) be poured on the stung area. This caused the tentacle to wither and gave a somewhat soothing, cooling effect. Jack also suggested the application of a tourniquet to the severely envenomated limb to prevent further absorption of venom. This treatment was later substituted by the application of a firm compression bandage. (In 1980, studies at James Cook University of North Queensland suggested that vinegar was preferable as a nematocyst inhibitor for box jellyfish stings [HARTWICK *et al* 1980].)

Barnes also suggested medical treatments for severe envenomation by both *Chironex fleckeri* and 'Irukandji' (BARNES 1966), which remain the basis for current in-hospital management of these marine envenomations today.

PROTECTIVE CLOTHING

As the medical adviser on marine envenomations to the SLSA, it was Jack Barnes who introduced the concept and use of pantyhose as an effective protective barrier against a serious *Chironex fleckeri* sting. The thickness of the nylon mesh was sufficient to prevent the discharged tubules of the tiny stinging organelles (nematocysts) on the tentacles from reaching through the keratin layers of the skin to the underlying sensitive and vascular epidermis and dermis (Plate 1.6).

It soon became commonplace in north Queensland to see surf life savers wearing pantyhose on patrol. One pair was worn to cover the legs and lower abdomen, and the other pair, with a hole cut for the head, worn upside down to cover the trunk and arms (Plate 1.7). This enabled patrolling lifesavers to enter the water safely while they dragged long 'fishing' nets through the shallows (Plate 1.8); this was not to clear the water of box jellyfish, but to see if *Chironex fleckeri* (or their natural prey of small fish and prawns) were present. It enabled the life savers to make a rational prediction of when the risk of a sting was sufficiently great to warrant the closing of the beach.

THE DEVELOPMENT OF *CHIRONEX FLECKERI* (BOX JELLYFISH) ANTIVENOM

In 1967 Barnes developed his innovative venom-collection technique, which led to production of the Commonwealth Serum Laboratories' (CSL) box jellyfish antivenom (BARNES 1967; BAXTER *et al* 1968; BAXTER & MARR 1974; SUTHERLAND 1983, 1994). Using a small vacuum flask with a human amniotic membrane (from a placenta 'borrowed' from the local hospital maternity ward) stretched across the neck of the flask (Plate 1.9a), he applied gentle suction to the flask through a side arm, which pulled the membrane slightly downwards. Into this centre dip were placed fresh, live *Chironex fleckeri* tentacles (Plates 1.9b and 1.9c). A small direct electric current was then passed across this membrane via electric terminals secured to the sides. In this way Barnes was able to induce the stinging capsules (nematocysts) to discharge through the membrane, enabling pure venom to be collected beneath the membrane.

The venom was then frozen, sent to CSL in Melbourne, and there injected into sheep, where antibodies were formed, enabling the first production in 1970 of the Commonwealth Serum Laboratories' box jellyfish (sea wasp) antivenom (SUTHERLAND 1983) {Plate 9.87}. It was the first specific treatment for severe *Chironex fleckeri* stings and remains a cornerstone of the management of any major *Chironex* envenomation (WILLIAMSON *et al* 1984b; FENNER *et al* 1989; BEADNELL *et al* 1992). Further research is only now suggesting some possible shortcomings to this method of extraction and ways to increase its undisputed clinical efficacy (ENDEAN & SIZEMORE 1988; BURNETT *et al* 1990; ENDEAN *et al* 1993).

CHIROPSALMUS SP. AUSTRALIA

Jack Barnes was also very interested in the other large 'box jellyfish' in his area and in 1965 identified another multi-tentacled cubozoan (BARNES 1965). To the untrained eye it looked the same as *Chironex fleckeri* (Plate 1.4), although it produced a milder sting pattern on the skin and milder symptoms. He produced a paper comparing and contrasting the morphology of *Chironex fleckeri* SOUTHCOTT, 1956 and *Chiropsalmus* sp. that inhabit the Australian coastline (BARNES 1965). He thus identified this other Australian box jellyfish (Plate 1.10) as *Chiropsalmus quadrigatus*, although, as he said 'not entirely conforming to original descriptions' (i.e. the classical descriptions, quoted by Barnes via Southcott as 'Haeckel 1879' [HAECKEL 1880] and MAYER 1910); Barnes stated that 'difficulty arises in reconciling Mayer's reference to cock's-comb-shaped gastric saccules', which Barnes regarded as typifying a growth phase of *Chironex fleckeri* (BARNES 1966). (A suggested new approach to the difficult question of the taxonomy of Indo-Pacific chirodropids is described by Dr J. Rifkin in Chapter 6, page 150 of this book).

THE 'IRUKANDJI'

After Flecker's death, Jack Barnes was asked to continue Flecker's quest to identify the jellyfish responsible for the 'Irukandji syndrome'. During the next four years, using detective work rivalling that of Sherlock Holmes, Jack deduced that the organism had to be a very small, quickly moving jellyfish that swam close to the surface. After estimating from his data the most likely time and place to catch the animal, he donned his scuba gear and lay on the bottom of the Palm Cove sea bed, in shallow water. After many hours his persistence was rewarded when he saw and caught a very small jellyfish swimming in front of his mask. Later another was caught when a small fish was observed swimming erratically; closer investigation showed that it was caught in the tentacles of another of these tiny jellyfish.

To discover if these were the jellyfish causing the 'Irukandji syndrome', Jack stung himself, his son Nick, and a surf life saver friend! After the characteristic 30-minute delay {see Chapter 9, page 249} all three developed Irukandji syndrome and required hospitalisation with severe back pain, muscle cramps, nausea, vomiting and headache. This experiment (guaranteed to give any modern Ethics Committee an attack of the terrors) was described in a classic paper in the *Medical Journal of Australia* (BARNES 1964). The experience prompted Jack to make one of his typically pungent and now most quoted remarks: 'I would rather be stung by a small *Chironex* than an "Irukandji", any day!'.

Dr Ron Southcott again closed a gap in the taxonomy of north Queensland cubozoans. He identified Barnes' little box-shaped jellyfish as a new genus and species, naming it, appropriately, *Carukia barnesi* (SOUTHCOTT 1967) {Plate 9.18} — '*Car*' from its order name 'Carybdeidae' and '*ruk*' from 'Irukandji', and Barnes' surname.

Barnes also believed {correctly — see Chapter 9, page 247} that there was more than one species of jellyfish that could cause the 'Irukandji syndrome' and referred to the others as 'pseudo-Irukandji' (KINSEY 1988). He was still working on this up to his untimely death. He died in 1985, a few short weeks after a fortuitous dinner held in Cairns in his honour by Queensland State Centre of SLSA, commemorating his great contribution to surf life saving and to water safety in northern Australia. One of the editors (JW), present on that memorable occasion, commented to Jack, who was always a man of few words, on his (expectedly) brilliant but (uncharacteristically) fulsome after-dinner speech. Quick as a flash, with his typical waspish grin, he shot back (perhaps unwittingly prophetically): 'It won't happen again!' With his death the SLSA lost a great friend and a medical adviser without parallel, and Australia lost a world respected scientist. Much of his original unpublished data have been faithfully collated since his death in two valuable publications by Dr Barbara Kinsey (KINSEY 1986, 1988).

Complete prevention of the sting of the 'Irukandji' and a completely adequate on-the-beach management regimen of its prostrating 'syndrome' have not yet been achieved. Research is continuing, including that by the present Marine Stinger Officer, Dr Peter Fenner (FENNER & WILLIAMSON 1993). Some progress has been made with medical understanding and in-hospital medical management (FENNER *et al* 1986, 1988; MARTIN & AUDLEY 1990) {see Chapter 9, page 249}.

SLSA STINGER SEMINARS

In 1967 the Queensland State Centre of Surf Life Saving became involved when several of its senior officers, including the late Dr Drury Clarke, who was then Superintendent, Deputy Superintendent Max Klinger and Secretary Alan Doig travelled to Cairns to confer with Jack Barnes and local government representatives. From this meeting came the prophetic quote:

> An intensive and coordinated effort is required to increase public and tourist awareness of the problem. But all such education must clearly recognise the fundamental limitation that it is only possible to reduce the incidence of fatal stinger attacks, (!) and not provide a blanket guarantee that attacks (!) from any stinger will not occur.

After the success of the preliminary meetings with Jack Barnes, Queensland State Centre held its first 'Stinger Seminar' on 31 July and 1 August 1971, at the James Cook University of North Queensland, under the chairmanship of Professor Cyril Burdon-Jones PhD, then Head of the University's Department of Biological Sciences. At this seminar all the known information was presented by Dr Jack Barnes and Maurie Mulcahy and distributed to the clubs in the north.

It proved so successful that a conference was convened less than a year later, on 19–20 February 1972, at the same venue. Chaired again by Professor Burdon-Jones, with much material presented by Jack Barnes, the conference was attended by the national office bearers Sir Adrian Curlewis, Harry Brown and Jack Dearlove. Also attending were the state officers Noel Heywood, Earle Smith, Dr Drury Clark and Jack McMaster. These guests shared the view that marine envenomation was not just a north Queensland problem, but one that the whole Surf Life Saving Association nationwide should address.

The most significant points that arose from this seminar were the importance of fast and efficient resuscitation, protective clothing, jellyfish 'spotting' from an inflatable rescue boat (IRB) {Plate 5.11} and netting of beaches (Plate 1.8). Also a drill procedure for the training of clubs in the rescue and treatment of marine stings was introduced by Board Secretary Maurie Mulcahy (Figure 1.3), which has remained largely unchanged to this day. The basis for an education film was also laid; it was

FIGURE 1.3
Maurie Mulcahy.

produced later that year and released under the title *Walk, Don't Run*.

In 1972 a Stinger Committee was also formed under the chairmanship of Earle Smith (Figure 1.4) with members Maurie Mulcahy, Peter Braddick and F. ('Tubby') Mathews. These people undertook preliminary training in the use of the IRB for jellyfish spotting at Bribie Island in south Queensland, with early input from Jack Barnes' skilled field colleague from Cairns and the North Queensland Branch, Ron Pollard (Figure 1.5).

The same year saw the establishment of the Queensland State Medical Panel under the chairmanship of Professor Tess Brophy (later Cramond). Dr John Williamson (Figure 1.6) later of *The Marine Stinger Book* fame, was also a foundation member of this Committee.

A third conference was held in 1973 at Cairns, and clubs were issued with booklets, warning cards and wall charts bought with a $7900 'Stinger Grant' provided by the Queensland State Government Minister for Primary Industries Mr Victor B. Sullivan. Six nets were also purchased and given to northern clubs for beach netting, and the booklet *The Ten Don'ts* (see Plate 1.11) was produced by Maurie Mulcahy and widely distributed to the general public. This contained advice on how to avoid jellyfish stings.

This state government Stinger Grant was to continue for many years and was of tremendous benefit to the SLSA and swimmers. It allowed the purchase of inflatable rescue boats for spotting, more nets, and the production of the first small edition (1974) of *The Marine Stinger Book*, initially named *Some Australian Marine Stings and Envenomations* (see Plate 1.12).

The fourth Stinger Seminar was organised by Queensland State Centre and the Department of Biological Sciences at James Cook University on 29-30 October 1977. It was again hosted by Professor Burdon-Jones, chaired by Earle Smith (Figure 1.4) and this time had the new faces of Dr Bob Hartwick (Figure 1.7) and Dr Vic Callanan (Figure 1.8), as well as the stalwarts Dr John Williamson and Dr Jack Barnes. The whole scene was again reviewed and updated where necessary. Box jellyfish reporting forms were introduced to gather information for the *Chironex* Research Program. More than 90 people attended the two-day conference; 82 of these even assembled on the Townsville Strand Beach at 6.30am on the second day for a display of netting techniques and other practical skills.

Further seminars were held at Picnic Bay on 10–11 October 1981 and 12 April 1986 at the Australian Institute of Marine Science (AIMS) under the chairmanship of Dr Joe Baker, OBE, the then Director of AIMS. This saw some of the leading authorities in Australia introduce the large amount of knowledge that had been gleaned during the previous decade. It is hoped that these seminars will be repeated in the coming years.

Summary of the Major Early Advances in Treatment

- The introduction of cardiopulmonary resuscitation (CPR) in 1964, with the combination of expired air resuscitation (EAR) and external cardiac compression (ECC) was a milestone in the treatment of *Chironex fleckeri* stings.

Figure 1.4
Earle Smith.

Figure 1.5
Ron Pollard.

Figure 1.6
John Williamson.

Figure 1.7
Bob Hartwick.

- The Queensland State Government Stinger Grant first came into regular operation in 1974. It consisted of $4000 for the first three years, increasing to $7000 in 1976, $10 000 in 1980, $11 770 in 1981, $14 400 in 1982, $15 700 in 1983 and $17 350 in 1984.
- Advanced Resuscitation Certificates (ARCs)* incorporating the training and use of positive pressure oxygen using the 'Oxy-Viva' Mark II, were introduced in 1974–75.
- Antivenom had been originally produced by the Commonwealth Serum Laboratories in 1970 (BAXTER *et al* 1968; BAXTER & MARR 1974; SUTHERLAND 1983, 1994) {see Chapter 21, page 420}. It was placed in First Aid Rooms in surf life saving clubs in 1979, along with a special 'injection kit'. It was available only for the use of surf life savers who held a current ARC and were trained in the technique of intramuscular injections. The first such injection courses were conducted in 1979 at the Townsville General Hospital and the Townsville Ambulance Centre under the supervision of Dr Vic Callanan and Dr John Williamson.
- The use of vinegar as a first aid treatment for *Chironex fleckeri* stings was introduced in 1980 by Dr Bob Hartwick. While experimenting with tentacles from *Chironex fleckeri* he observed that methylated spirits caused a mass firing of the nematocysts. Field research was carried out in Gove, NT, and Townsville, Qld, when volunteers (including surf life savers, and Drs Barnes, Callanan, Hartwick, Unwin and Williamson) were stung under controlled conditions — an experimental procedure not currently recommended by many scientists and all Ethics Committees! Various treatments were assessed in the laboratory and in the field, leading to the classic paper by HARTWICK and colleagues in the *Medical Journal of Australia* in 1980. This work demonstrated that 2–10% acetic acid in water (including household vinegar which is 4–6% acetic acid in water) irreversibly inhibited stinging capsule (nematocyst) discharge in *Chironex fleckeri* and a number of other species of box jellyfish. This would prevent further envenomation of the victim from remaining adherent tentacle material on the skin, the usual occurrence with *Chironex fleckeri* sting victims {Plate 9.76}. A paper by the same authors in the *Medical Journal of Australia* described a case where a pregnant woman was saved on a Townsville beach by EAR alone, without immediately available antivenom, after a severe *Chironex fleckeri* sting. These two papers exerted probably the greatest single influence on the Surf Life Saving Association at the time. After a meeting between the Queensland State Centre representatives, Dr Jack Barnes and the three authors of these two articles, the use of methylated spirits was abandoned and vinegar was adopted as the standard Australia-wide first aid treatment for inhibition of box jellyfish nematocyst discharge.

FIGURE 1.8
Vic Callanan.

- In 1980 there appeared the most important modern advance in the first aid management of envenomations of most kinds, with the Australian discovery of the compression/immobilisation bandaging technique (SUTHERLAND *et al* 1979; SUTHERLAND & DUNCAN 1980; SUTHERLAND 1983). This technique was shortly afterwards advocated for use in major *Chironex fleckeri* stings (WILLIAMSON 1981) {see Chapter 9, page 280, and Plate 5.15}, but although currently recommended and used (FENNER *et al* 1989b; BEADNELL *et al* 1992), experimental proof of its efficacy for jellyfish envenomation is still lacking.
- The Queensland Ambulance Service now has ready access to *Chironex fleckeri* antivenom and has used it successfully at sting sites on the beaches, on several occasions (FENNER *et al* 1989; BEADNELL *et al* 1992). As mentioned above, research is now taking place that may result in further improvements in marine antivenoms (ENDEAN *et al* 1993; COLLINS *et al* 1993) {see Chapter 21, page 420}.

* ARC also may stand for, and should not be confused with, the Australian Resuscitation Council.

STINGER BOOKS

In 1970 Dr John Williamson came to Townsville, in northern Queensland, and became associated with Earle Smith in Brisbane and the Queensland Surf Life Saving Association, particularly with regard to resuscitation of the near-drowned. A natural extension of these studies was in the area of marine stings and their treatment. An original idea of Maurie Mulcahy was brought to fruition by John and Earle with publication of the first edition of *Some Australian Marine Stings and Envenomations* by the Queensland State Centre in June 1974 (Plate 1.12).

To keep this publication in line with the increasing knowledge in the subject, it was updated by John Williamson in 1981, and entitled somewhat tortuously *Some Australian Marine Stings, Envenomations and Poisonings* (Plate 1.13). With the aid of the large Stinger Grant arranged through the Queensland State Centre of Surf Life Saving, the book was greatly enlarged and clearly illustrated with some of the best photographs available. It proved to be so popular that its stocks were quickly exhausted.

The third edition, again written by John Williamson and edited by David Exton, the Education Officer for the Queensland State Centre (and a contributor to this book), was released in September 1985. For simplicity, and at the perceptive suggestion of the SLSA rank-and-file beach patrol members, it was renamed *The Marine Stinger Book* (Plate 1.14). It has proven to be an up-to-date and well-used source on some aspects of marine stings, a publication of which Surf Life Saving Queensland can be justly proud. The third edition received financial assistance from the Rotary Foundation. John was awarded life membership of North Barrier Branch and then Surf Life Saving Queensland in 1995.

FIGURE 1.9
Peter Fenner.

Another landmark event occurred when Dr Peter Fenner (Figure 1.9) joined the SLSA marine envenomation research group in 1983 with his additional stimulus a period of intense activity for marine envenomation studies has occurred. A highlight was the development of collaboration between Surf Life Saving Queensland with its marine envenomation medical group, and the dynamic Professor J.W. (Joe) Burnett, MD (Figure 1.10) and his team in Baltimore, Maryland, USA. In 1983, following his letter to the *Medical Journal of Australia* concerning the possible role of verapamil in the acute management of *Chironex fleckeri* stings (BURNETT & CALTON 1983), Professor Burnett, whose internationally renowned research team had been taking increasing interest in marine toxinology worldwide, was contacted by John Williamson and Peter Fenner. Thus began a happy and fruitful international alliance, which continues unabated and has resulted in the publication of some 19 joint papers and with Dr Jacquie rifkin the medical editorship of this book. Burnett established the International Consortium for Jellyfish Stings (ICJS) {Chapter 23}, and all three colleagues have established international contacts with interested colleagues, a mailing list, a regular newsletter prepared in Baltimore, and an annual Sting Report Summary prepared in Australia (INTERNATIONAL CONSORTIUM FOR JELLYFISH STINGS 1993).

FIGURE 1.10
Joe Burnett.

Dr Jacquie Rifkin (Figure 1.11) began her research into the biological aspects of venomous and poisonous animals when she arrived in Australia in 1972, whereupon she was immediately sent to capture and study the box jellyfish (*Chironex fleckeri*) in north Queensland. During her 20 year career at the University of Queensland, Dr Rifkin was involved in the development and teaching of several practical and theoretical courses related to toxic marine and terrestrial animals. For many years her special interest lay in studying the biology of cnidae (especially nemato-

FIGURE 1.11
Jaquie Rifkin.

cysts) from a wide range of medically important cnidarians, including the box jellyfish. Some of this work has been published previously with Dr Robert Endean and is included in the bibliography. In addition, several previously unpublished observations are recorded in this text and many have important implications for first aid management of cnidarian stings.

Her assistance with research into first aid measures for the bluebottle (*Physalia*) stings was first sought in 1987 by SLSQ and some of the extensive work carried out on the beaches and in her home laboratory up to 1995 have involved:

- studies of nematocyst discharge in all classes of Cnidaria including the testing of a large number of reagents for use as potential inhibitors of further nematocyst discharge after cnidarian stings.
- research on analgesics appropriate for application after *Physalia* stings. After extensive testing of a variety of substances including local anaesthetics, hot packs and cold packs, she discovered that cold packs and ice were best to use on the beach after painful stings from the local bluebottle (*Physalia* sp). Further testing by SLSQ members was carried out and its results published, and cold packs are still in use today.
- research on taxonomic problems involving local carybdeid medusae (e.g. the morbakka and the little 'Darwin carybdeid'), bluebottles, and chirodropids of the tropical Indo-Pacific region. The preliminary results are summarised in the text (see Chapters 6, page 121, and 9, page 236).

In 1990 Surf Life Saving Queensland published the first edition of Peter Fenner's *Marine Stinger Guide* (Plate 1.15).* This further important publication for surf life savers treats *The Marine Stinger Book* material in a simplified and illustrated 'quick reference' style, aimed straight at the on-the-beach first aider. Subsequent editions of *The Marine Stinger Guide* will appear with the further evolution of *Venomous and Poisonous Marine Animals*.

The 'library' of SLSA documentation on marine envenomation continues to expand. In 1990, in a similarly simplified and clearly illustrated manner, the 'Marine Stinger' chart (Plate 1.16), which helps life savers to identify jellyfish, and the 'Stinger Treatment' chart (Plate 1.17) were each produced for Queensland Surf Life Saving by Peter Fenner, with the sponsorship of Rotary International (Brisbane), Hastings Deering and Castlemaine Perkins.

The first Marine Stinger Offi cer

In 1978 the first Marine Stinger Officer was appointed to the Queensland State Centre of the then Surf Life Saving Association of Australia. He was Ron Pollard of the Cairns Club (see Figure 1.5 page 48), who had a long and close association with Jack Barnes and had extensively assisted him with his field work into stinger research. Ron became one of the world's foremost authorities on the field skills of box jellyfish research, and his contribution to Dr Jack Barnes's successes was attested to again and again by Jack himself.

Under the guidance of Ron Pollard the 'Stinger Drill' charts were produced with the financial assistance of Lions International. He was also responsible for conveying to Queensland State Centre any progress in the field. Ron remained the Marine Stinger Officer until 1985 when he retired because of business commitments. For his extended work and dedication to his duties he was awarded life membership of the Queensland Surf Life Saving Association in 1986.

Stinger-resistant enclosures

Fixed swimming enclosures had been present in northern Queensland for many years. These consisted mainly of large posts embedded in the sand, supporting wire netting and usually

* NOTE by JWB and JFR: *The Marine Stinger Guide* (FENNER 1990a) reflects the dynamic and innovative influence of Peter Fenner on the practical aspects of marine envenomation, first aid treatment, and surf life saving. Peter Fenner has immersed himself in the surf life saving movement, is a practising, and senior examining life saver (and former St John Ambulance Australia examiner), is internationally known from his ever-increasing clinical publications and research in the field, and has developed special interests in the applied taxonomy of jellyfish.

replaced each year before the arrival of jellyfish in the summer months. A floating pontoon catwalk had also been tried in the Cairns area in the early 1970s. This enabled swimmers to walk from the beach to deeper water, thus avoiding the shallow waters where box jellyfish were known to be most prevalent (BARNES 1960). It is now appreciated that *Chironex* may be present in deeper waters, with the consequent danger to people jumping or diving on the unseen animal from a boat or jetty (INTERNATIONAL CONSORTIUM FOR JELLYFISH STINGS 1993).

Innovative research into suitably designed safe swimming enclosures, commenced by the late Professor Kevin Stark of the Department of Civil and Systems Engineering, James Cook University of North Queensland, led to the development of stinger-resistant enclosures (Plate 1.18). A trial net was set up at the Australian Institute of Marine Science (AIMS) in May 1982 and its effectiveness led to the purchasing of 13 nets now in use in northern Queensland. These were very effective in keeping out large specimens such as *Chironex fleckeri*, but the original designs quickly showed certain limitations, when 36 people were stung inside such a net at Trinity Bay, Cairns by the tiny Irukandji (*Carukia barnesi*)* on Christmas Day 1985 — something that was to happen in most other areas where the early design nets were installed.

The nets have now been modified by Kim Moss of the James Cook University to a smaller mesh size. They consist of a floating tube that can extend out into the water some 75 metres, parallel to the shore for 100 metres, and 75 metres back to shore; a fine mesh net hangs underneath the floating tube and is weighted to sit on the sea bottom (Plate 1.19). Unfortunately, initially this led to increased clogging of the nets with normal marine detritus, and they had to be defouled daily. This in turn led to an effective power-winch system (Plate 1.20) developed by the same Department at James Cook University, and the system now permits 'beaching' of the nets for cleaning. The winches, driven by electric or fuel engines, enable one person (a lifeguard) to operate and maintain the net. Such nets are now being developed internationally (BURNETT *et al* 1994).

STINGER SUITS

Following the lead given by Jack Barnes's pantyhose concept of protective clothing (described above), and conscious of the need for a more durable and appealing garment for life savers and swimmers in northern Australia, a small group met in Townsville in 1980 (Mike Ball, Vic Callanan, Bob Hartwick and John Williamson) to try to design a protective suit that would also prevent sunburn in northern Australia's fierce summer sun (WILLIAMSON 1981). A Townsville company specialising in ballet costumes made the first of what became known as 'stinger suits' to this group's specifications (Plate 1.21). These suits required the development of new sewing techniques, for the thin, two-way-stretch material. Eventually John Farmer of Sydney undertook the necessary practical research, and nylon/lycra stinger suits, with bright new designs, have become widely promoted and used for many water activities (Plates 1.22 and 1.23). The group never pursued a patent! {See also Chapter 5, page 103.}

Without some form of protective clothing, the survival of the competitive side of surf life saving in northern Queensland would have been threatened.

SUPPORT FROM THE QUEENSLAND STATE GOVERNMENT

At the same time as association members Dr Peter Fenner and Dr John Williamson and Marine Stinger Officers Ron Pollard and Maurie Mulcahy were invited to join the Queensland Government Task Force into Box-Jellyfish, the government granted $500 to

* *Carukia* was small enough to get through the initially somewhat-large gauge mesh of the nets {Plate 9.17}, as were sea snakes. After feeding well on trapped fish, the sea snakes became too big to fit back through the mesh to get out again and became trapped inside the nets. These events occasionally resulted in considerable unplanned activity among swimmers in the enclosures!

each North Queensland, North Barrier and South Barrier Surf Life Saving Clubs, to purchase lycra stinger suits for members on patrol (Plate 1.23).

To the present day...

Field and laboratory research continues and is becoming international, and this book contains testimony to such developments. Many surf clubs provide documented monthly information on jellyfish numbers and stings treated, and association members John Williamson and Peter Fenner, together with Professor Joe Burnett, collate data on stings from all round Australia and what continues to be available from the rest of the world. Informative cases are documented and frequently published in peer-reviewed scientific journals. All reports are now entered on a computer data base from which helpful information and trends can be collated (International Consortium for Jellyfish Stings 1993).

Surf Life Saving Queensland's Historical Calendar

Peter Fenner

1936 Formation of North Barrier Branch in Queensland, and surf life saving clubs from Mackay to Cairns.

1964 A 'special' North Barrier SLSA meeting to address problems of serious jellyfish stings and deaths. Dr Jack Barnes of Cairns enlisted to help SLSA;

1967 Queensland State Centre, Surf Life Saving Association (SLSA), formal involvement began.

1971 First 'Stinger Conference' which introduced:
- Wearing of pantyhose as protection
- 'Netting' of beaches for *Chironex fleckeri*
- Use of methylated spirits on jellyfish stings
- 'Spotting' technique to identify jellyfish from inflatable rescue boats (IRBs)

1972 Formation of Queensland State Centre's 'Stinger Committee'; formation of Queensland State Centre's 'Medical Advisory Committee'; second 'Stinger Conference'.

1973 Introduction of Queensland government 'Stinger Grant' which enabled issue of SLSA warning cards and posters about dangerous jellyfish. It also funded purchase of IRBs for 'spotting', as well as of drag nets, 'Oxy-vivas', brochures, books, wall charts and other audio-visual teaching aids on marine envenomation.

1974 First edition of Queensland SLSA's book *Some Australian Marine Stings and Envenomations* by Dr John Williamson, with the collaboration of Earle Smith and Jack McMaster, capitalising on the initial work of Maurie Mulcahy, Marine Stinger Officer, North Barrier Branch, SLSA.

1979 Introduction of Commonwealth Serum Laboratories' *Chironex fleckeri* antivenom into northern SLSA clubhouses, after teaching of 'injection courses' to senior SLSA holders of the Advanced Resuscitation Certificate.

1980 Volunteer life savers, medical officers and marine biologists in Townsville tested with *Chironex fleckeri* tentacle stings. This work, initiated by Dr Bob Hartwick of the James Cook University, together with Drs John Williamson, Vic Callanan and Marcus Unwin, led to the introduction of vinegar sting treatment to patrolled beach areas in northern Australia (Hartwick *et al* 1980).

1981 Second edition of *Some Australian Marine Stings, Envenomations and Poisonings* by John Williamson.

1985 Retirement of Ron Pollard of Cairns, Marine Stinger Officer, Queensland Surf Life Saving. He was presented with SLSA life membership for services rendered in the field.
Dr Peter Fenner of Mackay appointed as the new Marine Stinger Officer.
Publication (book launching by Professor Cyril Burdon-Jones and Mr Jack McMaster) of the third edition of *The Marine Stinger Book*, written by Dr John

Williamson and edited by David Exton; it was renamed *The Marine Stinger Book* at the request of practising Australian surf life savers, who coined this practical term.
Formation of a Queensland Government Task Force to inquire into Box Jellyfish, chaired by Dr J. T. (Joe) Baker OBE, then Director of the Australian Institute of Marine Science. Queensland SLSA. Medical Advisory Committee members Drs John Williamson, Vic Callanan and Peter Fenner, along with SLSA Stinger Officers Ron Pollard and Maurie Mulcahy, formed the largest single contingent in the Task Force.

1986 'Stinger Seminar' at Australian Institute of Marine Science chaired by its Director, Dr Joe Baker, who was later elected Vice-Patron of Surf Life Saving Queensland.
Introduction of patrols in 'Stinger-resistant enclosures'.
Queensland Government grant to northern surf clubs for five stinger suits per club for patrols.

1987 Acceptance by Queensland Education Department of Drs Fenner and Williamson's SLSA computer-assisted programme for teaching of 'Awareness, prevention and treatment of box jellyfish stings'.
Peter Fenner started trials of compressive bandaging for 'Irukandji' stings in the North Queensland Branch, SLSA. Further work on trials begun by Dr Jacquie Rifkin and David Exton using ice or cold packs for the skin pain of other stings.
Commencement of Marlin Coast Lifeguard Service (northern Queensland, Cairns area) operating jet boat patrols of stinger nets and of other beaches, with Lee Pardy as on-site officer;
Meeting of Surf Life Saving Queensland with James Cook University, and suggestion of sites of 'stinger-resistant enclosures' with a fully trained lifeguard to operate them.

1988 Withdrawal of *Chironex fleckeri* antivenom from north Queensland Surf Life Saving clubs because of on-going costs, and because of the demonstrable effectiveness of SLSA beach patrols there. No antivenom had been required to be administered by any ARC surf life saver during the previous 12 years.

1989 Publication by David Exton, Education Officer, and Drs Peter Fenner and John Williamson, Queensland State Centre, SLSA, of work initiated in 1987 by Dr Rifkin and Dr Robert Endean, on relief from skin pain of 'bluebottle' (*Physalia* species) stings using cold packs.
Queensland Surf Life Saving life membership awarded to Maurie Mulcahy for 20 years of distinguished work as Marine Stinger Officer, North Barrier Branch.
Marine envenomation section of SLSA Training Manual rewritten by Marine Stinger Officer, Dr Peter Fenner.

1990 Production of wall charts for 'Stinger Treatment' and 'Marine Stinger Identification', and the first edition of *The Marine Stinger Guide* by Dr Peter Fenner, Marine Stinger Officer, SLSA Queensland.

1991 Production of two teaching videos on (1) identification and treatment of marine stings and (2) marine envenomation, followed by the second Marine Stinger Set of slides by Dr Peter Fenner.
Presentation of SLSA Innovation Awards to Queensland Surf Rescue and to Dr Peter Fenner for *The Marine Stinger Guide*.

1992 Work on of *Venomous and Poisonous Marine Animals* commenced.

1993 Completion of a preliminary trial of compression bandages for Irukandji sting — no conclusive results were obtained. (Further research is planned.)

1995 Drs Vic Callanan and John Williamson awarded life membership of North Barrier Branch, SLSA. Dr John Williamson awarded life membership of Surf Life Saving Queensland.

1996 *Venomous and Poisonous Marine Animals* published.

REFERENCES TO CHAPTER 1

AUERBACH P S. 1987a. *A Medical Guide to Hazardous Marine Life.* Jacksonville, Florida. Progressive Printing Co. 47 pp.

AUERBACH P S. 1987b. Natural microbiologic hazards of the aquatic environment. In: (R M Mandojana, guest ed.) Aquatic dermatology. *Clinics in Dermatol* 5(3): 52–61

AUERBACH P S. 1995. Wilderness Medicine: *Management of Wilderness and Environmental Emergencies*, 3rd edn. (P Auerbach, ed.). St Louis, Missouri. Mosby. 1506 pp.

BARNES J H. 1960. Observations on jellyfish stingings in North Queensland. *Med J Aust* 2: 993–999.

BARNES J H. 1964. Cause and effect in Irukandji stingings. *Med J Aust* 1: 897–904.

BARNES J H. 1965. *Chironex fleckeri* & *Chiropsalmus quadrigatus* — morphological distinctions. *North Queensland Naturalist* 32: 13–22.

BARNES J H. 1966. Studies on three venomous cubomedusae. In: *The Cnidaria and their Evolution: Symposium of the Zoological Society of London. London.* Academic Press. No 16. pp307–332.

BARNES J H. 1967. Extraction of cnidarian venom from living tentacle. In: (F E Russell, P R Saunders, eds) *Animal Toxins: A Collection of Papers presented at the First International Symposium on Animal Toxins, Atlantic City, New Jersey*, 9–11 April, 1966. New York. Pergamon Press. pp115–129.

BAXTER E H, MARR A G M. 1974. Sea wasp (*Chironex fleckeri*) antivenene: neutralising potency against the venom of three other jellyfish species. *Toxicon* 12: 223–229.

BAXTER E H, MARR A G M, LANE W R. 1968. Immunity to the venom of the sea wasp (*Chironex fleckeri*). *Toxicon* 6: 45–50.

BEADNELL C, RYDER T, WILLIAMSON J, FENNER P. 1992. Management of a major box jellyfish (*Chironex fleckeri*) sting: lessons from the first minutes and hours. *Med J Aust* 156: 655–658.

BURNETT J W, CALTON G J. 1977. Review article: The chemistry and toxinology of some venomous pelagic coelenterates. *Toxicon* 15: 177–196.

BURNETT J W, CALTON G J. 1984. Box-jellyfish cardiotoxin and i.v. verapamil. *Med J Aust* 140: 50.

BURNETT J W, CALTON G J. 1987. Jellyfish envenomation syndromes updated. *Annals Emerg Med* 16: 1000–1005.

BURNETT J W, CALTON G J, FENNER P J, WILLIAMSON J A. 1988. Serological diagnosis of jellyfish envenomations. *Comparative Biochem Physiol* 91C: 79–83.

BURNETT J W, OTHMAN I B, ENDEAN R, FENNER P J, CALLANAN V I, WILLIAMSON J A. 1990. Verapamil potentiation of *Chironex* (box-jellyfish) antivenom. *Toxicon* 28: 242–244.

BURNETT J W, WILLIAMSON J A, FENNER P J. 1994. Mononeuritis multiplex after coelenterate sting. *Med J Aust* 161: 320–322.

CLELAND J B, SOUTHCOTT R V. 1965. Injuries to Man from Marine Invertebrates in the Australian Region. Canberra. National Health and Medical Research Council. Special Report series No 12. 282 pp.

COLLINS S P, COMIS A, MARSHALL M, HARTWICK R F, HOWDEN M E H. 1993. Monoclonal antibodies neutralising the haemolytic activity of box jellyfish (*Chironex fleckeri*) tentacle extracts. *Comp Biochem Physiol* 106B: 67–70.

COUTIÈRE H. 1899. Poissons venimeux et poissons vénéneux. *Thèse Agrég.* Paris. Carré et Naud.

CURRIE B. 1992. Jellyfish and top end waters. *Northern Territory Communicable Dis Bull* 1(3): 1–2.

CURRIE B, KHANH D M, ALDERSLADE P, WILLIAMSON J, FENNER P. 1991. Jellyfish envenomation in the Northern Territory of Australia. Programme and Abstracts, 10th World Congress on Animal and Plant Microbial Toxins, 3–8 November 1991, Singapore. (P Gopalakrishnakone, C K Tan, eds) Abstract No. 204: 256.

DORIT RL, WALKER WF, BARNES RD. 1991. *Zoology*. Saunders College Publishing. 1009.

EDMONDS C. 1989. Dangerous Marine Creatures. Sydney. Reed. 192 pp.

EDMONDS C. 1992. Dangerous marine creatures. In: (C Edmonds, C Lowry, J Pennefather, eds) *Diving and Subaquatic Medicine*, 3rd edn. Oxford. Butterworth-Heinemann. pp318–347.

ENDEAN R, MONKS S A, CAMERON A M. 1993. Toxins from the box-jellyfish *Chironex fleckeri*. *Toxicon* 31: 397–410.

ENDEAN R, SIZEMORE D J. 1988. The effectiveness of antivenom in countering the actions of box-jellyfish (*Chironex fleckeri*) nematocyst toxins in mice. *Toxicon* 26: 425–431.

FENNER P J. 1987. Report of Field Trip to Philippines, 1985. Queensland State Centre, Surf Life Saving Australia. Newstead Qld 4006. 20 pp.

FENNER P J. 1990a. *The Marine Stinger Guide*, 1st. edn. Brisbane. Queensland Surf Life Saving Association. pp57.

FENNER P J. 1990b. Medical frontiers in Australian seas. In: (J Pearn, M Cobcroft, eds) *Fevers & Frontiers*. Brisbane. Amphion Press (Department of Child Health Publishing Unit). pp73–84.

FENNER P J, WILLIAMSON J A. 1993. Acute chest pain on a tropical beach: myocardial ischaemia or 'Irukandji' syndrome? Scientific programme, 'The Spark of Life', Melbourne. Australian Resuscitation Council. 66 pp.

FENNER P J, WILLIAMSON J A, BLENKIN J A. 1989. Successful use of *Chironex* antivenom by members of the Queensland Ambulance Transport Brigade. *Med J Aust* 151: 708–710.

FENNER P, WILLIAMSON J, BURNETT J, COLQUHOUN D M, GODFREY S, GUNAWARDANE K, MURTAGH K. 1988. The 'Irukandji syndrome' and acute pulmonary oedema. *Med J Aust* 149: 150–156.

FENNER P J, WILLIAMSON J, CALLANAN V I, AUDLEY I. 1986. Further understanding of, and a new treatment for, 'Irukandji' (*Carukia barnesi*) stings. *Med J* Aust 145: 569–574.

FLECKER H. 1936. Cone shell mollusc poisoning with report of a fatal case. *Med J Aust* 1: 464–466.

FLECKER H. 1945. Injuries by unknown agents to bathers in North Queensland. *Med J Aust* 1: 98.

FLECKER H. 1952. Irukandji sting to North Queensland bathers without production of wheals but with severe general symptoms. *Med J Aust* 2: 89–91.

FLECKER H. 1956. Injuries from stone fish. *Med J Aust* 2: 371–373.

FREEMAN S E. 1974. Actions of *Chironex fleckeri* toxins on cardiac transmembrane potentials. *Toxicon* 12: 395–404.

GAILLARD C. 1923. Reserches sur les poissons représentés dans quelques tombeaux Égyptiens de l'ancien empire. *Mém Inst Français Arch Orient* 51:97–100, Figs 56–57.

GONZAGA R A F. 1985. Mordeduras e Picadas por Animais da Fauna Portuguesa. Porto, Portugal. *Prémio Bial de Medicina Clínica* 1984. 198 pp.

GUDGER EW. 1943. Is the stingray's sting poisonous? A historical resumé showing the development of our knowledge that it is poisonous. *Bull Hist Med* 14: 467–504.

HAECKEL E. 1880. System der Acraspeden. Zweite Halfte des Systems der Medusen. *Denkscher Med-Naturwiss Ges Jena* 1: 361–672.

HALSTEAD B W. 1988. *Poisonous and Venomous Marine Animals of the World* (2nd revised edition). Princeton, NJ. Darwin Press. 40 pp.

HALSTEAD B W, AUERBACH P S, CAMPBELL D. 1990. *A Colour Atlas of Dangerous Marine Animals*. London. Wolfe Medical Publications. 192 pp.

HARTWICK R, CALLANAN V, WILLIAMSON J. 1980. Disarming the box-jellyfish: nematocyst inhibition in *Chironex fleckeri*. *Med J Aust* 1: 15–20.

INTERNATIONAL CONSORTIUM FOR JELLYFISH STINGS. 1993. Some Australian and international marine envenomation reports: progress summary to October 31 1993 (P Fenner, J Williamson, J Burnett, eds.). Adelaide. Hyperbaric Medicine Unit, Royal Adelaide Hospital. 25 pp.

KINSEY B. 1986. Barnes on Box Jellyfish. Townsville. James Cook University of North Queensland, Sir George Fisher Centre for Tropical Marine Studies. 90 pp of tape transcripts.

KINSEY B. 1988. More Barnes on Box Jellyfish. Townsville. James Cook University of North Queensland, Sir George Fisher Centre for Tropical Marine Studies. pp48–59.

LIGHT S F. 1914a. Some Philippine Scyphomedusae, including two new genera, five new species and one new variety. *Philippine J Sci Ser* D9: 195–231.

LIGHT S F. 1914b. Dangerous jellyfish in Philippine waters. *Philipp J Sci* IX8(3): 291–295.

LUMLEY J, WILLIAMSON J A, FENNER P J, BURNETT J W, COLQUHOUN D M. 1988. Fatal envenomation by *Chironex fleckeri*, the north Australian box jellyfish: the continuing search for lethal mechanisms. *Med J Aust* 148:527–534.

MAGUIRE EJ. 1968. *Chironex fleckeri* ('sea wasp') sting. *Med J Aust* 2:1137–1138.

MARETIĆ Z. 1975. *Venomous Animals and Poisonous Animals of the Adriatic Sea*. J.A. Zagreb Univ.

MARSH L, SLACK-SMITH S, with GURRY D. 1986. *Sea Stingers*. Perth. Western Australian Museum. 133 pp.

MARTIN J C, AUDLEY I. 1990. Cardiac failure following Irukandji envenomation. *Med J Aust* 153: 164–166.

MAYER A G. 1910. *Medusae of the World*, Vol 3: The Scyphomedusae. Washington, DC. Carnegie Institution. No 109. i-iv: 499–735.

MEGLITSCH PA, SCHRAM FR. 1991. *Invertebrate Zoology*. Oxford University Press. 623 pp.

MEIER J, WHITE J. 1995. *Clinical Toxicology of Animal Venoms and Poisons*. Boca Raton, Florida. CRC Press. (in press).

PEARN J. 1994. Hugo Flecker's contributions to medicine and natural history. *Med J Aust* 161: 63–70.

PORTIER P, RICHET C. 1902a. De l'action anaphylactique de certains venins. *Compt Rend Soc Biol* 54: 170–172.

PORTIER P, RICHET C. 1902b. Noveaux faits d'anaphylaxie, ou sensibilisation aux venins par doses réitérées. *Compt Rend Soc Biol* 54: 548–551.

QUEENSLAND SLSA RECORDS. 1956. Surf Life Saving Queensland Inc. PO Box 2136, Fortitude Valley Qld 4006.

READ B E. 1939. Chinese materia medica. Fish drugs. *Peiping Nat Hist Bull*. 136 pp.

RICCI A. 1982. Sala absidata - Il pavimento. In: (A Carandini, A Ricci, M de Vos, eds) *Filosofiano la Villa di Piazza Armerina, imagine di un aristocratico romano al tempo di Costantino*. Palermo. S F Flaccovio. (Fig 157, page 260).

RIFKIN J, ENDEAN R. 1983. The structure and function of the nematocysts of *Chironex fleckeri* Southcott, 1956. *Cell Tissue Research* 233: 563–577.

RIFKIN J F, ENDEAN R. 1988. Arrangement of accessory cells and nematocysts bearing mastigophores in the tentacles of the cubozoan *Chironex fleckeri*. *J Morphology* 195: 103–115.

RUSSELL F E. 1975. *Poisonous and venomous marine animals and their toxins*. Ann New York Acad Sci 245: 57–60.

SAFAR P, ESCARRAGA L, ELAM J. 1958. A comparison of the mouth-to-mouth and mouth-to-airway methods of artificial respiration with the chest-pressure arm-lift methods. *New Eng J Med* 258: 671–675.

SCHMIDT H. 1974. Evolution in the Anthozoa. In: AM CAMERON, BM CAMPBELL, AB CRIBB, R ENDEAN, JS JELL, OE JONES, P MATHER, FH

TALBOT (eds)*Proc. Second International Symposium on Coral Reefs*. 533–563.

SOUTHCOTT R V. 1956. Studies on Australian cubomedusae, including a new genus and species apparently harmful to man. *Aust J Marine Freshw Res* 7: 254–280.

SOUTHCOTT R V. 1959. Tropical jellyfish and other marine stingings. *Military Med* 124: 569–579.

SOUTHCOTT RV. 1967. Revision of some Carybdeidae (Scyphozoa: Cubomedusae), including a description of the jellyfish responsible for the 'Irukandji syndrome'. *Aust J Zool* 15: 651–657.

STORER T I, USINGER R L. 1957. *General Zoology*, 3rd edn. New York. McGraw-Hill. 664 pp.

SURF LIFE SAVING QUEENSLAND, INC. 1963–1993. Annual Branch sting report summaries to 1993 (P Fenner). PO Box 2136, Fortitude Valley Qld 4006.

SUTHERLAND S K. 1983. *Australian Animal Toxins: The Creatures, Their Toxins and the Care of the Poisoned Patient*. Melbourne. Oxford University Press. 540 pp.

SUTHERLAND S K. 1994. Antivenom research in Australia. *Med J Aust* 161: 48–50.

SUTHERLAND S K, COULTER A R, HARRIS R D. 1979. The rationalisation of first-aid measures for elapid snakebite. *Lancet* 1: 183–186.

SUTHERLAND S K, DUNCAN A W. 1980. New first-aid measures for envenomation: with special reference to bites by the Sydney funnel-web spider *Atrax robustus*. *Med J Aust* 1: 378–379.

VIMPANI G, DOUDLE M, HARRIS R. 1988. Child accident-mortality in the Northern Territory, 1978–1985. *Med J Aust* 148: 392–395.

WILLIAMSON J. 1981. *Some Australian Marine Stings, Envenomations and Poisonings*, 2nd. edn. Brisbane. Surf Life Saving Association of Australia, Queensland State Centre. p. 23.

WILLIAMSON J A. 1983. Children and box-jellyfish in north Australia. In: (J Pearn, ed.) *Accidents to Children: Their Incidence, Causes and Effects*. Melbourne. Child Accident Prevention Foundation of Australia. Royal Australasian College of Surgeons. pp93–102.

WILLIAMSON J A. 1992. Current challenges in marine envenomation: an overview. *J Wilderness Med* 3: 422–431.

WILLIAMSON J A, CALLANAN V I, HARTWICK R F. 1980. Serious envenomation by the northern Australian box-jellyfish (*Chironex fleckeri*). *Med J Aust* 140: 13–15.

WILLIAMSON J A, CALLANAN V I, UNWIN M L, HARTWICK R F. 1984a. Box-jellyfish venom and humans. *Med J Aust* 1: 444–445.

WILLIAMSON J A, Le RAY L E, WOLFHART M, FENNER P J. 1984b. Acute management of serious envenomation by box-jellyfish (*Chironex fleckeri*). *Med J Aust* 141: 851–853.

CHAPTER 2

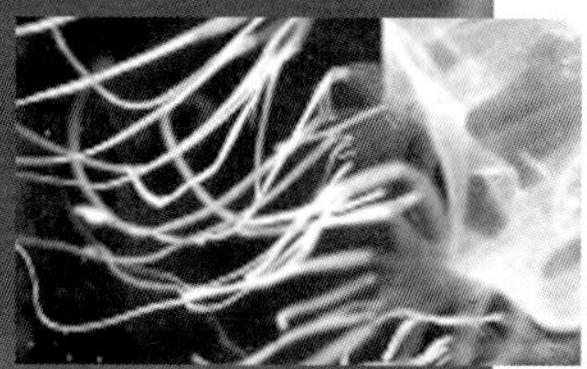

ANIMAL TAXONOMY

JACQUIE RIFKIN

While detailed knowledge of the classification of the animal kingdom is hardly a prerequisite for successful first aid and medical management of marine animal injuries, a broad appreciation of the taxonomy of marine animals helps us to recognise the patterns of injuries that may result and assists in the application of good medical management principles. Taxonomic knowledge also facilitates epidemiological understanding, which is important for injury prevention and environmental conservation.

This chapter does not pretend to cater for serious professional biologists, who are referred to specialised texts in their own areas of interest. This chapter and Chapter 6 serve as an introduction, (albeit containing some highly specialised terminology!) to get the interested beginner started. Some more detailed information is also presented at the beginning of certain other sections and chapters {e.g. class Cubozoa, page 236}.

TAXONOMY FOR BEGINNERS

Most of us are used to referring to animals and plants with a common name, such as 'garden snails', 'jellyfish', 'sea wasp' or 'sea horses'. Certain names, such as 'the garden snail', describe to some degree the specimen in question, even giving general details on where the animal is located, but others, such as 'jellyfish' and 'sea wasp' and 'sea horse' are quite unhelpful and even misleading. For example, the term 'sea wasp' could imply one is referring to a stinging marine insect {see page 236}. Common names are not useful as scientific tools; a scientist in one part of the world must understand exactly to which organism another scientist somewhere else is referring. It is unhelpful simply to use a term such as 'jellyfish' when there is a multitude of different types distributed in all parts of the world and when details of the signs, symptoms and medical treatment of their separate stings need to be communicated widely. In order to communicate biological ideas effectively it is essential to use accurate scientific names. Hence the name *Conus textile* describes only one type of cone shell out of several hundred possible types {Plate 13.5} whenever it is found.

Taxonomy is the science of classifying organisms, and it reflects the manner in which scientists have attempted to pigeonhole the great diversity of living organisms. Use of scientific names is heavily regulated. Not only must every name stand for one particular sort of organism, everywhere that it is found. It must be described with a scientific standard so that anyone can make reference to it.

Carl von Linné (1707–1778), a Swedish naturalist, professor and physician, whose name was Latinised as Carolus von Linnaeus, developed the binomial system (a scientific name in two parts) for naming living organisms. The first part is the **generic name**, or the name of the genus to which the species is referred; the genus is the equivalent of a surname. The second is the species or **specific name**, which is the equivalent of a given name. The generic and specific parts together comprise the name of the species (e.g. *Chiropsalmus quadrigatus*). Scientific names are always written in lower case (except for the first letter of the genus) and printed in *italics* (or underlined) in script. Throughout this text italics are used for names of species.

When multiple references are made to the same organism it is common practice to abbreviate the name of the genus, thus '*C. quadrigatus*' for *Chiropsalmus quadrigatus*, as long as reference is not made to another genus starting with the same initial letter 'C' in the same description. The specific name (e.g. '*quadrigatus*") is never abbreviated. Where a worker knows the genus but cannot identify a species, it is usual to write '*Chiropsalmus* sp.'. If several species are referred to collectively, it is common to write '*Chiropsalmus* spp.'.

In scientific writing, the first time a species is mentioned, the name of the person who first described it should be included, as in

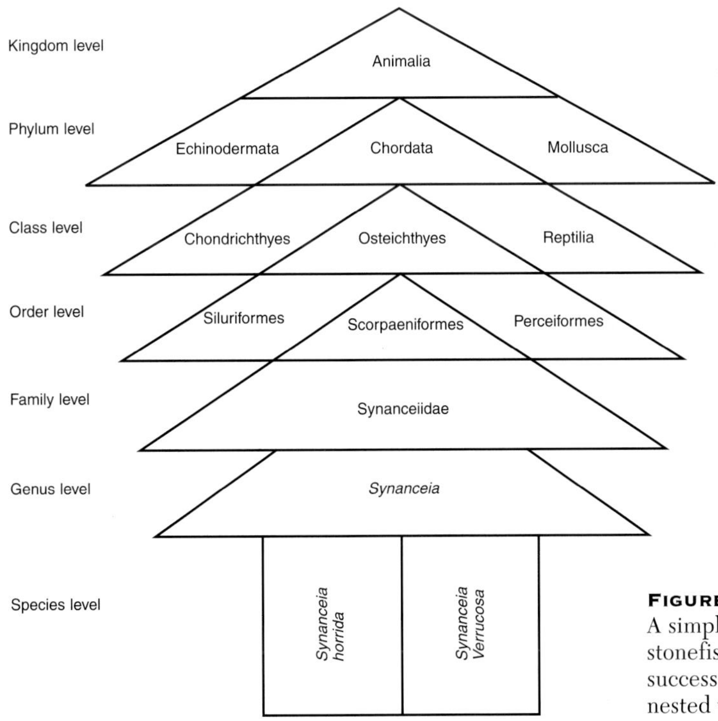

FIGURE 2.1
A simple taxonomic scheme for stonefish showing how each successive level of taxon is nested within the last one.

Chiropsalmus quadrigatus HAECKEL.* The date of publication is also sometimes included after the author's name, but this is optional. However, when the date is included, the author and the date are separated by a comma, as in *Chiropsalmus quadrigatus* HAECKEL, 1880. When the author's name appears in parentheses, as in *Gonionemus vertens* (AGASSIZ), this means that the author originally assigned the species '*vertens*' to some genus other than '*Gonionemus*' and '*vertens*' was subsequently reassigned to '*Gonionemus*'.

Classification of living organisms produces a hierarchical system, consisting of groups within groups (Figure 2.1, page 59). As one reads down the actual categories of names (the 'taxa') increasingly more information may be learned about a particular organism. The higher taxonomic categories are often used in discussions about animal relationships, and they help taxonomists to catalogue the animals with which they are working (Table 2.2).†

* Within the body of this text the following reference convention has been adopted for all authors, including persons who first described particular species: When their publication is listed in this book's bibliography, the surname only (except for identical surnames of different authors referenced for the same topic) is written entirely in UPPER CASE, followed by the year of the publication. Authors' names appearing in lower case within the body of the text (whether following species names or anywhere else) indicate that that particular publication is not contained in this text's bibliography.

† It is common for biologists to use the vernacular for taxonomic groups, e.g. 'cnidarians' when referring to members of the class Cnidaria, or 'chirodropids' for members of the order Chirodropidae.

Reading up the taxa, **species** that are closely related are grouped into **genera**. Genera that are closely related are grouped into **families**. In the Latinised manner of animal nomenclature each family name ends in (has the suffix) '-idae'. Similar families are in turn grouped into **orders**; these are based on stable and very general characters. Characteristics uniting orders into **classes** are often few and are usually generalised. The **phylum** (pl. **phyla**) is the next highest grouping. Each **phylum** represents a distinct line of evolution in animals and is characterised by very basic features. Some phyla are small, comprising only a few species while others are enormous. **Kingdom** is a taxonomic grouping of related or similar phyla, and represents the highest-level category in biological classification. The kingdom Animalia consists of multicellular organisms requiring complex organic substances for food, and the food is usually ingested.

This system of categorising living organisms by hierarchy is very efficient. Further, the system of classification based on binomial nomenclature is an effective means of allowing biologists to condense information.

CLASSIFICATION HIERARCHY

The hierarchical classification system used in taxonomic work, and the application of the binomial system for naming living organisms is outlined by an example in Figure 2.1, page 59. Table 2.1 lists the main categories (phyla) of marine animals known to be implicated in human injuries worldwide. Tables 2.2 and 2.3 respectively list examples of such invertebrate

TABLE 2.1
SOME TOXIC OR POTENTIALLY TOXIC MARINE ANIMALS

Phyla:	PROTOZOANS	PORIFERA	CNIDARIA	PLATYHELMINTHES	NEMERTEA	ANNELIDA	ARTHROPODA	MOLLUSCA
Examples:	dinoflagellates*	sponges	jellyfish	flatworms	ribbon worms	segmented worms	crabs	cones
Phyla:	ECHINODERMATA	CHORDATA						
Examples:	sea urchins	sea snakes						

* NOTE: Dinoflagellates are unicellular organisms with characteristics of plants as well as animals. DNA studies have revealed that they are derived from a variety of organisms, and a phylum name cannot by attributed to the group at this time (J.E.N. Veron 1995, personal communication).

TABLE 2.2
SOME INVERTEBRATES ASSOCIATED WITH HUMAN INJURIES

Protozoans

* Dinoflagellates e.g. *Gambierdiscus toxicus* (associated with ciguatera poisoning)

Phylum PORIFERA

Class Demospongiae: sponges

Phylum CNIDARIA (formerly Coelenterata)

Class Hydrozoa:
  Subclass Siphonophora
    Order Cystonectae: Portuguese man-o'-war, *Physalia*
  Subclass Hydroida
    Order Milleporida: fire corals, *Millepora* spp.
    Order Anthomedusae/Athecata: little red stinger, *Leuckartiara*
    Order Limnomedusae: e.g. the orange-striped jellyfish, *Gonionemus vertens*
    Order Leptomedusae/Thecata -
      Family Aglaopheniidae - stinging hydroids e.g. *Lytocarpus, Aglaophenia* *
Class Scyphozoa:
    Order Coronatae: e.g. thimble jellyfish, *Linuche unguiculata*
    Order Semaeostomeae: e.g. hair jelly, *Cyanea capillata*
    Order Rhizostomeae: e.g. cabbagehead jellyfish, *Stomolophus meleagris*
Class Cubozoa:
    Order Carybdeidae: e.g. *Tamoya haplonema, Carybdea* spp., *Tripedalia* spp., *Carukia* sp.
    Order Chirodropidae: e.g. box jellyfish *Chironex fleckeri*
Class Anthozoa:
    Order Actiniaria: sea anemones
    Order Ceriantharia: tube anemones
    Order Corallimorpharia: e.g. *Discosoma* spp.
    Order Zoanthinaria: zoanthids, *Palythoa* spp.

Phylum PLATYHELMINTHES

Class Turbellaria:	free-living flatworms

Phylum NEMERTEA (ribbon worms)

Class Anopla:	toxic, without venom apparatus
Class Enopla:	toxic, some with venom apparatus

Phylum ANNELIDA

Class Polychaeta:	marine sandworms, fire worms, bristle worms
Class Hirudinea:	leeches

Phylum ARTHROPODA

Class Merostomata:	horseshoe crabs
Class Malacostraca:	shrimps, lobsters, crayfish and reef crabs
Class Cirripedia:	barnacles

* See note in Table 2.1

TABLE 2.2 (CONTINUED)
SOME INVERTEBRATES ASSOCIATED WITH HUMAN INJURIES

Phylum MOLLUSCA	
Class Gastropoda:	whelks, cone shells, abalone, nudibranchs (e.g. *Glaucus* spp.)
Class Bivalvia :	bivalves (e.g. clams, oysters, mussels, scallops)
Class Cephalopoda:	octopus, squid, cuttlefish
Phylum ECHINODERMATA	
Class Asteroidea:	sea stars
Class Echinoidea:	sea urchins
Class Holothuroidea:	sea cucumbers

NOTE: Formidable difficulties presently beset the taxonomy of the chirodropids {see Chapter 6 page 150}. Other taxonomy was current at the time of publication, so far as the author was aware.

TABLE 2.3
SOME MARINE CHORDATES ASSOCIATED WITH HUMAN INJURIES

Phylum CHORDATA

- Subphylum Urochordata: sea squirts
- Subphylum Cephalochordata: lancelets
- Subphylum Vertebrata
  - Class Chondrichthyes: rays, sharks, skates
  - Class Osteichthyes: catfish, scorpionfish, surgeonfish, stonefish, mi mi, including certain fish that cause ciguatera poisoning
  - Class Reptilia: sea snakes
  - Class Mammalia: mammals
    - *subclass Prototheria: monotremes e.g. duck-billed platypus
    - subclass Theria: typical mammals e.g. whales, sea lions

*The duck-billed platypus, *Ornithorhynchus anatinus* frequents rivers and streams in Australia. It is not a marine animal.

(without backbones) and chordate (with backbones present at some stage of the life cycle) marine animals. This text does not address all phyla so listed, but it does consider phylum Cnidaria in some detail (Section II, page 236).

Acknowledgments: The author thanks Dr Richard Willan, Curator of Molluscs at the Museums and Art Galleries of the Northern Territory, for helpful comments on the manuscript, and for information concerning molluscan taxonomy. She also gratefully acknowledges Dr Paul Cornelius, Curator of Cnidarians at the British Museum of Natural History, London, for information on the current taxonomy of the class Hydrozoa, and Dr John ('Charlie') Veron, Senior Scientist at the Australian Institute of Marine Science in Townsville, for information concerning the taxonomy of the anthozoans and the dinoflagellates.

CHAPTER 3

The Injuries, their Incidence and the Toxins that Produce Them

John Williamson,
Peter Fenner and
Joe Burnett

Fatalities from Marine Toxins

In previous decades the emphasis on specific 'dangerous' marine animals has been determined by the number of human deaths attributed to them. Many of these attributions were anecdotal, undocumented, unwitnessed and repeated uncritically from text to text; not surprisingly more than a few of these 'deaths' evaporated when traced to their origins (FLECKER 1936; WILLIAMSON 1992). HALSTEAD (1965, 1967, 1970, 1988) and CLELAND & SOUTHCOTT (1965), in two of the most distinguished works in the marine envenomation literature, were among the first modern investigators to attempt rational documentation based on verified data.

Apart from its inherent scientific inaccuracy, this uncritical acceptance of data had two particularly undesirable effects: morbidity was relatively neglected (e.g. coral injuries {see Chapter 8, page 181} and ciguatera poisoning {see Chapter 15, page 346}); and public emphasis and funds (occasionally along with misguided human aggression!) were misdirected (e.g. sharks*). In keeping with this text's

* Non-venomous sharks are barely discussed in this book (see page 74). The reader is referred to the accurate Australian shark attack mortality data of Dr John West of Taronga Zoological Park, Sydney, and the equally accurate Australian crocodile attack mortality/morbidity data of Dr Charles Manolis of the Conservation Commission of the Northern Territory, Wildlife Operations (see below). Other works (COPPLESON 1958; ELLIS 1975; EDMONDS 1989, STEVENS (ed.) 1994), including the extensive South African experience with this subject (DAVIES 1964), are also worthy of consultation.

philosophy and attempts to record only *verifiable* facts, the following data have been carefully gathered. This approach will mean that some mortality and morbidity listings are likely to have been omitted. The editors would be grateful if any reader can identify such omissions (or errors of commission) based on verifiable information, and can advise them {see pages 13–16 for international contact details}. The hoped-for result will be the steady emergence of an accurate international picture of mortality and morbidity risk from marine animals.

Already this more disciplined approach has produced some useful effects:

- It is apparent that actual confirmed human death is much less common from some animals than was popularly believed (e.g. 'bluebottle', stonefish, cone shell, sea snakes), and more common from jellyfish stings worldwide than was previously realised {see Chapter 3, page 77}.
- It is no longer acceptable to us (ie JW, PJF and JWB) as referees/editors for *Medical Journal of Australia*, *Journal of Wilderness Medicine* and *Toxicon* respectively, for authors to claim mortality on the basis of hearsay alone.
- It is no longer acceptable to most medical scientists for authors to claim mortality on the basis of unconfirmed hearsay alone.
- The emphasis is shifting towards the significant and much more common morbidity {see Chapter 3, page 77}.

Disciplined mortality records of specific marine envenomations are now being kept in a number of nations, such as those for tetrodotoxic fish poisoning (Japan), ciguatera poisoning (some Pacific Island nations and protectorates, and Australia), jellyfish stings (Australia, China, Italy, Malaysia, Oman, Pakistan, USA) and stingray injuries (Australia, USA, Fiji). THE INTERNATIONAL CONSORTIUM FOR JELLYFISH STINGS (1993) now compiles an annual 'Sting Report Summary', containing mainly Australian but increasingly International 'first-hand only' written accounts of significant (fatal and non-fatal) envenomations.

While no claim for completeness is made, it is now possible with reasonable certainty to verify

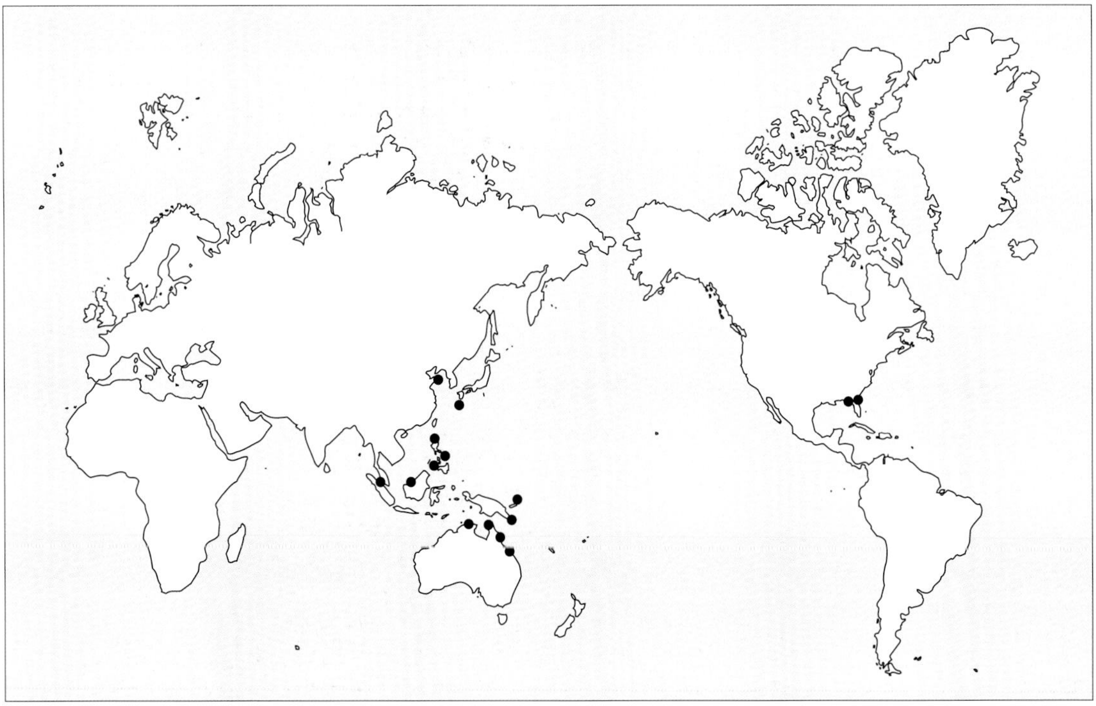

FIGURE 3.1
The known locations of confirmed human fatalities from jellyfish envenomations. Each one of these indicators is firmly documented and/or verified. Note the spread beyond tropical regions in the northern hemisphere.

the numbers of deaths, summarised in Tables 3.1 to 3.7. Australian deaths are listed separately here on page 70 because of the greater detail available to the editors.

JELLYFISH ENVENOMATION FATALITIES (EXCLUDING AUSTRALIA)

Figure 3.1, page 64, shows the locations of human fatalities from jellyfish envenomation, and Tables 3.1 and 3.2 below summarise the documentation of these fatalities.

MARINE ENVENOMATION FATALITIES OTHER THAN FROM JELLYFISH (EXCLUDING AUSTRALIA)

PUFFER FISH (TETRODOTOXIN) POISONING, '*FUGU*'

SOUTH AFRICA, JAPAN, CHINA, THAILAND

Although the number of deaths is unknown, it is undoubtedly significant (SHINCHO WEEKLY 1974;

TABLE 3.1
JELLYFISH FATALITIES WORLDWIDE (EXCLUDING AUSTRALIA)

Deaths	Jellyfish species	Country/region	References
100s	*Chirodropids various*	Indo-Pacific, Philippines	CLELAND & SOUTHCOTT 1965; FENNER 1987; AZILA & OTHMAN 1993;
2*		Borneo	FENNER & WILLIAMSON 1996, unpublished observations RIFKIN 1996 (this volume)
1		Japan: Okinawa	TOMIHARA 1992, unpublished observations FENNER & WILLIAMSON 1996
8	*Stomolophus nomurai*	Northern China:	
		1984 Qingdao 1	MINGLIANG ZHANG 1990,† 1992
		1986 Beidaihe, Hebei province 2	MINGLIANG ZHANG 1987, 1988
		1987 Beidaihe 3	MINGLIANG ZHANG 1987
		1987 Haiyang county, Shandong province 1	MINGLIANG ZHANG 1990†, 1991†
		1988 Longkou City, Shandong province 1	MINGLIANG ZHANG 1991† QIN SHEDE & MINGLIANG ZHANG 1992
3	*Physalia physalis* (Portuguese man-o'-war)	USA: Atlantic Coast	STEIN et al 1989; BURNETT & GABLE 1989
1	*Chiropsalmus quadrumanus* L. AGASSIZ (MÜLLER 1859)	USA: Texas	BENGSTON et al 1991

*Chirodropids resembling *Chironex fleckeri* have been reported from Sabah, Vietnam and Brunei (RIFKIN 1993, unpublished observations) and the Philippines (RIFKIN 1994, this volume). The reported deaths could thus perhaps have been caused by either *Chiropsalmus* or *Chironex* species in those areas (See chapter 6 this volume).

There have been two confirmed chirodropid sting deaths in Borneo (Sarawak and Sabah areas): a seven-year-old girl at Tutong beach 1989; and a ten-year-old boy on the north coast of Labuan 1992. A further four as-yet-unconfirmed deaths are reported (14-year-old British expatriate girl in 1983 on Tutong beach; a Malay man on Tutong beach in 1989; a Malay man in Sarawak in 1990; a Malay man in 1991, locality unknown). The details of these cases are currently being sought and prepared for publication (FENNER & WILLIAMSON 1996 unpublished observations).

† Mingliang Zhang's co-author was Professor Qin Shede.

TABLE 3.2
JELLYFISH FATALITIES IN THE INDO-PACIFIC REGION (EXCLUDING AUSTRALIA) DOCUMENTED BY CLELAND & SOUTHCOTT 1965

Date	Identity	Age(years)	Sex	Locality	Time	Survival Time
New Guinea:						
Indefinite	native	...	...	D'Entrecasteaux Island Group	...	...
1932–1933	native	12–13	M	Iwi Plantation, Bougainville I	...	...
Borneo:						
6 November 1945	...	ca. 20	M	East Labuan	late am	minutes
27 August 1945	D.K.P.	? 22	M	Gaard I, Brunei Bay	3 pm	few minutes
26 May 1956	A.C.K.	16 months	F	Bak-Bak, Kudat North Borneo	11.30 am	2–3 minutes
Malaya:						
?1946/47	fisherman	...	M	Telok Bahang, Penang I	...	...
Philippines:						
16 July 1907	Filipino	14	M	Olongapo	4 pm	3 hours
Indefinite	...	...	...	Palawan	...	...
? 1920s	Filipino	young	M?	Culion	...	few hours
21 September 1927	D.T.	19	M	Culion	...	2 minutes
Indefinite	persons	...	...	Busuanga	...	...
1927–1928	...	?	M	Leyte	...	minutes?
10 May 1928	fisherman	?	M	Sual, Pangasinan	...	3 hours
1 April 1957	J.V.	6	M	Camantilis, Pangasinan	...	<5 minutes
6 July 1958	...	15	F	Cebu	...	...

TSUNENARI *et al* 1980; SEUNAGA & KOTOKU 1980). In Japan alone during the period 1974–79, 60 deaths were reported (WORLD HEALTH ORGANISATION 1984). Halstead quotes KAWABATA (1965 unpublished observations) reporting a Japanese mortality rate of 59.4 per cent (HALSTEAD 1988); SEUNAGA and KOTOKU (1980) estimated a mortality of 51% from a series of 3000 puffer fish poisonings in Japan during a 21 year period, 1955–1975. KAKU and MEIER (1995) have obtained accurate Japanese data that documents 449 cases, with 49 deaths in Japan during the decade 1983–92; they suggest a current Japanese lethality rate of about 11 per cent. It seems that the case fatality rate, internationally, has declined from what was 80 per cent in the late nineteenth and early twentieth centuries (WORLD HEALTH ORGANISATION 1984). (See Chapter 26, page 442, for data recently obtained from Thailand.) Documentation of autopsy reports exists for at least seven fatalities in countries other than Australia (RICHARDSON J. 1861; RICHARDSON B.W. 1893; LARSEN 1925; FUKUDA 1951; BENSON 1956).

PARALYTIC SHELLFISH POISONING (PSP)

EUROPE, NORTH AMERICA, JAPAN, INDO-PACIFIC, VENEZUELA

(MEYER ET AL 1928; MEYER 1953; MCFARREN ET AL 1957; SMART 1995)

While the relationship between ciguatera and paralytic shellfish poisoning is being currently studied, documented PSP fatalities have occurred in Brunei, Sabah, New Britain, Papua New Guinea, Philippines, New Hebrides, Thailand, the United States (including Alaska), and Venezuela (McFARREN *et al* 1957; HALSTEAD 1988; SMART 1995). Autopsy results are referred to by

TABLE 3.3B
PREVIOUSLY UNDOCUMENTED CONE SHELL FATALITIES FROM JAPAN*

Date of sting	Species	Location	Original description
June 1944	*Conus geographus*	Okinawa, Japan	Tomihara (1994, unpublished observations)
August 1954	*C. geographus*	Okinawa, Japan	Tomihara (1994, unpublished observations)

* See also Chapter 20, page 415

WEEVERFISH (FAMILY TRACHINIDAE)

EUROPE AND AFRICA

Documented deaths from weeverfish envenomation are confined to the European and west African coastlines (EVANS 1943; HALSTEAD & MODGLIN 1958; RUSSELL & EMERY 1960; SKEIE 1962). However, at least three of these deaths were from secondary septicaemia (RUSSELL 1965).

CATFISH (SUBORDER SILUROIDEI)

TEXAS

(McKINSTRY 1993)

HALSTEAD (1988) states deaths have resulted from the stings of *Pterodoras granulosus* (Valenciennes), *Plotosus lineatus* and *Pimelodus clarias* (Bloch) but provides no details. One report of a diabetic person who died from septicaemia nine days after envenomation in Texas was found (McKINSTRY 1993).

ZEBRAFISH, SCORPIONFISH (BUTTERFLY COD)

PHILIPPINES

(HERRE 1952; HALSTEAD 1988)

Detailed documentation of fatalities is sparse, and deaths must be very rare. HALSTEAD (1988) lists five references, but some appear to lack specificity. Philippine deaths seem likely (HERRE 1952).

STONEFISH

SEYCHELLES, EAST AFRICA (SMITH 1957), THURSDAY ISLAND, JAPAN

Three specific documentations were found, reporting a total of three deaths. Two are attributed to envenomation by the stonefish *Synanceja verrucosa:* one at Mahé, Seychelles, and the other at Pinda, Mozambique (SMITH J.L.B. 1957; SMITH M.M. 1977). A third death occurred on Thursday Island in 1915, 'a few days' following envenomation (COOPER 1991), although doubt as to the causative animal exists (FLECKER 1936). There is a fourth death reported from Japan prior to 1989 of a male who was envenomed on the foot in 40 cm of sea water, trying to put a stonefish in a bag. He had four puncture marks and 'fell down and was drowned' (ARAKI & TOMIHARA 1992, unpublished observations).

Special note on needlefish fatalities (TOMIHARA *et al* 1994, unpublished observations) in Okinawa, Japan:

Needlefish (e.g. *Tylosurus crocodilus* {Plate 20.2}), are non-venomous but formidable fish, which have caused human fatalities, five deaths known to date. A recent injury occurred in 1992 {see Chapter 20, page 415), particularly in Japanese divers at night {Plate 20.3 and Figure 20.1, page 415}. Death has resulted from stab wounds from the strong needle-shaped jaws of the fish, which are plunged into the victim.

BLUE-RINGED OCTOPUS

SINGAPORE

(WILLIAMSON 1987)

Only one record of an actual human fatality outside Australia could be found; this was a child in Singapore, according to a newspaper report (*Straits Times*, 4 February 1986, page 63).

SEA ANEMONE ENVENOMATION

VIRGIN ISLANDS

(GARCIA *ET AL* 1994)

A 28-year-old caucasian male died of complications of fulminant hepatic failure nine days after a sting from a sea anemone of unknown species. The patient's serum tested positive to a dilution of 1:450 by ELISA against *Condylactis* species antigen, but negative to HIV and hepatitis A, B, and C profiles (GARCIA *et al* 1994).

BOX JELLYFISH DEATHS IN AUSTRALIA

These are detailed in Tables 3.4–3.7.

TABLE 3.4A
BOX JELLYFISH DEATHS IN QUEENSLAND AS REPORTED IN CLELAND & SOUTHCOTT 1965

Date	Initials	Age in years	Sex	Locality	Time	Survival time
30 December 1884	F.W.S.	11	M	Ross I, Townsville	7.30 am	3 minutes
16 January 1885	C.E.K.	6	M	Strand, Townsville	7.00 am	5–6 minutes
25 December 1910	F.L.G.	12	M	Cannonvale,Proserpine	...	within 2 hours
14 April 1911	A.G.W.	11	M	Cannonvale, Proserpine	...	2 hours (30 min utes)
5 March 1916	C.E.T.	17	M	Kissing Point, Townsville	ca.10.20 am	minutes
7 January 1923	G.M.	36	M	Mars Beach, Cannonville Proserpine	...	'almost instantaneous'
26 December 1934	S.C.	ca. 28	M	Googarra Beach, Tully	a.m.	minutes or up to 2 hours
20 January 1937	D.W.T.	19	M	Bramston Beach, near Babinda	7.00 am	5–10 minutes; other estimates 7–8 minutes
26 February 1939	R.H.	nearly 6	F	Googarra Beach, Tully	4.30 pm	'almost immediately'
20 December 1939	F.P.P.	3 1/2	M	Horseshoe Bay, Cannonvale	7–7.30 am	seconds or minutes
19 January 1941	J.C.	13	F	Townsville	1.50 pm	10–90 minutes
6 December 1941	R.E.D.	nearly 8	M	Townsville	ca.12.00	minutes
6 March 1943	W.J.H.	21	M	South Mission Beach, Tully	4.00 pm	3 minutes
18 December 1949	B.McN	10	M	North Mission Beach, Tully	4.30 pm	within 3 minutes
April 1951	J.G.	14	F	Kissing Point baths, Townsville	...	minutes
4 May 1953	J.E.L.	adult	M	Saltwater Creek, north of Townsville	1.00 pm	3–4 minutes
20 January 1955	D.J.M	5	M	Cardwell	9–9.35 am	few minutes
13 December 1956	T.W.S.	38	M	Pearce Creek, Ross River, Townsville	11.10 am	7–10 minutes
16 March 1957	R.W.T.	10	M	Bluff Beach, Yeppoon	8.00 am	80 minutes
13 December 1957	L.M.S.	11	F	North Mission Beach, Tully	10.50 am	0.5–10 mins
21 January 1962	R.V.B.	50	M	Yorkeys Knob, Trinity Bay	11.30 am	<3 minutes
31 December 1962	G.C.D.	6	M	South Mission Beach, Tully	11.00 am	few minutes
12 February 1963	M.D.H.	10	M	Bucasia Beach, Mackay	4.20 pm	2.5 minutes
North Queensland Torres Straits						
13 January 1962	J.R.K.	9	M	Bamaga, Cape York	1.00 pm	15 minutes

Table 3.4b
Box jellyfish deaths in the Northern Territory as reported in Cleland & Southcott 1965

Date	Initials	Age in years	Sex	Locality	Time	Survival time
North Qld Gulf						
December 1947	D	young	F	Appel Channel, Mornington I	am	'little later'
Indefinite ?2 deaths	...	children	...	Aurukun	...	...
NT. Darwin						
approx. 1894*	Master H.	15	M	Darwin baths	...	few minutes
12 March 1938	T.C.	11–12	M	Near Darwin baths	Mid-afternoon	few minutes
3 April 1941	...	ca. 20	M	Darwin	...	3 minutes
ca. March 1944	...	ca. 20	M	Darwin	pm	? 3 minutes
3 December 1950	W.C.	28	M	Mindil Beach	pm	few minutes
28 February 1954	C.W.C.	8	M	Larrakeyah Beach, Darwin	pm	2 minutes
NT. Islands						
15 December 1954	S.W.	4	F	Snake Bay, Melville I	am	<1 minute
25 December 1955	D.	young	M	Myilly Point, Bathurst I	1.00 pm	minutes
NT. Arnhemland						
1955	B.	6	M	Rose R	...	...
3 May 1959	D.N.	3	M	Maningrida, Liverpool R.	Late pm	...
16 December 1961	P.Y.	4	F	Maningrida, Liverpool R	Late pm	...

* 'A 15-year-old boy, Harwood, was stung to death by a jellyfish in the swimming baths' (Darwin). Reported by Ernestine Hill in *The Territory*. Angus & Robertson, 1951. Reprinted Walkabout Pocket Books–Ure Smith, Sydney (454 pp). 1970. p210.

Table 3.5
Box jellyfish deaths in the Northern Territory verified by Dr Bart Currie FRACP (Currie 1995, unpublished observations)

Date	Identity	Age in years	Sex	Locality	Time	Survival after sting
Northern Territory						
23 February 1964	*All cases identifiable*	6	M	Mandorah beach, Darwin	*No data in most cases*	*No data in most cases*
18 January 1967		7	F	Milingimbi, Arnhem Land		
28 March 1970		8	F	Shoal Bay, Darwin		
6 April 1974		2	M	Melville I		
2 February 1975		4	F	Bathurst I		
11 December 1975		4	F	Yirrkala, Arnhem Land		
15 January 1977		6	F	Snake Bay, Melville I		
23 January 1978		6	M	Maningrida, Arnhem Land		

continued ▶

TABLE 3.5 (CONTINUED)

Date	Identity	Age in years	Sex	Locality	Time	Survival after sting
26 December 1979		6	M	Maningrida, Arnhem Land		
20 September 1980		6	M	Snake Bay, Melville I	stung 4.30 pm	Died 9.00 pm
4 April 1983		8	F	Goulburn I		
14 December 1983		8	M	Milingimbi, Arnhem Land		
4 November 1984		8	F	Elcho Island, Arnhem Land		
2 December 1985		6	F	Milingimbi, Arnhem Land		
1 February 1988		9	M	Garden Pt, Melville I		Cardiac arrest in diastole
4 April 1988		13	M	Snake Bay, Melville I		2.3 m of tentacle contact; cardiac arrest in diastole

TABLE 3.6
BOX JELLYFISH DEATHS IN AUSTRALIA REPORTED BY BARBARA KINSEY (KINSEY 1986 & 1988)

Date	Initials	Age in years	Sex	Locality	Time	Survival time
21 January 1964		9	F	Palm Beach, Cairns		
January 1965		5	F	Mornington I		
31 January 1966		4	M	Half Moon Bay		
4 January 1967		10	M	Aurukun		
?1967		?	?	? Cardwell (to Tully Hospital)		
12 December 1970	K.C.	20	F	Forrest Beach Ingham		
24 November 1971		19?	F	North Mission Beach, Tully		
18 December 1980		5	M	Wongaling Beach, Tully		
20 January 1987*		5	M	Gladstone		

*(LUMLEY *et al* 1988)

The following fragmentary details of deaths following envenomation are subject to current investigation but cannot yet be either confirmed or rejected:
1950 South Mission Beach: 4-year-old male
1964 Cardwell: 12-year-old (approx) boy
1967 Cardwell: 9-year-old (approx) boy
1969 Tully: (name D.?) 12-year-old boy
late 1970s–early 1980s Milla: (name L.?) 6-year-old (approx) boy;
1980? Wongaling Beach

MARINE ENVENOMATION DEATHS (OVERALL) IN AUSTRALIA

TABLE 3.7
FATALITIES IN AUSTRALIA (TO JANUARY 1994)

Deaths	Cause	Region	References
61*	*Chironex fleckeri*	Tropical Australia	CLELAND & SOUTHCOTT 1965 WILLIAMSON 1985 CURRIE *et al* 1991; CURRIE 1994
2	*Hapalochlaena* spp. blue-ringed octopus	NSW Darwin	FLECKER & COTTON 1955 LANE & SUTHERLAND 1967 SUTHERLAND & LANE 1969
2	Stingray	North Queensland Melbourne, Vic.	FENNER *et al* 1989 WRIGHT-SMITH 1945
1	Cone shell	North Queensland	FLECKER 1936
1	Ciguatera	North Queensland	TONGE *et al* 1967
14	Puffer fish (tetrodotoxin)	Australia-wide: Autopsy:	7 BENNETT 1871; DUNCAN 1951; SOUTHCOTT 1975; SYDNEY GAZETTE 1821; COLONIAL TIMES 1831.
		Medical supervision:	5 THE WEST AUSTRALIAN 1950; DAILY TELEGRAPH 1950; SYDNEY MORNING HERALD 1951; WEST AUSTRALIAN 1988.
		No details:	2 SUTHERLAND 1983

* This figure differs from the 55 fatalities according to CURRIE, listed in the text below.

The absence of confirmed sea snake and stonefish fatalities from this list, at the time of preparation, should be noted.

CHIRODROPID DEATHS IN AUSTRALIA

CURRIE and colleagues (1990, 1991, 1992, 1994) compiled careful data relating to fatal chirodropid stings to 1993:

Worldwide	72
Australia	55*
Queensland	33
Northern Territory	28

* The differences between these two data groups (in this text and in Table 3.7) is being investigated, and illustrates the challenge in obtaining accurate fatality data for marine envenomations.

By comparison, from the above Tables the authors have calculated, to 1994:

Australia	63 (two of these lacking details)
Queensland	36
Northern Territory	27

A further six deaths in Queensland (referred to above in Table 3.6) remain unconfirmed.

PUFFER FISH (TETRODOTOXIN) POISONING IN AUSTRALIA

Documentation of 14 Australian cases is available, at least seven having an autopsy (BENNETT 1871; DUNCAN 1951; SOUTHCOTT 1975; SYDNEY GAZETTE 1821; COLONIAL TIMES 1831), another five under medical supervision (WEST AUSTRALIAN 1950; DAILY TELEGRAPH 1950; SYDNEY MORNING HERALD 1951; WEST AUSTRALIAN 1988), and two with no details (SUTHERLAND 1983).

MORTALITY FROM STINGRAY IN AUSTRALIA

Australian documentation records two stingray deaths (WRIGHT-SMITH 1945; FENNER *et al* 1989 Table 3.7, page 73). Both these Australian deaths were associated with stingray spine penetration of the heart. The first was presumed by those who conducted a careful autopsy (WRIGHT-SMITH 1945); the second was a delayed fatality in a 12-year-old boy following a freak accident in which stingray venom caused an insidious myocardial necrosis resulting in myocardial rupture and tamponade five days later (FENNER *et al* 1989).

COMPARATIVE FIGURES FOR SHARKS AND CROCODILES

MORTALITY FROM SHARK ATTACK IN AUSTRALIA

Since 1791 there have been 184 recorded deaths in Australia from shark attacks. Since the first confirmed documentation of a death in Australia from *Chironex fleckeri* envenomation, in Townsville, in 1884, there have been 163 shark attack deaths. (Data kindly supplied by Dr John West, Taronga Zoological Park, Sydney, November 1994). (Editors' Note: This certainly exceeds the mortality from chirodropid envenomation in Australia. However, a comparison of the morbidity figures would likely give a different picture.)

MORTALITY AND MORBIDITY FROM CROCODILE ATTACK IN AUSTRALIA

Dr Charles Manolis, Chief Scientist of the Conservation Commission of the Northern Territory, Wildlife Operations, kindly provided his database related to Australian crocodile attacks recorded since the 1870s. There have been 27 fatal and 31 non-fatal crocodile attacks since that time, all in the Northern Territory excepting nine of them (six fatal) in Queensland.

DOCUMENTATION SOURCES OF FATALITIES FROM MARINE TOXINS

AGASSIZ L. 1862. *Chiropsalmus*. *Contribution Natural Hist USA* 4: 174.

AZILA N, OTHMAN I. 1993. Cases of jellyfish envenomation in Malaysia and some cytotoxic studies on their toxins, in (N.H.Tan, S.L.Oo, V.Thambyrajah, N.Azila, eds.) Advances in Venom and Toxin Research. Proceedings of the Third Asia Pacific Congress on Animal, Plant and Microbial Toxins Malaysia, 27 June–1 July, 1993. Kuala Lumpur. *Malaysian Society on Toxinology*. 113–118.

BAGNIS R. 1970. Concerning a fatal case of ciguatera poisoning in the Tuamotu Islands. Clin Toxicol 3:579.

BAGNIS R, KUBERSKI T, LAUGIER S. 1979. Clinical observations on 3009 cases of ciguatera (fish poisoning) in the South Pacific. *Amer J Trop Med Hyg* 28: 1067–1073.

BENGSTON K, NICHOLS M M, SCHNADIG V, ELLIS M D. 1991. Sudden death in a child following jellyfish envenomation by *Chiropsalmus quadrumanus*. *JAMA* 266: 1404–1406.

BENNETT G. 1871. On the 'toad fish' (*Tetraodon hamiltoni*) of New South Wales. *Med Gaz* 1: 176–181.

BENSON J. 1956. Tetraodon (blowfish) poisoning: a report of two fatalities. *J Forensic Sci* 1: 119.

BURNETT J W, GABLE W D. 1989. A fatal jellyfish envenomation by the Portuguese man-o'-war. *Toxicon* 27: 823–824.

CLELAND J B, SOUTHCOTT R V. 1965. Injuries to Man from Marine Invertebrates in the Australian Region. Canberra. National Health and Medical Research Council. Special Report Series No 12. 282.

COLONIAL TIMES, Hobart. 1831. Coroner's inquest on three victims of toadfish poisoning (journalist unnamed). 29 March, 5 April.

COOPER N K. 1991. Historical vignette — the death of an Australian Army doctor on Thursday Island in 1915 after envenomation by a stonefish. *J Royal Army Med Corps* 137: 104–105.

CREVAUX J, Le JANNE E. 1882. Recit du troisième voyage dans l'Amérique équatoriale. *Arch Méd Nav* 37: 50–53.

CURRIE B. 1992. Jellyfish and top end waters. *Northern Territory Communicable Diseases Bulletin* 1(3): 1–2.

CURRIE B. 1994. Review article: Clinical implications of research on the box-jellyfish *Chironex fleckeri*. *Toxicon* 32: 1305–1313.

CURRIE B, ANSTEY N. 1990. Clinical observations of jellyfish envenomation. In: Annual Report 1989–1990, Menzies School of Health Research, PO Box 41096, Casuarina NT 0811, Australia. pp84–85.

CURRIE B, KHANH D M, ALDERSLADE P, WILLIAMSON J, FENNER P. 1991. Jellyfish envenomation in the Northern Territory of Australia. Singapore. 10th World Congress on Animal, Plant and Microbial Toxins, 3-8 November 1991, Singapore (P Gopalakrishnakone, C K Tan, eds). Abstract No 204, p256.

DAILY TELEGRAPH, Sydney. 1950. Boy and cat die from eating toado (journalist unnamed). 9 December.

DUNCAN C. 1951. A case of toadfish poisoning. *Med J Aust* 2: 673–675.

EDMONDS C. 1989. *Dangerous Marine Creatures.* Sydney. Reed. pp140–145.

EVANS H M. 1943. *Sting-fish and Seafarer.* London. Faber & Faber. 180 pp.

FENNER P J. 1987. Report of field trip to the Philippines by Dr P Fenner and Dr R Hartwick. Report to Queensland State Centre, Surf Life Saving Australia 1985. 20 pp.

FENNER P J, WILLIAMSON J A, SKINNER R A. 1989. Fatal and non-fatal stingray envenomation. *Med J Aust* 151: 621–625.

FLECKER H. 1936. Cone shell mollusc poisoning with report of a fatal case. *Med J Aust* 1: 464–466.

FLECKER H. 1956. Injuries from stonefish. Med J Aust 2: 371–373.

FLECKER H, COTTON B C. 1955. Fatal bite from octopus. *Med J Aust* 2: 329–332.

FUKUDA T. 1951. Marked increase of cases of puffer poisoning (in Japanese). *Clinics and Studies* 29(2).

GARCIA P J, SCHEIN R M H, BURNETT J W. 1994. Fulminant hepatic failure from a sea anemone sting. *Annals Int Med* 120: 665–666.

GILLESPIE N. 1987. Ciguatera poisoning. In: (J Covacevich, P Davie, J Pearn, eds) *Toxic Plants & Animals: A Guide for Australia.* Brisbane. Queensland Museum. pp160–169.

GLAZIOU P, CHINAIN M, LEGRAND A M. 1995. Clinical toxicology of ciguatera poisoning. In: (J Meier, J White, eds) *Clinical Toxicology of Animal Venoms and Poisons.* Boca Raton, Florida. CRC Press. pp 59–74.

GRANT C. 1801. *The History of Mauritius.* London. W Bulmer.

GUDGER E W. 1943. Is the stingray's sting poisonous? A historical résumé showing the development of our knowledge that it is poisonous. *Bull Hist Med* 14: 467–504.

HAECKEL E. 1880. System der Acraspeden. Zweite Halfte des Systems der Medusen. *Denkschr Med-Naturwiss Ges Jena.* 1: 361–372.

HALSTEAD B W. 1965, 1967, 1970. *Poisonous and Venomous Marine Animals of the World,* Vol I (1965), Vol II (1967), Vol III (1970). Washington, DC. US Government Printing Office.

HALSTEAD B W. 1988. *Poisonous and Venomous Marine Animals of the World* (2nd revised edition). Princeton, Darwin Press. 1168 pp; 288 pp plates.

HALSTEAD B W, MODGLIN R F. 1958. Weeverfish stings and the venom apparatus of weevers. Z *Tropenmed Parasitol* 9: 129–146.

HERRE A W. 1952. A review of the scorpaenoid fishes of the Philippines and adjacent seas. *Philippine J Sci* 80(4): 381–482.

INTERNATIONAL CONSORTIUM FOR JELLYFISH STINGS. 1993. Some Australian and International Marine Envenomation Reports: progress summary to October 31, 1993 (P Fenner, J Williamson, J Burnett, eds). Adelaide. Hyperbaric Medicine Unit, Royal Adelaide Hospital. 5: 1–25.

KAKU N, MEIER J. 1995. Clinical toxicology of Fugu poisoning. In: (J Meier, J White, eds.) *Clinical Toxicology of Animal Venoms and Poisons*. Boca Raton, Florida. CRC Press. pp 75–83.

KINSEY B. 1986. *Barnes on Box Jellyfish.* Townsville. James Cook University of North Queensland, Sir George Fisher Centre for Tropical Marine Studies. 76 pp; 81 pp appendices.

KINSEY B. 1988. *More Barnes on Box Jellyfish.* Townsville. James Cook University of North Queensland. Sir George Fisher Centre for Tropical Marine Studies. 109 pp.

KNIGHT J. 1989. Obituary: Andonis Neofitou, better known as Anthony Newly. *SPUMS J* 19: 197–198.

KOHN AJ. 1963. Venomous marine snails of the genus *Conus*. In: (H L Keegan, W B MacFarlane, eds) *Venomous and Poisonous Animals and Noxious Plants of the Pacific Region.* Oxford. Pergamon Press.

LANE W R, SUTHERLAND S K. 1967. The ringed octopus bite: a unique medical emergency. *Med J Aust* 2: 475.

LARSEN N P. 1925. Fish poisoning. *Queen's Hospital Bull* 2: 1–3.

LIGGINS J B. 1939. An unusual bathing fatality. *New Zealand Med* J 38: 27–29.

LUMLEY J, WILLIAMSON J A, FENNER P J, BURNETT J W, COLQUHOUN D M. 1988. Fatal envenomation by *Chironex fleckeri*, the north Australian box jellyfish: the continuing search for lethal mechanisms. *Med J Aust* 148: 527–534.

McFARREN E F, SCHAEFER M L, CAMPBELL J E, LEWIS K H, JENSEN E T, SCHANTZ E J. 1957. (Title unknown). In: Proc Nat Shellfisheries Assoc 47: 114, as reported in MURTHA E F. 1960. Pharmacology study of poisons from shellfish and pufferfish. *Annals New York Acad Sci* 90(3): 820–836.

MACGOWAN A J. 1887. Poisonous fish in China. *Bull US Fish Comm* 6: 130–131.

McKINSTRY D M. 1993. Catfish stings in the United States: case report and review. *J Wilderness Med* 4: 293–303.

MARINKELLE CJ. 1966. Accidents by venomous animals in Colombia. *Ind Med Surg* 77:988–994.

MEYER K F. 1953. Medical progress: food poisoning. *New Eng J Med* 249: 765–773, 804–812, 843–852.

MEYER K F, SOMMER H, SCHOENHOLZ P. 1928. Mussel poisoning. *J Preventive Med* 2: 365–394.

MORDECAI E R. 1860. Wound inflicted by a stingray. *New Orleans Med News Hospital Gaz* 7: 679–680.

MÜLLER F. 1959. Zwei neue quallen von Santa Catharina (Brasilien). *Abhandl Naturf Gas Halle* 5: 1–12.

QIN SHEDE (SHIDE), ZHANG MINGLIANG. 1992. *Nematocyst Dermatitis in China Sea.* Qingdao. Institute of Dermatology, Qingdao Medical College. 51 pp.

RATHJEN W F, HALSTEAD B W. 1969. Report on two fatalities due to stingrays. *Toxicon* 6: 301–302.

REID H A. 1956. Sea-snake bites. *Brit Med J* 2: 73–85.

REID H A. 1957. Antivenene reaction following accidental sea-snake bite. *Brit Med J* 2: 26–29.

REID H A, LIM K J. 1957. Sea-snake bite: a survey of fishing villages in northwest Malaya. *Brit Med J* 2:1266–1272.

RICHARDSON B W. 1893. *Fish poisoning and the disease 'siguatera'*. Asclepiad, London 10: 38–42.

RICHARDSON J. 1861. On the poisonous effect of the small portion of the liver of a *Diodon* inhabiting the seas of southern Africa. *J Linn Soc London* 5: 213–216.

RUSSELL F E. 1965. Marine toxins and venomous and poisonous marine animals. In: (F S Russell, ed) *Advances in Marine Biology*, Vol III. London. Academic Press. pp 255–384.

RUSSELL F E, EGEN N B. 1991. Ciguateric fishes, ciguatoxin (CTX) and ciguatera poisoning. *Toxin Reviews, J Toxicol* 10: 37–62.

RUSSELL F E, EMERY J A. 1960. *Venom of the weevers Trachinus draco* and *Trachinus vipera*. Ann NY Acad Sci 90: 805–819.

RUSSELL F E, PANOS T C, KANG L W, WARNER A M, COLKET T C. 1958. Studies on the mechanism of death from stingray venom: a report of two fatal cases. *Amer J Med Sciences* 235: 566–584.

SHINCHO WEEKLY. 1974. Report of 102 cases of tetrodotoxin poisoning of which 27 died (journalist unnamed). October 1974. p 141.

SILAS E G, BASTIAN FERNANDO A. ?1986. Turtle poisoning. *CMFRI Bull* 35: 62–74.

SKEIE E. 1962. Weeverfish toxin. *Acta Pathol Microbiol Scand* 56: 229–238.

SMART D. 1995. Clinical toxicology of shellfish poisoning. In (J Meier, J White, eds) *Clinical Toxicology of Animal Venoms and Poisons.* Boca Raton, Florida. CRC Press. pp 33–57.

SMITH J L B. 1957. Two rapid fatalities from stonefish stabs. *Copeia* 3: 249.

SMITH M M. 1977. *Sea and Shore Dangers: Their Recognition, Avoidance and Treatment.* Grahamstown, South Africa. J L B Smith Institute of Ichthyology, Rhodes University. pp16–17.

SOUTHCOTT R V. 1975. *Australian Venomous and Poisonous Fishes.* Adelaide. R V Southcott.

STEFANSSON V. 1944. *Arctic Manual.* New York. Macmillan.

STEIN M R, MARRACCINI J V, ROTHSCHILD N E, BURNETT J W. 1989. Fatal Portuguese man-o'-war (*Physalia physalis*) envenomation. *Annals Emerg Med* 18: 312–315.

SUENAGAK, KOTOKU S. 1980. Detection of tetrodotoxin in autopsy material by gas chromatography. *Arch Toxicol* 44: 291–297.

SUTHERLAND S K. 1983. *Australian Animal Toxins: The Creatures, their Toxins and the Care of the Poisoned Patient.* Melbourne. Oxford University Press. p 396 (venomous cone shells) and p 457 (fish poisonous to eat).

SUTHERLAND S K, LANE W R. 1969. Toxins and mode of envenomation of the common ringed or blue banded octopus. *Med J Aust* 1: 893–898.

SYDNEY GAZETTE. 1821. Coroner's inquest on man poisoned by toadfish at Parramatta (journalist unnamed). 21 April.

SYDNEY MORNING HERALD. 1951. Woman dies, family ill after eating toad fish (journalist unnamed). 6 March.

TONGE J I, BATTEY Y, FORBES J J. 1967. Ciguatera poisoning: a report of two outbreaks and a probable fatal case in Queensland. *Med J Aust* 2: 1088–1090.

TSUNENARI S, UCHIMURA Y, KANDA M. 1980. Puffer poisoning in Japan — a case report. *J Forensic Sci* 25: 240–.

WEST AUSTRALIAN. 1950. Two deaths, eight poisoned from eating a toado (journalist unnamed). Perth. 11 August.

WEST AUSTRALIAN. 1988. Seaman dies after eating blowfish (journalist unnamed). Perth. 14 September. p6.

WHITE J. 1995. Clinical toxicology of sea snake bites. In (J Meier, J White, eds) *Clinical Toxicology of Animal Venoms and Poisons.* Boca Raton, Florida. CRC Press. pp159–170.

WILLIAMSON J A. 1985. *The Marine Stinger Book,* 3rd edition. Brisbane. Queensland State Centre. Surf Life Saving Australia. pp19–38.

WILLIAMSON J A H. 1987. The blue-ringed octopus bite and envenomation syndrome. In: (R M Mandojana, guest ed) Aquatic Dermatology. *Clin in Dermatol* 5: 127–133.

WILLIAMSON J A. 1992. Current challenges in marine envenomation: an overview. *J Wilderness*

Med 3: 422–431.
WILSON B R, GILLETT K. 1980. *Australian Shells*. Sydney. A H & A W Reed.
WORLD HEALTH ORGANISATION. 1984. Aquatic (marine and freshwater) biotoxins. Environmental Health Criteria, No 37. Geneva. World Health Organisation. p15.
WRIGHT-SMITH R J. 1945. A case of fatal stabbing by a stingray. *Med J Aust* 2: 466–467.
YASHIRO H. 1939. Fatal bite of *Conus geographus*. *Venus IX* (3–4), October 1939 (in Japanese).
ZHANG MINGLIANG. 1987. Study on nematocyst dermatitis caused by jellyfish *Stomolophus nomurai*. *Acta Acad Med Qingdao* 4: 1–.
ZHANG MINGLIANG. 1988. Study of jellyfish *Stomolophus nomurai* stings on Beidehe. *Nat Med J China* 9: 499.
ZHANG MINGLIANG, QIN SHEDE. 1990. Nematocyst dermatitis in Huanghai and Bohai (The North China Sea). Proceedings 2nd China–Japan meeting of Dermatology. Shanghai. 50 pp.
ZHANG MINGLIANG, QIN SHEDE. 1991. The marine stinger in South China Sea. *Chinese J Marine Drugs* 4: 36.

MORBIDITY FROM MARINE TOXINS

Quantitatively speaking, the morbidity resulting from human encounters with venomous and poisonous marine animals dwarfs the mortality problem. Despite this, the accumulation of accurate morbidity data has been neglected and remains lacking worldwide. General collections show that the correlation between the lists of marine animals that produce death and injury, respectively, is a poor one (with the possible exception of jellyfish) (e.g. POPE 1947; HALSTEAD 1965 Vol I, 1967 Vol II, 1970 Vol III; CLELAND & SOUTHCOTT 1965; MINTON 1975; MARETIĆ 1975; SUTHERLAND 1983; and some more recent and specific ones, e.g. RUSSELL 1984; BURNETT & CALTON 1987; EDMONDS 1989; WILLIAMSON *et al* 1992; INTERNATIONAL CONSORTIUM FOR JELLYFISH STINGS 1993; McKINSTRY 1993; HALLEGRAEFF 1993; GLAZIOU *et al* 1994; KAKU & MEIER 1994; SMART 1994). (See Table 3.8.)

Coral cuts and abrasions are commonplace the world over {Plate 8.3} and seldom reported. When neglected they may produce protracted and sometimes serious morbidity {see Chapter 4, page 88, and Chapter 8, page 181}. From what little data exists it is clear that ciguatera poisoning is a major morbidity problem among certain peoples (LEWIS & SELLIN 1992; ADAMS 1993). (See also Chapter 26, page 445.) With the possible exception of South-East Asia (REID & LIM 1957), sea snake bite has been presumed to be uncommon globally; some doubt may now exist (ARAKI *et al* 1989; HIGA *et al* 1990) {see also Chapter 17, page 396, and Chapter 18, page 409 and Chapter 20, page 415}. The major oversight with respect to morbidity, however has been jellyfish stings. These injuries comprise a vast morbidity group worldwide (BURNETT & CALTON 1987; WILLIAMSON *et al* 1992; BURNETT 1992; FENNER *et al* 1993; CURRIE 1994) (Table 3.9, page 78) {see Chapters 8 and 9, pages 194 and 236, respectively and Chapter 26, page 442}, and their epidemiology is an important research frontier, which this text attempts to address.

TABLE 3.8
A COMPARISON OF THE RELATIVE FREQUENCIES OF MORTALITY AND MORBIDITY IN HUMAN VICTIMS FROM MARINE ANIMALS, IN DESCENDING ORDER OF FREQUENCY. THE RELATIVE MORBIDITIES ARE INFORMED ESTIMATES ONLY.

Mortality	Morbidity
Pufferfish* poisoning	Coral cuts, abrasions, stings (including wound infections)
Paralytic shellfish poisoning	Jellyfish stings (especially *Physalia* spp.)
Chirodropid stings	Ciguatera poisoning
Ciguatera poisoning	Spiny fish stings (including stingrays)
Turtle flesh poisoning	Other marine animal poisoning from ingestion
Sea snakes	
Spiny fish injuries	
Physalia stings	

*NOTE: Although pufferfish are poisonous and not venomous, they can be unpredictably aggressive. Some species grow to considerable size (e.g.50 cm long) and have been known to attack people in the water and inflict serious bite injuries.

TABLE 3.9
REPORT SUMMARY OF JELLYFISH STINGS ON AUSTRALIAN BEACHES OVER SIX SUCCESSIVE AUSTRALIAN SUMMERS, 1987–1994

Jellyfish	1987/88 summer	1988/89 summer	1989/90 summer	1990/91 summer	1991/92 summer	1992/93 summer	1993/94 summer
***Physalia* 'bluebottle'**							
Stings:	1960	1271	1579	1176	1224	539	2027
Sightings:	20000	9000	5000	6200	13857	1770	21745
***Cyanea* 'hair jelly'**							
Stings:	192	15	20	198	69	31	296
Sightings:	1500	204	1400	9200	2084	1	8023
***Catostylus* 'blubber'**							
Stings:	114	37	9	271	236	5	67
Sightings:	200	364	825	4100	4626	26	14742
***Chironex* 'box jellyfish'**							
Stings:	5	1	3	3	4	0	20
Sightings:	331	126	220	274	159	64	368
***Carukia* 'Irukandji'**							
Stings:	2	4	0	0	104	1	8
Nettings	0	38	14	1	0	0	1

Source: Based on the SURF LIFE SAVING QUEENSLAND 1992-93 63rd Annual Report.

DOCUMENTATION SOURCES OF MORBIDITY FROM MARINE TOXINS

ADAMS M J. 1993. An outbreak of ciguatera poisoning in a group of scuba divers. *J Wilderness Med* 4: 304–311.

ARAKI Y, TOMIHARA Y, HIGA H. 1989. Sea snake bite in Okinawa, Japan (1). In: Annual Report of antivenin study by Okinawa Prefectural Institute of Public Health, Okinawa, Japan. pp33–44. (in Japanese).

BURNETT J W. 1992. Human injuries following jellyfish stings. *Maryland Med J* 41: 509–513.

BURNETT J W, CALTON G J. 1987. Jellyfish envenomation syndromes updated. *Annals Emerg Med* 16: 1000–1005.

CLELAND J B, SOUTHCOTT R V. 1965. Injuries to Man from Marine Invertebrates in the Australian Region. Canberra. National Health and Medical Research Council. Special Report Series No 12. 282 pp.

CURRIE B. 1994. Review article: Clinical implications of research on the box-jellyfish *Chironex fleckeri*. *Toxicon* 32: 1305–1313.

EDMONDS C. 1989. *Dangerous Marine Creatures.* Sydney. Reed. 192 pp.

FENNER P, WILLIAMSON J, CURRIE B. 1993. Update of marine envenomation in tropical Australia. Presented at 'Tropical Medicine at the Top End'. Mirambeena Tourist Resort, Darwin, Northern Territory, 17–19 June, 1993. Townsville, Qld. Australasian College of Tropical Medicine Publication. 88 pp.

GLAZIOU P, CHINAIN M, LEGRAND A M. 1995. Clinical toxicology of ciguatera poisoning. In (J Meier, J White, eds) *Clinical Toxicology of Animal Venoms and Poisons.* Boca Raton, Florida. CRC Press. pp 59–74.

HALLEGRAEFF G M. 1993. A review of harmful algal blooms and their apparent global increase. *Phycologia* 32: 79–99.

HALSTEAD BW. 1965, 1967, 1970. *Poisonous and Venomous Marine Animals of the World,* Vol I (1965), Vol II (1967), Vol III (1970). Washington, DC. US Government Printing Office.

HIGA H, UEZATO H, ARAKI Y. 1990. A case of sea snake bite in Okinawa, Japan. *The Snake* 22: 100–105.

INTERNATIONAL CONSORTIUM FOR JELLYFISH STINGS. 1993. Some Australian and International Marine Envenomation Reports: progress summary to October 31, 1993 (P Fenner, J Williamson, J Burnett, eds). Adelaide. Hyperbaric Medicine Unit, Royal Adelaide Hospital. 5: 1–25.

KAKU N, MEIER J. 1995. Clinical toxicology of Fugu poisoning. In: (J Meier, J White, eds) *Clinical*

Toxicology of Animal Venoms and Poisons. Boca Raton, Florida. CRC Press. pp 75–83.
LEWIS R J, SELLIN M. 1992. Multiple ciguatoxins in the flesh of fish. *Toxicon* 30: 915–919.
MARETIĆ Z. 1975. *Venomous animals and poisonous animals of the Adriatic Sea*. Zagreb. J A Zagreb Univ.
McKINSTRY D M. 1993. Catfish stings in the United States: case report and review. *J Wilderness Med* 4: 293–303.
MINTON S A. 1975. Venomous marine invertebrates. In: (W T Hubbert, W F McCulloch, P R Schurrenberger, eds) *Diseases Transmitted from Animals to Man*, 6th edition. Springfield, Illinois. C C Thomas. 1031–.
POPE E C. 1947. Some sea animals that sting and bite. *Aust Museum Magazine* 9: 164–168.
REID H A, LIM K J. 1957. Sea-snake bite: a survey of fishing villages in northwest Malaya. *Brit Med J* 2: 1266–1272.
RUSSELL F E. 1984. Marine toxins and venomous and poisonous marine plants and animals (invertebrates). In: (J H S Blaxter, F S Russell, M Yonge, eds) *Advances in Marine Biol* 21: 60–194.
SMART D. 1995. Clinical toxicology of shellfish poisoning. In: (J Meier, J White, eds) *Clinical Toxicology of Animal Venoms and Poisons*. Boca Raton, Florida. CRC Press. pp 33–57.
SURF LIFE SAVING QUEENSLAND. 1992–1993. 63rd Annual Report. (Marine stingers — Dr Peter Fenner). Newstead, Qld. pp38–41.
SUTHERLAND SK. 1983. *Australian Animal Toxins: The Creatures, their Toxins and the Care of the Poisoned Patient*. Melbourne. Oxford University Press. pp338–468 (Fish poisonous to eat).
WILLIAMSON J A, FENNER P J, BURNETT J W. 1992. Some Australian and international marine envenomation reports: progress summary to March 31, 1992. Adelaide. Hyperbaric Medicine Unit, Royal Adelaide Hospital. pp1–22.

INTRODUCTION TO MARINE TOXINS

JOHN WILLIAMSON AND JOE BURNETT

The terms 'toxin', 'venom' and 'poison' are defined in the Glossary {see pages 37 and 38}. A venom is usually a 'toxin' or collection of toxins that gain entry to the body ('envenomation') by injection into the tissues through the skin (e.g. nematocyst tubule discharge, stonefish spine penetration). A 'poison' most commonly describes a toxin which gains entry into the body by absorption across intact body layers ('poisoning') (e.g. ciguatoxin or paralytic shellfish poison via the gastrointestinal tract). 'Toxin' is a non-specific term used to describe any substance harmful to living cells; thus venoms and poisons are both toxins. The toxin 'tetrodotoxin' is a venom when injected by a blue-ringed octopus bite, and a 'poison' when ingested from the flesh of the puffer fish (Plate 3.4) {see pages 331 and 83, respectively}; in both situations it is a 'toxin'. However, each of the three terms can unfortunately be loosely used at times, to imply one of the other two, leading to confusion. Throughout this text the three terms are used only with the specific respective meanings as defined above and in the Glossary, unless otherwise indicated.

The properties of marine venoms that are of significance in human envenomation include:

- toxicity (including lethality)
- lability
- molecule size (molecular weight)
- chemistry
- antigenicity
- physiological and biochemical actions

Other factors are also of clinical significance, including the volume of toxin involved in an envenomation or poisoning process, the physical size and condition of the victim (e.g.smaller body weight and/or frail persons are at greater risk), and the mechanism the animal uses to administer the toxin (e.g. venom deposited in a single site [sea snake bite], or spread over an area of the victim's body [jellyfish sting]).

SELECTED MARINE TOXINS OF MEDICAL INTEREST

JACQUIE RIFKIN AND JOHN WILLIAMSON

TOXINS OF PROTOZOANS (ONE-CELLED ANIMALS)

NOTE: The marine toxins (venoms and poisons) considered here are a selection only. For a more specific approach, the reader is referred to the respective sub-sections that follow, dealing with specific marine animals.

DINOFLAGELLATES

Dinoflagellates are one-celled organisms with characteristics of both plants and animals. Some are reported as toxic, and in large numbers as in a 'bloom' (also referred to as an 'algal bloom' or 'red tide'), they may become windborne in sea spray, causing respiratory tract symptoms in humans, such as rhinorrhoea and bronchospasm with a non-productive cough (PIERCE 1986).

When these dinoflagellates are fed upon by organisms such as molluscs, crabs (FOXALL *et al* 1979) and fish, these animals then harbour the toxins. This in turn may result in serious poisonings (including fatalities) in humans when the seafood is ingested (SMART 1995). Molluscs {see Chapter 13, page 327} such as mussels and oysters may contain paralytic shellfish poisons (PSPs). Crabs may contain PSPs, tetrodotoxin (TTX) and palytoxin (PTX) (see below), and ciguatoxins are found in fish {see Chapter 15, page 346}. These toxins are all very potent and exhibit toxic properties in humans, including occasional fatal effects (HALSTEAD 1988; HALLEGRAEFF 1993); they also exhibit thermostability (they are unaffected by cooking). No antivenoms, antitoxins or vaccines presently exist for this (presumably) heterogeneous group of toxins.

PARALYTIC SHELLFISH POISONS (PSPs)

1. PSPs causing toxicity in molluscs (phylum Mollusca)

PSP is a severe health risk and is one group of the most potent shellfish toxins. Blooms of dinoflagellates cause toxic 'red tides'. The dinoflagellates from these blooms are removed from the water by filtration by bivalves and other organisms. However, marine and freshwater (HUMPAGE *et al* 1993) dinoflagellates are only one source of PSPs. The best known such PSP is saxitoxin (STX), but others, all of similar general molecular structure, include neosaxitoxin, a series of gonyautoxins (I-VI and VIII), and decarbomyl saxitoxin (SMART 1995); of these the most toxic are gonyautoxin III and saxitoxin (GENENAH & SHIMIZU 1981). Shellfish poisons of public health significance linked with dinoflagellates are classified clinically as PSP, neurotoxic shellfish poisoning (NSP), diarrhetic shellfish poison (YASUMOTO *et al* 1984) (DSP — okadaic acid) (BADEN 1988; KUMAGAI *et al* 1986), and amnesic shellfish poisoning (SMART 1995) (PERL and colleagues [1990] suggest that this is caused by domoic acid and is better called 'excitatory encephalopathic shellfish poisoning').

Toxic dinoflagellates have been accidentally introduced into Australia from cold-water ports by ships' ballast water, causing toxic algal blooms. During 1986, and several years since, shellfisheries in Tasmania were closed because of the high levels of toxicity found in the shellfish. Monitoring programmes to alert oyster and mussel farmers to any presence of significant toxicity are now in place (JONES 1991). The introduction of toxic dinoflagellates from foreign warm-water ports in ballast water, into warm Australian waters (e.g. Moreton Bay, off Brisbane), is a real possibility and could have serious consequences, both medical and economic (R. LEWIS 1994 unpublished observations). †

2. PSPs causing toxicity in crab flesh (phylum Arthropoda)

The relationship between saxitoxin and its derivatives (known collectively as paralytic shellfish poisons or PSPs), palytoxin (and tetrodotoxin and its derivatives [KAO 1983], see phylum Mollusca {Chapter 13, page 327} and pufferfish, below), is not yet clear. However, all of these toxins have been found in reef crabs such as xanthids (class Xanthidae) and in portunids (class Portunidae) (LLEWELLYN & ENDEAN 1991) and have produced mortality and serious morbidity in the Philippines and in Japan (ALCALA 1983). The presence of toxicity may vary with the food source, and the geographical area in which the crabs are found. Unfamiliar crabs should always be considered to be potentially toxic and should never be eaten.

3. PSPs found in intertidal organisms, zooplankton and algae

According to JONAS-DAVIES & LISTON (1985), PSPs are also present in several intertidal organisms, zooplankton, and a red alga.

4. Dinoflagellates causing toxicity in fish (phylum Chordata)

Marine dinoflagellates (HOLMES *et al* 1991), in particular the benthic dinoflagellate *Gambierdiscus toxicus*, are probably a principal source of the very potent toxins of ciguatera poisoning

† A major marine environmental problem has been developing in the southern Tasmanian regions due to the accidental introduction of a voracious sea star from Japan which came there in ships' ballast water {see also Chapter 12, page 312}.

{see Vertebrate toxins, below, and Chapter 15, page 346}. Ciguatoxin chemical structures are shown in Figure 15.3 page 349.

Toxins of the phylum Porifera: venomous sponges (Class Demospongiae)

The toxicity of the slimy substances from marine sponges is impressive and can produce significant human morbidity, with contact dermatitis,* local oedema with pain and itch {see Chapter 14, page 342}, even progressing to peripheral vascular compression and compromise in susceptible individuals (Figure 3.2, page 83). A number of sponges that cause dermatitis are listed by KIZER (1984a). The skin-damaging (dermatopathic) agent has not been accurately identified. When large areas of skin surface are affected, systemic symptoms are described; anaphylaxis has been claimed (FISHER 1978). However, precise chemical or toxinological knowledge is still meagre. ENDO and colleagues (1986) have isolated a number of toxic (including cytotoxic) compounds from sponges, such as suberitine, halitoxin, nucleosides and macrolides. Sponges also contain antibacterial substances which may contribute to or be responsible for the biosynthetic processes (GARSON 1994).

Sponges also possess needle-like glassy spicules, which can pierce the skin when handled — alive or dead (MARSH & SLACK-SMITH 1986). These spicules may be retained in the skin, and sponge toxins compound the irritation by penetrating into the micro-traumatic lesions resulting from the spicules.

Some toxins of the phylum Cnidaria

The source of these toxins varies, but humans may be injected by coming in contact with stinging capsules or nematocysts (see Chapter 6, page 121), or poisoned when the flesh is consumed. These toxins are described below under the headings: Jellyfish toxins, sea anemone and corallimorpharian toxins, and palytoxin.

Jellyfish toxins

The majority of jellyfish toxins are complex mixtures of toxic and/or antigenic polypeptides and enzymes of high molecular weights, and pathogenic to humans, both locally and systemically (BURNETT & CALTON 1987a and 1987b). They are of the greatest medical significance, are labile (OLSON *et al* 1984; COMIS *et al* 1989; OTHMAN & BURNETT 1990; LONG-ROWE & BURNETT 1994), exhibit myotoxic effects upon the heart (BURNETT *et al* 1986a; ENDEAN & SIZEMORE 1988; LUMLEY *et al* 1988; MARTIN & AUDLEY 1990; MUSTAFA *et al* 1994), the pulmonary vasculature (FENNER *et al* 1988), the systemic vasculature (ADIGA KM 1984; ABU-NEMA *et al* 1988; WILLIAMSON *et al* 1988), and have dermatonecrotic and *in vitro* haemolytic effects (CRONE 1976; LONG-ROWE & BURNETT 1994a) {see Chapter 7, page 175}. They have been associated with many human deaths {see 'Fatalities', page 63}. Neurotoxicity also occurs (PRATAP CHAND & SELLIAH 1984; WILLIAMSON *et al* 1984a; BURNETT *et al* 1994b). They may also induce serious allergic phenomena (FISHER 1984; TOGIAS *et al* 1985; WILLIAMSON 1985). Thus antivenom production is feasible and already exists successfully for at least one jellyfish {see Chapter 21, page 420}. Serology permits increasingly specific identification of both stings and high-risk individuals (BURNETT *et al* 1988a) {see Chapter 7, page 176}.

Sea anemone and corallimorpharian toxins

These toxins are potent and may produce intense and sustained local skin reactions {see Chapter 8, page 186}. On the basis of *in vivo* animal studies, as well as anecdotal clinical experience {see Chapter 8, page 187}, these venoms contain neurotoxins, cardiotoxins, haemolysins and proteinase inhibitors (SHIOMI *et al* 1985).† Possible

* According to MANOWITZ and ROSENTHAL (1979), certain anemones live with and on sponges, and the toxins from these cnidarians {see Chapter 8, page 186} are responsible for a dermatitis and local necrotic skin reaction called 'sponge diver's disease' {see Chapter 14, page 343}.

† Water–ethanol extracts of sublittoral (>50 m) sea anemones from the Aleutian and Comandorskiye Islands (north Pacific Ocean), which were non-toxic to mice, exhibited inhibition of DNA and RNA synthesis *in vitro* in Ehrlich carcinoma tumour cells (MAHNIR *et al* 1993).

human hepatotoxicity is now reported (GARCIA *et al* 1994) (see also Chapter 26, page 448).

PALYTOXIN

Some *Palythoa* species, colonial members of the phylum Anthozoa, contain a potent toxin termed 'palytoxin' (SCHEUER 1964). Palytoxin (PTX) is a complex terpenoid which, in simple terms, may form 'pores' in cell membranes. It produces a powerful cell membrane depolarisation mainly due to an increase in the permeability of tetrodotoxin (TTX)-sensitive sodium ion (Na^+) permeability. This produces both excitatory and inhibitory effects, and it is possible that such 'PTX-activation' of Na^+ channels may influence regulation of ionic movement in many cells, through Na^+- and potassium ion- (K^+) adenosine tri-phosphatase (ATP) mechanisms (HIRATA *et al* 1988). Other organisms such as crabs may incorporate palytoxin into their flesh after ingesting it.

TOXINS OF PHYLUM PLATYHELMINTHES — FLATWORMS (CLASS TURBELLARIA)

Flatworms are often delicate but brightly coloured with conspicuous markings* — a characteristic of many toxic animals. There have been no reports of human intoxications, but the secretions of flatworms are distasteful to other animals. *Leptoplana tremellaris* Oersted has been reported as toxic (HALSTEAD & VINCI 1988), and TTX has been reported in the marine flatworm *Planocera multitentaculata* (MIYAZAWA *et al* 1986). However, little is known about flatworm toxins.

TOXINS OF THE PHYLUM NEMERTEA

Although some ribbon worms of the class Enopla have an impressive envenoming apparatus and apparently elaborate a venom capable of being delivered, there is no clear record of human envenomation to date. 'Amphiporine', the general term for toxins from the nemertean, was originally isolated from the nemertean worm called *Amphiporus lactifloreus* and thought not to be associated with the envenomation apparatus (BACQ 1937). More recent investigations have involved its major toxin 'anabaseine' (KEM 1971).

Anoplan toxins from the heteronemertean *Cerebratulus lacteus* are proteinaceous and have neurotoxic and cytotoxic activity. This material appears to be extruded through the integument of the worm (KEM 1988) and apparently no venom apparatus is present.

TOXINS OF THE PHYLUM ANNELIDA SEGMENTED WORMS (CLASS POLYCHAETA)

FAMILY NEREIDAE — RAGWORMS

Little is known about the toxins from segmented worms. The toxin 'nereistoxin' from the marine worm *Lumbriconereis heteropoda* has produced headache, respiratory difficulties, vomiting and itching in a victim who handled it (SAKAI 1970). Long thought to be a neurotoxin (NITTA 1934), nereistoxin chemistry is described by HASHIMOTO and OKAICHI (1960); it contains a cyclic disulphide and bears little resemblance to other marine toxins. It has been shown to act on the nicotinic receptor of the vertebrate neuromuscular junction (KEM 1988) (cf. mollusc toxins, below). It is also an effective insecticide (SAKAI 1970).

FAMILY GLYCERIDAE — BLOODWORMS

Other polychaete, *Glycera* spp. (the blood worms) also possess potent toxins delivered by their powerful jaws. These toxins affect neuromuscular transmitter release (KEM 1988).

FAMILY APHRODITIDAE — BRISTLE WORMS

Bristle worms have a world-wide distribution. Envenomation results from contact with the sharp bristles projecting from each body segment. Bristle worm envenomations produce local itching, burning and numbness, which some authors describe as lasting for some hours

*Aposematic colouration is the technical term for the characteristically bright colouration of some animals that advertises their toxicity. Also, see Glossary.

(CLELAND & SOUTHCOTT 1965; MARSH & SLACK-SMITH 1986). Some bristle worm species have hollow bristles, but the evidence that a venom is present in the bristles is weak (KEM 1988). More research is required. However, in one Australian (north Queensland) snorkeller, significant pain at two sting sites on his left elbow and forearm (Plate 3.2) only commenced six hours after the sting, when the victim was having a hot shower. He was readmitted to hospital with this local pain and with paraesthesia of the ipsilateral fingers. The latter persisted for 10 days. Venomous bristle worms include *Chloeia flava* of the Indo-Pacific including tropical Australia, and the 'sea mouse' *Aphrodita australis* of the cold temperate southern Australian coastline [MARSH & SLACK-SMITH 1986]. {See also Chapter 18, page 407}.

FAMILY AMPHINOMIDAE — FIREWORM, BRISTLE WORM

A South Australian patient was envenomated by a 'fireworm' on an ankle, while wind surfing (and falling off into waist-deep water where the sea bottom was covered with seaweed). She developed severe local ankle pain and oedema, then proximal leg pain, headache, nausea and vomiting and unilateral photophobia, with subsequent scattered paraesthesiae lasting 3-5 days. The worm was identified as the salmon-pink 'fireworm' *Eurythoë complanata* (Plate 3.3),

TOXINS OF THE PHYLUM MOLLUSCA

SNAILS, BIVALVES, OCTOPUS

Venomous and poisonous members of this phylum have a range of toxins that interfere with normal acetylcholine/receptor interaction, although not all are of importance in human toxinology. They include tetramine in whelks, quaternary ammonium compounds in cone shell venom, esters of choline (murexine) in muricid shellfish, choline in callistin shellfish, amines in poisonous turban shells, and a cholinergic poison in sea hares (HALSTEAD 1988). Thus

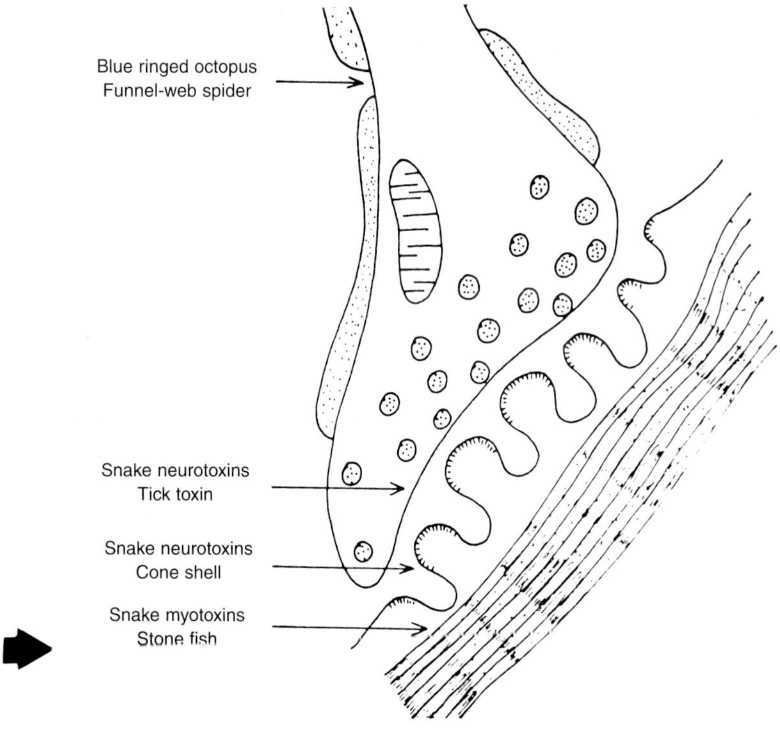

FIGURE 3.2
Schema of the human neuro-muscular junction, showing some sites of action of various animal toxins (after Sutherland). (Courtesy of Associate Professor V. Callanan, Townsville.)

disruption of neuromuscular transmission (Figure 3.2, page 83) characterises their effects. By contrast tetrodotoxin-like activity is obtained from the venom of the blue-ringed octopus. It is a neurotoxin, which specifically inhibits sodium channel function, thus producing presynaptic neural blockade (Figure 3.2). Tetrodotoxin itself (obtained from puffer fish) is a low molecular weight, non-antigenic molecule and (like many mollusc toxins) is also absorbed well via the gastrointestinal tract as a poison when ingested. The latter is known as puffer fish poisoning and it results from the ingestion of improperly prepared *'fugu'* or fillets of puffer fish, (phylum Chordata), (see below and Plate 3.4) (KAKU & MEIER 1995). Mollusc toxins thus may affect either cholinergic autonomic nerve function, and/or vascular smooth muscle (KAO 1966) and somatic (particularly respiratory) muscle contraction (SUN *et al* 1994). (See also PSPs, page 80, and Plates 13.2–13.9.)

TOXINS OF THE PHYLUM ECHINODERMATA

CLASS ASTEROIDEA — SEA STAR* TOXINS

The biological activity of poisons in the flesh of toxic sea stars is derived from highly surface-active saponins. These produce only vomiting upon ingestion, but are dangerously toxic if injected (HALSTEAD 1988). The venom of the spines of the crown-of-thorns sea star *Acanthaster planci* {see Chapter 12, page 315} includes saponins, a neurotoxin, vasoactive serotonin-like and histamine-like substances, and a protein toxin (BIRKELAND & LUCAS 1990). The venom exhibits allergenic properties, and the pain from the spine injuries is somewhat relieved by immersing the envenomated part in hot water {see also Chapter 5, page 98}.

Contact with other less injurious sea stars (e.g. the sun star, or rose star, *Solaster papposus*) may produce a pruritic, papulo-urticarial eruption (FISHER 1978).

* Sea stars are more commonly known as 'starfish'. However, this term may be confusing as they are not in any way related to vertebrate fishes. Thus the authors have chosen to use the term 'sea stars'.

TOXINS FROM THE CLASS HOLOTHUROIDEA — SEA CUCUMBERS

Work has been done on the saponins found in sea cucumbers. Sea cucumbers contain holothurins saponins which are potent cardiac–glycosides, capable of producing severe illness upon ingestion. The holothurins may produce contact dermatitis when the Cuvierian tubules of the animals are touched or even when a diver's mask is cleared in close vicinity to the evacuated tubules (AUERBACH 1988).

CLASS ECHINOIDEA — SEA URCHIN TOXINS

The venom mixtures of sea urchins contain a range of pharmacologically active substances, including steroid glycosides, haemolysins, proteases, serotonin and cholinergics (MANOWITZ & ROSENTHAL 1979). Sea urchins are nocturnal animals and hence may present a hazard to night divers {see Chapter 22, page 425}. Spines as well as seizing organs known as pedicellariae are present, and in some species, either one or both may be venomous. (Plates 12.10–12.15.)

The commonest injury is the penetration of tissue by a spine, producing instant pain. Multiple spine penetrations are claimed to produce serious systemic symptoms and signs (LINAWEAVER 1967) {see Chapter 12, page 323}.The local purplish discolouration from spine dye that may be left at the entry site in the skin from a sea urchin spine penetration (for example *Diadema setosum* or *Strongylocentrotus purpuratus*) can give the false impression of a retained spine (O'NEAL *et al* 1964). As with any marine wound, secondary infection and indolent ulceration can follow in neglected spine wounds. Spine penetration of a joint may also result in severe synovitis (CRACCHIOLO & GOLDBERG 1977).

Multiple stings of pedicellariae are stated to produce even more serious effects in those species in which they are venomous (see Chapter 12, page 321), with widespread systemic disturbance (FREYVOGEL 1972). The local effects of such envenomation from multiple pedicellarial bites may be short-lived (1–2 hours) but localised muscular weakness may persist for 4–6 hours, and cases of prolonged pain are recorded.

PHYLUM CHORDATA: SOME SIGNIFICANT MARINE VERTEBRATE TOXINS

CLASS CHONDRICHTHYES: STINGRAY VENOM

The venomous species of these magnificent animals envenomate by the avulsion and deposition, within the wound created by their spine, of the integumentary sheath that envelopes the spine {see Chapter 16, page 354}. The venom produces intense and immediate local pain, and leads to an inexorable local tissue necrosis (BARSS 1984; FENNER *et al* 1989b; BENDT & AUERBACH 1991; FENNER 1995). RUSSELL (1965) suggests the venom is a thermolabile mixture of protein and certain enzymes, which has marked effects on cardiac conduction and pacemaker tissue in animal studies (HALSTEAD 1988). There appears to be no action on neuromuscular transmission (SUTHERLAND 1983). Little else is known. Local pain is reduced acutely by immersing the injured part in hot water. Human deaths have occurred from trunk penetration followed by uncontrolled haemorrhage, or tissue necrosis in a vital organ that has been penetrated with spine venom deposition (FENNER *et al* 1989b) or by secondary infection {see Chapter 3, page 67}. No specific record could be found in Australia of human cardiac disturbances from the direct systemic effects of stingray venom, despite repeated claims of this effect in the literature (MARETIĆ 1988). There are no known antivenoms.

CLASS OSTEICHTHYES: THE VENOMS OF BONY FISH

Most knowledge in this area comes from studies of the venoms of weeverfish (e.g. MARETIĆ 1988), stonefish *Synanceia horrida* and *Synanceia verrucosa* (WIENER 1959a, 1959b; SAUNDERS 1959; CAMERON 1988) and scorpionfish including the 'zebrafish' e.g. *Pterois volitans* (SAUNDERS & TAYLOR 1959) {see Chapter 16, page 354}. They are an important cause of morbidity worldwide.

Stonefish venom is non-dialysable, heat labile, antigenic protein for which a successful Australian antivenom has been developed (WEINER 1959b; SUTHERLAND 1992). In animals, vascular smooth muscle depression seems prominent, leading to profound hypotension. *In vitro* cytolysis (cell destruction) and vascular permeability-increasing properties (including pulmonary oedema in animals) have been demonstrated (KREGER 1991). Human envenomation is characterised by local rather than systemic effects; however, a case of life-threatening acute pulmonary oedema from a stonefish (species unknown) envenomation has been reported (LEHMANN & HARDY 1993). Human fatalities may have occurred (SMITH 1957); their true incidence is the subject of controversy (FLECKER 1956; WILLIAMSON 1985; CAMERON 1988), but appear to be minimal worldwide {see Chapter 3, page 69}, although repeatedly subject to lurid media exaggeration. However, stonefish envenomation is a dreadful experience for any victim. Stonefish stings are undisputedly common (KIZER *et al* 1985; FENNER *et al* 1993a) and usually produce savage local pain. Zebrafish venom is similar. The pharmacology and chemistry of most of the remaining spine-armed venomous fishes remain largely unstudied (SUTHERLAND 1983; HALSTEAD 1988). The majority of these envenomations may be relieved by hot water immersion of the envenomated part {see Chapter 5, page 104}. Denaturation of the protein venom seems an unlikely complete explanation, as the pain usually returns upon withdrawal of the injured limb from the hot water.

NOTE: The interesting topic of the significance of ichthyocrinotoxic fishes is addressed only briefly by this text, but the reader is referred to GOSLINE & BROCK (1960) and HALSTEAD (1988). Their toxin, a glandular secretion (usually in the skin of the fish), which may be released into the surrounding water, is poisonous to other fishes but is not associated with a venom apparatus. Stonefish (genus *Synanceia*) have both a venom apparatus and tubercle glands in their skin which secrete a mixure of a water-soluble and a water-insoluble toxin.

CLASS OSTEICHTHYES: SOME IMPORTANT FISH POISONS

PUFFER FISH ('FUGU') POISONING

Flesh of puffer fish (Plate 3.4) is considered a delicacy in some countries, such as Japan, and the eating of the prepared flesh has produced much serious morbidity and many deaths (*fugu* poisoning) (KAKU & MEIER, 1995). The pure crystalline puffer poison is termed 'tetrodotoxin' (hence 'tetrodotoxication'). Tetrodotoxin-like substances are the principal components involved in envenomation from the bite of the blue-ringed octopus (phylum Mollusca). As explained above, tetrodotoxin is a selective sodium channel inhibitor on cell membranes (KAO 1966 & 1983), producing respiratory paresis and hypoxia.

(CIGUATERA POISONING) — TROPICAL FISH POISONING

'Tropical fish poisoning' is a somewhat non-specific term that can apply to poisoning from any fish caught in the tropics or subtropics and eaten (e.g. 'tropical' puffer fish). However, it is more customarily used to refer to ciguatera poisoning. Ciguatera poisoning is a major public health problem in many communities around the world both in the tropics and subtropical regions (GLAZIOU *et al* 1995), and is the subject of a special chapter in this text {see Chapter 15, page 346}. The powerful toxins involved may originate from dinoflagellates (HOLMES *et al* 1991). They are most concentrated in large, carnivorous fish situated high in the food chain, and these are most favoured for human consumption. The chemistry of at least the three major, closely related, polyether ciguatoxins has been determined {Figure 15.3, page 349}. These toxins bind firmly to sodium channels, causing depolarisation particularly of nerve cells. Deaths have occurred, but usually non-lethal gastrointestinal and peripheral neurological symptoms predominate {see Chapter 3, page 63}. No antivenoms or antitoxins are presently available. The toxins are remarkably potent and resistant to denaturation.

SOME OTHER FISH POISONS*

ICHTHYOCRINOTOXINS — from such fish as boxfish, triggerfish, trunkfish, and toadfish. These fish secrete skin toxins into the water, eliciting antibiotic activity ('antibiotic' being here a general term used to mean 'against life'). In these fishes, the poisonous slime may be eaten as in pufferfish, and dermal contact may cause dermatitis. The trunkfish contain substances which foam from the mouth or from the base of the fins when the fish are alarmed.

CLUPEOTOXINS — from plankton-feeding fish from tropical island areas, such as sardines, herrings, tarpon. It is thought (see HALSTEAD 1988) that it may be similar to if not the same as ciguatoxin.

SCOMBROTOXINS — originally thought to include only scombrid fish such as tunas and mackerel, but other fish may be implicated as well (AUERBACH 1988). Fish that are inadequately preserved or refrigerated contain bacteria that produce the toxin. The histamines ingested with the flesh by humans may produce a severe allergic-like reaction (HALSTEAD 1988)

ICHTHYOOTOXINS — from fish that contain toxin generally restricted to the gonads, and there is a definite relationship between gonadal activity and toxin production (see Palythoa, class Cnidaria, Chapter 8, page 180). Eggs of the marbled sculpin or cabezon, *Scorpaenichthys marmoratus*, from off the North American west coast have been responsible for severe poisonings.

ICHTHYOHEPATOTOXINS — some fish may cause fatalities after the liver is ingested. The deaths may be due to vitamin A poisoning (see class Mammalia, below).

* Halstead 1988 and Auerbach 1988.

CLASS REPTILIA: SEA SNAKE VENOMS

In keeping with its spectacular array of venomous terrestrial snakes (SUTHERLAND 1983), Australia possesses an extensive variety of venomous sea snakes in its waters {see Chapter 17, page 397}. South-East Asian sea snakes are also known for their danger to fishermen (REID & LIM 1957). Sea snake venom is predominantly neurotoxic and myotoxic in its effects, with the former capable of acting more rapidly. At least three neurotoxins are composed of some 60 amino acid residues with some disulphide bonds (HALSTEAD 1988), and induce a post-synaptic neuromuscular block of acetylcholine action (Figure 3.2, page 83). Myotoxins may lead to serious effects from myglobinuria.

Sea snake venom is antigenic, and (in sharp contrast to the venoms of many terrestrial snakes) most sea snake venoms can be effectively opposed antigenically and clinically by the antivenom preparation against the venom from the Australian beaked sea snake *Enhydrina schistosa* (SUTHERLAND 1983) {see Chapter 21, page 419}. Thus precise identification of the offending sea snake is unnecessary. In Australia the antivenom of the tiger snake *Notechis scutatis* is also effective against Australian sea snake envenomation, but larger volumes of antivenom are required, with the attendant increased risks (SUTHERLAND 1992).

CLASS MAMMALIA: TOXIC MAMMALS

SUBCLASS PROTOTHERIA — MONOTREMES, PLATYPUS VENOM

Chapter 25 describes human envenomation by platypus, a remarkable (non-marine) egg-laying mammal, which inhabits rivers and streams in eastern Australia. The toxinology of platypus venom is imperfectly understood. KELLAWAY and MESSURIER (1935) demonstrated coagulant effects, and more recent studies have been conducted by TEMPLE-SMITH (1978), and DEPLATER *et al* (1995) (see Chapter 26, page 452). Clinically it produces an immediate and intense local pain, which is mercilessly persistent without medical intervention. There does not seem to be any vascular compromise.

SUBCLASS THERIA — WHALES, SEA LIONS, CANINES

Several mammals, such as the sperm whale, the polar bear, Arctic seals and the husky — most of which inhabit cold temperate or Arctic regions — have been implicated in causing vitamin A poisoning in humans when their liver is ingested (CLELAND & SOUTHCOTT 1965; HALSTEAD & VINCI 1988). A similar phenomenon has been observed relating to mammals of the southern Australian and Antarctic regions (CLELAND & SOUTHCOTT 1969; SOUTHCOTT 1982b).

Acknowledgment: The authors thank Dr Richard Willan of the Museums and Art Galleries of the Northern Territory for helpful advice.

CHAPTER 4

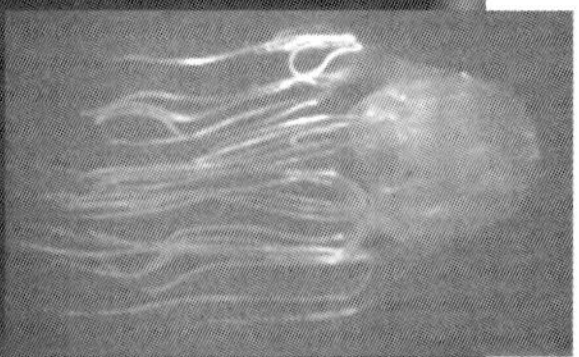

MARINE WOUND INFECTIONS

DR LES ASHDOWN AND

ASSOCIATE PROFESSOR VIC CALLANAN

DEDICATION TO DR LES ASHDOWN

This chapter serves as one memorial to the late Dr Les Ashdown, PhD, who died suddenly a few short months after its preparation. A full tribute to Dr Ashdown's life and important contributions to Australian and world microbiology and tropical medicine has been published elsewhere (AUSTRALASIAN COLLEGE OF TROPICAL MEDICINE 1993).

NOTE: This chapter is written in the language of the microbiologist. Some of the more basic terms only, appear in the Glossary {page 30}, but interested readers who are unfamiliar with microbiological and laboratory terms might also consult the standard microbiological literature or a microbiologist colleague.

Marine wounds are infected either by the direct introduction of microorganisms situated on the object causing the injury (BUCK *et al* 1984; KLONTZ *et al* 1993) or subsequently by contamination of the wound with sea water or by land-based microorganisms. The resultant clinical disease is protean and ranges from acute and fulminant, through subacute, to chronic recurrent suppurative, or chronic granulomatous stages. It may also be self-limiting with no serious sequelae.

AETIOLOGY

The number of organisms naturally resident in sea water and with the potential to infect marine injuries is substantial (AUERBACH *et al* 1987; INTERNATIONAL SYMPOSIUM ON DIVING IN HAZARDOUS

WATER 1991; MURPHEY *et al* 1992; MILLINGTON & WILHELM 1993); the more common pathogens are listed in Table 4.1. Land-based bacteria that are implicated in marine wound infections are usually endogenous to the person sustaining the injury, and are primarily pyogenic streptococci, *Staphylococcus aureus* or enteric bacteria. A brief description of only the more important bacteria complicating marine wounds will be given here.

The family Vibrionaceae consists of three genera whose members produce marine wound infections, namely *Vibrio*, *Plesiomonas*, and *Aeromonas* (BARON & FINEGOLD 1990; INTERNATIONAL SYMPOSIUM ON DIVING IN HAZARDOUS WATER 1991; MURPHEY *et al* 1992). These bacteria are motile by means of polar flagellation, unencapsulated, asporogenous, non-acid-fast, facultatively anaerobic Gram negative cells. They are seen singly or in pairs, from straight rods with rounded ends to coccoid in form, or may show a somatic curvature as in the comma-shaped cells of *Vibrio* species, and show considerable variation in size. Marine (halophilic) vibrios are members of the genera, *Vibrio* or *Photobacterium*, which are oxidase positive, ferment dextrose, do not grow in nutrient broth in the absence of added salt (NaCl) but will grow with the addition of 1 per cent salt. The vibrios are speciated in the laboratory by tests for salt tolerance, reduction of nitrate to nitrite, various carbohydrate fermentation reactions, and the presence or absence of arginine dihydrolase, lysine and ornithine decarboxylase. The mesophilic (optimal growth temperature, 20–45°C) aeromonads are speciated by using

TABLE 4.1
POTENTIALLY PATHOGENIC MICROORGANISMS PRESENT IN SEAWATER

Bacteria predominating in seawater:	
Vibrio alginolyticus	*Alteromonas espejiana*
Vibrio (Photobacterium) damsela	*Alteromonas haloplanktis*
Vibrio fluvialis	*Alteromonas macleodii*
Vibrio furnissii	*Delaya venustus*
Vibrio hollisae	*Pseudomonas marina*
Vibrio parahaemolyticus	*Erysipelothrix rhusiopathiae*
Vibrio vulnificus	*Mycobacterium marinum*
Bacteria predominating in estuarine or brackish water:	
Members of the family, *Enterobacteriaceae* including *Escherichia coli, Klebsiella-Enterobacter* species, *Salmonella* species and *Proteus* species	
Acinetobacter species	*Clostridium perfringens*
Aeromonas caviae	*Clostridium botulinum*
Aeromonas hydrophila	*Clostridium tetani*
Aeromonas sobria	Other clostridia
Chromobacterium violaceum	*Bacteroides fragilis*
Flavobacterium species	Other *Bacteroides* species
Moraxella catarrhalis	Actinomycetes
Pseudomonas aeruginosa	*Bacillus cereus*
Other pseudomonads	*Staphylococcus aureus*
Vibrio cholerae	Other staphylococci
Vibrio mimicus	Micrococci
Coryneform bacteria	Streptococci

the tests for esculin hydrolysis, growth in the presence of potassium cyanide, acid production from salicin, gas production from glucose, the Voges-Proskauer reaction and elastase production. *Plesiomonas shigelloides* is differentiated from *Vibrio* and *Aeromonas* by tests for DNase, and mannitol and inositol fermentation reactions (KUDOH 1988).

Erysipelothrix rhusiopathiae is a non-motile, asporogenous, microaerophilic, non-acid-fast, thin Gram positive rod, which has a tendency to form long filamentous forms (CHAN 1988). The bacterium grows between 5 and 44°C, and growth is more abundant with increased carbon dioxide tension. It is oxidase and catalase negative, and ferments only a small number of carbohydrates (dextrose, lactose, galactose and fructose). On blood agar medium, the bacterium may grow as either smooth or rough alpha-haemolytic colonies after 48–72 hours incubation.

Mycobacterium marinum (HARRIS *et al* 1991) (previously *M. balnei, M. platypoecilus*) is a photochromogenic (Runyon Group 1) acid-fast bacillus which grows optimally at 32°C but commonly not at temperatures of 35°C or higher (BARON & FINEGOLD 1990). It is a fast-growing, non-fastidious *Mycobacterium,* which does not reduce nitrate, produce catalase or Tween hydrolase.

Acinetobacter species, *Chromobacterium violaceum*, *Flavobacterium* species, *Pseudomonas aeruginosa* and other pseudomonads are Gram negative bacilli or coccobacilli. They are motile (except for *Acinetobacter* and *Flavobacterium* species), oxidase positive (except for *Acinetobacter* species) and are oxidative in their attack on carbohydrates.

The pyogenic streptococci encountered in marine wound infections are mainly
Streptococcus pyogenes (Group A),
Streptococcus agalactiae (Group B),
Streptococcus equisimilis (Group C),
Streptococcus anginosus, and Group G
Streptococcus. These are facultatively anaerobic, non-motile, asporogenous, non-acid-fast Gram positive cocci that grow in chains of varying length (BARON & FINEGOLD 1990). They are oxidase and catalase negative and frequently grow as beta-haemolytic colonies on blood agar. In the laboratory, they are identified by a number of tests including the detection of group specific (Lancefield) antigen. Staphylococci and micrococci are members of the family *Micrococcaceae*, and are also non-motile, asporogenous, non-acid-fast Gram positive cocci, but they grow in tetrads or grape-like clusters. Staphylococci and micrococci are differentiated by the inability of micrococci to attack carbohydrates anaerobically, while *Staphylococcus aureus*, the most pathogenic species in this family for humans, is identified essentially by morphology and coagulase or DNase production.

EPIDEMIOLOGY

The number and species of bacteria in sea water are dependent on several factors, including the impact of human habitation, the content and concentration of organic and inorganic substances, the degree of light, and the salinity, temperature and oxygenation of the water. Some bacteria, such as the marine vibrios, are natural residents of sea water. Other bacteria are regularly leached from terrestrial sources into the oceans either directly from soil into the river systems or indirectly from processes such as sewerage or storm-water effluents. Therefore in coastal waterways adjacent to large population bases, sea water is often heavily contaminated with a wide variety of enteric, terrestrial and marine bacteria. For example, it is well accepted that the primary habitat of *Aeromonas* species is non-faecal sewage, which in turn inoculates other water sources (JANDA & DUFFEY 1988). Although *Aeromonas* is found naturally in marine systems, particularly those that interface with fresh water, sea water trauma (unlike injuries contaminated with fresh water) is only occasionally associated with aeromonad infection. The marine vibrios are the predominant bacteria found in deep sea water which are pathogenic for humans, but they are present in larger numbers (together with a wider variety of bacteria) in estuarine and coastal waters. They may also be present and result in infections in wounds sustained in brackish or even freshwater habitats. Likewise, *S. aureus* is halophilic (one method of isolation in the laboratory is to grow the organism on nutrient agar with 7.5 per cent NaCl, which inhibits growth of most other bacteria) and is recovered readily from coastal waters but only rarely from deep sea water.

Marine bacteria have worldwide distribution and have been isolated from all oceans, with their presence being recorded both in cool climates and in the tropics. Their global distribution is due to their ability to survive and multiply under a wide variety of environmental conditions. Most marine wound infection and marine microbiological pathogen reports have come from the North American, European and Australian coastal regions (HAZEN *et al* 1978; INTERNATIONAL SYMPOSIUM ON DIVING IN HAZARDOUS WATER 1991; MURPHEY *et al* 1992; MILLINGTON & WILHELM 1993). There is a distinct seasonal distribution of cases of marine wound infections. In tropical regions, cases occur throughout the year, whereas in temperate climates, cases predominate in the warmer months of the year. This summer preponderance of cases probably reflects both increased numbers of bacteria in sea water, and more people using the marine environment for recreation during the summer months. Marine vibrios are mesophilic and are rarely encountered in sea water colder than 8°C.

The mode of transmission of bacteria occurs either directly by the transfer of bacteria from the marine object producing the injury, or by the wound being subsequently contaminated with sea water or sea-bottom sediment, or by land-based endogenous flora. Prior wounds subsequently contaminated with marine bacteria may also become infected. The incubation period can be as short as 8–24 hours in cases of *S. pyogenes*, aeromonads or marine vibrios or 2–8 weeks or longer in the case of *M. marinum* infections. Most marine wounds occur in young active men, with injuries to the extremities accounting for more than two-thirds of cases. The face is another common site of injury. Wounds caused by coral, sea shell or rock lacerations, surfboards, boating and fish hooks, fish or stingray spine punctures, and other trauma due to marine animals are the most common marine injuries encountered.

The frequency and extent of these types of injuries in northern Queensland are unknown {see Chapter 3, page 77}. Medical colleagues in general practice in this region indicate that recreational marine wounds occur far more often than injuries sustained during other forms of recreation, including contact sports such as football and hockey. Coral cuts are by far the most common marine wounds encountered in northern Queensland {see Chapter 8, page 181}. The wound is typically complicated by a non-infective cellulitis that often develops within 12 hours of injury. Many wounds are further complicated by retained coral fragments

TABLE 4.2
SEAWATER-CONTAMINATED INJURIES REQUIRING HOSPITAL ATTENTION IN NORTHERN QUEENSLAND DURING 1990–1991

Injury	No.	Non-infected	Number infected with:				
			Pyogenic *Streptococci*	*Staphylococcus aureus*	Enteric bacilli	Marine vibrios	Aeromonads
Coral cuts	25	4	13	14	8	5	0
Fish spines	2	0	0	1	2	2	1
Fish hooks	6	1	1	2	2	2	1
Boating mishap	3	0	1	1	2	2	1
Other injuries	5	0	4	2	3	1	1

and by growth of pathogenic bacteria (pyogenic streptococci, *S. aureus*, enteric bacilli, and marine vibrios) at the wound site (Table 4.2, page 91.)

In Australia, many marine puncture wounds are caused by stingrays and catfish. In one general medical practice in New South Wales about 1 per cent of all tourist consultations and 3 per cent of all traumatic wounds in tourists consisted of injuries of this nature (PACY 1962). In another study, at an island resort in south Queensland, over 100 stingray injuries in tourists were treated over a two-year period (FENNER *et al* 1989b). These injuries usually arise when persons are wading through shallow water or handling the animals (especially when unwisely pulling them into a boat) or when professional fishermen walk inside beach nets. Marine animal bites (moray eel, shark, dolphin, etc.) or contaminated fish hooks, crustacean shells or other such agents may also result in infected marine puncture wounds. Marine vibrios, aeromonads, enteric bacilli, pyogenic streptococci and *S. aureus* are the usual infecting bacteria, but *M. marinum* may also infect these types of wounds.

The epidemiology of marine wound infections induced by *E. rhusiopathiae* ('erysipeloid of Rosenbach', 'seal finger', 'whale finger', 'fish handler's disease' [AUERBACH 1987]) is obscure. The infection occurs almost exclusively on the hands and is mainly an occupational hazard of the seafood industry (CHAN 1988). The bacterium is found occasionally in fish slime, but there is no clear evidence that fish are naturally infected while alive. It is thought that fish are contaminated with the bacterium subsequently from trawler holds or fish boxes or from rats and mice on the wharves, and previous or fresh wounds on hands then become infected when people handle the fish. Also, some authors have reported *E. rhusiopathiae* as a cause of infected coral wounds, but we have never cultured the bacterium from these types of wounds, nor has the bacterium been isolated from slime from coral in this region (SUTTON 1992, unpublished observations). In addition, a computer search of the literature has failed to show documented cases of *E. rhusiopathiae* infected coral injuries.

CASE HISTORY

A recent chest infection in a 20-year-old female German tourist to north Queensland (December 1993) was caused by *Aeromonas hydrophilia* and *Serratia marscescens*, cultured purely by chance (LOWE 1994, unpublished information). As the former is mainly a marine organism, closer questioning of the patient confirmed that she had been in good health, was on no medications and had no previous relevant medical conditions. She had been snorkelling some 3–7 days prior to the infection and had otherwise not been in sea water. Thus the hired snorkel is one possible source of the infection. She began to improve clinically on oral cephalexin 500 mg three times daily but left the area before recovering completely; no follow-up details are available. There is little current data concerning cross-infection from hired snorkelling equipment, but careful cleaning between hirings or loanings would seem to be sensible.

PATHOGENESIS

Natural immunity to marine bacteria is high, and the majority of marine wound infections are self-limiting. Clinical infection develops because of impaired host defences, the virulence of the infecting bacteria, the size of the inoculum or a combination of these factors. As well, penetrating injuries may inoculate the marine bacteria deep in the dermis, thereby allowing clinical infection to establish more readily. Infections are frequently polymicrobial (MURPHEY *et al* 1992). Consequently, the role of each individual bacterium and of the interaction between bacteria isolated from marine wound sites in the pathogenesis of the disease is obscure. Nonetheless, pyogenic streptococci and *Staphylococcus aureus* are well-known causes of soft-tissue infection, and each Gram negative bacillus that has been recovered from marine wounds can effect soft-tissue infection in its own right. These bacteria regularly possess one or more virulence factors, including the ability to adhere to human cells by piliation or by other means, haemagglutinins, and various exotoxins. These exotoxins include cytotoxin, haemolysins, aerolysin, enterotoxins, proteases, collagenase, lecithinase, chitinase and others and correlate with invasiveness and pathogenicity. Severe or even fatal illness due to the marine vibrios and to the aeromonads may occur in persons with previous underlying disease or with immunosuppression (JANDA & DUFFEY 1988; MURPHEY *et al* 1992). In *Vibrio vulnificus* infection, increased levels of available iron appear to enhance the virulence of the bacterium, and this is reflected by the preponderance of clinical *Vibrio vulnificus* infections in patients with haemochromatosis and liver disease. The pathological changes of marine wound infections caused by marine vibrios and aeromonads are suggestive but not diagnostic.

Little is known of the pathogenic mechanisms of *E. rhusiopathiae* in humans. Pathogenicity may be related to surface antigen structure which probably influences adherence to hosts' cells, and to the production of neuraminidase, an enzyme that probably modifies host cell membranes and thereby induces cell lysis (SMITH 1983).

With *M. marinum* wound infection, the disease is almost always confined to the cooler superficial body tissues such as the extremities or the elbows (FADDA 1988) (Figure 4.2, page 94). Infection varies from frank suppuration to organised granulomas. Acid-fast bacilli in biopsy material are either present in very scant numbers or not seen at all.

CLINICAL MANIFESTATIONS

The manifestations of marine wound infections are protean and present most commonly as mild infection, cellulitis, erysipelas, ulceration, ecthyma, pyoderma gangrenosum, nodular lesions, vesicular or bullous eruptions (Figure 4.1) and gangrene (sometimes with gas). The most commonly encountered disease is a self-limiting infection, which generally requires domiciliary treatment only and is rarely seen in general practice or in hospitals.

The most serious marine infections encountered, which may result in limb amputation or even death, are necrotising cellulitis, fasciitis, or

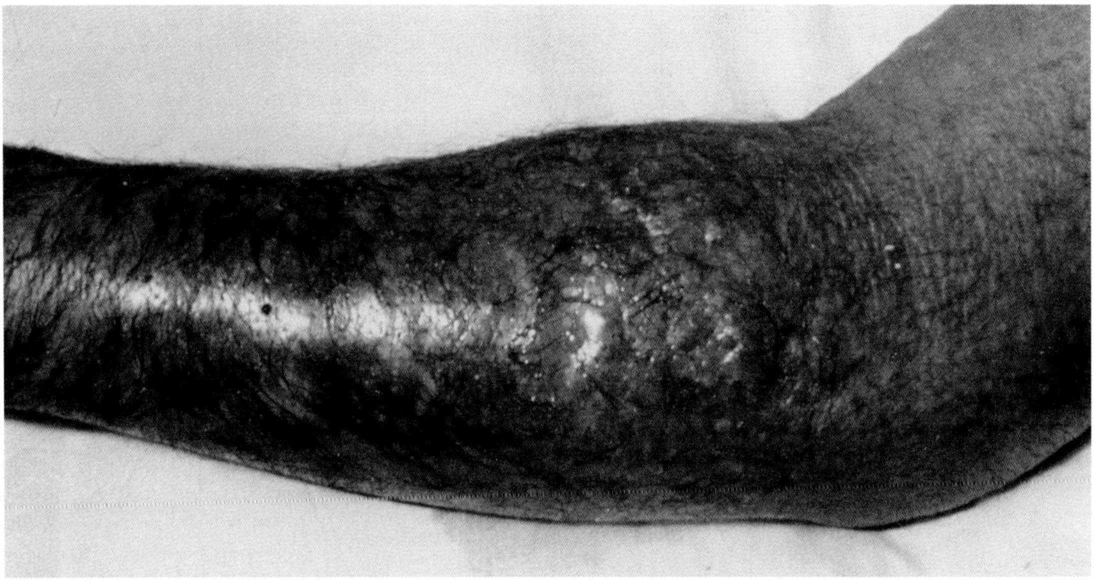

FIGURE 4.1
Haemorrhagic bullae of *Vibrio vulnificus* infection of the left arm of a 68-year old man following scratched sandfly bites sustained during a crabbing excursion. (Photograph courtesy of B.J. Russell)

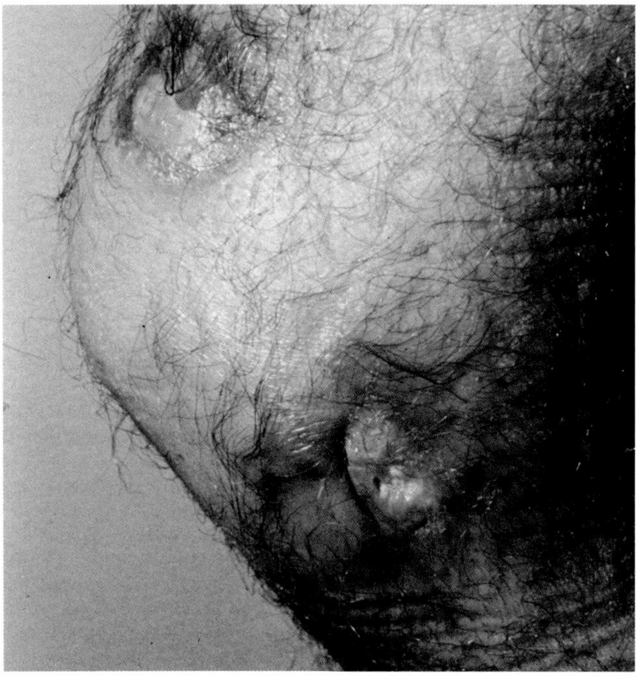

FIGURE 4.2
Chronic suppurative and granulomatous lesions caused by *Mycobacterium marinum* following occupational trauma to the elbow of a 47-year-old marine environmental officer. (Photograph courtesy of B.J. Russell)

myositis with septicaemia, caused particularly by *V. vulnificus* or *Aeromonas* (JANDA & DUFFEY 1988; JOHNSON 1992). Patients typically have underlying disease such as haemochromatosis, chronic liver disease, chronic renal failure, diabetes mellitus, or are immunosuppressed (LEVINE *et al* 1993). With *V. vulnificus*, the infected wounds typically begin as spreading erythematous lesions, which rapidly progress to haemorrhagic bullae (Figure 4.1, page 93), then to necrotic ulcers. These features are accompanied by systemic manifestations of high fever, bacteraemia, rigors, and hypotension due to septic shock. Death may occur in as many as 20 per cent of such patients with this syndrome. *Aeromonas* myonecrosis or pyomyositis is rare, and the manifestations are rapid onset (24–48 hours), rapid progression after trauma, localised oedema, and occasionally the presence of gas in the fascial planes of the muscles (JANDA & DUFFEY 1988). The patients usually are toxic, in considerable pain, with marked localised oedema, and spreading serosanguinous bullae, while bacteraemia is commonly present. Clinically, the infection may resemble clostridial gangrene.

Other marine bacteria, pyogenic streptococci, *S. aureus*, and enteric bacilli may provoke less serious but still severe wound infections, which may lead to septicaemia. The marine bacteria most commonly encountered in these cases are *V. alginolyticus*, *V. damsela* (*Photobacterium damsela*) and *V. parahaemolyticus* (MURPHEY *et al* 1992). There may be rapidly progressive localised infection without metastasis. In such cases the incubation period is short, one to two days, but occasionally it is as little as eight hours. Occasionally, there is a tendency for the bacteria to invade the deeper structures of muscle, tendon, fascia, joint and bone. Patients are febrile, commonly toxic and show leucocytosis in their blood. Occasionally leucopenia may be present. The infection may follow a protracted course, with patients returning for medical attention over many months.

The hands, feet or legs are most commonly involved. On examination, infected wounds are generally painful and typically show purulent drainage, necrosis of the wound margins, and spreading erythema. Crepitation or radiographic evidence of gas formation, and ulceration with associated regional lymphadenitis are rare. Occasionally, haemorrhagic or flaccid bullae at the wound and spreading along the limb follow the initial infection. Without appropriate treatment, mild cellulitis can develop into progressive fulminant disease.

Localised cutaneous infection with *E. rhusiopathiae* is characterised by throbbing pain, tingling, itching and burning with pain in the proximal joints, usually the finger, wrist, elbow or shoulder in 2–7 days of inoculation (CHAN 1988). The lesions are pathognomonic with erythema and dark purple-red areas, which are distinctly demarcated and sometimes elevated. The vesicles may be haemorrhagic. As the lesion progresses, it extends peripherally, and the centre becomes pale and desquamates about 3–5 days later. In about 20 per cent of cases, erythema spreads proximally from the hands to the arms, while fever and severe arthritis occur in about 10 per cent and 5 per cent of cases, respectively. In susceptible patients, dissemination may lead to *E. rhusiopathiae* endocarditis.

Clinical manifestations of *M. marinum* include superficial cutaneous lesions, which are solitary or multiple in a sporotrichoid distribution (FADDA 1988). The infection typically originates as a small papule that becomes slowly and progressively larger, developing a blue-purple hue. The lesions suppurate, sometimes with sinus formation, ulceration or granulation (Figure 4.2). Infection may extend to the tendon sheaths and periarticular tissue, but dissemination beyond these sites is very rare.

DIAGNOSIS

Except for *E. rhusiopathiae* or perhaps *M. marinum* infection, the signs of marine wound infections are not pathognomonic, and the aetiology of the infection cannot be made with confidence from the clinical features. A rapidly progressive necrotising cellulitis with septicaemia may furnish a best-guess diagnosis of *V. vulnificus* infection in a patient with a marine wound, but many other bacteria will produce this syndrome, and without the history of sea water exposure the definitive diagnosis may not be made. Obtaining a history of sea water trauma or exposure, and informing the laboratory of this finding, are crucial for diagnosis.

Localised cutaneous *E. rhusiopathiae* infection (erysipeloid) can be distinguished from erysipelas and cellulitis by the absence of pitting on pressure, the absence of suppuration, and the proximal but not distal spread of the lesions.

The definitive diagnosis of marine wound infections is made by culturing clinical specimens, usually blood, biopsy material (particularly for erysipeloid and *M. marinum* infection) and pus. The salinity of sea water is about 3.5 per cent compared to 0.85 per cent saline for physiological fluids. Hence, culture of specimens from marine wounds solely on media customarily used for non-marine specimens will recover only about 0.1–1.0 per cent of the total number of marine bacteria present. Consequently, for the optimal recovery of marine bacteria, the microbiology laboratory must be informed that the specimens are from a patient with sea water exposure and must then utilise specialised techniques for marine bacteria isolation and identification (MURPHEY *et al* 1992; LEVINE *et al* 1993). Culture media with a high salt content, such as 3 per cent saline–peptone enrichment broth and thiosulphate–citrate–bile salts–sucrose (TCBS) agar, a selective differential medium for *Vibrio*, as well as routine media should be used for primary isolation. Incubating the plates at both 35°C and at room temperature for two weeks will yield increased isolations, because some marine bacteria have low optimal growth temperatures and some grow slowly. *M. marinum*, if not overgrown by contaminant bacteria, can be recovered readily from inoculated blood agar plates that have been incubated at room temperature for 10–14 days. Collecting two sets of blood cultures, and increasing the salinity of one set to 3 per cent will also improve isolation rates of marine bacteria.

Even if growth occurs, marine bacteria may not be identified and consequently not be reported because they are frequently accompanied by other bacteria as a polymicrobial infection and may be disregarded as colonising bacteria. As well, marine bacteria can be easily misidentified with conventional routine microbiological tests. This is because of their many colonial and biochemical similarities to members of the *Enterobacteriaceae*, particularly in the early stages of the identification process. In the routine diagnostic laboratory, the most important features that lead to a presumptive identification of aeromonads and marine vibrios are a positive oxidase reaction, and the fermentative, not oxidative, attack on carbohydrates. Once the laboratory has reached this stage, it is then only a matter of using standardised methods to obtain a definitive identification.

PROGNOSIS

The signs and symptoms of most marine wound infections generally resolve even without antibiotic treatment, although relapses may occur after apparent cure. In the absence of chronic liver, renal or malignant disease, the response to appropriate antibiotic therapy and the outcome

TABLE 4.3
ANTIBIOTIC SUSCEPTIBILITY OF BACTERIA ISOLATED FROM SEA WATER INJURIES IN NORTHERN QUEENSLAND DURING 1990–1991

Bacteria isolated	No.	Percentage of bacteria isolated from marine wounds, which were susceptible to:							
		Flu.	Amox.	Aug.	Doxy.	SXT	Cip.	Gent.	Cefot.
Pyogenic *Streptococci*	19	100	100	100	95	95	100	0	100
Staphylococcus aureus	20	100	0	100	95	100	100	100	100
Enteric bacilli	17	0	35	70	82	65	82	100	100
Marine *vibrios*	12	0	25	42	100	100	100	100	100
Aeromonads	4	0	0	25	100	50	100	100	100

Flu. = flucloxacillin; Aug. = augmentin; Amox. = amoxycillin/clavulanic acid; Doxy. = doxycycline; SXT = co-trimoxazole; Cip. = ciprofloxacin; Gent. = gentamycin; Cefot. = cefotaxime.

of patients infected with marine bacteria is usually good, although those with necrotising myositis suffer severe infection that may lead to limb amputation. The mortality rate may be as high as 20 per cent with death occurring as rapidly as 3–4 days after onset of symptoms with *V. vulnificus* or *Aeromonas* disseminating from marine wounds (JANDA & DUFFEY 1988; JOHNSON 1992). Without specific therapy, infections with *M. marinum* typically progress to chronicity, but the lesions are usually localised, and infections disseminate rarely, and only with immunosuppression.

THERAPY

Marine wounds should be attended to as soon as possible after the trauma. Minor wounds should be given early and efficient toilet {see Chapter 8, page 182}; patients should seek medical attention if the wounds are more severe. Once the wound has been suitably cleaned, most patients only require domiciliary treatment with an antiseptic agent such as povidone iodine (Betadine). Serious injuries including large lacerations, major burns, deep penetrating wounds, and wounds with retained foreign bodies usually become infected and need prompt medical management. Optimal additional therapy of these wounds consists of early administration of appropriate antibiotics and anti-tetanus measures, appropriate surgery under antibiotic cover, and control of coexisting medical conditions.

Septicaemia with *V. vulnificus* or *Aeromonas* usually requires intensive care support to control multisystems failure. Wounds must be adequately irrigated with sterile normal saline or sterile water, be explored to remove foreign objects and to clean puncture wounds, and be debrided to remove crushed or devitalised tissue. Suturing of deep, sea-water-contaminated wounds should be avoided to prevent infection following primary closure. Loose approximation of the wound edges, or delayed wound closure, should be utilised if there is involvement of the deep tissues or if devitalisation is extensive.

Antimicrobial agents usually active against bacteria causing marine injuries are given in Table 4.3. If hospitalised, parenteral therapy should be commenced early with a broad spectrum antibiotic regimen including a tetracycline and an aminoglycoside, such as gentamycin, or an extended-spectrum cephalosporin, such as cefotaxime, to cover for pyogenic streptococci, *S. aureus*, enteric bacilli, and marine bacteria (AUERBACH *et al* 1987; JANDA & DUFFEY 1988; MURPHEY *et al* 1992; JOHNSON 1992). Specimens for culture should be collected on admission to identify the causative bacteria in case the infection is refractory to empirical treatment. Some bacteria likely to cause marine wound infections, such as the flavobacteria, are inherently resistant to the aminoglycosides, the extended-spectrum cephalosporins, and the tetracyclines, but are susceptible to co-trimoxazole and ciprofloxacin. Less serious infections can be treated with orally administered doxycycline or co-trimoxazole. The newer quinolones (ciprofloxacin, etc.) are active against most bacteria, but the efficacy of these drugs in the treatment of marine wound infections is largely unknown. Penicillin, amoxycillin, flucloxacillin, and erythromycin, although active against pyogenic streptococci and some *S. aureus*, have limited or no activity against enteric bacilli and marine bacteria.

If an antibiotic agent is required to treat erysipeloid, then penicillin is the drug of choice (CHAN 1988). Erythromycin or doxycycline is an alternative for patients who are sensitised to the penicillins. *E. rhusiopathiae* is inherently resistant to the glycopeptides, vancomycin and teicoplanin, and these drugs should not be used as specific therapy for complicated infections such as infective endocarditis. For *M. marinum* infection, both surgical debridement with adequate drainage and antibiotic therapy are essential (FADDA 1988). Doxycycline or co-trimoxazole should be administered orally for early or minimal disease, while rifampicin and ethambutol should be prescribed for advanced disease invading the deeper structures of the hand, wrist or elbow. Antibiotic treatment for 2–3 months or even longer may necessary.*

PREVENTION

There are few preventative measures against marine wounds becoming infected other than the prevention of the wound itself by appropriate in-water skills, the use of protective clothing and devices, and those procedures concomitant with good hygiene and early wound toilet. Wounds should be cleaned promptly to remove all foreign bodies as described previously, but fresh, clean drinking water, not seawater, should be used for irrigation. Further contact with the marine environment should be avoided until the wounds have healed.

If the patient is non-immune, tetanus toxoid prophylaxis should be offered.

Despite the potential of soft tissue infection to progress to life-threatening disease with certain marine bacteria such as *V. vulnificus*, prophylactic antibiotics are not warranted for healthy persons with non-infected marine wounds. However, it is advisable to use orally administered doxycycline or co-trimoxazole as prophylaxis for patients who are chronically ill, immunosuppressed, or suffer with chronic liver or renal disease, or diabetes mellitus.

NOTE: Many marine animal species themselves elaborate physiologically active substances, including some with antimicrobial activity (BAKER & WILLIAMSON 1986).

Acknowledgments: We are grateful to Mr Brian J. Russell, Medical Scientist, Bundaberg Base Hospital, for permission to use the photographs of the marine wound infections depicted in Figures 4.1 and 4.2, pages 93 and 94. We also thank Dr David Sutton, Sir George Fisher Centre for Tropical Marine Studies, James Cook University of North Queensland, for information about bacteria present in the marine environment in northern Queensland. Thanks are due in particular to Mr Peter Lowe, Central Queensland Pathology Laboratory, for his timely technical assistance with references.

* One editor (JWB) has found thrice daily 20 minute hot water (up to 40°) compresses or soaks are also effective.

CHAPTER 5

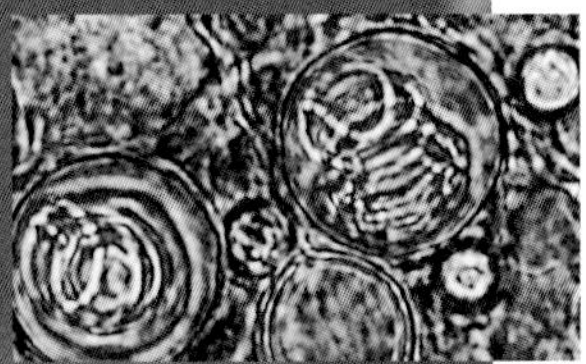

PRINCIPLES OF PATIENT CARE IN MARINE ENVENOMATIONS AND POISONINGS

JOHN WILLIAMSON,

PETER FENNER AND

JOE BURNETT

DIAGNOSTIC PRINCIPLES IN MARINE ENVENOMATION

Early and accurate diagnosis of any marine envenomation or poisoning is important in facilitating correct first aid and medical management. It is also of prognostic importance (BURNETT *et al* 1988a). Diagnosis is often possible, based on one or more of the following features:

1. Observation of the animal involved at the time
2. Geographic locality and climatic conditions at the site
3. Clinical features:
   history
   immediate effects
   delayed effects
4. Retrieval of nematocysts in cnidarian stings
5. Dermatologic and ophthalmologic manifestations
6. Serology and immunology

DIRECT OBSERVATION OF THE ANIMAL

After any marine envenomation it is unusual for the victim and bystanders to observe the animal involved. This may be due to the superb camouflage of some of the animals under their natural conditions (e.g. cubozoan jellyfish {see Plate 5.8} and stonefish {Plate 16.13}). More often the 'observers' do not observe because of the panic and pain of the moment and/or their unfamiliarity with the marine environment, especially if they are tourists. Furthermore, as has been repeatedly verified by experience (SUTHERLAND 1983; BURNETT, FENNER, WILLIAMSON 1993, unpublished observations), most people's identification of venomous animals generally is fraught with error; reliable identification requires expertise (LIMPUS 1987).

Consequently, with occasional exceptions, such as out-of-water envenomations by sea urchins or blue-ringed octopus, direct observation has so far proved a relatively unrewarding diagnostic tool in marine envenomation. It also seems to have improved little with public education. More needs to be done.

GEOGRAPHIC AND CLIMATIC CONDITIONS

It is a simple fact that specific venomous and poisonous animals favour certain regions of the world, and certain ecological habitats; for example, ciguatera poisoning from the flesh of a *freshwater* fish has not yet been recorded. It is impossible to sustain a chirodropid sting under natural conditions in the Antarctic Ocean (HARTWICK 1991b); and it is unlikely in tropical eastern Australia in the middle of the winter season, or out at sea in deep water (HARTWICK 1987, 1991b; FENNER *et al* 1993a). A spiny fish injury in Fiji is unlikely to be from a European weeverfish (HALSTEAD 1988). A blue-ringed octopus envenomation of a swimmer actually immersed in the water is most unlikely (SUTHERLAND 1983).

Thus, although geographic and climatic epidemiology of marine envenomations and poisonings has scarcely begun on an international scale, existing knowledge can provide useful diagnostic clues.

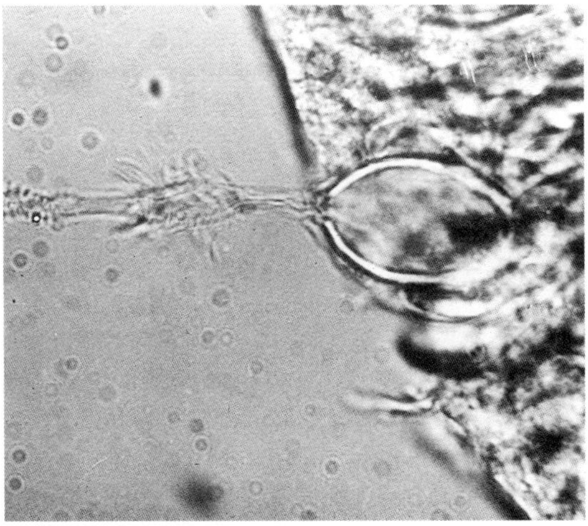

FIGURE 5.1
An unstained preparation of a single discharging *Carukia barnesi* {Irukandji—see Chapter 9, page 246} nematocyst, still *in situ* within the tentacle. Note the shaft spines close to the capsule, and the tiny helical spines visible on the distal tubule. Note also the elliptical shape. (Photograph courtesy of Dr Ron Southcott, Adelaide)

CLINICAL FEATURES

Clinical features provide the most important diagnostic aid, but usually they are only fully elicited retrospectively, by trained observers or experienced medical persons (BARNES 1960). It is common for on-site, enthusiastic but untrained persons (and occasionally self-styled 'experts') to jump to unwarranted conclusions about the cause of the envenomation.

HISTORY

As always, the history of what happened at the sting site and the medical history, taken systematically by a trained person, is invaluable. Many patients and bystanders do not realise the significance of the information obtainable from what they experienced or witnessed (although the potential value of any history degrades rapidly with the increase in time between the event and the history taking).

Objective facts such as 'it was a big jellyfish', or 'there were red tentacles on me', or even 'I saw nothing, even though I was being bitten' may form a piece of the puzzle which fits into place with information obtained from epidemiological and clinical details of the case.

It is particularly important to enquire if the animal (or pieces of the animal) was obtained, if other persons were envenomated in the area at the same time, or even if photographs were taken. The latter are of especial value, and tourists commonly carry cameras.

IMMEDIATE EFFECTS

Immediate effects provide excellent clues. All significant jellyfish and spiny-fish envenomations produce immediate and sharply localised skin pain. Most jellyfish tentacles cause acute inflammatory whip-wheals on the skin, the appearance of which may be absolutely diagnostic of the genus (BARNES 1960; CLELAND & SOUTHCOTT 1965; WILLIAMSON *et al* 1980) {Plate 1.2 and Plate 8.42}. One important cubozoan, 'Irukandji', leaves scarcely any skin mark {Plate 9.23}, itself a diagnostic clue.

The bite of the blue-ringed octopus, uncommonly the cone shell sting, and sometimes a sea snake bite, can be painless. Immediate (2–3 minutes) on-site life-threatening collapse from acute envenomation toxicity is confined virtually to chirodropid stinging (WILLIAMSON *et al* 1980; LUMLEY *et al* 1988), although an African report of two stonefish envenomations, and perhaps another from Japan {see Chapter 3, page 69}, describes such a sequence (SMITH 1957). Uncommonly, however, anaphylactic collapse from marine envenomation may occur just as rapidly (TOGIAS *et al* 1985).

A single, severely painful skin puncture wound on a foot or hand with rapidly developing marked local oedema and a blue discolouration surrounding the puncture site, occurring in someone on a tropical coral reef who did not see what caused the injury, is almost diagnostic of stonefish envenomation* {Plate 16.14}.

DELAYED EFFECTS

If the diagnosis has not already been made, delayed effects may reveal it. More commonly the delayed effects are anticipated by an acute medical diagnosis, and subsequently confirm it. A great range of delayed effects (30 minutes to several hours or days) have been documented — from progressive muscle paralysis (in serious sea snake, blue-ringed octopus, or the uncommon cone shell envenomation) and ascending muscle pains in 'Irukandji' envenomation {see Chapter 9, page 236}, to local tissue necrosis (stingray), ischaemia from vascular spasm (certain jellyfish) {see Plate 7.6} and peripheral neural paralysis (corallimorphorian sting [BURNETT *et al* 1994b]) {see Chapter 8, page 180}. Days or weeks later may bring dermatitis {Plate 11.3}, keloid formation {Plate 9.47}, pigmentation {Plate 7.8} and even visual disturbances (BURNETT & CALTON 1987a; GLASSER *et al* 1992; KOKELJ *et al* 1993).

Uncommonly, delayed effects can result from treatment (e.g. delayed serum sickness from antivenom administration) (SUTHERLAND 1992).

RETRIEVAL OF NEMATOCYSTS FROM THE SKIN IN CNIDARIAN STINGS

Superficial skin scrapings may be examined microscopically for nematocysts adhering to or penetrating the keratin layer. The gentle scraping of freshly stung skin in jellyfish stings, and the subsequent smearing of the scraped material onto a glass microscope slide, permits a low-power microscopic examination (x40 objective) for nematocysts from the offending cnidarian. Recently CURRIE & WOOD (1995) have developed an even simpler technique in which sticky tape is pressed on the skin in the sting area (after vinegar and ice application if necessary {see pages 111 and 167}) and then examined microscopically sticky-side up on a glass slide; this does not produce morphological distortion of the nematocysts (see below). However, this method does not result in a permanent microscopic record for subsequent study. It is desirable, once identification in this way is achieved, for investigators then to prepare a slide using a method which permits permanent and distortion-free preservation of the nematocyst sample. The possibility of preparing a permanent record from this simple sticky tape method of Currie and Wood, is being investigated by Dr Rifkin. (See also her detailed description in Chapter 7, page 167.)

The physical characteristics of either isolated or *in situ* cnidae may be helpful in identifying the animal responsible for a sting. However, only at times does simple low-power inspection of the nematocysts alone, either stained (Plates 5.1 and 5.2) or unstained (Plate 5.3), permit

* A recent report of a marine envenomation on a foot of a wader in north Queensland — fitting the above description precisely — failed to obtain pain relief from early, skilled and specific therapy for stonefish envenomation. Many unknowns remain in marine envenomation!

diagnosis of the actual species of cnidarian that caused the sting (BARNES 1960; CLELAND & SOUTHCOTT 1965; FENNER *et al* 1985; RIFKIN 1988a; LUMLEY *et al* 1988; SCHNADIG *et al* 1991; CURRIE & WOOD 1995). The information obtained may be of value for implementing more specific treatment and also for epidemiology.

MODIFIED METHOD OF NEMATOCYST PREPARATION FROM THE SKIN*

BART CURRIE AND JACQUIE RIFKIN

The Darwin group had earlier found an alternative method of preparation developed by Dr K.A.P. Lee, to be likely to give a positive result. The skin tentacle stings are vigorously scraped as usual with a scalpel blade, which is placed into a 50 ml sterile container (a urine container is ideal). Five to ten millilitres of distilled water are added to the specimen container, which is then sonicated,† thus removing the nematocysts and any cells and debris from the blade into the solution. The solution is syringed through a 13 mm Millipore filter so that the nematocysts are left on the Millipore filter paper. Then an eosin stain is syringed through the filter, and the filter paper is removed and placed on a glass slide, fixed and mounted with a cover slip for microscopic viewing.

Using this method, which results in a permanent identifying slide (Plate 5.4), the Darwin group are correlating clinical features of jellyfish stings with nematocyst identification. As well as *Chironex fleckeri* stings (Plate 5.3), they have thus identified stings from the newly recognised 'Darwin carybdeid' (Plate 5.4) (CURRIE & WOOD 1995), which can present like a mild *Chironex fleckeri* sting. In addition an as-yet unidentified carybdeid (Plate 5.5A) has produced a sting (Plate 5.5B) that has been associated with a severe Irukandji-like syndrome {see Chapter 9, page 249}. A more typical size and appearance of the bell imprint left by the 'Irukandji-type' carybdeid is shown by Plate 5.6.)

This simple retrieving of nematocysts from human skin is one of the most valuable (albeit sometimes imperfect) means of diagnosis in cnidarian stinging. It appears to be virtually unknown outside Australia, although it was described more than 30 years ago (BARNES 1960; CLELAND & SOUTHCOTT 1965). By virtue of their mechanical function, nematocysts nearly always remain attached to the victim's skin following cnidarian stings {see Plate 5.7}. Some may persist on the skin for hours or even days, despite the application of lotions, and/or rubbing and showering (BARNES 1960; STEIN *et al* 1989; WILLIAMSON & FENNER 1993 unpublished observations).

Dr Robert Hartwick, of the Department of Biological Sciences, James Cook University of North Queensland, has constructed a chart that displays the relative morphological features of the nematocysts from the important stinging jellyfish of northern Australia. This chart has been distributed to the Emergency Medicine Departments of many hospitals along Australia's northern tropical coastline, to facilitate the diagnosis of cnidarian stings. Similar efforts in other parts of the world, featuring the cnidarians com-

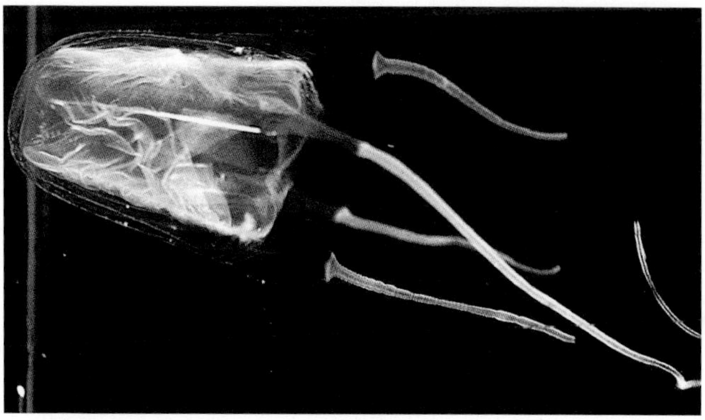

FIGURE 5.2
The 'Darwin carybdeid', photographed under artificially clear conditions. (Courtesy of Royal Darwin Hospital and the Menzies School of Health Research, Northern Territory, Australia)

* By Dr K.A.P.Lee, Forensic Pathologist, Royal Darwin Hospital.

† This method is known to produce distortion of carybdeid nematocysts, but not the larger, tougher microbasic mastigophores of chirodropids. The effects on other cnidarian nematocysts have not been studied by the Darwin group to date. Such distortion may make the detailed study of thus-affected nematocysts difficult.

mon to respective locations, are to be encouraged. Skin scraping microscopy should help clarify the jellyfish species associated with the many varied clinical presentations of stings occurring around the world.

OTHER METHODS OF NEMATOCYST IDENTIFICATION

Nematocysts may be identified by histological examination of autopsy skin biopsy samples (SOUTHCOTT & KINGSTON 1959; LUMLEY *et al* 1988; STRUTTON & LUMLEY 1988) (Plate 5.7).

Where the tentacles of an animal are available, BURNETT and colleagues (1992) have recently described a simple beachside technique of preparing a suspension of nematocysts (using cheap, readily available implements) for later examination.

We encourage people to send skin scrapings or 'sticky-tape skin specimens' (CURRIE & WOOD 1995) with the Marine Stinger Report Form to 'The Marine Stinger Officer', Surf Life Saving Queensland, Inc., P.O. Box 2136, Fortitude Valley, MAC Qld. 4006 Australia (Facsimile 61-(0) 7-32524511). The ideal sample is either left on a dry scalpel or wooden scraper, or smeared on a clean microscopic slide, and transported in a slotted specimen container which protects the nematocyst-bearing surface from abrasion. However, other 'blades' and other containers are acceptable.

DERMATOLOGIC AND OPHTHALMOLOGIC MANIFESTATIONS

Acute epidermal changes are seen in skin biopsy specimens during the early phases of the sting inflammation. Delayed and recurrent reactions to jellyfish stings exhibit a different pathologic picture with a dermal mononuclear cell infiltrate (BURNETT 1992a) {see Plate 7.5}. However, very few epidermal changes will be seen in many of these specimens.

Nematocysts may be detected in the stratum corneum of the skin or the cornea of the eye soon after envenomation (GLASSER *et al* 1993) (Plate 5.10). Nematocysts attached to the cornea may be detected a few hours after the stinging by direct or slit lamp examination if the light is bright and shining sideways onto the eye. Corneal epithelial oedema (Plate 5.10) and inferior punctate epithelial keratitis may also be found (GLASSER *et al* 1992) {see Chapter 11, page 310}.

SEROLOGY AND IMMUNOLOGY

BURNETT and colleagues have pioneered the serological diagnosis of jellyfish stings (RUSSO *et al* 1983; BURNETT *et al* 1983a & 1988b; BURNETT 1992c). Jellyfish immunology and its considerable significance is dealt with elsewhere in this text {see Chapter 7, page 176}. Envenomated victims develop immunoglobulin (Ig) M and G antibodies, within days, which are specific to the species venom injected from nematocysts. There is some cross-reactivity of resulting antibodies, but the antigens in the venom of the actual offending animal usually elicit the highest titre (BURNETT 1991), thus permitting diagnosis months or years later and, conversely, identification of unknown stinger parts from known convalescent serum antibodies.

Serum can be drawn for acute and convalescent IgG antibody titres, which are performed experimentally in some laboratories (e.g. International Consortium for Jellyfish Stings, Baltimore Laboratory, Maryland, USA). A few millilitres of serum can be sent by air mail after the cells of the blood are separated. Lyophilised serum (after measuring the original volume, so that it can be reconstituted) can be mailed overseas. As with other antibody tests a comparison to the acute and convalescent titres is made. The latter specimen becomes positive two to four weeks after the sting. Elevated antibody levels can persist for years (BURNETT *et al* 1988a).

Tests for detecting abnormalities in cell mediated immunity are available experimentally. In patients who have exaggerated, repeated or unusual manifestations of stings, the lymphocyte proliferation and natural killer (NK) cell enhancement assays have been found to be abnormal (BURNETT *et al* 1986c). These assays are more time consuming and require fresh cells.

Immunological advances, such as improvements in titre specificities, and the useful application of monoclonal antibodies (GAUR *et al* 1981, 1982; COLLINS *et al* 1993) will continue to occur.

PREVENTION OF MARINE ENVENOMATIONS

THE NEED FOR PUBLIC EDUCATION

{SEE ALSO CHAPTER 23}

'People will continue to do
not what they should,
but what they want!'
— plaque on the wall of an Australian Rotary Club

Prevention measures in marine envenomation have been repeatedly emphasised in the past because they are so relatively simple and effective (SOUTHCOTT 1959; BARNES 1973, in KINSEY 1986; SMITH 1977; WILLIAMSON 1981 & 1985a; MARSH & SLACK-SMITH 1986; WILLIAMSON *et al* 1987; EDMONDS 1989; BURNETT 1991; FENNER 1991; WILLIAMSON 1992b; WARRELL & FENNER 1993). Nevertheless, public response remains poor worldwide, and there is still a need for public and government* education (WILLIAMSON 1992).

It is encouraging to note that the number of deaths from *Chironex fleckeri* envenomation in Australia has decreased from at least 1 per year for the past century (CLELAND & SOUTHCOTT 1965), to zero in recent years (see Chapter 3, page 70), although the steady destruction of coastal habitats by 'developers' may have contributed to this. Eventual acceptance of the problem by tourist operators in Australian tropical areas has allowed increased promotion of the dangers, especially by Surf Life Saving Queensland, Inc.

BASIC PRINCIPLES OF PREVENTION

Specific details of prevention as they apply to individual venomous or poisonous marine animals, are discussed in detail in specific chapters of this book. However, some basic important principles of prevention (listed in Table 5.1) are common to all situations.

* The Queensland State Government has recently (June 1994) supported publication of information that addresses both the attractions and the avoidable hazards of a holiday in North Queensland.

SAFETY CONSIDERATIONS FOR RESCUERS AND BYSTANDERS

GENERAL CONCEPTS

No marine animal (with the doubtful exception of the shark) has unprovoked malice of forethought towards human beings in the water; humans are not part of their normal food range. Encounters between venomous marine animals and humans not infrequently result in injury or death to the animal. Venomous marine animals will usually attempt to escape from the risk of such an encounter, given the opportunity. It follows that human envenomation is invariably the result of either intentional or accidental interference with the animal or its territory, or of actual entrapment of the animal. The animal reacts purely defensively. Consequently desires to seek 'vengeance' against animals are illogical and should be resisted.

RESUSCITATION

Any resuscitation effort must begin promptly, but only after the victim and rescuers are removed from any on-going danger, and after verifying the need for resuscitation, as described later ('**DRABC**': **D**anger, **R**esponse, **A**irway, **B**reathing, **C**irculation). It is necessary to remember that an envenomated would-be rescuer will not be much help to the initial victim. The techniques of rescue and resuscitation of marine envenomation victims are considered in detail later in this chapter.

PROTECTIVE CLOTHING

As described in Chapter 1 (page 45), it is sensible to wear protective clothing and gloves when entering the water to rescue an envenomated victim. Most significant jellyfish stings cannot occur through close-weave clothing,* even

* There have been at least two unusual, but confirmed occasions when cnidarian stings have occurred directly through the material of an intact stinger suit: a corallimorphorian (see Chapter 8, page 187) and a carybdeid (see Chapter 9, page 242). This should not detract from the protective value of these suits, which offer excellent protection against the vast majority of cnidarian stings; indeed a fatal envenomation from a chirodropid is impossible through a correctly worn, intact stinger suit.

TABLE 5.1
PRINCIPLES OF PREVENTION

1. Know the risk areas, risk seasons (e.g. summertime in the tropics, especially for jellyfish, as shown in Table 5.2) and likely local conditions for possible animal encounters (e.g. shallow sloping sandy bottom for stingrays and chirodropid jellyfish).
2. Do not make 'duck dive' entries into unprotected waters in risk areas. Most people are stung when they move quickly into trailing tentacles of *Chironex* as it swims along in shallow water over a sandy bottom, 'fishing' for its food. Even looking at the water first may be of little help, as the bell and most of the tentacles of even full-grown *Chironex* are translucent and very difficult to see under natural conditions (Plate 5.8). Walk, don't run.† Do not allow children to frolic unsupervised in previously unchecked beach water areas.
3. Choose to swim on beaches where stinger enclosures are present {Plate 1.19}, and/or skilled water supervision is present {Plates 5.9 and 5.11}. Always obey the instructions of beach patrols and lifeguards.
4. Do not handle nor allow children to pick up strange-looking 'rocks' or unfamiliar marine structures. Do not tease or provoke marine animals; observe them, but leave them alone. Do not attempt to capture a live or dead marine animal with which you are unfamiliar, or if you have no experience in such matters.
5. Always wear protective clothing (e.g. a 'stinger suit' {see Plate 1.22} when wading, swimming, snorkelling or scuba diving in isolated or unfamiliar marine environments. When walking on coral reefs or marine rocks, wear stout footwear. (It is appropriate here to emphasise that tramping carelessly over living marine structures wearing boots or shoes is damaging to the marine environment.)
6. If going to an isolated beach or marine area, ensure that you have means of communication in event of an emergency. Scuba divers and snorkellers in the Australasian region can call a 24 hour telephone service in an emergency {see also Chapter 22, page 423}:

   Within Australia (free call):

   DIVERS EMERGENCY SERVICE (DES) 1 800 08 8200

   From outside Australia (user pays):

   International DES Australia 61 8 8373 5312
7. Carry with you three rolls of broad compression bandages {see Figures 5.3A and 5.3B, page 110, and Plate 5.15} and, in box jellyfish sting risk areas, 2 litres of household vinegar {see Chapter 5, page 108}. Where possible check that hot water is immediately available, for the severe pain from many spine injuries from marine animals can be eased considerably by immersion of the envenomated part into hot water.
8. Educate yourself and responsible members of your family in the manual skills of bare-handed cardiopulmonary resuscitation (CPR). Maintain your competence by annual attendance at authorised CPR classes. In many countries of the world, St John Ambulance, Red Cross, Royal Life Saving Societies, Surf Life Saving Organisations and certain scuba-diving training organisations conduct such classes on a regular basis. The life you save may be a member of your own family!

† *Chironex* (and all other cubozoans) have rudimentary 'eyes', enabling them to distinguish light from shadow {see Figures 6.7 and 6.8, page 127}; they can also sense vibration. They will swim away from large, dark objects (such as humans) that may signify danger to them, and will avoid turbulence (vibration) from large objects such as humans moving into the water. Thus, a *Chironex* sp., given the opportunity, will turn and swim actively away from any human near it. This is obvious when one tries to catch specimens for research. This knowledge leads to the advice 'walk don't run' when entering the water. And never dive.

intact pantyhose (BARNES 1971 in KINSEY 1986). However, stinger suits, gloves and thin shoes will not necessarily protect against injury from venomous spines {see Chapter 16, page 354}. Nevertheless, gloves are recommended for the sorting process on trawlers.

CAPTURE OF SPECIMENS

No attempt should be made to capture or handle marine animals thought to be responsible for human envenomation, unless there is some point in doing so, and then only if the person attempting it is skilled and experienced in the

TABLE 5.2
THE RISK PERIOD FOR *CHIRONEX FLECKERI* STINGS IN TROPICAL COASTAL AUSTRALIA, ACCORDING TO THE THE GEOGRAPHICAL LOCATION. NOTE THE INVERSE CORRELATION BETWEEN THE RISK PERIOD AND THE LATITUDE.

Latitude Sth	Place	July	Aug.	Sept.	Oct.	Nov.	Dec.	Jan.	Feb.	March	April	May	June
12° 27′	Darwin		<——					risk period					——>
16° 55′	Cairns					<——							——>
19° 16′	Townsville						<——						——>
21° 09′	Mackay						<——					——>	
23° 22′	Rockhampton						<——			——>			

task. However, parts of venomous animals still attached to the victim (e.g. jellyfish tentacles, stingray barb [Plate 5.12]) should accompany the patient and be preserved for later examination by an expert.

Capture of jellyfish

This should be attempted only by people who know what they are doing and who are wearing protective clothing.

Nematocysts are present on the bell of chirodropids (RIFKIN 1993, personal communication) though they do not appear to be arranged in warts {shown in Plate 9.7}; however, it appears that they may not be present in large numbers. Thus far, specimens of *Chironex fleckeri* and *Chiropsalmus* spp. have been apparently safely handled by the *bell* using bare hands.* Notable species that should not be handled like this are the Irukandji *Carukia barnesi*, the mauve stinger *Pelagia noctiluca* and many large carybdeids {see Plates 9.6A and 9.6B}. Such animals may only be safely handled wearing gloves.

If specimens are to be lifted out of the water, a firm grip is taken on the bell. The tentacles will usually be very long and very 'sticky' catching on everything they touch. Care must be taken that no one is standing close down-wind of the tentacles, as the wind can swing the free end of the tentacles onto their skin.

* Care should be exercised. One of the editors (JWB) has experienced palmar pain after handling *Physalia physalis* and *Chrysaora quinquecirrha*, even though the float and bell, respectively, of these animals appeared 'clear' to cursory inspection.

Specimens to be preserved should be placed in formalin sea water; one part of formalin is mixed with nine parts of sea water. Care must be taken not to inhale the formalin vapour, as it is an irritant to mucosa; and not to touch it without rubber gloves, as it may cause serious skin irritation (EDWARDS & CAMPBELL 1984). Fresh specimens can also be placed directly into a freezer, either immersed in sea water or dry; some damage occurs at times with ice crystals affecting the jellyfish structures, but this is a simple second-best method (RIFKIN 1993, personal communication).

Jellyfish specimens captured should have the name of the collector with contact details; the date, time, place and depth at which it was captured; the weather and wind conditions; the solution it was preserved in; and any other information considered relevant.

PHOTOGRAPHS AND WRITTEN RECORDS

Almost all the photographs reproduced in this book are by observers who were present at the time of envenomation and/or sighting of the animal. Such photographs are of inestimable value when made on the spot (or immediately afterwards), with details of when the photographs were taken in relation to the envenomation. However, such efforts must never be allowed to interfere with resuscitation efforts or endanger the victim, rescuers or bystanders.*

* Surf Life Saving Queensland, Inc., a totally voluntary organisation, would be grateful to receive photographs, which will be duly acknowledged.

PRINCIPLES OF FIRST AID IN MARINE ENVENOMATION

PETER FENNER, JOHN WILLIAMSON, JOSEPH BURNETT AND JACQUIE RIFKIN

DRABC: **D**anger, **R**esponse, **A**irway, **B**reathing, **C**irculation

The principles of rescue and resuscitation in marine envenomation are the same as for any other situation requiring first aid assistance.

First:

1. Remove victim from danger (e.g. surface the victim if submerged and remove from the water as soon as possible). Protect yourself and others from **D**anger.
2. Assess conscious state; 'shake and shout' to check the victim's **R**esponse.
3. Care of **A**irway **B**reathing **C**irculation (**ABC**) (See the flow chart below.)

Then:

4. Prevent increased envenomation (e.g. do not rub the area of a jellyfish sting).
5. Minimise the effects of venom already injected or ingested (e.g. reduce the 'muscle pump' effect by lying the victim down).
6. Summon medical/ambulance assistance where indicated, if this has not already been done. (See the flow chart below).
7. Provide some pain relief when possible (e.g. cold packs for the skin pain of jellyfish stings, and hot water immersion for spiny fish injuries).

These step-wise actions are specific, yet first-aiders can become confused about the priorities. Steps 1, 2, 3 are the familiar '**DRABC**' of resuscitation. *Resuscitation, in the order described, takes absolute priority at all times.*

Table 5.3 lists the most likely marine envenomations in which elements of resuscitation may become necessary. Not uncommonly, only the first steps of the resuscitation sequence will be required — i.e. **D**anger, **R**esponse and protection of the **A**irway — in a spontaneously breathing victim.

TABLE 5.3
MARINE ENVENOMATIONS WHERE RESUSCITATION MAY BE NECESSARY

Box jellyfish *Chironex fleckeri* (north Australian species)
Indo-Pacific and Philippine *Chiropsalmus* spp.
Chiropsalmus quadrumanus
Blue-ringed octopus *Hapalochlaena* spp.
Sea snakes e.g. beaked sea snake *Enhydrina schistosa*, *Astrotia stokesii*
Cone shell *Conus geographus*
Portuguese man-o'-war *Physalia physalis* (Atlantic species)
Severe Irukandji syndrome
Stingray barb injury to trunk (chest or abdomen)
Stonefish *Synanceia* spp.
Respiratory effects from ingestion of poisonous marine flesh (e.g. fugu, shellfish poison)
Anaphylactic reaction to envenomation

Resuscitation may be used in a number of emergencies, but always follows the sequence described in the basic 'flow chart', below. These principles are taught around the world, using resuscitation manikins (AUSTRALIAN RESUSCITATION COUNCIL 1989).

PREVENTION OF INCREASED ENVENOMATION

1. RESTRAIN THE PATIENT'S OR BYSTANDERS' HANDS FROM RUBBING THE FRESH STING AREA.

Although the desire to rub or massage any acutely painful skin area is natural (almost reflex), such action will only serve to discharge more nematocysts adherent on the skin in any cnidarian sting, thus increasing pain and envenomation. Restrain such well meaning attempts, pending the more appropriate first-aid measures, as described below.

2. REMOVE ADHERENT TENTACLE MATERIAL IN JELLYFISH STINGS

Adherent tentacles, avulsed from the bell of the jellyfish by the struggles of the victim in the water, are usually present on the skin (together with discharged and undischarged nematocysts)

FLOW CHART FOR CARDIOPULMONARY RESUSCITATION (CPR)

Send others for ambulance/medical assistance.
Remain with your patient.
Check for **D**anger; move if necessary.
Assess consciousness by shaking and asking simple commands.

UNCONSCIOUS:
Place on side.

CONSCIOUS:
Make comfortable.
Monitor ABC.

CLEAR AIRWAY:

ASSESS BREATHING:

BREATHING ABSENT:
Five deep breaths.

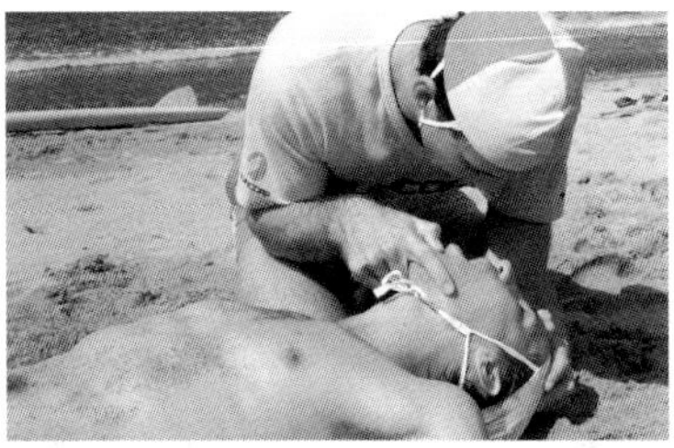

Check pulse.

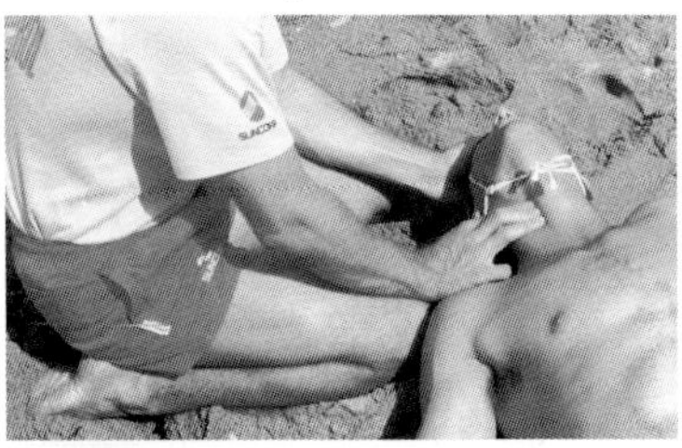

BREATHING PRESENT:
Lateral position.

Monitor ABC.

PULSE PRESENT:
Continue expired air resuscitation.

PULSE ABSENT:
Commence cardiopulmonary resuscitation.

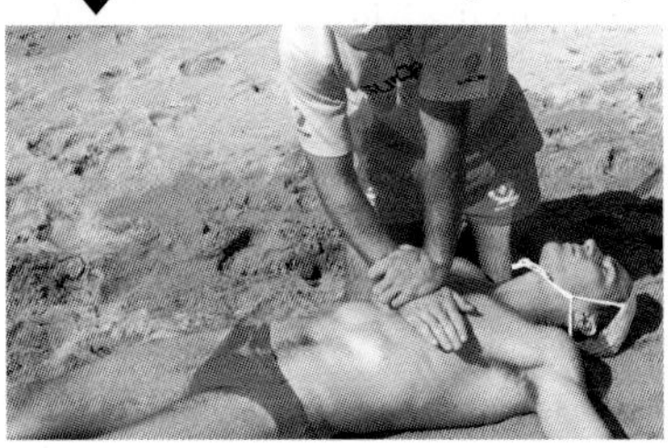

Reassess after the first minute, then every two minutes. Move patient into the lateral position only after successful resuscitation.
Continue to monitor airway and breathing

following *Chironex* stings and *Chiropsalmus stings* {see Chapter 6, page 163}. Actual tentacles are not often present on the skin after other jellyfish stings, but may be seen after envenomation by *Morbakka* {see Chapter 9, page 244} or *Physalia* {see Chapter 8, page 198}; however, nematocysts are always left on skin that has been stung (BARNES *et al* 1960). The longer the fresh adherent tentacle remains in contact with the skin, the greater the risk of further envenomation.

Some nematocysts in jellyfish tentacles are 'adapted' to act as adhesors {see Chapter 6, page 121}. Others work like grappling irons, designed to draw the natural prey (e.g. prawns, small fish) closer to the tentacle as the prey continues to struggle (RIFKIN & ENDEAN 1983). It is therefore important to remove any remaining tentacles to prevent further envenomation. If other methods are unavailable, and in the case of cubozoan stings, *only after treatment with vinegar* {see below} should you pick tentacles off the skin using fingers.

NOTE: The skin on the pads of the fingers usually has keratin that is thick enough to prevent the penetration of the nematocyst tubule {see Chapter 6, page 173} as far as the sensitive deeper epidermal layers, or the vascular dermis. Although envenomation (passage of venom into the skin) happens, no pain or other sequelae occur, although a harmless 'prickling' sensation may be felt. In contrast, the skin of the back of the hand has thin keratin and is easily stung!

3. Inhibition of further discharge of nematocysts

This is an important aspect of first aid for cnidarian stings and is the subject of intense current research (RIFKIN 1993; RIFKIN *et al* 1993: FENNER *et al* 1993b). It seems that the extent of nematocyst-discharge inhibition may be at least partially a species-specific matter (varying from species to species). A summary of current understanding is set out below:

Vinegar (4–6% acetic acid in water)

The original work (HARTWICK *et al* 1980a, 1980b) showed that 2–10% acetic acid in water, when poured on the envenomated skin area in large quantities for a minimum period of 30 seconds, irreversibly inactivated the nematocysts of some jellyfish species including the dangerous northern Australian box jellyfish *Chironex fleckeri*.

Fortunately, since that work, vinegar has been shown to inactivate the discharge mechanism of all other known 'box' jellyfish (cubozoan) nematocysts, including Morbakka (FENNER *et al* 1985), 'Irukandji' (FENNER *et al* 1986b) and *Carybdea rastoni* (FENNER & WILLIAMSON 1987). Vinegar thus remains the most effective method of preventing further envenomation during the first aid management of a fresh sting in all box jellyfish (cubozoan) species. It assumes special importance in serious chirodropid (multi-tentacled cubozoan) envenomations, e.g. *Chironex fleckeri* (WILLIAMSON 1985a; CURRIE 1994), *Chiropsalmus* sp. (Australia) and *Chiropsalmus quadrumanus* (BENGSTON *et al* 1991).

- Vinegar or acetic acid should never be splashed into or on the open eyes, even when they are involved in any sting {see Chapter 5, page 102 and Chapter 9, page 280}.

Pending further studies, vinegar is not currently recommended for the treatment of stings by other jellyfish such as *Cyanea* spp. (FENNER & FITZPATRICK 1986) {see Chapter 8, page 235}, *Chrysaora* (BURNETT *et al* 1983b) {see Chapter 8, page 228}, *Pelagia* {see Chapter 8, page 230} or *Physalia* species (FENNER *et al* 1993b) {see Chapter 8, page 197}. Nor is vinegar recommended for the first aid management of the sting from the stinging hydroid *Lytocarpus philippinus* (RIFKIN *et al* 1993a).

Despite a clear distinction in the original research between 'relief of sting pain' and 'inhibition of nematocyst discharge' (HARTWICK *et al* 1980a & 1980b), confusion persists among some first aiders, and even some researchers. The distinctions are restated here:

- Vinegar does not reverse any effects of venom already injected;

° One study with *Physalia* sp. stings in Sydney, NSW did suggest a weak acute analgesic benefit from vinegar application (TURNER *et al* 1980). Since no subsequent objective confirmation of this observation has appeared with any cnidarian stings, this may be a placebo effect.

- Vinegar prevents further stinging (and thus may also assist in preventing worsening of pain) by inhibiting further nematocyst discharge*

Slurry of baking soda

BURNETT and colleagues (1983b and 1987a) initially recommended inhibiting the discharge mechanism of *Chrysaora* nematocysts with a 'slurry' of baking soda (sodium bicarbonate). As *Pelagia noctiluca* is a close relative, they suggested it might be suitable for this species also. However, further field work by the editors in 1991 indicated that neither baking soda nor vinegar inhibited all *Chrysaora quinquecirrha* nematocyst discharge (Plates 5.13A and 5.13B). Each test substance was applied for a minimum of 30 seconds to fresh tentacles placed on healthy human skin. The treated tentacle pieces were subsequently placed on the thin mucous membrane of the lips and inside the wrists. Mild lip pain, indicating nematocyst discharge, occurred with tentacles treated with either substance.* This may be explained by only partial inhibition of nematocyst discharge, or incomplete cnidom discharge, or 'roofing-over' (HARTWICK *et al* 1980) in which more deeply placed nematocysts are not reached by the vinegar and can subsequently cause envenomation. Microscopic observation of nematocysts was not performed in these field studies.

It should be noted that remaining, adherent tentacles are not often present in *Chrysaora quinquecirrha* stings; inactivation may not then be an essential step in the first aid management of envenomation by this cnidarian.

Other substances

Aluminium sulphate (20 per cent suspension) (HENDERSON & EASTON 1980) in two separate studies did not cause nematocyst discharge in the tentacles of *Chironex fleckeri* (HARTWICK et al 1980), *Physalia* spp. or *Cyanea* sp. (FENNER 1986 unpublished observations); neither did it inhibit nematocyst discharge in *Chironex fleckeri* (HARTWICK *et al* 1980a & 1980b).† However, its actions in other cnidarian or animal and plant stings were not evaluated in these tests.

Papain contributes to inhibition of nematocyst discharge of *Chrysaora quinquecirrha* (BURNETT et al 1983b).

There are many other folk-lore recipes advocated and used for treatment of stings including hot sand, ammonia preparations, extracts of onion, diesoline, petrol, kerosene, paw-paw, aloe vera, wet tobacco and urine. None of these has been shown in published controlled tests to date to be effective for nematocyst discharge inhibition, pain relief or reversal of venom effects. More studies are necessary. It is noteworthy that during recent field trials in the Chesapeake Bay area of the United States with living *Chrysaora quinquecirrha* tentacles, two newly marketed jellyfish sting substances failed to affect either the tentacle stinging ability or the sting pain. Hot or warm water also made no difference, and in fact was subjectively unpleasant on the fresh sting (BURNETT & WILLIAMSON 1991, unpublished observations) (Plates 5.13A and 5.13B).

Recently HEEGER and colleagues (1992) suggested that a layer of water-resistant sunscreen lotion applied to human skin blocked *Cyanea* nematocysts from penetrating the skin, thus preventing envenomation. Other scientists are also studying this approach with other marine stings, but results are unclear at present (BURNETT 1994, unpublished observations).

Water or moisture on dried nematocysts

It has been previously shown that dried undischarged nematocysts from different cnidarians

---

* The editors are aware that self-experimentation with marine venoms is not recommended. However advances in understanding have been achieved quickly by such means (HARTWICK *et al* 1980; BURNETT & CALTON 1987b; FENNER & WILLIAMSON 1987; see also Temple-Smith's work in Chapter 25, page 436). No reader is advised to adopt this method of investigation.

† After the report of the above results relating to the actions of aluminium sulphate suspension, opposing views were expressed by a representative of the company marketing the product (EASTON 1980) and the developer of the product (HENDERSON 1980). Despite a subsequent report (HENDERSON & EASTON 1980), no published report, then or since, has caused modification of the original findings.

in vitro are made to fire by the application of moisture: the sea anemones *Anemonia sulcata* and *Actinia equina* (WEILL 1934), an Atlantic *Physalia* sp. (LANE 1960a), *Chironex fleckeri, Cyanea capillata, Lytocarpus philippinus, Cerianthus sp. and Stichodactyla haddoni* (RIFKIN 1988a) and again *Chironex fleckeri* (ENDEAN *et al* 1991). Workers who have handled jellyfish nets with sweaty hands, after months of storage, have sustained stingings (NORTH BARRIER BRANCH, SURF LIFE SAVING QUEENSLAND 1988–1993, unpublished observations). It is common knowledge among researchers that undischarged nematocysts can remain on unwashed hands for hours (WILLIAMSON and colleagues 1980–1990, unpublished observations), and even inanimate laboratory surfaces for many hours (LANE 1960b; WILLIAMSON and colleagues 1980–1986, unpublished observations), only later to cause stinging on contact (e.g. rubbing an eyelid or touching lips). Either the keratin of the palms is too thick or there may be another modifying factor (BURNETT 1994, unpublished observations) for stinging to be felt there. {See note, page 108}.

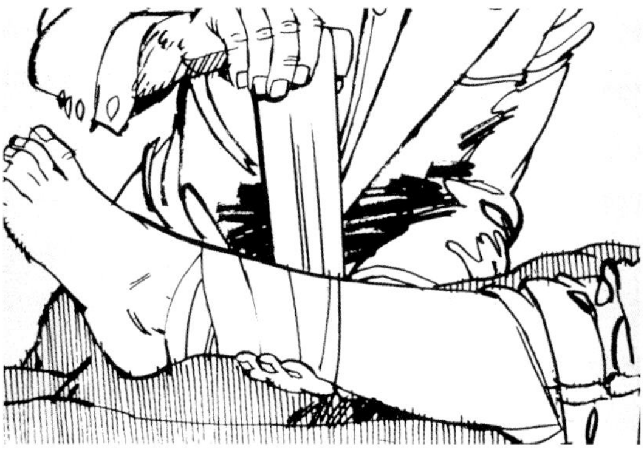

FIGURE 5.3A

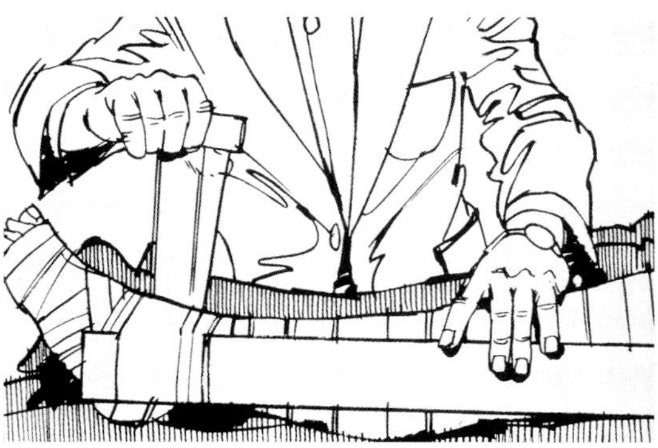

FIGURE 5.3B
Technique of compression/immobilisation bandaging. (Courtesy of Associate Professor S. Sutherland and Commonwealth Serum Laboratories, Melbourne)

BARNES (1966; and in KINSEY 1986) was of the firm opinion that showering after a major *Chironex fleckeri* sting was contraindicated because of the likelihood of further envenomation; he quotes a case [in which methylated spirits but not vinegar was applied initially] where cardiac arrest occurred upon showering.

On these grounds it seems advisable to avoid exposure of dried-off cubozoan stings to any subsequent moisture, prior to the application of vinegar. Once vinegar has been adequately applied (preferably immediately after the stinging, as outlined above), there need be no further concern relating to exposure to moisture; but note that at our present levels of understanding this applies only to cubozoan stings.

In dried off non-cubozoan cnidarian stings, every attempt should be made on the beach to remove tentacles by gently picking off any tentacle fragments (preferably with gloved hands, or forceps if available) and to remove as many undischarged, isolated nematocysts as possible by blotting the area with a clean, dry, disposable tissue if available. Wrapped ice can then be used for analgesia (see RIFKIN Chapter 6, page 121 of this volume.)

MINIMISATION OF VENOM EFFECTS BY FIRST-AIDERS

COMPRESSION/IMMOBILISATION BANDAGING

The entrapment of peripheral venom in limb tissues using a compression bandage is a firmly vindicated first aid measure for people bitten by venomous Australian snakes (SUTHERLAND *et al* 1979) and the funnel web spider (SUTHERLAND & DUNCAN 1980; SUTHERLAND 1983). It is also of the greatest importance as an early first aid measure in the following marine envenomations:

- Sea snake bite (FULDE & SMITH 1984) {see Chapter 17, page 396, and Plates 17.3 and 17.4};

- Blue-ringed octopus bite (SUTHERLAND 1983; WILLIAMSON 1987);
- Cone shell envenomation (SUTHERLAND 1983; WILLIAMSON 1985a).

The splinting of the bandaged limb is to reduce muscle contraction and the resultant 'muscle pump' effect {see Glossary}. It does not matter whether the bandage is applied from below upwards, or above downwards in envenomation, as long as it is first applied directly over the envenomated area, as soon as possible (but after vinegar dousing, following a major chirodropid sting). However, recent studies suggest that the application pressure range should be neither too loose nor (surprisingly) too tight, to impede lymphatic flow of toxins following envenomation. The recommended pressure ranges are 5.3–9.3kPa (40–70mmHg) for an upper limb, and 7.3–9.3kPa (55–70mmHg) for a lower limb (HOWARTH *et al* 1994). This may at first glance present some practical difficulties for first aiders in the field, but application of the bandages so that they are comfortable for the patient (neither constricting nor loose) will produce the correct pressure. (See Figures 5.3A and 5.3B page 110.)

The place of compression/immobilisation bandaging in the acute first-aid management of major chirodropid stinging requires further evaluation; it is not uniformly adopted (CURRIE 1994) but is currently recommended in northern Queensland (WILLIAMSON 1985a; FENNER *et al* 1989a). There is now some circumstantial evidence that it is associated with a favourable clinical outcome (FENNER *et al* 1989a; BEADNELL *et al* 1991). It is not recommended in this situation for other than a *major* chirodropid sting (Plates 5.14 and 5.15).

A major *Chironex* envenomation {see Chapter 9, page 273} is defined as a sting which:

- occupies a total surface (skin) area equal to or more than one half of one limb (Plate 5.14)
- and/or is associated with
  impaired consciousness
  impaired breathing
  impaired systemic circulation
  (e.g. hypotension)
  irregular heart beat (pulse) (not pre-existing)

Compression bandaging should not be used for venomous fish such as stingray envenomations (WILLIAMSON 1985) {see Chapter 16, page 354}.

SPECIFIC ANTIVENOMS

Specific antivenoms provide the most effective counter to the effects of venom already present in the body (SUTHERLAND 1983; WILLIAMSON *et al* 1984b; SUTHERLAND 1992), but there are only a few specific marine antivenoms currently available {see Chapter 21, page 418}, and their usage must be preceded by accurate diagnosis and appropriate precautions, matters that are still under examination (SUTHERLAND 1992; TIBBALLS 1994). Thus their use by first-aiders is strictly limited to special circumstances (FENNER *et al* 1989; BEADNELL *et al* 1992) {see below}.

SPECIFIC DRUG THERAPY

Drugs, correctly chosen and correctly administered, may reduce some of the effects of an envenomation. A good example of this is the monitored use of alpha receptor blocking drugs in the Irukandji syndrome, to reduce some of the severe systemic signs and symptoms (FENNER *et al* 1986b; MARTIN & AUDLEY 1990) {see Chapter 9, page 254}. However, first-aiders' use of such methods is hardly appropriate.

PAIN RELIEF

Although first-aiders are seldom able to prescribe drugs for the relief of pain, fortunately there are two simple techniques that may offer significant pain relief (analgesia) when correctly applied to specific marine envenomations: cold packs or wrapped ice; and heat.

COLD PACK OR WRAPPED ICE TREATMENT

Cold therapy has long been used for the early treatment of acute sporting injuries (DIROX *et al* 1988) and is emerging as a most useful local analgesic for a wide range of localised pain. It is currently the first aid treatment of choice for most jellyfish stings for mild to moderate skin pain (EXTON *et al* 1989). Ice wrapped in a towel is equally effective. (Despite the reservations outlined above concerning fresh water on jellyfish stings the wetness of towels wrapped around ice does not pose a significant hazard. An additional precaution of placing a dry cloth between the ice pack and the skin to prevent ice 'burns' is shown in Plate 5.16.)

In spite of earlier doubts (WILLIAMSON 1985a), ice packs or cold packs (Plates 5.16 and 5.17) are surprisingly of considerable analgesic help in the treatment of even large *Chironex fleckeri* stings (FENNER 1990; CURRIE 1992, unpublished observations), and more recently were very effective on severe localised *Chironex* stinging on lifeguards (INTERNATIONAL CONSORTIUM FOR JELLYFISH STINGS 1993). Cold packs must never be used on top of cubozoan tentacles that have not been first treated with vinegar, and must not be allowed to distract from resuscitation measures.

The local application of cold has also been shown to be effective in other marine envenomations (MULLINS *et al* 1957) and has been successfully used for the pain from a stonefish wound when heat failed to bring benefit (FENNER 1993, unpublished observations).

BURNETT (1990, unpublished observations) observed that ice 'delayed' the pain in stings of the sea nettle *Chrysaora* sp. and the American *Physalia* Portuguese man-o'-war but did not fully alleviate it. This observation was subsequently supported by further field tests by the editors (BURNETT *et al* 1991, unpublished observations).

The practical application of cold therapy:
1. Warn the patient that ice or cold packs may be uncomfortable at first! Wrap cold pack or ice in a cloth and apply directly over the sting site (apply only to the stung area to avoid systemic hypothermia, especially in small children).
2. Reassess pain after 15 minutes, and reapply the cold pack or ice if necessary.
3. Reassess pain after second application.
4. Send for medical assessment if cold fails to relieve pain, or if other symptoms develop.
If this simple treatment is not effective, the patient is referred to medical help.

Large stings: If a sting covers a large area, particularly if the victim is cold or wet, using a large number of cold packs (Plate 5.17), or packing the area with ice may cause hypothermia (low body temperature). In these cases 'ice massage' seems to help.

Ice massage: A small block of ice is held with a glove or grasped with a cloth, and rubbed by the first-aider all over the stung area for 10–20 minutes or until pain relief occurs. At the same time, the rest of the body can be kept warm with blankets or clothing.

'Icy Cup': If water is frozen in a paper cup, the top part of the cup can be torn back to expose the ice while the first aider's hand remains insulated (Plate 5:18).

HEAT TREATMENT

Heat treatment is recommended (WILLIAMSON 1985a; FENNER *et al* 1989b) and has proven effective (FENNER *et al* 1993c) in the treatment of penetrating injuries caused by venomous barbs or spines {see also Chapter 16, page 361}, including:
- stingray spines
- stonefish spines
- bullrout spines
- other fish armed with venomous spines, including 'happy moments', butterfly cod, and weeverfish {see Chapter 16, page 354}

Heat may be useful with penetrating injuries from echinoderms including the crown-of-thorns sea stars, but further evaluation is required {see Chapter 12, page 312}. Heat application is contra-indicated in the treatment of *Physalia* ('bluebottle') stings (BURNETT & CALTON 1977).

The practical application of heat therapy:
1. First, test the temperature of the water with your own limb, and aim for a temperature that is tolerably hot (e.g. up to 40°C {104°F}. Test this yourself; do not ask the victim to do it. Because of the pain of envenomation, the victim may be unable safely to discriminate 'hot' from 'too hot' with the affected limb (or any part of the body!), and a hot water burn could result.
2. After the water temperature has been so tested, immerse the patient's affected limb in the hot water until the pain is reduced, or relieved. It may be necessary to change the water or 'top it up' with more hot water several

times. Test the new temperature each time, prior to reimmersion of the patient's limb.

OTHER METHODS OF PAIN RELIEF

Papain (meat tenderiser)

In an uncontrolled trial, papain has been recommended as a treatment to relieve the skin pain of jellyfish stings (ARNOLD 1979). Its use is based on the theory that protein venom will be denatured by the papain. However, subsequent studies failed to show any pain relief benefit in stings from *Physalia physalis* or the North American sea nettle *Chrysaora quinquecirrha* (BURNETT *et al* 1983b), although in these studies papain was shown to inhibit the nematocyst discharge of *Chrysaora quinquecirrha*.

Entonox

Entonox is a gaseous mixture of 50 per cent oxygen and 50 per cent nitrous oxide, usually administered from a demand flow apparatus by facemask, to a spontaneously breathing, co-operative patient. It has some well-established, useful but limited analgesic properties (STEWART 1988). It is carried by many Australian, British (and other) ambulance services and is widely used for first aid analgesia generally. It has provided only minor relief of pain from the muscle spasms of the Irukandji syndrome (FENNER *et al* 1986b), but documented experience with its analgesic effects for the pain from other marine envenomations is limited (BEADNELL *et al* 1992). Entonox (or any nitrous oxide, such as may be used in general anaesthesia) should not be administered to scuba divers immediately after diving, or for at least six weeks after an episode of decompression illness (DCI), because of the risk of re-inducing symptoms of DCI (ACOTT & GORMAN 1992).*

'Penthrane' (methoxyflurane) inhaler

'Penthrane' is the volatile liquid analgesic agent, methoxyflurane, which was initially used in general anaesthesia until its dose-dependent toxicity to the kidneys, when used in high doses, was appreciated. It is a powerful analgesic in low doses, when used by a fully conscious patient from a self-administered, disposable, single-use, draw-over vaporiser fitted with a simple mouthpiece.† Used correctly in this way, and preceded by trained instruction, toxic doses are not possible; the small, hand-held vaporiser can actually be thrown to a victim who is trapped on a rocky shore, immediately inaccessible to rescuers, and in pain. However, its onset and offset are rather slow (it has a high blood/gas solubility partition coefficient); it may be unsafe in a drowsy patient; and prolonged use may cause clouding of consciousness and perhaps result in an unprotected upper airway. The vapour has a characteristic (not unpleasant) fruity odour. The inhaler may be used with either air or oxygen. It is also carried by some Australian ambulance services.

SPECIAL NOTE ON EYE STINGS

{See also Chapter 5, page 102 and Chapter 11, page 310}

Jellyfish stings directly to the eyes require immediate medical referral. Apart from the removal of obvious pieces of tentacle on the face, there should be no first aid attempts to treat eye stings, other than simple irrigation with clean saline sea water, followed by pad cover of the eye(s) and immediate medical consultation (GLASSER *et al* 1992). Do not douse or irrigate the eyes with vinegar. Also, at the present level of understanding, fresh water irrigation may be best avoided (see above) unless there is absolutely no alternative.

FOLLOW-UP MEDICAL MANAGEMENT

Follow-up medical management will be determined on an individual basis, dictated by the condition of the patient, the experience of the

* Nitrous oxide enters any nitrogen (or any other breathing gas) 'bubbles' remaining in the diver's tissues, simply because the partial pressure of nitrous oxide in the bubbles is normally zero (Dalton's law of partial pressure of gases). This will result in the (re-)expansion of the bubbles and the (re-)production of symptoms.

† Experience with its use for marine envenomation pain is limited. It has given only minor relief of pain in the muscle spasms of an Irukandji syndrome (FENNER *et al* 1986b).

medical officers and the facilities available. Medical management can be considered under the following broad headings: Advanced life support, analgesia, antivenom administration, management of peripheral vascular compromise, other specific drug therapy or measures, long term management.

ADVANCED LIFE SUPPORT

Intermittent positive pressure ventilation (IPPV), cardiovascular and renal support, pain relief and/or prevention of secondary infection may variously be necessary following serious envenomation by chirodropids (WILLIAMSON *et al* 1980), carybdeids (FENNER *et al* 1988), sea snakes (FULDE & SMITH 1984), blue-ringed octopus and cone shell (SUTHERLAND 1983). Such support may also be required for extreme cases of ciguatera poisoning {see Chapter 15, page 346}, stingray spine injury (FENNER *et al* 1989; KNIGHT 1989), and uncommonly in a variety of other marine envenomation situations (TOGIAS *et al* 1985; BURNETT *et al* 1994a). In serious Irukandji stings producing cardiovascular complications (FENNER *et al* 1988; MARTIN & AUDLEY 1990), IPPV will correct hypoxia and assist with control of pulmonary oedema.

ANALGESIA

Many marine envenomations are accompanied by acute and protracted severe pain: e.g. spiny fish injuries (SUTHERLAND 1983; WILLIAMSON 1985a), large chirodropid stings (WILLIAMSON *et al* 1980; BEADNELL *et al* 1992) and Irukandji syndrome* (FENNER *et al* 1988). Effective analgesia is important.

NARCOTIC ANALGESIA

The severe pain of spiny fish injuries, and the severe muscle pain of the Irukandji syndrome, responds in the early stages to:

- Morphine 0.1 mg/kg (to an adult dose of 5 mg) as an IV bolus, and thereafter 2.5 mg increments as necessary every 5 minutes until pain relief is adequate.
- Pethidine 1.0 mg/kg (to a maximum adult dose of 50 mg) as an immediate intravenous (IV) bolus dose, and followed by an infusion of 30 mg per hour for adult patients.

Such therapy should be conducted with respiratory resuscitation skills and facilities immediately available.

LOCAL ANAESTHESIA

This may be dramatically effective for pain localised to one site (such as in stonefish or weeverfish spine injuries), especially as a regional nerve block (WILLIAMSON 1985; FENNER *et al* 1992). Local infiltration of the anaesthetic around the wound may also be very effective (FENNER *et al* 1989b; BENDT & AUERBACH 1991). Such measures are best conducted by experienced medical persons under safe conditions of patient monitoring.

ANTIVENOM ADMINISTRATION

{See Chapter 21, page 418.}

There is little therapy of a specific nature in most advanced life support measures for a serious marine envenomation, the outstanding exception being the administration of specific antivenoms when the identity of the animal is known with certainty (SUTHERLAND 1992). It is important also to appreciate that the successful use of a specific antivenom in a painful marine envenomation may confer excellent analgesia, and may reduce scarring as well (SUTHERLAND 1983 & 1992; WILLIAMSON *et al* 1984b; FENNER *et al* *1993b*). However, the use of antivenom for analgesic or cosmetic indications alone must be based upon expert advice.

Antivenoms available for marine envenomations in Australia are listed in Table 5.4.

TABLE 5.4
MARINE ANTIVENOMS AVAILABLE IN AUSTRALIA

Northern Australian box jellyfish *Chironex fleckeri* antivenom
Stonefish *Synanceia horrida* (= *S. trachynis*) and *S. verrusosa* antivenom
Beaked sea snake *Enhydrina schistosa* antivenom

*BARNES suggested that oral aspirin was very effective for Irukandji syndrome muscle pain after the initial severe stage had passed (KINSEY 1988).

The use of specific antivenoms in marine envenomation is discussed in detail in Chapter 21, page 418. In general, specific antivenom administration for a marine envenomation, when available, should be considered in the following circumstances:

- a life threatening envenomation
- unrelieved severe pain
- to minimise scarring when the cosmetic outcome may be poor (e.g. a massive sting [Plate 1.2], or significant facial stings)

Antivenom should be administered only in the presence of expert medical (or selected paramedical) support, preferably slowly intravenously,* and in a dilute form (SUTHERLAND 1992). However, intramuscular use by trained paramedics outside hospital is effective (FENNER *et al* 1989a; BEADNELL *et al* 1992).

MANAGEMENT OF PERIPHERAL VASCULAR COMPROMISE

Early rehydration and maintenance of good fluid balance, aggressive analgesia and the administration of oxygen (preferably 100%) are essential.

Care should be exerted to help preserve local limb blood supply by keeping the patient warm. (For most spiny fish injuries, e.g. stingray and stonefish, and for some cnidarian envenomations [WILLIAMSON *et al* 1988], compromise of perfusion in the area of the sting, rather than systemic absorption of venom, is a greater concern.) Peripheral pulses over the involved extremity should be monitored by direct physical examination and/or Doppler measurements. Any circulatory compromise should be aggressively treated with local measures such as intra-arterial urokinase (50 000 i.u.) infused over a period of 30 minutes and repeated if necessary. Systemic heparin, dextran, corticosteroids, intra-arterial reserpine and intravenous pentoxifylline have been ineffective. In one patient a cervical dorsal sympathectomy produced no improvement. Tolazoline intra-arterially was also of little apparent benefit.

* Stonefish antivenom is more commonly administered intramuscularly.

Judicious anticoagulation and intravenous infusion of lignocaine, vasodilators (such as papaverine), prostaglandin E_2 or antiplatelet aggregation agents (i.e. prostacyclin) should be employed, with careful monitoring. Surgical intervention and/or hyperbaric oxygen therapy should be reserved for unresponsive cases, but decisions to employ these interventions should be made early rather than late in the clinical course.

OTHER SPECIFIC DRUG THERAPY OR MEASURES

α-ADRENERGIC RECEPTOR BLOCKADE IN THE IRUKANDJI SYNDROME

BARNES (1964) was the first to document the use of alpha-adrenergic receptor blocking drugs for the control of the severe systemic hypertension that may be part of the Irukandji syndrome. The efficacy of this therapy has subsequently been confirmed and appears also to reduce the general duration and severity of the whole syndrome (FENNER *et al* 1988; MARTIN & AUDLEY 1990).

Catecholamine-like effects in the Irukandji syndrome may produce sweating, pilo-erection, anxiety state, restlessness, headache, nausea, hyperventilation, tremor, pallor, peripheral cyanosis, oliguria, tachycardia, extrasystoles, hypertension.

Management: Phentolamine (a short-acting α-adrenergic receptor blocking drug) is given intravenously (IV) as a loading dose of 5 mg, followed by 1.0 mg IV, repeated when necessary until there is satisfactory control of the blood pressure, sweating, anxiety and tremor (FENNER *et al* 1986b, 1988). MARTIN & AUDLEY (1990) suggested 2.5 mg was sufficient to control these symptoms. Careful blood pressure monitoring is essential using this regime. The use of longer acting α-adrenergic receptor blocking drugs, such as phenoxybenzamine, requires intensive care facilities for safety.

ANTIBIOTICS

According to microbiological culture and clinical assessment, specific antibiotic therapy may be indicated {see Chapter 4, page 96}, especially in cases where secondary infection becomes established (Figures 4.1, page 93 and Plates 5.20A, B and C).

ANTI-TETANUS PROPHYLAXIS

Tetanus has caused death after a penetrating marine wound (stingray [RATHJEN & HALSTEAD 1969]).* All significant marine wounds should be given a tetanus toxoid injection unless the patient has proven current immunity. Any deep puncture wound (e.g. stonefish, stingray) or any marine wound with retained foreign body fragments (e.g. coral cuts and abrasions) should alert the medical officer to the risk of tetanus; appropriate tetanus immunisation should be ensured, although it should never replace careful wound exploration and primary wound toilet, under appropriate anaesthesia. Primary wound closure under such circumstances also carries the risk of infection from other dangerous clostridial organisms. Suspicious or heavily contaminated wounds that are difficult to clean and explore should be X-rayed, and left open pending further skilled assessment.

PENETRATING WOUNDS

Such wounds can also prove difficult to heal (BENDT & AUERBACH 1991), although the following treatment has been found successful for the management of penetrating injuries of stingray envenomation:

1. If necessary inject local anaesthetic (1% plain lignocaine) directly into the wound and around the site of envenomation to ease the pain.†
2. X-ray the wound area for traces of foreign material.
3. Explore the wound carefully, following the full length of any tract left by the penetrating spines. Remove all traces of broken barb or integumentary sheath and clean the wound using a suitable non-toxic tissue antiseptic; use an operating microscope if available.
4. Leave the area open to granulate, or suture loosely to allow drainage from deeper tissues.
5. Give tetanus vaccine or 'ADT vaccine', if indicated.
6. Consider using antibiotics if the injury is more than six hours old and/or is extensive (check for any drug allergy or intolerance).
7. If you think that the wound has penetrated the abdomen or chest, immediately refer to a major medical centre for further assessment.
8. Follow up all wounds until they are completely healed.

EYE STINGS

{See Chapter 5, page 102, and Chapter 11, page 310} (HERCUS 1944; MITCHELL 1962; CLELAND & SOUTHCOTT 1965; GLASSER *et al* 1992)

Corneal stings may produce eye pain, foreign body sensation, tearing, photophobia and decreased visual acuity. Medical care is a specialist ophthalmological matter, but pending this, topical anaesthesia will ease the unpleasant corneal pain.

Specialist slit lamp examination, removal of impacted nematocyst bodies if necessary, and empirical therapy directed towards control of intra-ocular pressure, inflammation and secondary infection are advised (GLASSER *et al* 1992).

LONG-TERM MANAGEMENT

SCARS AND COSMETIC CONSIDERATIONS

Troublesome itching in the healing scar is not infrequent after cnidarian stinging (WILLIAMSON & CALLANAN 1981, unpublished observations; BEADNELL *et al* 1992). Although this is self-limiting with time, interim symptomatic treatment may be required.

Delayed and recurrent inflammatory skin reactions are also well documented (BURNETT 1991; KOKELJ *et al* 1993) and are discussed elsewhere in this text {see Chapter 9, page 177}.

Disfigurement from scarring {Plates 9.47 and 9.48} and keloid formation (BURNETT 1991) may

* Non-fatal tetanus has occurred in a wound from a terrestrial snake bite (CALLANAN 1980, unpublished observations).

† During a study of 100 cases of small stingray envenomations it appeared that when 1% plain lignocaine was used to anaesthetise the painful wound prior to treatment, its analgesic effect seemed to take effect much more rapidly than usual, suggesting either enhanced susceptibility of venom-affected peripheral sensory nerves to lignocaine, or some possible direct inactivation of the venom in the wound; or both (FENNER *et al* 1989b).

be of particular concern following chirodropid (multi-tentacles) and other selected cnidarian stings. This is especially in view of the high incidence of intimate tentacle–skin contact with such stings in the relatively hairless limbs of children and females (WILLIAMSON 1983 & 1985a; VIMPANI *et al* 1988) (Plate 5.19A). It is also important in the uncommon event of facial stings.

Apart from the life-saving indications, the appropriate use of a specific Commonwealth Serum Laboratories' antivenom for any *Chironex fleckeri* sting for which a long-term cosmetic problem can be anticipated, prevents or significantly reduces long-term disfigurement (WILLIAMSON *et al* 1980; SUTHERLAND 1983; WILLIAMSON *et al* 1984b; BARNES in KINSEY 1986) (Plates 5.19A and 5.19B).

The prevention or aggressive management of secondary infection in such stings (Plate 5.20A) {see also Plates 9.65 and 9.66}, and in puncture wounds such as stingray injuries, will also assist with a better cosmetic result. Rarely plastic and reconstructive surgery of scarring may be required, more likely for chirodropid stings (Plates 5.20B and 5.20C).

PSYCHOLOGICAL AND STRESS REACTIONS

Apart from life-threatening envenomations, the savage pain, panic and distress in victims, parents and even rescuers can result in 'post-traumatic' stress reactions (McFARLANE 1989, 1991; BEADNELL *et al* 1992; WILLIAMSON 1992, unpublished observations). This has been a relatively neglected area but should be anticipated and searched for tactfully over ensuing days and weeks by attending physicians, in non-fatal cases as well as fatalities. The advent of such sudden distressing events affecting children during what is commonly a family holiday, can be especially psychologically disruptive in the absence of counselling. Long-standing phobias concerning the ocean and the beach, affecting several family members, have resulted. Skilled counselling of post-traumatic stress reactions is usually very successful (McFARLANE 1991; MARMAR *et al* 1993).

SECTION TWO

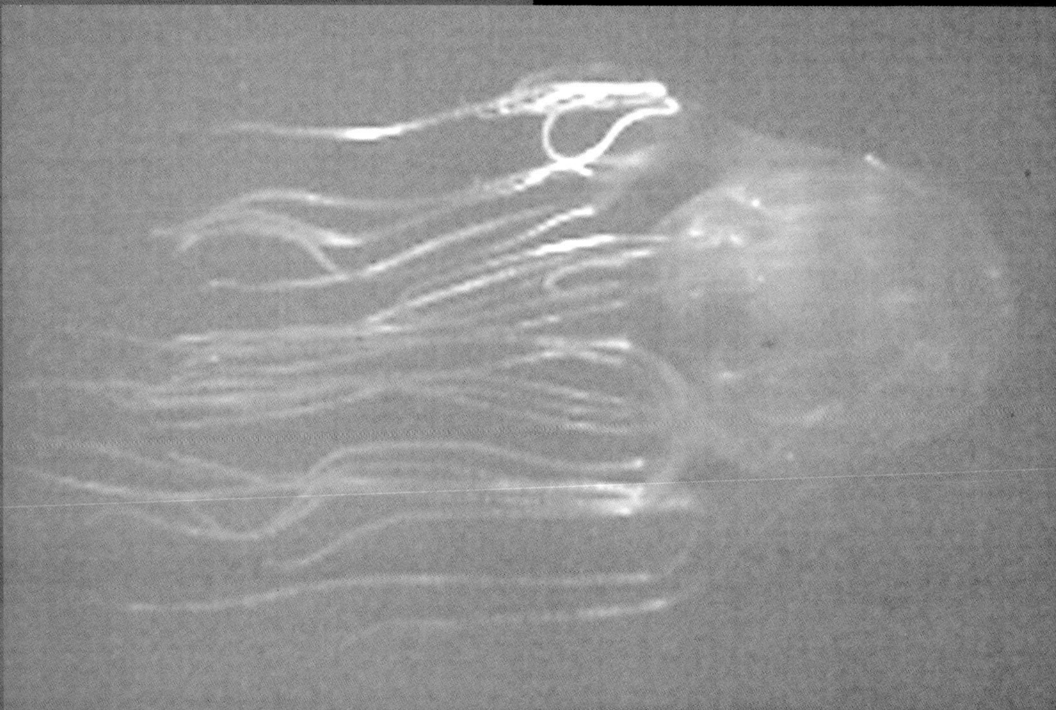

CNIDARIANS: ANIMALS WITH ENDOGENOUS CNIDAE

INTRODUCTION

JACQUIE RIFKIN

Biologists currently place 'Cnidaria'[1] at the taxonomic level of phylum. This term is derived from the Greek word for 'thread'[2] and nearly all species possess a myriad of minute capsules called cnidae, containing threadlike tubules.[3] The old phylum Coelenterata once contained gelatinous animals called ctenophores[4] or 'comb jellies' as well as jellyfish and their relatives. The ctenophores were later removed from this group and placed into a phylum of their own (Ctenophora), leaving the jellyfish, sea anemones, corals and other organisms as cnidarians. The term 'coelenterates' is still used occasionally to refer to both phyla.

[1] The 'c' is silent in such terms as 'Cnidaria', 'cnidae', 'cnidom', 'ctenophore' and 'Ctenophora'.

[2] Other groups of invertebrates including protozoans possess different kinds of 'thread organelles'. These are adapted for anchoring the organism during feeding, defense and prey capture.

[3] Cnidae is the general term for the three categories of capsular organoid found within cnidarians. They are nearly ubiquitous throughout the phylum and include nematocysts (stinging capsules), spirocysts and ptychocysts, the latter two comprising non-stinging capsule categories exclusive to the class Anthozoa (e.g. corals, sea anemones).

[4] Ctenophores are gelatinous marine organisms that are mainly pelagic. They possess bands of cilia called combs which propel them through the water. Some ctenophores resemble medusae. Stinging cells are absent, though they have special adhesive cells adapted for prey capture.

[5] Stinging cells are absent in *Mastigias* a rhizostome jellyfish in the Palau lakes. These animals cannot sting and are totally dependent upon photosynthetic processes of symbiotic algae to obtain their nutrients.

Cnidarians harmful to humans fall into all four taxonomic classes:

- class Hydrozoa (plume-like hydroids, 'fire corals', jellyfish with small bell-shaped bodies called medusae, and members of the subclass Siphonophora which may be buoyed up by gaseous floats.)
- class Scyphozoa (jellyfish, such as the 'hair jelly', *Cyanea* spp.)
- class Cubozoa (box jellyfish, such as the deadly northern Australian species, *Chironex fleckeri*)
- class Anthozoa (many body forms such as sea anemones, hard and soft corals, sea whips, sea fans)

Some of the phylum's major characteristics include the exhibition of polymorphism (wide varieties of form presented by a single species) and alternation of generations (where both a medusoid [or 'jellyfish'] form and a polyp form are present during the course of the life cycle). Indeed, the morphological variation found among representatives of this entire phylum is enormous. Nearly ubiquitous within the phylum are the cnidae (Figure 6.43, page 168) which are employed for prey capture and for defence.

However, there are some cnidarians without nematocysts[5] (e.g. the rhizostome medusa *Mastigias* in the Palau lakes).

Cnidarians are the only animals that manufacture nematocysts, but other aquatic creatures sequester them from various cnidarians to use them for their own protection. These 'stolen nematocysts', or 'kleptocnidae', are thus found in such extrinsic sources as sea slugs, freshwater turbellarian flatworms, comb jellies, and sponges. Although there are many similar examples of the use of living cnidarians for defence, some species of crab carry nematocyst-laden anemones on their shells or in their claws {see Plate 8.7} and use them as weapons against potential predators.

A description of the more important physical characteristics of the phylum follows.

CHAPTER 6

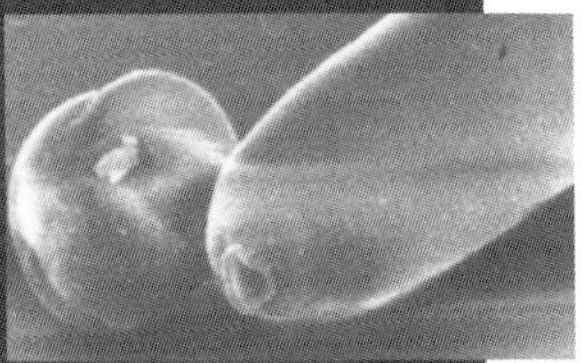

JELLYFISH MECHANISMS

JACQUIE RIFKIN

JELLYFISH ANATOMY AND BEHAVIOUR

Three of the four classes of phylum Cnidaria are represented under the broad but familiar heading of 'jellyfish'. Mainly marine inhabitants, the members of the classes Hydrozoa, Scyphozoa and Cubozoa all develop 'medusa', or jellyfish stages in their life cycles. The fourth class, Anthozoa, includes the sea anemones, true corals and other benthic (sea-bottom) life forms. In this class, only the polyp form is present.

Jellyfish of the three classes of Cnidaria in which the medusa stage is present will be termed 'hydromedusans', 'scyphomedusans' or 'cubomedusans', respectively. Bluebottles and other hydrozoans that exist as colonies of individuals are also commonly regarded as 'jellyfish' by the general public, but will be given separate consideration as is usual in zoological texts.

The basic cnidarian life cycle usually involves an asexually reproducing polyp stage (Figure 6.1a, page 122), alternating with a sexually reproducing medusoid stage (Figure 6.1b, page 122). The polyp is sessile and usually found attached to a substrate such as rock or seaweed, whereas most medusae are free-living, dispersing widely to areas that are ideal for survival and reproduction. As most medusae move about to survive, they are structurally and behaviorally more complex than polyp forms.

Hydrozoans include both freshwater and marine species. They may be either solitary or colonial in form. The medusae are usually small and inconspicuous, reaching no more than a few centimetres in diameter.

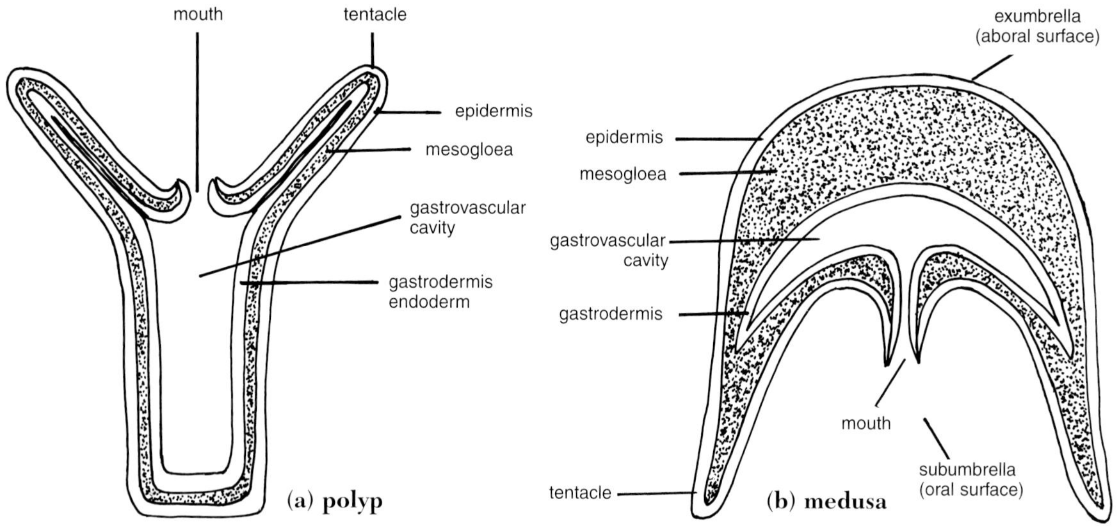

FIGURE 6.1
The basic cnidarian body plan of (a) the polyp and (b) the medusa. (After HICKMAN, HICKMAN and HICKMAN 1974, redrawn by Jan Hynes.)

All **scyphozoans** are marine, and the medusa stage is dominant in the life cycle. They are usually quite conspicuous because of their size and colouration and are often found washed up on the beach after onshore winds. Although they usually range from 2 to 40 cm in diameter, AGASSIZ (1865) recorded a specimen of the jellyfish *Cyanea capillata* in Massachusetts Bay that measured 2.3 metres across the bell.

Cubozoans are marine and are highly inconspicuous, being almost invisible in the water {see Plates 5.8}. The body or bell of the deadly northern Australian box jellyfish *Chironex fleckeri* is cube-shaped and may grow rather larger than a human head {see Plate 1.4}.

Most jellyfish come to the attention of bathers only when they cause a painful or irritating stinging sensation. However, at present the first aid measures to be applied to a human victim of a hydromedusan sting are not necessarily the same as those to be applied to the victim of a scyphomedusan or a cubomedusan sting (HARTWICK *et al* 1980; FENNER *et al* 1985; FENNER & FITZPATRICK 1986; FENNER & WILLIAMSON 1987; FENNER *et al* 1993). Thus knowledge of the major features of the three classes of jellyfish of medical importance will assist in the administration of appropriate first aid. (See the Glossary for additional definitions of the terminology used below.)

OUTLINE OF ANATOMY

Every medusa resembles more or less an inverted umbrella, saucer-shaped disc, or a bell (Figures 6.1b, above, and 6.6a–d, page 126). Structures situated on the outside of the umbrella are thus associated with the **exumbrella**; the exumbrella surface is also termed the **aboral** surface. The **subumbrella** refers to the area on the underside of the umbrella; this area is also called the **oral** surface.

Cnidarians typically follow a radial symmetry (Figure 6.2), in which there are many equal portions that radiate outwards like the spokes of a wheel. Most jellyfish exhibit tetramerous radial symmetry in which the animal can be divided into four symmetrical parts about a central axis (Figures 6.3 and 6.5, page 125).

The various anatomical structures found in the jellyfish are arranged with reference to particular vertically distributed radial planes (see Figures 6.3 and 6.4, pages 123 and 124). Four radial planes or radii forming right angles (90°) to each other and corresponding to the 12:00, 3:00, 6:00 and 9:00 positions on a clockface are called the perradii (Figures 6.3 right, 6.4 page 124, 6.27 page 149 and 6.29 page 151). They indicate the corners of the mouth. The **interradial planes** or **interradii**, also forming right angles to each other, are found halfway between the four perradii, and they bisect the four quadrants of the jellyfish (Figures 6.3 right, and 6.4 page 124). **Adradial planes** (Figures 6.3

and 6.4) occur halfway between the perradii and the interradii.

HYMAN (1940) states that cnidarians are at the 'tissue grade of organisation' (i.e. evolution), where different cells of different functions are gathered together to form tissues but are not themselves organised into organs. However, this idea is presently being debated (e.g. see FAUTIN & MARISCAL 1991). Evidence (see later in this chapter) suggests that within cubomedusans at least, the structure and functioning of the lens eye [Figure 6.8, page 127] are complex, and challenges HYMAN's original concept. In this text the cnidarians are not strictly classified anatomically on the basis of their organ systems. However, anatomical differentiation of medusae commonly permits recognition of a central or 'vegetative system' and a marginal or 'sensory system'.

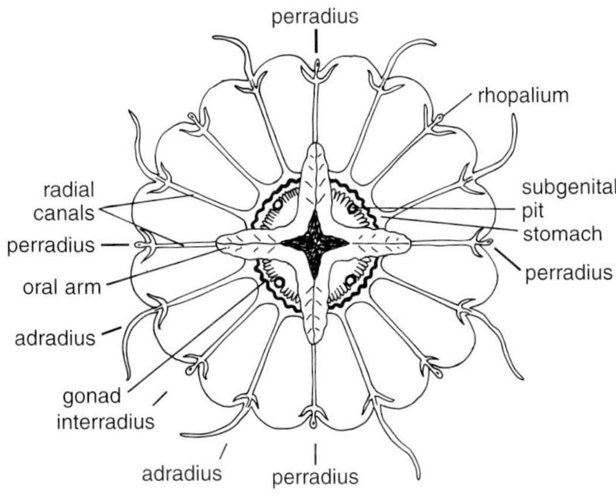

FIGURE 6.3
Diagrammatic view of a semaeostome scyphomedusa, from the oral end. (After Hyman 1940, redrawn by Jan Hynes.)

THE CENTRAL 'SYSTEM'

In hydrozoan and most scyphozoan medusae, the **manubrium** or food tube hangs from the subumbrella. The **mouth**, which is located at the distal end of the manubrium, is generally bounded by **lips** and leads to the **gastrovascular cavity** (Figure 6.1b). This cavity and the various tubes leading from it serve for digestion, distribution of food, and for circulation. In hydromedusans (Figure 6.5, page 125), four perradially-located gastrovascular tubes called the **radial canals** emanate from the simple **stomach** to the bell margin where they fuse with the **ring canal**. In scyphomedusans the gastrovascular cavity is present, as are the manubrium, mouth, oral (mouth) appendages and gonads; the manubrium is usually drawn out to form four frilled, perradial **oral arms** (e.g. Figures 6.3 above, 6.6b, page 126 and 6.23, page 144), and these are used in capture of prey when tentacles are absent, and for ingestion of prey and sometimes for hatching larvae. The medusa forms of the semaeostomes and rhizostomes (members of two orders in the class Scyphozoa) possess radial canals, which may be simple or branched (Figures 6.3 and 6.4, page 124), and many also possess a **circular** or (ring) **canal** (part of the marginal system, described below and shown in Figure 6.4c, page 124) as part of the gastrovascular system. In some groups of adult scyphozoans, the gastrovascular cavity is divided by four main **septa** into four perradial sub-cavities termed the **gastric pouches** or **pockets** (Figure 6.4b, page 124). Gastric septa are absent in adult semaeostome and rhizostome medusae. Radial septa divide the marginal area of the gastrovascular cavity into pouches. Both gastric and radial septa are present in coronate medusae, and **gastric ostia** open into the pouches (Figure 6.4a, page 124). Other pouch types may be present when septa are absent.

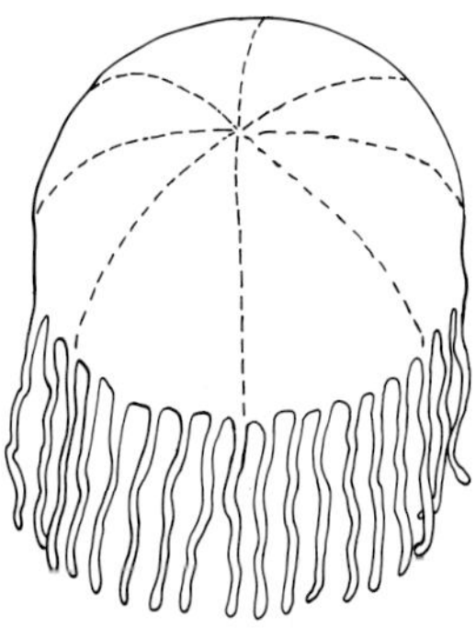

FIGURE 6.2
Diagram of a jellyfish exhibiting radial symmetry. Note that any plane passing vertically through the centre of the organism divides the organism into equal symmetrical halves. (After Curtis and Barnes 1989, redrawn by Jan Hynes.)

In the cubomedusans (Figures 6.25, 6.26, 6.27 and 6.29, pages 148–151) the gastrovascular cavity is also divided by four gastric septa into flattened, lateral gastric pouches or pockets. The ostia are considerably narrowed by the enclosure of funnels by the septa in the corners of the bell (Figures 6.25 and 6.26, page 148) LESH-LAURIE & SUCHY 1991; THOMAS & EDWARDS 1991). In chirodropid medusae, evaginations of the pockets called **gastric diverticula** hang into the subumbrellar cavity, covering the superior gonads (Figures 6.27–6.30, pages 149–151 and 153). Many scyphomedusans also have four interradial external openings, **funnels, peristomal pits** or **subgenital pits** (Figures 6.6b and c, page 126) leading into the **septa** (Figures 6.4a and b). In scyphomedusans and cubomedusans the internal layer of the gastrovascular cavity may extend down in fine, finger-like **gastric cirri** or **filaments** (Figures 6.6d, page 126 and 6.25, page 148). The filaments contain digestive enzymes and nematocysts.

In cubomedusans, a canal leads from each gastric pocket into the rhopalium; there also is a connection from each gastric pocket into the ring canal. In addition, The **pedalial canal** (Figures 6.6d and 6.25, pages 126 and 148) leads from each gastric pouch into the pedalium. If the tentacle is hollow, this canal extends down into its tentacle cavity.

MARGINAL STRUCTURES

In hydromedusans the marginal structures include the **ring canal**, which runs around the bell margin (Figure 6.5). An **inner and outer nerve ring** run along the ring canal (Figure 6.5). **Tentacles** first form at the ends of the radial canals. **Ocelli** (eye spots) and **statocysts** (balance 'organs') may form at tentacle bases. A shelf-like **velum** (Figure 6.5) is directed in towards the centre of the bell from the bell margin.

In scyphomedusans the marginal features include the **lappets, marginal tentacles** (where present), club-shaped sensory centres called **rhopalia** (Figures 6.6a, page 126 6.19–6.24, pages 141–146), and **pedalia** in the coronate medusae (Figures 6.6c, page 126, 6.19, page 141 and 6.20, page 142). The margin of the umbrella is usually scalloped, and each indentation bears a pair of **lappets** (Figures 6.6a, b and c, page 126), containing branches of the gastrovascular cavity (Figure 6.23, page 145). Between each pair lappets is a rhopalium. A velum is absent from the umbrella of scyphomedusans (Figures 6.6a, b and c, page 126).

Cubomedusans (Figures 6.6d, page 126 and 6.25–6.30, pages 148–153) possess marginal structures including interradial **pedalia** and **tentacles**, and a **velarium** (a velum-like structure containing velar canals and supported by four perradial **frenula** [folds]). **Rhopalia** occur above the margins at the perradii in **sensory niches**. As in scyphomedusans, these rhopalia contain balance organs and eyespots, but in cubomedusans some of the latter are complex and lens-bearing.

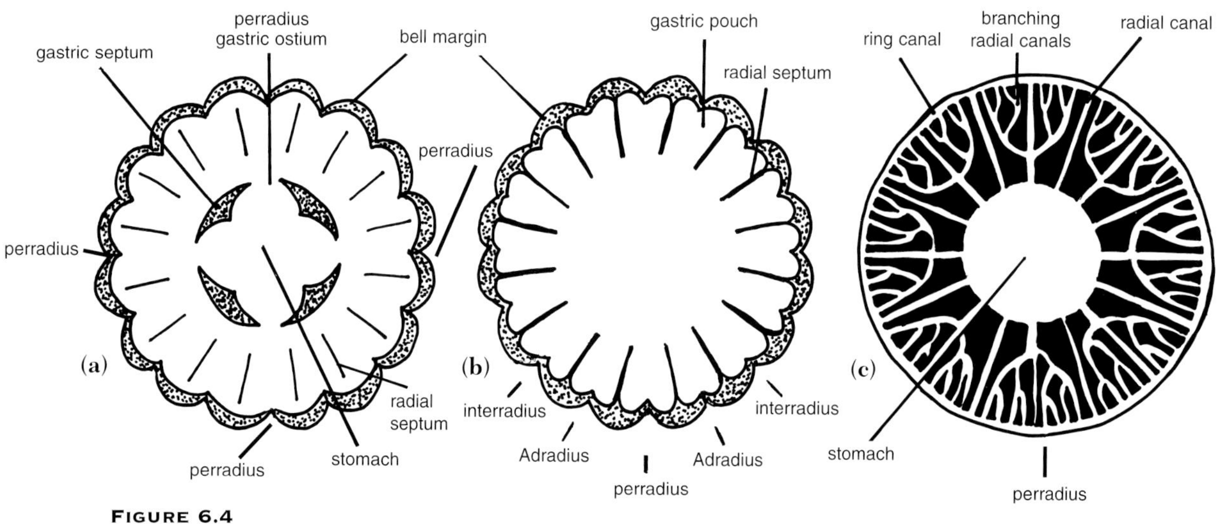

FIGURE 6.4
Diagrammatic view of transverse sections of the gastrovascular system in scyphomedusans: (a) coronate, (b) semaeostome, (c) semaeostome. (After RUSSELL 1970, redrawn by Jan Hynes, Brisbane.)

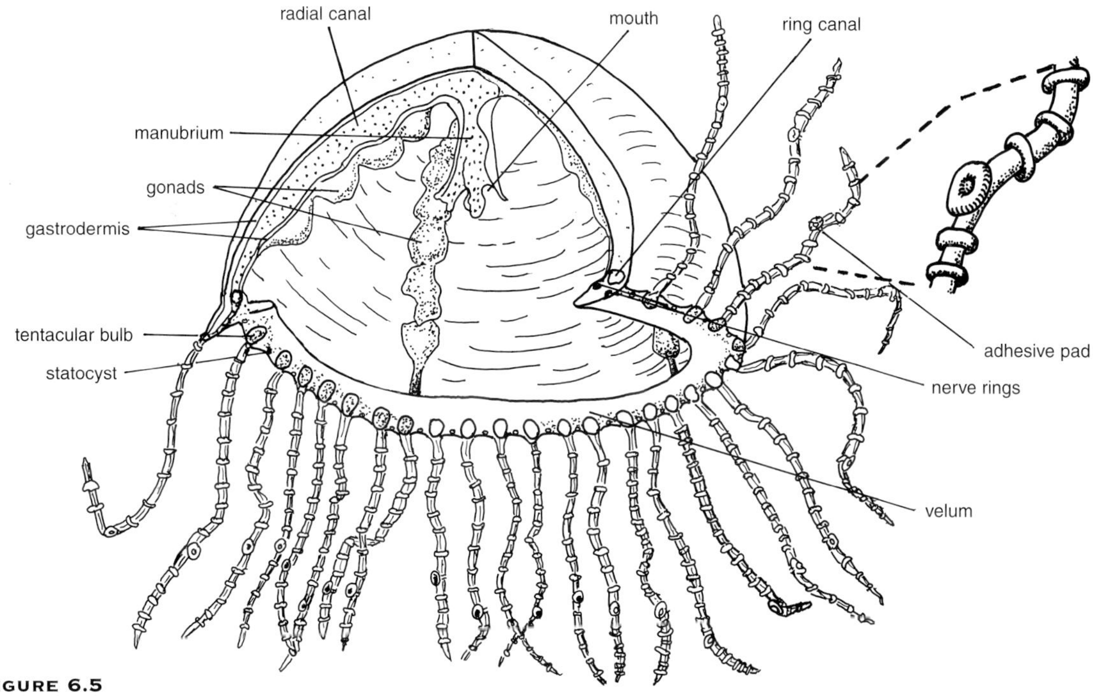

FIGURE 6.5
Diagram of a hydromedusan (*Gonionemus* sp.), showing the typical tetramerous arrangement and body parts. On the right, the enlargement shows a portion of tentacle with an adhesive pad and ridges of nematocysts. (After HICKMAN, HICKMAN and HICKMAN 1974, redrawn by Jan Hynes, Brisbane). (See also Figure 8.8, page 202.)

Table 6.2 on page 147 summarises the structures that are present in the cubomedusans and which are shared by medusae of the classes Hydrozoa and Scyphozoa.

OTHER ANATOMICAL/FUNCTIONAL FEATURES

Cell layers

Cnidarians possess only two cell layers: the outer ectoderm and the inner endoderm (or entoderm) (see Figure 6.1, page 122). The adult tissues will hereafter be termed **epidermis** and **gastrodermis**.* The epidermis covers the exumbrella and the subumbrellar surface, including the stomach, mouth and tentacles. Gastrodermis lines the inside of the stomach, radial canals, ring canals and tentacles. Between these two layers lies the jelly-like **mesogloea**. (See Figure 6.1, page 122.)

* The gastrodermal layer consists of a simple epithelium, which is composed of cells that participate directly in absorption, digestion and egestion.

The mesogloea, containing fibres and cells in scyphozoans, gives the bell a firm consistency. Among other functions, it provides a restoring force once the bell has contracted and also provides buoyancy and protection against buffeting.

SENSE 'ORGANS'

The cnidarians' sensory 'organs', which are intrinsically associated with the nervous 'system', are generally poorly developed, perhaps reaching a zenith in the cubozoans. However, they are far more evolved in the medusae than they are in the sessile polyps.

Generalised neurosensory cells of several types are present largely in the epidermis of cnidarians, detecting touch, chemicals and light. However, most medusae develop specialised aggregations of different cell types, loosely termed 'organs' to detect individual stimuli. The functional complexity of these structures may not bear a close relationship to their macro-anatomical complexity.

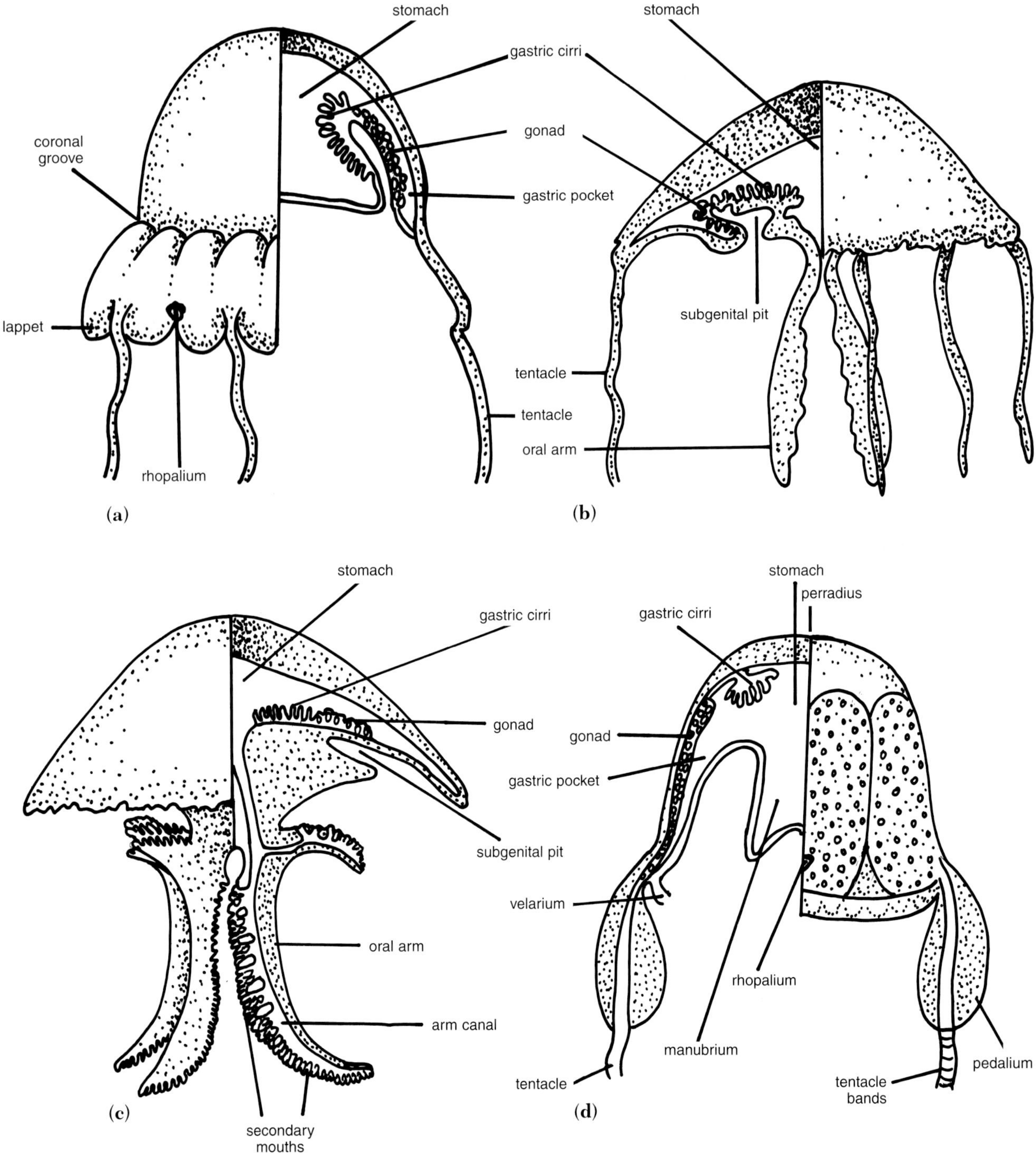

FIGURE 6.6
General anatomical comparison of scyphomedusans and cubomedusans: (a) a coronate scyphomedusan (order Coronatae), (b) a semaeostome scyphomedusan (order Semaeostomeae), (c) a rhizostome scyphomedusan (order Rhizostomeae), (d) a carybdeid cubomedusan (order Carybdeidae). (After LARSON 1976B, redrawn by Jan Hynes, Brisbane.)

OCELLI OR EYESPOTS

Some jellyfish are attracted to light; others may actively avoid it. Generalised or dermoptic light sensitivity is very common among cnidarians. Although there is strong evidence for photoreceptor function of ocelli, (Figures 6.7 and 6.8), medusae that lack ocelli (e.g. *Gonionemus*) are capable of responding to altered light intensity by changing their swimming rates.

The eyespot in hydromedusans may be simple (Figure 6.7) or complex, with a lens similar to that for the cubomedusan shown in Figure 6.8. In hydromedusans, ocelli occur at the bell margin near structures called **tentacular bulbs** (Figure 6.5, page 125). Ocelli are also present in scyphozoan and cubozoan medusae. Simple ocelli may form retinas; and complex ocelli contain lenses, retinas and corneas (see below). Cubozoans have both simple and complex ocelli.

Statocysts

Many jellyfish have balance 'organs' called **statocysts** (Figure 6.9, page 128). These are variable in structure, but they all have hair cells called **lithocytes** containing heavy particles called **statoliths**. This construction gives position information to cnidarians, orientating the jellyfish with respect to gravity. When the bell tilts, gravity pulls the statolith down against the **hair cells**. This stimulates the nerve fibres and signals the animal to right itself (Plate 6.6). In hydrozoan medusae (Figure 6.5, page 125), these structures occur on the outside surface of the bell (HORRIDGE 1969; SINGLA 1975). In scyphomedusans and cubomedusans they are found towards the outside of the bell, embedded in club-shaped sensory centres called rhopalia. In scyphomedusans and cubomedusans the statoliths are composed of gypsum ($CaSO_4$), while in hydromedusans, they are composed of $MgCaPO_4$ (CHAPMAN 1985).

Rhopalia

The rhopalium (Figure 6.10, page 129) is a club-shaped structure containing sense organs and is found in both scyphomedusans and cubomedusans. Although the rhopalia in the medusae of both classes are structurally different, they both contain ocelli and statocysts.

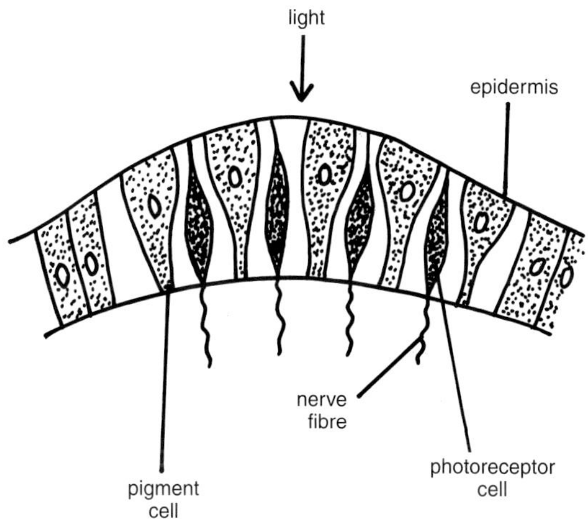

FIGURE 6.7
Cross-section of a simple ocellus (eyespot). (After CURTIS & BARNES 1989, redrawn by Jan Hynes, Brisbane.)

Rhopalia in the scyphomedusans are located at the bell margin between the lappets.

There are four rhopalia in the cubomedusans, one situated on each of the four sides at the perradius, above the bell margin within a sensory niche (Figures 6.6d and 6.25, pages 126 and 148). HARTWICK (1991a) observed that cubomedusan rhopalia are not fixed in one position but are controlled by fine muscles so that they can

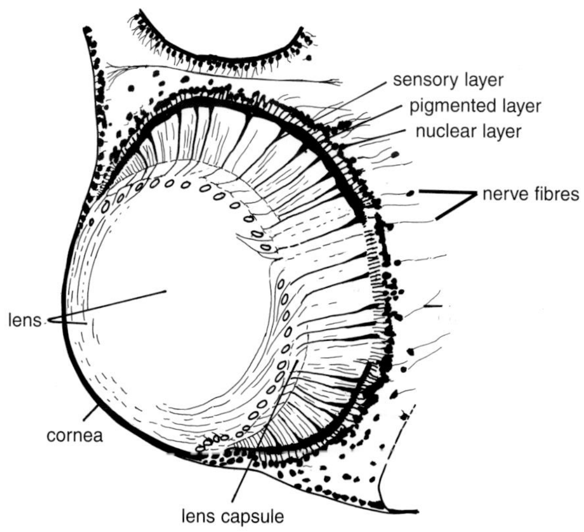

FIGURE 6.8
A complex lens-bearing eye of a cubomedusan, *Carybdea* sp. (After PEARSE & PEARSE 1978; redrawn by Jan Hynes, Brisbane.)

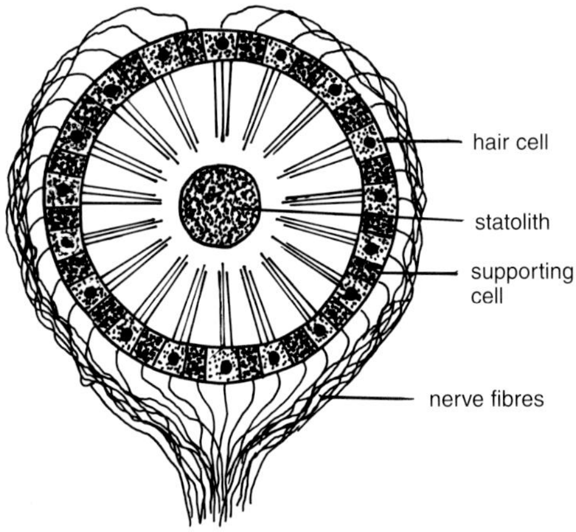

Figure 6.9
A statocyst. (After CURTIS AND BARNES 1989, redrawn by Jan Hynes, Brisbane.)

rotate, allowing the jellyfish 'to see' both outside and inside the bell.

The complex ocellus or lens eye of cubomedusans

BERGER (1900) described in detail the histology of the complex ocellus of *Carybdea* sp. (Figure 6.8). Much later YAMASU and YOSHIDA (1976) examined the ultrastructure of the different cell types in the cornea, the spheroidal cellular lens, the lens capsule and the three-layered retina from the lens eye of *Tamoya haplonema* (*T. bursaria*). The presence of all of these structures implies that the complex ocellus is an anatomically advanced photoreceptor. Curiously, in this same study YAMASU and YOSHIDA showed that only one kind of ocellar cell is appropriately structured to receive light.

Furthermore, not all work supports the assumption that cubomedusans can actually 'see'. As yet the function of the jellyfish lens is not known, but PIATIGORSKY and colleagues (1989) suggested that an absence of a space between the retina and lens of cubomedusan complex ocelli could mean that the jellyfish lens does not resolve an image onto a retina. Moreover, LARSON (1976), in observing the behaviour of cubomedusans, including their reaction to lights at night, stated that although *Carybdea marsupialis* possess (complex) ocelli, his observations suggested that 'they neither swam toward prey nor avoided obstacles'. He thus suggested that 'these ocelli show only directional sensitivity toward light'. HARTWICK (1991a) discusses evidence for coordinated vision in *Carybdea sivickisi* and its possible use during pair recognition.

BARNES (1966) stated that *Chironex fleckeri* is very light sensitive and will orient itself to the light of a match as much as 1.5 metres away even if the match is extinguished before orientation is complete. However BARNES in KINSEY (1986) and HAMNER & DOUBILET (1994) showed that *Chironex* will actively swim away from dark objects such as human beings, piers and jetties. In addition, BARNES (in KINSEY 1986) observed that this avoidance behaviour was particularly obvious when the objects were neither directly in front nor directly behind them. These observations have important implications for the initiation of preventative measures against stings (see page 135).

It is not yet understood how medusae can process visual information without a brain, but complex eyes found in some cubomedusans may play a role in effecting highly developed behaviour such as pairing and the spermatophore (gamete package) transfer described by BARNES in KINSEY (1986) and by HARTWICK (1991b) and others.

Jellyfish Behaviour

The topic of jellyfish behaviour and the factors that affect it may appear at first to be purely of interest to the biologist. However, preventive and first aid measures initiated from an understanding of jellyfish behaviour (BARNES 1960 & 1966) have contributed to the reduction in the rate at which stings, and indeed fatalities, occur in Australia.

Swimming behaviour

Jellyfish can aggregate due to waves or currents near shores, in bays or inlets or in the lee of islands. This gathering is useful in ensuring feeding and reproductive success (LARSON 1991). In addition, many species follow daily migration patterns. Many jellyfish alternate between an active swimming upwards in the water column and a slow sinking with tentacles outspread to capture prey items (MILLS 1984). *Carybdea rastoni* moves toward the surface at night to feed and returns to the sandy ocean bottom during the day (MATSUMOTO 1995, personal communication). Other jellyfish such as *Aurelia* and *Mastigias* are capable of horozontal migration. *Mastigias* migrates from one side of the Palau lakes to another, some using the sun as a compass cue (HAMNER & HAURI 1981). The swimming strokes consist of a series of rhythmic pulsations of the bell margin (GLADFELTER 1973) with the same basic opening and closing action as a rain umbrella. The rate of these pulsations depends on such conditions as the presence of food, the concentration of light and the temperature.

Basically, when the diameter of the subumbrella aperture decreases (Plate 6.2B), the force of the water jet increases, driving the jellyfish forward in the direction opposite to that in which water is expelled. The mesogloea provides the force that passively restores the bell to its original shape.

In hydromedusans, contraction of the subumbrella and velar musculature act to fill the bell cavity. The velum (Figure 6.5, page 125) bulges out with the force of the water jet, having narrowed the bell aperture considerably. After contraction it is sucked loosely inwards as water fills the cavity. In cubomedusans, this process is carried out with the velarium (Figures 6.27, 6.29 pages 149 and 151), a structure which fulfils an equivalent role of a velum.

Scyphomedusans, which lack either a velum or a velarium, depend mainly on the powerful circular (coronal) and radial muscles of the subumbrella (Figure 6.23, page 144) and to some extent on the action of the marginal lappets (Plate 6.1B) to reduce the diameter of the bell cavity (Plates 6.2B and 6.7). Saucer-shaped bells such as those found in *Cyanea* may be provided with hinges, which allow the bell to become concave upon contraction of the swimming muscles (Plates 6.1B and 6.7).

Turning* by jellyfish can be observed during the righting response described above (Plate 6.6). When the hydromedusan *Gonionemus* changes direction or descends, the velar and adjacent bell musculature contract more

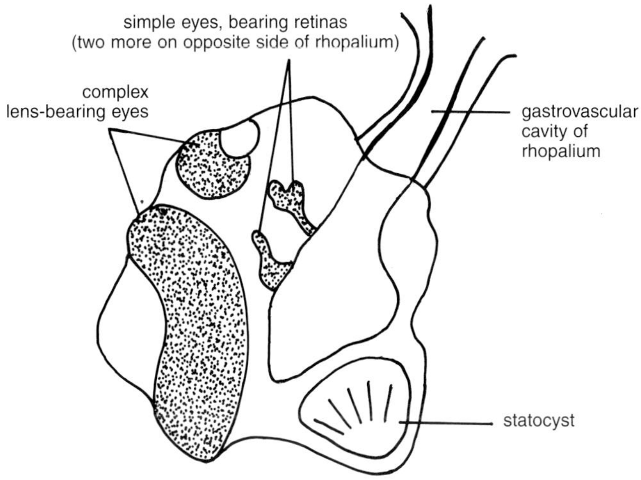

FIGURE 6.10
A rhopalium of a cubomedusan, *Carybdea* sp. (After CONANT 1898, redrawn by Dr Jacquie Rifkin.)

strongly on one side than the other, forcing water out asymmetrically. Similarly, the box jellyfish, which is highly manoeuvrable, uses asymmetric contraction of the velarium and bell musculature for adroit steering. According to BARNES (in KINSEY 1986), box jellyfish swimming and turning occur rapidly (see below).

Jellyfish move downwards either actively or passively. Rapid active descent, with the aboral end facing the sea floor, constitutes one form of escape response employed by medusae (HORRIDGE in discussion following BARNES 1966; MILLS *et al* 1985; LARSON *et al* 1991). This has been observed in jellyfish such as *Cyanea* (e.g. RUSSELL 1970 and Plate 6.7). BARNES (in KINSEY 1986) observed this behaviour when *Chironex* escapes from shadows or other offending stimuli. However, BARNES also observed *Chironex* passively sinking to the bottom and apparently 'playing dead' when disturbed.

Hydromedusans and scyphomedusans have been observed to move horizontally passively, mainly as a result of wave and current action, although active horizontal swimming with the aboral end anterior has been recorded in some scyphozoan jellyfish (e.g. RUSSELL 1970 and Plates 6.2A and B). In contrast, horizontal swimming is common behaviour among cubomedusans (Plate 6.8).

BARNES (in KINSEY 1986) estimated that the swimming speed of the box jellyfish *Chironex fleckeri* was very rapid, (1.02 m/s — 2.04m/s

* In the scyphomedusan *Cyanea*, turning is accomplished when the two sides of the bell contract sequentially and to slightly different degrees (GLADFELTER 1970 & 1972).

[2–4 knots]) and that when alarmed it can reach a speed of 2.55 m/s (5 knots). BARNES's estimations of these unusually high swimming speeds have been disputed e.g. by HEEGER 1995, personal communication.

Cubomedusans appear to surpass other jellyfish in the power and complexity of their swimming behaviour. They have good control over speed and manoevrability. The efficiency and speed of the box jellyfish steering mechanism is best demonstrated while watching the jellyfish turn, and according to Heeger (1995, personal communication), they can turn 'on the spot'. Cubomedusans have also frequently been observed to swim against the current (e.g. BARNES in KINSEY 1986). Since cubomedusans live entirely below the water level and manoevre so efficiently, they are not as easily trapped in the shallows and washed ashore as, for example, *Physalia* or slow-moving rhizostome scyphomedusans (HEEGER 1995, personal communication).

Physiology of jellyfish swimming behaviour

Scyphomedusans and hydromedusans have swimming musculature that is similar in myofibrillar arrangement, function and location (LESH-LAURIE and SUCHY, 1991). The main gross anatomical difference is that in the scyphomedusans, a velum is absent. In scyphomedusans, bell contractions occur in the circular and radial fibres of the subumbrella.

In hydromedusans, concentrations of bipolar **neurones** (nerve cells) that conduct stimuli in any direction are densely situated along the radial canals and the margins of the bell. In the epidermis of the bell margins, neurones form an inner and an outer nerve ring, each lying on opposite sides of the mesogloea (Figure 6.5, page 125). Giant photosensitive neurones, which are also called swimming motor neurones, connect with the subumbrella swimming musculature (SPENCER 1981).

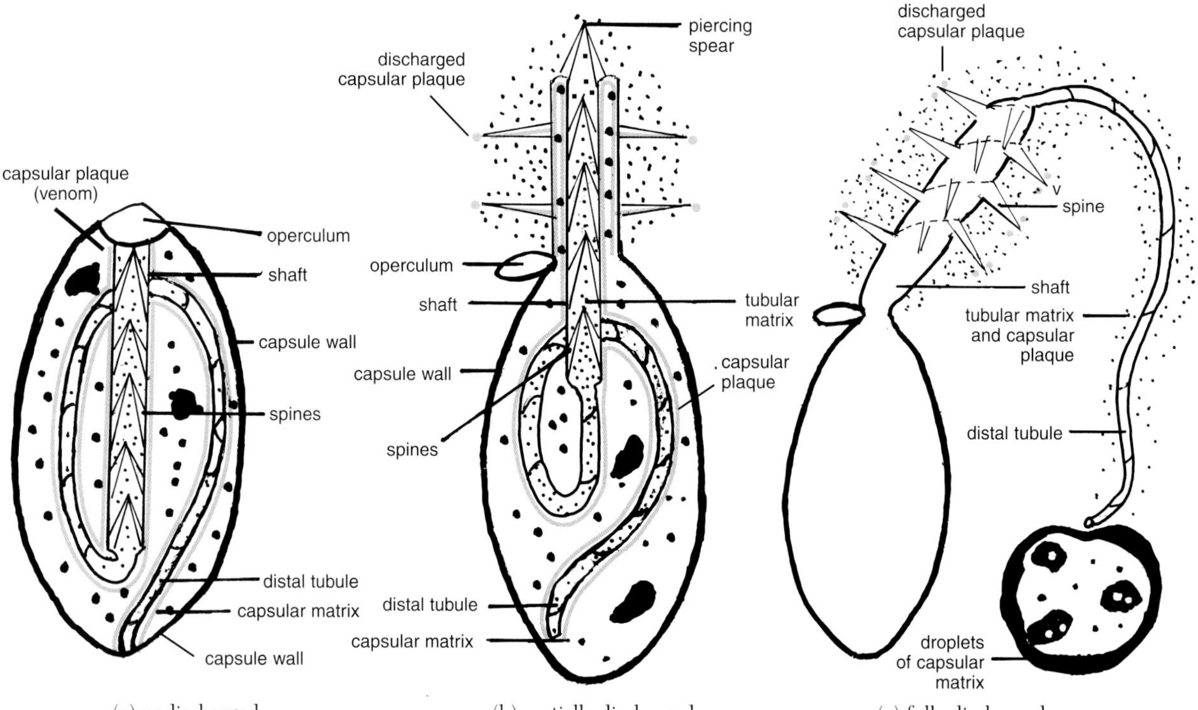

FIGURE 6.11
Schematic diagram of stages in the discharge of a nematocyst, showing (a) undischarged (b) partially discharged (c) fully discharged. Also showing the release of venom from 3 possible sites within the undischarged capsule. See also Plate 6.10. (Drawn by Dr Jacquie Rifkin.)

In hydromedusans swimming function is controlled by the inner nerve ring. It serves the balance organs and swimming muscles. Pacemaker activity, which initiates the rhythmical swimming beat, also occurs. The outer nerve ring connects to the ocelli, tentacles and tentacular bulbs. (Thus in hydromedusans we have a rudimentary form of 'nervous system organisation' with motor [inner nerve ring] and sensory [outer nerve ring] functions.)

Scyphomedusans have two epidermal nerve nets that interact with one another in marginal centres or ganglia, but there are no nerve rings as in hydromedusans. There are three parts of the 'nervous system' in scyphomedusans that are associated with their swimming action; these are reviewed in LESH-LAURIE and SUCHY (1991) and briefly summarised here:

- A motor neuron net that conducts impulses from the marginal ganglia to the swimming muscles.
- A diffuse nerve net, wherein neurones extend throughout the subumbrellar epidermis, the oral arms, and tentacles and sometimes innervate the subumbrella muscles. The net also serves the marginal ganglia.
- Marginal centres including a marginal ganglion, and rhopalia. These marginal centres are responsible for control of pacemaker activity, which in turn controls the rate of swimming pulsations (PASSANO 1982).

According to PASSANO (1982), in cubomedusans the marginal centres consist of groups of nerve cells or marginal ganglia, a rhopalium with statocysts, ocelli and sensory pits or niches. In cubomedusans there are well-developed nerve rings around the bell margin (Figure 6.25, page 148), where at the perradii and interradii each ring enlarges to form a ganglion; at the perradius, the ganglion is near the site of the rhopalium. Although the cubomedusan rhopalium is different from that of the scyphomedusans, in both the scyphomedusans and cubomedusans impulses generating swimming movements are initiated in these structures. Impulses travel then to the nerve ring (in cubomedusans) and to the subumbrellar swimming musculature. Swimming will cease if the rhopalia are removed, though severing the nerve ring does not initiate the same response (CONANT 1898; SATTERLIE 1979; SATTERLIE & SPENCER 1979; PASSANO 1982). Pacemaker activity in the rhopalia has been recorded in cubomedusans and scyphomedusans and this controls and coordinates swimming behaviour. However, PASSANO (1982) suggested that cubomedusans possess complex forms of behaviour because the 'nervous system' of the members of this class may be more highly organised than is usual for cnidarians.

FIGURE 6.12
Scanning electron micrograph of a microbasic mastigophore nematocysts from an anemone, penetrating a prawn carapace. (Photograph by Dr Jacquie Rifkin.)

FEEDING BEHAVIOUR

Some jellyfish may adopt solitary behaviour, whereas others, for example *Chironex fleckeri*, may be found in large numbers swimming in amongst swarms of small crustaceans in the shallows.* Many jellyfish tend to aggregate at temperature- or salinity-discontinuity layers in the ocean, feeding on zooplankton, which also concentrate at these locations.

Generally medusae feed as their tentacles are moved continuously through the water by active swimming or by 'sinking behaviours' described

* Although solitary specimens of jellyfish can cause problems for bathers, many difficulties arise when humans encounter jellyfish feeding in swarms. Good examples of this are *Pelagia noctiluca* in the Mediterranean Sea (see Chapter 10, page 287), and the thimble jellyfish, *Linuche unguiculata* in the Caribbean (see Chapter 8, page 221).

above (MILLS 1984). Although it is generally thought that prey is caught by chance encounter, some species actively create currents bringing food into range of the tentacles and/or mouth arms (e.g. see *Cassiopea* below). Prey that come in contact with these structures are immobilised by myriads of minute structures called nematocysts.

Nematocysts

Nematocysts (Figures 6.43 I–XXXI, page 168), which are stinging organoids or capsules, are one category of cnida or capsular organoid. Of all three cnida categories, only nematocysts are found amongst jellyfish of medical importance. The other two categories, the **spirocysts** (Figure 6.43B, page 168) and the **ptychocysts** (Figure 6.43C, page 168) are found exclusively among anthozoans.

Nematocysts may be distributed on all parts of the jellyfish, including the tentacles, the oral and stomach regions and the exumbrellar epidermis. They may stick to, ensnare or penetrate prey, but they are used for defence as well as feeding. Venom may be contained within the undischarged capsules, in specific locations, (ENDEAN *et al* 1969; ENDEAN & RIFKIN 1975; ENDEAN & RIFKIN 1983; RIFKIN & ENDEAN 1983; RIFKIN 1988, RIFKIN et al 1993) and translocation of venom from one compartment to another appears to occur during discharge (LOTAN 1995). Discharging nematocyst tubules are capable of turning inside out at great speed, penetrating and paralysing prey, the eversion process itself being completed within 3 milliseconds. An acceleration power of 40 000 XG enables the tubule to penetrate even the carapace of a crab (e.g. HOLSTEIN & TARDENT 1984). This power is comparable to that of a missile penetrating the superstructure of an aircraft (BURNETT 1995, unpublished). Subdued prey is conveyed by tentacles up to the mouth and gastrovascular cavity (Figure 6.14 and Plate 6.3).

Nematocysts are arranged in a specific pattern on the tentacles of *Cyanea* (as well as other cnidarians). When a human victim is accidentally stung by tentacles, the pattern of the weal on the skin reflects the way in which the nematocysts are organised (BARNES 1966) {see Plate 9.91A}.

Rhizostome scyphomedusans

Rhizostome scyphomedusans (Figures 6.6C page 126 and 6.24 page 146, and Plates 6.2A and 6.2B), do not have a central mouth; they have thousands of tiny secondary mouths in the oral arms. These rhizostome medusae are usually microplankton feeders; in plankton-rich waters, massive numbers of organisms collide with the arms as the jellyfish swims along and are devoured by these small mouths. However, the remains of fish have been found in the stomach of certain rhizostomes (RUSSELL 1970) — apparently these fish may be paralysed and held against the mouths. Enzymes are secreted into the prey, and digestion takes place outside the body. Minute partially-digested food particles are swept into the mouths by ciliary action. The jellyfish *Cassiopea* (the upside down jellyfish) {Figure 6.24 page 146 and Plate 9.91A} has symbiotic algae from which it indirectly obtains part of its nutrients by photosynthesis. The remainder is obtained as the jellyfish lies on the bottom, upside down, using the pulsations of the bell to bring food into reach of the oral arms.

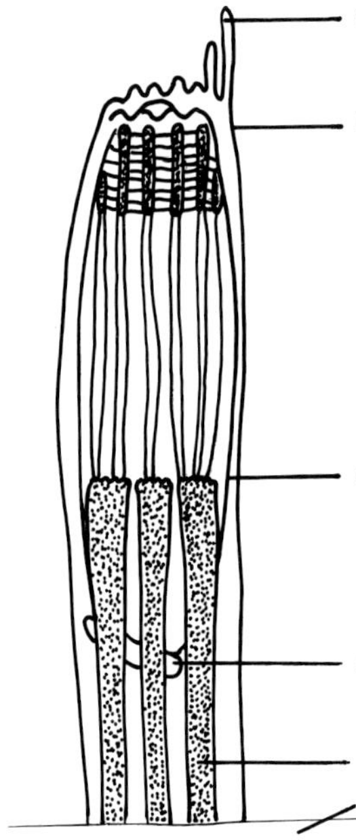

FIGURE 6.13
Diagram of a microbasic mastigophore nematocyst from *Chironex fleckeri* within its nematocyte. Radial fibres help to anchor the whole nematocyte to the mesogloea. (After RIFKIN & ENDEAN 1988, redrawn by Jan Hynes, Brisbane.)

Cubomedusans

Feeding behaviour has been described in detail for the carybdeid cubomedusan *Carybdea mar-*

supialis (Figure 6.14). According to LARSON (1976), nematocysts of the tentacles penetrate prey; the tentacles contract, and adhering prey is brought closer to the mouth by the flexing of the pedalia. The prehensile lips of the manubrium grasp the food, and the food is engulfed. Then cilia and muscular contractions carry the mucus-laden food up the central, interradial part of the manubrium to the gastric cirri which also produce digestive enzymes. Wastes can move simultaneously down the perradial stomach furrow, down the manubrium and out of the body. LARSON stated that it takes only three to four hours to digest a 20 mm fish. HAMNER (1994) described similar feeding behaviour in a chirodropid cubomedusan, the box jellyfish *Chironex fleckeri* (see also Plates 6.3, 6.9).

BARNES (1966) also described a form of feeding behaviour in which the *Chironex* he observed finally sank to the bottom after prey capture, apex downward, with its tentacles draped over the subumbrellar aperture, ingesting all the prey caught on the tentacles. After this, it resumed swimming. Finally, BARNES (in KINSEY 1986) reported a remarkable hunting behaviour that is carried out by pairs of *Chironex* early in the season; he observed that the pair co-operate with one another in herding their crustacean prey by encircling them. A similar co-operative herding behaviour was observed in the carybdeid cubomedusan, *Carybdea rastoni* in South Australian waters (GOWLETT-HOLMES 1995, personal communication).

In *Chironex fleckeri*, as in other jellyfish, the tentacles that are ready to feed are extended behind them while fishing {Plates 9.2, 9.8 and 9.42} so that all nematocysts on the tentacle are potentially available for use {Plate 9.68}. A mere touch by a human hand will immediately cause contraction of the tentacle (HEEGER 1995, personal communication). When this occurs, many thousands of potent nematocysts are effectively 'roofed over' beneath the uppermost layers of tentacle tissue (Plate 9.41 and Figure 6.42 page 162).

REPRODUCTION

Most jellyfish have separate sexes apart from a few species such as *Chrysaora* in which male and female gonads often occur within the same individual. Fertilisation may be either internal or external. The development of the free-swimming, ciliated planula larvae after fertilisation is an important characteristic of members of the phylum Cnidaria. In some cases the female broods developing stages of the new generation in pouches.

In free-living hydromedusans, germinal cells develop from interstitial (multi-potent) cells that migrate to specific sites on the bell surface.

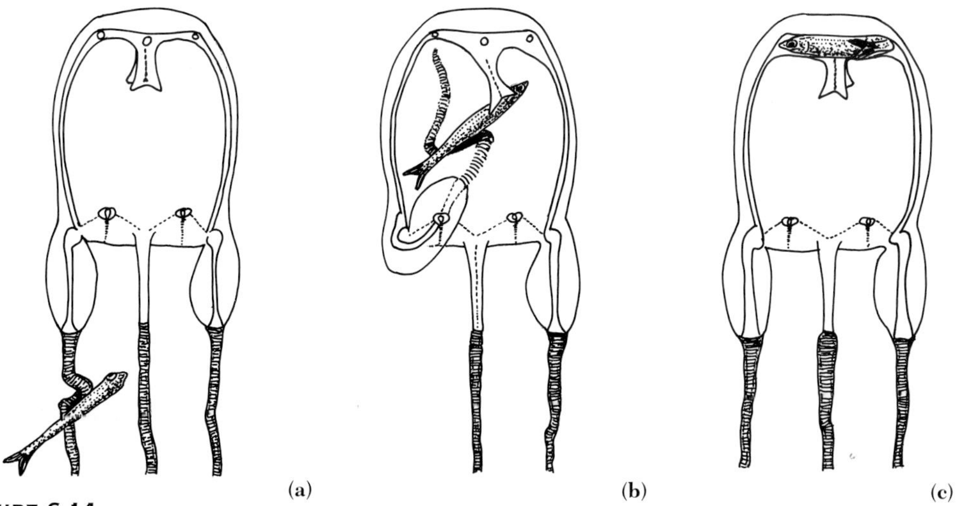

FIGURE 6.14
Stages of prey transfer in the cubozoan *Carybdea marsupialis*. (After LARSON 1976a, redrawn by Jan Hynes, Brisbane.) [labels: (a) Fish attached to contracted tentacle. (b) Pedalium flexed, prey grasped by prehensile lips. (c) Fish 'swallowed' via mouth into stomach cavity.]

Gonads develop from pouches of epidermis on the sides of the manubrium, beneath the radial canals (Figure 6.5, page 125) or anywhere on the subumbrellar surface. Gametes (sex cells) are shed when the overlying epidermis ruptures (THOMAS & EDWARDS 1991).

In the typical life cycle of hydrozoans, small medusae release either eggs or sperm directly into the water, where fertilisation occurs. In some species, only sperm are released and fertilisation occurs on or within the female; the fertilised eggs develop via planulae into sessile polyps that reproduce asexually to form colonies which often resemble plants.

In scyphozoans, germinal cells develop from amoeboid cells. The gonads in adult scyphomedusans are always associated with the gastrodermis, and these are usually located on the floor of the gastric pouches (Figures 6.4 and 6.6, pages 124 and 126). Gametes are released into the stomach cavity and generally emerge from the mouth.

The majority of scyphozoans have well-developed asexual or polyp stages, but the medusa stage is the dominant form of the life cycle. Fertilisation generally takes place in the open sea outside the animal or in the gastric pouches of the female. The fertilised eggs form planula larvae, each of which develops into an asexual polyp called a **scyphistoma**. This undergoes a process called **strobilation** by which miniature medusa-like structures are formed, often one on top of the other, resembling a stack of dinner-plates (a **strobila**). These are then budded off as small medusa-like forms (**ephyrae**) which eventually develop into adult medusae.

In cubomedusans, the gonads are located gastrodermally, on either side of the interradial septa (Figures 6.25 and 6.26, page 148). Superior gonads, present in chirodropids, are described on page 153, Figures 6.27–6.30 pages 149 – 153). Unusual aspects of medusan fertilisation and development are discussed below. The cubozoan life cycle does not include a strobila, and the medusa forms directly from a polyp, which undergoes metamorphosis. The life cycle of *Chironex fleckeri* is described in more detail in Chapter 9, page 236.

Specialised aspects of reproductive behaviour in cubomedusans

The process of internal fertilisation has been observed in several cubomedusans (ARNESON 1976). However, in the carybdeid cubomedusan *Tripedalia cystophora* {Plate 10.28} WERNER (1973) observed a very specialised behavioural phenomenon. Male and female *Tripedalia* locate and hook onto one another with their tentacles. They then engage in a form of copulation in which the male and female medusae become entwined and the male ejaculates bundles of sperm called spermatophores through its extensible, tube-like mouth onto a tentacle of the female. Apparently the spermatophores are then transferred to the mouth of the female and these bundles travel to the stomach where individual sperm fertilise the ova. Within the female, the fertilised eggs develop into planulae and these are later released to continue development outside the body. HARTWICK (1991a) recorded evidence for similar behaviour in the minute sexually dimorphic cubomedusan *Carybdea sivickisi*, as well as reporting several significant differences. (See also note on page 173.)

External fertilisation is believed to occur in the chirodropid *Chironex fleckeri* after unfertilised gametes are released into the water (YAMAGUCHI & HARTWICK 1980). However, BARNES (in KINSEY 1986) has reported seeing *Chironex* travelling in pairs. He also observed the mutual passage of tentacles from one specimen into the bell cavity of the other, behaviour similar to that recorded by WERNER (1973) for *Tripedalia*. BARNES (in KINSEY 1986) observed this behaviour both in the laboratory as well as in the field. Further fortuitous observations by RIFKIN and ENDEAN (unpublished) revealed the presence of unidentified larvae within the bell cavity of a mature *Chironex*. The polyps that developed from the planulae resembled those in published works by YAMAGUCHI & HARTWICK (1980) and HAMNER, (1994) and RIFKIN (unpublished observations). This circumstantial evidence suggests that in the chirodropid *Chironex fleckeri* fertilisation may be either external or internal. Further research is necessary.

JELLYFISH BEHAVIOUR AND AVOIDING JELLYFISH STINGS

Since box jellyfish will avoid large, and especially dark objects if given the opportunity, BARNES (1966) suggested that bathers walk slowly into the water, wearing dark clothing, giving the box jellyfish a chance to avoid them (also see Chapter 9 page 236). Obviously running or diving into the water is dangerous; small children splashing in the shallows are particularly at risk.

It is also helpful to be aware that the tentacles of most jellyfish are many times longer than the bell length (tentacles of some stretch out up to 30 metres in length in the relaxed state) and the absence of a bell in the immediate vicinity is no indication that the area is necessarily safe {see 'Prevention' in Chapter 5, page 104}.

As people work or swim in the marine, estuarine or even fresh water environment they will continue to come in contact with jellyfish. It is important to remember that however dangerous they may be, jellyfish do not actively attack humans. They become a problem only when a victim blunders into them. By understanding their behaviour patterns and bearing in mind such factors as their geographical distribution, favourable environmental conditions, and the seasons and times of day in which they are likely to be found, painful stings may be avoided. It is fortunate that preventative measures such as stinger-resistant enclosures and stinger suits have been developed and implemented (also see Chapter 5, page 98 and Chapter 9, page 236) to ensure that those who make their living in or on the water or who enjoy water sports can do so in relative safety.

Higher taxa to which jellyfish of medical importance belong may be identified by the following key. The key also serves as a summary of major class and order characteristics. Each

A CLASSIFICATION KEY FOR MEDICALLY IMPORTANT JELLYFISH

1 (a)	Jellyfish, rounded or square, without float, and capable of diving or swimming below the surface	(2)
1 (b)	Jellyfish, with a gas-filled float (pneumatophore) existing above the water surface..	(7(c))
2 (a)	Bell of generally cubical form with umbrella margin built inwards as a muscular velarium. Tentacles arise from blade-like or hand-like structures (pedalia) placed at the (interradial) corners of the bell. Four perradial rhopalia present; septa present...	class Cubozoa (3)
2 (b)	Bell rounded or circular	(4)
3 (a)	Four single, paired, or trebled interradial tentacles each arising from single, simple pedalia and 4 stomach pouches without superior gonads.......................................	order Carybdeidae (Figures 6.6d, page 126, 6.25 and 6.26, page 148)
3 (b)	Four interradial tentacle clusters arising from complex pedalia; with 4 stomach pouches, and four pairs of superior gonads..	order Chirodropidae (Figures 6.27–6.30, pages 149-153, Plates 6.8 and 6.9)
4 (a)	Tetramerous symmetry, 4–32 rhopalia; septa present, velum absent, gastric filaments present............................	class Scyphozoa (5) (Plates 6.3 and 6.9)
4 (b)	Mouth of bell provided with a clear, circular shelf or velum which partly occludes the open end of the bell; usually small and simple medusae	class Hydrozoa (7a or b)

A CLASSIFICATION KEY FOR MEDICALLY IMPORTANT JELLYFISH

5 (a)	Margin separated from bell by a circular furrow distinguishing the upper and lower portions of the bell............	order Coronatae (Figures 6.6a, page 126, 6.19, and 6.20, pages 141 and 142)
5 (b)	Margin not separated from bell by circular furrow. Tentacles if present, not borne on pedalia	(6)
6 (a)	Margin of bell scalloped. Corners of mouth elongated into four long, frilly mouth lobes or mouth arms. Gastric septa absent in adults; 8 or 16 rhopalia, gonads in folds of endodermal wall of subumbrella, tentacles present	order Semaeostomeae (Figures 6.3, page 123, 6.6b, page 126, 6.21, page 143, 6.22, page 144, 6.23, page 145, and Plates 6.1A, 6.1B, 6.6)
6 (b)	Margin of bell scalloped. The oral lobes are joined into a mass, obliterating the mouth, which is replaced by numerous small mouths distributed over the mouth lobes. Marginal tentacles absent (in a few cases extensions of the bell margin may appear as pseudo-tentacles)..	order Rhizostomeae (Figures 6.6c, page 126, 6.24a and b, page 146 and Plates 6.2A and 6.2B)
7 (a)	Medusae generally hemispherical or flattened. Gonads present on stomach alone or on stomach and a little way along radial canals only; ocelli absent, statocysts present or absent, but when present they are in the form of sensory clubs with endodermal axes	order Limnomedusae (Figures 6.5, page 125, 6.18, page 140)
7 (b)	Medusae generally deep bell-shaped. Gonads present only on stomach, or occasionally both on stomach and extending for a short distance along radial canals; statocysts absent	order Anthomedusae/Athecota
7 (c)	Members are comprised of colonies of organisms highly specialised for secreting gas, for reproduction, for feeding and for defence	subclass Siphonophora, order Cystonectae (Figures 6.15–17, pages 138–139, Plates 6.4, 6.5)

entry line in the key concludes with a number which directs the reader to the next appropriate descriptor in the key.

The term 'jellyfish' is used here in its loosest sense. It refers largely to the motile medusa stage in the life cycle of the class Hydrozoa (also encompassing hydrozoan colonies that are buoyed up by means of gaseous floats), as well as medusoid representatives of the classes Scyphozoa and Cubozoa.

ANATOMY OF SELECTED JELLYFISH OF MEDICAL IMPORTANCE

The text on the following pages describes taxonomic features of selected jellyfish species from the classes Hydrozoa, Scyphozoa and Cubozoa, with brief comments on distribution and medical significance. Where possible anatomical features are summarised in point form for convenience. Most anatomical information on jellyfish from the classes Hydrozoa and Scyphozoa is derived from MAYER (1910), TOTTON (1960, 1965), KRAMP (1961), RUSSELL (1970) and SOUTHCOTT (1982). Some of the information on the section on the cubomedusans is based on a summary of recent research carried out by the author on specimens from Australia and the remainder of the Indo-Pacific region.

CLASS HYDROZOA

SUBCLASS SIPHONOPHORA

Siphonophores are not solitary organisms like medusae, but colonies of animals, and some individuals form spectacular displays in the water (DYBAS 1992). They may be either free swimming or floating; some have floats and some do not. Most siphonophores are capable of diving; many have no stinging powers on humans. The genus of dominating medical importance is *Physalia*. Other less common siphonophores* causing discomfort to humans include *Agalma okeni* which stings severely with pain and erythema, and members of the genus *Rhysophysa*, specifically *R. filiformis* and *R. eysenhardti*, which cause painful injuries to Japanese fishermen handling seines (fishing nets equipped with floats and weights) in the Pacific (SOUTHCOTT 1987b). More recently, in South Australian waters, a species of the genus *Förskalia* caused severe pain and whealing for several hours on the face of a diver, who swam unwittingly into the elongated, diaphanous colony (GOWLETT-HOLMES 1995, personal communication).

ORDER CYSTONECTAE

Genus *Physalia* Lamarck, 1801

The 'bluebottle' *Physalia utriculus*, 'Portuguese man-o'war' (Atlantic *P. physalis*) and 'Pacific man-o'war' (Pacific *P. physalis*) {see also Chapter 8, page 192, and Plates 8.32–8.36 and Chapter 10, page 302, and Figure 10.9, page 303}.

* *Velella* {Plate 8.36} and *Porpita*, jellyfish which also live at the air-sea interface, were once thought to be siphonophores but they are now considered by different workers to be either colonial or solitary members of the 'order Hydroida'. FIELDS and MACKIE (1971) suggest that they are highly specialised solitary tubularian zooids which float upside-down at the surface.

Physalia is an open-water hydrozoan colony which lives at the air–sea interface. The genus is easily recognised by its prominent blue or purplish gas–filled sac (**pneumatophore**) with its erectile sailing crest (Figures 6.15–16, pages 138–139). The pneumatophore floats on the surface as a bladder above the tentacles and zooids. Following onshore winds, it is common to find them blown up onto beaches in swarms sometimes called 'navies' or 'armadas'. Left- and right-handed colonies are present, enabling *Physalia* to sail at an angle of 40 degrees to the left or right of the wind direction (TOTTON 1965, FRANCIS 1987).† This characteristic would apparently ensure that all specimens are not blown up on the beach. The zooids, including the tentacles act as stabilisers while sailing.

Distribution and taxonomic considerations

Physalia is widely dispersed throughout all ocean regions but there are no arctic or antarctic forms. Swarms are very common all around the Australian coastline; on the east coast of Australia it is usually found from October to March. There is current interest in the taxonomy of *Physalia*, as it is debated (but not known at this time) whether two distinct species of bluebottle exist — namely *Physalia utriculus* and *Physalia physalis* — or whether there is only one cosmopolitan species of *Physalia* (TOTTON 1960; FENNER *et al* 1993). Recently FENNER *et al* (1993) suggested that both *P. physalis* and *P. utriculus* should be retained {see Chapter 8, page 192}.

TABLE 6.1

Main anatomical differences between the two morphs (Plates 6.5 and 8.35) *Physalia physalis* and *Physalia utriculus*:

Physalia physalis	***Physalia utriculus***
bearing several long, thick fishing tentacles with numerous shorter, finer tentacles of varying lengths	bearing only one main fishing tentacle with numerous shorter tentacles usually found just beneath the flout

† Dr T. Heeger (1994, unpublished observations) currently questions this observation.

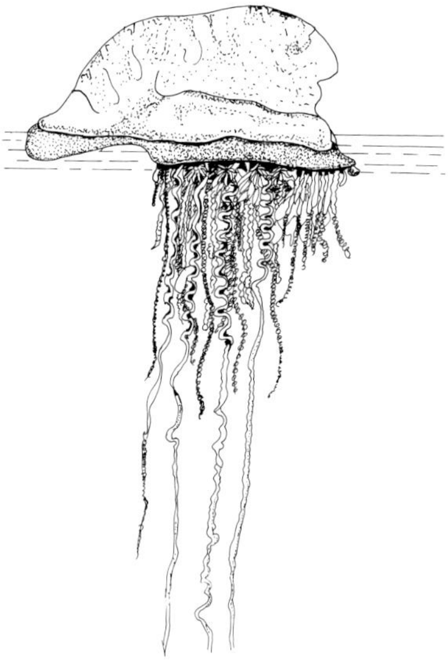

Figure 6.15
Drawing of *Physalia physalis* (Portuguese man-o'-war), a siphonophore colony. (After TOTTON 1965, redrawn by Rebekah Lewis, Brisbane.)

Morphological features of the genus *Physalia*

The *Physalia* colony consists of highly modified medusa and polyp types called 'zooids', each of which is integrated to act as a single individual. Each zooid may also be considered as specialised, serving as an organ of this 'super-organism'. Different individuals in the colony take on specialised functions such as feeding, defence and reproduction (see below). Future generations of individuals bud from the float and hang suspended in the water; up to a thousand zooids may be found in a single colony.

Variously modified zooids are organised into clusters called **cormidia** (Figure 6.22, page 144) consisting of:

- **Gastrozooids** which are feeding polyps with a mouth. During feeding, the fishing tentacles carry the attached, subdued prey upwards, and the mouths of the feeding polyps attach to it (Figure 6.17). The prey is engulfed by one or more gastrozooids; or if the prey is large, many feeding polyps spread out over it and begin to digest it. Partially digested food is distributed by a series of internal canals of the gastrovascular cavity to all other colony individuals. Indigestible remains are ejected through the mouths of the feeding polyps.
- **Dactylozooids** (also called palpons or feelers), which are the defensive polyps. These look like the gastrozooids but they do not have mouths. They also possess tentacles, often called **fishing tentacles**, one or more of which may be long; the remainder are usually shorter. Each carries bean-shaped, **stinging buttons** along its length (Figure 6.17). These buttons or 'batteries' are replete with nematocysts, which are responsible for paralysing prey. (The nematocysts are also responsible for causing considerable morbidity [and uncommonly mortality] to humans who accidentally become entangled in the trailing tentacles.)
- **The pneumatophore** or float which is believed to have expanded from the original larval polyp. Each is comprised of a double walled chamber lined with chitin (Plate 6.4), and houses a gas gland, with 15–20% oxygen, 0.5–13% carbon monoxide, usually around 70% nitrogen, and a very small amount of carbon dioxiode and argon (TOTTON, 1969 and

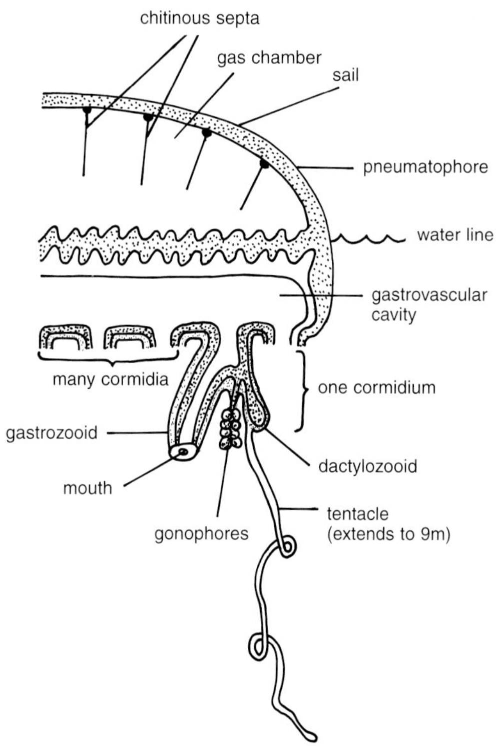

Figure 6.16
An individual *Physalia* cormidium, which consists of a cluster of different zooid types. (After BAYER & OWRE 1968, modified by Jan Hynes, Brisbane.)

MACKIE, 1974). This large, thick-walled float appears not to be capable of allowing the colony to expel enough gas to sink below the surface except in the earliest growth forms (MACKIE *et al* 1987). Mackie and colleagues state further that gas is secreted from the gas gland (small in relation to the size of the float) and continues to secrete as the float grows, and that diffusional losses are also replaced by the gas gland. Mackie and colleagues also argue that gas secretion will be easier at atmospheric pressure (on the surface). (See also Chapter 8, page 193.)

- In Australia the usual float length of the common *Physalia utriculus* ranges between 3 and 6 cm but specimens with a float length of 10 cm or greater are found (RIFKIN 1995, unpublished observations), and in the less-common *Physalia physalis* (Pacific man-o'-war), float lengths of up to 15 cm are found (FENNER *et al* 1993). Floats lengths of up to 30 cm have been recorded for specimens of the *P. physalis* from the Atlantic Ocean.
- **Reproductive zooids**: Medusoid individuals include the **gonophores** (Figures 6.16 and 6.17), which are tiny sacs containing gametes of each sex.
- Because nematocysts remain active, beached specimens and broken tentacle fragments can still sting. Usually the float may be safely handled, although in certain parts of the world (e.g. Bermuda) the float itself may produce severe irritations (HEEGER 1994 unpublished observations).

It is also possible to be stung by animals that feed upon *Physalia* and incorporate nematocysts into their tissues. The pelagic sea slug *Glaucus* {Plate 11.6} of the phylum Mollusca is capable of causing irritating stings similar to, but less extensive than those sustained by victims of *Physalia* itself (THOMPSON & BENNETT 1970; SLSQ Gold Coast, personal communication to RIFKIN 1990-1992; FENNER 1991, unpublished observations).

ORDER ANTHOMEDUSAE/ ATHECATA

One species of the order, *Leuckartiara gardineri*, although not a serious stinger, bears a morphological similarity and a distribution similar to a significant stinger, the carybdeid jellyfish *Carukia barnesi* {page 246}. To facilitate the differential recognition of the two species, the morphological characteristics of *Leuckartiara gardineri* are listed here.

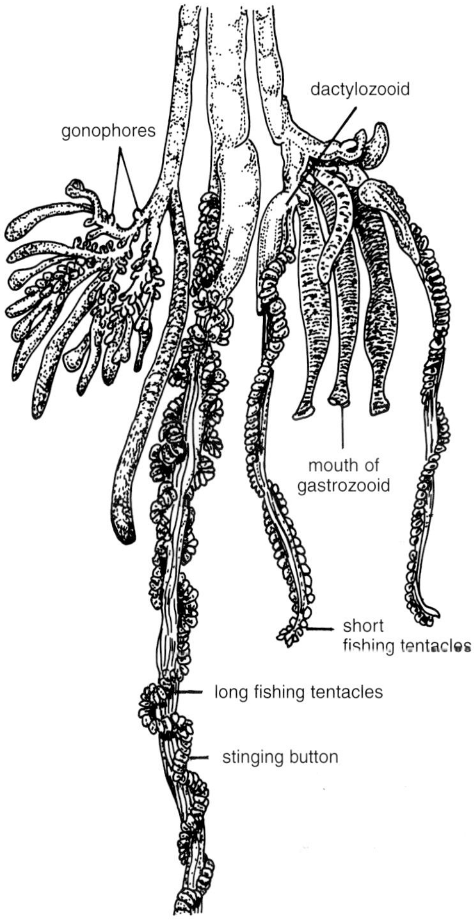

FIGURE 6.17
Detailed drawing of zooids found in *Physalia physalis*. (After LANE 1960, redrawn by Jan Hynes, Brisbane.)

Leuckartiara gardineri is a small jellyfish also known as the 'little red stinger'. It is found in the Indian Ocean and North Queensland and will probably be found on the north coast of Western Australia (MARSH & SLACK-SMITH 1986). This jellyfish is about the same size and bears the same number of long tentacles as *Carukia barnesi*, the carybdeid causing the Irukandji syndrome {Chapter 9, page 236 and Figure 9.1, page 238}. However unlike *C. barnesi*, it is a relatively harmless stinger, producing a very sharp sting pain which subsides rapidly, leaving no local effects (BARNES in CLELAND & SOUTHCOTT 1965). The medusae of this order are deep and bell-shaped. The gonads are present only on the stomach and may extend part way down the radial canals; statocysts are absent.

Leuckartiara gardineri Browne, 1916

Bell shape:	• Conical
Bell size:	• Height 10 mm • Diameter 5 mm
Marginal tentacles:	• 4 perradial, pink colour • About 25 mm long • Many additional tiny, thin tentacles between perradii
Manubrium:	• Very large, folded
Gonads:	• Very large, filling the entire subumbrellar cavity

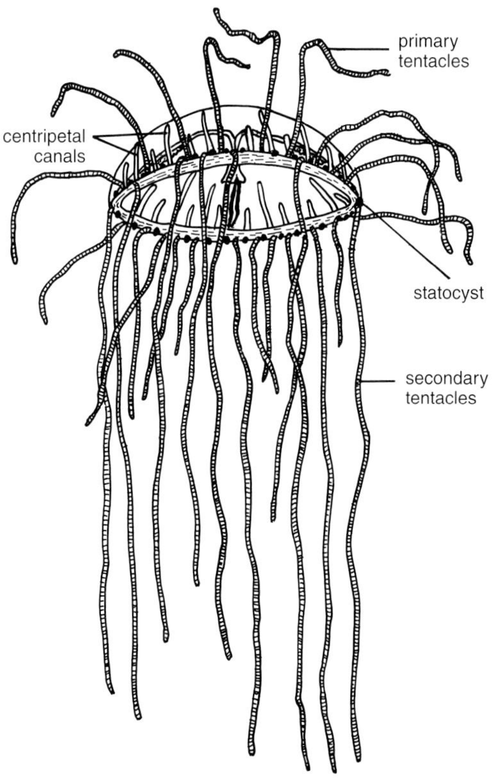

FIGURE 6.18
The limnomedusan jellyfish *Olindias* from Bermuda. (After HYMAN 1940, redrawn by Jan Hynes, Brisbane.)

ORDER LIMNOMEDUSAE

Only a few species of limnomedusans are known to cause problems for humans. *Gonionemus vertens*, commonly known as the orange-striped jellyfish {Figure 8.8, page 202} is found in north temperate Sea of Japan, and Pacific and Atlantic Oceans and the Mediterranean Sea. However, numerous serious stingings have also been reported along on the eastern coast of Russia and China (YAKOVLEV & VASKOVSKY 1993 and see Chapter this volume).

Olindias singularis BROWNE, 1905 (Figure 8.9, page 205) is distributed throughout the warm waters of the Indian Ocean, West Pakistan, the Iranian Gulf, the Philippines, north Queensland (KRAMP 1961) and in sheltered waters off South Australia (SOUTHCOTT 1982). These stings can be painful.

Olindioides formosa GOTO, 1903 {Figure 8.11, page 208} is a stinging jellyfish found in Japanese waters. There have been severe generalised symptoms reported after stings from this jellyfish (AOKI 1922) {see Chapter 8, page 208}.

Gonionemus vertens Agassiz, 1862

Gonionemus vertens (Figure 6.5, page 125) {see also Chapter 8, page 201 and Figure 8.8, page 202} is a small hydromedusan, which swims about, or clings to seaweed by means of adhesive pads found on its tentacles. *Gonionemus* exhibits its typical tetramerous radial symmetry in the four radial canals and the quadrangular mouth. The effect on humans is documented in Chapter 8, page 203.

Bell shape:	• Hemisphere or flatter
Bell size:	• 1.5–2.5 cm in diameter
Bell margin:	• Smooth
Statocysts:	• Large number, similar to number of tentacles
Tentacles:	• Numerous (60–80) • Sharply bent near their distal ends with adhesive pads • Nematocysts borne on ridges
Mouth:	• With four short lips
Gonads:	• Arranged along most of length of radial canals

Olindias singularis Browne, 1905

(Figure 8.9, page 205, also see *Olindias* sp., Figure 6.18}:

Bell shape and consistency:	• Circular and saucer-like • Clear and firm
Bell size:	• 13-36 mm in width • Height about half diameter
Manubrium:	• Central, with four radial canals
Centripetal canals:	• Short and unbranched • 4–12 in each quadrant
Primary tentacles:	• Situated at bell rim • 28–86 in number • Rudimentary adhesive clubs at tips
Secondary tentacles:	• Longer than primaries • 16–50 in number
Statocysts:	• One at the base of each primary tentacle
Gonads:	• Frilly and pink in colour • Run along part of the length of the radial canals

CLASS SCYPHOZOA

There are three orders of the class Scyphozoa to which most jellyfish of medical importance belong. These include the Coronatae, the Semaeostomeae, and the Rhizostomeae.

ORDER CORONATAE (THE GROOVED MEDUSAE)

Coronate medusae of medical importance include *Linuche unguiculata* {Plates 8.73–8.77} and *Nausithoë punctata* {Plate 8.80}. Both dwell in warm waters in the shallows near the surface. Great swarms of the jellyfish *Linuche unguiculata* may occur in the tropical Indo-Pacific and also in the tropical Atlantic (SOUTHCOTT 1987b; BURNETT *et al* 1995). *Nausithoë punctata* Kölliker is found in all oceans, but the medusoid form of *Nausithoë* does not appear to be implicated in stings. Its polyps often live in association with sponges (JARMS 1991). Contact with these irregularly branched, plant-like colonies results in pain, itching and fever that may last for several weeks (SOUTHCOTT 1987b). The presence of the medusoid form of *Nausithoë* {Figure 6.20, page 142 and Plate 8.80} as described below, heralds outbreaks of stings from early growth stages of this organism.

Taxonomic features of coronate medusae

The jellyfish of this order are mainly found in deep water and possess either a conical, dome-shaped or flattened bell each with the typical coronal groove extending around the exumbrella. This results in the bell margin being separated from the rest of the bell. Below the coronal groove, or circular furrow, is a circle of fleshy pedalia, most bearing one solid tentacle. Below the level of the pedalia, the bell margin is formed into lappets that alternate with the pedalia. Rhopalia are carried in the niches of some of the lappets (from 4–32 rhopalia are found). A mouth is present. This leads into the manubrium, which in turn leads into the stomach. The stomach is supported by fleshy septa containing subumbrellar funnels. The edges of the septa bear gastric cirri.

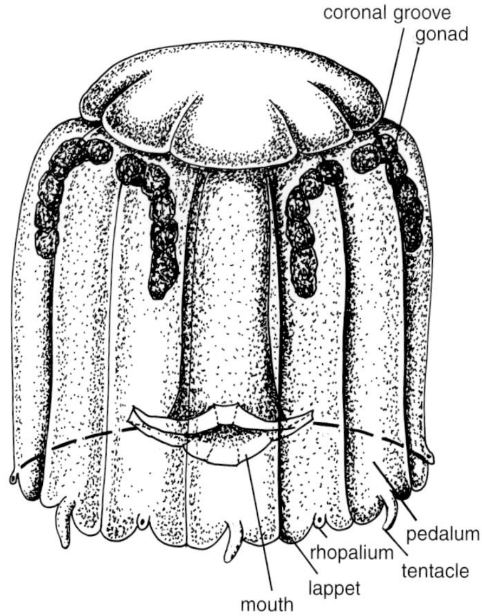

FIGURE 6.19
The thimble jellyfish *Linuche unguiculata*, a coronate medusa. (After HYMAN 1940; redrawn by Jan Hynes, Brisbane.)

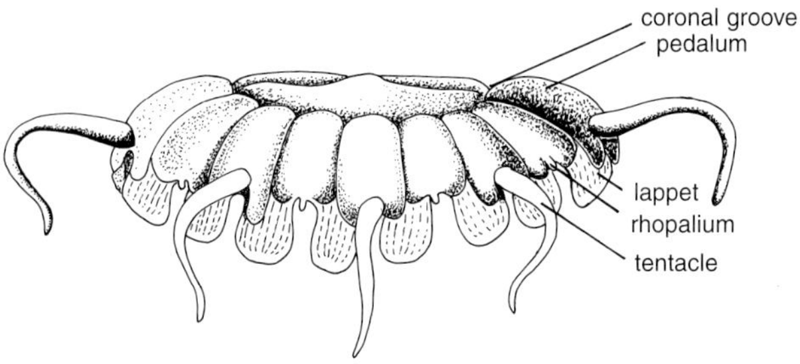

FIGURE 6.20
Nausithoë, a coronate medusa. (After MAYER 1910, redrawn by Jan Hynes, Brisbane.)

Linuche unguiculata (Schwartz, 1788)

(Figure 6.19, page 141) {Plates 8.73–8.77}

Bell:	• Thimble-shaped • Flat apex • Brownish in colour due to the presence of symbiotic algae
Bell size:	• 13 mm high, 16 mm wide • Bell width slightly greater than depth • Total length up to 100 mm, with tentacles extended
Rhopalia:	• 8
Lappets:	• 16
Pedalia:	• 16
Tentacles:	• 8 stumpy
Subumbrella:	• With three circles of hollow protuberances arising from radial stomach pouches
Gonads:	• 8, which form 4 cleft crescents

Nausithoë punctata Kölliker, 1853

(Figure 6.20):

Bell:	• Bowl shaped • Central portion thick and raised with a 'finely dotted appearance' • No radiating furrows
Bell size:	• 9–15 mm wide, rarely to 40 mm wide
Bell colour:	• Clear to red-brown • Small red spots on exumbrella
Rhopalia:	• 8 hanging from pedalia
Lappets:	• 16 marginal
Pedalia:	• 16
Tentacles:	• 8 situated adradially, between the lappets
Gastric cirri:	• Not grouped in clusters — 7 filaments per row in each quadrant
Stomach pouches:	• 16 radiating, simple
Gonads:	• Large, adradial, red-brown to bright yellow in colour

ORDER SEMAEOSTOMEAE (THE SCALLOPED MEDUSAE)

{See Chapter 8, page 223}

The semaeostome medusae are very common in temperate waters and include some of the largest specimens known. Species of medical importance include *Chrysaora quinquecirrha* {Plate 8.82}, *Chrysaora helvola*, *Chrysaora hysoscella* {Plates 6.6 and 8.81}, *Cyanea capillata* {Figure 6.23, page 144 and Plates 6.1A, 6.1B, 6.7}, *Pelagia noctiluca* {Plate 8.87A} and possibly *Aurelia* sp. (BURNETT *et al* 1988) {Figure 8.13, page 224 and Plate 10.1}. *Chrysaora quinquecirrha* is found in the Atlantic, Indian and Pacific Oceans; the species has not been positively identified in Australian waters (CLELAND & SOUTHCOTT 1965; MARSH & SLACK-SMITH 1986). *Chrysaora quinquecirrha* is an important stinger along the US east coast, and a special section {page 226} is devoted to it. In Western Australia another species of *Chrysaora* is considered to be a severe stinger, and stings produce general and local symptoms. *Cyanea capillata* ('snotty', the 'hair jelly' or the 'lion's mane jellyfish') (Plates 6.1A, 6.1B) {see also Plates 8.57, 8.90 and 8.91} is found in cold as well as temperate waters of Australia and the world, and numbers have at times reached plague proportions. Recently *Cyanea* species off the east coast of Australia have been identified as *Cyanea nozakii* (HEEGER 1994, unpublished observations) and *C. mjöbergi* (CORNELIUS 1994, unpublished observations).

Pelagia spp. are found in warm and temperate oceanic waters worldwide {see Chapter 8, page 228, and Chapter 10, page 287}. As the scientific name implies *Pelagia noctiluca* is luminescent at night. This species has attained plague proportions in the Mediterranean Sea, where it is a very familiar stinging jellyfish {see Chapter 10, page 287}. Swarms of this jellyfish were reported in the northern Adriatic coast of Yugoslavia (MARETIĆ *et al* 1980) where four or five of the luminous specimens per square metre of water were counted. An estimated 250,000 people were stung in one summer. In Australia, species of *Pelagia* have been found off the eastern and western coasts, also usually occurring in large swarms. Another stinger, *Pelagia flaveola*, has been recently reported (CORNELIUS 1994, unpublished observations) off the northern coast of Papua New Guinea {Plates 8.84A and B}.

Taxonomic features of semaeostome medusae

The semaeostome jellyfish possess a flat, saucer or bowl-like bell (Figures 6.21, 6.22 and 6.23, pages 143–4), the edge of which is scalloped and fashioned into eight or more lappets. Usually, from 8–16 club-shaped rhopalia (sense 'organs') are present in the niches between the lappets. Tentacles are present between the rhopalia, either on the lappets themselves or between them in the niches, or on the subumbrella. Tentacles may also be arranged along the bell margin or grouped in U-shaped or V-shaped bunches, as in *Cyanea* (Figure 6.23, page 144). The mouth is drawn out into four oral arms. It leads into the large stomach. Stomach pouches and septa are not present in adults. In *Chrysaora* and *Aurelia*, septa and pouches are present in the larval stages only, as in adults, a fusion of the upper and lower layers of gastrodermis at the periphery of the stomach occurs; these form the radial canals. Gastric filaments are present. Semaeostome medusae such as *Cyanea* possess four horseshoe-shaped gonads which lie on the floor of the periphery of the stomach.

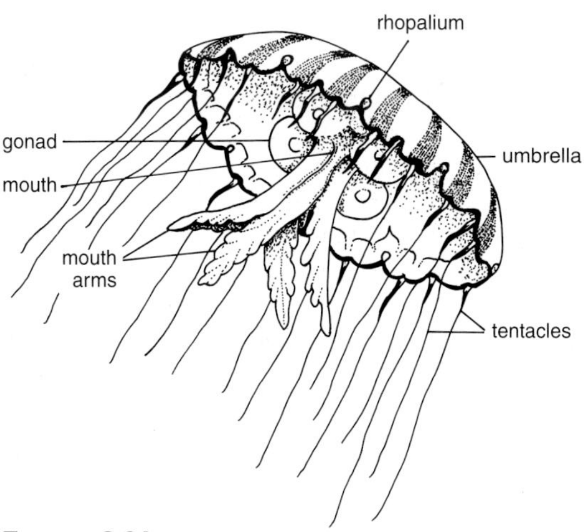

FIGURE 6.21
The semaeostome medusa *Chrysaora* sp., shown with some tentacles removed. (Composite drawing by Jan Hynes, Brisbane.)

Genus *Chrysaora* Péron & Lesueur, 1809

(Figure 6.21 and Plates 6.6, 8.81 and 8.82)

Bell shape:	• Somewhat saucer-shaped
Bell colour:	• White, pink, rusty red or light brown
Marginal tentacles:	• 24 or more in total
Marginal lappets:	• 32–48
Rhopalia:	• 3–9 with several tentacles in between • Marginal in positional
Stomach:	• Central • Gives rise to 16 separated unbranched pouches • Gastric septa absent • Gastric filaments present • Ring canal absent
Nematocyst warts:	• Present on exumbrella

Genus *Pelagia* Péron and Lesueur, 1809

Type species: *Pelagia noctiluca* (Forskäl, 1775) (Figure 6.22, page 144 and Plate 8.87A) — similar in appearance to *Chrysaora* in Australia.

Bell shape:	• Somewhat saucer-shaped
Bell size:	• 100 mm in diameter (generally smaller than *Chrysaora sp.* in Australia)
Bell colour:	• Highly variable, from brownish-yellow to pink or purple particularly on the tentacles, gonads and nematocyst warts

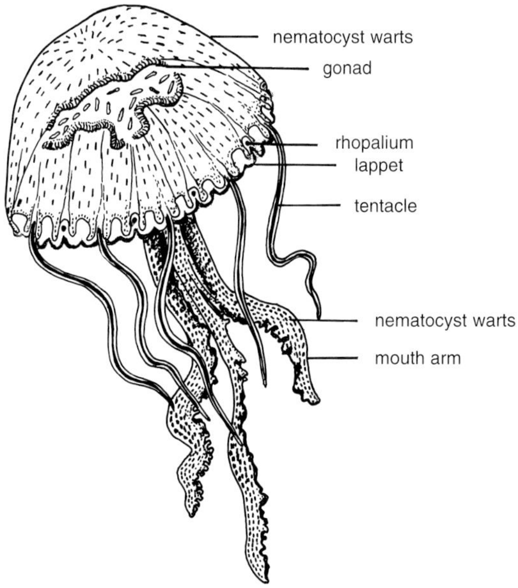

Figure 6.22
The semaeostome medusa *Pelagia noctiluca*. Note the eight marginal tentacles, the long mouth arms and the nematocyst warts distributed over the exumbrella and mouth arms. (After MAYER 1910, redrawn by Jan Hynes, Brisbane.)

Stomach:	• Gastric septa absent • 16 gastric pouches each ending in a pair of canals entering lappets • Gastric cirri present
Tentacles:	• 8 alternating with 8 sense organs at bell margin • Colour variable
Mouth arms:	• 4 simple mouth arms with frilled lips
Lappets:	• 16
Nematocyst warts:	• Present on exumbrella and oral arms • Variable in size, shape and number
Bell:	• Bioluminescent (gives off a green light), especially from the nematocyst warts near the bell margin

Cyanea capillata (Linné, 1758)

(Figure 6.23 and Plates 6.1A, 6.1B, 6.7)

Bell shape and texture:	• Saucer-shaped • Smooth exumbrella
Bell diameter:	• Ranges from 15–20 cm to 2.3 metres in diameter
Colour:	• Variable, from yellowish-brown to reddish-brown to light yellow
Bell margin:	• With lappets
Rhopalia:	• 8
Tentacles:	• Eight adradial tentacle clusters • Grouped in U- or V-shaped bunches, away from bell margin • Each tentacle thin, highly extensile
Subumbrella:	• Possesses both radial and circular muscles, • Divided into 16 parts by radial septa

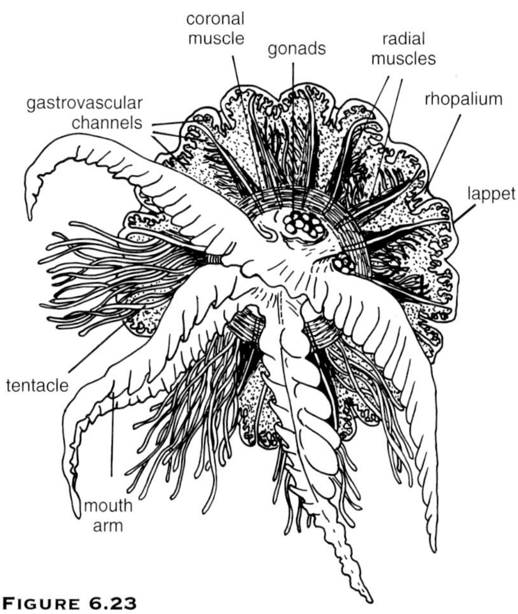

Figure 6.23
Subumbrella region of the semaeostome medusa *Cyanea* sp. Note that some of the tentacles have been removed. (After HYMAN 1940, redrawn by Jan Hynes, Brisbane.)

Mouth arms:	• 4
Stomach:	• Central, leads to gastric pouches then to many branching canals in the lappets • Gastric septa absent • Gastric cirri present
Gonads:	• Interradial, on stomach floor at its periphery

ORDER RHIZOSTOMEAE

Although rhizostome medusae are often considered to be relatively mild stingers (MARSH & SLACK-SMITH, and HEEGER 1994, unpublished observations) several species may cause very severe stings.

Stomolophus nomurai {Plate 8.62} is a potentially lethal stinger off the coast of China {see Chapter 8, page 214}, causing much morbidity and several confirmed deaths (ZHANG MINGLIANG 1987, 1988, 1990, 1991, 1992).

Members of the genus *Cassiopea* {Plate 8.60} are commonly called 'the upside-down jellyfish' because of their habit of floating upside-down in shallow waters, allowing debris and other food material to float down into the mouth arms. *Cassiopea* has been found in the Pacific, Atlantic and Indian oceans and several species are known to be severe stingers (SOUTHCOTT 1987b). *Cassiopea* sp., from waters around the Philippines, has caused pain, whealing, vomiting and joint pains.

On the east coast of Australia, *Catostylus mosaicus* {Plate 8.61} is very common. Contact with great swarms of these jellyfish causes annoying irritation for at least 30 minutes {see Case Report, page 213}. Other rhizostomes of medical importance include *Rhizostoma pulmo* of the Atlantic Ocean and the Mediterranean Sea (Figure 10.4, page 288}, *Lobonema smithii* (Plate 8.69) recently photographed on Seringatapam Reef, Western Australia (ALLEN & STEENE 1994), and in Darwin Harbour, Northern Territory (CORNELIUS & ALDERSLADE 1994, unpublished observations), and *Acromitoides purpurus*, from the Philippines (SOUTHCOTT 1987b).

Recent field evidence from Penang implicates the rhizostome genus *Thysanostoma* (Plates 6.2A and 6.2B) in possible human stings (BURNETT & CALDER 1991, unpublished observations). In Western Australia, the most common rhizostome jellyfish is *Phyllorhiza punctata* (Plate 8.70), although it is also found in eastern Australia, the Philippines and Japan. This jellyfish gives a mild sting to sensitive areas of the human body. MARSH and SLACK-SMITH (1986) reported prolonged morbidity from eye stings. HEEGER (1995) related personally a case of stinging by the rhizostome medusa *Versuriga anadyomene*, by which he himself was stung. This medusa caused a painful sting which lasted for 6 hours, and caused ulcers. Scar tissue and marks were evident three weeks after the stinging episode. *Pseudorhiza haeckeli* {Plate 8.72} is found in waters from off the Western Australian coastline to Victoria and it has been reported to give mild stings that persist for half an hour (MARSH & SLACK-SMITH 1986).

Taxonomic features of rhizostome medusae

Rhizostome jellyfish are similar in structure to those of the semaeostomes. However, they differ in the absence of marginal tentacles and the manner in which the oral arms are formed. The bell is usually quite firm and may be saucer-shaped or bowl-shaped. It is usually cut into numerous intricate scallops along the margin, and 8–16 rhopalia are present. During development of the organism, the early four-lobed mouth margin grows out and splits into eight thick oral (mouth) arms. These mouth arms may branch again. The grooves present in the mouth arms are closed over, forming a brachial (arm) canal or duct. The mouth opening is obliterated. Small, filamentous appendages are used to catch and paralyse prey, which is then carried to the inner surfaces of the arms and the small secondary mouths. These may be surrounded by five tentacles. Food passes from the small mouths into brachial canals in turn leading to the central stomach (Figures 6.24A and B).

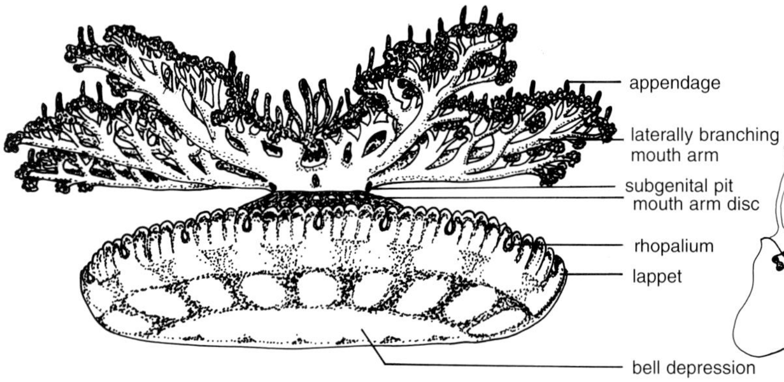

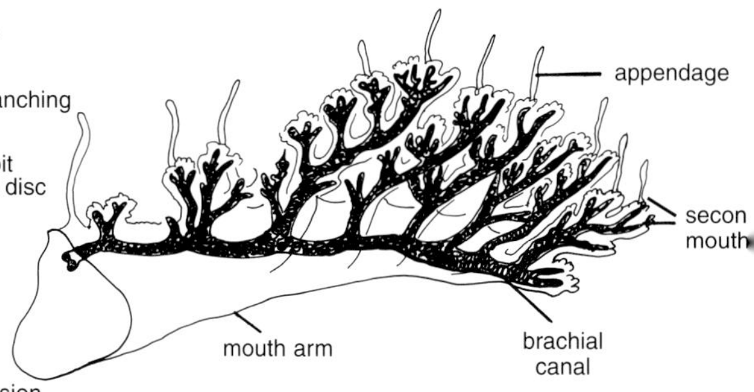

FIGURE 6.24A
Cassiopea xamachana ('upside down jellyfish') showing laterally branching mouth arms and concave bell apex. (After MAYER 1910, redrawn by Jan Hynes, Brisbane.)

FIGURE 6.24B
Branching mouth arm of *Cassiopea xamachana* showing gastrovascular canals and the small secondary mouths. (After MAYER 1910, redrawn by Rebekah Lewis, Brisbane.)

Cassiopea xamachana R.P. Bigelow, 1892

(Figures 6.24A and B and Plate 8.60)

Bell shape:
- Disc-shaped, flat, rounded edges, usually about 150 mm in diameter

Exumbrella:
- Distinctive concavity present in middle of exumbrella

Rhopalia:
- From 11 to 23, usually 16

Lappets:
- 5 short lappets present between each successive pair of sense organs

Mouth arm disc:
- Projects from subumbrella centre
- Mouth arms arise from this disc

Mouth arms:
- 4 pairs of adradial mouth arms arise from arm disc
- 1.25 x length of bell radius
- Triangular in cross-section
- Each gives rise to 10–15 alternate primary branches, which themselves give rise to secondary branches

Mouths:
- Mainly present on oral sides of primary and secondary branches of mouth arms
- Mouths fringed with fine tentacles

Appendages:
- Ribbon-like filaments are present in association with primary branches and on the oral surface of the mouth arm disc
- Appendages associated with mouth arm disc are relatively long
- Short, club-shaped nematocyst-bearing structures are present among the mouths

Stomach:
- Cruciform
- Ducts from 8 mouth arms empty into central stomach

Radial canals:
- Lead to the bell margin

Gonads:	• Sac-like, found at the interradial sides of stomach
Colour:	• Variable, generally greenish grey-blue
Pattern:	• Exumbrella patterned with a circle and radial stripes

CLASS CUBOZOA

Basic characteristics, as compared with other cnidarian classes

In this book we follow WERNER (1975) and regard the cubozoans as a class even though this is not a universal convention. Characteristics of the medusoid forms of the Cubozoa are summarised in Table 6.3 and features shared with medusae of

TABLE 6.2
SUMMARY OF FEATURES DIAGNOSTIC OF CUBOMEDUSANS, SOME OF WHICH ARE SHARED WITH SCYPHOMEDUSANS AND HYDROMEDUSANS

Hydromedusans	Cubomedusans	Scyphomedusans
+*	cuboidal bell*	+*
–(velum),–	velarium, velarial canals	–,–
–	pedalia	+ (in order Coronatae)
+(2)	marginal nerve ring	–
–	nematocysts in rings on tentacles in most species	–
–,–	rhopalia, in sensory niches	+, –
+,+(rarely)	ocelli, with lenses	+, –
–	statolith composed of gypsum ($CaSO_4$)	+
–	septa present in adults	+
+	development of polyp to medusa without strobilation	–
	present + absent –	

*NOTE Many hydromedusans such as *Leuckartiara* spp. do appear box-like as the four tentacles are equidistant, arising from the bell margin. Some scyphomedusans such as the coronate medusae *Pericolpa* spp. are 4-sided individuals in which the tentacles arise from pedalia. As such they superficially resemble box jellyfish.

TABLE 6.3
IMPORTANT DIAGNOSTIC DIFFERENCES BETWEEN JELLYFISH OF THE TWO EXISTING ORDERS OF THE CLASS CUBOZOA

Carybdeidae	Chirodropidae
4 single, paired or trebled* interradial tentacles arise individually from separate, blade-like pedalia	4 interradial tentacle clusters arise from complex pedalia. Each tentacle hangs from a gelatinous claw which branches from the main pedalial trunk.

*Most carybdeid species have only one tentacle at each corner, but *Tripedalia cystophora* has 3 separate tentacles in these sites, each tentacle arising from a separate, simple pedalum (Figure 6.25, page 148). Similarly, *T. binata* has 2 tentacles arising interradially from 2 separate, simple pedalia. Neither of these cubomedusans appears to be medically important (See Chapter 9, page 236).

Carybdeidae	Chirodropidae
4 stomach pouches containing lateral (sidewall) gonads but without superior gonads*	4 stomach pouches containing lateral gonads and with 4 pairs of perradial, superior gonads**

* In carybdeids, the gonads consist basically of paired 'leaflike' sheets that arise at the interradii and extend laterally into the stomach pouches, growing towards the perradii.
** Chirodropids (Figures 6.27–6.30, pages 149–153) have similar lateral gonads as carybdeids, though they differ markedly in the presence of paired superior gonads covered by the gastric diverticula, situated on either side of the perradii.

the classes Hydrozoa and Scyphozoa are compared with them. Although some taxonomists still consider that class Cubozoa is intermediate in position between the classes Hydrozoa and Scyphozoa, the data outlined in Table 6.2 suggest that the Cubozoa in general appears to be a sister group to the Scyphozoa. There are 2 existing orders in the class Cuboza (see Table 6.3).

ORDER CARYBDEIDAE

Some carybdeid medusae presently known and formally named (see Chapter 9 for their known distribution):

Order Carybdeidae (Figure 6.6d, page 126)
Genus *Carukia* SOUTHCOTT, 1967
 C. barnesi SOUTHCOTT, 1967 (Plate 9.2)
Genus *Carybdea* PÉRON & LESUEUR, 1809
 C. sivickisi STIASNY, 1926 (Plate 9.46)
 C. marsupialis LINNÉ, 1758
 C. alata REYNAUD, 1830 (Plate 9.8)
 C. alata var. grandis MAYER, 1910 (Plate 9.15)
 C. rastoni HAACKE, 1886 (Plate 9.27)

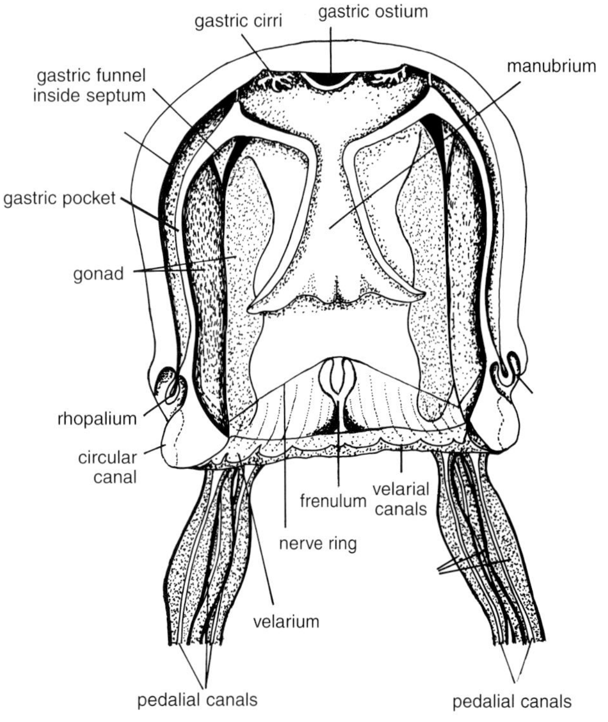

FIGURE 6.25
Perradial longitudinal section of *Tripedalia*, a carybdeid cubomedusan. (After CONANT 1898, redrawn by Jan Hynes, Brisbane). See also Figure 6.6d, page 00.

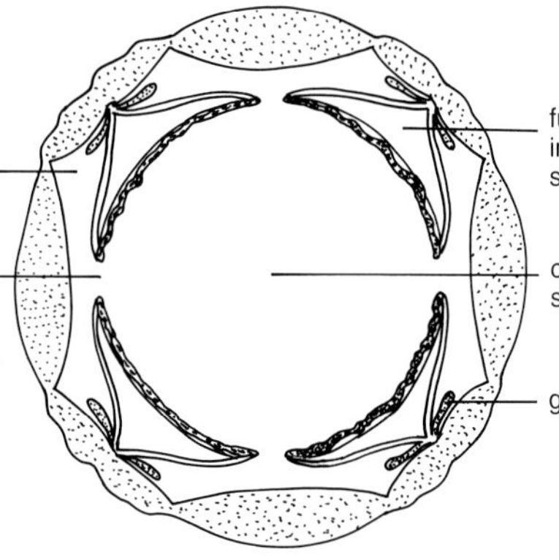

FIGURE 6.26
Transverse section through the upper part of the bell of a carybdeid cubomedusan. (After CONANT 1898, redrawn by Jan Hynes, Brisbane)

Genus *Tamoya* MÜLLER, 1859
 T. haplonema (MÜLLER, 1859) (Figure 10.6, page 300)
 T. virulenta KISHINOUYE (1910) (Plates 9.4, 9.7)
Genus *Tripedalia* CONANT, 1897
 T. cystophora CONANT, 1897 (Figure 6.25)
 T. binata MOORE, 1988

The general anatomy of carybdeids was discussed earlier in this chapter, while further details differentiating between the individual species listed above are included in Chapter 9. The severity of stings received from individual species worldwide varies according to species and ranges from minor to serious, and detailed accounts of the stings also appear in Chapter 9. In Australia, victims of the tiny carybdeid *Carukia barnesi* have sustained the most severe symptoms and produced the Irukandji syndrome (see Chapter 9, page 249), though other carybdeids, and in fact members of other jellyfish classes, are also capable of producing severe Irukandji-like reactions (see below and Chapter 9).

TAXONOMIC STUDIES OF THE MORBAKKA (MORETON BAY CARYBDEID)

T. virulenta KISHINOUYE (1910) (Plates 9.4, 9.7). 'The morbakka' is the common name coined by SOUTHCOTT (1987) for one of a group of Australian carybdeids under current investigation by the author. Various morphologically distinct members of the group are geographically isolated from one another. These carybdeids closely resemble the well-described *Tamoya haplonema* in appearance. The latter carybdeid is characterised by the presence of vertical rows of gastric phacellae and cirri situated in the interradial portions of the stomach.

Other group members in the Indo-Pacific specimens examined resemble it except that the phacellae and cirri are absent. All similar specimens from the Indo-Pacific have a bell taller than it is wide, with 4 tentacles, 4 rhopalia, a sensory niche with a covering scale above and below, a stomach supported by 4 mesenteries and complexly forked velar canals. However there are several significant variations within the species including the size and shape of the bell. The author suggests that morphs of the carybdeid *Tamoya virulenta* constitute a species complex in the Indo-West Pacific region.

Tamoya virulenta from Moreton Bay has been known to cause serious stings, with Irukandji-like characteristics (see below and Chapter 9, page 244).

Dr Frank Carroll, a general practitioner from Stradbroke Island, off Brisbane, recorded and treated the serious systemic effects of an Irukandji-like reaction from a 4-tentacled carybdeid found in the Stradbroke Island area. The offending jellyfish was caught and tentatively identified as the morbakka. The case history as related personally, taken on 3.2.87 is as follows: The victim presented a 10 cm sting on the right hand. The wheal was 5mm wide at its widest point, but 1–2 mm wide at its narrowest. The symptoms included respiratory distress, throbbing back pain localising in the lumbar areas. The wheal resembled a first degree burn with no superficial necrosis. Treatment was with 1 ml of promethazine (25 mg). The tentacles fragmented, and were wrapped around crab pot ropes. Ten cm pieces, 4 mm wide were obtained and portions were sent to the author.
Dr Carroll also kindly donated photographs of Morbakka stings, some of which are included in Chapter 9, pages 244 and 246–7.

ORDER CHIRODROPIDAE

1. Genus *Chironex* SOUTHCOTT 1956
C. fleckeri SOUTHCOTT, 1956 (Australia) (Figures 6.27, 6.28, and 9.8A, page 264)
C. fleckeri, *Chironex*-like specimens (northern Indo-West Pacific, Plate 10.17)

2. *Genus Chiropsalmus* L. AGASSIZ 1862
C. quadrigatus HAECKEL, 1880 (northern Indo-West Pacific)

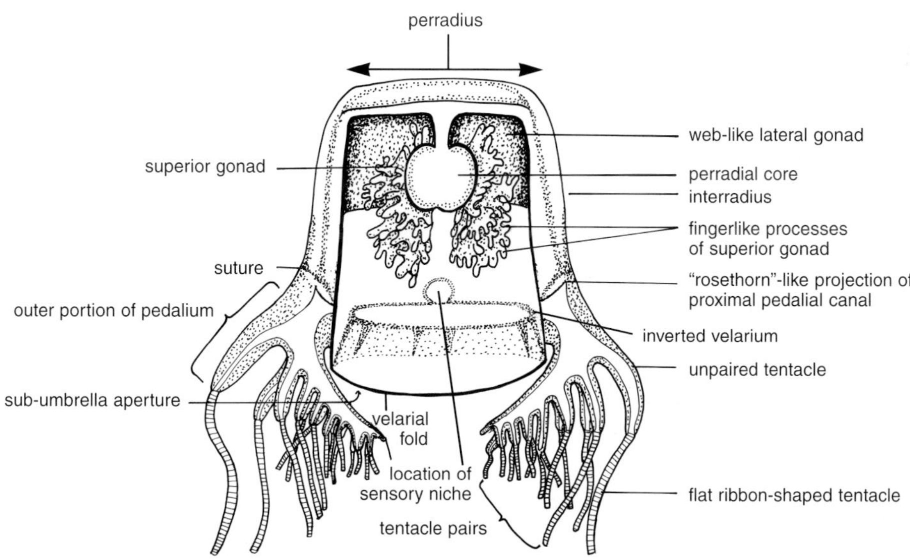

FIGURE 6.27
The chirodropid *Chironex fleckeri* from Australia during swimming, with bell and velarium in the relaxed state, showing structure of pedalium, the 'rosethorn'-like proximal pedalial canal; mature superior gonad with digitations; lateral gonad; ribbon-like tentacles; kidney-shaped perradial cores. (Drawing based on photographs and preserved specimens, by Jan Hynes.)

C. quadrigatus (Philippines) (Plate 9.35)
C. quadrigatus (Japan) (Plates 9.38A and 9.38B)
C. quadrigatus sensu BARNES (Australia) (Figure 6.29 page 151 and Plates 1.10 and 6.9)
C. quadrumanus (MÜLLER 1859) (Atlantic /Caribbean) (Plate 9.36)

3. Genus *Chirodropus* HAECKEL, 1880
C. gorilla HAECKEL, 1880 (West African coastline) (Figure 9.9 page 268)
C. palmatus STIASNY, 1922

4. Genus *Chiropsoides* SOUTHCOTT, 1956 = *Chiropsalmus* L. AGASSIZ
Chiropsoides buitendijki (HORST, 1907) (Plates 9.9A, 9.9B, 9.60 and 9.61)

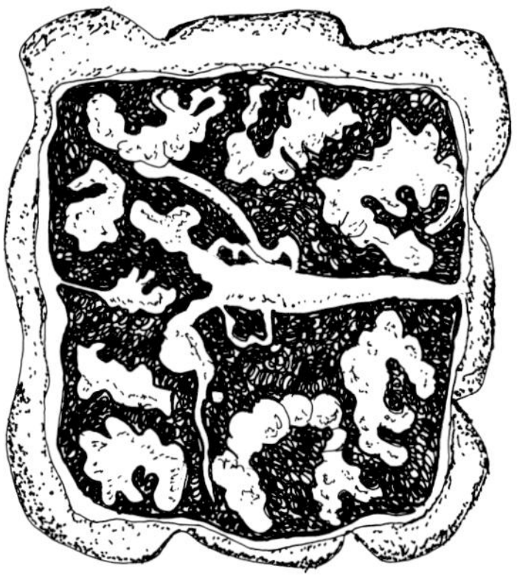

FIGURE 6.28
Chironex fleckeri, looking into the bell from below, showing the distinctive 'cockscomb'-like superior gonads. Tentacles have been removed. (After BARNES 1965, redrawn by Jan Hynes, Brisbane.)

ASPECTS OF CHIRODROPID TAXONOMY IN THE INDO-PACIFIC REGION

Australia's most common chirodropid species are *Chironex fleckeri* SOUTHCOTT, 1956 (Plate 9.3) and *Chiropsalmus quadrigatus* HAECKEL, 1880 *sensu* BARNES 1965 (from now on referred to simply as *Chiropsalmus* sp. [Australia]) (Plate 6.9). They have been named and described in some detail by SOUTHCOTT (1956) and BARNES (1965, 1966) respectively. These species are readily identifiable in Australia (see Tables 6.4A and 6.4B and also Chapter 9, page 258), though *Chiropsalmus* sp. (Australia) has not yet been formally identified. Moreover, the identity of the chirodropids of the Indo-Pacific region north of Australia is still problematic.
For a summary of the description of chirodropids presently known and listed above, and their world distribution see Chapter 9, page 258.

Brief history of chirodropid taxonomy in the Indo-Pacific region:

Ernst Haeckel identified an Indo-Pacific chirodropid in 1880 caught off the coast of Burma, and named it *Chiropsalmus quadrigatus*. Unfortunately, his specimen was spirit preserved, damaged and immature and his resulting description is inadequate for identification purposes. This has resulted in confusion and in the attribution of several potentially different species to the name *Chiropsalmus quadrigatus*. Specimens to which the name *C. quadrigatus* have been attributed have been collected from all over the Indo-Pacific region including Australia, and northwards to Japan (Okinawa) (see Chapter 9, and Plates 9.38A, 9.38B and 9.39, and see list above). MAYER (1910) redescribed *C. quadrigatus* according to the appearance of specimens he caught in waters off the Philippines, and several authors appear to have followed his example. His descriptions resembled those of SOUTHCOTT (1956) for *Chironex fleckeri* of Australia. BARNES (1965, 1966) later used the name *Chiropsalmus quadrigatus* for a second morphologically distinct chirodropid commonly found in Australian waters. Another distinctive chirodropid medusa is found in the Indo-Pacific region north of Australia, to which the author attributes the name *Chiropsoides buitendijki*, according to suggestions made by SOUTHCOTT (1956).

Up to the present time, names have not been attributed with any certainty to chirodropid specimens in the tropical Indo-Pacific and distributional and epidemiological data taken so far are in doubt. Outside Australia, few collections of the cnidarians actually causing stings have been carried out. Those few include one *Chironex*-like specimen identified by the author collected in the area in which a victim was stung in waters off Brunei (unpublished observations 1993, and see Chapter 9, page 264). It is a fact that unless animals responsible for stings are collected at the precise time of envenomation, it is not possible to attribute stings or fatalities to specific species. Furthermore, pharmacological work cannot be carried out with confidence until species are properly defined. The author is presently carrying out taxonomic research in this area with specimens collected from Australia and from the remainder of the Indo-Pacific region. Much of the research centres around the interspecific and intraspecific variation present among cubozoan specimens in general. At times variation appears to be attributable to stages in the maturation of the medusa, while at others it is not.

Chirodropid Anatomy

A comparison of distinctive features of carybdeid and chirodropid medusae were presented above. For a summary of the anatomical features present in all chirodropids described up to the present time, see Chapter 9, page 258.

Anatomy of Australian chirodropids

As stated above, most of the anatomical features that distinguish between Australia's chirodropids of medical importance have been studied, and morbidity and mortality studies following stingings from both morphs have been well documented. Up to the present time, deaths have been attributed only to *Chironex fleckeri* (see Chapter 9, page 259).

In Australia, *Chiropsalmus* and *Chironex* appear similar until detailed anatomical studies are carried out. A comparison of the similarities and differences are presented below in the form of 2 tables (Tables 6.4A and 6.4B). Difficult concepts summarised in the tables are described in more detail in the text below. Most of the data is from BARNES, 1965, 1966 and BARNES in KINSEY (1986) and all has been confirmed by the author. Other data is by RIFKIN (unpublished).

Further information on major morphological differences between *Chironex fleckeri* of Australia and *Chiropsalmus* sp. (Australia)

(SOUTHCOTT, 1956, BARNES 1965, 1966, RIFKIN unpublished)

1. The pedalium and the pedalial canal

There are four strong and outstanding symmetrical pedalia, each of which originates at one of the four interradial corners of the bell. Each consists of a proximal portion, which branches into a trunk from which paired claws emanate. The claws join with the tentacles. A single unpaired claw branches from the outside of each 'hand'.

The pedalium contains a branched canal which is an extension of the gastrovascular system and which connects with each tentacle cavity. The proximal portion contains the part of the canal that is of the most interest taxonomically. In *Chironex fleckeri*, it contains a sharply

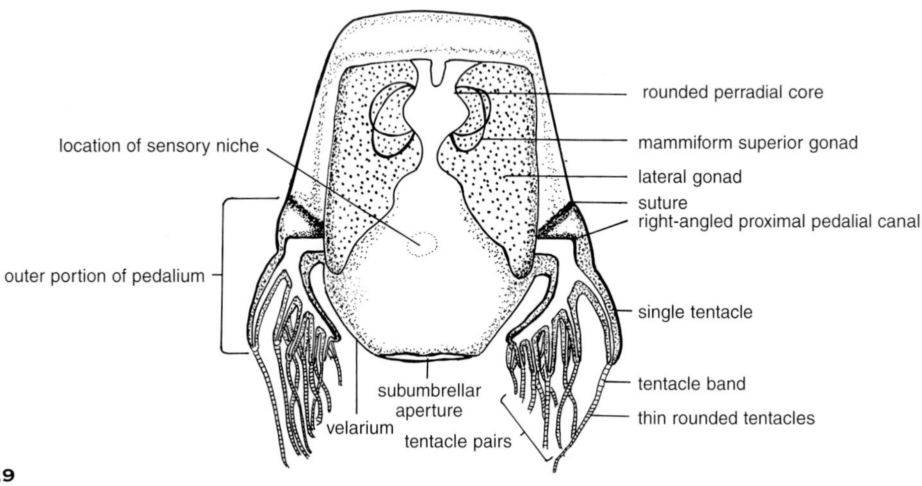

FIGURE 6.29
The chirodropid *Chiropsalmus quadrigatus* with bell and velarium in fully contracted state during swimming, right-angled bend in proximal pedalial canal, rounded shape of perradial cores, mammiform superior gonad; lateral gonad. (Drawing based on photographs and preserved specimens by Dr Jacquie Rifkin.)

TABLE 6.4A
BASIC SIMILARITIES BETWEEN *CHIRONEX FLECKERI* (AUSTRALIA) AND *CHIROPSALMUS* SP. (AUSTRALIA)

Bell:	box-shaped
Pedalium:	located at each of the 4 interradii (corners) of the bell hand-like shape
pedalial canal:	connects gastrovascular cavity proximally with tentacle cavities distally proximal portion of pedalium:contains pedalial canal with distinctive angulation
distal portion of pedalium (main trunk):	with paired sets of gelatinous claws. Single, unpaired claw outermost
Tentacles:	several paired sets single, oldest tentacle outermost Nematocysts arranged in bands or rings along length.
tentacle buds:	located between the last paired (youngest) sets of tentacles.
Gastrovascular cavity:	with 4 gastric pockets or pouches with diverticula. gastric cirri present.
Gonads:	
superior gonads:	4 pairs associated with gastric diverticula
lateral gonads:	sheet or web-like, originating at the interradius and extending into the gastric pouches

TABLE 6.4B
SUMMARY OF MORPHOLOGICAL DIFFERENCES BETWEEN *CHIRONEX FLECKERI* (AUSTRALIA) AND *CHIROPSALMUS* SP. (AUSTRALIA) (DATA FROM SOUTHCOTT 1956; BARNES 1965, 1966; RIFKIN UNPUBLISHED 1995).

Feature		*Chironex fleckeri*	*Chiropsalmus* sp.
Bell size		up to 200mm + in width	up to 100 mm in width*
Proximal pedalial canal		rose-thorn-like	rounded to acutely-angled (upwards)
Tentacles		flat in cross-section	rounded in cross-section
		relatively robust	relatively fine
		up to 15 per pedalium	up to 9 per pedalium
Lateral gonad		fine, weblike	sheets
Superior gonad:	1. perradial cores	kidney-shaped	rounded
	2. perradial eminences	papillated → 'cockscomb' → tiered bunches of fingers	mammiform
	3. reproductive tissue	covers perradial eminences	covers perradial eminences
Gastric diverticula		envelops perradial eminences	envelops perradial eminences

* The largest specimens recorded by one of the editors is 60 mm from apex to the rhopalar niche

defined 'rosethorn-shaped' corniculum or spur, which is directed upwards (Figure 6.27, page 149), whereas in *Chiropsalmus quadrigatus* this part of the canal is curved (knee-shaped), right-angled, or angled upwards parallel with the suture joining the pedalium with the bell (RIFKIN, unpublished observations) (Figure 6.29, page 151).

2. Tentacles

The manner in which the tentacles are arranged on the pedalia is shown in Figures 6.27, page 149 and 6.29, page 151. *Chironex*, however, acquires almost twice the number of tentacles as *Chiropsalmus quadrigatus*. Also, tentacles of *Chironex* are flattened (Plate 1.5), whereas in *Chiropsalmus quadrigatus* they are finer and rounder in cross section (Plate 1.12) and contain correspondingly fewer nematocysts.

3. Gonads (SOUTHCOTT 1956, BARNES 1965, 1966 and RIFKIN, unpublished observations)

In *Chironex fleckeri*, the web-like gonad tissue arises from the upper portion of the interradial septum and migrates towards kidney-shaped perradial cores.* The latter are paired mesogloeal formations formerly called perradial nuclei by BARNES 1965. They appear to form the foundations from which the superior gonads arise. In the early stages, the perradial cores develop papillae, which soon develop into cockscomb-like structures (Figure 6.28, page 150), and these later become whorled. Reproductive tissue becomes associated with these structures. The subumbrella of the most mature stages of *Chironex* contains a branching series of stubby, finger-like projections on either side of the perradius. These are sometimes likened to bunches of grapes (Figure 6.27, page 149). Complex tiered diverticula, which arise from the upper portion of the gastric pouches on either side of the perradii, conform to these projections like a glove on a hand.

In *Chiropsalmus quadrigatus*, smooth sheets of reproductive tissue (gonad) originating at the central portion of the interradial septum but eventually occupying its entire length, grow towards rounded perradial cores† (Figure 6.29, page 151). These cores develop into still larger rounded eminences which, covered in gonad sheets, hang into the subumbrellar cavity (Figures 6.29, page 151, 6.30 and Plate 6.9). Simple diverticula arising from the upper portion of the gastric pouches on either side of the perradii envelop these eminences.

Other species sighted in Australian waters

While *Chironex fleckeri* and *Chiropsalmus quadrigatus* are the most common chirodropids found in Australia, there have been early isolated reports of other chirodropid species here, including an immature and damaged *Chiropsalmus quadrumanus* specimen identified by by STIASNY (1926) (its identity is now disputed) and several specimens of *Chiropsoides buitendijki* (HORST, 1907) (=*Chiropsalmus buitendijki* HORST) (Plates 9.60 and 9.61) by POPE in 1953, but the paucity of these sightings prompted BARNES (1966) virtually to discount these last 2 species as major causes of stings in Australia.

* These structures have been renamed by RIFKIN as the term 'nuclei' is suggestive of other structures.

† The most mature forms present at least three tiered bunches of stubby 'fingers'.

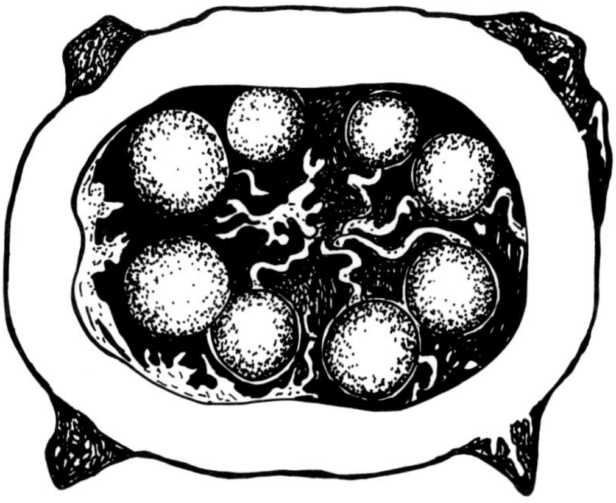

FIGURE 6.30
Chiropsalmus quadrigatus looking into the bell cavity from below. Tentacles removed. Note the eight mammiform superior gonads. (After BARNES 1965, redrawn by Jan Hynes, Brisbane.)

Chirodropids found in the remainder of the Indo-Pacific region

In the Indo-Pacific region north of Australia, chirodropids are also responsible for very serious stings and many fatalities (CLELAND & SOUTHCOTT 1965; Chapter 3, page 66). Two species of chirodropids have been reported; *Chiropsalmus quadrigatus* and *Chiropsoides buitendijki* (KRAMP 1961). However, early reports of specimens of *Chiropsalmus quadrumanus* by, for example KRUMBACH (1925) from Sumatra and by RAO (1931) from the Indian Ocean, suggest that the situation is complicated.

The author has identified both *Chiropsalmus*-like and *Chironex*-like specimens in waters north of Australia based on comparisons with Australian specimens and those found overseas. She has identified *Chironex*-like specimens from the Philippines, Vietnam and Indonesia (Borneo), and Malaysia and *Chiropsalmus*-like

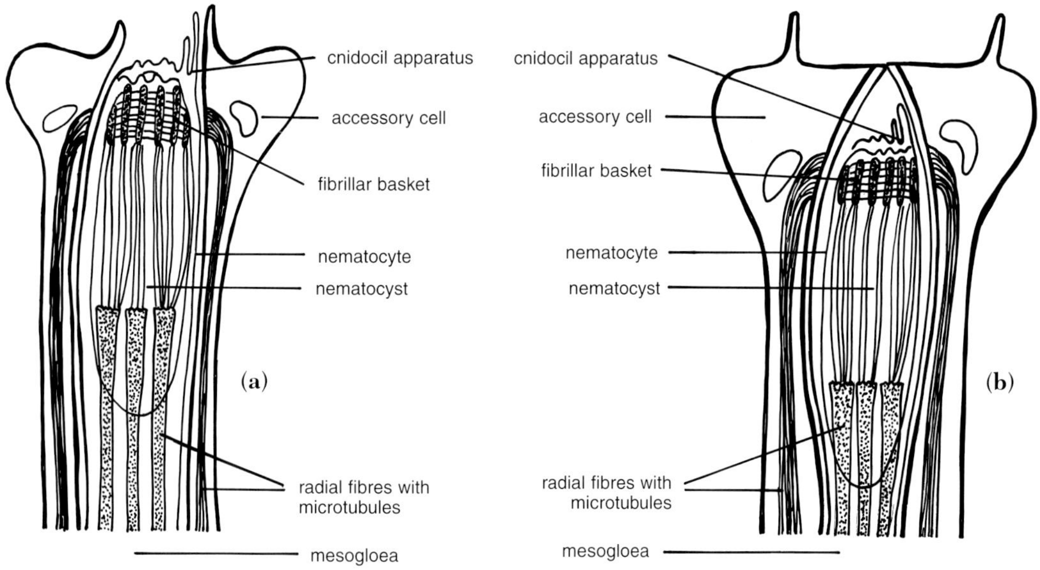

FIGURES 6.31a AND b
Chironex fleckeri mastigophore nematocyte *in situ* in the tentacle:(a) in the 'fire-ready' position; (b) in the retracted or 'resting' state. (After RIFKIN & ENDEAN 1988, redrawn by Jan Hynes.)

specimens from Borneo and the Philippines. Specimens examined from these areas were preserved in various stages of development of the superior gonad. These developmental stages are similar to those described earlier for *Chironex fleckeri* and *Chiropsalmus* sp., both from Australia.

Although studies by the present author suggest that that it is not yet possible to attribute deaths in these areas to specific species (see above), work is continuing.

The taxonomy of cubozoan specimens presents particular difficulties

Jellyfish are often damaged in fixatives, especially alcohol, and it is important that artifacts not be interpreted as species characteristics. A suggested narcotising, fixing and preserving regime, follows.

- Narcotisation: If required, specimens may be narcotised prior to fixation by placing them in sea water to which a solution of MS222 has been added. Alternatively, specimens may be relaxed in sea water to which a solution of 7.5% magnesium chloride in distilled water is added in equal parts.
- Fixation: Alcohol should not be used for fixation or preservation as it causes gross shrinkage of the mesogloea (including that of the superior gonad). The author's recommended preservative for jellyfish is Perspex box preservative fluid (20 L distilled water, 4L glycerine, 400ml formalin [conc.], 400ml camphor and 800g sodium acetate) to be used after 24 hours fixing in 10% neutral-buffered formalin in sea water (9 parts sea water to 1 part formalin conc., buffered in lithium carbonate). It contains a reduced amount of formalin and is more easily tolerated by the researcher than a solution consisting entirely of formalin. However, if this is not available, specimens should be fixed as well as preserved in 10% neutral-buffered formalin sea water. Preserved specimens should always be handled with gloves in a well-ventilated place as formalin is toxic. Specimens can be frozen for later use although ice crystals cause damage to delicate structures.

Recommendations for obtaining reproducible length and width data in cubozoan jellyfish

NOTE: The reader will be aware that there is little consistency amongst authors with respect to the recording of length and width measurements in cubozoans (even in the present text). In some cases, bell width measurements only are recorded, in others only the bell length is recorded, while in still others, both length and width measurements are taken. However, because of the presence of special features such as the velarium, and the pedalium, confusion often arises amongst researchers as to how actually to proceed with the measuring task. Indeed, it is not adequate to record only the length or width, as the medusae may not be perfectly square. In order to standardise these measurements, and allow for damage and shrinkage of the velarium, the author suggests (following BARNES in KINSEY 1986) that measurements be taken in the following manner (see Figures 6.25, 6.27, 6.29, pages 148–151).

- Height measurements of the medusae of cubozoans should be recorded from the top of the bell to the rhopalar niche, and rhopalar niche to the bottom of the velarium; width measurements of the bell should be taken along one side of the bell from one interradius (corner) to another. Measurements are best taken just above the level of the pedalia.

Acknowledgments: The author is indebted to Professor Thomas Heeger of the Philippines, Associate Professor Jack Greenwood and Dr George Matsumoto, both of Australia, and Dr Gerhard Jarms of Germany for critically reading this manuscript. The author is especially grateful to Dr Paul Cornelius of the British Museum of Natural History for advice concerning the taxonomy.

CNIDA STRUCTURE AND FUNCTION

JACQUIE RIFKIN

Cnidae are among the most highly complex intracellular secretory products known (Figures 5.1, page 99, 6.11, page 130, Plates 6.10, 6.11 and Figures 6.31, 6.39 and 6.43). These remarkable non-living structures and their surrounding living cells, the **cnidocytes** (Figure 6.43, page 168) characterise the phylum Cnidaria. 'Cnida' is the collective term used to encompass the three main categories of capsular organoid; namely **nematocysts**, **spirocysts** and **ptychocysts** (Figure 6.40, page 159). 'Cnidocyte' is a general term which refers to the cell in which the mature nematocyst is housed. If the category of cnida is known, the specific terms 'nematocytes', 'spirocytes', or 'ptychocytes' should be used.

Nematocysts (Figure 6.43 I–XXXI, page 168), to which the greater part of this discussion is devoted, are the only cnidae that are known to penetrate the skin. This penetration is carried out by a complex explosive drilling action {decribed below}.

However, all categories of cnidae discharge in a similar manner. **Spirocysts**, Figure 6.43b, page 168 found exclusively amongst the anthozoans, may differ morphologically from one taxonomic group to another (RIFKIN 1991). They contain sticky tubules that act as adhesors. **Ptychocysts** (Figure 6.43c, page 168) are elongate capsules that are used in the construction of protective tubes for certain burrowing anemones (class Anthozoa, order Ceriantharia).

Mature **nematocytes** function as both sensory and effector cells that may or may not be associated with a 'nervous system'. Their contained nematocysts, which are themselves capable of discharging in the isolated state (a fact that has important first aid implications) {see Chapter 5, page 108}, are used primarily as tools for capturing prey or repelling predators.

There are many types of nematocyst, which are morphologically distinct (Figure 6.43, page 168). Although most nematocysts penetrate and immobilise prey, some are adapted as adhesors, or as miniature 'lassoes' or 'grappling hooks'.

The microbasic mastigophore nematocyst (Figures 6.33 and 6.34, page 156) is a penetrant found in many cnidarians of medical importance. These nematocysts may be structurally similar from one cnidarian class to another, but their contained venoms differ (HESSINGER 1988). In most cases this reflects the intended use of

FIGURE 6.32
Scanning electron micrograph of outer surface of tentacle epidermis of *Chironex fleckeri*, showing cnidocil and cnidocil-like receptors, all surrounded by microvilli. The prey walking on the surface touches the cnidocil, stimulating the nematocyst (buried below) to discharge. (Photo by Dr Jacquie Rifkin.)

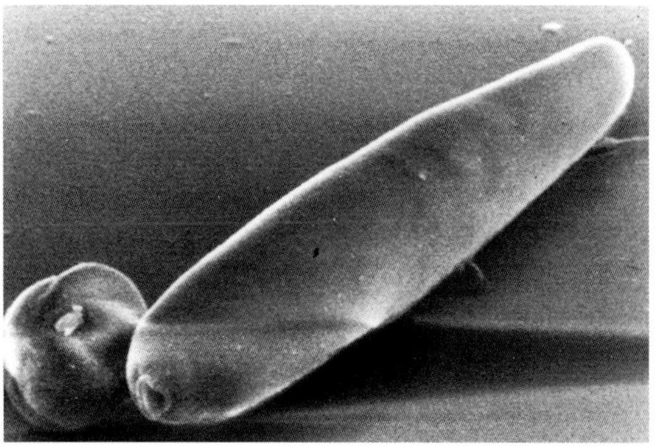

FIGURE 6.33
Scanning electron micrograph of an undischarged microbasic mastigophore capsule from *Chironex fleckeri* tentacles. Thses are present in two functionally different sized classes. Note operculum on capsule in foreground. (Photograph by Dr Jacquie Rifkin.)

the venom by each individual cnidarian species upon the usual prey. A sea anemone (class Anthozoa) feeds largely on small crustaceans, whereas the box jellyfish, *Chironex fleckeri* (class Cubozoa) which also feeds on crustaceans, immobilises large fish, since the capture of large fish could otherwise damage the jellyfish during the fish's capture. Rapid immobilisation of such vertebrate prey is required, and the cardiotoxic, myotoxic (and possibly neurotoxic and haemolytic) actions take effect rapidly (ENDEAN *et al* 1969).

Examples of nematocysts found in medically important cnidarians include microbasic mastigophores, a type found in every class of cnidarian (RIFKIN 1988, 1996, in press) although previously thought not to exist in the class Scyphozoa. The mastigophore has received much publicity as it is associated with fatalities in human victims of *Chironex fleckeri* stingings (HARTWICK *et al* 1980; ENDEAN & SIZEMORE 1988). However, all cnidarians of medical importance have nematocysts consummately adapted for penetrating human skin. These types may or may not have spines, and they may or may not have basal shafts, (Figure 6.43, page 168); though all nematocysts appear to operate in the same basic manner as that described below for *Chironex fleckeri*.

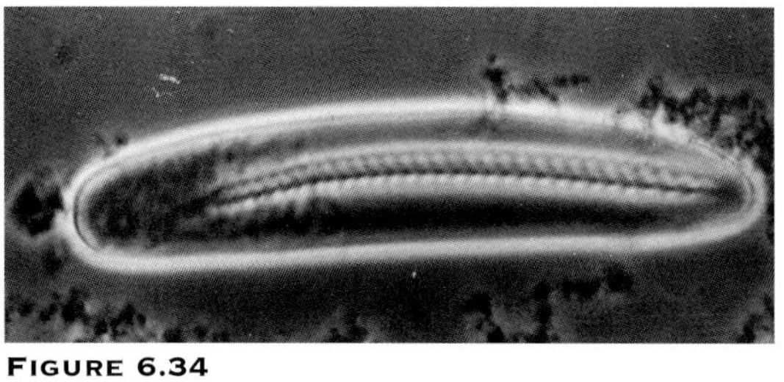

FIGURE 6.34
Light micrograph of an undischarged microbasic mastigophore capsule from a sea anemone, showing the central rod-like shaft and the distal tubule coiled up at the bottom of the capsule. (Photograph by Dr Jacquie Rifkin.)

THE NEMATOCYSTS OF THE BOX JELLYFISH *CHIRONEX FLECKERI*

There are at least six different types of nematocysts found on the tentacles of the cubozoan *Chironex fleckeri*. These include microbasic mastigophores in two functionally different sized classes (e.g. Figures 6.33, 6.35, 6.36, 6.38), atrichous isorhizas, holotrichous isorhizas and two kinds of trirhopaloid (formerly identified as 'heterotrichous microbasic euryteles') (ENDEAN *et al* 1969; RIFKIN & ENDEAN 1983).

The venom found within the elongate, penetrating, microbasic mastigophore nematocysts (Figure 6.43, page 168 [XIV–XIX] and Figures 6.33 and 6.35) has been responsible for more than 60 human fatalities in the past 100 years, and for painful morbidity in countless many more {see Chapter 3, page 77, and Chapter 9, page 266}. The atrichous isorhiza is also very abundant in the tentacles of *Chironex fleckeri*. These nematocysts do not carry spines, but are filled with a sticky secretion which appears to help adhere the tentacle onto the skin or cuticle of a victim. Holotrichous isorhizas which are coiled upon discharge, may entangle small prey (Figure 6.43, page 168). The large trirhopaloid (Figure 6.43) is a penetrant that appears to act as a grappling hook, dragging the integument of the prey into closer contact with the tentacle (RIFKIN & ENDEAN 1983).

FACTORS CAUSING CNIDA DISCHARGE

Cnidae *in situ* in the tentacle are known to discharge in response to 'contact–chemical stimuli' (PANTIN 1942a & 1942b). These are, in essence the usual chemical and mechanical stimuli that are

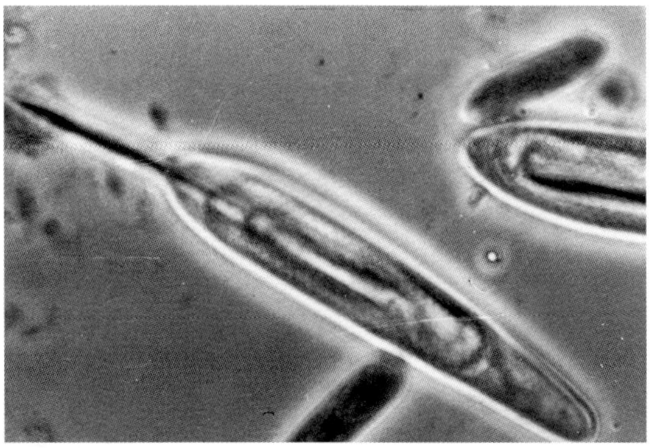

FIGURE 6.35
Light micrograph of a partially discharged large microbasic mastigophore shaft, showing piercing tip and drops of discharged tubular matrix and possibly capsular plaque at left-hand margin of photo. (Photograph by Dr Jacquie Rifkin.)

at hand when the cnidarian responds to the presence of prey. It is probable that such a response triggers the envenomation process when humans come into contact with the tentacles. Most reports suggest that the discharge of cnidae *in situ* under natural conditions requires both chemical and mechanical stimuli.

The cnidocil (the hair trigger) has long been known to be directly associated with nematocyst discharge (MARISCAL 1974) (see Figures 6.31, page 154, and 6.32 page 155). Sensitising agents in prey activate specific chemoreceptors. In their studies with sea anemones (class Anthozoa) {see Chapter 8, page 181}, THORINGTON and HESSINGER (1988) found two groups of naturally occurring substances in the feeding tentacles of *Aiptasia pallida* that chemosensitise cnidae to discharge. Although anthozoan nematocytes do not have cnidocils, they suggest that chemoreceptors in the epidermis detect such specific substances from potential prey, and that the cnidocytes are thus sensitised to discharge their cnidae in response to mechanical stimuli originating from the struggling prey. The cnidarian's nervous system may or may not be active in the nematocyst discharge process (CARRÉ & CARRÉ 1980). Other stimuli (including electrical stimuli) have also been shown to elicit nematocyst discharge (BARNES 1967; RIFKIN 1982).

Recent research by LOTAN *et al* (1995) has shown that the discharge of nematocysts is driven by the capsule's high internal hydrostatic pressure of 15 megapascals (=150 atmospheres). This pressure causes eversion to occur in 3 milliseconds, with a maximum velocity of 2 metres per second generated, corresponding to an acceleration up to 40,000 G* (HOLSTEIN and TARDENT [1984]). In comparison, the explosive device in an airbag detonates in 30 ms (Ford Motor Company, Australia). Because of its exceedingly small size

*G = 9.8 m/s/s; it is the acceleration produced by gravity; force = mass X acceleration.

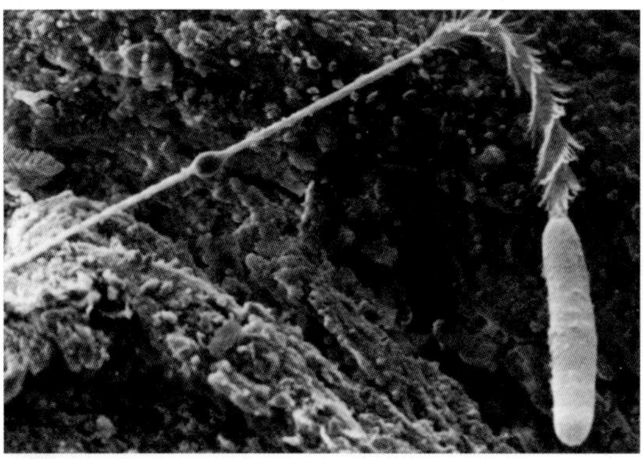

FIGURE 6.36
Scanning electron micrograph of a fully discharged microbasic mastigophore nematocyst from a sea anemone. Elongate capsule in lower right-hand side is joined to a spine shaft, and the long thin distal tubule contains a droplet originating from within the capsule. The droplet is in the process of passing down the tubule length. (Photo by Dr Jacquie Rifkin.)

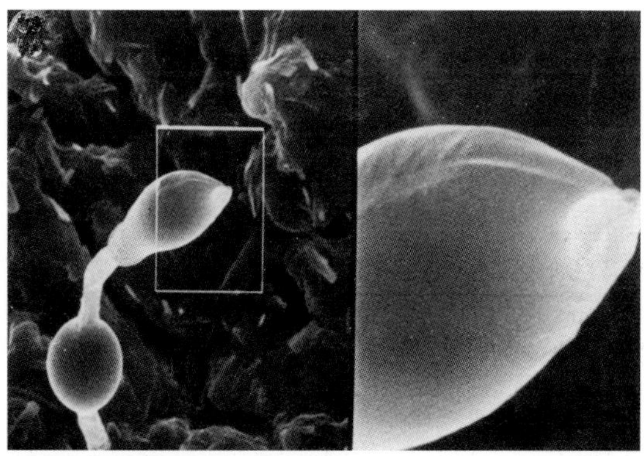

FIGURE 6.37
Scanning electron micrograph showing capsular matrix droplet arrested during nematocyst discharge, at the closed tip (see box) of the nematocyst tubule. Another droplet (shown) joins the terminal one, and pressure causes the nematocyst tip to rupture, releasing the droplets. (Photo by Dr Jacquie Rifkin.)

(1-2μm), the force generated by a penetrating tubule would be considerable.

STRUCTURE AND FUNCTIONING OF THE MASTIGOPHORE NEMATOCYTE FROM *C. FLECKERI*

In order to understand the manner in which discharge occurs in the nematocysts of box jellyfish, it is helpful to examine the structure of the cell (nematocyte) in which the nematocyst is contained (Figure 6.39).

MICROANATOMY OF THE MASTIGOPHORE NEMATOCYTE

The cnidocil is a whip-like structure called a **flagellum**, surrounded by a ring of microvilli (Figures 6.31 and 6.32, pages 154 and 155). Fibrils emanating from the basal end of the microvilli are connected distally and proximally respectively, to the **operculum** and the **fibrillar basket** surrounding the collar of the nematocyte (Figure 6.31, page 154).

The fibrillar basket consists of a collar of interwoven fibrils. The longitudinally-running fibrils are called **supportive rods** and emanate from the small microvilli; other fibrils run in a cross-wise direction.

The fibrous bundles are longitudinally-arranged bundles running from the fibrillar basket and are orientated parallel with the long axis of the nematocyst capsule. These fibrous bundles cover most of the nematocyst and are directly connected to thick, microtubule-laden radial fibres.

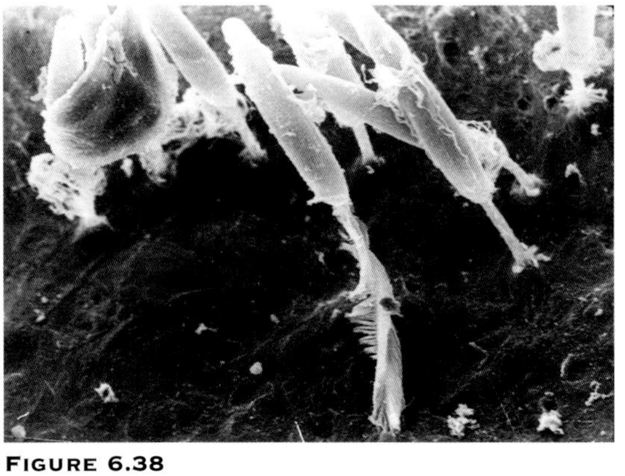

FIGURE 6.38
Scanning electron micrograph of discharged microbasic mastigophore nematocysts from *Chironex fleckeri*, penetrating the carapace of a prawn (bar = 20μm). Note mastigophore in foreground is anchored by the distal tubule only. (Photo by Dr Jacquie Rifkin.)

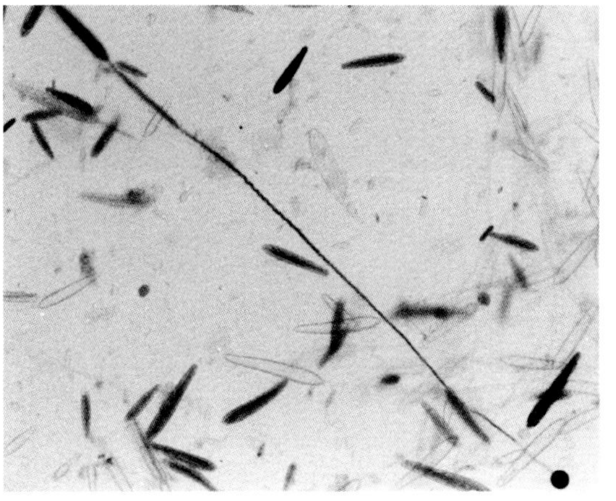

FIGURE 6.39
Light micrograph showing a field of mixed nematocyst types isolated from the tentacles of the box jellyfish *Chironex fleckeri* and dyed using neutral red dye. Some of the nematocysts are visible in various stages of discharge. One nematocyst, a microbasic mastigophore has discharged completely and a droplet of capsular matrix is visible in the lower right-hand corner of the photograph. Early studies demonstrated that microbasic mastigophores comprise about 42% of the cnidom of the tentacles of *Chironex fleckeri* (Endean *et al* 1969).

Some of the radial fibres belong to the nematocyte itself and others belong the neighbouring accessory cells. The radial fibres are firmly anchored in the mesogloeal (connective tissue) layer.

Accessory cells consist of an apical region filled with secretory granules and a basal portion consisting of a radial fibre filled with microtubules.

OPERATION OF THE NEMATOCYTE-ACCESSORY CELL COMPLEX

RIFKIN and ENDEAN (1988) suggest that nematocytes and accessory cells work antagonistically. When the nematocyte is in the resting state (Figure 6.31 page 154), it is removed from the firing line by the microtubules found within the radial fibres supporting the nematocytes, and is usually found below the surface of the epidermis, covered over by the apices of the neighbouring accessory cells. When the nematocytes are to be placed in the 'fire-ready' position, the apical regions of the neighbouring

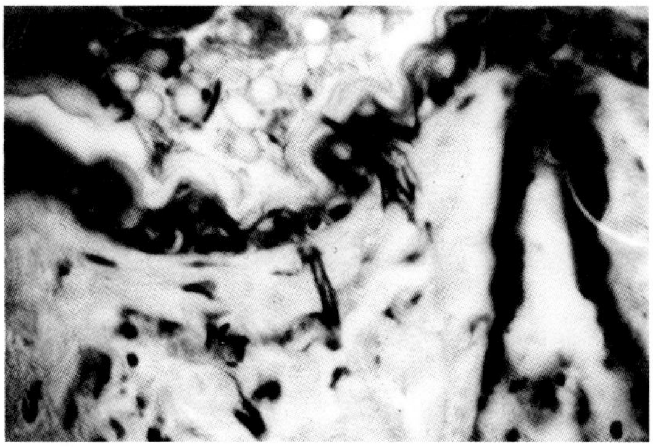

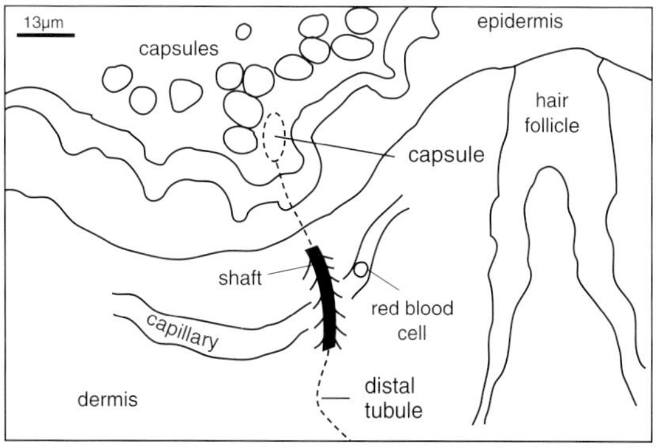

FIGURE 6.40
Serial section of biopsy of stung mouse skin, showing some oedema and degeneration in venom-damaged cell layers. Also visible is the transfixion of a capillary by discharged microbasic mastigophore tubules. (See diagram left.) Note the transverse sections of the discharged nematocysts on the skin surface, the epidermis, the dermis and the capillary. Dotted lines indicate the entire nematocyst. (Photograph by Dr Jacquie Rifkin; accompanying diagram Jan Hynes.)

accessory cells are pulled away, and the fibres supporting these accessory cells pull them proximally into the epidermis. This exposes the cnidocil to the sea water and its attendant mechanical and chemical stimuli (Figure 6.31, page 154).

It is possible that this mechanism would allow the box jellyfish to 'turn off' the triggering mechanism of its nematocysts to prevent discharge under undesirable circumstances. It is also possible that the entire process is under the control of the nervous system or special chemical inhibitory substances that help keep the cnidocils enclosed and dormant when near rocks, or when the jellyfish is fed to repletion, conserving mature nematocysts (e.g. RIFKIN 1988a). The phenomenon of inhibition of nematocyst discharge in fully fed cnidarians was first reported in sea anemones by SANDBERG KANCIRUK AND MARISCAL (1971).

DESCRIPTION OF THE MICROBASIC MASTIGOPHORE NEMATOCYSTS

UNDISCHARGED CAPSULES FROM *C. FLECKERI*

These are generally elongated but slightly broader at their apical ends (Figure 6.33, page 156). They contain a coiled inverted tubule with internal spines (Figures 6.33 and 6.35, pages 156 and 157). The capsule is smooth apart from a cloverleaf-shaped operculum situated at its apical end (Figure 6.33, page 156). A homogeneous, granular, electron-dense matrix is found within the tubule, bathing the spines. Similar matrix is found within the undischarged capsule, clinging to the outside of the tubule (Figure 6.11, page 130 and Plate 6.10).

THE MECHANISM OF NEMATOCYST DISCHARGE

Several events occur in the discharge process of the microbasic mastigophore nematocyst of *Chironex fleckeri*:

1. The cnidocil is stimulated mechanically and chemoreceptors chemically by stimuli given off by prey or a human victim.

2. The operculum is tripped, and the basal portion of the tubule appears (Figures 6.11 and 6.35 pages 130 and 157, and Plate 6.10).

3. The coiled tubule found within the undischarged capsule twists as it everts, and spines in the shape of a triangle appear (Figure 6.11, page 130 and Plate 6.10). These spines initially form a piercing spear which opens outward, with further spines continually presented as the nematocyst everts.* When the tentacle and prey

* The 'piercing tip' of the spear is continually renewed as the nematocyst everts, so that the nematocyst may become anchored at any point in the dischage process, and irrespective of whether the process proceeds to completion. Thus only the distal tubule is buried in the integument in the nematocyst in the foreground of Figure 6.38, page 158 (RIFKIN 1994, unpublished observations).

are closely adhering, this action anchors the shaft (proximal portion of the tubule) within the tissues of the prey or a human victim (Figure 6.38, page 158 [in background]).

4. Granular, electron-dense material found inside the tubule is released during the entire eversion process. Plaque is released at about the same time, possibly through hollow tubules (Figures 6.11, page 130; 6.35, page 157 and Plate 6.10)*.

5. As the spines continue to emerge they form three helices about the shaft (Figure 6.38, page 158).

6. Then the distal tubule appears. This carries spines which course down the remainder of the tubule in three similar helices. The distal tubule spines are smaller than those on the shaft portion of the tubule (Figures 6.11 and 6.43, pages 130 and 169, and Plate 6.10).

7. Once the entire tubule has everted, the capsular matrix flows down the length of the hollow tubule in the form of small droplets (Figures 6.36 and 6.37, page 157). This causes the end of the tubule to swell and burst, releasing the capsular matrix from the end of the tubule that is embedded in the integument (Figures 6.12 and 6.38, pages 131 and 158).

Note: Research into the mechanical, chemical and ionic events associated with nematocyst discharge continues (WATSON & HUSON 1994). Studies on isolated nematocysts from the sea anemone *Calliactis polypus* suggest that Ca^{++} ions influence the osmotic properties of the intracapsular fluid (see below) rather than the mechanical properties of the capsule wall (HIDAKA & AFUSO 1994). Activation of *Pelagia noctiluca* {see Chapter 8, page 229} nematocytes is Ca^{++}-dependent, and *in situ* activation of these nematocysts is powerfully blocked by gadolinium (Gd^{+++}) (SALLEO *et al* 1994).

KINGSTON and SOUTHCOTT (1960) and STRUTTON and LUMLEY (1988) all noticed certain pathological changes in the skin in fatal cases of jellyfish (*Chironex fleckeri*) envenomation. Histological sections showed that microbasic mastigophore nematocysts were present on the surface of the skin, and the tubules had penetrated into the dermis (Plate 5.8). The stratum corneum appeared oedematous, the Malpighian layer had degenerated and pyknotic nuclei were present.

* Published pictures of electron-dense plaque as well as "hollow" (channelled) spines appear in Figures 9–12 and Figure 17 in RIFKIN and ENDEAN 1983.

CURRENT THEORIES ON THE PHYSICAL CAUSES OF NEMATOCYST DISCHARGE

Much has been written about the general mechanism of nematocyst discharge. The most popular current theories are: the contraction hypothesis, the osmotic hypothesis and the tension hypothesis. A fourth, an osmotic/tension hypothesis, has recently been advanced (ENDEAN *et al* 1991).

The contraction hypothesis

Contractile material in the capsule wall, or contractile elements in the cell surrounding the capsule, are activated and thus increase intracapsular pressure (YANAGITA & WADA 1954 & 1959; CORMIER & HESSINGER 1980).

The osmotic hypothesis

This theory suggests that intracapsular pressure is rapidly raised by an osmotic inflow of water or ions, bringing on discharge. This inflow must occur either just before the moment of discharge, or simultaneously with it. Uptake of water could occur either because of an altered permeability of the capsule wall, or by entry of water into the capsule when the operculum has been dislodged. The tubule is then forced out by eversion (PICKEN & SKAER 1966).

The tension hypothesis

CARRÉ (1980) proposed that the energy for eversion of the tubule is obtained when the tubule twists as it invaginates during the formation of the nematocyst. The energy is thus stored in the highly coiled and pleated tubule within the mature nematocyst.

The osmotic/tension hypothesis

In 1991 ENDEAN and colleagues proposed an osmotic/tension hypothesis for microbasic mastigophore nematocyst discharge. In this theory, the apical ends of accessory cells are pulled away from the nematocyte prior to discharge. With the way clear, the cnidocil of the nematocyst protrudes and is thus in the 'fire-ready' position (Figures 6.31a and b, page 154).

Once the cnidocil is triggered mechanically and appropriate chemical stimuli are received,

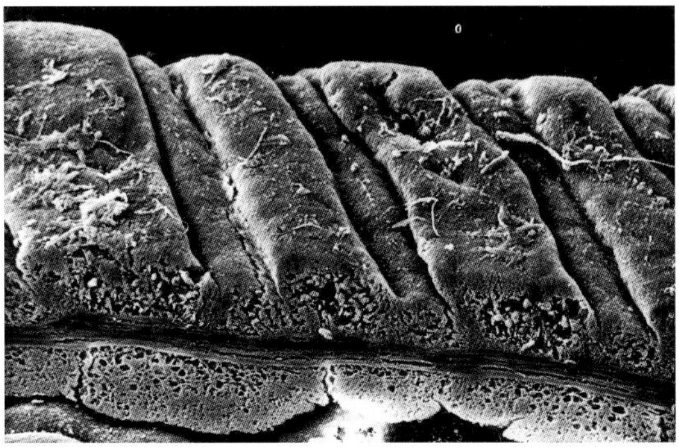

FIGURE 6.41A
Scanning electron micrograph of a relaxed tentacle, with only slight elevation of epidermis at the band sites wherein lie the mature nematocysts. Primary bands are highest and widest, secondary bands are narrower and tertiary bands are narrowest. No roofing over is apparent on the tentacles. The rings of mature nematocysts account for the ladder-like pattern imprinted on the skin of a victim (see Plate 9.91). Note the thread-like discharged nematocysts on the surface of the epidermis. (Photo by Dr J. Rifkin.)

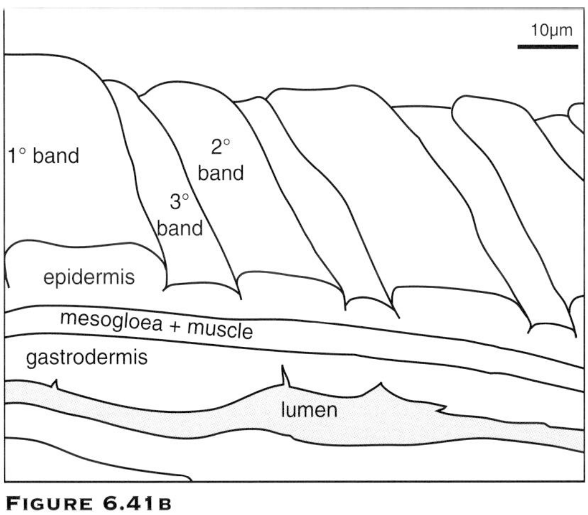

FIGURE 6.41B
Drawing of Figure 6.41A with the relevant features labelled. Visible in the longitudinal section (slice) and labelled here are: the outer epidermis, the mesogloeal layer containing muscle fibres, and the gastrodermis. (Drawn by Jan Hynes.)

the operculum that covers the apical end of the nematocyst is tripped {study Figure 6.11, page 130, and Plate 6.10}. Subsequently, two important events occur very rapidly:

1. In nematocysts *in situ* there is tension applied by the fibrous basket to the capsular wall.

2. The capsular matrix, which normally consists of homogeneous material in unfired nematocysts, is polymerised, forming clusters of regularly spaced electron-lucent granules arranged in hexagonal patterns. The theory suggests that following this reorganisation (or polymerisation?) of precursor material within the capsular matrix to form an electron-lucent network, water rushes into the capsule.

This results in rapid pressurisation of the capsular matrix, thereby inducing an explosive discharge of the tubule and the tubular matrix, followed by the capsular matrix, through the open apical end of the capsule. The pressure on the capsular matrix is sustained throughout the eversion of the entire tubule.

LOTAN et al (1995) suggest further that the high internal hydrostatic pressure supplies the necessary propulsion required, and that the tensile strength necessary to withstand this pressure is provided by the fibre-like structure of the inner wall of the capsule and of the tubule.

The postulated osmotic inflow of water may be illustrated practically when nets, previously used for catching live box jellyfish, are handled after several months, and stings result (RIFKIN 1972, unpublished observations; NORTH BARRIER BRANCH, SURF LIFE SAVING QUEENSLAND 1988–1993, unpublished observations). The ever-present moisture on the hands may activate the nematocyst discharge process in tentacles that have been trapped and dried on the mesh of the nets. This has been proven microscopically (e.g. RIFKIN 1988a; ENDEAN *et al* 1991). (When the tentacles are dead and the nematocysts are operating as 'independent effectors' the mechanical and chemical receptors usually required to produce discharge *in vivo* are not operational.) Such a propensity for water to fire dried-out, undischarged nematocysts still on the skin (BARNES 1966 and in KINSEY 1986; RIFKIN et al 1993, unpublished observations) has first aid implications, which are discussed in Chapter 5, page 110 and below.

MODE OF ENVENOMATION IN *CHIRONEX FLECKERI*

ENDEAN *et al* (1969) suggested that the venom of the box jellyfish may need to be injected directly into the bloodstream (Plate 6.11 and Figure 6.40, page 159) if systemic effects are to be elicited within the time observed clinically in life-threatening human stings {see Chapter 9, page 236}. No measurable systemic effects are elicited when toxic material from crushed isolated nematocysts of *Chironex fleckeri* tentacles is injected subcutaneously or intra-peritoneally into test animals (ENDEAN *et al* 1969).

Ultrastructural studies of *Chironex fleckeri* nematocysts as well as those found in other species of cnidarian showed there is granular or amorphous material present within the undischarged nematocyst tubule, in addition to the material in the capsule (RIFKIN 1988a). A plaque of capsular matrix clings to the outside of the undischarged tubule (Figures 6.11 and 6.41, pages 130 and 161). Studies have also shown that upon discharge the intra-tubular matrix is released progressively, external to the tubule, as the tubule everts (Plate 6.10 and Figure 6.11, page 130). Once each tubule is fully everted, the intracapsular matrix is then released (Figure 6.39, page 158). Histochemical tests, pH indicator tests, and histological studies showed that the matrix in the tubule differs from that found within the capsule (RIFKIN 1988a). KLUG and colleagues (1988) performed haemolytic tests using nematocysts of *Hydra attenuata*, *Physalia physalis*, *Chrysaora quinquecirrha* and *Anemonia sulcata*. In that study, red blood cell cell lysis apparently occurred along the length of fully discharged nematocysts, but did not occur at the very end of the fully discharged tubule where the capsular matrix would be expected to be deposited.

ENDEAN and RIFKIN (1982) suggested the possibility that the toxin was in the tubule and KLUG *et al* 1986 demonstrated the toxicity of matrix released during discharge experimentally (see above). However, recent studies by LOTAN et al (1995) demonstrated phospholipase A_2 toxin associated with the outer wall of the undischarged tubule within the intact capsule (See Figures 6.13, page 132 and 6.41, page 161). Apparently this toxin is translocated to the inside of the tubule through hollow spines during the discharge process. When the spines emerge during the eversion process, this material would emerge at about the same time. Although this latest research suggests that translocation of the venom from the capsule to the tubule is required before envenomation can occur*, the fact remains that venom is delivered via the tubule and is released gradually as eversion proceeds as suggested originally by ENDEAN and RIFKIN (1982). Endean and Rifkin suggested that the capsular matrix may act as a propellant during the discharge process, but whether it is in itself toxic is as yet unclear.

Note: It is clear that envenomation of phospholipase A_2 toxin occurs in the manner described in nematocysts of *Rhopilema nomadica*. However, variations in the intrinsic structure of spines and in the distribution of matrices within the capsules and tubules of different nematocyst types in this and other species, may well produce differences in the manner in which venom is delivered (RIFKIN 1995, personal observation).

If systemic effects attributable to this venom are to occur as rapidly as they clearly do (WILLIAMSON *et al* 1980; LUMLEY *et al* 1988), the venom is most probably introduced directly into blood vessels. The longest tubules are capable of penetrating to a depth of more than 550 µm in mammalian skin, which will carry the distal tubule to the vascular deep epidermal or dermal tissues in humans {Plate 5.8}.

A proportion of these numerous small blood vessels (microvasculature) in the human dermis would be unlikely to escape penetration by at least some of the many thousands of everting tubules of the mastigophores. (Dr Robert Hartwick, PhD, has calculated that the tentacles of an adult *Chironex fleckeri* contain up to some 3000 million nematocysts [HARTWICK 1982, unpublished observations]). Nor is it likely that many of the numerous tips of fully discharged tubules would be deposited neatly in the lumen of a blood vessel, as would be necessary if the intracapsular matrix is introduced directly into

FIGURE 6.42
Scanning electron micrograph of a contracted tentacle of *C. fleckeri* showing the wide, primary bands on the outside, and other bands underneath. A sting from a length of contracted tentacle produces a continuous wheal. (Photograph by Dr J. Rifkin.)

the vascular system. However, if toxic material matrix is released progressively as the tubule everts, as shown by LOTAN *et al* (1995) in *Rhopilema nomadica* nematocysts, it could enter the blood stream as the everting tubules transfix the blood vessels (Plate 6.11 and Figure 6.40, page 159). In fact, studies show that some of the tubules of mastigophores penetrating mouse skin do indeed transfix blood vessels (Figure 6.40, page 159). This would account for the rapid onset of cardiovascular effects, as well as contributing to the full-thickness skin necrosis observed in *Chironex fleckeri* stingings (ENDEAN *et al* 1969; WILLIAMSON *et al* 1980) {plate 5.14}.

However, the bulk of all three matrices present passes into extravascular spaces, resulting in the well-described skin pain, damage and acute inflammatory reaction at the sting site.

Potential envenomation by all nematocyst types

It is possible to observe the passage of dyed capsular matrix passing down the length of the discharged tubule and being released in the form of droplets near the tubule tip (Figure 6.37, page 157). On the basis of WEILL's (1934) belief that an orifice must be present at the distal end of nematocyst tubules for the passage of the 'venom', nematocysts that were expected to penetrate were termed 'stomocnides' (Plate 6.10). Droplets were not seen outside the relatively wide tubule tips of the 'astomocnides' and thus they were not thought to be capable of injecting venom. However, studies by RIFKIN (1988a) in *Chironex fleckeri* (Figure 6.37, page 157) and later by HEEGER (1992) in *Cyanea capillata* show that the tips of at least some fully discharged stomocnide nematocysts are closed. Pressure released during the discharge process causes the tubule tip to rupture, releasing the droplets of capsular matrix (Figures 6.36, 6.37 and 6.39, pages 157–158). Material released progressively as the tubule everts, which has been shown to be toxic in other species (see above) is released prior to the discharge of capsular droplets (Figure 6.35, page 157). Since the release of venom as proposed by LOTAN et al does not depend on the rupture of the tubule tip, it is possible that venom is released from all nematocyst types, whether or not they are fully discharged. See also: THOMASON, J.C. 1983. Cnida discharge and the mechanism of venom delivery in *Anemonia viridis* (Cnidaria, Actinaria Hydrobiologia 216/217: 649654.)

Tentacle adhesion and nematocyst discharge

In serious stingings from *Chironex fleckeri*, the morbakka, and the multitentacled *Physalia*, tentacles are usually found adhering to the skin of the victim. As observed by the author, most discharged nematocyst tubules under which tentacles were attached are fully buried within the skin. However, such strong adhesion is not chateracteristic of all or even most cnidarian stings reported or observed. The author has observed that in situations where tentacles do not adhere to the skin, they move around, exposing new surfaces to contact-chemical stimuli. In these cases, firing may be initiated where no skin target is available for all everting tubules. However, since piercing spear tips form continuously during the eversion process, penetration is possible at virtually any point along the tubule length. For example, Figure 6.38, page 158, shows only the distal portion of a nematocyst tubule buried within a crustacean carapace. Adherence of tentacles to the skin is not always an indicator of a serious sting. For example, the tentacles of the sea anemone *Stichodactyla haddoni* will sometimes cling strongly to the fingers of an observer without causing any irritation whatsoever.

Arrangement of nematocysts on tentacles

Nematocysts are abundant on jellyfish tentacles and are usually grouped into arrangements such as warts, ridges, rings, spirals and terminal swellings. In many cases, the pattern of the wheal arising from the sting of certain dangerous cnidarians reflects the manner in which nematocysts are arranged on tentacles. For example, in stingings involving the jellyfish *Cyanea capillata*, BARNES (1960) remarked that the wheal resembled the zig-zagging teeth of a saw, which corresponds to the spindle-like swellings present on the *Cyanea* tentacle. In contrast, the wheal from the tentacles of

Chironex fleckeri often presents a ladder-like pattern {see Plate 9.91}, corresponding to the transverse bands of nematocysts found on cubozoan tentacles {see Plate 9.43}.

Over most of the tentacle of *Chironex fleckeri*, nematocysts are arranged on primary, secondary, and tertiary bands, corresponding with the width and height of each respective band, creating a 'pseudo-stratified' histological appearance (Plate 6.12 and Figure 6.41, page 161). Wheals carried by human victims of *Chironex fleckeri* stingings are usually described as darkly coloured ladder-like bands alternating with lighter coloured bands {see Plate 9.43} (BARNES 1960; WILLIAMSON *et al* 1980). The discharge of large numbers of mature mastigophores and the large trirhopaloids from the primary bands on the tentacle may be responsible for the darkly-coloured bands, while discharge of smaller numbers of mature penetrants on the secondary and tertiary bands may be responsible for causing the light-coloured wheal (ENDEAN & RIFKIN 1988, unpublished observations). Furthermore, the light and dark pattern on the wheal would be more likely to arise from contact with the relaxed or slightly-contracted tentacle (Figure 6.41, page 161). In the fully-contracted tentacle it is mainly the primary bands that are exposed to the outside of the tentacle, producing a continuous unsegmented wheal (Figure 6.42, page 162) (RIFKIN 1988a; 1996). Jellyfish, having recently fed, will have fewer mature penetrants (RIFKIN 1996), so clinical severity of the sting could vary independently of the length of tentacle contacting the skin.

Thus the gross morphology of the tentacle and the arrangement of both mature and immature cnidae on the tentacles have a direct bearing on the severity of a sting received by a victim.

ACTIVATION AND INHIBITION OF NEMATOCYST DISCHARGE

Several studies have been carried out to assess nematocyst discharge and inhibition in response to a variety of stimuli (YANAGITA 1960a, 1960b & 1960c; HARTWICK *et al* 1980). Both nematocysts *in situ* in the tentacle and nematocysts isolated from the tentacles are capable of discharge. Experiments have shown that nematocysts *in situ* in tentacles discharge in response to different stimuli from those causing discharge in isolated nematocysts. For example, methylated spirits induces a powerful discharge response in *Chironex fleckeri* nematocysts *in situ* (HARTWICK *et al* 1980). (On this basis methylated spirits was abandoned in favour of vinegar use in first aid after *Chironex fleckeri* envenomation.) However, the same response is not elicited in the same nematocysts when they are isolated from the tentacle (ENDEAN & RIFKIN 1983, RIFKIN *et al* [1993]). In order to qualify as a suitable discharge inhibitor for use in first aid, a substance must deactivate the nematocyst discharge mechanism in response to whatever mechanical or chemical stimulus is subsequently applied to the tentacles. Since both nematocysts *in situ* as well as nematocysts dislodged from the tentacle may be present on the skin of a sting victim, inhibition of nematocyst discharge in both presents a challenge.

Nematocysts *in situ* in the tentacle and nematocysts isolated from the tentacles — implications for first aid

First aid for cubozoan stings: After any serious stinging from a cubozoan medusa, the victim should be treated at once (see Chapter 5, page 98). Pouring vinegar immediately onto cubozoan tentacles has a 3-fold effect: to inhibit the discharge mechanism of nematocysts *in situ* so that no further venom can be injected into the skin (HARTWICK *et al*, 1980), to denature all remaining venom in the tentacles apart from that which has already been injected into the bloodstream, and to loosen the stinging tentacles thereby facilitating their removal (RIFKIN 1995, unpublished observations and see Chapter 5, page 98). Any isolated cubozoan nematocysts containing active venom are inactivated upon contact with vinegar (ENDEAN and RIFKIN 1983).

First aid for non-cubozoan stings: Although first aid measures have been devised for cubozoan jellyfish stingings (see Chapter 5, page 98), the same nematocyst inhibitors cannot be employed consistently for the same task after scyphozoan, hydrozoan and anthozoan stings (FENNER 1987, RIFKIN 1988a and 1996). This is possibly because the nematocysts of these different classes, and even different members of the same classes, are responsive to different discharge stimuli. Thus the task remains to provide first-aid treatment for nematocysts from non-cubozoan stingings, when they are *in situ* in the tentacle and when they have been dislodged from it.

On the premise that a physical barrier between tentacles and skin usually provides adequate protection against stings (e.g. CLELAND & SOUTHCOTT, 1965 and BARNES in KINSEY, 1986) the author suggests using a stick or other appropriate object found on the beach, disposable forceps, or disposable gloves to remove adhering tentacles. The fingers should be used as a last resort for reasons explained below.

Neither fresh nor sea water should be used to remove tentacles. Although initial *in vitro* studies by the author showed that water did not cause discharge in *C. fleckeri* or *Physalia* nematocysts, nematocysts within detached tentacles are still very reactive to the contact-chemical stimuli (mechanical and chemical stimuli) associated with a victim or rescuer who grabs at tentacles and stretches them, and whose own skin is a rich source of chemical stimuli. To remove tentacles with a stream of water is to risk further envenomation as tentacles roll off the skin, exposing new, undamaged surfaces to the stimuli required for further discharge. As well, water spreads the mass of nematocysts isolated from the tentacles to sites remote from the initial sting site where they may discharge later (JR unpublished observations, and see below).

The keratinised pads of the fingers are safe for removal of tentacles in most cases although the author has occasionally sustained stings in the palm of the hand, between the fingers and near the nail bed. However, touching tentacles usually dislodges more of their nemotocysts both discharged and undischarged and adds to the pool of organoids already dislodged during the initial envenomation. These isolated nematocysts become lodged in calloused tissue, near the nails and other places once the tentacle is discarded.

Unfired nematocysts that become isolated whenever tentacles are touched are capable of lying dormant for many months in a dehydrated condition, until moisture is added, thereby producing discharge (e.g. LANE 1960) The phenomenon was explained decade ago (e.g. by WEILL 1934), and the author has personally observed this discharge *in vitro* in smear preparations of nematocysts from every class of cnidarian. Isolated nematocysts may sometimes act as independent effectors because the living cnidocils and sensory receptors usually necessary for discharge in nematocytes *in situ* in the tentacles are absent.

The author has personally found that when inadvertently scratching oneself or accidentally wiping the eyes or other mucous membranes, undischarged nematocysts are transferred to these locations, where they may discharge immediately, or where they may await hydration for subsequent discharge. She has handled nets used a year before to catch box jellyfish, and received slight stings. In addition, she has handled tentacles from the apparently innocuous scyphozoan medusa *Cyanea nozakii*, washed her hands with soap and dried them on towels. Several hours later she used the same towel and received irritating stings. ALDERSLADE (1995, personal communication) reported receiving irritating stings after wearing clothing used previously for collecting box jellyfish, but only after the clothing got wet. The author has accidentally rubbed the side of her face, felt nothing at the time, only to experience irritation, swelling and erythema hours later during a light rain shower. Localised irritations also occurred at other times after collecting jellyfish while the author was perspiring. When collecting box jellyfish it is helpful to rinse fingers and gloves and collecting equipment in vinegar (freely available) to reduce the subsequent occurrence of these irritating stings.

For non-cubozoan stings, the author suggests that the area in the vicinity of the immediate sting should be lightly blotted with a clean, dry, disposable tissue to remove many of the nematocysts present. Cold packs may offer all the analgesia required in stings from the relatively innocuous non-cubozoan jellyfish.

The author suggests taking adequate preventative measures against stings, such as wearing protective clothing or swimming within jellyfish-resistant enclosures where they are available. (Plates 1.19 and 1.22)

FUTURE NECESSARY RESEARCH

An organised approach to the study of the treatment of cnidarian stings is required. Firstly, the structure and ultrastructure of tentacles of representative cnidarians should be studied to determine whether structural elements are involved in modifying the severity of stings received. As previously stated, in *Chironex fleckeri* the radially-orientated fibres associated with each nematocyte and neighbouring accessory cell appear to be capable of retracting or lifting the nematocyst and its associated cnidocil apparatus to enable, or prevent discharge (Figures 6.31, 6.32, pages 154–155). It is possible that radially-orientated stalks associated with the nematocysts of other species act in the same manner and affect the severity of stings administered.

Secondly, studies are needed to find suitable general nematocyst discharge inhibitors for all cnidarians, particularly if these could be applied on the skin before people go in the water; research is underway (HEEGER *et al* 1992). A more complete identification of substances that are contraindicated for this purpose is equally desirable. The apparent variation in discharge responses to given chemical stimuli among cnidarians presents a major research challenge {see also Chapter 23, page 430}.

A NOVEL CLASSIFICATION SYSTEM FOR CNIDAE

Figure 6.43, page 168 illustrates the cnida types that have been discovered so far. They have been organised into a new key which will allow the researcher to distinguish between cnidarians on the basis of their cnidae. Cnidae may be useful in helping to distinguish one cnidarian from another, and if so they make good taxonomic tools. After envenomation by cnidarians, many cnidae isolated from the tentacles remain clinging to the skin of the victim (BARNES 1960; CLELAND & SOUTHCOTT 1965; CURRIE and RIFKIN this volume; CURRIE and WOOD 1995). Samples of these may be taken from the skin by scrapings (BARNES 1960; CLELAND & SOUTHCOTT 1965; SCHNADIG *et al* 1991); the sticky tape method is described in detail in Chapter 5 and below. However, they will rarely allow the interpreter to distinguish species on their own. Cnidarians are most effectively diagnosed (i.e. identified with precision) using a combination of gross anatomical features and microscopic features described earlier such as nematocysts.

The keys presented below will allow diagnosis of cnidae from all classes of cnidarians.

With today's popularity of water sports, a first aider may never know with certainty whether a sting was sustained in a reef or open water environment, especially if there is a delayed response such as is characteristic of the Irukandji syndrome. A knowledge of the types of nematocyst present in cnidarians of medical importance may be of practical significance both in helping determine the cause of the stings and in deciding upon the most appropriate first aid to be administered after stings (HARTWICK *et al* 1980; FENNER *et al* 1993).

Several nematocyst classification systems have been invented over the years, but the most famous and most widely used is that of WEILL (1934). However, despite the additions and alterations made throughout the years, problems with the use of revised versions of his system still persist. The classification system included here is presented in the form of two keys and although it has been somewhat simplified by virtue of its new format, it is designed primarily to alleviate many of the problems previously encountered by persons particularly interested in determining the formal diagnosis of cnidae from all classes of cnidarian.

Characteristics of this classification system:

Historically, nematocysts have been divided into two groups depending on whether or not they possess openings at the tips of the fully discharged tubules (see above). The stomocnides were thought to be the only penetrants, as the relatively broad-tipped astomocnide tubules (Figure 6.43 I–V, page 168) appeared to be incapable of allowing passage of droplets of capsular matrix, (once thought to be the venom). It is now known that passage of venom is possible whether or not the tubule tip has ruptured (see above). The description in the classification system presented below takes this information into account. In addition:

- It is arranged into a system of two keys — the first being a key to the classification of the

3 major cnida categories, i.e. nematocysts, spirocysts and ptychocysts; the second being a key to types of nematocyst.

- It retains the technical names for the nematocysts so that scientific work may be carried out with direct reference to previous work done on the subject.
- It improves upon the definitions of the major descriptors of the keys. Changes to the basic definitions of the heteroneme and haploneme categories (see below) are adapted from SCHMIDT (1974) whose approach the author believes is the simplest yet devised (see key descriptors). This is an especially important consideration for accurate diagnosis in the interpretation of skin scrapings where in the author's experience, artefacts are produced during the processing required for the microscopic slide preparations.
- It contains several new nematocyst types recently discovered by the author. For example the trirhopaloid nematocyst and the basitrichous microbasic mastigophore are included. The microbasic mastigophore from *Chironex fleckeri* has been redefined, as have microbasic p-mastigophores.

Basic instructions for making a simple nematocyst preparation from living or preserved cnidarian tissue:

1. Obtain a very small (1mm) portion of tentacle and place it on a microscope slide. Bear in mind that the cnidom may differ from one portion of the organism to another.*
2. Using a coverslip, squash the sample by gently tapping on the surface of the coverslip using the point of a pencil until the preparation has flattened.
3. Place a drop of sea water or 0.9% saline at the edge of the coverslip and allow the liquid to spread under the coverslip by capillary action. It is helpful to mix dye such as methylene blue (1%) with the sea water.
4. If no dyes are available, the phase contrast microscope or interference contrast microscope should be used to differentiate the necessary anatomical features including capsules, tubules and spines.
5. A high power (x40) objective lens will usually be required to differentiate the parts required, and the oil immersion objective lens will be required where nematocysts are particularly small. At times the scanning electron microscope may be necessary for confirmation of different nematocyst types when certain diagnostic features (e.g. spines) are not otherwise visible.
6. Depending upon the cnidarian examined and the anatomical portion sampled, the cnidom may consist of one cnida type, or several.

The sticky tape method of retrieval of nematocysts from the skin

The most efficient method for retrieval of nematocysts from the skin involves the use of sticky tape (CURRIE and WOOD 1995) and see Chapter 5, page 98. Nematocyst samples should be taken after discharge inhibitors and ice have been applied.

- Press a piece of sticky tape to the skin in the sting area (after vinegar and ice application if necessary).
- Remove the tape.
- Attach one of the portions of tape to a glass slide, sticky side up, using additional pieces of tape to hold the preparation in place.
- Place sticky side up in a dry, dust proof container.
- Examine microscopically, sticky-side up, on the same day if possible.

Basic instructions for diagnosing cnidae in nematocyst preparations using the keys provided

1. In the squash preparations or the tape preparations described above, nematocysts will usually be present in both the discharged as well as the undischarged states. Both states of each type should be examined if possible to obtain an accurate diagnosis. Length and width data of the capsule, shaft and distal tubule are required for records, and drawings or photographs of each cnida type should be obtained.
2. Diagnosis of the type examined can be made by comparing the morphology of a targeted

*The author has examined skin scrapings provided by Dr Bart Currie of Darwin, from the victim of an apparent Irukandji sting. Nematocysts retrieved coincided with those characteristically found on the bell rather than the tentacles.

cnida in the preparation with the diagrams and the descriptors in the keys provided. There are usually several morphologically different types of cnidae present in a tentacle, and each should be examined in the same manner.

3. The targeted cnidae retrieved from a victim's skin need to be compared with the cnidae present in the cnidarian thought to be causing the sting.

At times the presence of distinctive capsule shapes and types helps in differentiating a member of a higher taxonomic group, but not usually to species level. For example, the presence of spirocysts always identifies an anthozoan specimen; the siphonophore *Physalia* is characterised by spherical nematocysts of 2 size classes (Figure 5.1, page 99); mastigophores in all chirodropid species so far described, are elongate capsules that are wider at the apical end (Figure 6.33, page 156) and vary greatly in size from one species to another (JR, 1995, personal observation). Some Cubozoan nematocysts (such as in Figure 6.43 this page) have a short, bare area between shaft and distal tubule. Capsule shapes are of little value, however in formal nematocyst classification.

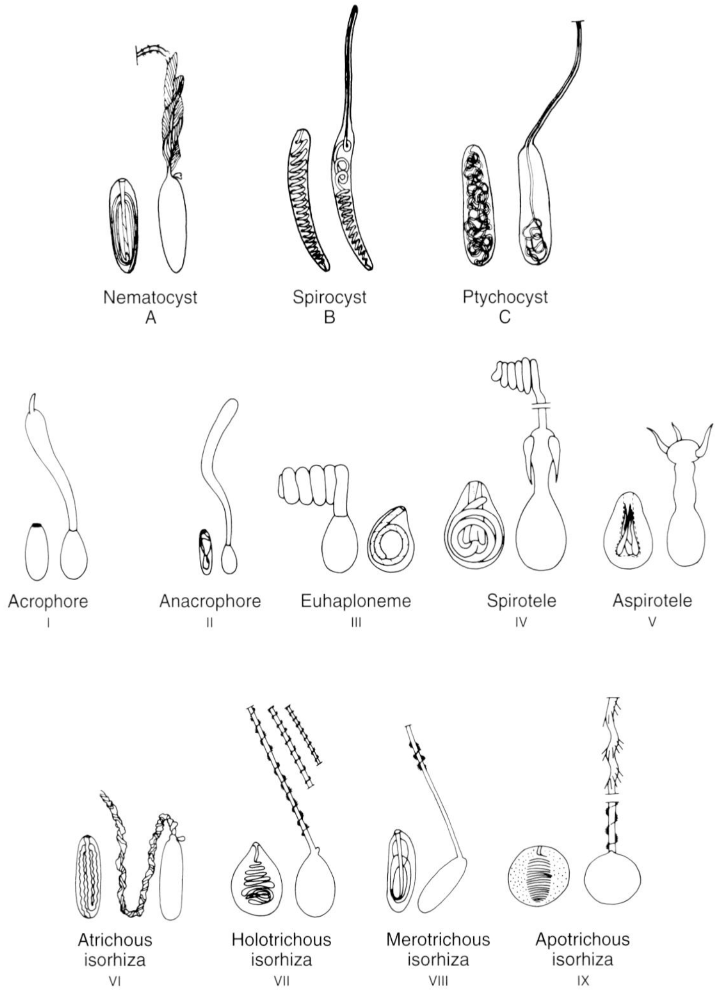

FIGURE 6.43
The various cnidae described to date. See the keys for description and classification of the various cnida types. (Figs I—V, VII—XIII, XX, XXI, XXIII, XXIV, XXVI—XXXI (with permission) after Mariscal 1974; Figs A, VI, XXII, XVI (with permission) after RIFKIN & ENDEAN 1983; Fig. XIX (with permission) after England 1987 by courtesy of the British Museum (Natural History); Fig. XXV (with permission) after Bouillon, Gravier-Bonnett, Boero (1986). Composite by Dr Jacquie Rifkin, redrawn by Dr Matthew Sheehy)

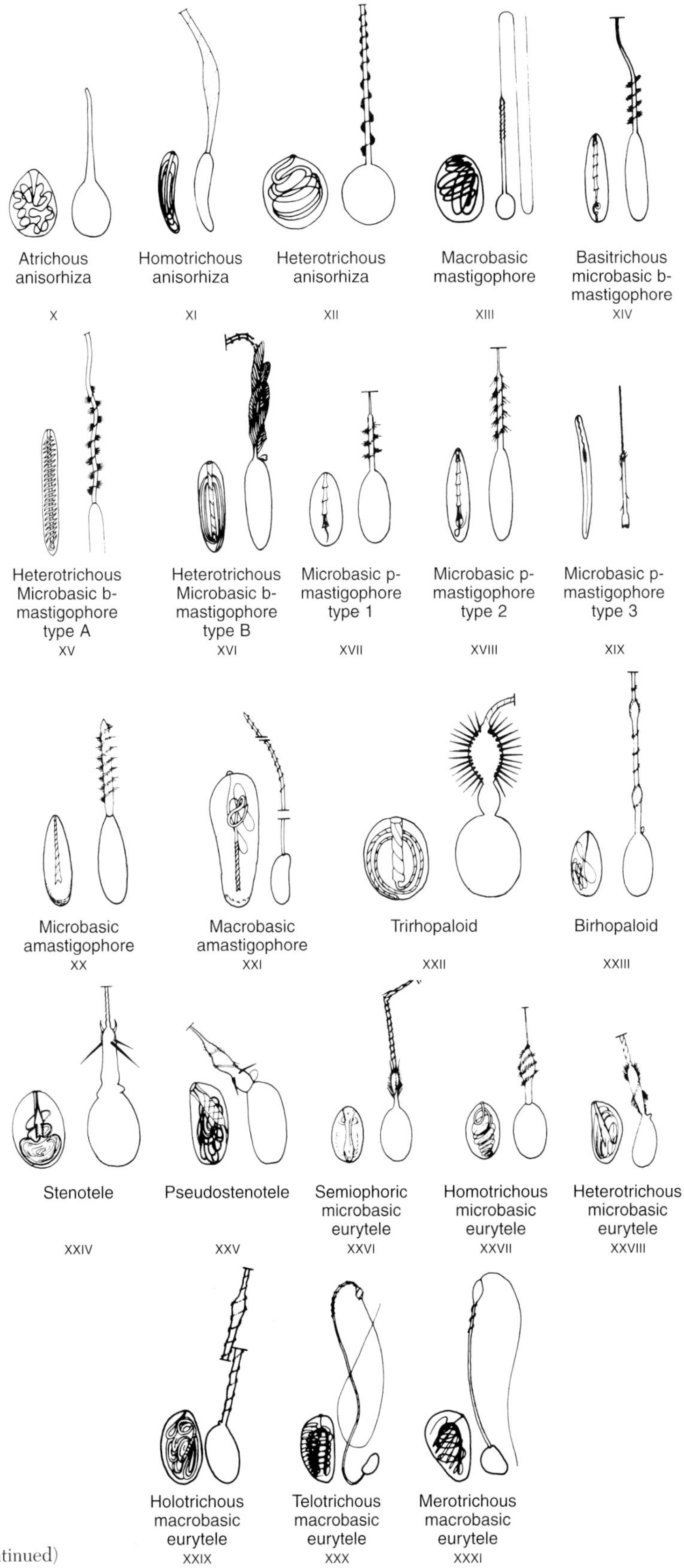

FIGURE 6.43 (continued)

A KEY TO CATEGORIES OF CNIDAE

(SEE ALSO FIGURE 6.43, PAGES 168–169)

Each entry line in the key concludes with a number that directs the reader to the next appropriate descriptor in the key.

1(a)	Undischarged capsule containing a densely packed tubule; discharged tubule lacking helical folds but bearing longitudinal folds; no spines or filaments present on discharged tubule	Ptychocyst (Figure 6.43[C])
1(b)	Undischarged capsule containing helically folded tubule; discharged tubule bearing filaments or spines (sometimes spines are absent over the whole or part of the tubule)	(2)
2(a)	Undischarged capsule containing helically coiled tube wall, with or without a basal shaft, with or without spines, and no longitudinal folds in tubule wall	Nematocyst (Figure 6.43[A])
2(b)	Undischarged capsule, thin-walled, containing tubule which is itself coiled like a spring; bearing hollow filaments which emanate from discharged tubule	Spirocyst (Figure 6.43[B])

KEY TO TYPES OF NEMATOCYST

(SEE ALSO FIGURE 6.43, PAGES 168–169)

Each entry line in the key concludes with a number that directs the reader to the next appropriate descriptor in the key.

1(a)	Discharged tubule permanently closed at tip	Astomocnide (2)
1(b)	Tubule ruptures at tip upon discharge	Stomocnide (7)
2(a)	Astomocnide with tubule club-shaped and much greater in volume than capsule	Rhopaloneme (4)
2(b)	Astomocnide without club-shaped tubule	(3)
3(a)	Astomocnide tubule forming a spiral coil distally	Spironeme (5)
3(b)	Astomocnide with tubule not spirally coiled distally	(4)
4(a)	Discharged tubule bearing apical projection	Acrophore (Figure 6.43[I])
4(b)	Discharged tubule without apical projection	Anacrophore (Figure 6.43[II])
5(a)	Spironeme bearing tubule without well developed shaft and with corkscrew-like coil	Euhaploneme (Figure 6.43[III])
5(b)	Spironeme bearing tubule with well-defined shaft (=Eurytele; =Rhopaloneme)	Euheteroneme (6)
6(a)	Discharged euheteroneme tubule extends beyond shaft	Spirotele (Figure 6.43[IV])
6(b)	Discharged euheteroneme tubule does not extend beyond shaft	Aspirotele (Figure 6.43[V])
7(a)	Stomocnide tubule without a well-defined shaft and in which changes in the tubule diameter and in the size and sometimes the shape of spines are gradual	Haploneme (8)
7(b)	Discharged stomocnide tubule possessing an abrupt change in size and shape of spines between proximal shaft and distal tubule	Heteroneme (14)

8(a)	Discharged haploneme tubule of the same diameter throughout or narrowing slightly toward distal end....	Isorhiza (9)
8(b)	Discharged haploneme tubule slightly dilated proximally....	Anisorhiza (12)
9(a)	Discharged haploneme tubule bearing spines....	(10)
9(b)	Discharged haploneme tubule without spines....	Atrichous isorhiza (Figure 6.43[VI])
10(a)	Discharged haploneme tubule completely covered in similarly-sized spines throughout; spines may be smaller towards distal end....	Holotrichous isorhiza (Figure 6.43[VII])
10(b)	Discharged haploneme tubule covered in similarly-sized spines in certain areas....	(11)
11(a)	Discharged haploneme tubule bearing spines on intermediate tubule portion only....	Merotrichous isorhiza (Figure 6.43[VIII])
11(b)	Discharged haploneme tubule bearing spines on distal tubule portion only....	Apotrichous isorhiza (Figure 6.43[IX])
12(a)	Discharged anisorhiza tubule bearing spines....	(13)
12(b)	Discharged anisorhiza tubule without spines....	Atrichous anisorhiza (Figure 6.43[X])
13(a)	Discharged anisorhiza tubule spiny throughout; spines all of equal size....	Homotrichous anisorhiza (Figure 6.43 [XI])
13(b)	Discharged anisorhiza tubule spiny throughout; spines larger at base of tubule....	Heterotrichous anisorhiza (Figure 6.43[XII])
14(a)	Shaft of discharged heteroneme tubule of constant diameter....	Rhabdoid (15)
14(b)	Shaft of discharged heteroneme tubule of unequal diameter....	Rhopaloid (23)
15(a)	Discharged rhabdoid tubule continues beyond shaft	Mastigophore (16)
15(b)	Discharged rhabdoid tubule does not continue beyond shaft....	Amastigophore (22)
16(a)	Discharged mastigophore shaft short, less than three times capsule length....	Microbasic mastigophore (17)
16(b)	Discharged mastigophore shaft long, more than four times capsule length....	Macrobasic mastigophore (Figure 6.43[XIII])
17(a)	Discharged shaft of microbasic mastigophore tapers gradually into distal tubule and is not associated with an abrupt change in tubule diameter....	Microbasic b-mastigophore (18)
17(b)	Discharged shaft of microbasic mastigophore always tapers abruptly into distal tubule....	Microbasic p-mastigophore (20)
18(a)	Discharged shaft and distal tubule of microbasic b-mastigophore spiny, but spines unequal on either portion....	Heterotrichous microbasic b-mastigophore (19)

18(b)	Discharged distal tube of microbasic b- magistiphore without spines	Basitrichous microbasic b-mastigophore (Figure 6.43[XIV])
19(a)	Smooth transition of undischarged shaft with distal tubule (no V-shaped notch visible at base of undischarged shaft); spines occupy entire length of discharged tubule	Heterotrichous microbasic b-mastigophore type A (Figure 6.43[XV])
19(b)	V-shaped notch present at base of undischarged shaft; short length of tubule between discharged shaft and distal tubule devoid of spines	Heterotrichous microbasic b-mastigophore type B (Figure 6.43[XVI])
20(a)	Spines of discharged shaft of microbasic p-mastigophore orientated at right angles to capsule ..	Microbasic p-mastigophore type 1 (Figure 6.43[XVII])
20(b)	Spines of discharged shaft of microbasic p-mastigophore not orientated at right angles to capsule	(21)
21(a)	Spines of discharged shaft of microbasic p-mastigophore orientated towards capsule	Microbasic p-mastigophore type 2 (Figure 6.43[XVIII])
21(b)	Spines of discharged microbasic p-mastigophore shaft short and orientated towards the capsule (proximally); some longer spines point away from it (distally)	Microbasic p-mastigophore type 3 (Figure 6.43[XIX])
22(a)	Discharged amastigophore tubule with shaft short, less than three times capsule length	Microbasic Amastigophore (Figure 6.43[XX])
22(b)	Discharged amastigophore shaft long, more than four times capsule length	Macrobasic Amastigophore (Figure 6.43[XXI])
23(a)	Discharged shaft of rhopaloid dilated at one point (distally) along length	Eurytele (27)
23(b)	Discharged shaft of rhopaloid dilated at more than one point along length	(24)
24(a)	Discharged shaft of rhopaloid dilated at 2 points along its length	(25)
24(b)	Discharged shaft of rhopaloid dilated at more than 2 points along its length; largest (=middle) swelling bearing spines	Trirhopaloid (Figure 6.43[XXII])
25(a)	Discharged shaft of rhopaloid dilated at 2 points along its length; without 3 distinctive, well-developed spines between the 2 swellings but bearing regular spines	Birhopaloid (Figure 6.43[XXIII])

25(b)	Discharged shaft of rhopaloid dilated at 2 points along its length; with 3 large spines between swellings	(26)
26(a)	Discharged shaft separated by a constriction; basal portion of shaft usually longer than distal, spinous portion; large spines present at point of constriction. Capsule rounded or egg-shaped....................................	Stenotele (Figure 6.43[XXIV])
26(b)	Discharged shaft separated by a constriction; basal portion of the shaft always shorter than distal, spinous portion; distal portion of shaft possesses more spines than stenotele; large spines present at point of constriction. Capsule oblong..	Pseudostenotele (Figure 6.43[XXV])
27(a)	Discharged shaft of eurytele short, less than 3 times capsule length	Microbasic eurytele (28)
27(b)	Discharged shaft of eurytele long, more than 4 times capsule length	Macrobasic eurytele (30)
28(a)	Discharged tubule of microbasic eurytele bent and whip-like, with large, flat spine medially	Semiophoric microbasic eurytele (Figure 6.43[XXVI])
28(b)	Discharged tubule of microbasic eurytele neither bent nor whip-like; with several spines	(29)
29(a)	Spines of discharged shaft of microbasic eurytele all the same size	Homotrichous microbasic eurytele (Figure 6.43[XXVII])
29(b)	Spines of discharged shaft of microbasic eurytele of unequal size..	Heterotrichous microbasic eurytele (Figure 6.43[XXVIII])
30(a)	Shaft of macrobasic eurytele spiny throughout entire length...	Holotrichous macrobasic eurytele (Figure 6.43[XXIX])
30(b)	Shaft of macrobasic eurytele spiny only along part of length...	(31)
31(a)	Spines confined to distal portion of macrobasic eurytele shaft..	Telotrichous macrobasic eurytele (Figure 6.43[XXX])
31(b)	In macrobasic eurytele, spines not distal but confined to shaft area of uniform diameter, proximal to terminal swelling..	Merotrichous macrobasic eurytele (Figure 6.43[XXXI])

Acknowledgments: The author is grateful to Professor Thomas Heeger of the University of San Carlos in the Philippines and to Dr Richard Mariscal of the University of Southern Florida for helpful comments concerning the manuscript.

CHAPTER 7

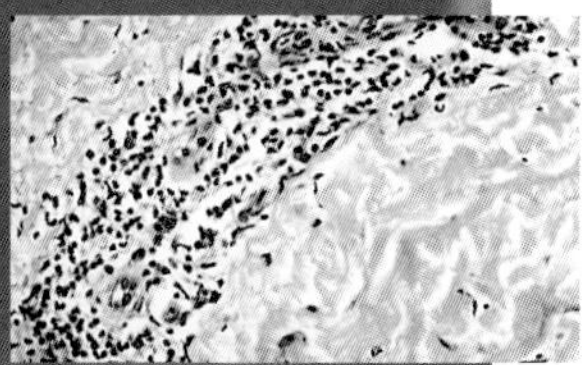

CNIDARIAN ENVENOMATION

JOSEPH W. BURNETT, JACQUIE RIFKIN AND JOHN WILLIAMSON

THE ENVENOMATION PROCESS OF CNIDARIANS: CLINICAL SIGNIFICANCE

JOHN WILLIAMSON AND JACQUIE RIFKIN

An appreciation of the efficiency of cnidarian envenoming can be gained by comparing it with that of snake bite. The snake injects its venom through its usually paired fangs into the victim's tissues in either a single deposit or, in the case of multiple 'strikes' in several closely adjacent deposits (Plate 7.1). This limited-surface-area venom depot then acts as a reservoir from which absorption may occur systemically, via the lymphatic system in particular, over the ensuing minutes and hours (SUTHERLAND 1983). Virtually none of this venom normally gains direct intravascular entry. The actual onset of symptoms may take at least 15 minutes but usually an hour or more (unless absorption is arrested by a compression bandage {see Chapter 5, page 98}).

In contrast, the cnidarian has its venom dose divided into millions of micro-doses delivered from multiple discharging nematocysts {see Chapter 6, page 121}. Upon such widespread nematocyst discharge* (e.g. tentacle contact), these multitudinous micro-doses are injected simultaneously into the epidermal and dermal tissues {Plate 5.7} in closely

* The mechanical events and micro-anatomical structures concerned with cnidarian nematocyst discharge are discussed more fully in Chapter 6.

adjacent but discrete areas, distributed in serious stings over a significant portion of the body surface area (Plate 7.2). The speed of this envenomation is remarkable; for example the discharge time of nematocysts of *Hydra attenuata* was measured at 3 milliseconds (e.g. HOLSTEIN & TARDENT 1984). As explained above, the speed of the clinical onset of symptoms is explicable only by direct intravascular deposition of venom. In animal experiments the extravascular deposition of *Chironex fleckeri* venom failed to elicit any cardiac effects (ENDEAN *et al* 1969; ENDEAN & RIFKIN 1983). The discharging tubules may transfix micro-vessels (Plate 6.11) (ENDEAN & RIFKIN 1983), and it has been demonstrated that venom exodus occurs along the full length of the discharging tubule (Plates 6.10 and 6.11), leading to the likelihood of such direct intravascular deposition of venom (ENDEAN & RIFKIN 1983, LOTAN *et al* 1995).

The remaining majority of the venom dose that is not directly deposited intravascularly is subdivided into many thousands of mini-doses and presented over a wide area of vascular bed (i.e. healthy skin) thus offering a large surface area of venom for local toxic damage of tissues. This is invariably accompanied by sudden, sharply localised and sometimes savage skin pain. Such tissue damage can be expected to affect adversely the integrity of micro-vasculature, with enhanced egress of inflammatory exudate and perhaps even enhanced ingress of some venom components. The speed and vigour of the local acute inflammatory reaction to cnidarian envenomation has few parallels in clinical medicine {see Plates 1.2 and 5.19A}. The clinical impact of the postulated absorption of some of this venom across damaged capillary endothelium may be mitigated by the known thermolability of cnidarian venoms (OTHMAN & BURNETT 1990) {see below}.

In nature, the practical result of such an envenomation process is the rapid immobilisation of cnidarians' struggling prey, e.g. a prawn or fish for chirodropid jellyfish. (On occasions the fish may be nearly as large as the bell of the chirodropid itself.) It makes teleological sense, mindful of the relative frailty and limited mobility of many cnidarians. It is also clear from direct observation that cnidarians may employ not massive but selective nematocyst discharge to capture small prey {Plate 6.9}. Humans have intruded upon this dynamic biological event. The massive discharge that may accompany a human envenomation by a chirodropid reflects a 'survival' response by the animal, which nonetheless commonly perishes following such encounters, with most of its tentacles avulsed and even disruption of its bell from thrashing human limbs. This and other observations, such as variation in the severity of sting effects in the face of significant tentacle contact (ENDEAN 1988a; RIFKIN 1988a), has led to the suggestion that some cnidarians may in part exercise a form of 'central' control over their nematocyst discharge. This was referred to intriguingly by BARNES, concerning *Chironex fleckeri*, as 'the changeable "mood" of the animal' (KINSEY 1986).

CNIDARIAN VENOMS: CURRENT UNDERSTANDING 1996

JOSEPH W. BURNETT

The nematocyst venoms of cnidarians are groups of polypeptides that stimulate toxic and immunological reactions (BURNETT & CALTON 1987a; BURNETT 1991; BURNETT 1992b). This complement of polypeptides, each several thousand kilodaltons in size, acts concomitantly to produce the clinical sting in humans. The immunological reactions may involve either the humoral or cellular immune systems (BURNETT 1992a) {see below}. Serious reactions such as anaphylaxis (TOGIAS *et al* 1985) are extremely rare, but the expected immunological reactions seen in humans exposed to foreign proteins are usual.

Multiple different toxic reactions to cnidarian venoms are most likely produced by the polypeptide side chains. A membrane effect that increases ion permeability, particularly to sodium and calcium, is among the most important (BURNETT & CALTON 1977, 1983, 1987b; COBBS *et al* 1983). Once this membrane-pore effect of the venoms is initiated, a cascade of actions alter both the structural and the electrical (polarisation) integrity of the envelope. These abnormalities may occur in cells of several organs simultaneously, thus broadening the spectrum of sting-induced clinical disease.

The fact that the venom is dislodged from the nematocyst as it penetrates the epidermis and dermis (Plate 5.7), and possibly transfixes the dermal micro-vasculature en route (Plate 6.11) (ENDEAN & RIFKIN 1983), means that damaging toxins can be placed intravascularly and absorbed immediately, i.e. within a circulation time (approximately 20 seconds in healthy persons); the remainder of the toxin deposited is available in a reservoir. As discussed above, the vast numbers (millions) of nematocysts that fire with tentacle–skin contact enable the harmful toxin dose to be increased rapidly to supra-threshold levels for tissue damage.

Cnidarian venoms are hydrophobic, relatively thermolabile (OTHMAN & BURNETT 1990) and have the ability to aggregate or deaggregate, especially under *in vitro* conditions. The range of potency between venoms of different cnidarian species is quite remarkable. The venom components vary in antigenicity, with some of the more active compounds capable of binding antibody within a very few minutes. Cross-reactivity of epitopes present in the polypeptides of different venoms is quite common, but currently cross-protection of polyclonal antibody preparations against lethal factors of crude venom from different species is limited.

IMMUNOLOGY OF JELLYFISH ENVENOMATIONS

JOSEPH W. BURNETT

NOTE by JAW, PJF AND JFR: This is a relatively new and developing sub-branch of science. The author of this segment and his colleague G.J. Calton, PhD, of Baltimore, are the world pioneers.

NORMAL RESPONSE TO ENVENOMATION

Immunocompetent individuals respond acutely to jellyfish envenomations with an increase in serum concentration of specific serum immunoglobulin M (IgM) antibodies. A few days or weeks later this response abates and is gradually replaced by a rising concentration of specific immunoglobulin G (IgG) antibodies which persist for many years. The presence of these circulating serum proteins can be used as an immunodiagnostic marker to identify previous exposure to the venom of specific stinging animals (BURNETT *et al* 1988). ELISA (Enzyme-Linked Immuno-Sorbent Assay) IgG titres of greater than 1:300 have been rarely detected in unexposed healthy people (false positives), and antibodies to one cnidarian venom frequently cross-react to toxins of other jellyfish. In general, the more severely envenomated patients subsequently develop higher titres of specific IgG.

Normal people also respond to the sting by increasing the activity of their cellular immune system within several days following the sting. The event can be demonstrated by noting the appearance of lymphokines (products released by sensitised T cells) or by measuring the suppression of natural killer (NK) cell enhancement after these cells were exposed *in vitro* to jellyfish venom proteins (BURNETT *et al* 1986; BURNETT *et al* 1987; WACHSMAN *et al* 1991).

ABNORMAL IMMUNOLOGICAL REACTIONS FOLLOWING STINGING

HUMORAL MEDIATED IMMUNOLOGICAL REACTIONS

The most serious abnormal reaction following cnidarian stings mediated by the humoral antibody system is anaphylaxis (TOGIAS *et al* 1985). This adverse complication has been demonstrated in dogs after exposure to sea anemone venoms (class Anthozoa) and in a patient after a clinical sting from *Pelagia noctiluca*. Anaphylaxis in this patient was confirmed by experimentally exposing her basophils *in vitro* to venom of the North American sea nettle *Chrysaora quinquecirrha*. Thus this serious sequela can occur after exposure to cross-reacting jellyfish. Yet, anaphylaxis is very unusual after jellyfish stings, possibly because the priming doses that people usually receive are small and infrequent, thus resulting sensitisation may be difficult. It should be stressed that the clinical manifestations of anaphylaxis can mimic those of severe acute toxic reactions following jellyfish envenomation, and that the latter are far more common.

CELL MEDIATED IMMUNOLOGICAL REACTIONS

Clinical abnormalities following jellyfish envenomation that are thought to result from cell mediated hypersensitivity reactions (BURNETT & CALTON 1987) include: persistent skin eruptions, including urticaria; papular urticaria; recurrent eruption at the site of envenomation; and distant reactions.

1. Persistent lesions following jellyfish envenomation appear as erythematous, indurated nodules or inflammatory plaques lasting from a few days to several months (Plate 7.3). Examination of biopsy specimens taken from the involved skin reveals granulomatous infiltration of lymphocytes. These lesions are usually corticosteroid-responsive but require prolonged systemic therapy (REED *et al* 1984).

Cases of jellyfish envenomation have been complicated by persistent urticaria lasting up to one month. Symptomatic treatment with systemic antihistamines has been effective. The pathogenesis of this manifestation is unknown, although both humoral and cell mediated antibodies may be involved.

2. Some of the patients who have jellyfish envenomations may demonstrate papular urticaria, an eruption characterised by symmetrical papules and urticarial lesions on areas of the body unassociated with the original tentacular contact (BURNETT *et al* 1987; BURNETT & CALTON 1987). Spontaneous resolution of these lesions requires several days. Treatment for this complication is systemic corticosteroids and topical mentholated emollients.

3. Recurrent eruptions can appear at the site of the original envenomation in 5–30 days (WILLIAMSON 1985; BURNETT *et al* 1983 and 1987) (Plate 7.4). In contrast to the initial stings which are painful, recurrent episodes are pruritic. The clinical appearance of the lesions with each successive episode is identical except that there may be exudation. The interval between successive episodes expands, whereas the duration of each episode shortens with each exacerbation. Envenomations producing this complication have occurred worldwide appear following stings by many different jellyfish* species, and have a 4:1 prevalence for females (see Plate 7.4). NK cell and T cell (lymphocyte) functions are active and persistent in some of these patients (BURNETT *et al* 1987). Examination of biopsied lesions shows an abundance of helper T cells in the perivascular dermal mononuclear cell infiltrate (Plate 7.5) {see also Plate 8.83}.

4. A localised but distant eruption has been reported in one woman who exhibited oral lesions several hours after a sting on the lower extremities (MATUSOW 1980). Whether this reaction was a manifestation of an immunological response to envenomation remains speculative.

ASSORTED LESIONS, POSSIBLY IMMUNOLOGICALLY MEDIATED

Six cases of localised arterial vasospasm underlying the sting site (see Figure 7.1, page 178), with ischaemic muscular and ischaemic/necrotic cutaneous lesions distal to the site of the vascular compromise, have been reported (DRURY *et al* 1980; ADIGA 1984; WILLIAMSON *et al* 1988; ABU-NEMA *et al* 1988;). Five were females (e.g. Plate 7.7) and the sixth was a young male teenager (Plate 7.6). All patients sustained limb stings near to or in the compromised area. Most patients had residual contractures, dysfunction and scarring.

Other unusual reactions to jellyfish envenomations possibly resulting from immunological abnormalities include mononeuritis multiplex in which the neurological loss was adjacent to or at the sting site (FILING-KATZ 1984; PEEL & KANDLER 1990; MOATS 1992). Two further such cases (one female, one male) from a Penang jellyfish and a Papua New Guinea corallimorphorian sting respectively, have been recently reported

* A Red Sea coral contact dermatitis producing recurrent pruritic papular eruptions has recently been reported (CAMARASA *et al* 1993); further information is being sought.

(BURNETT *et al* 1994) (Plates 7.7 and 7.8). Another young female stung by a Papua New Guinea jellyfish in 1994 brings the total neuropathy cases to six. A seventh is mentioned in Chapter 26 (page 448).

In 1992 a reliable report was received of temporary (three or four days duration) delayed (two weeks later) aphonia (loss of voice) in three male scuba divers (two Thai, one American), each stung on a separate prior occasion, and in differing body sites, by unidentified jellyfish in Hua Hin waters near Bangkok (Lt Col. A.B. RHODES, Bangkok 1992, unpublished observations).

In 1984 a curious reversible, month-long parasympathetic dysautonomia resulted from a jellyfish envenomation in a Chinese fisherman; the envenomation commenced with a pattern of cardio-respiratory symptoms and progressed to produce parasympathetic dysfunction of the gastro-intestinal, genitourinary, cardiovascular and lachrymal systems (PRATAP CHAND & SELLIAH 1984).

Prolonged blurred vision (BURNETT & BURNETT 1990), decreased visual acuity (GLASSER *et al* 1992), arthralgias that persist significantly beyond the duration of the cutaneous pain (BURNETT & BURNETT 1990; WEINBERG 1988), fat atrophy (BURNETT & CALTON 1987), and possible granuloma annulare (localised death of dermal tissue) (MANDOJANA 1990) have followed various jellyfish envenomations. Episodes of recurrent cutaneous herpes simplex at the injury site can follow a sting (BURNETT & CALTON 1987); it has been hypothesised that this infection was caused by a decrease in local immune defences.

IMMUNOPROPHYLAXIS AND IMMUNOTHERAPY

Toxoids for *Chironex fleckeri* and *Chrysaora quinquecirrha* (the North American sea nettle) stings have been developed experimentally but not commercially (BAXTER & MARR 1975; BURNETT & CALTON 1977). The protective potency was low for the latter preparation but adequate for the former (against only *Chironex fleckeri*). However, the practical obstacles to the former's clinical application (e.g. public compliance) inhibit its pharmaceutical production even as an orphan drug.

Experimental hyperimmune serum has been produced against crude venoms of the same two jellyfish (BAXTER & MARR 1974; BURNETT & CALTON 1977; SUTHERLAND 1983). In addition, commercial hyperimmune serum (ovoid antivenom) against *Chironex fleckeri* venom

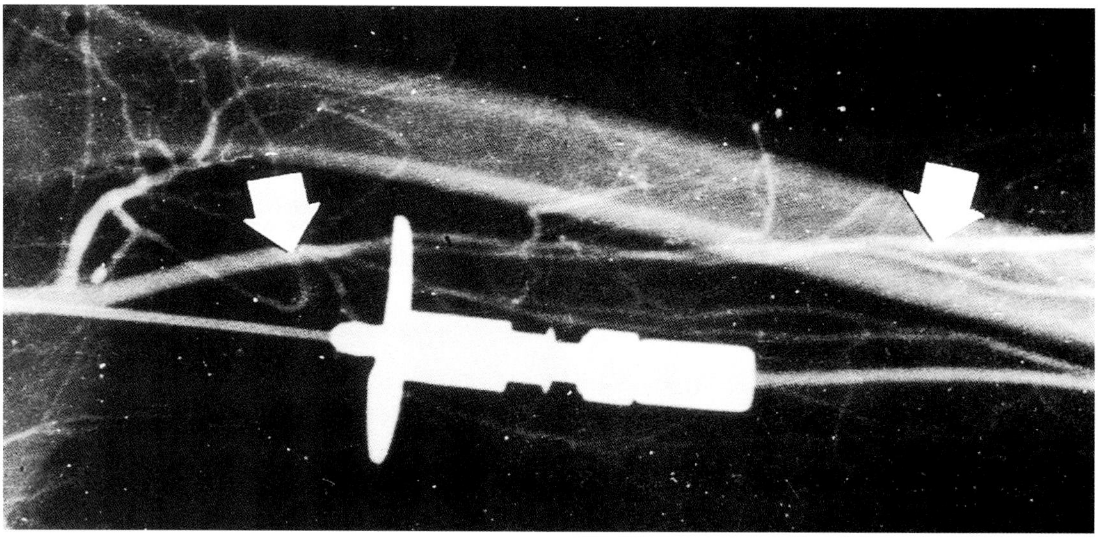

FIGURE 7.1
A brachial arteriogram on a boy following an Indian *Physalia* sp. sting on the overlying upper arm, which resulted in serious hand ischaemia. Note the marked arterial spasm (narrowing) between the arrows. (Copyright © Medical Journal of Australia 1984; 140: 180–181)

antigens is made by the Commonwealth Serum Laboratories in Melbourne. This preparation has been used repeatedly and successfully to alleviate the savage pain and skin damage and to help counteract the lethal action specifically of the box jellyfish *Chironex fleckeri* venom (SUTHERLAND 1983; WILLIAMSON *et al* 1984; BEADNELL *et al* 1992; INTERNATIONAL CONSORTIUM for JELLYFISH STINGS 1993). This product may have slight effect, but is not clinically useful for Irukandji stings (FENNER *et al* 1986); it is ineffective experimentally against the venoms of the Atlantic *Physalia* and *Chrysaora* (BAXTER & MARR 1974).

EXPERIMENTAL IMMUNOLOGY

MONOCLONAL ANTIBODIES

Monoclonal antibodies against *Physalia*, *Chrysaora* and *Chironex* have been produced (GAUR *et al* 1981, 1982; NAGUIB *et al* 1988; COLLINS *et al* 1993). They exhibit considerable interspecies cross-reactivity even to non-cnidarians (OLSEN *et al* 1985). Unfortunately none of these reported antibodies offers significant protection against the lethal actions of the above venoms in experimental animals, but they have been successfully employed to partially purify active principles of the venoms (COBBS *et al* 1983).

ULTRAVIOLET IRRADIATION EFFECTS

Recent experimental work has demonstrated suppression of normal immune responses in mice immunised with the venoms of *Chrysaora quinquecirrha* (the North American sea nettle) or the Atlantic *Physalia* sp. (Portuguese man-o'-war), by either prior or subsequent ultraviolet (UV) irradiation of the skin of the intact animal (MIURA *et al* 1993). Confirmatory results have recently been obtained experimentally in humans (MIURA *et al* 1995 in press). This work complements an already significant body of evidence indicating a major influence of UV irradiation upon immune responses to a cutaneously applied antigen. Such observations have significance in view of the fact that natural human contact with jellyfish venoms occurs usually together with exposure to strong sunlight.

CONTACT DERMATITIS

Contact with jellyfish can produce an initial toxic eruption, but also, evidence is being compiled that a true contact dermatitis may be evolving (OHTAKI *et al* 1990; KOKELJ *et al* 1992). Thus the interplay of toxic and immunologic mechanisms involved in the clinical jellyfish sting is not fully understood and remains a focus for future investigations.

CHAPTER 8

ANTHOZOANS, HYDROZOANS AND SCYPHOZOANS

JACQUIE RIFKIN
(BIOLOGICAL ASPECTS)
JOHN WILLIAMSON AND PETER FENNER
(MEDICAL ASPECTS)

INTRODUCTORY COMMENTS

In Chapters 8 and 9 the cnidarians are considered in their four classes: Anthozoa, Hydrozoa, Scyphozoa and Cubozoa. Because there is an emphasis on first aid and medical the animals producing the somewhat 'simpler' injuries', both conceptually and therapeutically (e.g. class Anthozoa: coral cuts, abrasions and stings), are discussed first.

Worldwide, an accurate appreciation of the medical significance (human morbidity and mortality) of the cnidarians continues to be lacking, and with a few distinguished exceptions (e.g. MAYER 1910; LIGHT 1914a, 1914b, 1921; UCHIDA 1929, 1970; RUSSELL 1953, 1970; CLELAND & SOUTHCOTT 1965; MARETIĆ 1983; SHOKITA 1986; HALSTEAD 1988; AUERBACH 1995) their disciplined study has been somewhat neglected. One of the aims of this book is to help redress this imbalance.

Section II (Chapters 6–11) is extensive and consequently we have separated cnidarians into two chapters, with the cubozoans (box jellyfish) assigned to their own chapter (Chapter 9) because of their medical importance. Thus the sequence is:

Chapter 8: Class Anthozoa
Class Hydrozoa
Class Scyphozoa
Chapter 9: Class Cubozoa

Of necessity, the following pages contain biological aspects as they are seen to bear upon medical understanding. Particulars of jellyfish anatomy, and behaviour in jellyfish of medical importance, with accompanying diagrams, are included in Chapter 6 (see RIFKIN page 121). Chapter 8 in particular draws on Chapter 6 for some anatomical descriptions. Chapters 8 and 9 are illustrated by colour photographs and drawings. However, the reader is referred back to Chapter 6 for further assistance with anatomical features and to the Glossary for definitions of terms.

CLASS ANTHOZOA

JACQUIE RIFKIN

Anthozoa is by far the largest class of the phylum Cnidaria and exhibits the widest diversity of form and colour.* It includes the coral polyps (responsible for the construction of the world's massive coral reefs and atolls) and the sea anemones. In all anthozoans, the medusa or jellyfish stage is absent in the life cycle.

The subclass Zoantharia ('true' corals, sea anemones, zoanthids and corallimorpharians) has marine toxinological significance. A terminology confusion exists in the layman's description of 'stinging corals'. The classic 'stinging coral' or 'fire coral' is a hydrozoan colony {class Hydrozoa, page 189} with an external, hard skeleton. However 'hard' or 'stony corals', which may also be equipped with nematocysts injurious to humans, are colonial or solitary individuals, also with hard external skeletons, and these belong to the class Anthozoa.

ORDER SCLERACTINIA: TRUE, STONY OR HARD CORALS

These pleomorphic animal colonies may occasionally produce human injury by the use of stinging nematocysts (e.g. *Goniopora*,† *Plerogyra*, *Physogyra* [VERON 1986; MARSH & SLACK-SMITH 1986]) and/or much more commonly by direct mechanical abrasion or laceration (e.g. *Acropora* [Plates 8.1, 8.2 and 8.3]). This they do with their hard, calcareous and usually sharp skeleton, as snorkellers and divers and fossickers on the reef brush against them.

Their mode of envenomation by nematocysts is typical of all cnidarians and is described in Chapter 6, page 121.

THE INJURIES

Hard coral stings

The envenomation resulting from the relatively few hard corals that sting {see above} is seldom serious and usually affects only those regions of the body where the skin is thin (VERON 1986), but it can produce significant pain and discomfort locally. The theoretical possibility of an allergic reaction to any antigenic marine venom exists (BURNETT *et al* 1987a), and suggestive symptoms are recorded (CLELAND & SOUTHCOTT 1965), but no conclusive reports of this complication from coral stings were found in the scientific and medical literature.

It is most uncommon to encounter a pure sting from a hard coral (scleractinian), as the injury is usually a combination abrasion and sting, with the former predominating in clinical significance (Plate 8.3).

Coral cuts and abrasions

These injuries are tremendously common {Chapter 3, page 77}, and many fortunately heal well (CAMPBELL 1980). Consequently they may be neglected initially, with sometimes serious and even disastrous results {see Chapter 4, page 92}. Their commonness also partly explain the paucity of detailed documentation. Every coral abrasion or laceration includes one or more of the following features:

- direct mechanical injury to soft tissues resulting from the force of the impact
- haemorrhage from disrupted blood vessels in the wound
- deposition of proteinaceous (including mucoid) coral polyp material and other foreign bodies into the depths of the wound

* For an expert treatise on the taxonomy of corals, combined with glorious and detailed photographic illustrations, the reader is referred to the definitive text by Dr John Veron, of the Australian Institute of Marine Science (VERON 1986).

† An example of toxinological research involving hard corals includes the isolation of polypeptides with cardiotoxic properties on rabbit myocardium from the coral *Goniopora* (FUJIWARA *et al* 1979).

- introduction of coral venom {see above}
- contamination of the wound with potentially pathogenic marine or terrestrial microorganisms {see Chapter 4, pages 93–96}, with the possible development of serious secondary infection
- production, uncommonly, of an antigen/antibody (allergic) reaction locally or systemically (generally) {see above}

SYMPTOMS AND SIGNS OF INJURIES FROM HARD CORALS

Excepting major lacerations or acute haemorrhage from such injuries, it is an unfortunate fact that many non-stinging coral cuts and abrasions either pass unnoticed for some hours (Plate 8.3), or are actually intentionally neglected because of their initial trivial appearance (EDMONDS 1989), lack of symptoms and/or distraction with activities of the moment. This is a serious mistake, which may result in protracted morbidity {see Chapter 4, page 92} (WILLIAMSON 1985; HALSTEAD 1987). Such complications are more likely in victims with underlying illness such as diabetes mellitus, peripheral vascular disease, chronic liver disease, or immunosuppression (e.g. steroid dependency) {see Chapter 4, page 92}.

Within hours such neglected lesions, including seemingly trivial ones, may become itchy, swollen and painful (Plate 8.3). A clear discharge, turning purulent-looking, may then begin. Regional lymph glands are commonly enlarged and tender. Localised erythema will develop early, and pain and itch increase. If secondary infection becomes established (within 6–10 hours), the patient may complain of malaise, rigors and headache, and become pyrexic. Local oedema may extend.

MANAGEMENT

A sting from hard coral

This should be dealt with as for a sting from a stinging hydroid (see below, and page 190).

FIRST AID WOUND TOILET

{see also Chapter 5, page 106}

Nowhere in the field of outdoors medicine is the value of immediate wound toilet better illustrated than with fresh coral cuts or abrasions. This may result in rapid, primary healing with no morbidity. The fresh wound, following pressure to control usually 'welling-type' (rapid oozing) venous or capillary haemorrhage, should be 'compound scrubbed' using a soft bristle brush, and a bowl of fresh water (not sea water) containing a mild, non-irritant antiseptic/detergent. The depths of the wound must be cleansed, and there may be some patient discomfort and further bleeding from this procedure. Visible foreign material should be removed. Large or deep wounds, and wounds in small children, may require more expert medical attention for appropriate anaesthesia or analgesia. This should not be delayed longer than three hours — sooner if possible.

Thorough and early wound toilet should be followed by bandaging over a soft clean dressing, particularly with the use of the new seaweed-derived compounds (e.g. kaltostat) that allow rapid haemostasis, yet diffusion of potentially infective exudate. Steroid creams should not be applied to wounds receiving delayed or incomplete wound toilet. The injured part should be elevated and the patient rested, pending referral to medical attention. Oral aspirin or paracetamol is suitable for pain relief. The patient should not enter the sea again prior to receiving medical attention.

Wounds receiving delayed toilet or which are already showing signs of infection require urgent medical referral.

Medical management

Perform a thorough 'compound scrub', if this has not been done. Consider tetanus prophylaxis, and appropriate pain relief. Radiographic examination of deep wounds for retained foreign bodies may be indicated.

Established infection requires immediate and vigorous management, e.g. appropriate anaesthesia and analgesia, wound toilet, wound and perhaps blood culture, and antibiotic sensitivity testing of discharge (consider salinated culture medium {Chapter 4, page 95}), antibiotic therapy (consider marine pathogens), tetanus prophylaxis, and rest with hospital admission where indicated.

CASE HISTORY:

A healthy male scuba diver sustained a minor coral cut to his left forearm on Nedipus Reef in Palau. No wound toilet was performed. Initial symptoms of discomfort and a localised erythema were not apparent at the time of injury. The next day the patient felt well but noticed a red streak up the left forearm. Microbiological investigations were impossible. Following discussion with a medical colleague, a delayed wound toilet and careful selection of antibiotics (including oral clarithromycin) resulted in the lymphangitis subsiding over the next week. No systemic symptoms developed.

ORDERS ACTINIARIA (SEA ANEMONES), CERIANTHARIA (TUBE ANEMONES), CORALLIMORPHARIA (CORALLIMORPHARIANS), AND ZOANTHINARIA (ZOANTHIDS)

BIOLOGICAL FEATURES AND APPEARANCE

Whereas many other anthozoans such as corals are either solitary or colonial in form, actinarian sea anemones consist of solitary marine polyps. Ceriantharians, commonly called the tube anemones, are also solitary, and they do not possess a basal disc. The corallimorpharians include both solitary and colonial organisms (SOUTHCOTT 1987c); some experts consider that the latter anemone-like organisms (Plates 8.21A, 8.21B and 8.23) are corals without skeletons (FAUTIN & ALLEN 1992). Zoanthids are small polyps (a few centimetres tall), which are joined together partially or completely to form carpet-like colonies; they are linked together and attached to the bottom by tubes called stolons and do not have basal discs. They are non-venomous to humans but are very poisonous (MARSH & SLACK-SMITH 1986) {see below}.

Sea anemones, sometimes called 'flowers of the sea' are found in many beautiful colours and in a wide range of sizes ranging from a few millimetres to a metre in diameter [Plate 8.4]). There are about 1000 species. The anemone polyp has a stout muscular column-shaped body without a skeleton. The **column** has nematocysts and these may be concentrated in 'warts'. The upper portion of the column is expanded into an **oral disc** with a central **mouth** surrounded by relatively stubby, hollow, blunt or finger-like **tentacles**, arranged in radial rows or in several circlets (Plate 8.5). Other tentacle shapes are common (FAUTIN & ALLEN 1992) (Plate 8.6). Upon irritation these tentacles are capable of rapid withdrawal into the contracting body. The tentacles are usually full of stinging nematocysts and spirocysts {see Chapter 6, page 121}, used in capturing food and transferring it to the mouth. However these nematocysts may also be used in defence against predators by other organisms. For example, sea anemones are utilised by the boxer crab, *Lybia tessellatus*; it grasps a pair of stinging anemones, one in each claw, which it brandishes and uses both to immobilise its prey and to defend itself (Plate 8.7). Many anemones have sticky acontial filaments laden with potent nematocysts (Figure 8.1, page 184) shot out through the mouth as a defensive mechanism when the anemone is touched (Plate 26.3A).

The basal or lower end of the body forms a smooth, slimy **pedal disc** on which the anemone can slide about slowly or hold vigorously onto rocks or the sea floor (see arrow in Figure 8.5, page 195). The pedal end may be adapted for digging into sediments.

A muscular **gullet** hangs from the mouth down into the gastrovascular cavity where it is connected with the body wall by a series of vertical radiating partitions called '**mesenteries**' (Figure 8.1). The partitions are double sheets of gastrodermis supported by a central layer of jelly called mesogloea, and they increase the digestive surface of the gullet. Thus the animal can digest a large animal such as a fish, sea urchin, and various crustaceans, though many anemones feed upon small particles of organic matter. The free edges of the mesenteries form digestive filaments which bear the gland cells that secrete digestive enzymes. Digestion is completed by specialised gastrodermal cells lining the gastrovascular cavity.

Anemones possess unicellular algae on their tentacles and oral disc, which photosynthesise sugars for their use and supplement their nutrient supply. These sugars may be the anemone's major source of energy.

The gullet has a groove called a **syphonoglyph** running from top to bottom, which is lined with long hair-like cilia. When these special cilia beat downwards, they bring water and organic detritus into the gastrovascular cavity and provide the internal parts of the anemone with oxygen. And when they beat upwards they create an outward current of water that takes with it carbon dioxide and other wastes. When prey or food material is detected by the tentacles, the cilia of the gullet reverse their upward beat, and the food is swept into the digestive cavity.

Reproduction

Anemones sometimes reproduce asexually by pulling apart longitudinally into two parts. Each part grows larger and in turn may divide. A genetically identical group of individuals or

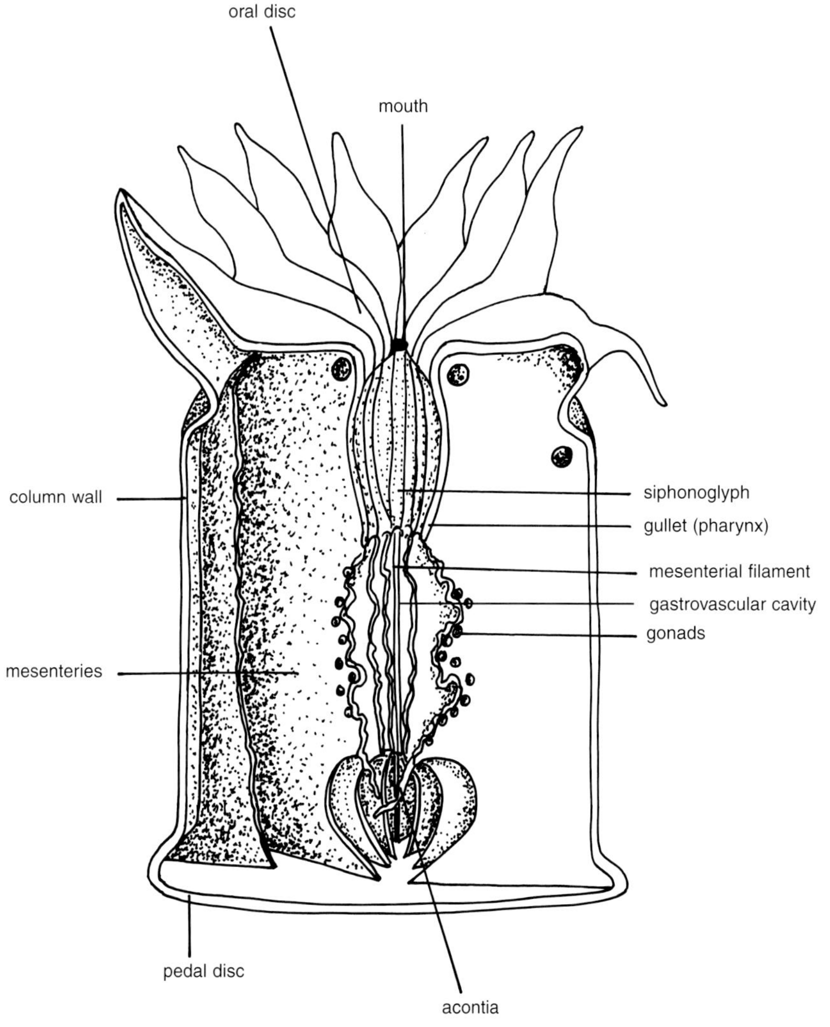

FIGURE 8.1
Diagrammatic representation of a coronal section through an actiniarian sea anemone, showing internal structure. (After Bayer & Owre 1968, redrawn by Jan Hynes, Brisbane.)

clone is formed from all descendants of the original sexually-produced anemone. In certain species, fragments of an injured anemone may regenerate into tiny anemones.

In sexual reproduction, the eggs or sperm form on the partitions of the gastrovascular cavity and are ejected through the mouth. In some species the eggs stay in the body cavity and are fertilised there by sperm that enter the mouth along with oxygenated water. If the zygotes grow inside the parent, they are eventually released as tiny anemones but, in most cases, the fertilised egg develops into a planula, which finally settles in a rocky crevice and grows into a single anemone.

Motility

Anemones may move about if environmental conditions are not favourable, or if they are being pursued by predators. They slide about on their pedal discs, driven by the basilar muscle of the disc, and may thus move only a few millimetres per day. However, they may actually detach altogether and roll to their destination. Some species are capable of swimming, albeit jerkily, through the water, away from predators (SOUTHCOTT 1987c).

Symbiosis with clown fish

Ten species of anemones have a well-known symbiotic relationship with small fish, usually called clown fish (Plate 8.9) (SCARR in EDMONDS 1989). These fish live among the tentacles and are not affected by the nematocysts on the tentacles, so long as they remain in continuous exposure to the tentacles. In this relationship the anemone and the clown fish share captured meals of immobilised prey.

However, should the clown fish have a period of time away from such contact, its 'acquired immunity' fades, and it becomes vulnerable to the nematocysts of its own anemone. In such an event the fish, in a remarkable demonstration of acquired pharmacological symbiosis, backs cautiously into the anemone, exposing only its tail and then pectoral fins, with brushing actions across the anemone tentacles. This apparently induces once again the clown fish's protective immunity to the anemone nematocyst venom, and the fish resumes its place unharmed among the tentacles.

Distribution of anemones

The anemones are abundant seashore creatures. Beyond these littoral zones they are found to depths of more than 5000 metres. Although they are most common in tropical seas, some species occur even in Arctic oceans (HALSTEAD 1988). Anemones may abound on coral reefs, where they blend perfectly with the reef environment and may not be seen by the unwary snorkeller or diver until contacted (Plates 8.10 and 8.11 and Case Reports 1 and 2 below) (NICHOLLS 1992).

Some sea anemones, tube anemones and corallimorpharians of documented stinging significance

Australia:

(See Figures 8.4 and 8.5, pages 194–195, 8.13, page 221 and Case Report 1 below)

Actinia arboreum (CLELAND & SOUTHCOTT 1965).

Actinodendron plumosum (SAVILLE-KENT 1893; SOUTHCOTT 1987c), nicknamed 'Hell's fire sea anemone' (HALSTEAD 1988).

Corynactis australis Jewel anemone (CLELAND & SOUTHCOTT 1965; SOUTHCOTT 1987c), a corallimorphorian {see Case Report 2 below}.

Dofleina armata Armed sea anemone (COLEMAN 1972).

Phlyctenactis tuberculosa (SOUTHCOTT 1987c).

Stichodactyla = *Stoichactis* spp., some known as 'giant anemones' (MARSH & SLACK-SMITH 1986).

Anthothoë albocincta (white-striped sea anemone) (family Sagartiidae) (GEORGE & GEORGE 1979).

Fiji:

Radianthus ritteri (NICHOLLS 1992) (also found in Australian waters).

Mediterranean Sea:

'Maladie des plongeurs' or sponge divers' disease (ZERVOS 1903; CLELAND & SOUTHCOTT 1965) is a sting/irritation caused by sea anemones growing on the sponges collected by bare-skinned divers (HALSTEAD 1988). The identity of these anemones is disputed; MANSON-BAHR (1960) identified them as genera *Hellenopolypus* and *Aktinion*; FISHER (1978) attributes the problem to *Sagartia*; and HALSTEAD (1988) says they are 'probably' *Sagartia* sp. or *Actinia* sp. The resulting lesions can be severe and protracted; resulting erythema and vesicles can become 'darkened purple' (FISHER 1978) [Editors' note: ?haemorrhage into lesions] and produce permanent scarring {see below}. No resistance is conferred by repeated exposure (CLELAND & SOUTHCOTT 1965). Also *Anemone sulcata* (MARETIĆ & RUSSELL 1983) produces Mediterranean stings with considerable oedema, urticaria and morbidity.

New Caledonia:

Cerianthus sp. (CATALA 1964).

New Guinea:

Actinia arborea (QUOY & GAIMARD, 1833). {see also Case Report 2, below}

Japan (Okinawa Island):

Phyllodiscus semoni (Plates 8.14, 8.15, 8.16, 8.16, 8.17 and 8.19) (NISHIHIRA & TOMIHARA 1994 unpublished observations).

Red Sea:

Triactis producta Klunzinger (LEVY *et al* 1970; MANDOJANA 1987).

Virgin Islands and the Caribbean region:

Condylactis spp. (GARCIA *et al* 1994).

Alicia mirabilis Johnson (berried anemone). This is a night-active venomous anemone with long tapering tentacles and prominent warts on the column. The warts contain batteries of stinging cells, which are probably used to defend the exposed columns from predators at night. During the day it withdraws into a tube in the sand (also found in the Indo-Pacific) (GEORGE & GEORGE 1979). Never touch with bare hands.

Lebrunia danae (GEORGE & GEORGE 1979).

The sting

(See also Case Report, page 448.)

Sea anemone stings may be very severe (CLELAND & SOUTHCOTT 1965; MARSH AND SLACK-SMITH 1986). It is the opinion of some observers that species equipped with wart-like 'vesicles' on the column produce particularly severe stings in humans. The warts or vesicles may be simple or compound projections of the column, and they occur in tropical anemones such as members of the families Aliciidae (e.g. *Alicia* and *Phyllodiscus* [see Plates 8.14 to 8.19]) and Phyllactidae. The toxicity of some anemone venoms and the extended morbidity that may result from sea anemone envenomation (GARCIA *et al* 1994) (Plate 8.20) still remains relatively unappreciated (see Case Reports below).

Although there is considerable variation in the immediate symptoms from an anemone sting (CLELAND & SOUTHCOTT 1965; MARSH & SLACK-SMITH 1986; HALSTEAD 1988; NICHOLLS 1992), when there are severe effects these are mainly localised to the contact site. The severe burning pain at the contact site is usually immediate but may be delayed by some 30–60 minutes. The skin at the sting site reacts swiftly with blanching, followed by wheal formation, surrounding erythema and oedema, and often multiple papule formation (Plate 8.12); subsequently vesicles may form (Plates 8.12–8.13). Vesicles may form even on occasions when local pain is mild. A rash may appear (Plates 8.11 and 8.16) (NICHOLLS 1992). Subsequent local skin necrosis and ulceration are repeatedly described (Plates 8.17 and 8.18) (MARSH & SLACK-SMITH 1986; MANDOJANA 1987; HALSTEAD 1988). Following the sting, the pain not uncommonly persists for hours and local soreness for days or weeks {see Case Report, page 448 and Plate 8.11} with notably delayed healing (CLELAND & SOUTHCOTT 1965; MARSH & SLACK-SMITH 1986; SOUTHCOTT 1987c; EDMONDS 1989) and scarring (Plate 8.19). Secondary infection of the ulcerated wounds, with radiating limb pain and regional lymphadenopathy are claimed (HALSTEAD 1988; EDMONDS 1989; NICHOLLS 1992), but no specific case descriptions could be found. Infection risk is claimed particularly following the stings of certain species (*Sagartia, Actinia, Anemonia, Actinodendron and Triactis*) (LEVY *et al* 1970; HANSEN & HALSTEAD 1971; MARETIĆ & RUSSELL 1983). Similarly, generalised symptoms of 'malaise, fever, chills, delirium, headaches, thirst, shock', etc., are repeated in the literature, but no specific case documentation or medical records could be found.

Variation in stinging reaction appears to be related to the specific toxicity of the venom, such that *Actinia* specimens produce painful urticarial lesions, but *Anemonia* specimens induce paresthesiae, oedema and erythema (FISHER

1978). In some cases variation in stinging reaction may also be related to the season in which the sting is sustained (RIFKIN 1994, unpublished observations). Recent reports and review of the clinical literature (BURNETT *et al* 1994), together with long-suspected effect on mammalian nerve conduction (WELSH 1956), suggest a propensity for sea anemone stings to produce a toxic peripheral neuritis in humans {see Case Report 2 below}. A recent report of fatal fulminant hepatic failure following a sea anemone sting has appeared (GARCIA *et al* 1994) (Plate 8.20).

One form of a disease pattern, known as 'seabather's eruption', has followed envenomation by a juvenile form of a sea anemone *Edwardsiella lineata* which was associated with a ctenophore *Mnemiopsis leidyi* (phylum Ctenophora) found off Long Island and elsewhere along the east coast of North America (FREUDENTHAL 1991; FREUDENTHAL & JOSEPH 1993) {see Chapter 11, page 307}.

CASE REPORT 1

(courtesy of Bob and Dinah Halstead, Alotau, Papua New Guinea):

1 A healthy 19-year-old female scuba diver, wearing a swimming costume but no wet suit, sat on the anemone identified as *Cryptodendrum adhaesium* (Figure 8.11, page 208), at a depth of 9 metres, in Torres Strait, north Queensland. Local irritation began after surfacing, and the sting area principally over the left buttock and extending around the left upper thigh (Figure 8.11, page 208) became increasingly painful and papulo-erythematous, with subsequent blister formation. Localised pain and itching increased in severity over the ensuing days to the point of severe distress, accompanied by nausea. The sting region was severely hyperaesthetic to touch. No other systemic symptoms and signs were described. There was no response to oral antihistamines and analgesics or local applications of various creams. No secondary infection supervened. Some nine days following the sting (on medical advice from JAW) the patient was administered a three day course of systemic and oral steroids (intramuscular hydrocortisone 100 mg, followed by three days of oral prednisolone), combined with simple wound dressing of the sting, with immediate (next day) commencement of symptomatic improvement. The wound remained tender to touch for a further two to three weeks, and over the four weeks following the steroid therapy the lesions faded to slight blemishes where blistering had been. No significant permanent scarring occurred. It is not possible to say whether the steroid therapy contributed to the recovery.

CASE REPORT 2

(courtesy of Bob and Dinah Halstead; and the Editors, Medical Journal of Australia)

A very fit male scuba diver, taking underwater photographs, was protected by a lycra stinger suit {see Plate 1.32}. He kneeled unwittingly (Plate 8.23) upon an undescribed species of the corallimorpharian genus *Discosoma* (Ruppell and Leuckart, 1828) (Plates 8.21A, 8.21B and 8.22) (kindly identified by Dr J.C. den Hartog of the Netherlands). He developed itchy, non-painful papulo-erythematous lesions over his elbow, knee (Plate 8.23) and ipsilateral shoulder. During the ensuing days and weeks there developed incapacitating pain with paraesthesiae of the left arm and shoulder with marked left upper limb paresis. Treatment for decompression illness was to no effect! This complication of 'mononeuritis multiplex' is reported elsewhere (BURNETT *et al* 1994) and in this patient took 18 weeks for complete recovery.

SOME POISONOUS ANTHOZOANS

Certain anemone species such as *Rhodactis howesii* SAVILLE-KENT, 1893, and zoanthids such as *Palythoa* spp. have been used in the Pacific for food, with some casualties (CLELAND & SOUTHCOTT 1965; MARSH & SLACK-SMITH 1986).

The corallimorpharian *Rhodactis howesii* in American Samoa is known as 'matamalu', and its ingestion raw has resulted in death (MARTIN 1960). However, cooking renders it harmless, and cooked matamalu is a traditional food there (CLELAND & SOUTHCOTT 1965).

The zoanthid *Palythoa* is highly poisonous; in earlier times, the native Hawaiians used *Palythoa* to poison-tip their spears. However, species of the genus *Palythoa* are found worldwide, including in the Caribbean and Australia where they are also poisonous (P. Sandstrom 1995 personal communication). 'Palytoxin' was first extracted from the Hawaiian *Palythoa vestius* (SCHEUER 1964; MOORE & SCHEUER 1971) and it has proven to be one of the most toxin substances known (BÉRESS *et al* 1983; MARSH & SLACK-SMITH 1986) (MOORE & SCHEUER 1971) {see Chapter 3, page 82}. Even the inhalation of dust from dried specimens is risky.

TOXINS FROM SEA ANEMONES, CORALLIMORPHARIANS AND ZOANTHIDS

The study of the actions of both sea anemone and *Physalia* venoms led to the first identification in medicine of anaphylaxis, by PORTIER, RICHET (with Perret) (1902a, 1902b). RICHET (1903) termed the toxin he obtained from sea anemone tentacles 'congestin'.

Sea anemone toxins have featured in the earliest toxinological studies of cnidarian venoms, and continue to do so, especially as tools for physiological and pharmacological research (e.g. see 'ATX II' below). Because of the large size of some sea anemone nematocysts (e.g. *Corynactis australis* Haddon and Duerden, 1896) and their ready availability, their nematocysts have also proved suitable for studies on nematocyst mechanisms (GODKNECHT & TARDENT 1988; WATSON & HESSINGER 1989).

It appears clear that the toxins are proteinaceous mixtures, which are polypeptide in nature (FARBER & LERKE 1965; MEBS & GEBAUER 1980; SUN *et al* 1993) but probably contain low molecular weight substances such as histamine, serotonin and tetramethyl-ammonium hydroxide (a ganglion blocker). Of interest, sea anemone species contain polyvalent proteinase inhibitors (BÉRESS & BÉRESS 1971; HALSTEAD 1988). The sea anemone toxin from *Anemonia sulcata*, ATX II (along with the other naturally occurring toxic substances: veratridine, aconitine, batrachotoxin, saxitoxin and tetrodotoxin), is a specific modifier of sodium channels (HALSTEAD 1988), which are essential to the generation of action potentials in mammalian muscle and nerve. Recently, ANERIOS and colleagues (1993) have demonstrated a peptide toxin with a unique chemical structure from the sea anemone *Bunodosoma granulifera*, which competes with dendrotoxin I. The latter acts as a probe for voltage dependent potassium channels, so that this anemone toxin exerts a potassium channel blocking action. It facilitates acetylcholine release at avian (bird) neuromuscular junctions.

Sea anemone toxins are likely to be thermolabile (MARTIN 1963). Sea anemone toxins exert paralysing effects on skeletal muscle and can depress vascular smooth muscle or block cardiac arrhythmias in experimental animals (NOVAK *et al* 1973). The venoms of some sea anemones, for example *Aiptisia pallida* and *Radianthus ritteri* are haemolytic *in vitro* to animal red cells (HESSINGER & LENHOFF 1974; BÉRESS *et al* 1983; SUN *et al* 1993) {see comments on the clinical significance in humans of such *in vitro* haemolysis in animal red cells, on page 20}. Toxin from the anemone *Rhodactis howesii* is antigenic in rabbits (MARTIN 1962, unpublished observations).

Palytoxin is also haemolytic *in vitro* to animal red blood cells (see above); it depolarises all excitable animal tissues so far investigated, initiating muscle contracture, and is the most potent coronary vasoconstrictor known (KAUL 1981).

Despite all the above evidence, the most important clinical effect in human sea anemone stings appears to be the dermatotoxic and dermatonecrotic actions (Plates 8.13 and 8.18). It is also suggested by experience (POPE 1963) and some studies (CSABA GY *et al* 1983, unpublished observations) that sea anemone venom may have particular allergenic and sensitising propensities and could perhaps present increased risks in patients with pre-existing 'allergic diatheses' such as asthma, atopic conditions and allergic eczema (cf. mauve stinger *Pelagia noctiluca* envenomation, page 228). Certainly ATX II is antigenic (MAIER & RATHMAYER 1982). The recent report of an apparently toxin-induced fatal hepatic necrosis following a sea anemone sting in the Virgin Islands (GARCIA *et al* 1994) justifies further careful scrutiny of sea anemone toxins.

CLINICAL MANAGEMENT

First aid

The editors could find no published studies on nematocyst discharge inhibition in sea anemone stings, but see RIFKIN 1996. Thus accurate recommendations concerning substances to apply to a fresh sea anemone sting to achieve this end are not currently (1996) possible. It is recommended, pending further research, to remove any adhering tentacles with fingers or forceps and then blot the area with a disposable tissue {see Chapter 5, page 108}; apply ice packs or cold packs (EXTON *et al* 1989) to the area. Simple oral analgesia (paracetamol or acetyl salycilic acid tablets) may also help with pain relief. Uncontrolled pain and/or systemic signs or symptoms require referral for medical management, especially in patients with a history of allergy {see above}.

Medical management

1. Appropriate analgesia as indicated.
2. Management of local allergic reactions: local antihistamines and anaesthetics have proved disappointing for this purpose. Systemic antihistamines or even steroids may assist {see Case Report 1 above}
3. Systemic allergy or unrelieved inflammatory reactions require vigorous treatment, e.g. parenteral steroids. However, steroids must not be used in the presence of, or to treat secondary infections of, the sting area {see above}. Anaphylaxis (TOGIAS *et al* 1985) to an anemone sting has not been described to date, but in such an event, systemic adrenalin is essential (EXECUTIVE EDITORIAL BOARD 1989).
4. Secondary infection is managed with topical and/or systemic antibiotics, using culture and sensitivity results as guidance.
The management of sea anemone poisoning is currently non-specific supportive therapy, along the same lines as for other potent marine toxins {see Chapter 5 page 114, and Chapter 15, page 346}.

CLASS HYDROZOA

The medically significant members of this diverse class include:
- colonies of delicate plant-like growths ('feather hydroids', 'stinging hydroids' or 'fireweeds', e.g. *Aglaophenia* (Plate 8.24)
- stinging or fire 'corals' in the order Milleporina, e.g. *Millepora* (CAMPBELL 1980; VERON 1986) (Plate 8.25), which may superficially resemble the 'true corals' (class Anthozoa)
- jellyfish of the order Limnomedusae, e.g. *Gonionemus* and *Olindias* {see Chapter 6, page 140}
- free-swimming, floating organisms of the subclass Siphonophora, e.g. *Physalia* spp. {also see Chapter 6, page 137}

Technically, any member of the class Hydrozoa may be referred to as a 'hydrozoan', the vernacular name.

SUBCLASS HYDROIDA
ORDER LEPTOMEDUSAE/THECATA

FAMILY AGLAOPHENIIDAE (STINGING HYDROIDS)

DISTRIBUTION

These hydrozoans are found worldwide, occurring in greatest profusion in tropical waters (Plate 8.26). They favour sheltered areas below the low-water mark along the mainland coasts, growing on rubble on reef flats, on piers and pilings, in caves and beneath overhangs, where they are commonly encountered by recreational snorkellers (Plate 8.27) and scuba divers (Plate 8.29) (RIFKIN *et al* 1993) {see Chapter 22, page 424}. They have an apparent predilection for coral reef edges and channels (MARSH & SLACK-SMITH 1986).

Several species in the family Aglaopheniidae cause discomfort to humans. *Lytocarpus philippinus* (Kirchenpauer) (RIFKIN *et al* 1993a) has been recorded in the tropical and sub tropical Indo-Pacific region and in the Atlantic Ocean and Mediterranean Sea. *Aglaophenia cupressina* Lamouroux (Plate 8.26) has been recorded throughout the Indo-Pacific from Zanzibar to northern Australia and north to the Philippines (HALSTEAD 1988).

Because of their appearance and their sting, *Lytocarpus* and especially *Aglaophenia* are often referred to as 'fireweed' (Plate 8.24) (MARSH & SLACK-SMITH 1986).

Australian hydroids

Hydroids are found along the coasts of Queensland, the Northern Territory and Western Australia as far south as Fremantle. *Lytocarpus philippinus* and *Aglaophenia cupressina* are both very common. *Lytocarpus philippinus* has been found as far south as Moreton Bay, near Brisbane, Queensland. *Aglaophenia cupressina* occurs along the subtropical coastline of Queensland as far south as Heron Island. Other stinging hydroids found in Australia include *Lytocarpus pennarius* (Linné), *L. phoenecius* (Busk), and *Thecocarpus divaricatus* (Busk). The family Haleciidae is represented by *Halecium beani* (Johnston).

APPEARANCE

These animal colonies vary in length from a few millimetres to many centimetres. Although there is considerable diversity of appearance, most have a horny skeleton. Many hydroids, including *Aglaophenia* and *Lytocarpus*, have feather-like branches lined with nematocyst-bearing polyps (Plate 8.28, see RIFKIN *et al* 1993), giving them their seaweed-like appearance (Plate 8.24). Long tentacles are absent, but the delicate fronds contain many nematocysts (RIFKIN *et al* 1993) that can cause immediate, burning skin pain.

HYDROID STINGS

Hydroid stings have long been recognised, and respected, among reef workers (SAVILLE-KENT 1893; De OREO 1946; CLELAND & SOUTHCOTT 1965) for their impressive stinging propensities. The response to a sting may be both acute and protracted. CLELAND and SOUTHCOTT (1965) document a trial with stings from the hydroid *Halecium* sp., in which a rapid urticarial reaction was followed 4–12 hours later by a papular, 'haemorrhagic and zosteriform' reaction locally (De ORIO 1946). They suggest elements of 'sensitisation' with eosinophilia in some of the skin biopsy lesions.

Certainly the sting is acutely painful following even simple brush contact on bare skin (Plates 8.27 and 8.29). The skin reaction is not immediate, developing itchy, painful whealing within 30 minutes, progressing to local urticaria (Plate 8.30). Usually the wheals last up to a week before fading and leave no remaining skin staining or scarring (RIFKIN *et al* 1993). In more severe reactions, angry blistering may follow with crusting over several days. Generalised symptoms are described (which include skin rash) (CLELAND & SOUTHCOTT 1965) but these are not common. Regional lymphadenitis may occur. No deaths are documented.

MANAGEMENT

Stinging hydroid or ('fireweed') stings

First aid treatment should commence with blotting the stung area gently with a clean dry tissue. Vinegar was proven to inhibit discharge of any remaining undischarged nematocysts in *Lytocarpus philippinus* (RIFKIN *et al* 1993a), but further studies are necessary for other species.

Pain relief is best obtained with the direct application of ice or ice packs (EXTON et al 1989). Itching may respond to application of ultrapotent steroid or mentholated creams. Exclude actual skin abrasion or puncture by any horny hydroid skeleton, with the attendant risks {see above} in cases of forcible contact. Systemic or unrelieved local symptoms require medical referral.

Medical management of more serious cases should proceed along symptomatic lines {see Chapter 5, page 113}.

ORDER MILLEPORIDA (FIRE 'CORALS')

Millepora alcicornis Linné is found in the Mediterranean Sea and in the tropical Pacific (HALSTEAD 1988). *Millepora tenera* BOSCHMA, 1949 (Plate 8.25) and *M. platyphylla* Linné, 1758, occur in Thailand waters (CAMPBELL 1980), and *M. platyphylla* is also known in north-western Australia (MARSH & SLACK-SMITH 1986). *Millepora dichotoma* Forskäl is found in the

Red Sea, the Gulf of Aden and the tropical Indo-Pacific. *Millepora complanata* Lamarck and *M. platyphylla* Hemprich & Ehrenberg have also been located there (RUSSELL 1984).

APPEARANCE

Fire corals resemble hard corals (Plate 8.25) and can appear innocuous to the unwary scuba diver. They are not the true corals of the class Anthozoa, but nevertheless they are important parts of the reef-building community. Fire corals may be distinguished from stony corals (Class Anthozoa) by their relative smoothness. Species of *Millepora* are usually differentiated by their skeletal structure (BOSCHMA 1949) which possesses millions of pores, each of which houses a polyp. Two types of polyp project from these pores: the dactylozooid or defensive polyp, and the gastrozooid or feeding polyp. Usually a gastrozooid is surrounded by five to seven dactylozooids. Both polyp types contain nematocysts capable of immobilising prey.

FIRE CORAL STINGS

Contact with any fire coral (Plate 8.25) produces an immediate burning sensation, which may range from mild to intense pain (BACHAND 1983). Mainly soft skin areas (e.g. backs of the hands, insides of the arms, the face) are susceptible (VERON 1986). Generally, a skin lesion develops 1–24 hours after the sting. Although the lesion may initially look like a series of mosquito bites, it soon progresses to raised erythematous urticarial wheals (AUERBACH 1987). Infrequently the lesion may progress to infected pus-containing sores; necrosis of the skin can also occur but is uncommon. Such a lesion can last for weeks depending on the extent of the envenomation (number of nematocysts discharged) and the sensitivity of the victim.

SAGI *et al* (1987) reported the case of a 20-year-old girl who was diving in the Red Sea when her calf brushed against a colony of the hydrocoral *Millepora dichotoma*. Immediately following the dive the calf became erythematous, and an itching and burning pain ensued. The pain gradually subsided, but during the next two days most of the reddened areas became necrotic. Because the necrotic area was painless, the girl did not seek medical assistance for 20 days following the incident. According to SAGI *et al* (1987), 'the skin burns were well-demarcated as two lenticular necrotic skin areas. An excision of both areas was performed with primary closure of the wound. As the wound was somewhat tightly closed, bed rest with the leg elevated continued to avoid dehiscence. Wound healing progressed uneventfully.' No human deaths have ever been reported.

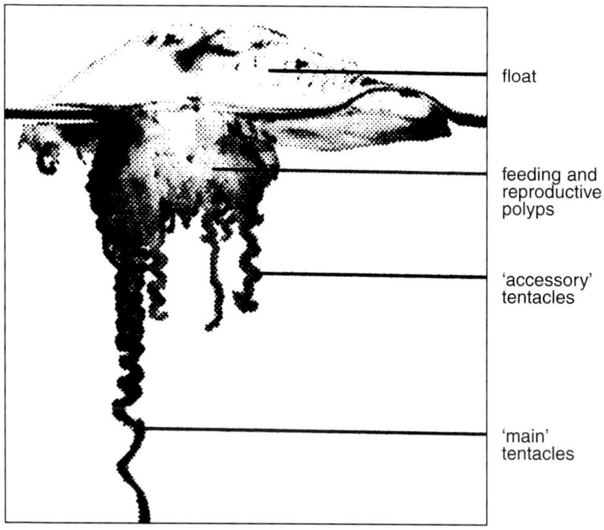

FIGURE 8.2
Diagram of *Physalia* sp. showing the main morphological features.

It is unusual, in the authors' experience, for a fire coral sting to persist much beyond the acute painful phase, and blistering following this sting is uncommon.

MANAGEMENT OF FIRE CORAL STINGS

These painful stings (AUERBACH 1987) can be managed in the same way as a non-dangerous jellyfish sting {see Chapter 5, page 106}. Following gentle clean, dry tissue dabbing of the freshly stung skin, application of cold packs or ice* to the painful area, with rest and elevation of the envenomated part, will usually ease the pain. Itching

* AUERBACH (1987) specifically advises against the application of ice to fire coral stings, but the editors have not encountered any contraindications.

may predominate as the pain fades (usually within hours). Local application of ultrapotent steroid creams (contraindicated in the presence of infection) or mentholated creams may ease the itching. Local antihistamines and local anaesthetic creams have proved disappointing.

SUBCLASS SIPHONOPHORA

BIOLOGICAL DETAILS — JACQUIE RIFKIN

The siphonophores are exclusively pelagic, free-swimming colonies consisting of several kinds of zooid (individuals) often supported by a float mechanism (CLELAND & SOUTHCOTT 1965) (Plates 8.31, 8.38) {see also Figure 6.15, page 138}. The float can act as a sail, influencing a colony's direction of travel (Figure 8.3, page 193).*

The siphonophore of dominating medical significance, *Physalia*, is one of the most important 'jellyfish' in the world, because it produces the most common marine sting presently known (INTERNATIONAL CONSORTIUM for JELLYFISH STINGS 1993). It is present in the majority of oceans and seas from cold temperate to tropical; and because of its ubiquity and its 'familiar' appearance to the general public, it has been falsely and repeatedly blamed for almost every type of marine sting (and some bites!) in the past (CLELAND & SOUTHCOTT 1965). However, the large, 'multi-tentacled'† Atlantic species is certainly responsible for some dangerous and fatal envenomations (STEIN *et al* 1989; BURNETT & GABLE 1989; WILLIAMSON *et al* 1993; BURNETT *et al* 1994a). Surprisingly, the taxonomy of this 'familiar' animal remains unresolved. For all these reasons the genus *Physalia* is deserving of the closest study.

* An excellent example of a float mechanism and a sail also exist in the non-stinging {usually — see Chapter 18, page 404} species *Velella velella* (Plates 8.37A and 8.37B) (BROWNLEE 1987). However, this animal (along with *Porpita* sp.) is now not considered to be a siphonophore (FIELDS & MACKIE 1971) {see Chapter 6, page 121}.

† All *Physalia* specimens have thick and thinner tentacles. Classically, those with one long thick fishing tentacle have been known as *P. utriculus* while those with more than one long thick tentacle have been termed *P. physalis*. *Although the* 2 species were combined by Totton in 1960 into one cosmopolitan species, namely *P. physalis*, FENNER *et al* 1993 reintroduced the 2 species for differentiating between the two different morphs found on Australian beaches. For the purposes of the following analysis, the terms 'single-tentacled' and 'multi-tentacled' will be used to describe the number of long tentacle(s) of the two distinct morphs.

GENUS *PHYSALIA*

MEDICAL DETAILS — PETER FENNER

The genus *Physalia* was first established by Lamarck in 1816 and named after its distinguishing sail. There is currently disagreement whether there is more than one *Physalia* species, and uncertainty as to how the multi-tentacled and single tentacled *Physalia* specimens should be classified.

FENNER and colleagues (1993) suggested that there are at least two and possibly more species of *Physalia*; they reintroduced the name *Physalia utriculus* for the single-tentacled variety (Plates 8.31 and 8.37), which is very common in the Pacific Ocean and elsewhere, and *Physalia physalis* for the multi-tentacled variety (Plates 8.33 and 8.35), which has long been associated with, but is now known not to be exclusive to, the Atlantic Ocean. To differentiate between the 'Pacific' and 'Atlantic' forms, they also suggested the name 'Pacific man-o'-war' for the smaller, Pacific multi-tentacled variety (Plate 8.35), retaining the name 'Portuguese man-o'-war' for the larger, potentially lethal Atlantic variety (Plates 8.33 and 8.43).

NOTE: Historically CHUN (1897) and BIGELOW (1911) held the view that there were two species, one in the Atlantic and one in the Pacific. Bigelow stated that all Pacific specimens had only one tentacle. TOTTON (1960) reported only one species (multi-tentacle), and he regarded the single-tentacled *Physalia* as immature specimens of the multi-tentacled species. Although this remains the current view of many biologists, one of the editors (JFR) has collected a 10 cm long *P. utriculus* (i.e. much larger than the usual *P. utriculus*) in Australia; this animal's toxicity was measured by JWB as very high (RIFKIN & BURNETT 1992, unpublished observations). In addition, the editors have in their possession immature multi-tentacled specimens as well as mature single-tentacled specimens.

1. *Physalia physalis*, the multi-tentacled variety: including the large, lethal Portuguese man-o'-

war found in the Atlantic (Plate 8.33, 8.34 and 8.43), and the smaller but multi-tentacled Pacific man-o'-war (Plate 8.35) (FENNER *et al* 1993)

2. *Physalia utriculus*, the single-tentacled variety (Plates 8.31 and 8.37): commonly found in Australia and worldwide (including the coast of China [QIN SHEDE & ZHANG MINGLIANG 1994, unpublished observations)

NOTE: The marine biologist editor (JFR), from specimens in her possession, is confident that the differences between the two varieties of Pacific *Physalia* are great enough to have two different species (FENNER *et al* 1993). The term 'Pacific *Physalia*' is employed in the following discussion to include these two different species.

Physalia morphology

(Plates 8.31–8.35, 8.37–8.41 and 8.43) {see also Plates 6.4, 6.5 and 10.26}

Physalia are hydrozoan colonies rather than solitary jellyfish. Each colony consists of four groups of animals living together symbiotically, each group performing and sustaining a different, specialised function for the common good of the whole colony {see Chapter 6, page 137, for a precise biological description}:

1. the float and gas production – which controls the buoyancy of the colony and acts as a sail

2. the long fishing tentacles, armed with nematocysts, that immobilise prey on contact

3. gastrozooids (feeding polyps) that digest the captured food

4. the reproductive organs

The reproduction of this complicated creature involves direct budding to form more polyps asexually, but each colony also has male and female reproductive organs. The exact reproduction process and life cycle are currently unknown.

The float mechanism

Physalia have large gas-filled floats (Plates 8.31–8.35, 8.38, 8.43 and Figures 8.2 and 8.4, pages 191 and 194), which may reach 25 cm in length. A gas gland, which increases in size as the colony gets larger, secretes the gas that inflates the float. The gas resembles atmospheric air, except curiously it contains some 8 per cent by volume of carbon monoxide (a lethal concentration to mammals!). (Normal air should contain less than 10 parts per million [i.e. < 0.001%] of carbon monoxide.)

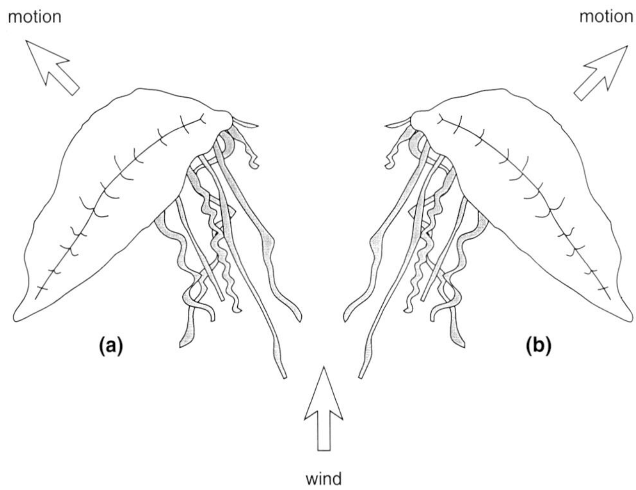

FIGURE 8.3
Diagrammatic representation of (a) right-handed and (b) left-handed float-sails of Physalia sp., illustrating the directional sailing characteristics of the two different varieties.

De QUATREFAGES (1854) is the only documented observer claiming to have seen a *Physalia* float deflate and inflate; but it did not sink below the water surface.*

Muscle fibres present in the float appear to control the float orientation on the surface. When the wind blows, the muscular fibres lift the crest into the wind, rather like a sail being hoisted (TOTTON 1960). This 'sail' causes *Physalia* to be blown by the prevailing wind, rather than propelled through the water by any contractions of the float (Plates 8.31–8.33 and Figure 8.3).

In calm conditions the muscle fibres also adjust gas pressures in the float, altering its buoyancy and causing 'rolling'. The animal 'contracts', lifting one end of the float out of the water and forcing itself to roll over. Rolling appears to occur only in calm weather and is thought to be a simple method of keeping the float cool in hot, still weather.

Right- and left-handed varieties of *Physalia* are recognised (TOTTON 1960; SHANNON & CHAPMAN 1983) (Figure 8.3, page 193) (as they are in

•See note in Bibliography under HUXLEY 1859

FIGURE 8.4
A preserved specimen of an Atlantic *Physalia* sp. from the waters of Uruguay. The photograph suggests one long tentacle. Living adult specimens do not sink below the water surface. (Courtesy of Dr H. Mianzan and Dr F. Ramírez.)

Velella velella [Plate 8.36] [L. Francis in BROWNLEE 1987]). The animals are efficient 'sailors'. They do not run directly before the wind, but right- and left-handed specimens move in opposite directions, about 40 degrees off the wind* (Figure 8.3, page 193), thus preventing all the specimens in an 'armada' being blown on to a lee shore, ending in their destruction. The tentacles, gastrozooids and other appendages act as sea anchors (Plates 8.39 and 8.40 and Figure 8.5, page 195) {see also Plate 6.5}.

Nematocyst assembly

Physalia nematocysts {Plates 5.1 and 8.41; Chapter 6, page 167} are grouped in 'buds' (which are not complete rings) around the circumference of the tentacle (Figure 8.6, page 195). No nematocysts are found on the float. Replacement nematocysts are produced in the proximal portion of the tentacle and migrate down a central tentacle canal to be 'slotted in' to gaps created by previously discharged nematocysts, or into areas where new ones are needed (TOTTON 1960). This functional anatomical arrangement is common in other jellyfish species (RIFKIN 1993, unpublished observations).

Tentacles are very retractile and may contract up to one-seventeenth of their extended length (TOTTON 1960). Tentacles 'to a depth of 30 m' are claimed in Atlantic species of *Physalia* (HALSTEAD 1988). In the smaller, single-tentacled Pacific specimens, tentacle length may reach 2–3 metres (WILLIAMSON 1985; FENNER *et al* 1993b).

DISTRIBUTION

Distribution of the genus is virtually worldwide, although occurrence in Arctic and Antarctic seas is undocumented. (The Mawson Antarctic collection [1929–1931] located in South Australian museum does not include *Physalia*.)

Atlantic Ocean

The multi-tentacled, sometimes lethal Atlantic *Physalia physalis* (STEIN *et al* 1989; BURNETT & GABLE 1989) is found in the Caribbean Sea and on the 'southern' (i.e. North Carolina) Atlantic coastline of North America, where it is particularly prominent in winter after south-east winds. Several hundred thousand envenomations can occur annually in south-eastern Florida during the Christmas holiday period (BURNETT 1992a).

This multi-tentacled species is apparently also found along the coast of South America as far south as Uruguay and northern Argentina. The single-tentacled *Physalia* has also been found on the coast of Brazil† (MIANZAN 1990, unpublished observations). A mass stinging in January 1994 of 295 persons in the State of São Paulo (250 in the city of Guarujá, 30 in the city of Praia Grande, 15 in the city of Santos and 128 persons in Rio de Janerio) has been confirmed to have been caused by an Atlantic *Physalia physalis* (unpublished data of the Hospital Municipal Lourenço Jorge, 12 February 1994; *Jornal do Brasil*, 15 and 24

* Some doubts have recently been expressed concerning the details of this 'sailing' behaviour of *Physalia* (Heeger 1994, unpublished observations).

† Thus it seems that the single-tentacled and the multi-tentacled varieties occur in both the Atlantic and the Pacific Oceans as stated previously by Chun

February 1994) (courtesy of Dr Hermes MIANZAN, Mar del Plata, Argentina, and Dr Rossana H. PITA VIRGA, University Cat. Santos, Brazil).

South Africa

South Africa has a problem with *Physalia* envenomations, with reports of several hundred stings at beaches of many coastal cities (SHANNON & CHAPMAN 1983; City of East London Municipal Directorate 1991, unpublished observations). Large numbers of these multi-tentacled medusae were beached on the south-western Cape Province (South Atlantic Ocean) in 1982–83 (SHANNON & CHAPMAN 1983). There is a discrepancy between the description of these specimens and the illustrations (photographs and line diagrams) of multi-tentacled *Physalia physalis* seen in BRANCH & BRANCH (1981). All photographs supplied to one of the editors (PJF) of the specimens from South Africa appear to be single-tentacled *Physalia* (*P. utriculus*?).

FIGURE 8.5
A preserved specimen of a *Physalia* sp. captured on a Karachi (Arabian Sea) beach in 1993 by Junaid Mahmood Alam, PhD. Note several long tentacles, three of which have been cut for microscopic examination. {Compare with Figure 10.9, page 303.} (Prepared by and reproduced with the courtesy of Junaid Alam, Rashida Qasim and colleagues, Department of Pathology, Basic Medical Sciences Institute, Jinnah Postgraduate Medical Centre, Karachi.)

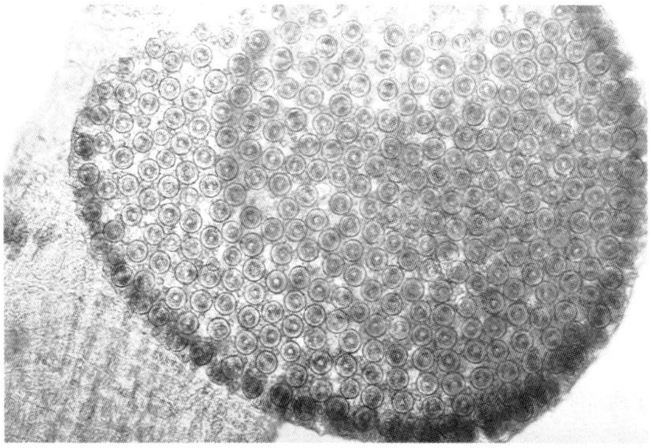

FIGURE 8.6
A single tentacular 'bud' of nematocysts of *Physalia physalis*. Note the bud does not completely encircle the tentacle. (Photograph courtesy of Professor Thomas Heeger, Philippines.)

Indian Ocean

The eastern shores of the Indian Ocean (Western Australia) have an equally plentiful supply of *Physalia* spp. (MARSH & SLACK-SMITH 1986). Only the single-tentacled *Physalia utriculus* has been reported so far. In April 1991 this *Physalia utriculus* interfered with the World Life Saving ocean swim race off Perth.

Arabian Sea

Physalia with apparent multiple long tentacles* (Figure 8.5) are found on the shores of the Arabian Sea in Pakistan, where they cause hundreds of stings each year (ALAM & QASIM 1991; J. M. ALAM [Karachi] 1993, unpublished observations). The sting of these Arabian Sea hydrozoans is severe, causing systemic side effects, muscle pain, sweating and anxiety {see Chapter 10, page 303}.

Pacific Ocean

Multi-tentacled specimens can injure swimmers during late summer on the Pacific beaches along the eastern coast of Honshu, Japan. Pacific

* Note that recent toxinological studies from these investigators refer synonymously to '*Physalia* (Portuguese man-o'-war)' (ALAM *et al* 1992) and '*Physalia utriculus*' (ALAM *et al* 1993). It thus remains uncertain whether two differing *Physalia* colony forms (FENNER *et al* 1993b) exist in Pakistani waters.

Physalia stings are also well known throughout the Pacific islands and Hawaii (KIZER 1984; HALSTEAD 1988). In the eastern Pacific many small *Physalia* sp. medusae are sighted in the Gulf of California (formerly known as the Sea of Cortes [Cortez]) from June to August (Dr Thomas WEBSTER, Arizona, 1994, unpublished observations).

On the eastern coast of Australia, where vast 'armadas' of *Physalia* spp. appear from time to time (CLELAND & SOUTHCOTT 1965; INTERNATIONAL CONSORTIUM for JELLYFISH STINGS 1993), approximately 5000 stings occur each year from the smaller, single-tentacled Pacific variety (CLELAND & SOUTHCOTT 1965; FENNER 1993). The beaches have to be closed by life savers, mainly in the summer months when thousands of what FENNER *et al* term '*Physalia utriculus* ("bluebottles")' are blown onto the beaches. In the New Year's summer of 1994 a huge influx of *Physalia* (mainly single-tentacled, but with some multi-tentacled specimens) blew in upon the beaches of Queensland's Gold Coast and northern New South Wales, causing thousands of stings on the first day alone, and the closing of many beaches for several days (SURF LIFE SAVING QLD 1994, unpublished records).

THE STINGS

Despite the large numbers of stings in Australia, there has never been a documented or reported fatality from the Pacific species. Multiple large stings from *P. utriculus*, the single-tentacled variety (Plate 8.42), usually cause only skin and regional lymph gland pain.* The subjective association of 'breathing difficulty' with the more extensive stings from this single-tentacled animal, is infrequent, and probably inaccurate. Such reports of 'breathing difficulty' (actually 'painful breathing' caused by intercostal muscle pain) are probably due to *P. physalis* stings (FENNER *et al* 1993). An early report suggested life-threatening symptoms after a *Physalia* sting, (GOLLAN 1968), where mention of one large tentacle was made.

Physalia stings causing severe toxic symptoms with back pain, breathing problems and anxiety {cf. Irukandji syndrome, Chapter 9, pages 251–2} appear to be more associated with the 'previously unrecognised', multi-tentacled 'Pacific man-o'-war' (FENNER *et al* 1993). Further studies are required.†

Following recognition of the severe stinging characteristics of this multi-tentacled Pacific variant in 1991 and the concomitant capture of the specimens, further specimens were captured throughout Queensland's coastline, and more recently in New South Wales. However, reports of multi-tentacled Australian *Physalia* are found in the literature as early as 1965 in CLELAND and SOUTHCOTT. Further studies are required.

FIRST AID AND MEDICAL MANAGEMENT

{See also Chapter 5, page 106.}

1. Using seawater (FENNER *et al* 1993), wash off any remaining tentacles on the skin, but only if the sting site has not dried.†† This step is not

* During the January 1994 Queensland Gold Coast epidemic, several patients were treated in hospital for pain, and one female case required 'EAR' on the beach following a post-envenomation syncopal attack. However, subsequent expert clinical assessment of the patient judged this to be a vaso-vagal episode. The same epidemic 'moved' southwards a few days later and caused similar and equally extensive beach problems in the Sydney area, again resulting in several hospital attendances.

† It is of historical interest that several undocumented but puzzling accounts of much more clinically severe envenomations from Pacific 'bluebottles' were known by medical clinicians in Australia for many years (CLELAND & SOUTHCOTT 1965; EDMONDS 1989). This apparent contradiction tended to be ascribed to variation in individual patient susceptibility to the envenomation (e.g. allergy)! Such occasional severe reactions also contributed to the initial confusion in Australia between the stings of the then unidentified *Chironex fleckeri* SOUTHCOTT 1956 and the Pacific *Physalia* sp. The brilliant field and analytic work of Flecker, Southcott and Barnes, which identified *Chironex fleckeri* as the culprit (thus exonerating the bluebottle), is chronicled in the pages of CLELAND & SOUTHCOTT (1965). In hindsight it is possible that some of these severe envenomations may have been from unrecognised multi-tentacled *Physalia*.

†† The use of water on dried tentacles containing nematocysts has been shown to cause discharge of unfired nematocysts, in several species of cnidarians (WEILL 1934; LANE 1960; BRANCH & BRANCH 1981; RIFKIN 1988a; ENDEAN *et al* 1991; RIFKIN 1993, unpublished observations), a phenomenon clinically recorded by BARNES in relation to fresh water (1966) {refer to Chapter 5, page 110}.

necessary if there are no adherent tentacles. (Editors' Note: We are sharply aware of the uncertainty concerning *Physalia* spp. nematocyst provocation and inhibition.* See below.)
2. Apply cold packs or wrapped ice for relief of skin pain for 5–15 minutes, then reapply if necessary.
3. Seek medical aid for symptoms not responding to cold packs.

The current puzzle of *Physalia* nematocyst inhibition

TURNER *et al* (1980) recommended topical application of dilute (4–6%) acetic acid (household vinegar) as a first aid treatment to ease the skin pain from Sydney Pacific *Physalia* (reported by TURNER *et al* [1980] as '*Physalia physalis*, bluebottle') envenomations. No corroborating experiments have since been reported, and it is difficult to understand how vinegar penetrated the skin to counteract instant skin pain produced by the injected venom. Earlier, *in vitro* experiments by HARTWICK *et al* (1980) suggested that vinegar also inhibited the discharge of nematocysts *in situ* in freshly separated north Queensland Pacific *Physalia* tentacle material, but did not provide any analgesia. FENNER *et al* (1993) found that vinegar appeared to inhibit the discharge of *P. utriculus* (their chosen name) nematocysts, but caused discharge of some of the nematocysts in the Pacific man-o'-war (which they called '*P. physalis*'). This offered one possible explanation for previous baffling data that showed that almost 40 per cent of nematocysts of some southern Queensland *Physalia* (undifferentiated at that time) may be discharged by dousing

* Although *in vitro* tests have shown that nematocysts *in situ* in the tentacle do not necessarily fire when exposed to sea or fresh water, the situation may differ when they are next to the skin, where receptors on the tentacle will be exposed to chemical stimuli required for discharge.

TABLE 8.1:
COMMON SUBSTANCES THAT CAUSE NEMATOCYST DISCHARGE IN THE PORTUGUESE MAN-O'-WAR

Recent field studies upon Florida Atlantic *Physalia* involved testing a range of substances for nematocyst discharge in living tentacles. Tests were carried out with live animals with float sizes ranging from 2 to 5 cm, and tentacles were examined with x100 magnification microscope, for evidence of nematocyst discharge, following previously described principles (HARTWICK *et al* 1980). The following estimated percentages of nematocyst discharge were observed *in vitro* (BURNETT 1993 unpublished observations).

Substance tested	Tentacle region tested		
	Proximal	Middle	Distal
4% methyl alcohol in ethyl alcohol	0-20%	nil	up to 100%
Household vinegar	nil	nil	nil
Acetone	nil	nil	nil
Baking soda slurry	nil	nil	nil
Urine	nil	nil	nil
$MgSO_4$ slurry or paste	nil	nil	nil
Formalin	up to 20%	nil	nil
Aluminium hydroxide suspension	nil	nil	nil
Cocoa Cola	slight	nil	nil
Hydrogen peroxide	nil	nil	nil
5% boric acid	nil	nil	nil
Insect repellant	nil	nil	nil
Sting remedy used in Penang	nil	nil	nil

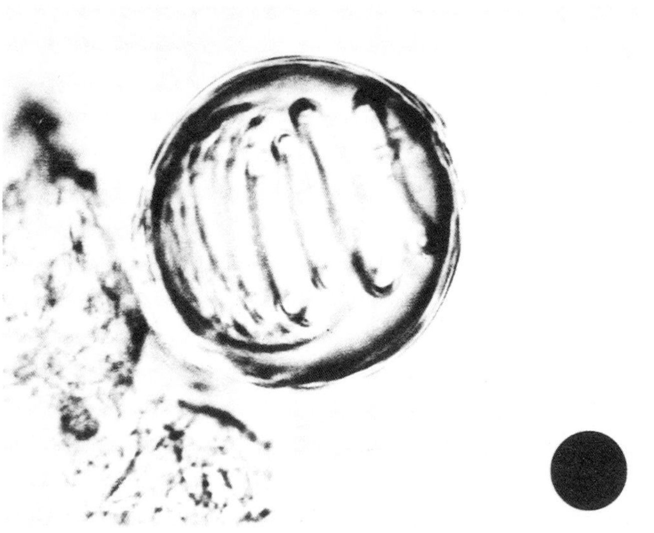

FIGURE 8.7
A solitary, undischarged, spherical Pacific (north Queensland) *Physalia utriculus* nematocyst. The small black circle is the approximate size of a human red blood cell (≈7.5 of µm) (Courtesy of Dr R.F. Hartwick, PhD, James Cook University of North Queensland.)

with vinegar (EXTON *et al* 1988). Specimens used in these experiments were later examined by one of the author/editors (JFR), and shown to be the multi-tentacled (*P. physalis*) variety.

It appears that most Pacific *Physalia* (especially bluebottle) nematocysts *in situ* can be inhibited from firing by topical vinegar applied *in vitro*. However, when the vinegar-doused tentacles are placed on the lip, pain will still be perceived. These data indicate that topical vinegar will probably be of some benefit, but it does not totally inhibit discharge, and so will not offer complete protection. Finally, a study of multi-tentacled *P. physalis* specimens caught in north Queensland demonstrated that when vinegar was applied to nematocysts *in situ*, mass discharge occurred in some specimens, but not in others. Because some of the stings from these animals were severe (including a modified Irukandji syndrome) and beach differentiation between the various types of *Physalia* stings is not a practical possibility for the general public at present, it is currently recommended that vinegar not be applied to any *Physalia* stings, pending further nematocyst discharge inhibition research (FENNER *et al* 1993).

PHYSALIA PHYSALIS (PORTUGUESE MAN-O'-WAR)

JOSEPH BURNETT AND PETER FENNER

APPEARANCE

Physalia physalis is a large blue-purple animal whose float may measure from 2 to 250 mm in length (Plate 8.43). Multiple, long tentacles descend from the centre of the float up to a depth of many metres (e.g. 32 m according to HALSTEAD [1988]) (Plates 8.33 and 8.35) {also Plate 10.22}.

The cnidom of this species includes two sizes of spherical isorhizas {see Chapter 6, page 139} (LANE & DODGE 1958; HULET *et al* 1974; BURNETT et al 1986). In the opinion of HULET and colleagues (1974), working with Florida specimens of *Physalia physalis*, 'isolated nematocysts as well as those observed in sections of fishing tentacles are of one structural type and two sizes...' They noted that cnidocils were present on these nematocysts.

THE STING

Contact with these tentacles usually results in some tentacle tissue remaining adherent to the patient. A characteristic imprint may appear on the skin (Plate 8.44), and instant severe pain is perceived which may persist for many hours. A linear erythematous eruption with localised sweating and piloerection follows. Localised muscular cramps are common. Stings around the eyes may produce small corneal ulcerations and visual disturbances.

Systemic symptoms claimed include nausea, vomiting, tender lymphadenopathy, syncope, convulsions and muscular cramps (HALSTEAD 1988; BURNETT *et al* 1994).* Respiratory acidosis

* HALSTEAD (1988, Plate IV:9) gives a dramatic account of a near death in a (breath-hold?) diver in the Caribbean Sea off Isla de Providencia (east of Nicaragua) who surfaced under an Atlantic *Physalia* sp., suffered excruciating acute pain which lasted overnight, with an extensive vesiculo-necrotic shoulder lesion, and subsequently was suspected of developing tetanus. Recovery was thought to have occurred. A fatal (BURNETT & GABLE 1989) and a near-fatal (BURNETT et al 1994) Atlantic *Physalia* sting under similar circumstances involving scuba divers {see Chapter 22, page 424} are documented.

from hypopnoea due to chest pain has resulted (KIZER & PIEL 1982). Recurrent eruptions following single envenomations have been reported (BURNETT *et al* 1987). *Physalia physalis*, Atlantic species, caused a recent mass stinging of 295 persons in Sâo Paulo in January 1994 (250 in Guarujá, 30 in Praia Grande, 15 in Santos) Many people required medical attention, and two children developed respiratory arrest, responding to resuscitation (translation taken from the Argentinian newspaper *'Diario La Capital'*, Thursday 20 January 1994) (MIANZAN 1994, unpublished observations).

In another report a distant reaction in the mouth was noted in one patient who was stung only on her feet (MATUSOW 1980). Serious envenomation from an Indian Ocean (Goa) jellyfish was reported, in which *Physalia* was suspected, on the basis of serological titres, of producing localised necrosis, vasospasm and gangrene (Plates 8.45 and 8.46) (WILLIAMSON *et al* 1988). However, the identity of the jellyfish was not confirmed.

Mononeuritis multiplex has been reported in a patient envenomated off the American mid-Atlantic coast (FILLING-KATZ 1984). Reactive arthralgia of significant duration can follow the sting (WEINBERG 1988). Clinical effects severe enough to cause acute renal failure and gross haematuria have been reported (SPELMAN et al 1982, GUESS et al 1982).

Recent stings in Australia (1991–92) caused severe local skin pain and within 20 minutes were followed by a modified Irukandji reaction {see below}. This included localised sweating; cramping muscle pains in all four limbs; cramping abdominal pain; respiratory muscle pain on inspiration (causing an inspiratory 'grunt'); and an anxiety state. Hospital admission was needed for control of pains with systemic analgesia. Inhalation of 'Entonox' (50% nitrous oxide, 50% oxygen) for pain relief (STEWART 1988) was ineffective (FENNER *et al* 1993).

Fatalities

{see Chapter 3, page 64}

Three fatalities from *Physalia* envenomation in the Atlantic, on the south-east coast of the United States, have been documented in the medical literature (STEIN *et al* 1988, BURNETT & GABLE 1989) {see also Chapter 22, page 424}.

The venom

Physalia physalis nematocyst venom is antigenic, haemolytic, dermatonecrotic (Plate 8.47) and potentially or actually lethal to both humans (Plates 8.48A and 8.48B) and smaller animals (Figure 8.55). One fragment of 240 000 molecular weight has been detected, but several other proteins of lower molecular weight, in particular a chick embryo cardiotoxin of 30 000 daltons, are present (TAMKUN & HESSINGER 1981; BURNETT *et al* 1986). The cross-reactivity with other jellyfish venoms has been noted (BURNETT *et al* 1988).

Treatment

Systemic analgesics should be promptly instituted because topical first aid measures (antihistamines, topical anaesthetics, corticosteroids, papain or aluminium suspension preparations) to decrease the severe skin pain are usually ineffective, in the experience of the authors. The local application of heat does not offer pain relief (BURNETT, FENNER, WILLIAMSON 1992, unpublished observations) and carries the possibility of increased venom uptake in the vasodilated tissues. Cold packs as a pain-relieving measure in *Physalia physalis* apparently only postpone the severe skin pain of victims in the United States (BURNETT 1991 unpublished observations), although they are reported as being of benefit to victims with skin pain from specimens of *Physalia* in Pakistan (ALAM 1992, unpublished observations). The toxinology of the Pakistani animal (Figure 8.5, page 145) is currently being studied (ALAM & QASIM 1991) and shows dermatonecrotic action and toxic effects on kidney, liver, and cardiac tissues in albino rats (ALAM *et al* 1992). (See also Chapter 26, pages 452–3.) As mentioned above, from the terminology used in recent reports for the Pakistani *Physalia* sp.(ALAM *et al* 1992 & 1993), it is uncertain whether the studies were conducted on the single tentacled or the multi-tentacled variety.

Systemic collapse should be treated in the usual resuscitative and supportive manner {see Chapter 5, page 114}. Early experimental animal results indicate that verapamil is effective against the cardiotoxin of this animal (BURNETT *et al* 1985). More work is required before human use can be recommended.

PHYSALIA UTRICULUS (BLUEBOTTLE)

PETER FENNER, JACQUIE RIFKIN AND JOHN WILLIAMSON

APPEARANCE

(Plates 8.31 and 8.37). The floats are smaller than those of the Atlantic *Physalia physalis*. Its pale blue float measures from 2 to 100 mm (usually 20–80mm) in length. A single, thickened, dark blue fishing tentacle descends from the centre of the float and extends to a length of more than a metre through the water (WILLIAMSON 1985) (Plate 8.37). The cnidom of this species also includes two sizes of isorhizas {see Plate 5.1}. Nematocysts are either spherical or oval (RIFKIN 1993, unpublished observations) and thick-walled (Figure 8.7, page 198) (see also CLELAND & SOUTHCOTT 1965 and HALSTEAD 1988).

THE STING

Contact with the tentacles can result in the tentacles adhering to the patient, producing a characteristic imprint on the skin. Instant skin pain occurs which may persist for hours.

- The fresh sting is usually a linear raised wheal (Plate 8.50), with prominent blanched, collinear, elliptical 'beading' effect (Plate 8.42). Multiple wheals may occur with mass stinging, which can result when a swimmer encounters 'armadas' of the animals in the water (Plate 8.42). Although this increases the area of skin pain, systemic effects are usually not troublesome in adults.
- Systemic symptoms may occur with extensive stinging in small children and include nausea, vomiting, and tender regional lymphadenopathy (WILLIAMSON 1985).

As described above {page 196}, more severe systemic effects such as breathing difficulty, muscular cramps, and even probable anaphylaxis and near death (GOLLAN 1968) have been reported in the past. However, it can now be speculated that these effects may have resulted from the previously unrecognised multitentacled Pacific man-o'-war.

THE VENOM

To date, only preliminary work has been undertaken on *Physalia utriculus* venom (BURNETT 1993, unpublished observations, in FENNER *et al* 1993).

Venom toxicity comparisons

Two Pacific man-o'-war specimens showed a similar LD_{50} in laboratory mice. Similar comparative LD_{50} results have recently been obtained by ALAM *et al* (1993) in Pakistan, using both single and multiple-tentacled *Physalia* specimens from the Karachi coast in the Arabian Sea (northern Indian Ocean) {see Figure 8.5, page 195 and Plate 8.41, and Chapter 10, page 303}. Identifying specifically a Pakistani '*Physalia utriculus*', these workers correlate Ca^{2+} accumulation (suppressible by verapamil) with the multiple organ histopathology resulting in experimental guinea pigs from injected *Physalia* venom (ALAM *et al* 1993). See also Chapter 26, page 453.

TREATMENT

{See Chapter 5, page 106.} Cold packs are effective as a pain-relieving measure for the skin pain of single-tentacled *Physalia* stings in Australia (EXTON 1989) (Plates 5.16 and 5.17). Systemic analgesics are rarely needed, at least in Australia, because of the effectiveness of this topical first aid measure.

For extensive stings in children with systemic symptoms, simple oral analgesics in appropriate dosage (e.g. paracetamol, soluble aspirin) are adequate.

SUBCLASS HYDROIDA ORDER LIMNOMEDUSAE

Two limnomedusan genera are considered here: genus *Gonionemus* and genus *Olindias*. Recent data indicate that some species from both these genera are of medical significance.

GENUS *GONIONEMUS*

There are two medically significant species of this hydrozoan genus {see Figure 6.5, page 125}, about which information is currently available.

1. *Gonionemus oshoro* UCHIDA, 1929, in Japan (Plate 8.52)
2. *Gonionemus vertens* AGASSIZ, 1862, on the south-eastern (far eastern) Russian coast (Figure 8.8, page 202)

GONIONEMUS OSHORO UCHIDA, 1929, IN JAPAN

The following information, based on Japanese publications, research and observations, has kindly been provided by Dr Yasuhiro Tomihara and colleagues.

SYNONYMS

The Japanese *Gonionemus* spp. include '*Gonionemus vertens*' and '*Gonionemus depressum*', although it is possible that these are one and the same species (ARAKI 1992, unpublished observations). (The proximity of the Japanese and Russian locations for *Gonionemus* raises the additional possibility that these may be the same species in the Sea of Japan — see the note in the text on *Gonionemus vertens*, below.)

DISTRIBUTION

Gonionemus oshoro is known with certainty in Japan from Niigata on the north-west coast of Honshu (latitude 38° north) (OTSURU *et al* 1974) (Plate 8.51), northwards to Hokkaido (SHIN KUSUNOKI 1970 [approx.]). This area forms part of the western shores of Japan, directly opposite the coastline of eastern Russia and Vladivostok.

OTSURU and colleagues were stimulated to study the Japanese animal in the Niigata region because of the numerous and serious summertime stingings (mid-June to late August), sustained by young female seaweed collectors in that region. *Gonionemus oshoro* UCHIDA was subsequently located in numbers among the seaweed *Sargassum confusum* Agardh, which grows plentifully along northern Japanese coastlines.

APPEARANCE

(See Plate 8.52) *Gonionemus oshoro* is a small jellyfish measuring 5–15 mm in diameter, with a circular bell, around the entire edge of which are attached many tentacles. Nematocysts encircle the surface of each tentacle for its full length; nematocysts are also located on the surface of the manubrium (oral lip). The transparent bell displays a symmetrical, right-angled cross (OTSURU *et al* 1974) {Plate 8.52}.

THE STING

Details of the Japanese species will be considered along with the Russian species, as no practical differences between the two species of any medical significance appear to exist.

GONIONEMUS VERTENS AGASSIZ, 1862

ON THE SOUTH-EASTERN (FAR EASTERN) RUSSIAN COAST

YU M. YAKOVLEV AND V. E. VASKOVSKY

Gonionemus is a small hydrozoan 'jellyfish' (Figure 8.8, page 202). World wide, excepting on the eastern Russian coast, it has not been considered to be a particularly toxic medusa (HALSTEAD 1988), but in the waters around Vladivostok in the Sea of Japan, envenomation from *G. vertens* causes distressing systemic symptoms (AZNAURIAN 1958, 1964; YAKOVLEV & VASKOVSKY 1993), somewhat similar to the Irukandji syndrome {see Chapter 9, page 246}. However, there are no reliably described fatal envenomations from *G. vertens* stingings (YAKOVLEV & VASKOVSKY 1993); those previously reported (HALSTEAD 1978) have not withstood close investigation.

A similar hydrozoan, *G. oshoro,* occurs in the seas of the northern rocky coast of Japan on the opposite side of the Sea of Japan to Vladivostok. It too causes an 'Irukandji-like' syndrome (OTSURU *et al* 1974).

The precise taxonomy of the genus *Gonionemus* has been a matter of uncertainty for decades. PICARD (1951) recognised only one species of the genus *Gonionemus*, the various 'species' being variants of a single species. NAUMOV (1955) suggested that the nine species of the genus described in the literature at that time be reclassified into three valid species: *G. vertens* AGASSIZ,

Gonionemus suvaensis AGASSIZ et MAYER, and *Gonionemus gennifora* (Kishinoye) (YAKOVLEV & VASKOVSKY 1993). He united five of the species — *Gonionemus murbachi, G. oshoro, G. vertens, G. depressum* and *G. agassizii* — into a single species *Gonionemus vertens*. More precise identifications are hoped to follow advances in isozyme research (MATSUOKA *et al* 1992).

Active *G. vertens* medusae generally emerge in June in Japan, and during early July around Vladivostok. They disappear in late August or early September, and during the rest of the year they exist as the non-dangerous polyp stage.

Around the Vladivostok coastline *Gonionemus* spp. appear to have a 'periodicity', as hundreds of stings are recorded in some years, but few or no stings in other years (AZNAURIAN 1964, MIKULITCH & NAUMOV 1974; YAKOVLEV & VASKOVSKY 1993). Such periodicity has not been reported with the Japanese *Gonionemus* species (OTSURU *et al* 1974). This periodicity, as well as the clinical syndrome *Gonionemus* produces, resembles that observed with *Carukia barnesi* {see Chapter 9, page 246}).

On 17 July 1966, on Amur Bay beaches (Vladivostok), more than 1000 people were stung by *Gonionemus* sp. (MIKULITCH & NAUMOV 1974). On one Saturday in 1986, 76 patients were similarly stung and required admission to a toxicological centre and hospital (YAKOVLEV & VASKOVSKY 1993). In 1991, 1992 and 1993 there were generally fewer such medusae stings in the vicinity of Vladivostok: respectively, 74, 52 and 95 stung patients were received at the toxicological treatment centre.

These medusae dwell at shallow depths of 0–10 m in quiet, sandy bays, usually hanging on to seaweeds and grasses by suckers which are

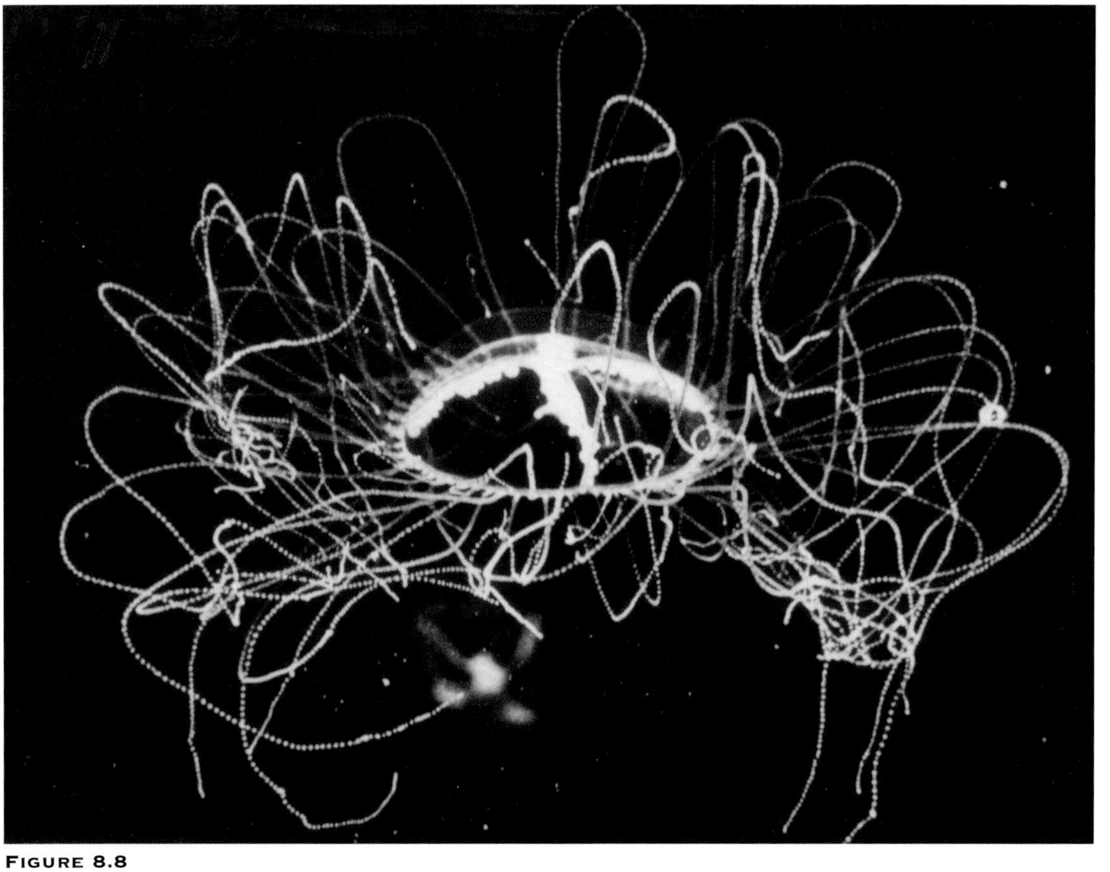

FIGURE 8.8
The medusa *Gonionemus vertens* AGASSIZ 1862 of Far Eastern Russian seas, photographed by A. Golubev at a depth of 1.5 m in the Sea of Japan near Vladivostok.

attached to the tentacles {Figure 6.5, page 125}. They become active on calm, warm days, usually early in the morning or in the evening, when they detach themselves and swim in search of food. The tentacles and the bell are contracted and they rise to the surface. Then they turn over with the exumbrella facing down, unfurl the tentacles (Figure 8.8) and descend slowly {cf. the rhizostome jellyfish *Cassiopea* sp., page 212}. It is at this stage that stings to humans often occur. (Because of this behaviour AGASSIZ (1862) gave the medusa the species name '*vertens*'). MILLS (1983) observed that the medusa was much more active at night.

Gonionemus nematocysts inject their venom in the usual way (PIGULEVSKY & MICHALEFF 1969) {see Chapter 6, page 155}. WESTFALL (1970), who studied nematocysts in *G. vertens*, suggested discharge was controlled by a nervous system {see Chapter 6, page 157}.

DISTRIBUTION

Gonionemus spp. are found in a number of regions of the world (EDWARDS 1976); they have a wide distribution in the northern hemisphere oceans (TAMBS-LYCHE, 1964).

On the Asian side of the Pacific Ocean they occur northwards from Zhejiang (ZHENG *et al* 1989) through the Sea of Japan to the Olga Bay, along the coasts of northern Japanese islands and southern Kurile Islands and southern Sakhalin (NAUMOV 1960, ZELICKMAN 1976). Along the west coast of North America they occur from the Aleutian Islands (NAUMOV 1960; BRINCKMANN-VOSS 1974; MILLS 1983) to southern California (TODD *et al* 1966). In European waters they occur in the Mediterranean Sea and north along the Atlantic coast (KRAMP 1959), in warmer years reaching Norway (GULLIKSEN 1971). Other species of the genus *Gonionemus* inhabit the tropical zone of the Atlantic, Pacific and Indian oceans (WEILL 1938; KRAMP 1959, 1968 & 1970).

Representatives of this genus toxic to humans apparently have been encountered only in eastern Russia, around Vladivostok (HALSTEAD 1988, MIKULITCH & NAUMOV 1974), and along the northern shores of Japan (UCHIDA 1929; OTSURU *et al* 1974). MIKULITCH and NAUMOV (1974) separated these toxic medusae into a toxic subspecies, *Gonionemus vertens vertens*, and refer the apparently non-toxic Atlantic medusae to the subspecies *Gonionemus vertens murbachi*.

APPEARANCE

The bell of *Gonionemus vertens* is usually less than 20 mm in diameter, although individuals of up to 40 mm in diameter have been found (IVANOV & STRELKOV 1949; RUSSELL 1953). The bell of *Gonionemus oshoro* has a diameter of only 5–15mm (OTSURU *et al* 1974) {Plate 10.12}.

The bell of both species is transparent, sometimes light greenish-yellow in colour. It is edged with up to 90 tentacles, each having suckers in the medium or distal portion {see Figure 6.5, page 125} (suckers are unusual for medusae). Throughout the length of the tentacle are ridges with stinging cells.

Four fine radial canals are often visible (Figure 6.5, page 125, and Plate 8.52), which form a noticeable cross (the common Russian name for the animal is 'krestovik', which means 'cross-shaped'). The gonads are yellow or red-brown and they are arranged along most of the radial canals in *G. vertens*. The number of statocysts usually equates with the number of tentacles present.

Some of the major distinguishing features of *Gonionemus vertens* (AGASSIZ, 1862) are detailed above and in Chapter 6, page 140.

ENVENOMATION

The typical syndrome caused by the medusa was first described by a Vladivostok physician, A.E. Bury (BURY 1922). It was later described in more detail by AZNAURIAN (1958), who included sting statistics (cf. the Irukandji syndrome {page 246}):

'Thirty-seven patients were treated between the months of July and September, during the years 1949–1957, who were stung in the waters off Vladivostok. All patients were males aged 20–24. A burning pain occurred on envenomation, followed a few minutes later by erythema covered sometimes with vesicles. The erythema lasted 2–3 days, then faded to scarcely visible pigmented zones. Eleven stings were on the hands, thorax 3, breast 3, face 1, legs 7, genital area 7, neck 3. Two patients had no visible sting marks.

Fifteen to 20 minutes after the initial envenomation the symptoms were: toxicity, weakness, fatigue, low crying with pain, and thrashing around with pain in the lower back, lower legs, joints and stomach. The distribution of the symptoms among the patients was:

> 'General fatigue and lower body pain 29, low back pain 15, cough 32, chest pain 21, abdominal pain 11, headache 1, chill 5, spasm 3, vomiting 2, throat pain 1, rash 1. Seventeen were classified as "heavy", 16 "middle" and 4 "light". Twenty-four patients were pyrexic during 2–3 days (from 37.2 to 38.5°C). Twelve had a leucocytosis (from 8500 to 20 000). Eosinophils were low or absent in "heavy" group. 31 had lymphopenia, 19 had moderate monocytosis. All had a normal ESR. Urinalysis showed transient albuminuria and microscopic haematuria mainly in 'heavy' group. Only 1 patient had bradycardia. Blood pressure remained normal; there were no other cardiovascular changes. The patients had hard hurried breathing, some of them had râles. There were no disturbances in defaecation or urination. Reflexes may have been increased.'

(*Translated from Russian by the authors, YuMY and VEV*)

The syndrome was also described by PIGULEVSKY and MICHALEFF (1969):

> 'The toxin of *G. vertens* causes pain and other localised symptoms, although neuropsychiatric symptoms may also occur. Sting marks are most often observed on the buttocks, chest, abdomen and legs. At the moment of stinging, there is a burning sensation with indistinct swelling and hyperaemia. Thirty to forty minutes after envenomation, patients develop flabbiness accompanied by acute muscular and joint pains. Relation of the upper and lower limb muscles can lead to the loss of the ability to move or even stand. Many reflexes are weakened. There is localised burning or, reversely, coolness on the skin. Pains in the fingers and toes may increase. Patients are usually excited, restless, failing to find a comfortable posture. To all appearances, envenomation subsequently involves the central parts of the nervous system and the clinical picture is changed. Patients develop pains in the internal organs, convulsive muscular contractions. Stupor and knee-jerk reflexes can occur. At this time, temporary weakening of sight and hearing, depression and hallucinations are possible.'

Extensive material from clinical records of *Gonionemus* stings is provided by the works of MIGAS and POPOVA (1988). This material was gathered at the Toxicological Centre in Vladivostok — the major medical unit at Primoriye, where patients with the medusa stings are treated. Over the period from 1974 to 1986, 648 cases of stings were observed (427 males and 221 females aged from three to 60 years). Three forms of intoxication are distinguished: painful, respiratory and mixed.

> 1. 'In the painful form (37 per cent of cases), the symptoms are severe muscular, joint, chest and loin pains which persist for 2–3 days. There is general motor anxiety. No deviations from the norm are observed in the internal organs.
> 2. In the respiratory form (44 per cent), patients develop allergic catarrhal symptoms in the upper respiratory tract. Besides, there is skin hyperaemia, rhinitis, lachrymation, hoarseness, cough, dyspnoea. These symptoms persist from several hours to 2 days.
> 3. The mixed form was observed in 19 per cent of patients. In addition to severe joint and muscular pains, patients have cough, bronchospasm, tickling in the throat, rhinitis and lachrymation. Some patients have tachycardia (up to 100 beats p.m.) and increased arterial pressure (up to 150/100 mm Hg).'

MIGAS and POPOVA (1988) also described different degrees of clinical severity.

> 'In mild cases only the skin is affected at the sting site. Moderately affected patients develop the above described forms of intoxication. Severe stings documented by physicians are accompanied by psychic dysfunction. Over the recent 10–13 years severe stings have not been reported.'

The venom does not appear to affect routine laboratory animals, although the sting can kill frogs (BREKHMAN & MINUT-SOROKHTINA 1951). PAVLENKO and colleagues (1990, unpublished observations) isolated toxic high-molecular-weight glycoproteins containing lipids. BERDYS-

HEV and colleagues (1991) suggested that a platelet activating factor may play a significant role in the toxicity of the venom from this medusa.

In Japan the stings mainly occur underneath bathing costumes. The syndrome is described (OTSURU *et al* 1974) as causing:

> 'severe pains in the various groups of muscles, oppression of chest, coughing, lachrymation, serious rhinorrhea, nausea and vomiting etc... In general recovery occurred in 1–2 days. In severe cases the victims were sent to hospital with severe pains and dysphasia and had to receive approximately one week's medical attention.'

It is of interest that a similar syndrome occurs in northern Japan following ingestion of a raw seaweed, *Nemacystus decipiens* Suringar. As *Gonionemus oshoro* lives among this seaweed it is presumed the syndrome results from ingestion of the hydrozoan (OTSURU *et al* 1974). Anaphylaxis is also suggested by Otsuru, on clinical grounds only, in some of his cases.

FIRST AID

{See also Chapter 5, page 106.} Russian sources suggest:

Assist the victim to the shore, wash the sting site with fresh water and then wet with an alcohol solution. The patient should be referred to the nearest medical unit. Rest and high fluid intake are recommended (MIGAS & POPOVA 1988).

Other remedies suggested in 'folk lore' include drinking alcohol(!) (AZNAURIAN 1964), and application of raw potato to the sting site. There is no reliable evidence for the effectiveness of the latter method.

WESTFALL (1970) studied the cnidom of *Gonionemus vertens*. No published controlled studies on nematocyst discharge inhibition for *Gonionemus* spp. are known.

MEDICAL MANAGEMENT (FROM RUSSIAN SOURCES)

Original treatment was systemic 5% ephedrine and antihistamines (AZNAURIAN 1958). According to more recent 'Instructions on rendering medical assistance to persons stung by a poisonous krestovik-medusa from the Sea of Japan' (MIGAS & POPOVA 1988), the following medications are used, irrespective of the severity of sting:

1. Antihistamine drugs: pipolphen (2.5%, 2 ml); chloropyramine (2%, 2 ml); or diphenylhydramine hydrochloride (1%, 2 ml) by the intramuscular route.

2. Sedative drugs: 'relanium' (local name) 0.5%, 2 ml either intramuscularly or intravenously (in 10 ml of 40% glucose).

Adrenergics (ephedrine and adrenalin) are not indicated.

Patients receiving such medical treatment at hospital are usually discharged without complication in two or three days.

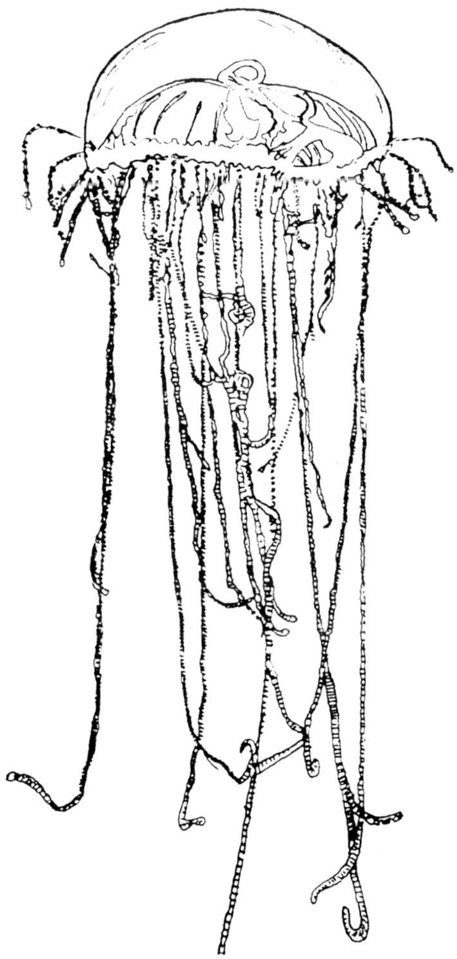

FIGURE 8.9
Drawing of a living specimen of *Olindias singularis* BROWNE, 1905, captured at Seacliff, Australia, in 1963. The bell diameter is about 1.5cm. (After CLELAND & SOUTHCOTT 1965.)

PREVENTION OF STINGS

Similar to Australian warnings on *Chironex* {see Chapter 5, page 103}, are Russian warnings:

> 'To date, the only reliable way to avoid stings from *Gonionemus vertens* has been not to bathe in the areas of its concentration.'

Hence, it is important to keep the public informed about the appearance of the medusae. In recent years, information leaflets have been posted up in the recreation zone of Vladivostok, and warnings have been made on TV and in the local press. These measures appear to have some effect. For example, in the Vladivostok area alone, 293 ambulance calls were recorded in 1987, 214 in 1988 and merely 163 in 1989, the year when *Gonionemus* abundance was greatest. (During 1990–1993 there were practically no medusa in the vicinity of Vladivostok.) If no reliable information on the development of *Gonionemus* is available, bathers are recommended to avoid places with growth of sea grasses and seaweeds.

Neither the Russian authors nor the editors of this book are able to find documented details of first aid or medical treatments for stings from the Japanese medusa.

GENUS *OLINDIAS*

{See also Chapter 6.}

Three species are considered below:

1. *Olindias singularis* BROWNE, 1905
2. *Olindias sambaquiensis* MÜLLER, 1861, along the Atlantic coast of South America
3. *Olindioides formosa* Goto, 1903

OLINDIAS SINGULARIS BROWNE, 1905

PETER FENNER

DISTRIBUTION

Indian Ocean, Pakistan, Iranian Gulf, Philippines, north Queensland (KRAMP 1961), South Australia (CLELAND & SOUTHCOTT 1965).

APPEARANCE

The bell is circular and saucer-like, with a diameter of 13–36 mm; the bell height is usually half its diameter. There are between 28 and 86 primary tentacles hanging from the bell rim, each with rudimentary adhesive clubs at their tips. Secondary tentacles numbering 16–50, all longer than the primary tentacles, are also present. There are 4–12 centripetal canals in each quadrant. A single statocyst is situated at the origin of each primary tentacle. Gonads are present on the radial canals (KRAMP 1961; CLELAND & SOUTHCOTT 1965) (Figure 8.9, page 205). {See also Chapter 6, page 141 and Figure 6.18, page 140}

ENVENOMATION

Southcott described a mildly painful sting to the back of the hand on his daughter in 1962 on a beach near Adelaide, South Australia (CLELAND & SOUTHCOTT 1965). The wheal was short and narrow with local erythema. Local stinging lasted some 30 minutes, and the wheals lasted up to two hours. He considered it to be a minor sting and suggested that many stings frequently attributed to 'sea lice' might in fact be *Olindias* {See Chapter 11, page 309}.

FIRST AID

No data exists. However, the mild skin pain could be expected quickly to respond to cold packs {see Chapter 5, page 00, First Aid}.

OLINDIAS SAMBAQUIENSIS MÜLLER, 1861

HERMES W. MIANZAN AND FERNANDO C. RAMÍREZ

Olindias sambaquiensis MÜLLER, 1861 (class Hydrozoa, order Limnomedusae, family Olindiadidae) (Plate 8.53) is a neritic jellyfish endemic in the area south of Buenos Aires, Argentina (Figure 8.10, page 207) (KOKELJ *et al* 1993). Common names for this jellyfish include *agua viva* in Argentina and Uruguay, and *mae d'agua* and *relojinho* in Brazil.

LIFE CYCLE

The species has a conspicuous medusa stage. The release of juveniles occurs during October, with individuals of 0.55 mm size, which reach 70 mm size during February and March. About this time the gonads become mature and release the gametes. The medusa then becomes smaller, losing the manubrium and other structures before dying. A few individuals survive during autumn and winter months (MIANZAN & RAMÍREZ 1994, unpublished observations). In the family Olindiadidae the polyp form is very primitive and rudimentary (NAGAO 1973; ZAMPONI 1987), making it difficult to find. The documentation of its life cycle is in preparation (MIANZAN & RAMÍREZ 1994, unpublished observations).

DISTRIBUTION

The animal occurs in South American sub-tropical and temperate coastal waters from latitude 23° to 42° south. It is common along the south-west Atlantic Coasts of South America in Brazil, Uruguay and Argentina (see Figure 8.10) (VANNUCCI 1951, VANNUCCI & TUNDISI 1962; KRAMP 1970, MIANZAN 1989).

Stings cause problems for tourists swimming at southern localities of the Buenos Aires district (MIANZAN 1988; MIANZAN & RAMÍREZ 1994, unpublished observations). Official records show 500–1000 bathers are stung in the Monte Hermoso area each season, and many others are probably not recorded. This problem occurs along the whole coastal area from Monte Hermoso to Necochea (Figure 8.10).

SEASONAL OCCURRENCE

Though originally described as abundant in Brazil during winter (VANUCCI 1951), recent records show the species is common during spring and summer in Uruguay and Argentina (MIANZAN & RAMÍREZ 1994, unpublished observations) and in southern Brazil (MIANZAN 1989, unpublished observations). *Olindias sambaquiensis* currently swarms annually in Argentina during January and February.

Climatic influence on its distribution

On the Monte Hermoso coast of Argentina (Figure 8.10), during continental north winds the chance of finding specimens of *Olindias* in shallow water increases by up to 65 per cent (MIANZAN & ZAMPONI 1988).

APPEARANCE

Olindias sambaquiensis is a conspicuous jellyfish with a flat bell, usually up to 10 cm diameter; a 21 cm bell diameter has been observed (MIANZAN 1991, unpublished observations). The non-coloured, transparent umbrella has 60–70 reddish blind radial channels. When the gonads are ripe they are yellowish and easily

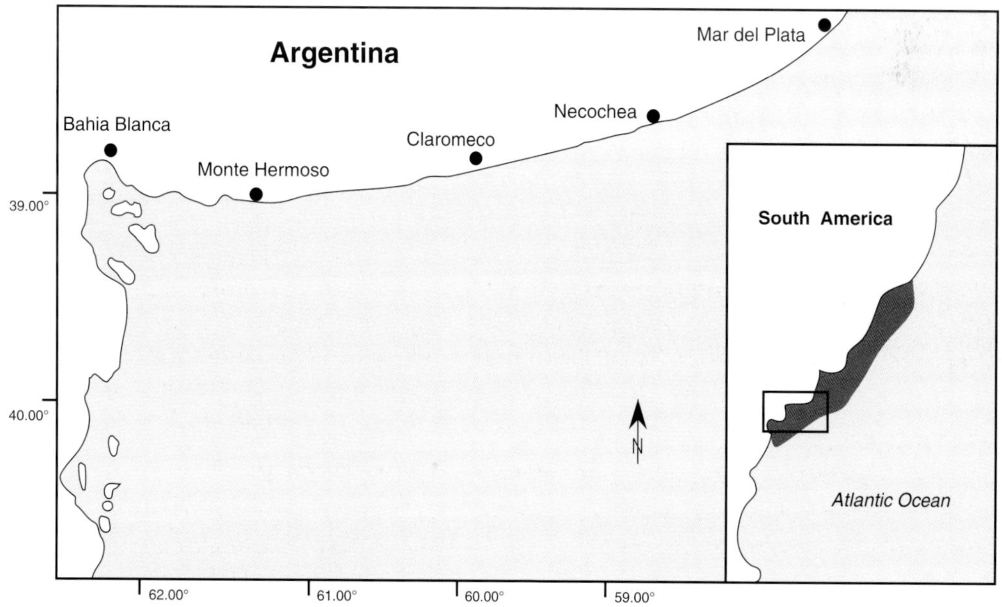

FIGURE 8.10
Distribution of *Olindias sambaquiensis* in the south-western Atlantic Ocean, along the coasts of Brazil, Uruguay and Argentina (shaded area of inset) and occurrence of numerous human stingings. The coast from Monte Hermoso to Necochea is particularly affected.

seen through the umbrella. Each medusa has about 380 tentacles of two types: primary tentacles are short and reddish brown; secondary tentacles are up to 50 cm long, thin, and yellow or pink coloured (KRAMP 1961) (Plate 8.53).

THE STING

Stings usually occur on the back and extremities of human victims. Instant burning skin pain occurs followed by the typical appearance of red zig-zag lines, or irregular white welts with a red surrounding (Plate 8.54). Pain can persist for several hours. In severe stings, fever and difficulty with breathing may be observed (KOKELJ *et al* 1993). Clinical skin signs can persist for up to a week (Plates 8.55A & 8.55B). There have been no deaths reported to date.

FIRST AID AND MEDICAL MANAGEMENT

In Argentina the methods used are to wash off with seawater any adherent tentacle {see Chapter 5, page 110}, and to apply a solution of 20% ammonia in water, to reduce the initial pain. However, this appears to afford very poor results (MIANZAN & RAMÍREZ 1994, unpublished observations) {see Chapter 5, page 109}. In patients who develop systemic symptoms (e.g. dyspnoea, fever) the medical practice in Argentina is to inject dexamethasone, with usually good results.

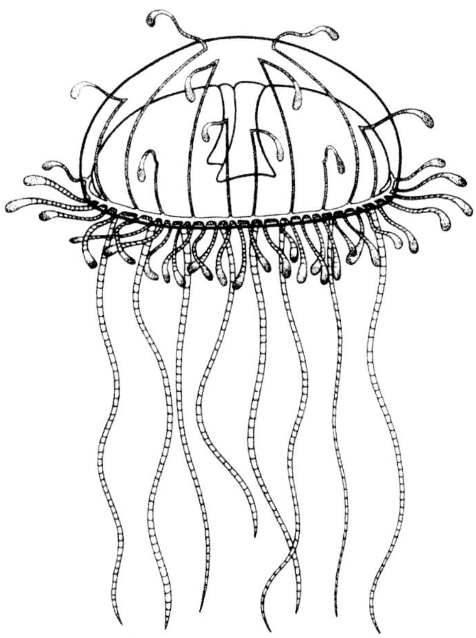

FIGURE 8.11
Drawing of *Olindioides formosa*, Goto, 1903. (From Halstead 1965, courtesy United States Office of the Assistant Secretary of Defense, Washington, DC.)

CURRENT ARGENTINIAN ACTIVITIES AND RESEARCH NEEDS

- The development of mechanical barriers to prevent entry of jellyfish into swimming areas (BURNETT *et al* 1994) {see, for example, Plate 1.19}
- The search for more effective medical methods to relieve the initial effects of the sting.
- A description and assessment of the polyp stages of *Olindias sambaquiensis* and their distribution, leading to possible methods of controlling development.
- Ecological investigation related to the impact of the species on the zooplankton, including the pelagic stages of benthic and nektonic communities.

Another cnidarian envenomation documented from Argentina (Playa Bristol, Mar del Plata) involved probably *Lychnorhiza lucerna* (spelling according to KRAMP 1961); the jellyfish was not positively identified (MÅNSSON *et al* 1985). This envenomation on both thighs and one wrist of a 27-year-old female produced severe acute localised and generalised symptoms, followed one week later by a spontaneous local reappearance of painless but intensely pruritic erythematous lesions at the sting site. These took a further week to resolve {see Chapter 7, page 177}.

Acknowledgments: The authors wish to thank Mr V. Dimartino (National Science Museum, Monte Hermoso, Argentina).

OLINDIOIDES FORMOSA GOTO, 1903

A report of serious systemic effects from envenomation by this Japanese species (HALSTEAD 1988) (Figure 8.11), appears in the German literature from AOKI (1922, written in Japanese). Halstead quotes FAUST (1924) as another source. STRONG (1942) and again CLELAND and SOUTHCOTT (1965) refer to Aoki's report. The editors have been unable to find more recent documentation or first-hand information concerning this species.

PLATE 1.1
A jellyfish, possibly a *Pelagia noctiluca*, illustrated in the mosaic floor of the Villa di Piazza Armerina in central Sicily. This was the home of a Roman magistrate in the 4th century AD. (From Figure 157 in, A. Carandini, A. Ricci, M. De Vos. 1982. Filosofiana la Villa di Piazza Armerina, immagine di un aristocratico romano al tempo di Costantino. Palermo. S F Flaccovio)

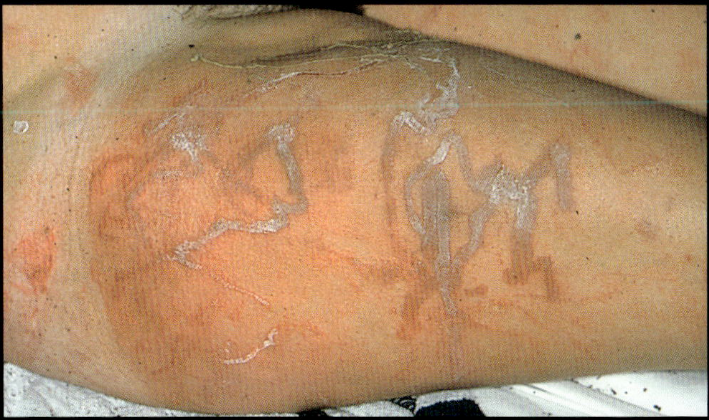

PLATE 1.2
Massive *Chironex fleckeri* sting on the lateral thigh of a young woman (with hair-free skin), who was sitting in knee-deep water close to the shore (photographed 30 minutes after envenomation).

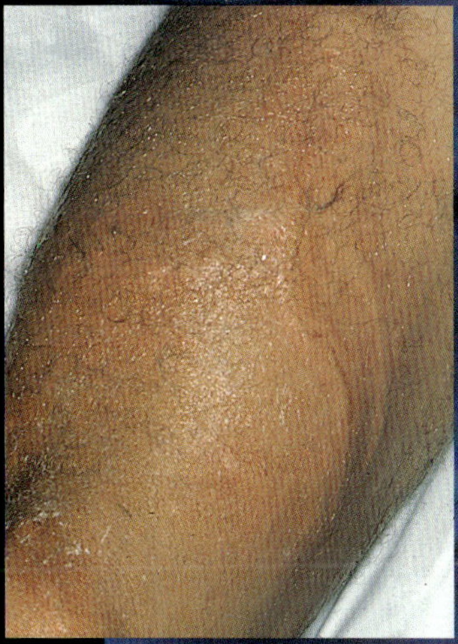

PLATE 1.3
Thigh of the husband of the patient in Plate 1.2, stung by the same animal at the same time, showing the protective effect of his skin hair which prevented the tentacles making intimate contact with the skin (photographed 30 minutes after envenomation).

PLATE 1.4
The new species, named *Chironex fleckeri* in 1956 by Dr Ron Southcott, after its discoverer, Dr Hugo Flecker. (Photograph courtesy of Dr Bob Hartwick, James Cook University of North Queensland, Townsville).

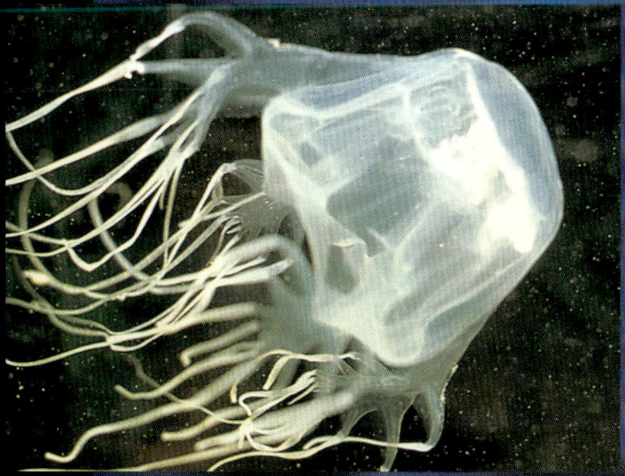

PLATE 1.5
Dr Jack Barnes OBE, a doyen of marine envenomation researchers, at work in his laboratory, which he built under his house in Cairns.

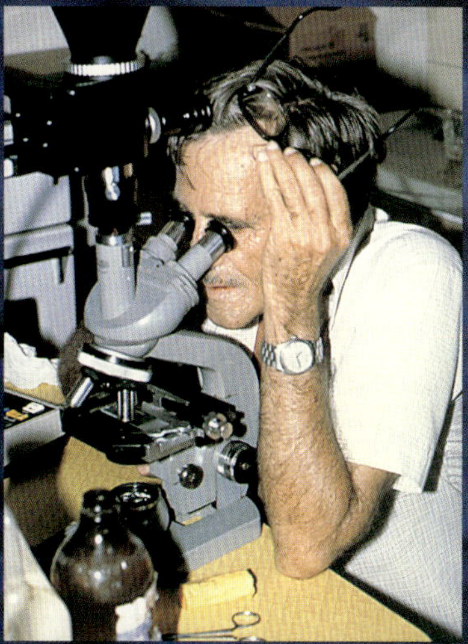

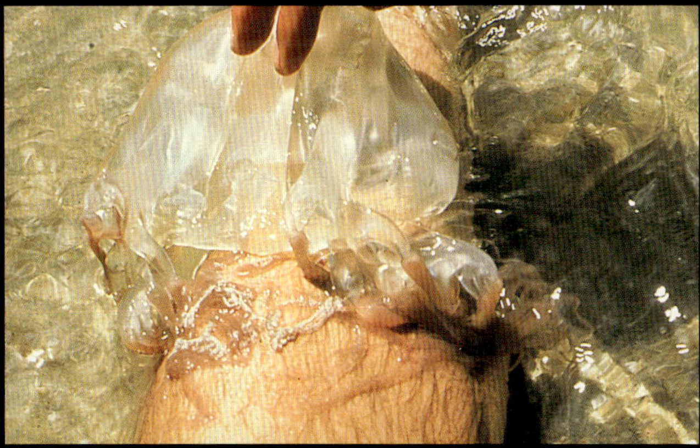

PLATE 1.6
Dr Jack Barnes demonstrating on himself the protective effect of pantyhose against stinging from live adult *Chironex fleckeri* tentacles. Dr Barnes' thigh is covered with a single layer of pantyhose.

PLATE 1.7
Protective pantyhose worn by north Queensland life savers on patrol in 1978.

PLATE 1.8
Life savers wearing pantyhose, checking the water with nets.

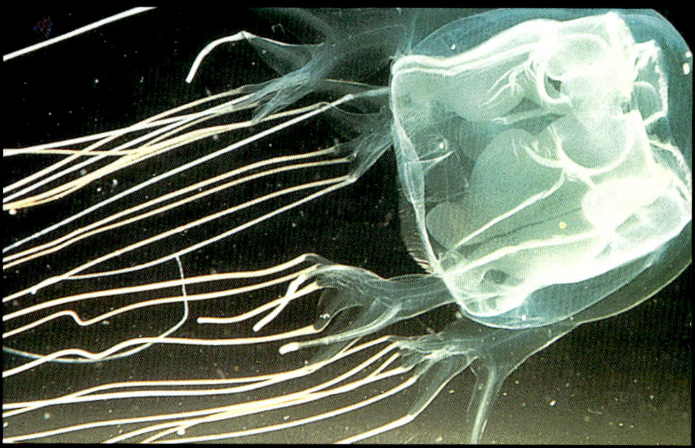

PLATE 1.10
Chiropsalmus quadrigatus (Australia), the species originally described by BARNES (1965). The taxonomy of chirodropids is being reconsidered {see page 150}.

PLATE 1.9A
A replica of Dr Jack Barnes's vacuum flask, sealed across the top with stretched human amnion, and wired to conduct an electric current through the membrane.

PLATE 1.9B
Dr Barnes then lowered the undisturbed fresh tentacles of a living *Chironex fleckeri* onto the membrane.

PLATE 1.9C
After cutting the tentacles free, Dr Barnes passed a current through them on the amnion, causing massive discharge of tentacular nematocysts. Venom was discharged through the membrane into the flask, to which a small suction was applied. The venom thus collected in the flask was snap-frozen.

PLATE 1.11
Second edition (September 1975) of Maurie Mulcahy's booklet, *The Ten Don'ts*, one of the earliest stinger publications by Queensland Surf Life Saving.

PLATE 1.12
First edition of *The Marine Stinger Book* (1974), then known as *Some Australian Marine Stings and Envenomations*.

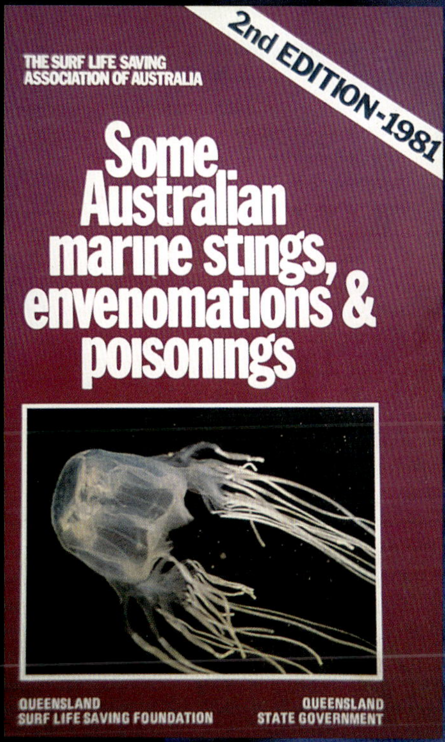

PLATE 1.13
Second edition of *The Marine Stinger Book* (1981), then known as *Some Australian Marine Stings, Envenomations and Poisonings*.

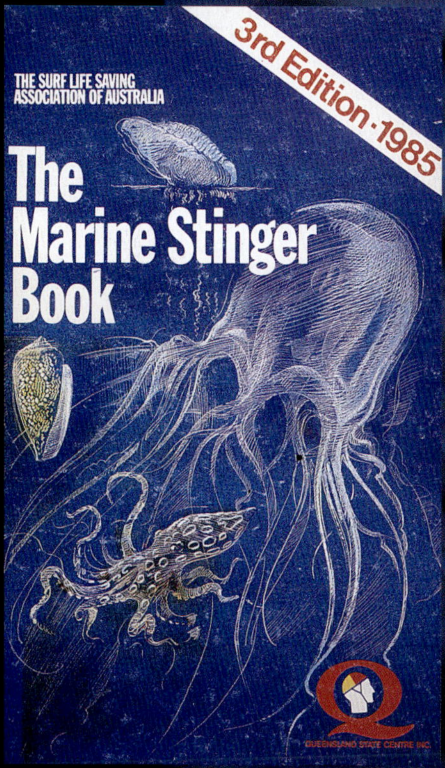

PLATE 1.14
Third edition of *The Marine Stinger Book* (1985), known as just that!

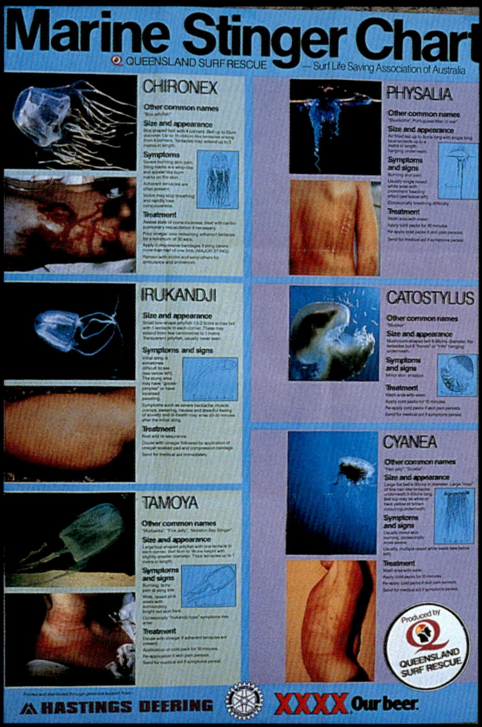

PLATE 1.16
Dr Peter Fenner's 1990 'Marine Stinger Chart' to assist with identification of the animals.

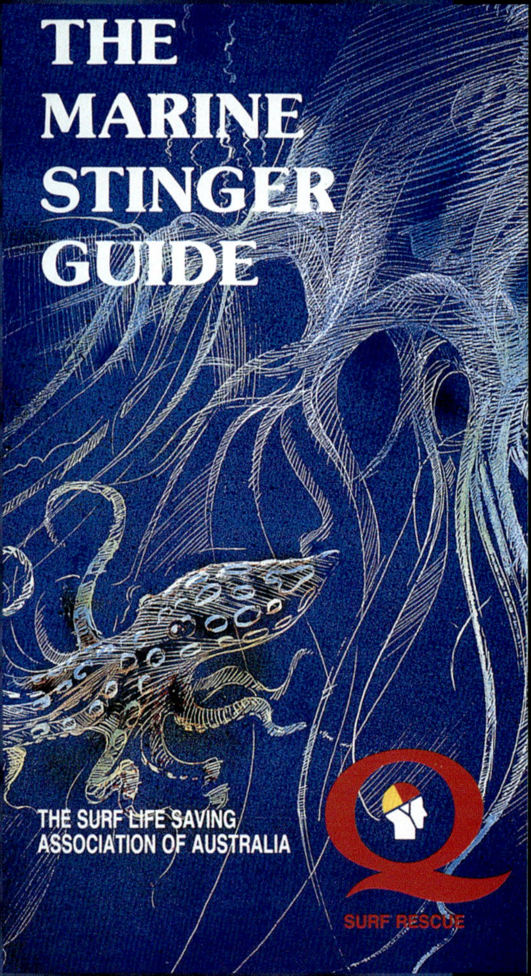

PLATE 1.15
Dr Peter Fenner's *The Marine Stinger Guide*, first edition, 1990.

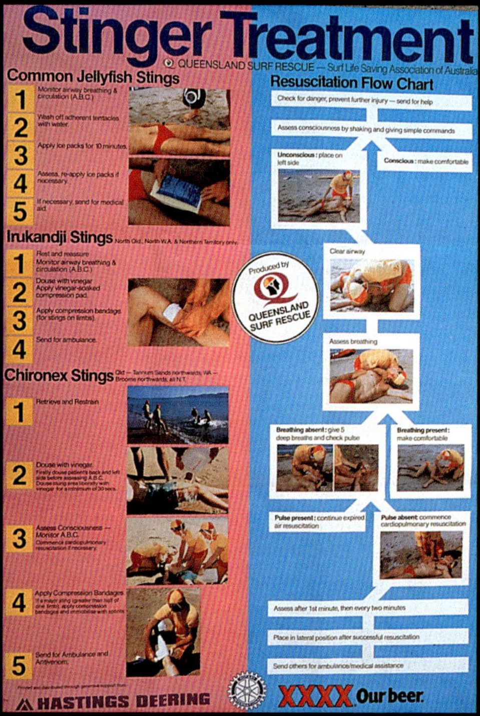

PLATE 1.17
Dr Peter Fenner's 1990 'Stinger Treatment' Chart.

PLATE 1.19
Modern stinger-resistant enclosure with a mesh net of finer gauge.

PLATE 1.20
Winch system for beaching the stinger enclosure for cleaning.

PLATE 1.21
The first Townsville stinger suit (on the left!).

PLATE 1.18
An early stinger-resistant enclosure.

PLATE 1.23
The colour-specific stinger suits of Surf Life Saving Queensland.

PLATE 1.22
The modern stinger suits in action. Their value for the protection of women and children cannot be over emphasised.

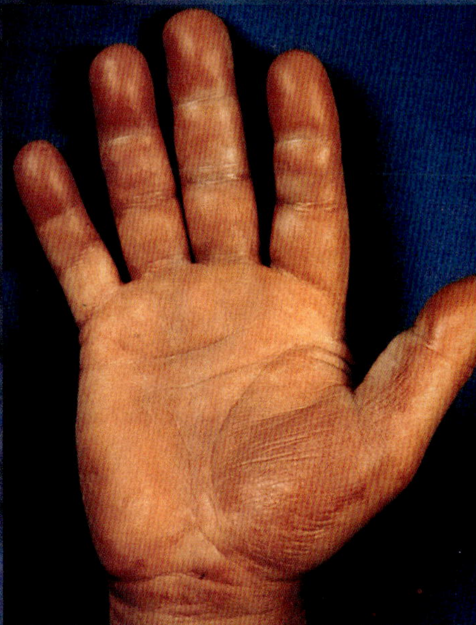

PLATE 3.1
Hand of a scuba diver who handled a toxic sponge underwater while wearing only one glove, showing general oedema of the unprotected hand, with dusky discolouration of all fingertips, indicating mild vascular compromise. The gloved hand was unaffected. (Photo taken during recovery on short-term systemic steroid therapy, some five days after the envenomation.)

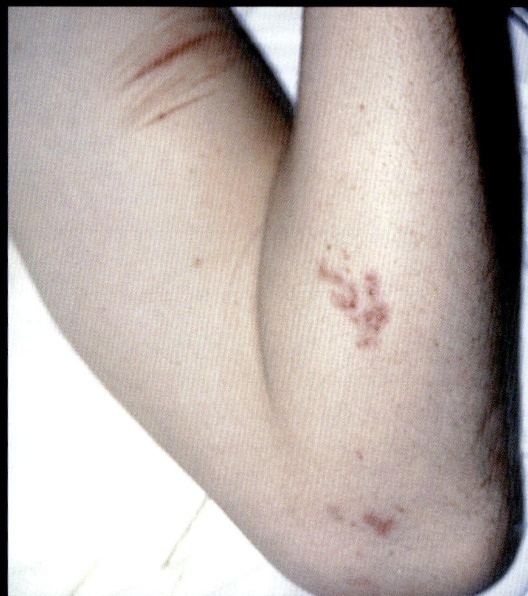

PLATE 3.2
Two-day-old stings from a bristle worm on the left forearm and elbow of a young snorkeller in north Queensland. Note the healing coral abrasions on the upper arm. (Photograph courtesy of Dr Kevin King, Mater Hospital, Townsville)

PLATE 3.3
The salmon-pink tropical bristle worm *Eurythoë complanata*. Also descriptively called 'fireworm'.

PLATE 3.4
A aggressive puffer fish (*Torquigener pleurogramma)* competing with the photographer for perceived food under rocks. (Photograph taken in 2–3m of water at Cockburn Sound, Western Australia, by Karen Gowlett-Holmes, Collection Manager, Marine Invertebrates, South Australian Museum, Adelaide).

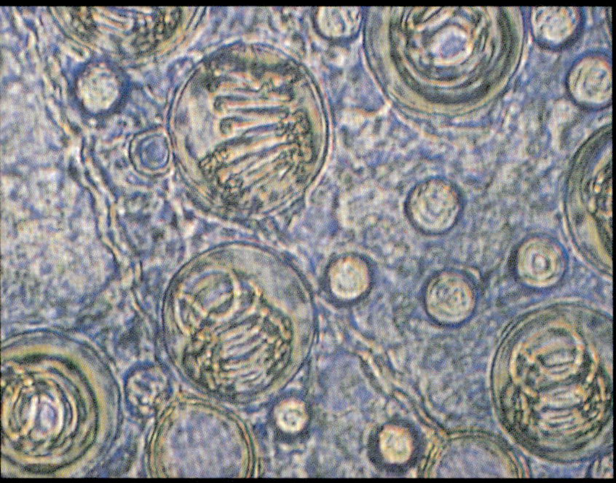

Plate 5.1
Microsection showing nematocysts of a Pacific *Physalia* sp.

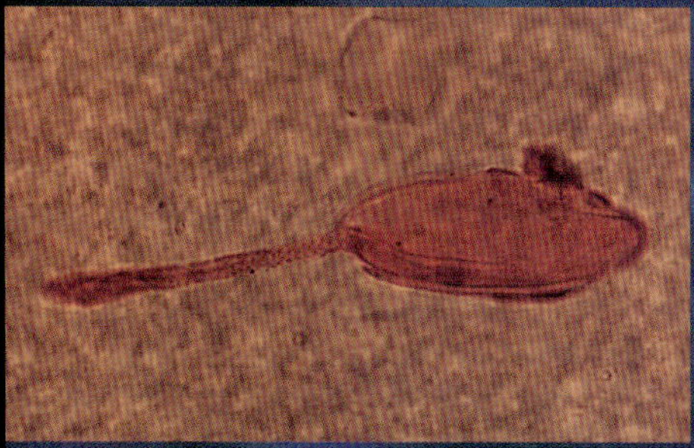

Plate 5.4
A partially discharged nematocyst taken from the skin of a sting by the 'Darwin carybdeid' (prepared by the K. A. P. Lee technique, described in the text).

Plate 5.2
A fresh skin scraping from an 'Irukandji' sting in the Northern Territory, Australia, prepared according to the technique of Dr K. A. P. Lee of the Royal Darwin Hospital, showing in centre field a completely discharged nematocyst from the bell of *Carukia barnesi*. Part of the everted tubule is missing. Note the more spherical shape of the bell nematocyst, compared with the more elliptical shape of the nematocyst shown in Figure 5.4.

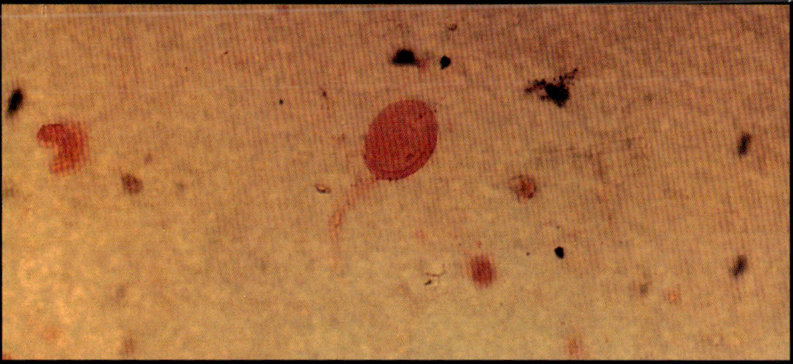

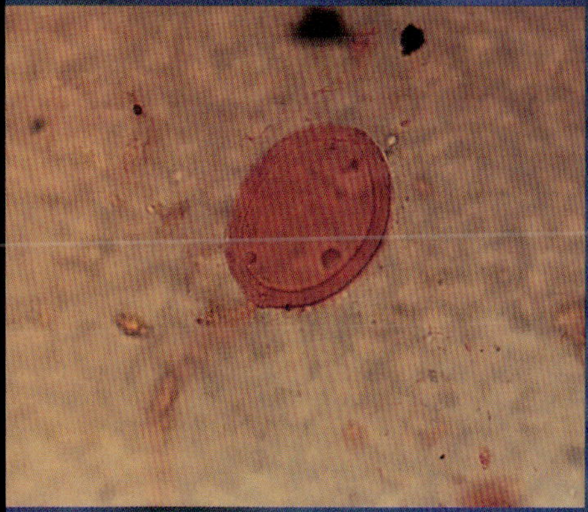

Plate 5.5a
A single discharged nematocyst from the Northern Territory sting in Plate 5.5b, which produced a clinical syndrome resembling an Irukandji syndrome. Compare with Plate 5.2.

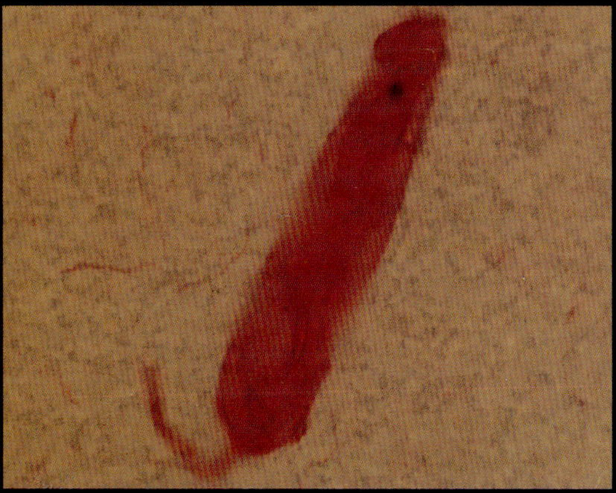

Plate 5.3
A single, partially discharged, microbasic mastigophore nematocyst taken from the skin of a mild *Chironex fleckeri* sting in the Northern Territory and prepared according to the technique developed by Dr K. A. P. Lee of the Royal Darwin Hospital (see text).

Plate 5.5b
Appearance of the upper arm sting (which was followed by an Irukandji-like syndrome), on the patient from whom the nematocyst in Plate 5.5a was obtained. Note the bell-like imprint possibly left by the nematocyst-bearing body of the animal. The size of this imprint is larger than that of the usually described Irukandji carybdeid (e.g. *Carukia barnesi*).

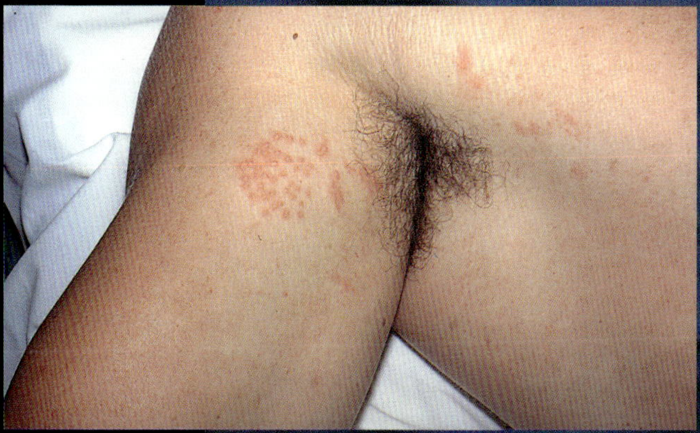

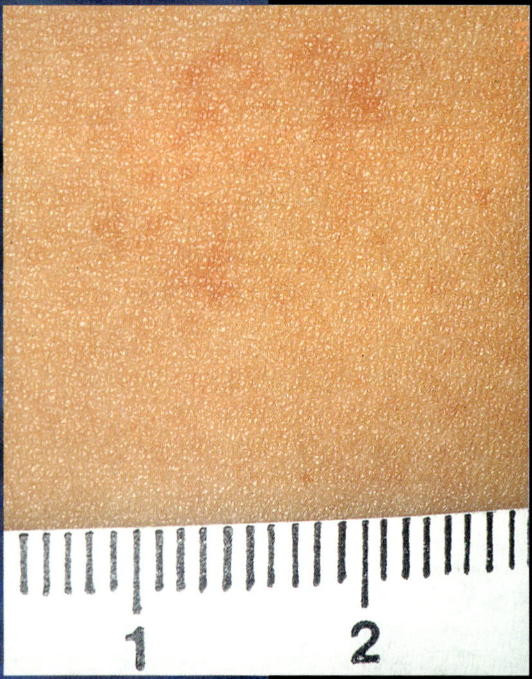

PLATE 5.6
The more diffuse, erythematous bell mark of a small carybdeid, more typical of the size (1-2 cm diameter) of the 'Irukandji' carybdeid. This sting occurred on the abdomen of a three-year-old child in Darwin in 1991. Compare with Plate 9.24.

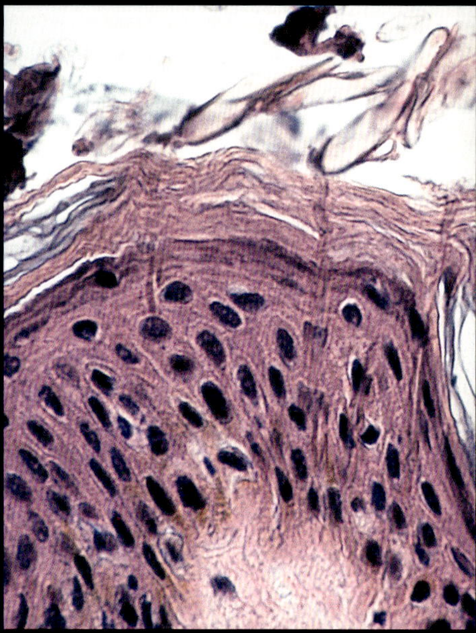

PLATE 5.7
Skin biopsy taken after a fatal *Chironex fleckeri* sting in Queensland (LUMLEY *et al* 1988). Note the empty nematocyst capsules anchored to the keratin layer by the penetrating discharged tubules. (Compare nematocysts with Plate 5.3)

PLATE 5.9
North Queensland life savers on beach patrol against stingers. Note the 'spotter' in the tower with field glasses, viewing the clear inshore waters for any suspicious signs of *Chironex fleckeri* jellyfish moving about. The inflatable rescue boat (IRB) in the foreground is used for systematic spotting beyond the swimming line, before the beach is opened to the public. The seated lifesavers, ready for an immediate rescue, are wearing protective pantyhose over their legs.

PLATE 5.8
A large adult *Chironex fleckeri* occupies the centre field of the photo. Even under such 'ideal' conditions for spotting (clear sunlit calm seawater), the animal is still difficult to see. (Courtesy of Dr Bob Hartwick, James Cook University of North Queensland.) (See also Plate 9.56.)

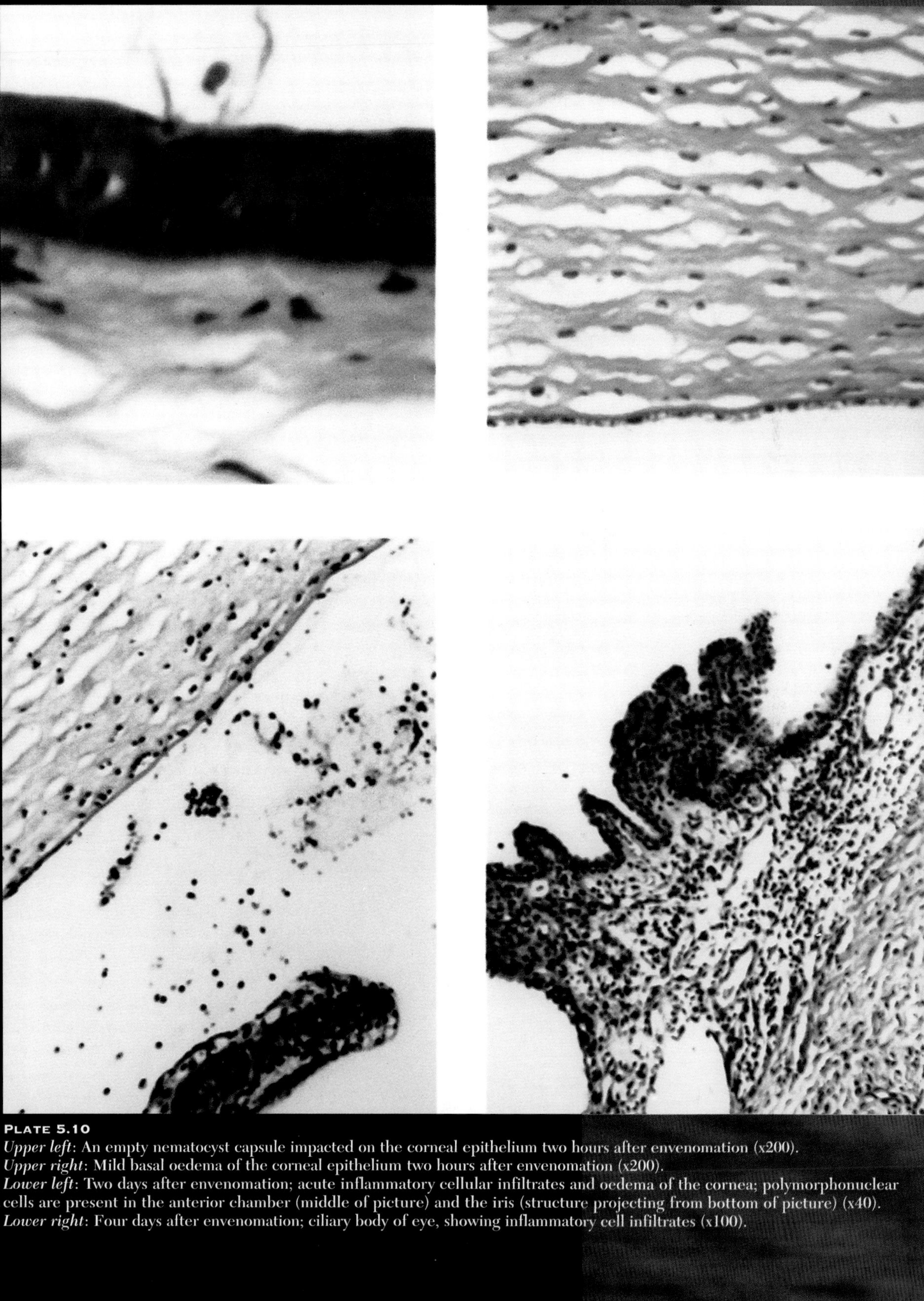

PLATE 5.10
Upper left: An empty nematocyst capsule impacted on the corneal epithelium two hours after envenomation (x200).
Upper right: Mild basal oedema of the corneal epithelium two hours after envenomation (x200).
Lower left: Two days after envenomation; acute inflammatory cellular infiltrates and oedema of the cornea; polymorphonuclear cells are present in the anterior chamber (middle of picture) and the iris (structure projecting from bottom of picture) (x40).
Lower right: Four days after envenomation; ciliary body of eye, showing inflammatory cell infiltrates (x100).

PLATE 5.11
The spotting technique from an IRB is here demonstrated by two of its innovators: Dr Jack Barnes does the spotting in the bow by scanning the still waters, wearing polaroid sunglasses; his famous field colleague, Ron Pollard, wearing pantyhose, slowly steers the IRB at very quiet, low throttle.

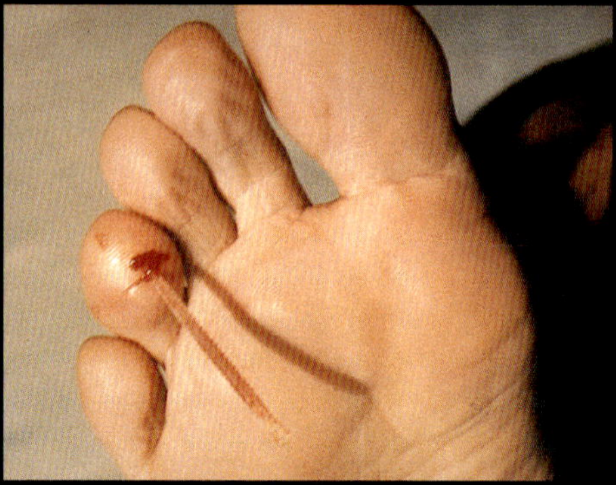

PLATE 5.12
A broken-off stingray barb still embedded in the fourth toe of the victim. If removed in the field, this barb piece should accompany the patient to medical attention.

PLATE 5.14
A major *Chironex fleckeri* sting, i.e. one in which the sting occupies a total skin area of more than half of one limb. Note the full-thickness skin death behind the right knee.

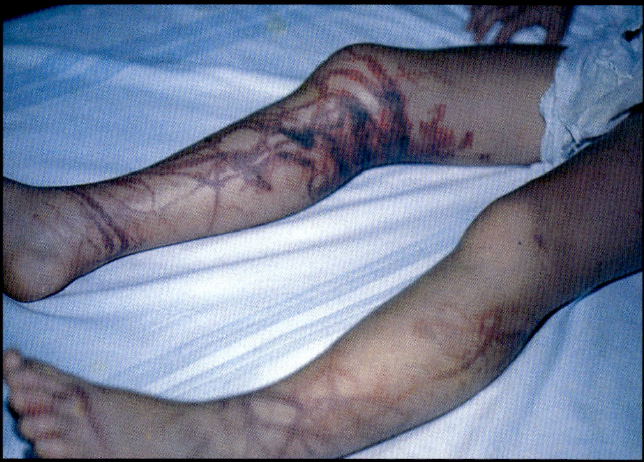

PLATE 5.15
A team of north Queensland lifesavers practising a full scale beach rescue of a major *Chironex fleckeri* sting victim. Note the application of 'stinger vinegar' followed rapidly by compression bandaging; resuscitation continues without interruption.

PLATE 5.13A
Two of the editors experimenting with *Chrysaora quinquecirrha* specimens fresh from the waters of Chesapeake Bay, Maryland, USA, in 1991.

PLATE 5.13B
One of the editors conducting further Chesapeake Bay experiments with *Chrysaora quinquecirrha*, in 1991.

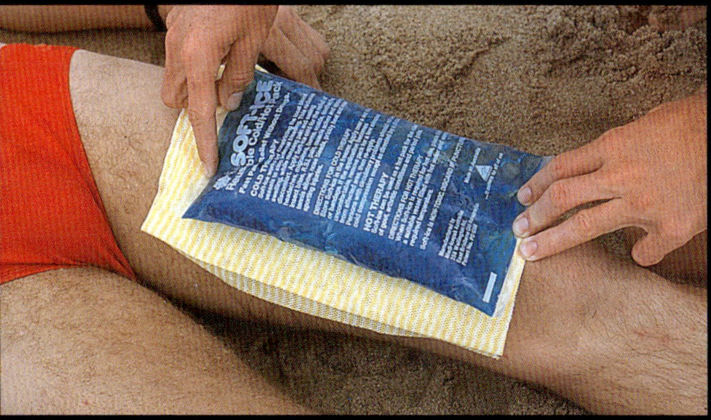

Plate 5.16
A cold pack applied through a dry cloth, which has been placed over a jellyfish sting on the thigh.

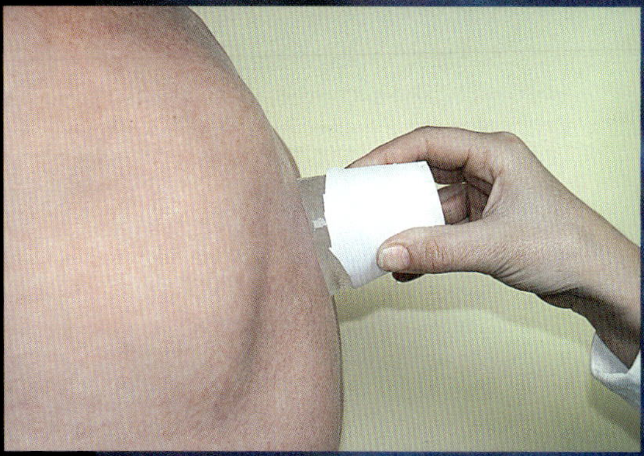

Plate 5.18
'Icy Cup' cold treatment of a fresh sting. The first-aider's hand is insulated from the piece of ice by the paper cup.

Plate 5.17
A large, fresh *Physalia* sting being treated with direct application of multiple cold packs. General (systemic) hypothermia must be avoided.

Plate 5.19a
A massive *Chironex fleckeri* sting {shown in Figure 1.2, at 30 minutes after envenomation in a healthy young female}, photographed eight hours later. Note the wide tentacle marks (indicating a very large animal) and the threatening full-thickness skin death.

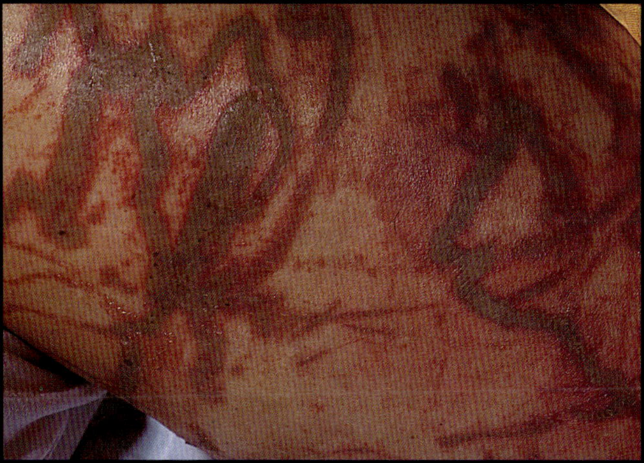

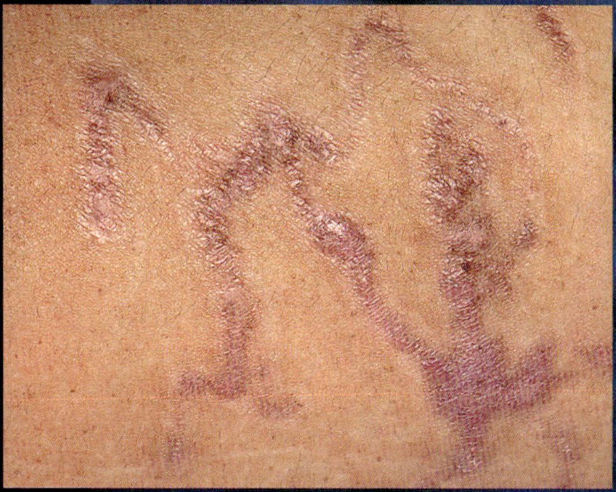

Plate 5.19b
The same sting nine weeks later. This patient received optimal IV dosage of specific CSL *Chironex fleckeri* antivenom, commencing within 30 minutes of her sting.

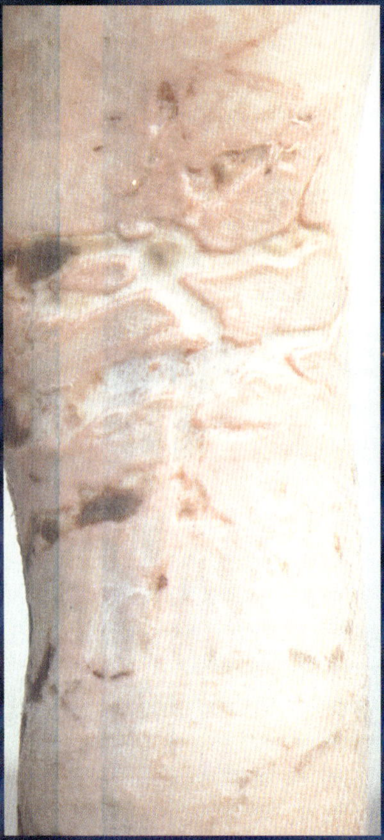

PLATE 5.20A
Secondary infection five days after a major *Chironex fleckeri* sting {the case shown in Plate 5.14}. The wound cultured *Staphylococcus aureus*, responded to systemic antibiotic therapy and wound dressings, but resulted in severe scarring.

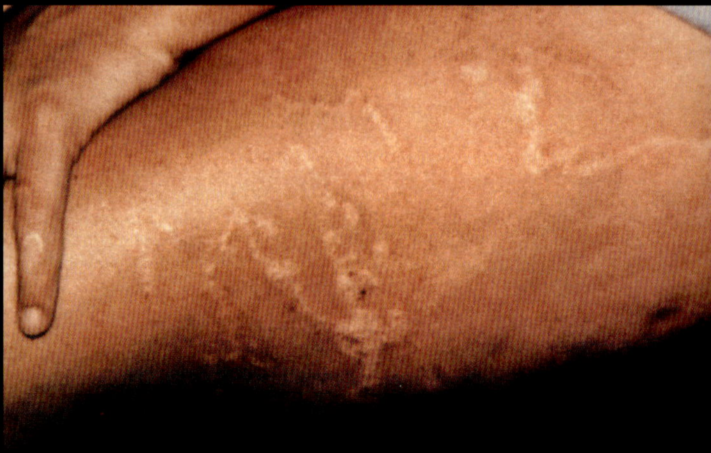

PLATE 5.19C
The same sting appearance, 18 months later. Further reduction in scarring occurred spontaneously over the next two years.

PLATE 5.20B
The same sting as shown in Figures 5.14 and 5.20A, photographed six weeks after the envenomation.

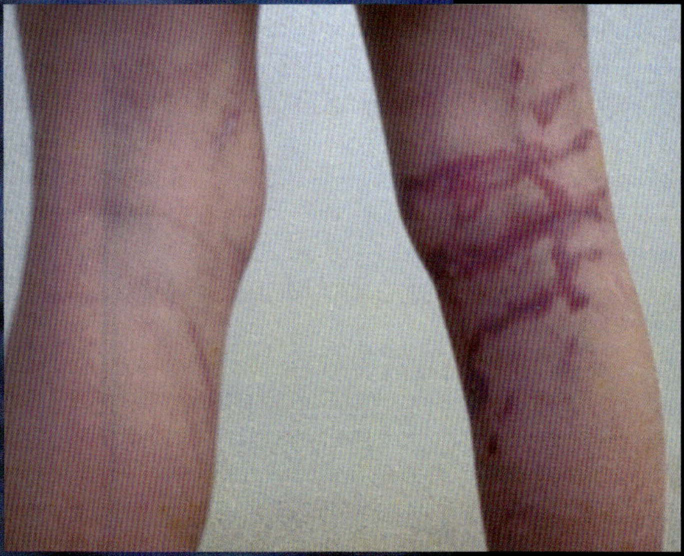

PLATE 5.20C
The scarring, ten years later.

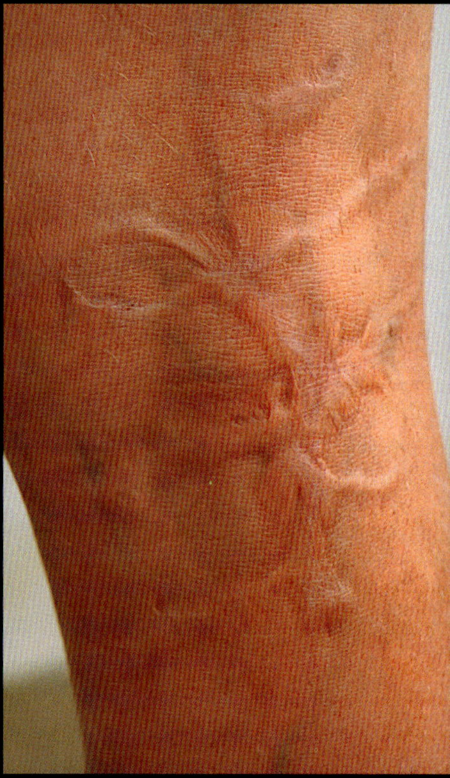

Plate 6.1a
The semaeostome scyphomedusan *Cyanea*, with the bell in the relaxed state; umbrella flat; radial swimming muscles visible under marginal lappets of bell. Note also large whitish clumps of gonads covering central mouth arms, and string-like tentacles.

Plate 6.1b
Cyanea in the contracted state; swimming muscles are contracted; and the size of the subumbrella is decreased. (Photograph courtesy of Professor Thomas Heeger, Philippines.)

Plate 6.2a
The rhizostome scyphomedusan *Thysanostoma thysanura* with bell cavity expanded allowing inflow of water; actively travelling in the horizontal plane. (Photograph courtesy of Professor Thomas Heeger.)

Plate 6.2b
Thysanostoma thysanura travelling in the horizontal plane with swimming muscles contracted, reducing the size of the subumbrellar aperture. Note the elongated mouth arms. (Photograph courtesy of Professor Thomas Heeger.)

PLATE 6.3
A view through the transparent bell of a Philippines chirodropid showing a partially digested fish, 16 cm in length, in the stomach. Note the yellowish clumps of gastric cirri. (Photograph courtesy of Professor Thomas Heeger.)

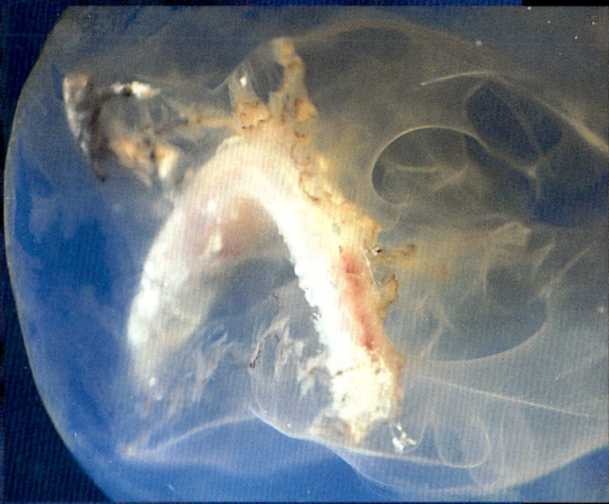

PLATE 6.6
Chrysaora hysoscella in a righting response. Note the asymmetric contraction of the bell. (Photograph courtesy of Professor Thomas Heeger).

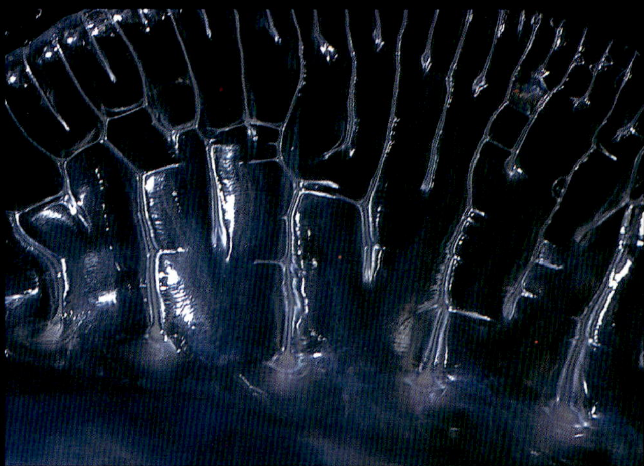

PLATE 6.4
Erect sail of a pneumatophore (float) from *Physalia physalis* showing partitions or septa of the crest with air pockets in between. (Photograph courtesy of Professor Thomas Heeger)

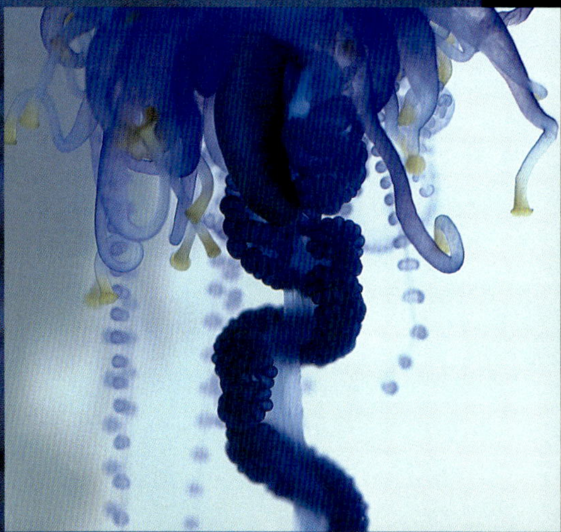

PLATE 6.5
The zooid mass under the pneumatophore (not shown) of the siphonophore *Physalia utriculus* showing one large dactylozooid bearing a single, large, highly contracted and twisted fishing tentacle with a line of closely packed stinging buttons; smaller dactylozooids with smaller, more relaxed tentacles bearing stinging buttons situated further apart in a line; and the yellow-tipped feeding polyps. (Photograph courtesy of Professor Thomas Heeger)

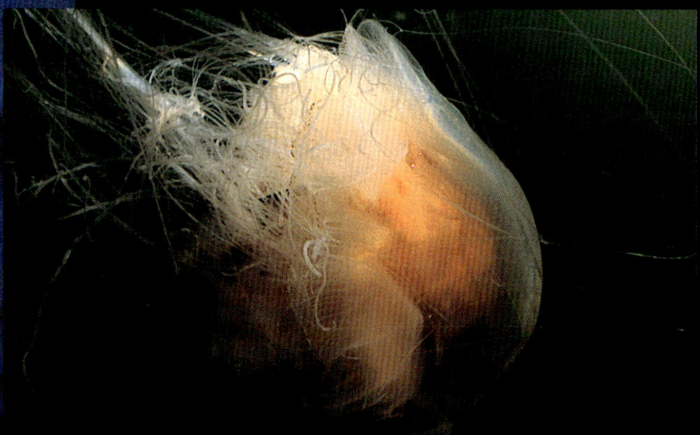

PLATE 6.7
The scyphomedusan *Cyanea capillata* exhibiting an escape response, in an active dive, bell fully contracted, aboral end downwards (cf. Plate 6.1B). (Photograph courtesy of Professor Thomas Heeger)

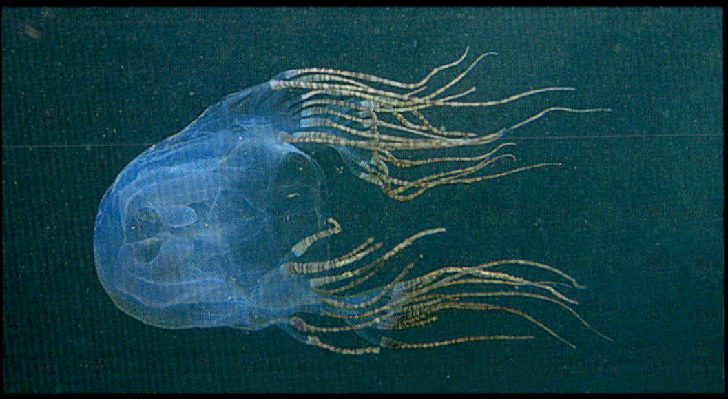

PLATE 6.8
A Philippines chirodropid, *Chiropsalmus sp.*, swimming in horizontal plane with aboral end anterior; with bell and velarium in systole. (Photograph courtesy of Professor Thomas Heeger)

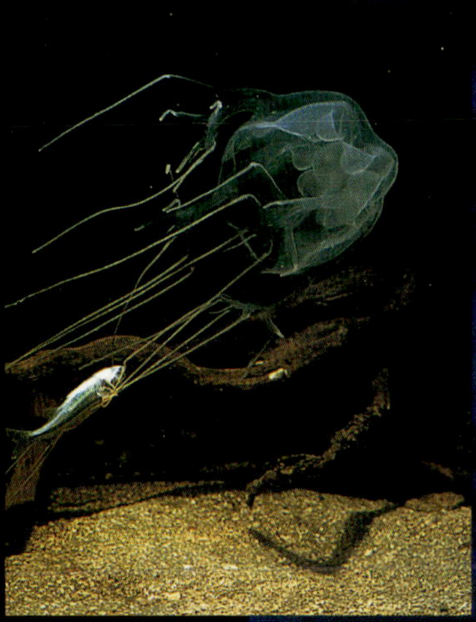

PLATE 6.9
A north Queensland *Chiropsalmus quadrigatus sensu* BARNES 1965 in the process of stinging a fish. Note the characteristic mammiform superior gonads, and the side-wall gonads of the medusa. Note also the selective adhesion to the fish of just the four tentacles concerned with its capture, and the non-involvement of the remainder of the tentacle armament. (Photograph courtesy of Ben Cropp Productions, Port Douglas, Qld)

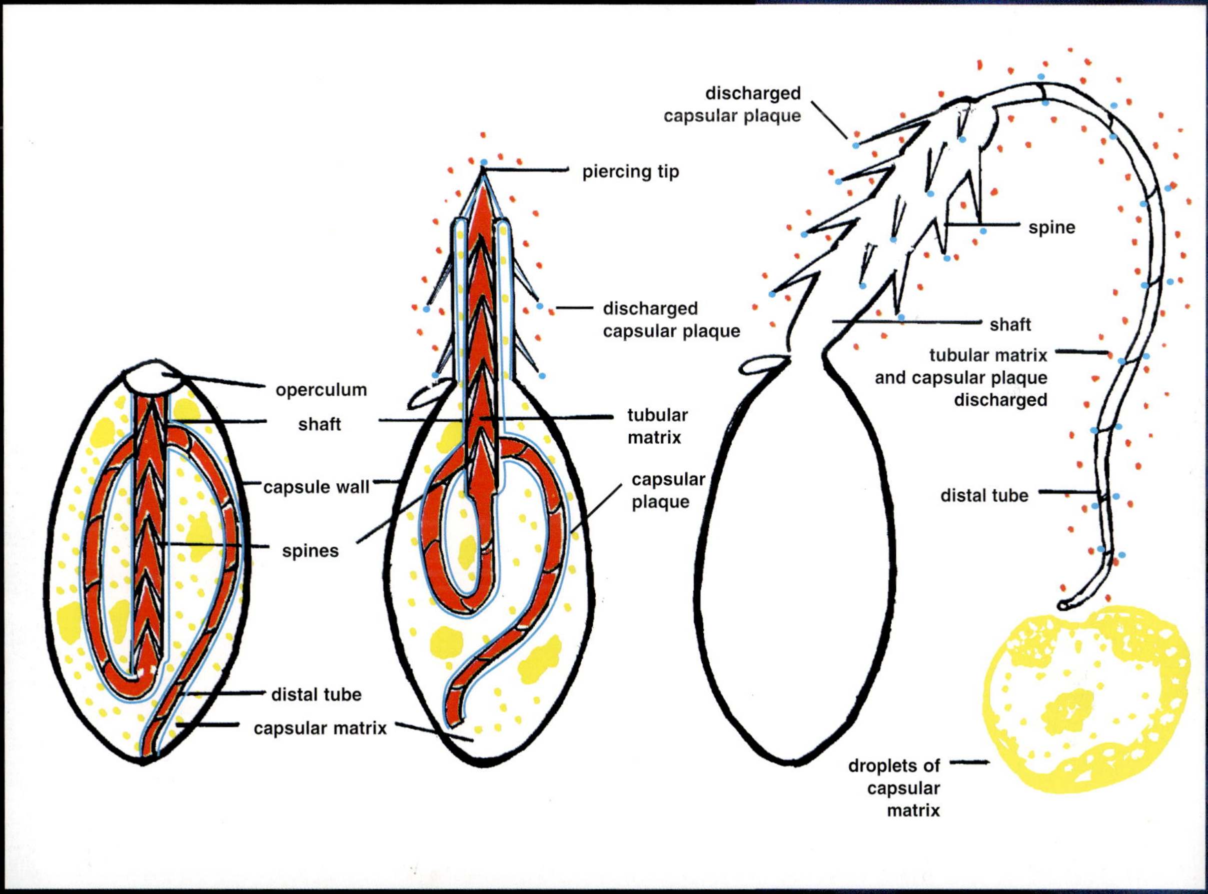

PLATE 6.10
Undischarged, partially discharged and fully discharged nematocyst, showing the distribution of tubular matrix, capsular matrix and capsular plaque in the discharge process.
Capsular matrix in yellow
Capsular plaque in blue
Tubular matrix in red

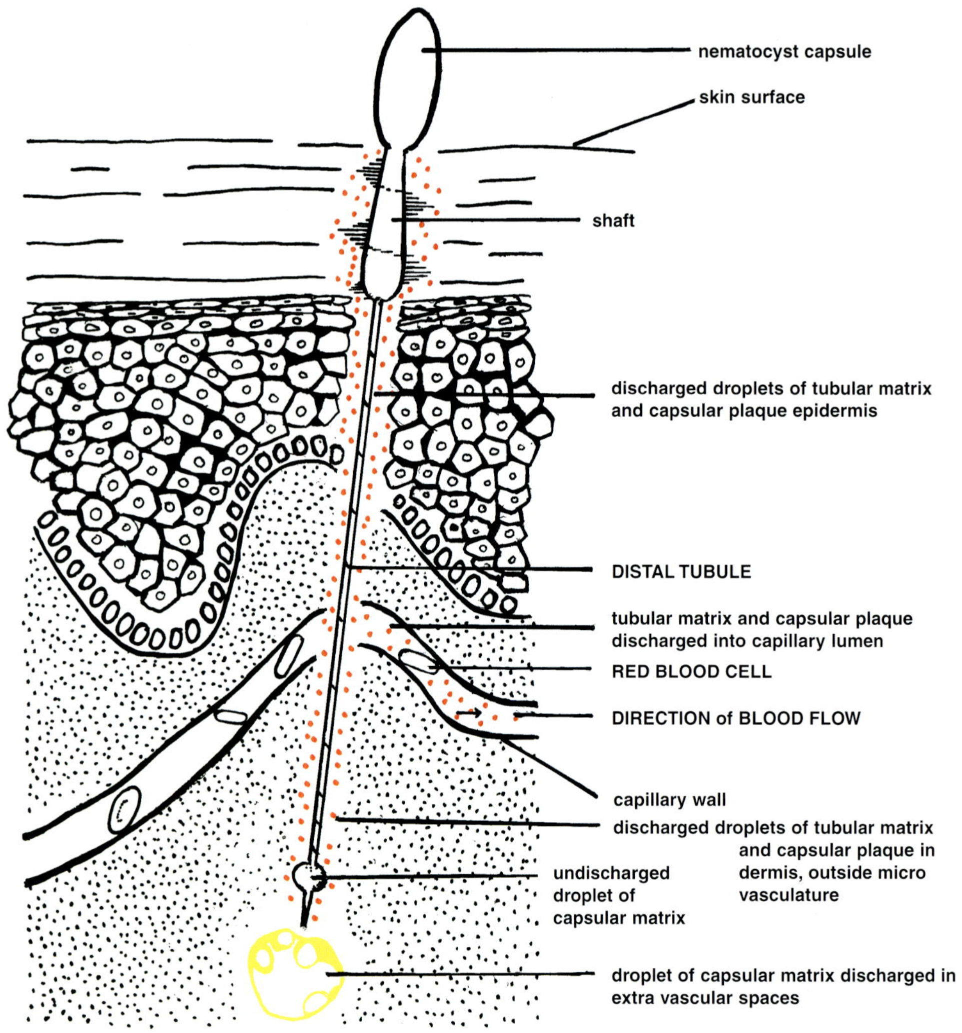

PLATE 6.11
Drawing of a fully discharged nematocyst penetrating the layers of mouse skin, showing the fate of tubular matrix and capsular plaque (red) and capsular matrix (yellow). Drawn by Dr. Jaquie Rifkin

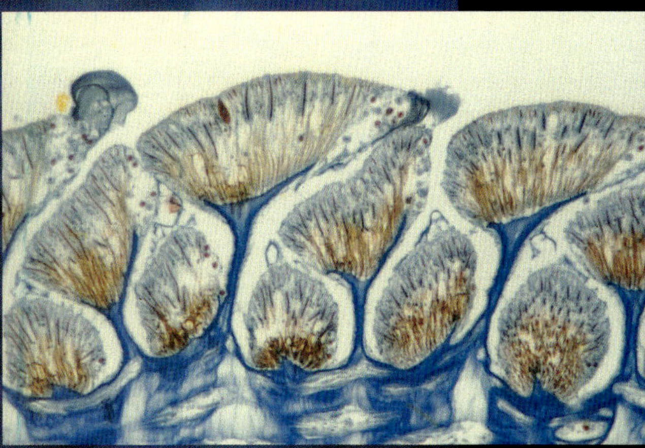

PLATE 6.12
A histologically stained light micrograph of a longitudinal section of contracted *Chironex fleckeri* tentacle. Note the three 'layers' of bands (primary, secondary and tertiary) of mastigophore-bearing epidermis. (Copyright © *The Medical Journal of Australia* 1980; 1:15-20. Reprinted with permission, and courtesy of Dr R. F. Hartwick, PhD, James Cook University of North Queensland)

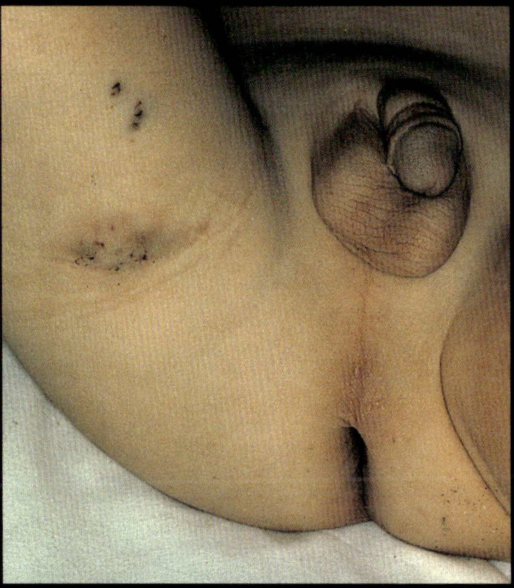

PLATE 7.1
A fatal Taipan snake bite on the right under-thigh of a small child in north Queensland; the multiple strike pattern is clearly visible. The patient died within about 30 minutes, which is the most rapid death from Australian snake bite so far recorded. (Photograph courtesy of Townsville General Hospital)

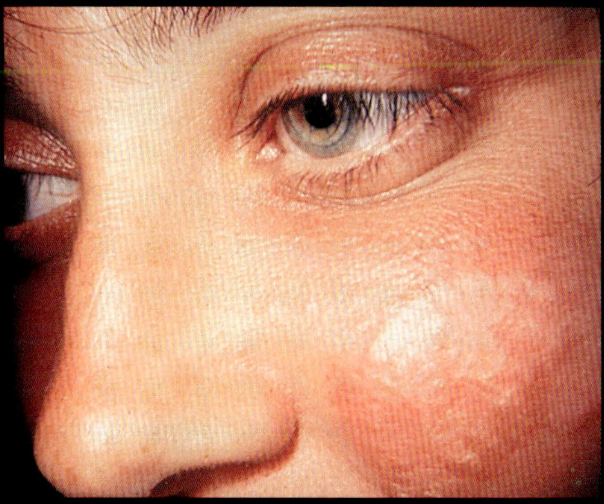

PLATE 7.3
A persistent granuloma on the left cheek following a *Pelagia* sp. sting. (Reproduced from Reed *et al*, J Am Acad Dermatol 1984; 10: 463, with permission of the publishers)

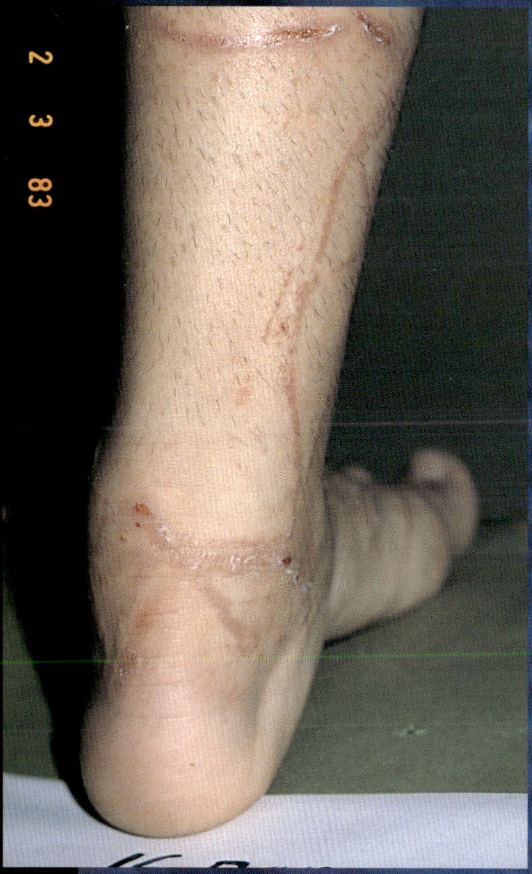

PLATE 7.4
A spontaneously recurrent, pruritic eruption at the sting site, 16 days after a central Queensland *Chironex fleckeri* sting on the leg of a patient with a past medical history of an allergic diathesis. The sting site was clear and asymptomatic from day 3 to day 15. Note the distinct reappearance of the tentacle marks.

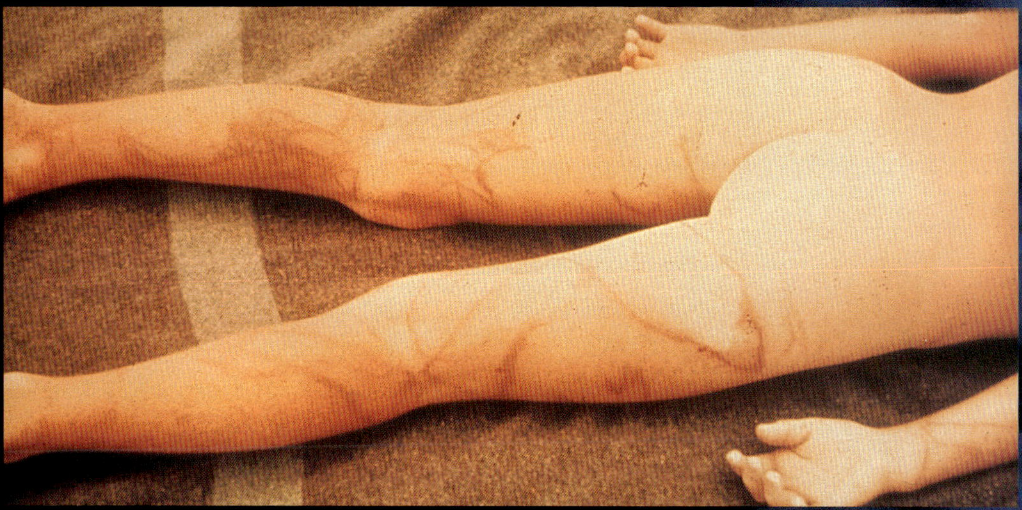

PLATE 7.2
A fatal northern Australian *Chironex fleckeri* sting in a small boy of comparable size to the patient shown in Figure 7.1. Loss of consciousness occurred within three minutes, followed quickly by circulatory arrest. It is this rapidity of demise that explains the apparent paucity of inflammatory skin reaction (cf. Figure 1.2).

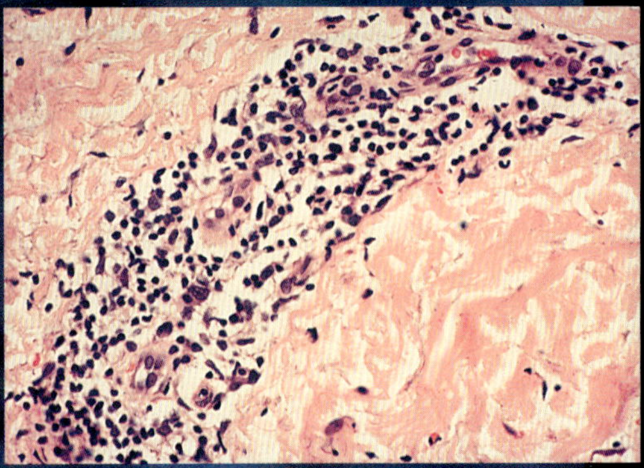

PLATE 7.5
Biopsy of a recurrent skin lesion from a ***Stomolophus meleagris*** sting, showing a dense perivascular dermal infiltrate of mononuclear cells. Numerous helper T cells can be found among this population.

PLATE 7.6
Ischaemic hand with cyanosed fingers of a young female stung on the same wrist by an unidentified jellyfish in Penang waters. There was residual hand weakness for two years (Copyright ©Medical Journal of Australia 1988; 149: 698–701)

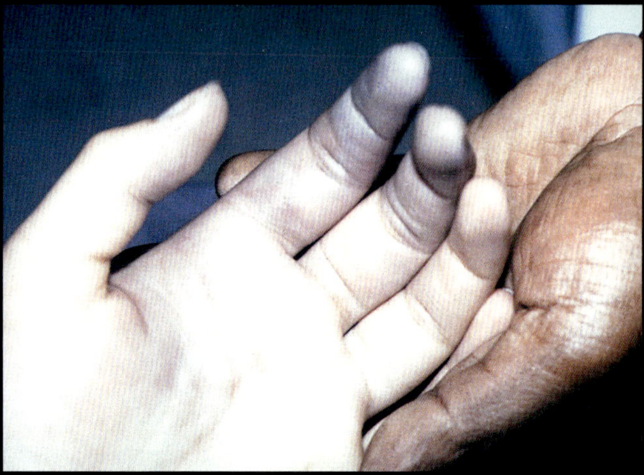

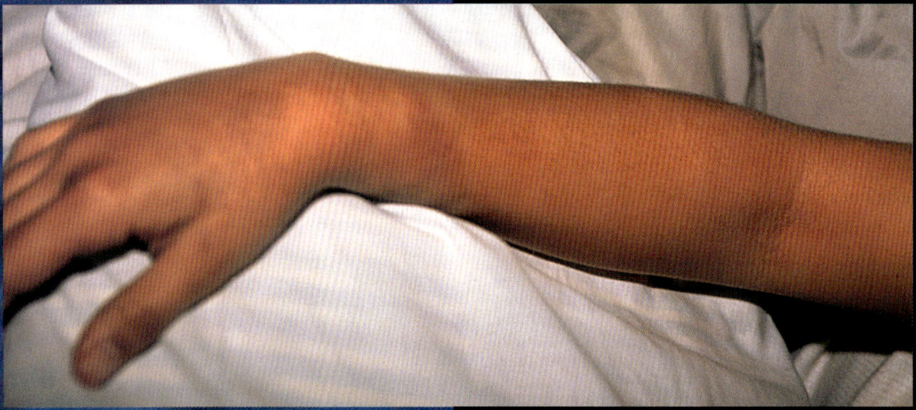

PLATE 7.7
Paresis of the right forearm of a young female, 2 days after a sting from an unknown jellyfish in Penang waters. The sting was over the dorsum of the right wrist; paraesthesiae and muscle weakness affected the entire right forearm and persisted for several months. (Copyright © Medical Journal of Australia 1994; 161: 320–322)

PLATE 7.8
Right hand of the patient in Plate 7.7, photographed several weeks after the sting. Note the hyperpigmentation of the sting area and the residual impairment of finger coordination. (Copyright © *Medical Journal of Australia* 1994; 161: 320–322)

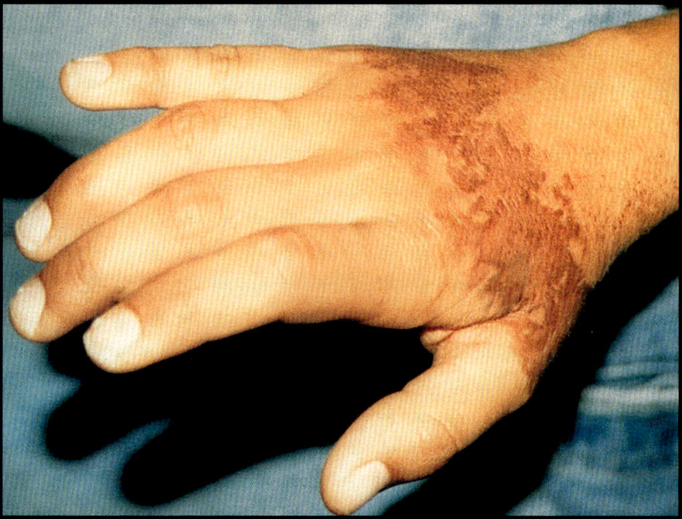

PLATE 8.1
Hard coral *Acropora* sp. — one of the world's most pleomorphic and successful hard corals [VERON 1986]) — on Australia's Great Barrier Reef. Most scleractinian corals do not deliver a sting, but even brush contact can abrade or lacerate skin.

PLATE 8.2
A close view of the coral *Acropora* sp., showing the extended polyps. Genus *Acropora* Oken, 1815, seldom stings but is associated with many coral abrasions because of its international reef occurence.

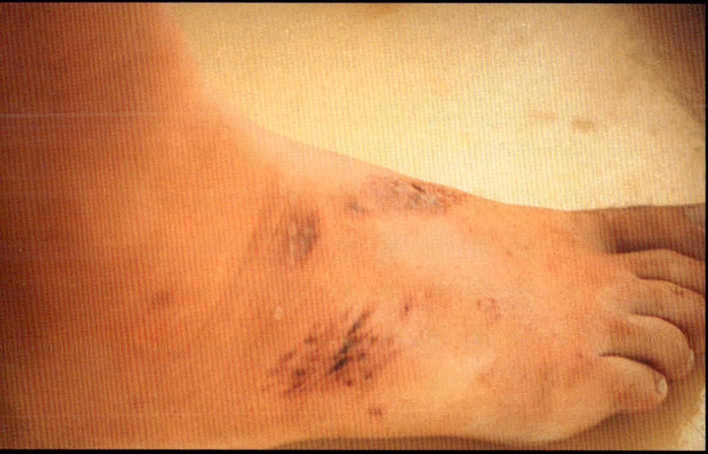

PLATE 8.3
A painful coral abrasion on the dorsum of a foot, 18 hours after the injury. No first aid was administered and infection is already evident.

PLATE 8.4
The flower-like sea anemone *Condylactis* sp. (see Plate 8.20)

PLATE 8.5
Sea anemone *Haliplanella luciae*. Arrow indicates location of pedal disc. (See Plate 8.12)

PLATE 8.6
Sea anemone *Bolocera tuediae*. Note the longer tentacles

PLATE 8.7
The boxer crab *Lybia tessellatus*, brandishing a live sea anemone in each claw. With remarkable adaptation it uses the stinging properties of the tentacles of these anemones for defence and for the capture of its prey. (Photograph courtesy of Karen Gowlett-Holmes, Collection Manager, Marine Invertebrates, South Australian Museum, Adelaide)

PLATE 8.8
Stinging sea anemones of the order Actiniaria, in 3 m of sea water at Point Turton, South Australia. Note the warty protruberances on the column of the anemones; these, along with the tentacles, are filled with powerfully stinging nematocysts. (Photograph courtesy of Karen Gowlett-Holmes, Collection Manager, Marine Invertebrates, South Australian Museum, Adelaide)

PLATE 8.9
A sea anemone with its clown fish. Note the impunity with which the clown fish swim against the nematocyst-laden anemone tentacles.

PLATE 8.10
The sea anemone *Cryptodendrum adhaesivum*, at 9m depth on the Great Barrier Reef, north Queensland. (Photograph courtesy of Bob Halstead, Cairns)

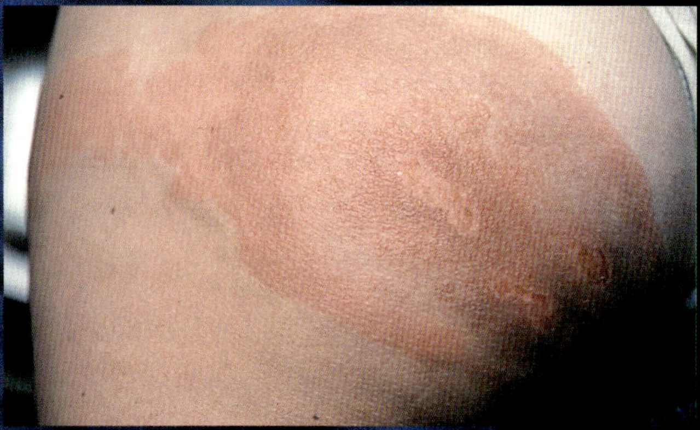

PLATE 8.11
A troublesome sting on the buttock of a scuba diver who contacted the sea anemone *Cryptodendrum adhaesivum*; photographed on day 7. (Courtesy Bob and Dinah Halstead, Cairns)

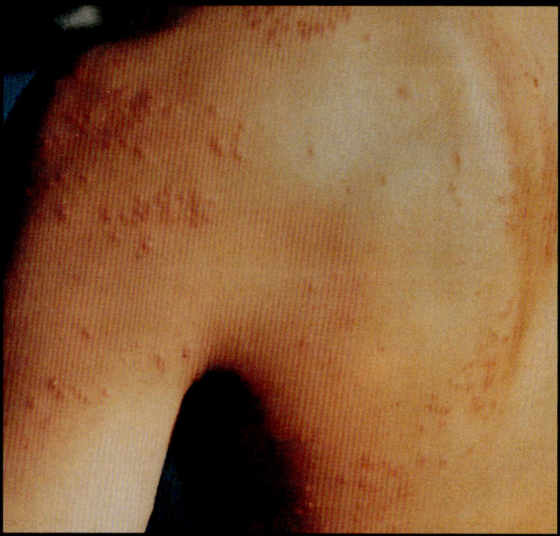

PLATE 8.12
Recent sting on the back and shoulder from the sea anemone *Haliplanella luciae*. Note the multiple, discrete papular lesions. (See Plate 8.5)

PLATE 8.13
Vesicle formation on the wrist in a sting lesion from the sea anemone *Haliplanella luciae*. Compare with Plate 8.5.

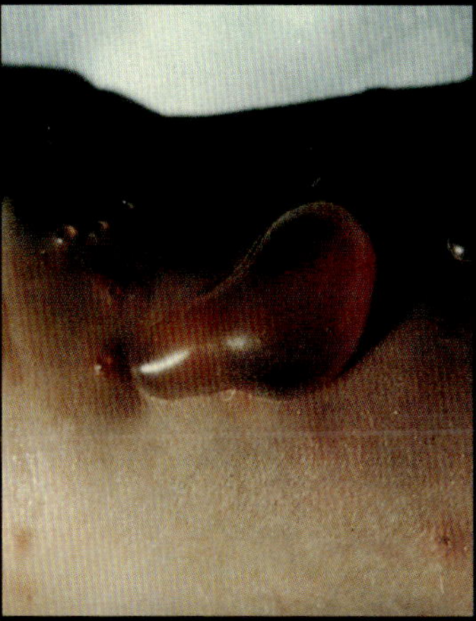

PLATE 8.14
The venomous sea anemone *Phyllodiscus semoni* in Okinawa Island, Japan. (Photograph courtesy of Dr Yasuhiro Tomihara and colleagues, Okinawa Prefectural Institute of Public Health, Japan)

PLATE 8.15
The nematocyst-bearing tentacles of the Okinawan sea anemone *Phyllodiscus semoni*, photographed in August 1989. (Photograph courtesy of Drs Nishihira, Yasuhiro Tomihara and colleagues, Okinawa Prefectural Institute of Public Health, Japan).

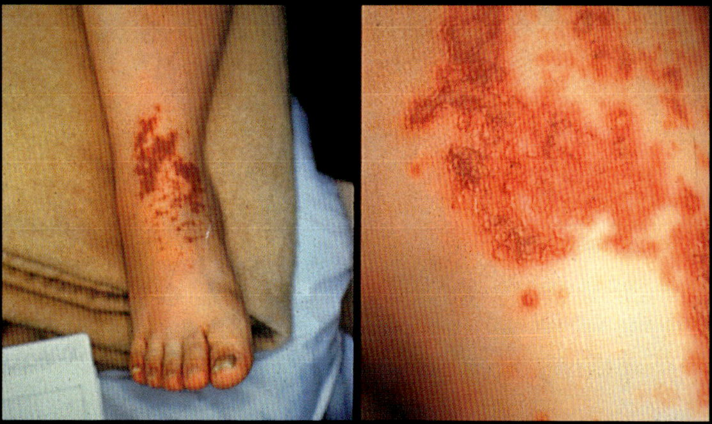

PLATE 8.16
A recent sting on the leg of a Japanese swimmer from the sea anemone *Phyllodiscus semoni*, in Okinawa, Japan. The right hand field shows a close-up of the vesiculo-erythematous lesions. (Photograph courtesy of Dr Yasuhiro Tomihara and colleagues, Okinawa Prefectural Institute of Public Health, Japan).

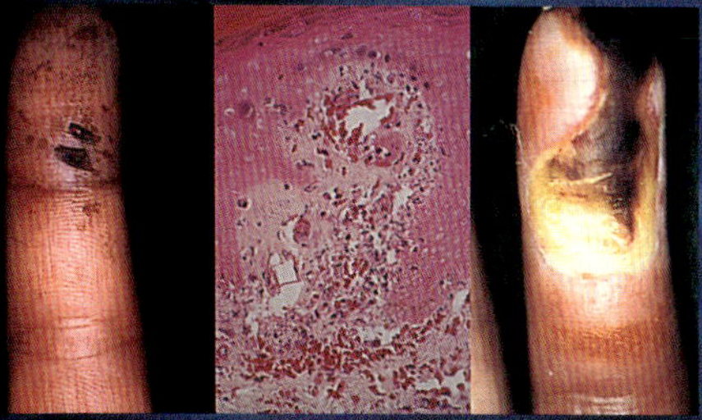

PLATE 8.17
Full-thickness necrosis on the palmar surface of fingers from contact with the Okinawan sea anemone *Phyllodiscus semoni*. The central section of the picture shows the histology of necrosis and cellular inflammatory infiltrate. (Photograph courtesy of Dr Yasuhiro Tomihara and colleagues, Okinawa Prefectural Institute of Public Health, Japan)

PLATE 8.18
Deep leg ulcers from the sting of the Okinawan sea anemone *Phyllodiscus semoni*. (Photograph courtesy of Dr Yasuhiro Tomihara and colleagues, Okinawa Prefectural Institute of Public Health, Japan)

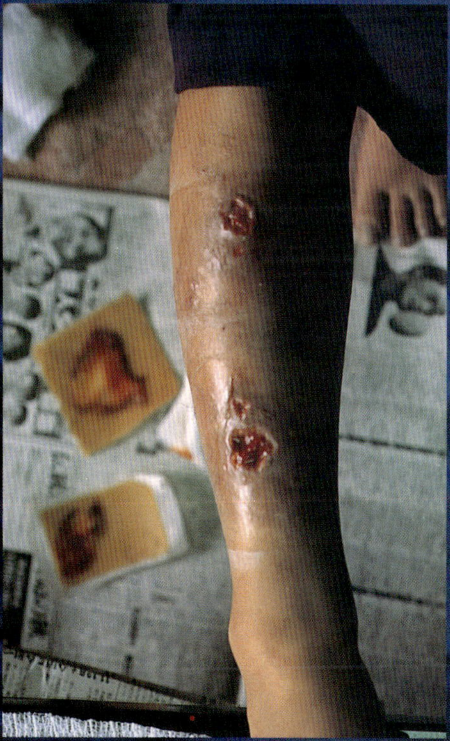

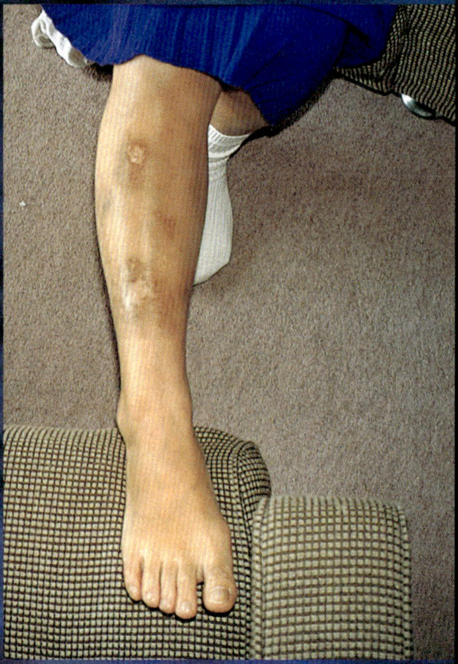

PLATE 8.19
The same patient as in Plate 8.18, showing scarring after healing from the lesions produced from the sting of the Okinawan sea anemone *Phyllodiscus semoni*. (Photograph courtesy of Dr Yasuhiro Tomihara and colleagues, Okinawa Prefectural Institute of Public Health, Japan)

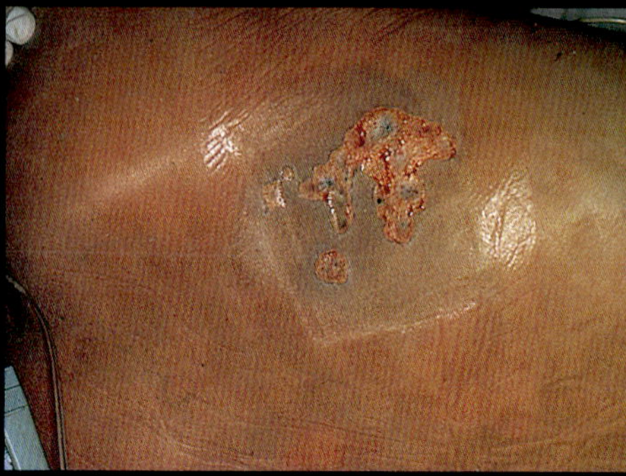

PLATE 8.20
Sea anemone *Condylactis* sp. sting on the back of a breath-hold diver, who became jaundiced, and died from fulminant hepatic failure nine days after envenomation. See also Plate 8.4. (Reproduced with permission, from P.J. Garcia, R.M.H. Schein, J.W. Burnett. Fulminant hepatic failure from a sea anemone sting. *Ann Intern Med* 1994; 120: 665–666.)

PLATE 8.21A
A corallimorpharian of the genus *Discosoma* (Ruppell and Leuckart 1828), shown with tentacles withdrawn.

PLATE 8.21B
Discosoma corallimorpharian with tentacles extended. (Photographs courtesy of Bob Halstead, Cairns)

PLATE 8.22
The reef referred to in Case Report No. 2 page 187, on which the diver knelt to take photographs. The reef is covered with numerous *Discosoma* corallimorpharian stingers, which blend superbly with their environment. (Photograph courtesy of Bob Halstead, Cairns)

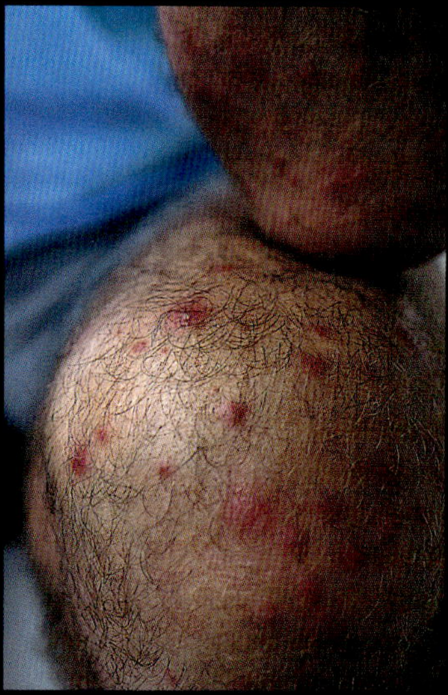

PLATE 8.23
The papular lesions on the left elbow and knee of the patient described in Case Report No. 2 page 187, photographed on day 2 after contact with the corallimorpharian. (Photograph courtesy of Bob Halstead, Cairns)

Plate 8.24
The feather hydroid (possibly *Aglaophenia whiteleggei*).

Plate 8.27
Wrist sting on a snorkeller from the hydroid *Aglaophenia cupressina*, showing erythema and discrete papules. (Photograph courtesy Neville Coleman)

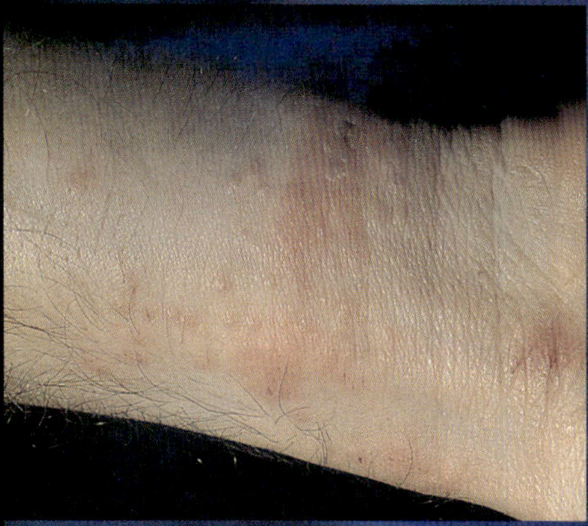

Plate 8.25
A species of the genus *Millepora* (possibly *M. tenera* BOSCHMA 1949), a stinging coral.

Plate 8.26
The stinging hydroid *Aglaophenia cupressina*, the 'cypress sea fern'. (Photograph by the Australian Institute of Marine Science)

Plate 8.28
The feather-like appearance of a nematocyst-armed branch of a species of tropical stinging hydroid (possibly *Aglaophenia cupressina*).

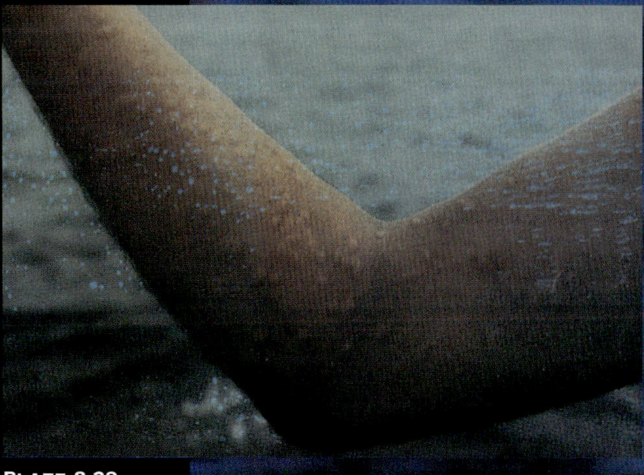

Plate 8.29
Fresh hydroid (*Aglaophenia cupressina*) sting on the arm of a female scuba diver following simple brush contact. (Photograph Neville Coleman).

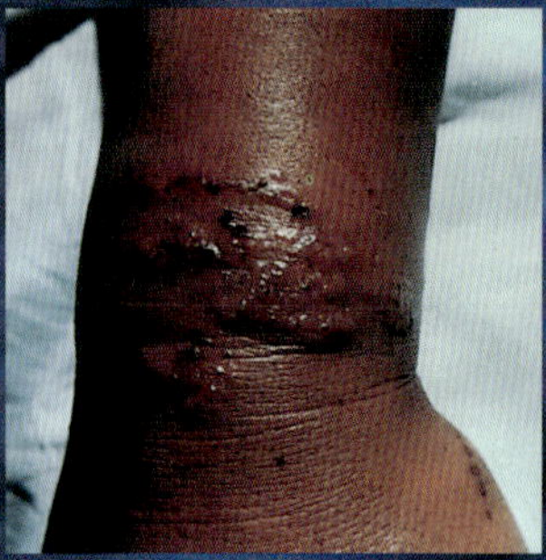

PLATE 8.30
Sting on the wrist from the hydroid *Aglaophenia whiteleggei* (see Figure 8.24), showing angry wheal-ing, vesiculation and early crusting. Note the concealment of erythema by the pigmented skin.

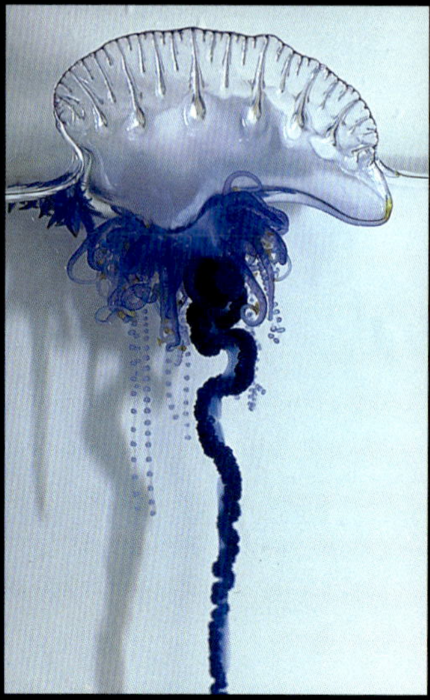

PLATE 8.31
The Pacific *Physalia utriculus* ('bluebottle'), captured from the Pacific Ocean in the Philippines. Note the single long tentacle (cf. Plate 8.37). (Photograph courtesy of Professor Thomas Heeger, University of San Carlos, Philippines)

PLATE 8.32
An Australian specimen of a Pacific *Physalia utriculus* photographed in captivity. Note the contracted state of the tentacles. (Photograph courtesy of Keith Gillett, Sydney)

PLATE 8.34
A Bermuda specimen of the Atlantic *Physalia physalis* displayed on beach sand. Note the multiple long tentacles. Compare with Plate 10.22. (Photograph courtesy of Professor Thomas Heeger, University of San Carlos, Philippines)

PLATE 8.35
The Pacific *Physalia physalis*. This specimen was captured in north Queensland. Note the multiple long tentacles.

PLATE 8.33
The Atlantic (USA) *Physalia physalis* (Portuguese man- o'-war). Note the large pneumatophore (gas-filled float; cf Plate 6.4), and the multiple long tentacles. (From Halstead 1965, courtesy of United States Office of the Assistant Secretary of Defense, Washington, DC)

PLATE 8.36A AND B

The 'harmless jellyfish' *Velella velella*, specimens captured in the Mediterranean waters of the Greek Islands. Note the obliquely positioned, transparent 'sail' on the upright specimens, and the mostly non-stinging subumbrella of the single inverted specimen. Each elliptically shaped animal is approximately 5 cm wide (short axis). The concentric elliptical markings on the top surface of the upright specimens are gas-filled flotation tubes. (L. FRANCIS, in BROWNLEE 1987; photograph courtesy of John Lippmann, Melbourne)

PLATE 8.37

A scuba-diver's-eye-view of the Pacific *Physalia* sp. ('bluebottle') floating on the surface, showing the down-current, trailing single long tentacle which characterises this species (see text).

PLATE 8.38

A large float of a *Physalia* sp. specimen photographed in Bermuda. (Courtesy of Professor Thomas Heeger, Philippines)

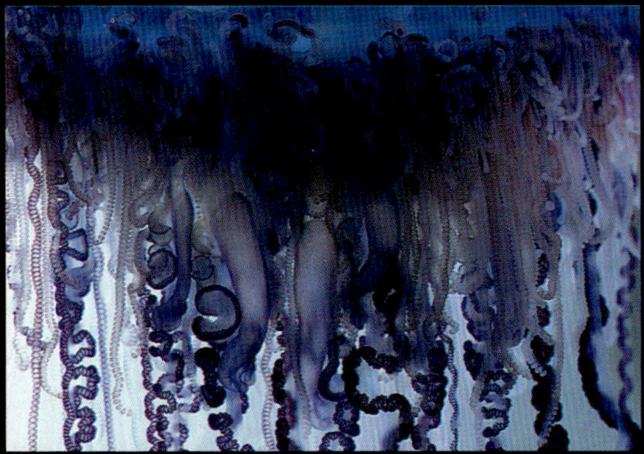

PLATE 8.39

The massed tentacles and zooids immediately under the float of *Physalia* sp. (Photograph by Professor Thomas Heeger, Philippines)

PLATE 8.40

Physalia dactylozooids. (Photograph by Professor Thomas Heeger, Philippines)

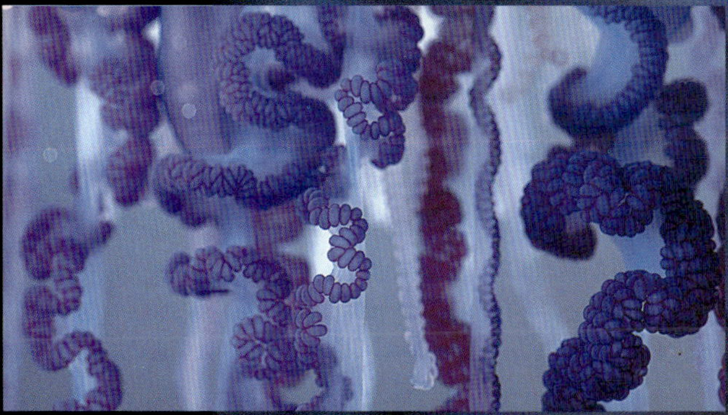

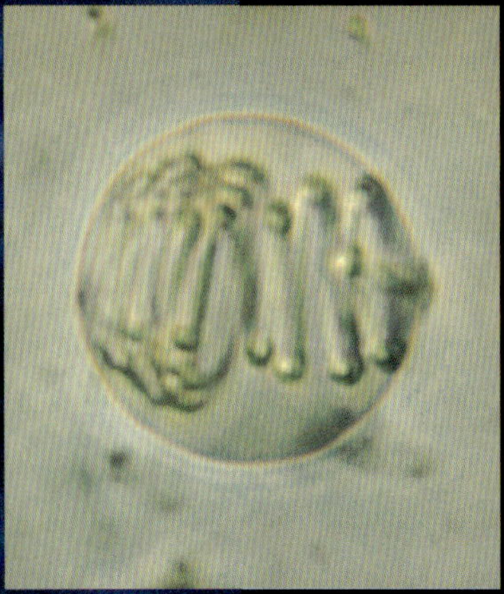

PLATE 8.41
A photomicrograph (x1400) of a solitary, spherical, undischarged nematocyst from one of the long tentacles cut from the animal pictured in Figure 10.9, page 303 (Prepared by and reproduced with the courtesy of Junaid Alam, Rashida Qasim and colleagues, Department of Pathology, Basic Medical Sciences Institute, Jinnah Postgraduate Medical Centre, Karachi). The bar is not visible.

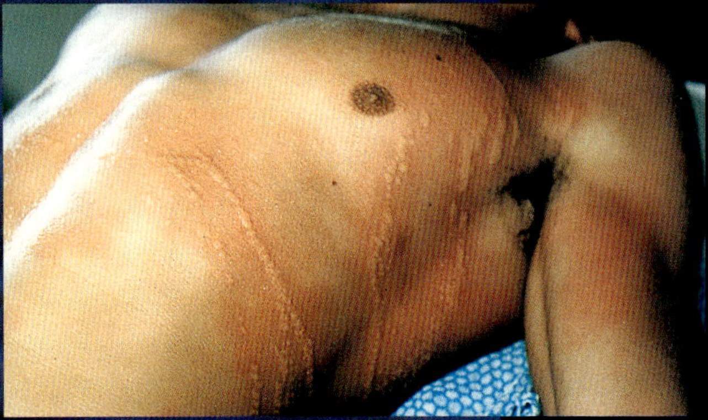

PLATE 8.42
Multiple Pacific *Physalia* sp. ('bluebottles', single long tentacles) stings on the torso of a champion Australian lifesaver, sustained on the south Queensland coast during a competition surf swim that took competitors unwittingly through a swarm ('navy') of bluebottles. Full recovery resulted. Note particularly the diagnostic beaded appearance of each line of tentacle contact (compare with Figure 6.5).

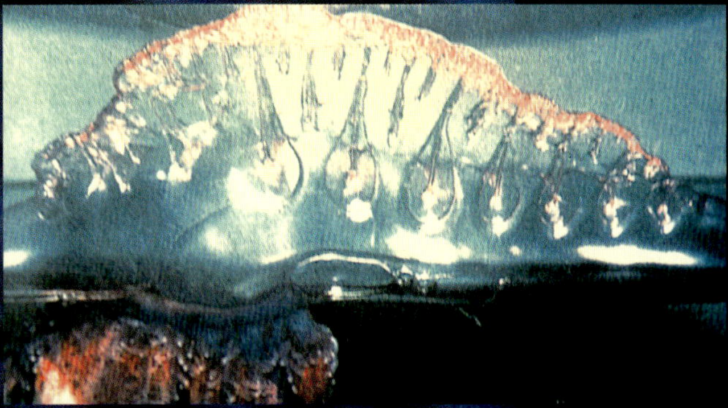

PLATE 8.43
The float and proximal tentacle mass of a large, multi-long-tentacled, potentially lethal Atlantic *Physalia physalis* ('Portuguese man-o'-war') (Photographed under artificial, specially lit conditions). (Reproduced from B.W. Halstead, *Poisonous and Venomous Marine Animals of the World*, Vol I 1965, with kind permission of the US Library of Congress, Washington, DC)

PLATE 8.44
Atlantic *Physalia physalis* sting, showing 'characteristic' imprint of linear wheal, with somewhat irregular 'beading' along the line of tentacle contact.

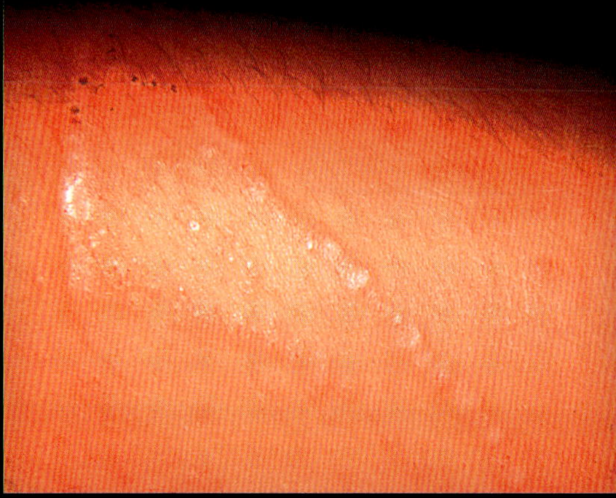

PLATE 8.45
The massive acute response to a serious envenomation from an unknown jellyfish in the Indian Ocean.
The patient was a healthy young female, and eight weeks later her serum gave a 1:1800 dilution positive ELISA test against *Physalia physalis* venom (WILLIAMSON *et al* 1988).

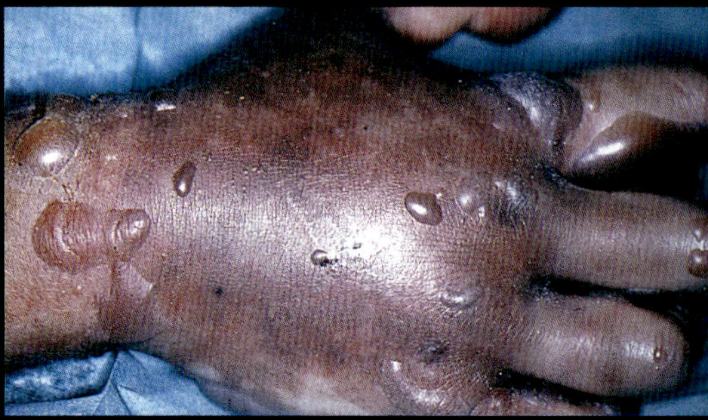

PLATE 8.46
The same patient's hand as Plate 8.45, showing gangrene of the finger tips. Both upper limbs were affected, and the patient recovered with partially flail upper limbs. (Copyright ©*Medical Journal of Australia* 1988; 149: 698–701)

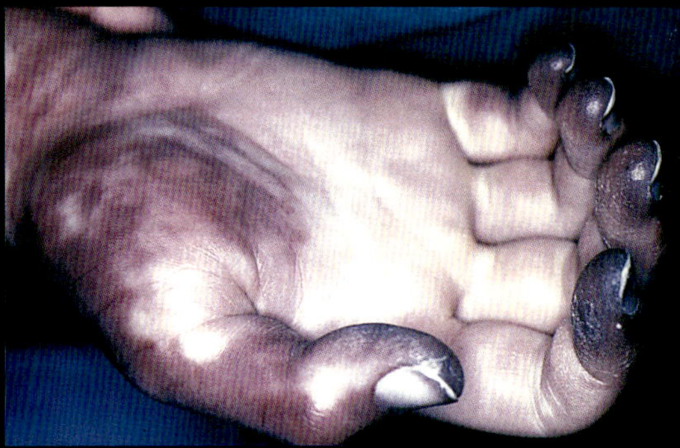

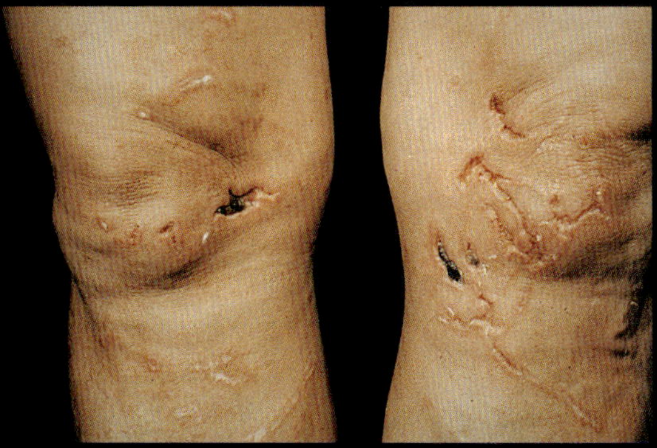

PLATE 8.47
Partially healed scars on the knees of an adult female stung weeks previously by a Florida *Physalia physalis*. Note the full-thickness skin destruction.

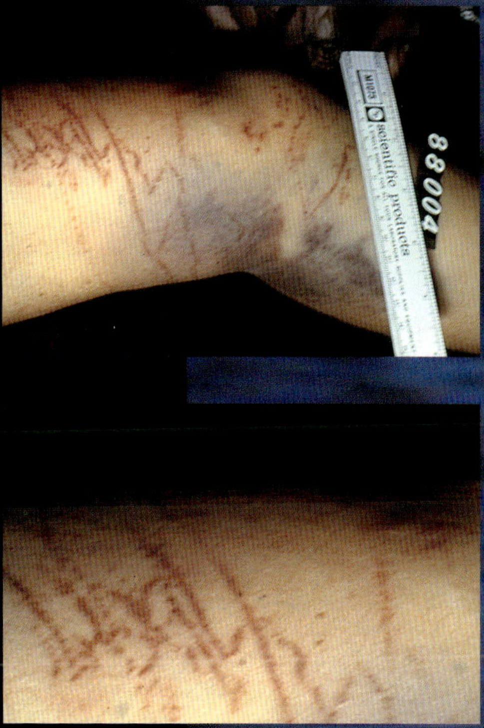

PLATES 8.48
Atlantic *Physalia physalis* (Portuguese man-o'-war) sting marks (24 hours post mortem) on a human fatality from massive envenomation on a Florida beach. Coma occurred within 30 minutes and death on the 5th day. (See *Annals Emerg Med* 1989; 18: 312–315)

PLATE 8.49
A dead 'Atlantic salmon' netted in an ocean-exposed Tasmanian fish farm and bearing the scar from a large *Physalia* tentacle. (Courtesy of Dr A. Clark, Margate, Tasmania)

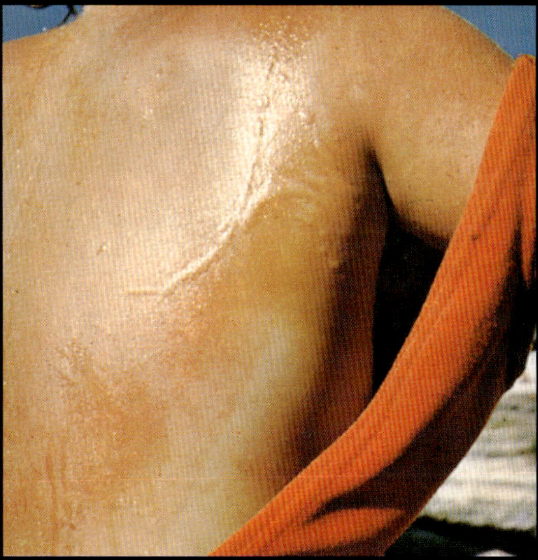

PLATE 8.50
A few minutes after a north Queensland bluebottle (*Physalia utriculus*?) sting, showing the distinct blanched linear wheal with a short segment of elliptical beading (see Plate 8.42 and text). (Courtesy of Mr Charles Fitzpatrick, Townsville)

PLATE 8.51
The typical habitat of *Gonionemus oshoro* UCHIDA, 1929, in Niigata Prefecture, on the north-west coast of Honshu (latitude 38° north) (Photograph courtesy of Dr Ohtsuru, University of Ryukyus, Japan)

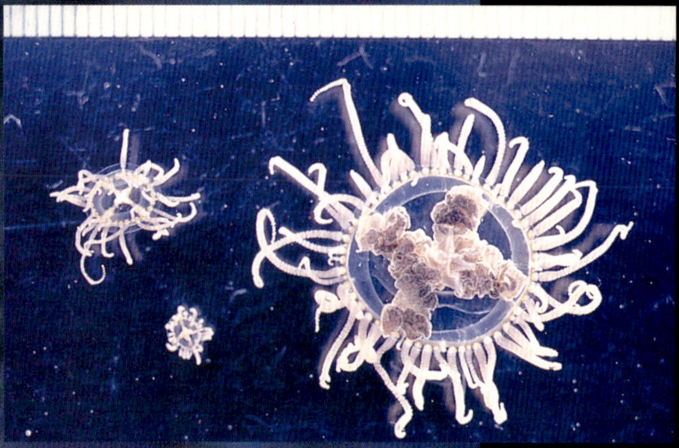

PLATE 8.52
The tiny stinging Japanese medusa *Gonionemus oshoro* UCHIDA, 1929. Note the numerous rim tentacles. Photographed in a petri dish; rule indicates millimetres. (Photograph courtesy of Dr Ohtsuru, University of Ryukyus, Japan)

PLATE 8.53
The neritic stinging Atlantic South American medusa *Olindias sambaquiensis* MÜLLER, 1861, whose distribution in Argentina is shown in Figure 8.10, page 207. Note the short and long tentacle types and the numerous radial channels in the bell.

PLATE 8.55A
The appearance of an *Olindias sambaquiensis* ('agua viva') sting after one week. Note the faint skin hyperpigmentation.

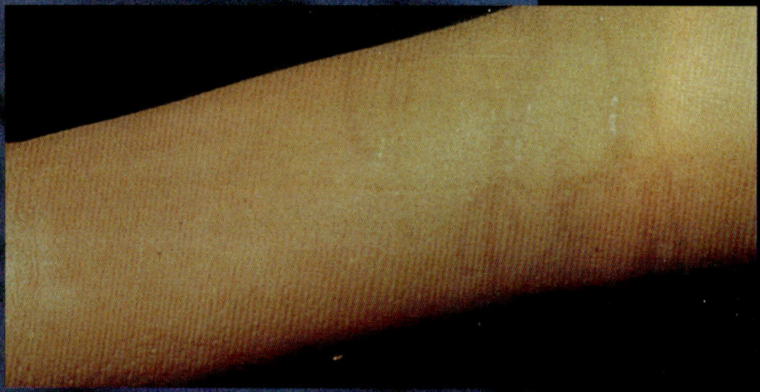

PLATE 8.54
A fresh (30 minutes old) sting on an upper arm from *Olindias sambaquiensis* in Argentina. Note the irregular white welts on the soft inside skin of the arm.

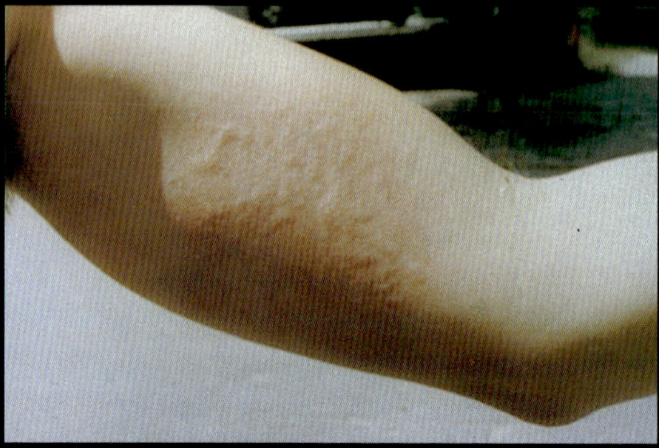

PLATE 8.56
A surface swarm of *Cyanea* spp. photographed during winter, in Chesapeake Bay, on the east coast of the USA. The floating white object is a diaper pail! For detail of a single species see Plate 8.57. (Courtesy of Mike Reber, Solomons, MD, USA)

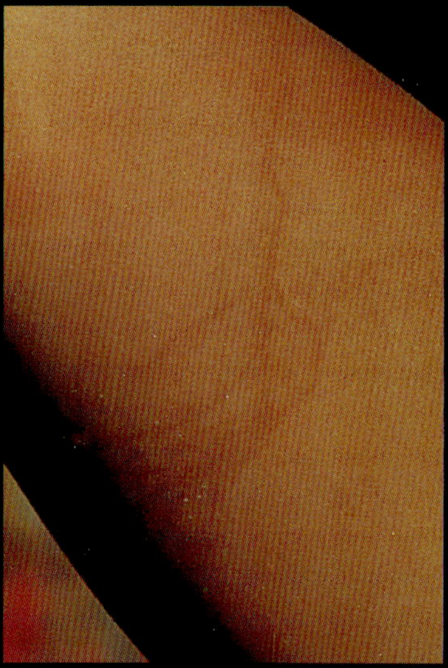

PLATE 8.55B
An *Olindias sambaquiensis* ('agua viva') sting after one week.

PLATE 8.57
Cyanea sp., commonly large (here about 80 cm across the bell), ocean-going and solitary. Note the tentacles — widely spread and of great length.

PLATE 8.58
A *Rhopilema* species. Note the smooth white hemispherical dome of the bell. (Photograph courtesy of Dr Amit Lotan, Tel Aviv, Israel).

PLATE 8.59
Rhopilema hispidum photographed on the surface, north of Roxas, Palawan, Philippines, in March 1994, by Dr Paul Cornelius. The bell is approximately 15 cm in diameter.

PLATE 8.60
Cassiopea xamachana at Girl's Bank, Harbour Island, Bahamas. This animal did not produce a sting on one of the editors (JWB).

PLATE 8.61
A beached specimen of the medusa *Catostylus mosaicus* (Quoy and Gaimard 1824) from Queensland waters. (Photograph courtesy of Dr Richard Willan, Darwin)

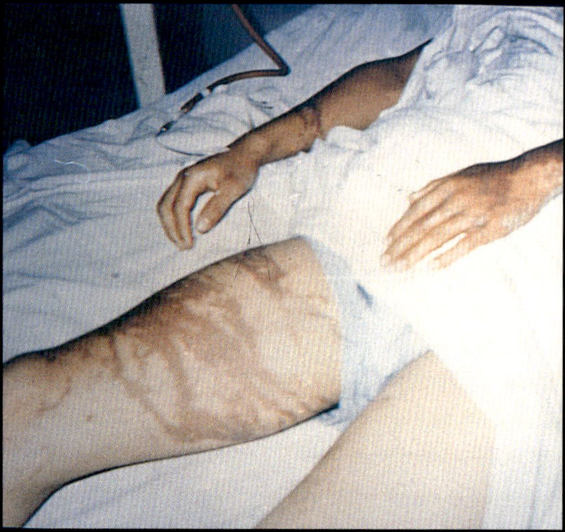

PLATE 8.64
A severe *Stomolophus nomurai* sting on a young adult male swimmer at Beidaihe Beach, China. The patient developed acute pulmonary oedema two hours after the envenomation and took two weeks to make a full recovery. (Photograph courtesy of Dr Zhang Mingliang and Professor Qin Shede)

PLATE 8.62
The dangerous medusa, *Stomolophus nomurai* in Chinese waters. The animal is being supported on the surface with a paddle. (Photograph courtesy of Dr Mingliang Zhang and Professor Qin Shede)

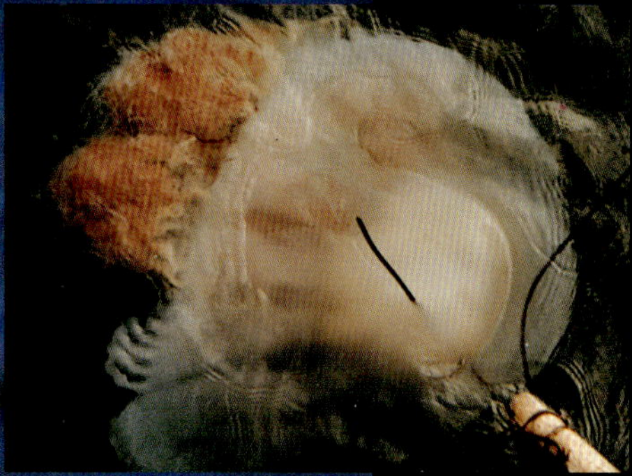

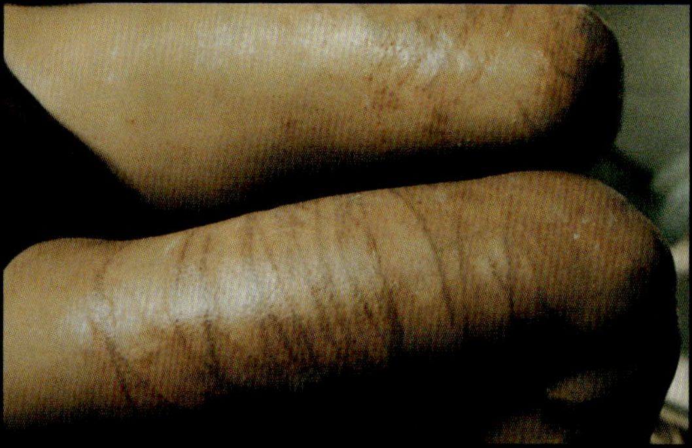

PLATE 8.65
The whip-like wheals of a fatal *Stomolophus nomurai* sting on a previously healthy Chinese male, on a beach east of Beijing, China (MINGLIANG ZHANG 1988). Within two hours of envenomation the patient developed acute pulmonary oedema. He died 24 hours post-envenomation in unresponsive cardiac arrest. (Photograph courtesy of Dr Mingliang Zhang and Professor Qin Shede)

PLATE 8.63
Experience with *Stomolophus nomurai* in China. (Top left) specimen enmeshed in a fisherman's net; (bottom left) a Chinese fisherman cuts off pseudo-tentacles of a large specimen; (top right) some of the oral arm nematocysts; (bottom right) pieces of the severed oral arms lying on the sand; these can still sting. (Photograph courtesy of Dr Mingliang Zhang, and Professor Qin Shede)

PLATE 8.66
A specimen of *Stomolophus meleagris* captured from the sea near Duck, North Carolina, USA.

Plate 8.67
Stomolophus meleagris. One can appreciate the 'white beach ball' description by a stung patient (see text).

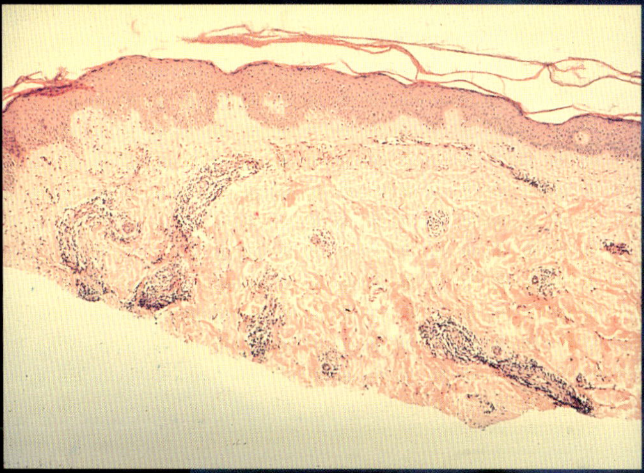

Plate 8.68
Skin biopsy from a recurrent lesion following a sting from the North American *Stomolophus meleagris* (BURNETT & CALTON), showing a mononuclear perivascular infiltrate.

Plate 8.69
Lobonema smithii in Philippine waters. (Photograph courtesy of Professor Thomas Heeger)

Plate 8.70
Phyllorhiza punctata photographed in a large petri dish.

Plate 8.71a
The scyphozoan *Phyllorhiza pacifica*, observed in Shark Bay, Western Australia in January 1994. Specimens were seen in a wide variety of colours. (Photographed by Dr Paul Cornelius).

Plate 8.71b
Phyllorhiza pacifica. The dense blue colouration in some of the specimens may have been due to diet and is evidently not present in all individuals. (Photograph Dr Paul Cornelius)

PLATE 8.72
Pseudorhiza haeckeli. (Photograph courtesy of Neville Coleman)

PLATE 8.73
The bell of *Linuche unguiculata* in profile, showing the typical 'punched out' appearance and the stumpy 'tentacles' (Photograph courtesy of Dr Ronald Larson, US Fish & Wildlife Service, Brunswick, Georgia)

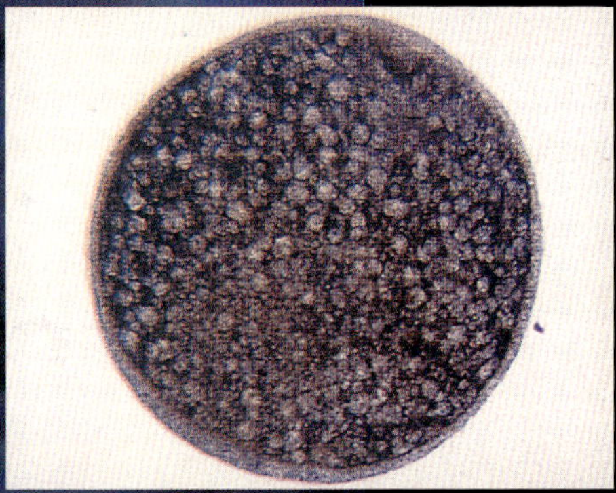

PLATE 8.76
Larval forms of *Linuche unguiculata* (x40). (Courtesy *J Amer Acad Dermatol* 1994; 30: 399–406)

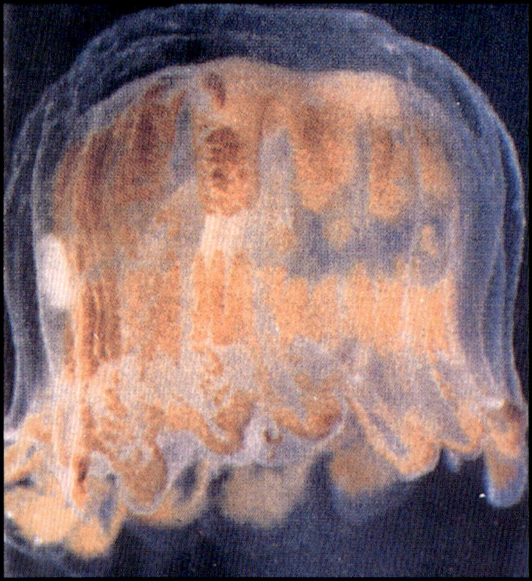

PLATE 8.75
The bell of *Linuche unguiculata* showing a typical brown-pinkish colour due to the presence of symbiotic algae. (Courtesy *J Amer Acad Dermatol* 1994; 30: 399–406)

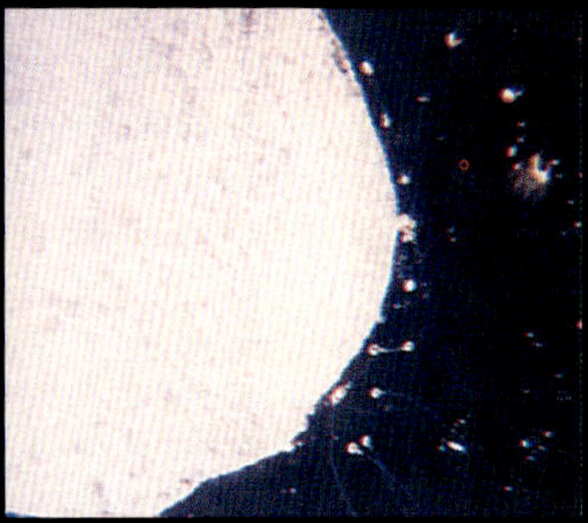

PLATE 8.77
Linuche unguiculata shown with nematocysts firing (x5). (Courtesy *J Amer Acad Dermatol* 1994; 30: 399–406)

PLATE 8.74
The bell of *Linuche unguiculata* from below. (Photograph courtesy of Dr Ronald Larson, US Fish & Wildlife Service, Brunswick, Georgia)

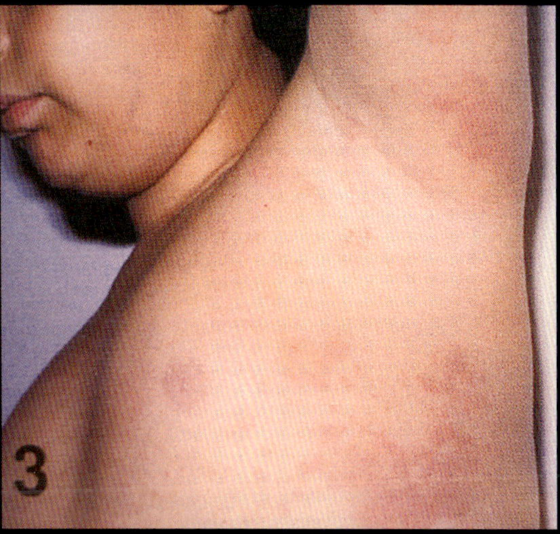

PLATE 8.78
Urticarial eruption from 'seabather's eruption' sting, showing axillary sparing in a prepubescent male. (Courtesy *J Amer Acad Dermatol* 1994; 30: 399–406)

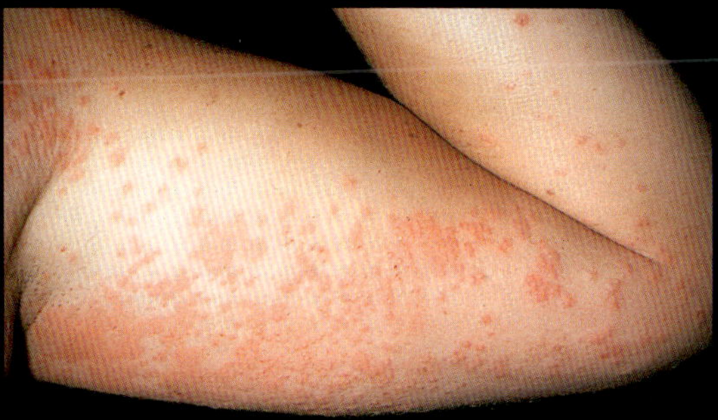

PLATE 8.79
A rash on an arm from *Linuche unguiculata* envenomation on a North American bather. (Courtesy of Dr Ronald Larson, US Fish & Wildlife Service, Brunswick, Georgia)

PLATE 8.80
The medusa form of *Nausithoë punctata*. (Courtesy of Dr Ronald Larson, US Fish & Wildlife Service, Brunswick, Georgia)

PLATE 8.81
A lovely specimen of *Chrysaora hysoscella* (also spelt '*hyoscella*') photographed well submerged. Note the typical brown bell markings. (Courtesy of Mr Ennio Vio, Dipartimento di Biologia, Universita Degli Studi di Trieste, Italy)

PLATE 8.82
The Chesapeake Bay (Maryland, USA) *Chrysaora quinquecirrha*. The bell diameter is about 12 cm. (See also Plate 26.5.)

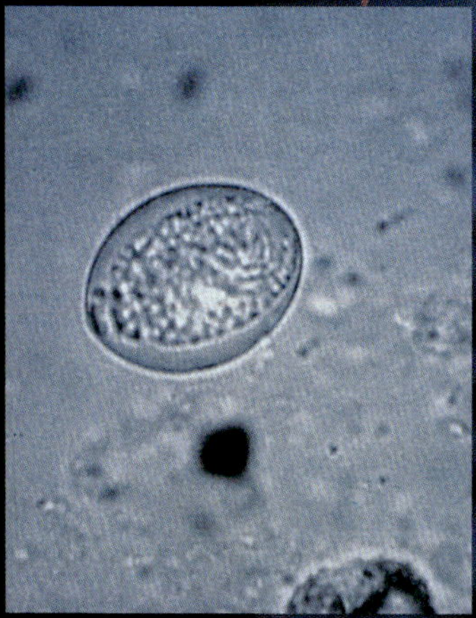

PLATE 8.83
An undischarged, solitary isorhiza nematocyst from the cnidom of the Chesapeake Bay *Chrysaora quinquecirrha*. The nematocyst measured 15 x 25 μm (the diameter of 2–3 human erythrocytes).

PLATE 8.84A
Specimen of *Pelagia flaveola* (original description *P. tahitiana*) after Agassiz & Mayer 1902 (from MAYER 1910). Note the prominent exumbrella warts, taller than wide. (Courtesy of Dr Paul Cornelius, Natural History Museum, London)

PLATE 8.84B
A 25 mm specimen of *Pelagia flaveola* caught at Laing Island, near Madang, Papua New Guinea, on 23 February 1994. (Photograph courtesy of Dr Paul Cornelius, Natural History Museum, London)

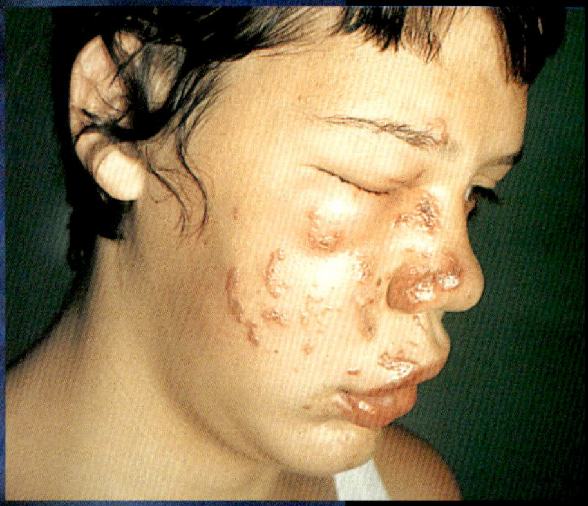

PLATE 8.86B
A 24-hour-old facial sting from the Mediterranean species of *Pelagia noctiluca*, showing large papular lesions and facial and adjacent palpebral oedema. (Courtesy of Dr Franco Kokelj, Trieste)

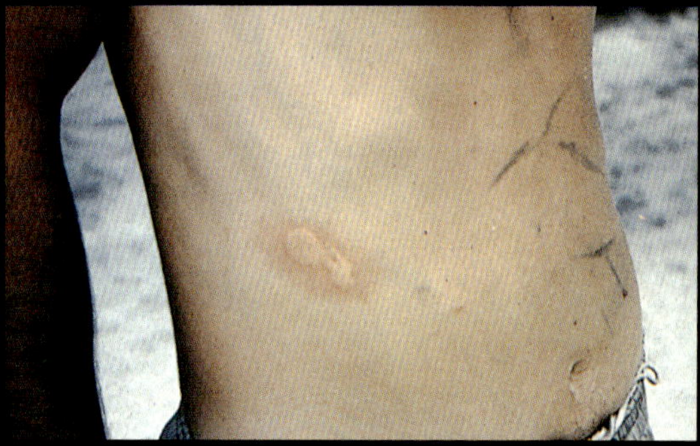

PLATE 8.85
A fresh "mauve stinger" sting in Australia. (Courtesy of Dr Jack Barnes, Cairns)

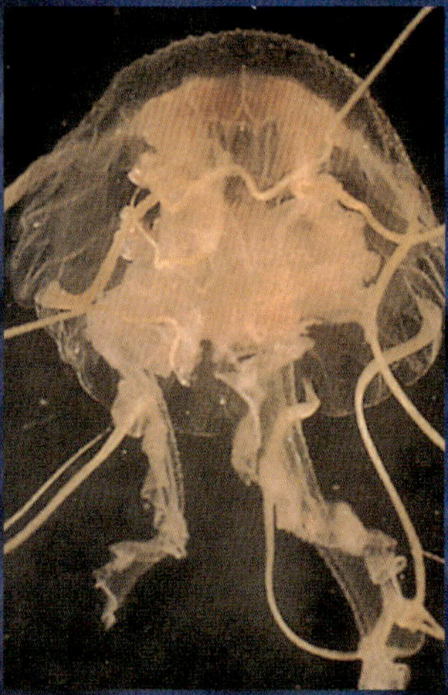

PLATE 8.86A
A pinkish Australian *Pelagia noctiluca medusa* ('mauve stinger'); the colour varies in different parts of Australia. Note the frilly oral arms inside the outer circle of whip-like 'fishing' tentacles. (See also Figure 10.3, page 287.)

PLATE 8.88
A drawing of *Sanderia malayensis* from life by VANHÖFFEN, 1902. Note the flat top bell and the shallow 'skirt' giving attachment to 16 tentacles. The bell may attain a diameter of 100 mm or more. The real life colour of the animal may not be as yellow as predicted here. (Courtesy Dr Paul Cornelius, Natural History Museum, London)

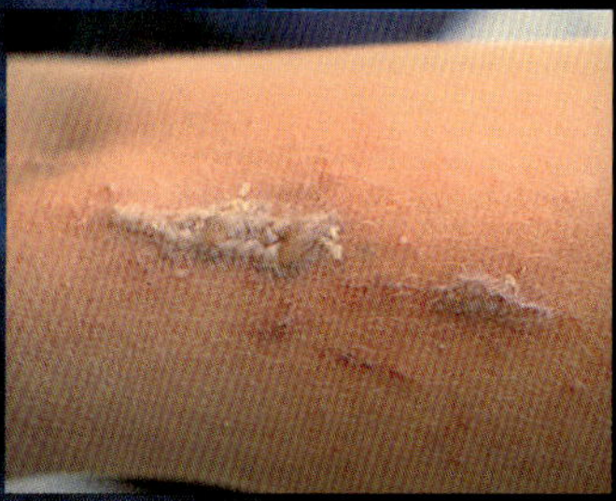

PLATE 8.87
An American *Pelagia noctiluca* sting on a limb after several days. Note the crusting and the adjacent less severe lesions.

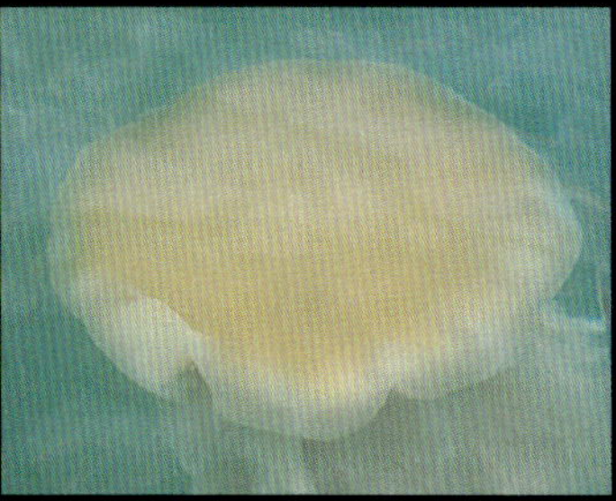

PLATE 8.89
A *Cyanea* sp. in the sea at Repulse Bay, north Queensland.

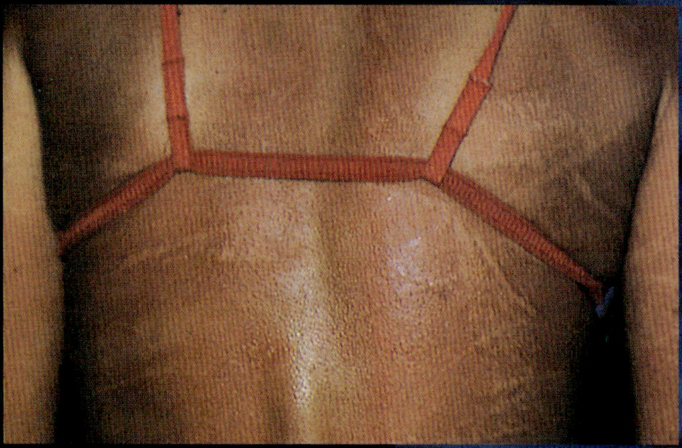

PLATE 8.92
A 1-hour-old Cyanea sting. The patient had only mild discomfort.

PLATE 8.90
A 'Hair jelly' *Cyanea capillata* photographed in Tasmania (latitude 43° south). (Courtesy of Neville Coleman)

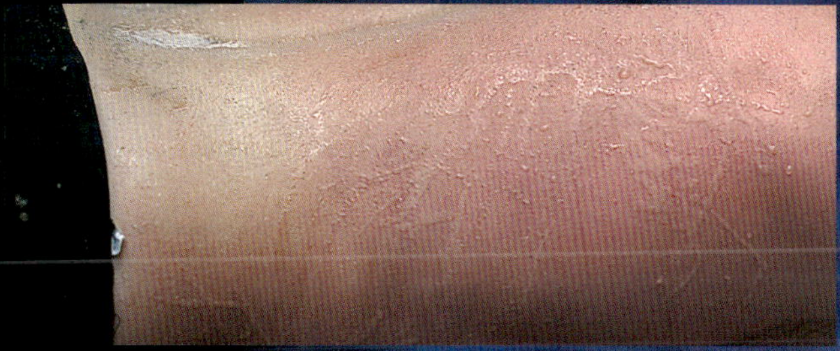

PLATE 8.93
A 90-minute-old *Cyanea* sp. sting on the lateral chest of a healthy Queensland lifesaver. This was treated effectively with ice for pain during this period, which may have contributed to the impressive surrounding erythema. Blanched whealing is still visible.

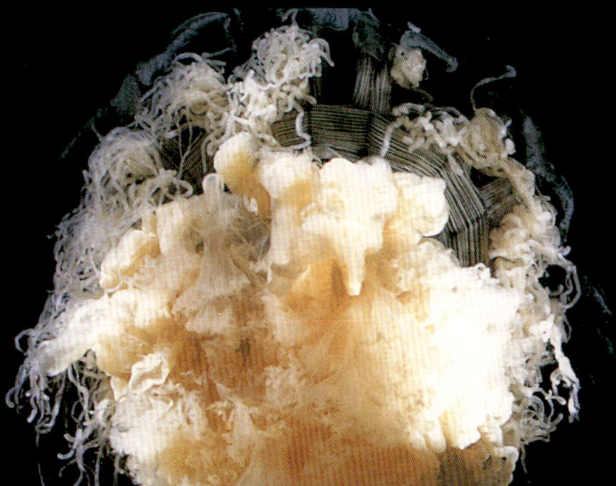

PLATE 8.91
A specimen of *Cyanea capillata* viewed from below. Compare with Plate 6.01.

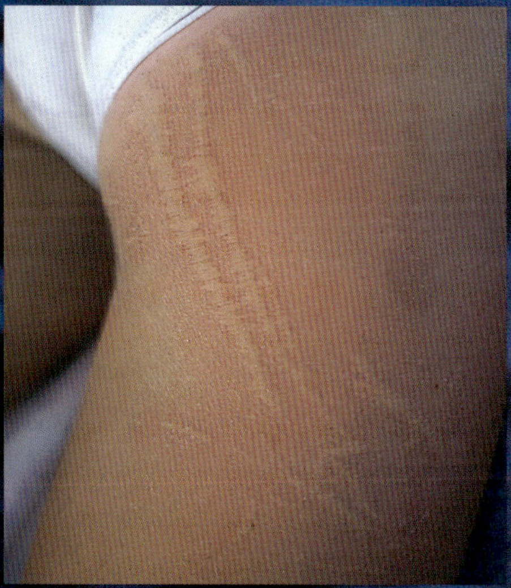

PLATE 8.94
Fresh wheals from a *Cyanea* sp. sting on a 17-year-old female in north Queensland. The characteristic 'saw-tooth' appearance of the wheals is plainly shown. The surrounding red flare is less evident.

PLATE 9.1
In-water photograph of the chirodropid *Chironex fleckeri* SOUTHCOTT, 1956, taken from directly above, showing the typical square-shaped body. Note the four pedalia (hands), each bearing tentacles, and arising from each of the four corners of the trailing edge of the bell. (Courtesy of Dr Bob Hartwick, James Cook University of North Queensland)

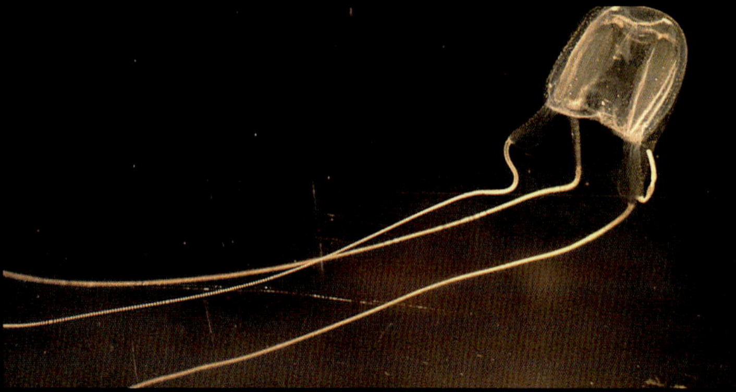

PLATE 9.2
The carybdeid medusa *Carukia barnesi* captured in Alma Bay, Magnetic Island, March 1981, immediately after a serious human Irukandji stinging. One of the four tentacles has been broken off during capture. The bell was about 1 cm in diameter. (Courtesy of Dr Bob Hartwick, James Cook University of North Queensland)

PLATE 9.3
An adult *Chironex fleckeri* medusa captured in Townsville, north Queensland, and photographed under artificially clear conditions by Mr Charles Fitzpatrick of Townsville, a great field pioneer of chirodropid capture and preservation {see Acknowledgments page 25}.

PLATE 9.4
A large carybdeid medusa captured in northern New South Wales. The bell measured 13 cm in height.

PLATE 9.5
An unidentified (at the time) large carybdeid captured in Queensland. The glove provides protection against stings from nematocysts on the bell, a common carybdeid feature.

PLATE 9.6A
An unidentified large stinging carybdeid photographed at Black Rock, Western Australia (latitude 22°46' south).

PLATE 9.6B
The carybdeid shown in Plate 9.6A. Contact with the specimen's bell produced a severe syndrome (see Case History 1 page 242). Its vertical height approached 50 cm. (Courtesy of Pat Baker, Western Australian Maritime Museum).

PLATE 9.7
An adult large carybdeid medusa (thought to be a species of *Tamoya*, or 'fire jelly'). The specimen was captured in north Queensland waters and photographed in a specially lit seawater tank. Note the nematocyst 'warts' covering the large bell (see Plate 9.16), and the four wide, ribbon-like tentacles (see text). (Photograph courtesy of Ben and Eva Cropp, Port Douglas, Qld.).

PLATE 9.8
A specimen of *Carybdea alata* photographed in Hawaiian waters in June 1993. The animal is visible in the lower half of the field, partially obscured by surface reflections. Compare with Plates 9.10 and 10.20. (Courtesy of Mike Nolan, Hawaii)

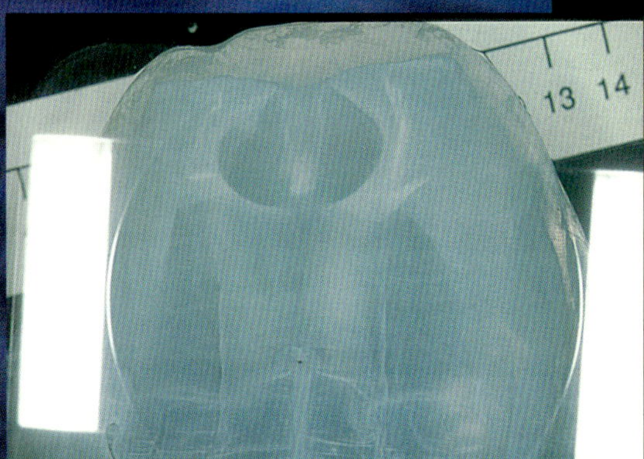

PLATE 9.9A
Chiropsalmus buitendijki = *Chiropsoides buitendijki* captured by a diver in Sri Lanka in 1991. Specimen viewed from the side. (Photograph courtesy Dr Malik Fernando, Colombo, Sri Lanka)

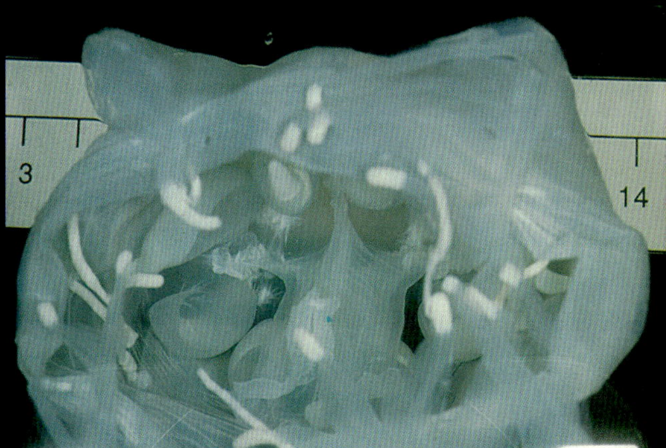

PLATE 9.9B
The specimen shown in Plate 9.9A, viewed from below. (Photograph courtesy of Dr Malik Fernando, Colombo, Sri Lanka)

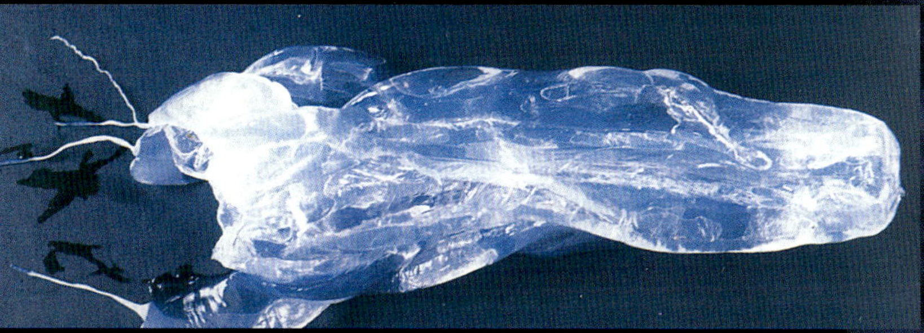

PLATE 9.10
A specimen captured in Oman in August 1991, of either *Tomoya* sp. or *Carybdea alata grandis*, with a bell 30cm tall. See also Plates 18.2. (Photograph courtesy of Professor Reginald Victor, Sultan Qaboos University, Muscat).

PLATE 9.14
A 'laddering' pattern on the skin, left by the Morbakka sting shown in Plate 9.13 and reflecting the arrangement of the nematocyst batteries on the tentacle. (cf. Plate 9.78) (Photograph courtesy Dr Frank Carroll, Dunwich, Stradbroke Island).

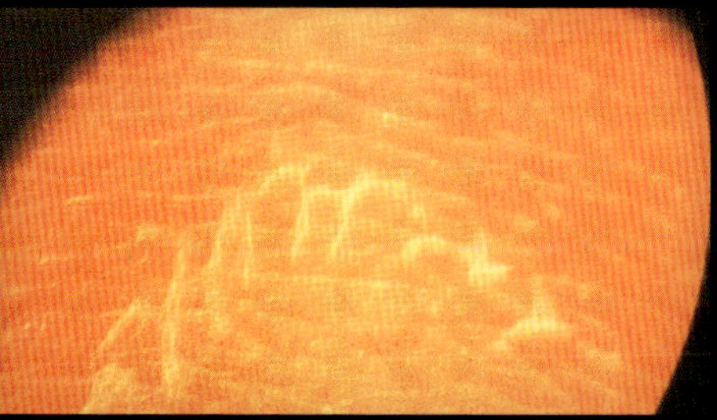

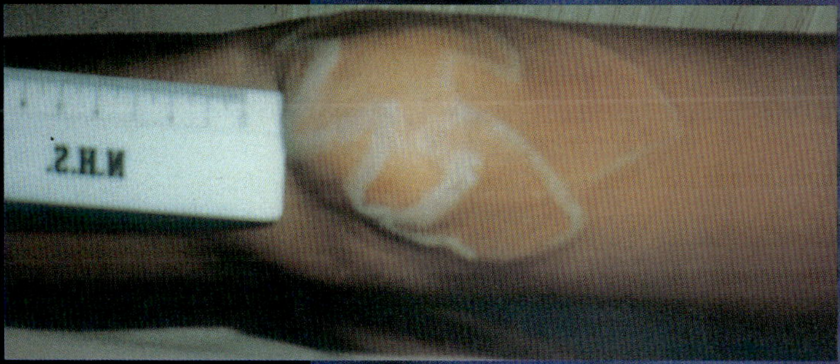

PLATE 9.13
A fresh Morbakka sting on the knee of a swimmer on Stradbroke Island, south Queensland, photographed in January 1987. Note the white whealing, the 'frosted' appearance and the tapering of the tentacle marks. (Courtesy Dr Frank Carroll, Dunwich, Stradbroke Island)

PLATE 9.12
The same sting as shown in Plate 9.11, photographed three days later. (Photograph courtesy of Dr Ralph Skinner, Queensland)

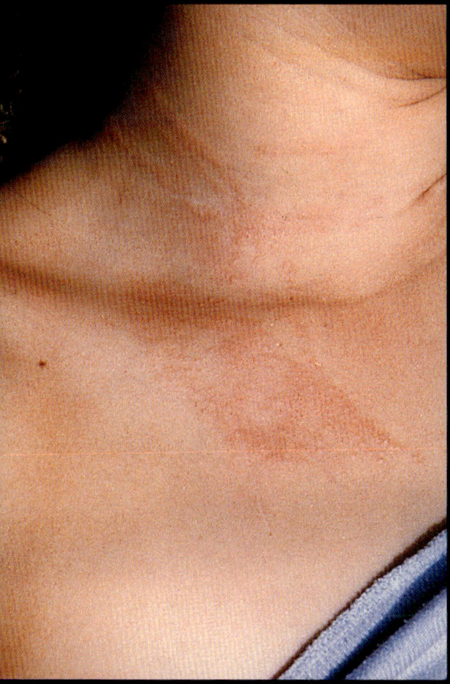

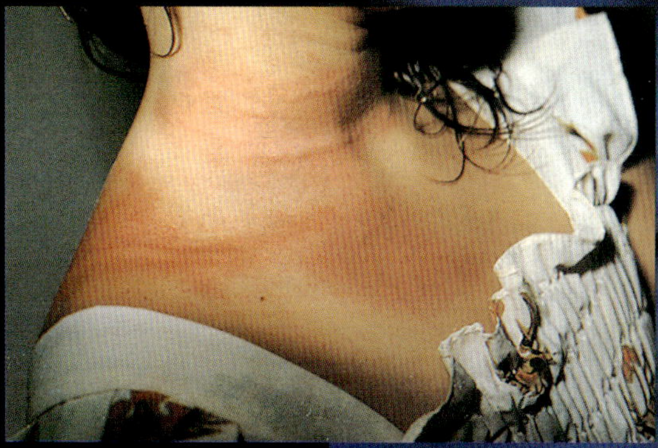

PLATE 9.11
A fresh (15 minutes old) 'fire jelly' sting which occurred in Moreton Bay ('Moreton Bay stinger'), south Queensland. Mild to moderate pain accompanied this sting. (Photograph courtesy of Dr Ralph Skinner, Queensland)

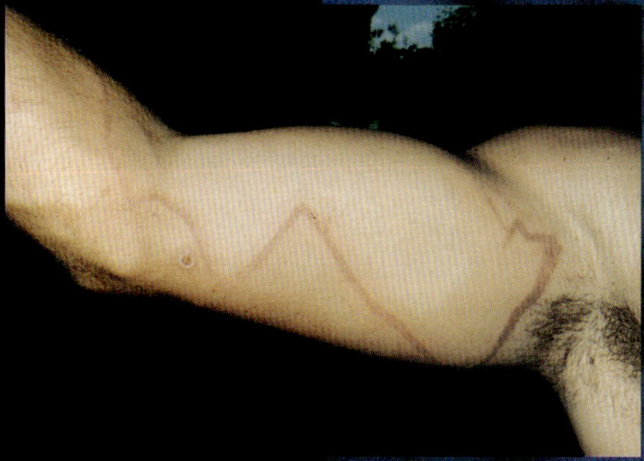

PLATE 9.15
A significant Morbakka sting from Bribie Island, south Queensland, photographed at day 3, showing early skin necrosis. Note the tapering tentacle mark (cf. Plate 9.13). This patient developed acute 'Irukandji-like' systemic symptoms (see text).

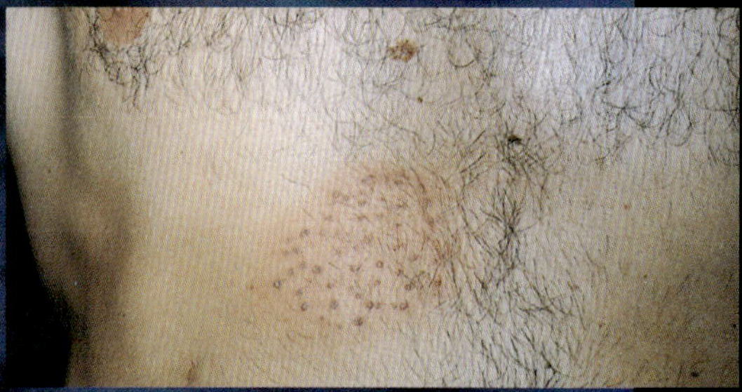

Plate 9.16
The clear imprint of a largish, nematocyst-bearing bell on the right chest of the patient pictured in Plate 9.15. Note the multiple punctate lesions within the erythematous area of bell contact, corresponding to the distribution of nematocysts on the bell.

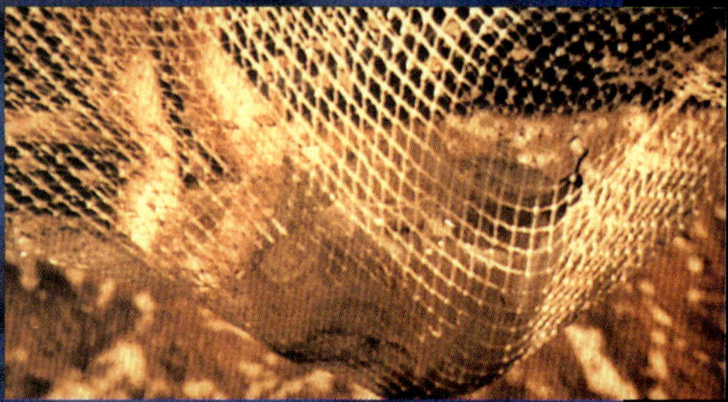

Plate 9.17
The original mesh size of the Mark I Stinger nets, which were completely effective in keeping out chirodropids (shown here captured within the net), but allowed some tiny Irukandji jellyfish to pass through, producing stings on bathers within the enclosure. This problem has now been controlled by the Department of Civil and Systems Engineering at James Cook University of North Queensland.

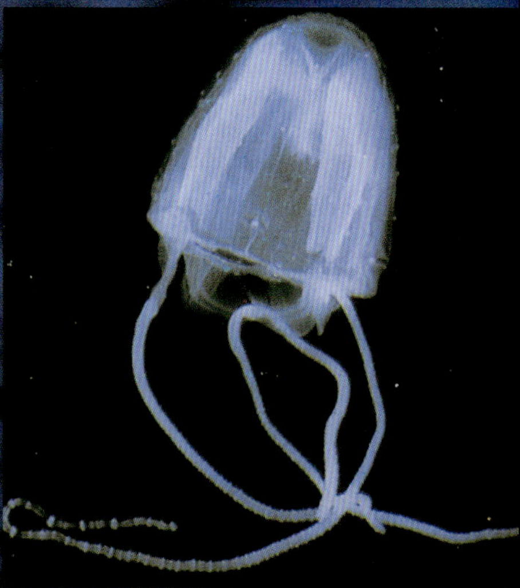

Plate 9.18
Carukia barnesi, the 'original' Irukandji jellyfish species, photographed by Dr Jack Barnes in Cairns.

Plate 9.19
Magnified photograph of a relaxed *Carukia barnesi* tentacle, showing the 'batteries' of undischarged nematocysts located on the series of encircling rings along the tentacle length.

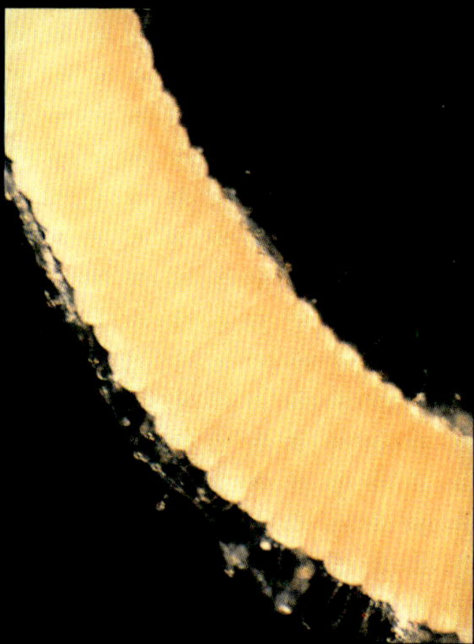

Plate 9.20
As in Figure 9.19, but showing discharged nematocysts. Note that the tentacle 'fired' by exposure to methylated spirits has also contracted in length. (Courtesy of Dr Bob Hartwick, James Cook University of North Queensland)

Plate 9.21
Photograph showing the 'pocket-handkerchief-like' tails as in Plate 9.22 (Photograph courtesy of Dr Ron Southcott, Adelaide).

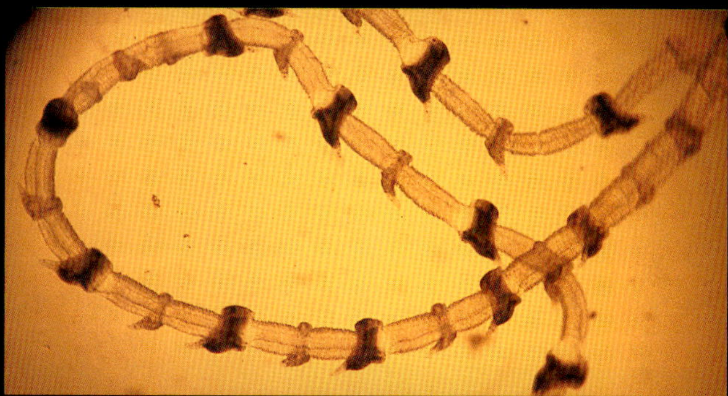

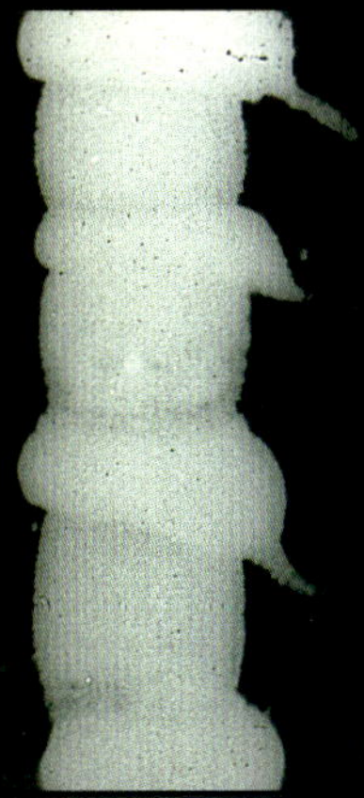

PLATE 9.22
As in Plate 9.21. (Photograph courtesy of Dr Ron Southcott, Adelaide)

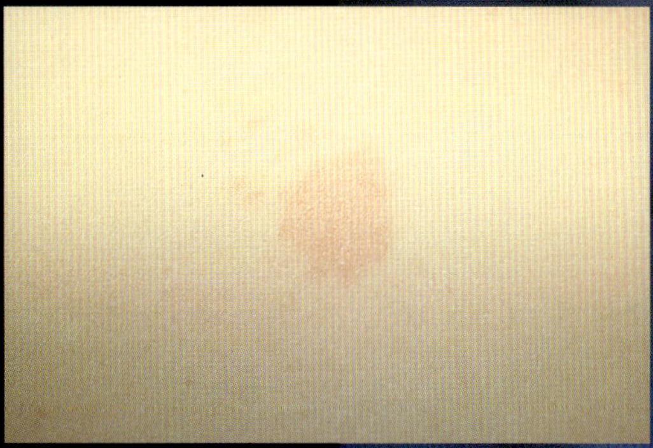

PLATE 9.24
The small imprint of the bell (but no tentacle marks) of an Irukandji jellyfish sting in a young female, taken 36 hours after envenomation.

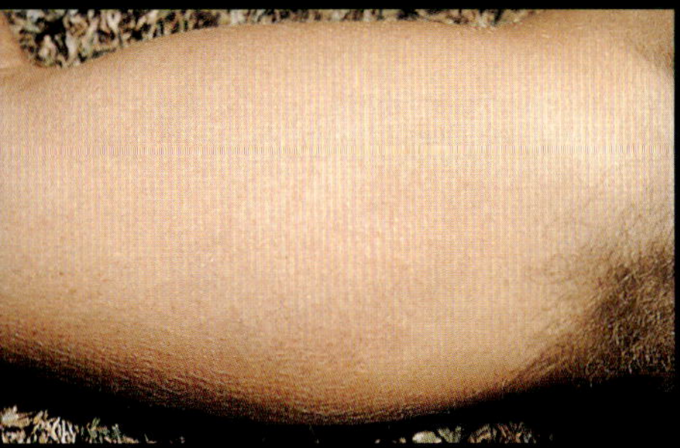

PLATE 9.23
A fresh *Carukia barnesi* (Irukandji) sting in North Queensland. There is the faintest area of erythema only, on the inside upper arm; however, a local 'goose-pimple' reaction is obvious. The patient developed a full-blown Irukandji syndrome some 30 minutes later. (Photograph by Dr Jack Barnes, Cairns)

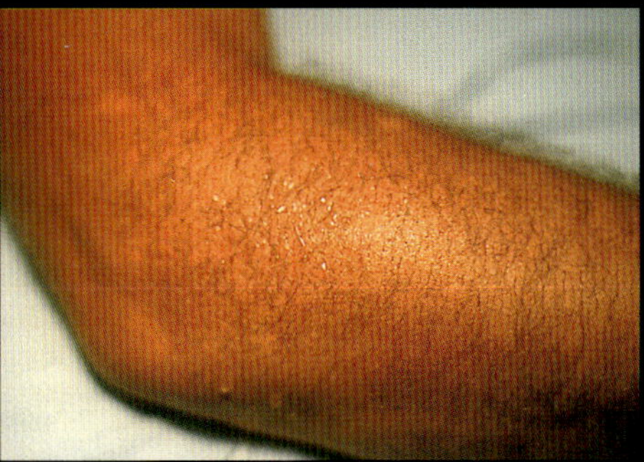

PLATE 9.25
Photograph of a victim of Irukandji syndrome showing localised sweating on right forearm. The patient had been stung on the right upper arm six hours previously.

PLATE 9.26
Carybdea rastoni ('Jimble'). This small carybdeid commonly measures 3 cm across the bell and may appear in swarms in the shallows of South Australian and Western Australian waters. Note that no mammillations are visible on the bell. (Photograph by Neville Coleman)

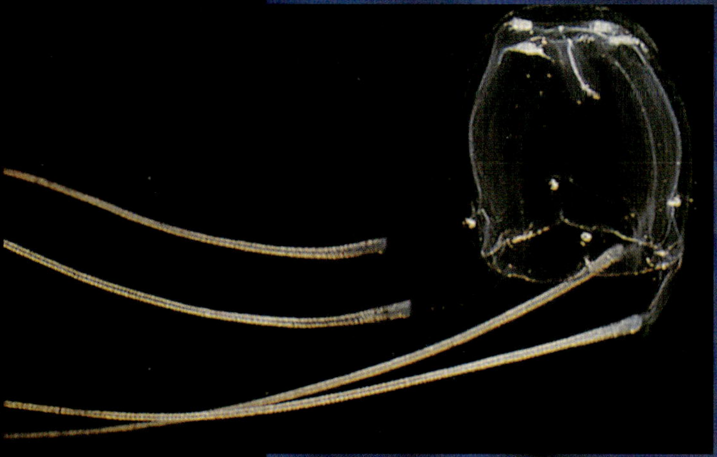

PLATE 9.27
Carybdea rastoni photographed in 2 m depth of water in Edithburgh, South Australia. Note apparent mammillations visible on the bell. Stinging from contact with the bell does not normally occur. (Photograph courtesy of Karen Gowlett-Holmes, Collection Manager: Marine Invertebrates, South Australian Museum, Adelaide).

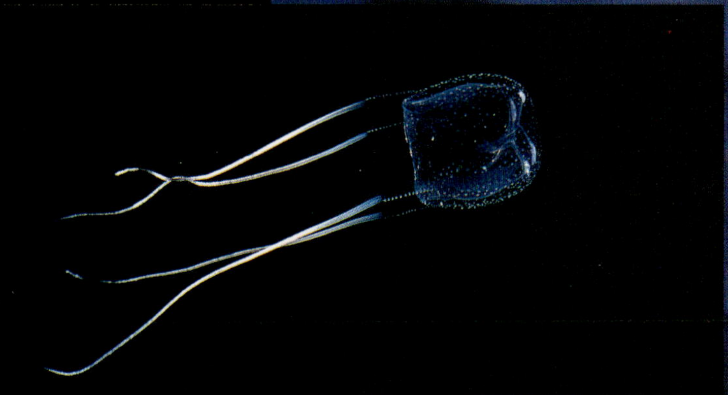

PLATE 9.28
Carybdea rastoni again photographed at Edithburgh, South Australia, with no 'mammillations' visible on the bell. However, clearly visible low on each of the four sides of the bell is an oscellus (eye). The animal exhibits clear light/dark perception under natural conditions, actively avoiding shadows. (Courtesy of Karen Gowlett-Holmes, Collection Manager, Marine Invertebrates, South Australian Museum, Adelaide.)

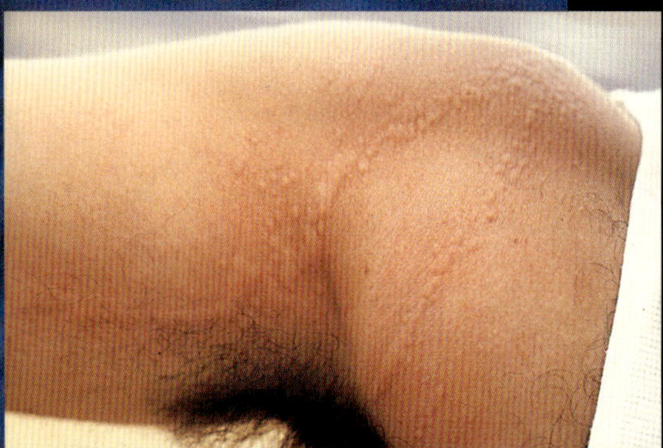

PLATE 9.30
A fresh minor sting from a jimble (*Carybdea rastoni*) on a lifesaver, on Cottesloe Beach, Perth, in April 1987.

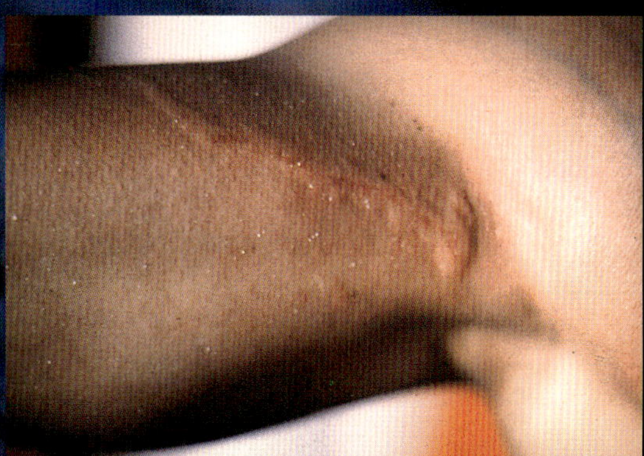

PLATE 9.31A
A fresh, more severe sting from a jimble *Carybdea rastoni* on a West Australian lifesaver on Mullaloo Beach, Perth.

PLATE 9.29
A South Australian jimble (*Carybdea rastoni*) digesting a captured fish within its bell. (Photograph courtesy Karen Gowlett-Holmes, Collection Manager, Marine Invertebrates, South Australian Museum, Adelaide)

PLATE 9.32
A specimen of small carybdeid jellyfish captured and photographed in Sydney Harbour by Dr Carl Edmonds. Species identification remains uncertain. (See Plate 9.33.)

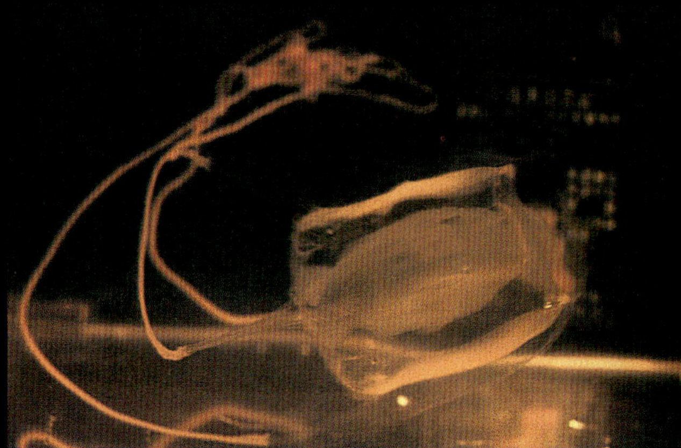

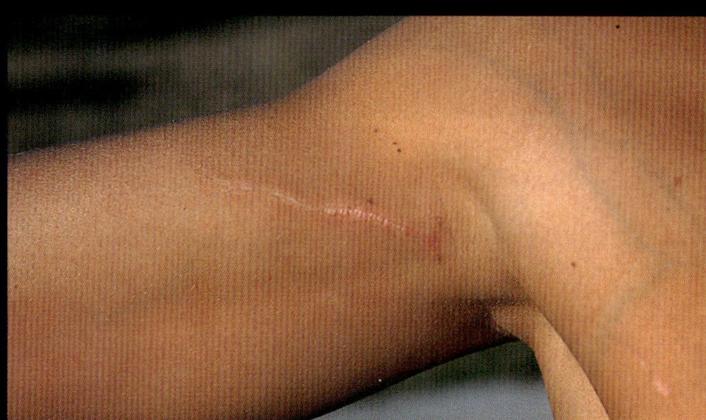

PLATE 9.31B
The scarring resulting from the sting shown in Plate 9.31A, pictured three months later

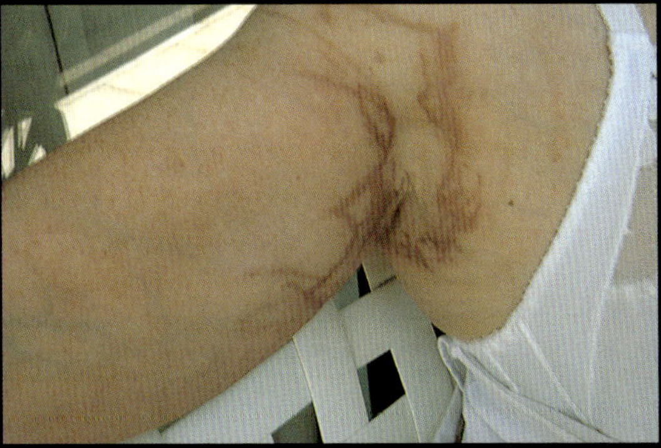

PLATE 9.33
A seven-day-old carybdeid jellyfish sting, which occurred in Sydney Harbour waters during a particularly virulent 'epidemic' of small carybdeid stingings in October 1993. Note the superficial skin necrosis (Photograph courtesy of Dr Carl Edmonds, Sydney).

PLATE 9.34
The northern Australian box jellyfish *Chironex fleckeri*, one of the world's most venomous creatures. (Photograph courtesy of Dr Bob Hartwick, James Cook University of North Queensland).

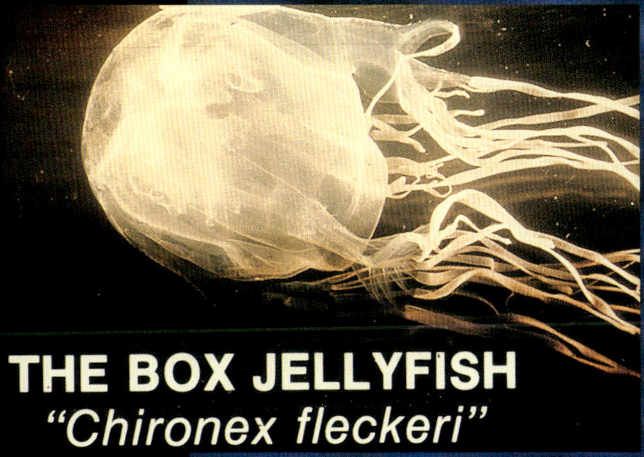

PLATE 9.35
A mature Indo-Pacific *Chiropsalmus* sp. from the Philippines, swimming horizontally with tentacles in differing states of relaxation, following stimulation from touching by the photographer. Note the purple and white banding visible particularly on the more contracted tentacles, which are rounded in cross-section. (Photograph courtesy of Professor Thomas Heeger, University of San Carlos, Philippines)

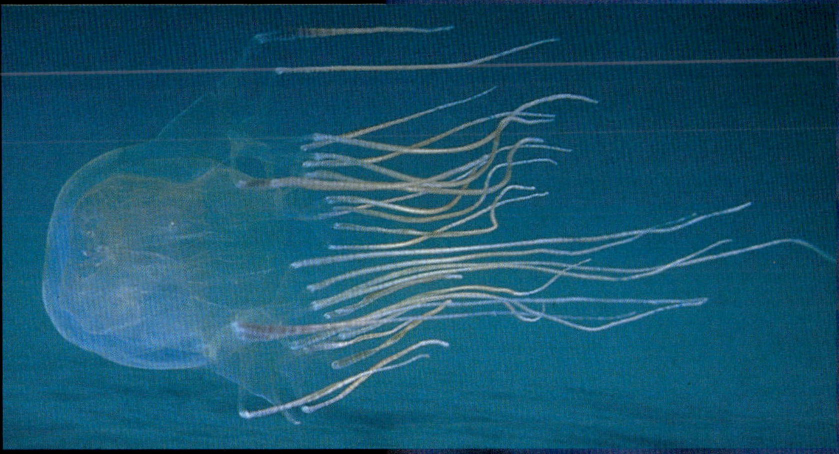

PLATE 9.36
The dangerous Atlantic/Caribbean chirodropid, *Chiropsalmus quadrumanus*. (Photograph courtesy of Professor 'Chuck' and Mrs Bertha Cutress, Puerto Rico)

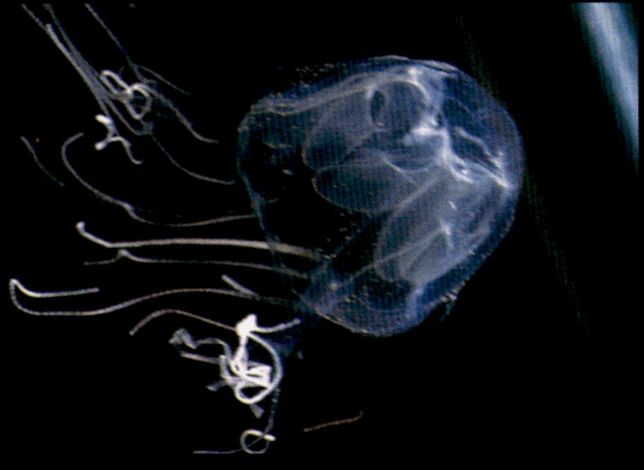

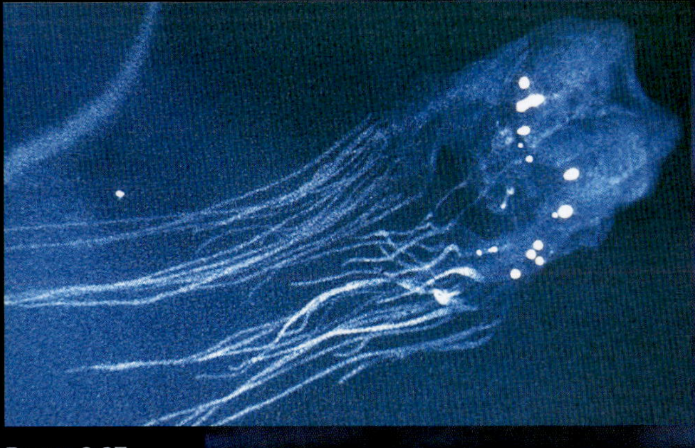

PLATE 9.37
A remarkable night photograph by a diver, of an unidentified chirodropid medusa swimming in the Andaman Sea in 1983. (Courtesy of Lt Col. (Ret.) A.B. ('Dusty') Rhodes, Bangkok, Thailand)

PLATE 9.38A
The Japanese '*Habu-kurage*' (see text) photographed at Chatan, Okinawa Island. The precise identity of this animal is being investigated, but it has gross similarities to *Chiropsalmus quadrigatus sensu* HAECKEL. (Courtesy of Dr Yasuhiro Tomihara, Okinawa Prefecture Institute of Public Health, Japan).

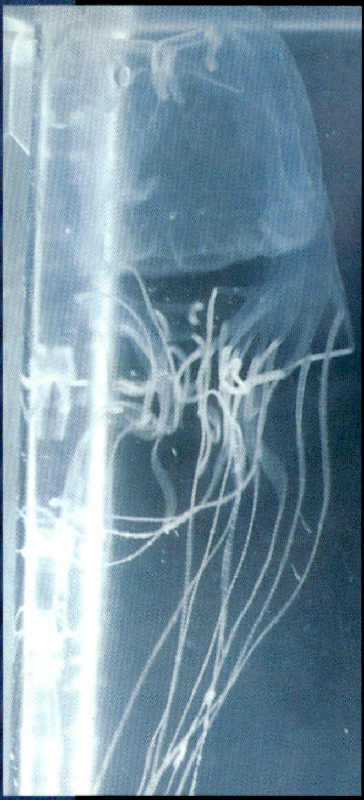

PLATE 9.38B
'*Habu-kurage*', as in Plate 9.38A. (Courtesy Dr Yasuhiro Tomihara, Okinawa Prefecture Institute of Public Health, Japan)

PLATE 9.40A
Preserved chirodropid specimen captured in Brunei in 1993. Precise identification is difficult, but studies are continuing.

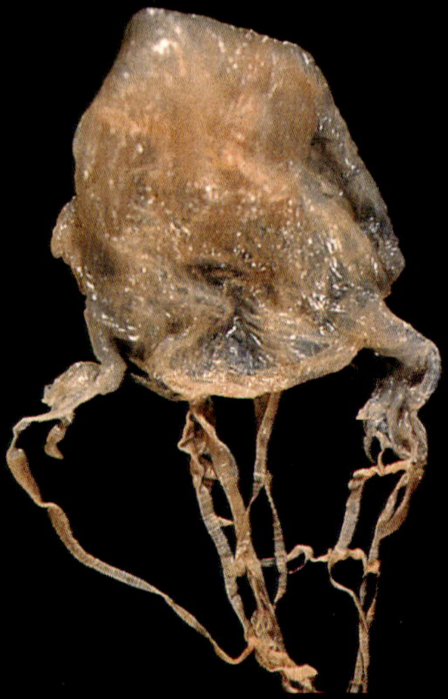

PLATE 9.40B
Preserved chirodropid specimen captured in Brunei in 1993.

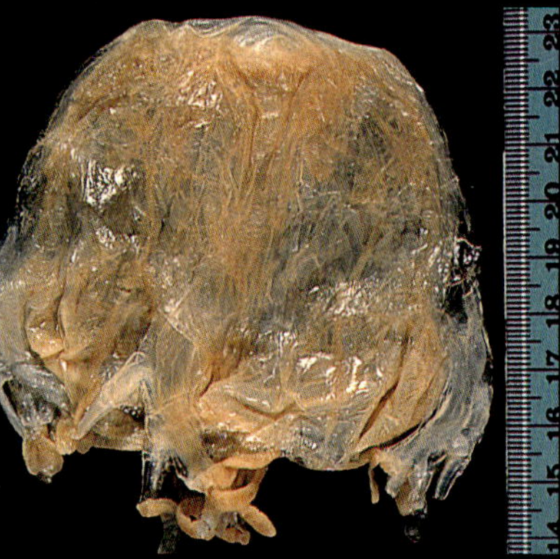

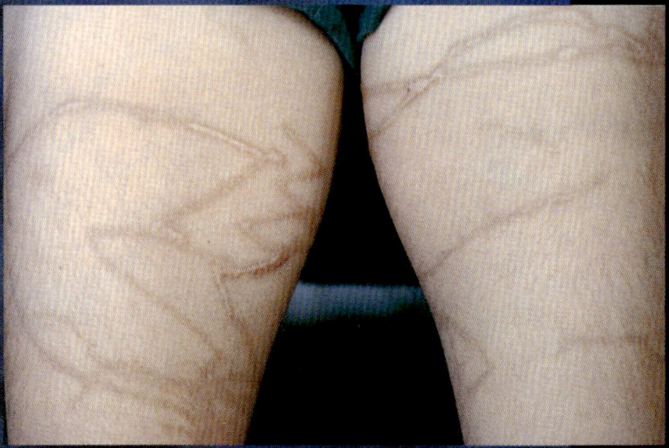

PLATE 9.39
A seven-year-old Japanese girl bearing two-year-old scars from a chirodropid sting (probably *Chiropsalmus quadrigatus sensu* HAECKEL) sustained at Shioya Bay, Okinawa Island, on 13 August 1989. Note the patch of keloid formation on the right thigh. (Compare with Plate 5.19C) (Courtesy of Dr Yasuhiro Tomihara, Okinawa Prefecture Institute of Public Health, Japan)

PLATE 9.41
An adult *Chironex fleckeri* SOUTHCOTT, 1956.

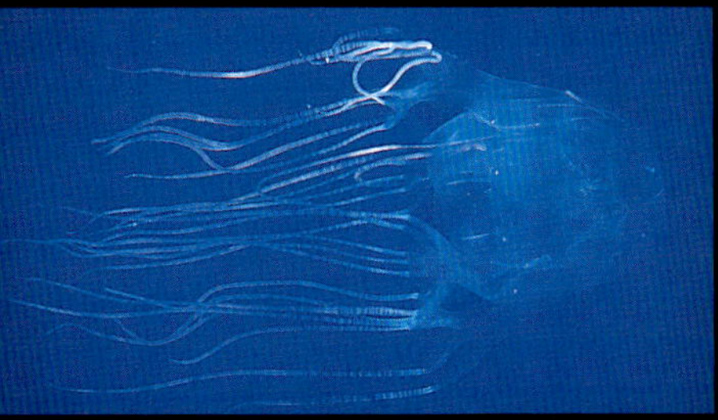

PLATE 9.42
A chirodropid, photographed in Philippine waters by Professor Thomas Heeger, University of San Carlos, Philippines.

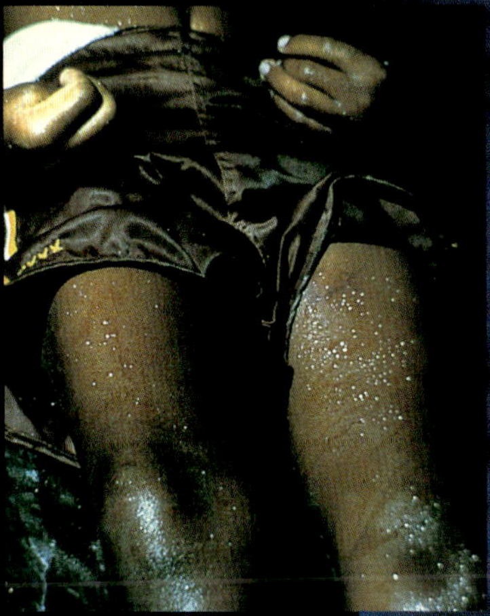

PLATE 9.45
Fatal massive *Chironex fleckeri* sting in a northern Australian aboriginal child (see text, and VIMPANI *et al* 1988). Note the lack of acute inflammatory response (cf. Plates 1.2, 5.14, 5.19H and 9.44). This is partly because of the rapid demise (approximately 5 minutes) of this unresuscitated patient, and partly, contributed to by the dark skin colour. Extensive tentacle marks are still visible all over both legs, indicating massive envenomation. (The flexural deformity of the hands was partially pre-existing.)

PLATE 9.43
A magnified close-up photograph *in vivo* of a portion of the tentacle mass of a large adult *Chironex fleckeri*, showing the robust ribbon-like shape. Note the transverse barring pattern on the tentacle surfaces (see text). (Courtesy of Dr Bob Hartwick, James Cook University of North Queensland).

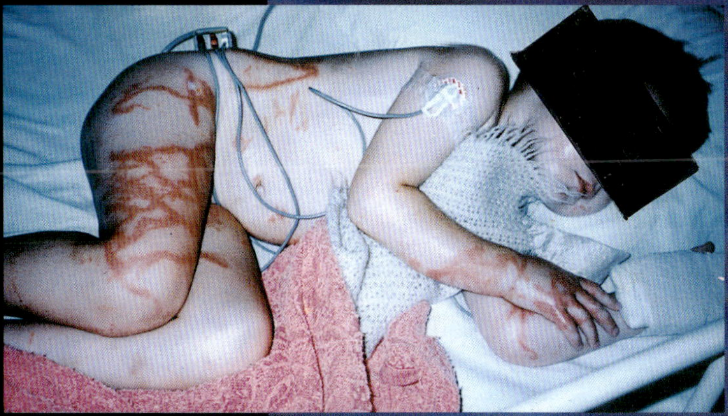

PLATE 9.46
A life-threatening chirodropid envenomation in a three-year-old child in the waters of Pulau Sapi, Labuan, in 1993. The patient survived following resuscitation from a cardio-respiratory crisis with loss of consciousness. (Photograph courtesy of Joehann Angkie of the *Sabah Times*, and Associate Professor Nor Azila, Universiti Malaya)

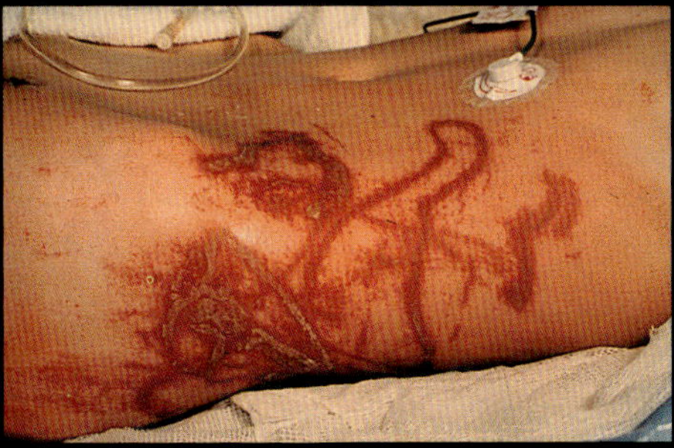

PLATE 9.44
A massive, near-fatal *Chironex fleckeri* envenomation in north Queensland; the young male patient survived a cardiac arrest. Note the full-thickness skin death in most areas of tentacle contact. 'Frosting' (see text) is visible over the lower left hip. Note also the large (wide) tentacle marks indicating a large adult animal. (Photograph by Dr Jack Barnes, Cairns)

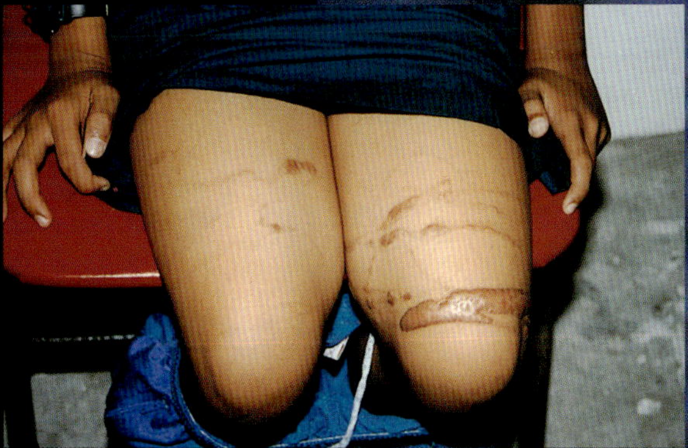

PLATE 9.47
One-year-old scars on the legs of an 11-year-old Labuan Island (North Borneo) survivor of a serious chirodropid sting in 1993 (species identity uncertain). This patient arrived at hospital unconscious. No antivenom was available. Note the vigorous keloid formation. (Courtesy of Associate Professor Nor Azila and Dr Iekhsan Othman, Universiti Malaya)

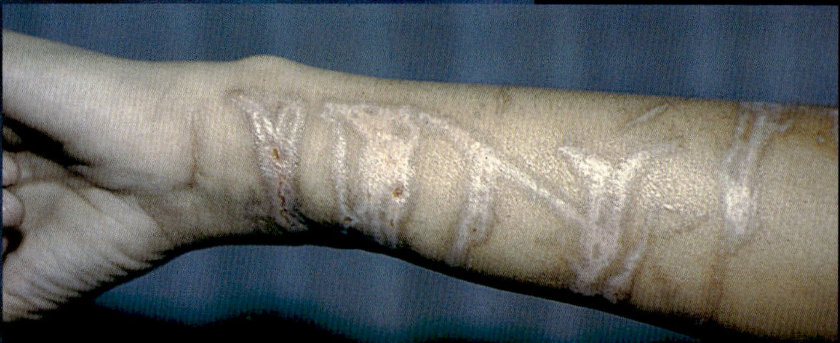

PLATE 9.48
Extensive scarring, with some residual ulceration, on the arm of a young Thai fisherwoman, stung in the waters of the Gulf of Thailand in 1980, and photographed several months later. The patient received no specific acute medical treatment at the time of the sting. The pattern is compatible with, but not diagnostic of a chirodropid sting. (Courtesy of Dr P Soontornpalin, Division of Dermatology, Mahidol University, and Dr K Pranich, Dermatologist, Bangkok)

PLATE 9.50
An example of a north Queensland stinger warning sign erected right on the beach. Tourists have been known to hang their clothing on such signs while they go swimming!

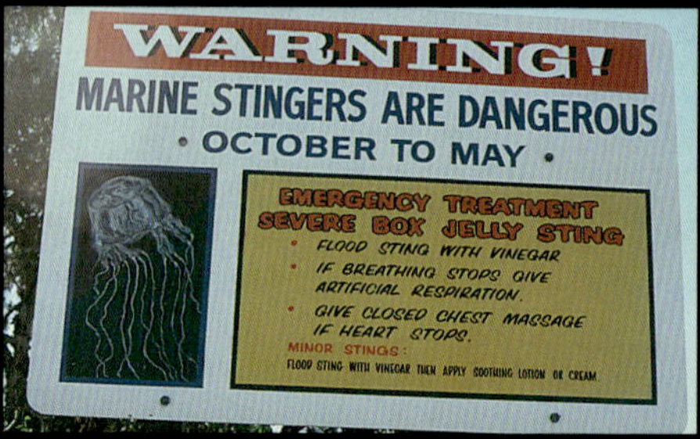

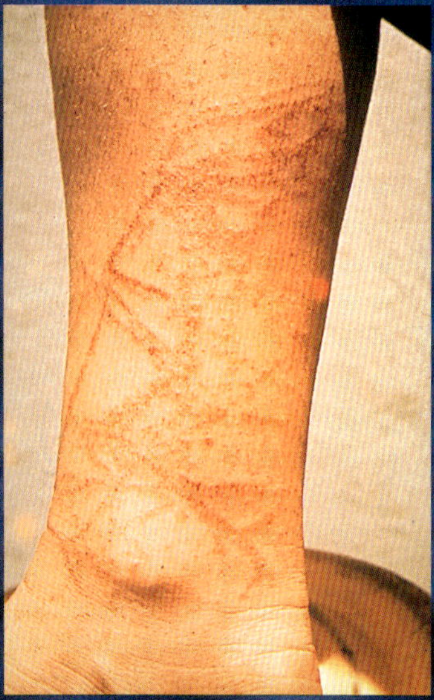

PLATE 9.49
A 24-hour-old sting on the leg from the Australian *Chiropsalmus* sp. Note the thin tentacle marks and the less vigorous inflammatory response.

PLATE 9.51
Dr Bob Hartwick, with Dr Jack Barnes and Ron Pollard (in the boat), three of the world's great chirodropid envenomation pioneers {see Chapter 1}, 'chirodropid-hunting' under the ideal shoreline and climate conditions of hot weather, calm water and gently sloping sandy beach, in Nhulunbuy (Gove), Northern Territory, Australia.

PLATE 9.52
The polyp stage of *Chironex fleckeri*. (Courtesy of Dr Bob Hartwick, James Cook University of North Queensland)

PLATE 9.53
The free-swimming juvenile stage of *Chironex fleckeri*. (Courtesy of Dr Bob Hartwick, James Cook University of North Queensland)

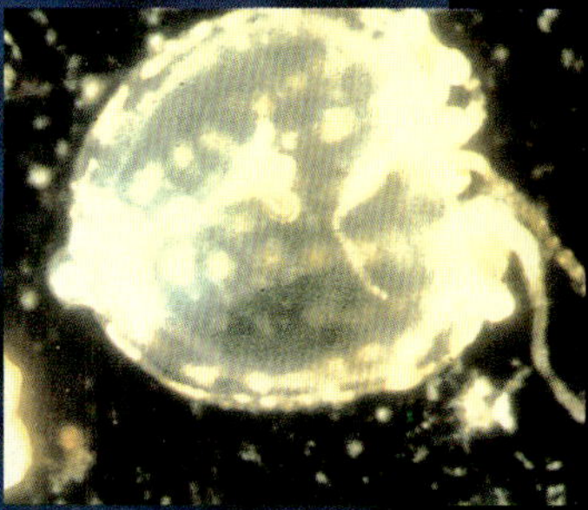

PLATE 9.54
A captive hawksbill turtle eating an adult *Chironex fleckeri*. (Photograph courtesy of Ben Cropp Productions, Port Douglas, Queensland)

PLATE 9.55
A butterfish photographed in the act of eating a young (inverted) box jellyfish *Chironex fleckeri*. (Courtesy Ben Cropp Productions, Port Douglas, Queensland)

PLATE 9.56
A large adult *Chironex fleckeri* occupies almost the entire field of the photo, swimming from left to right. Even under such 'ideal' conditions for spotting (clear sunlit calm seawater) the animal is still difficult to see. Compare with Plate 5.8. (Courtesy of Dr Bob Hartwick, James Cook University of North Queensland).

PLATE 9.57
Microscopic section (x400) of an unfired *Chironex fleckeri* tentacle piece, stained to show the large, plum-coloured, cigar-shaped, dangerous-venom-bearing microbasic mastigophore nematocysts. (Compare with Plate 6.12) (Photograph courtesy of Dr Bob Hartwick, James Cook University of North Queensland)

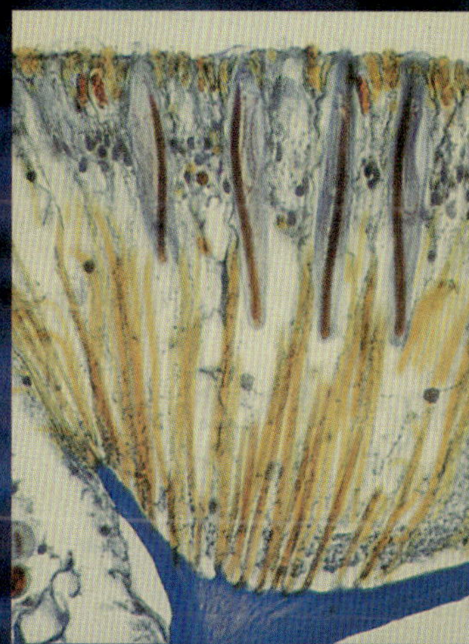

PLATE 9.58
A major Northern Territory *Chironex fleckeri* sting (in which specific *Chironex fleckeri* antivenom was administered), with a less common sting distribution involving the lower back, as well as the posterior upper thighs. Protection afforded by the swimming trunks is evident. Note the vesiculation on the welt just above the natal cleft. (Courtesy of Menzies School of Health Research, Darwin)

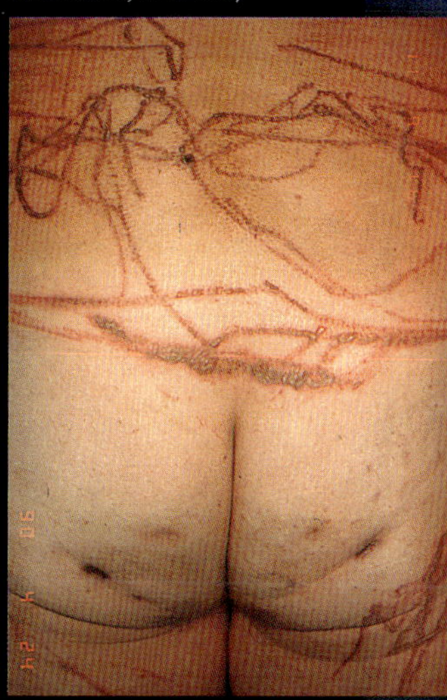

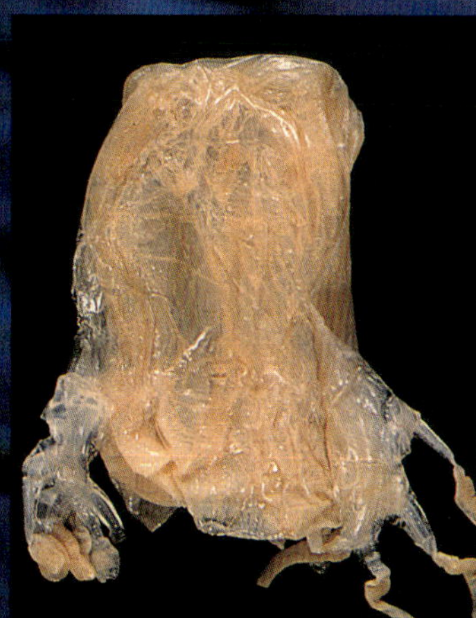

PLATE 9.59
A preserved specimen of *Chironex fleckeri* captured in Brunei in 1993. (Specimen courtesy of Major [Dr] R. Hooper, Labuan, Sabah)

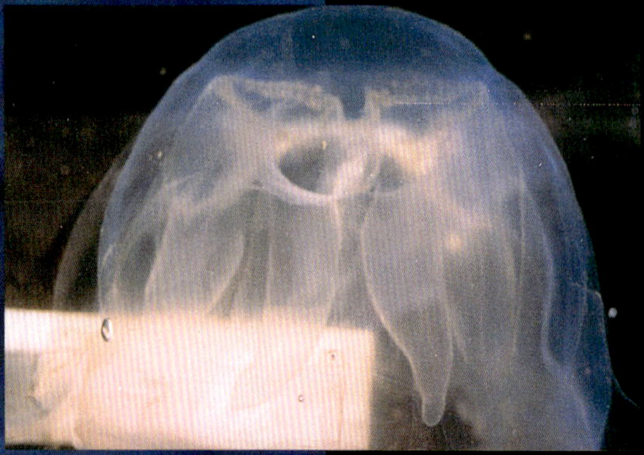

PLATE 9.60
The bell of a specimen of *Chiropsalmus buitendijki*, captured in deep water in Sri Lanka in 1993, showing some of the eight elongated gonads. (Courtesy Dr Malik Fernando, Colombo)

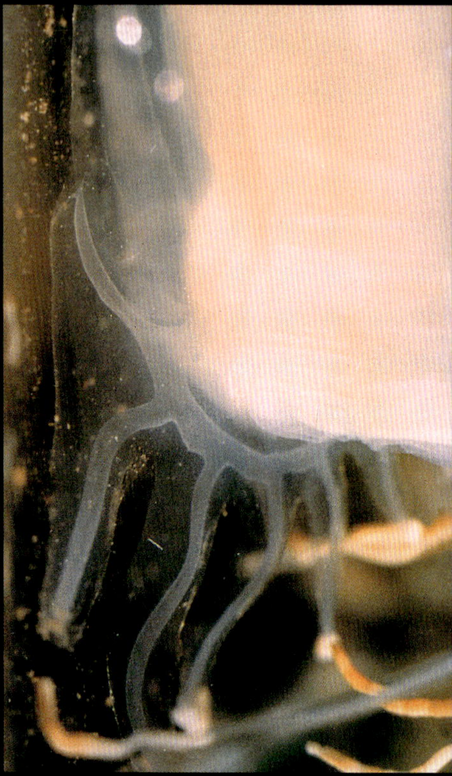

PLATE 9.61
A pedalium of *C. buitendiki* showing the characteristic linear arrangement of the tentacles and the upward extension of the pedalial canal. (Courtesy Dr Malik Fernando, Colombo)

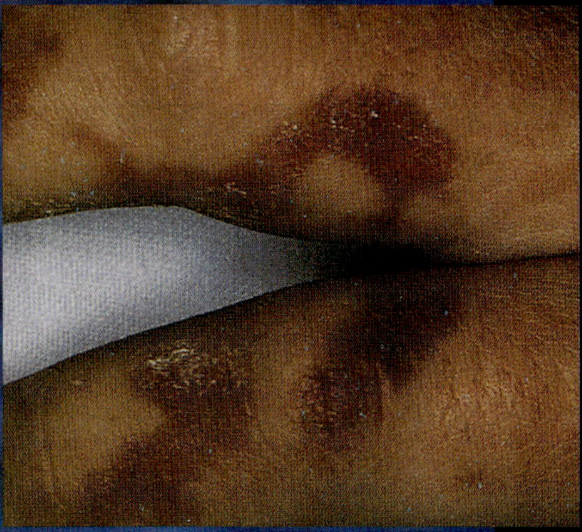

PLATE 9.62
A minor, but painful *Chironex fleckeri* sting on the fingers, photographed 20 hours after the sting.

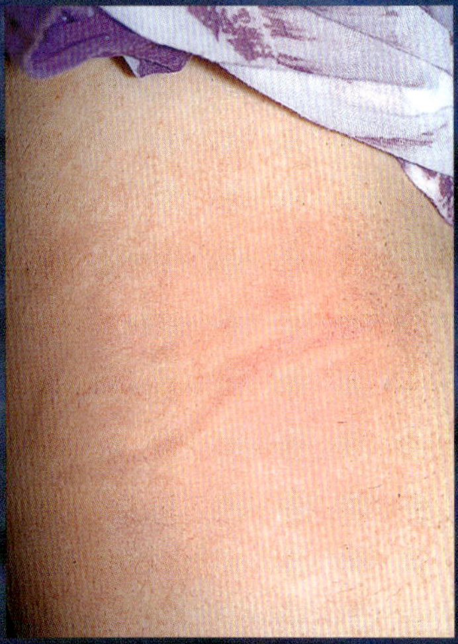

PLATE 9.63A
A small Northern Territory *Chironex fleckeri* sting on the thigh which occurred in October 1991, photographed two hours after envenomation (Courtesy Assoc. Prof Bart Currie)

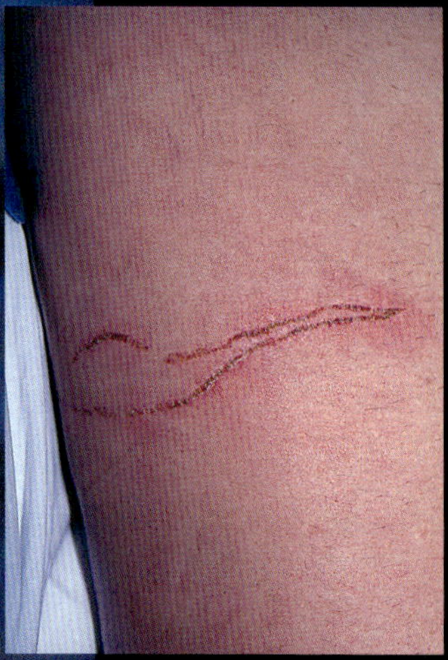

PLATE 9.63B
The *Chironex fleckeri* sting shown in Plate 9.63A six days after envenomation.

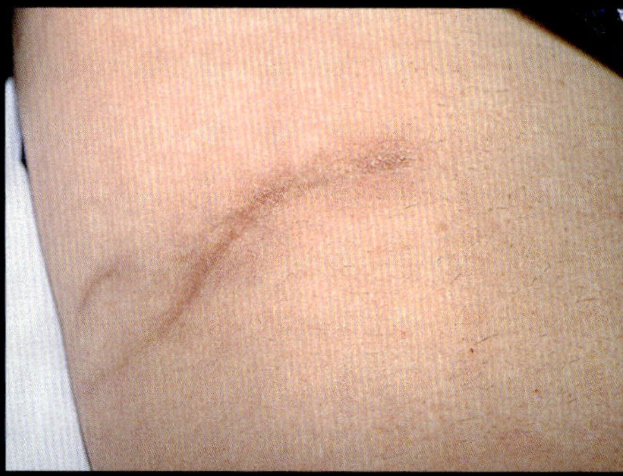

PLATE 9.63C
The *Chironex fleckeri* sting shown in Plate 9.63A eight weeks after envenomation. Good clean healing has occurred, with minimal scarring.

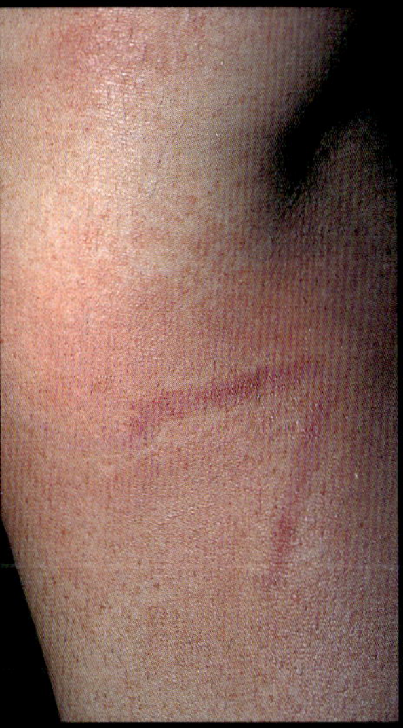

Plate 9.64a
A small Northern Territory *Chironex fleckeri* sting on the lower leg, which occurred in October 1991, photographed two hours after envenomation.

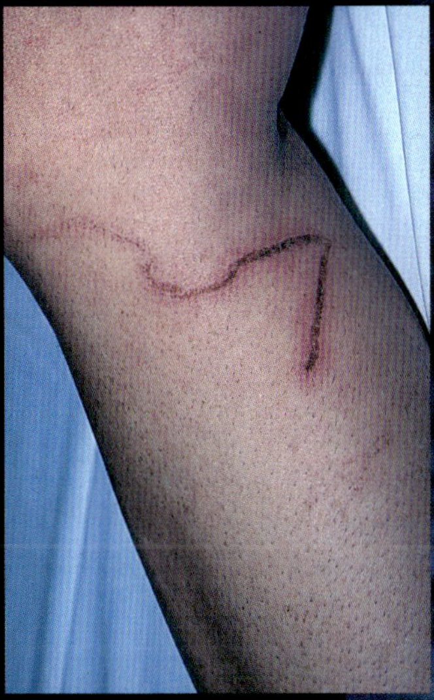

Plate 9.64b
The *Chironex fleckeri* sting shown in Plate 9.64a, photographed five days after envenomation.

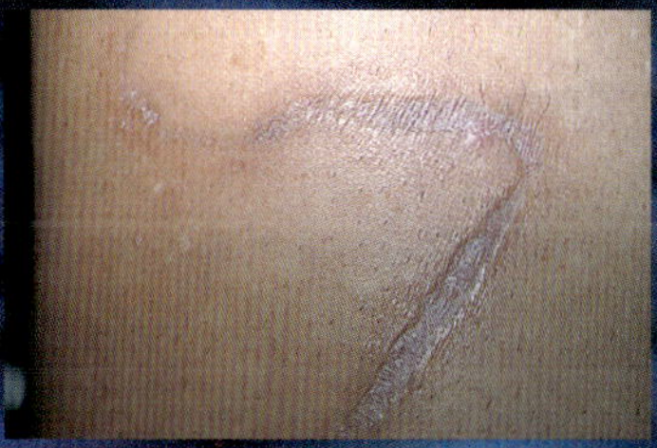

Plate 9.64c
The *Chironex fleckeri* sting shown in Plate 9.64a, photographed eight weeks after envenomation. This small sting has produced full-thickness skin death; and although infection has been avoided, scarring in the line of tentacle contact has resulted.

Plate 9.65
Secondary infection in a *Chironex fleckeri* sting. Note that the 'cross-hatched' appearance of the fresh sting (see text) has been retained into the healing phase, possibly enhanced by the infection process.

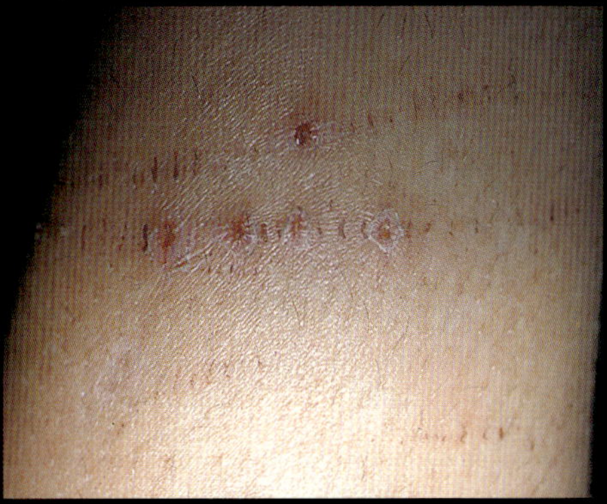

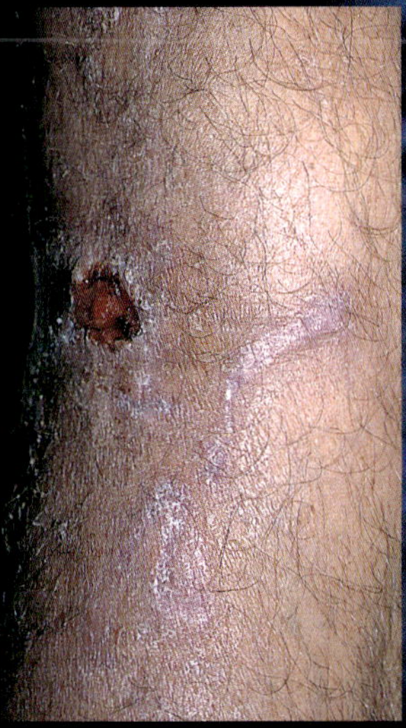

Plate 9.66
Secondary infection in a *Chironex fleckeri* sting. Note that the skin damage is greater (deeper) than in the skin of the patient shown in Plate 9.65.

Plate 9.67
The hairless leg of a child showing extensive and intimate tentacle contact during an Australian *Chironex fleckeri* sting. The photograph was taken 24 hours after the envenomation and shows full-thickness skin damage and vesiculation. (Courtesy of Dr Ian Audley, Cairns)

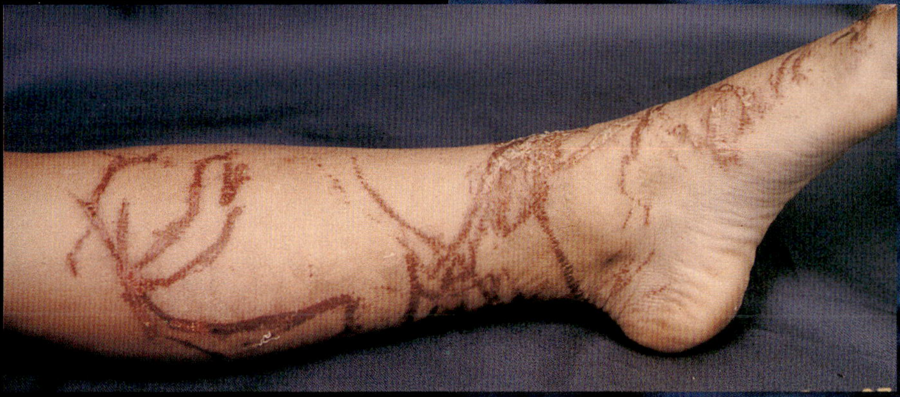

PLATE 9.68
A magnified (x100) longitudinal section of *Chironex fleckeri* tentacle relaxed in $MgCl_2$ solution in seawater, then immersed in methylated spirits prior to fixation and staining. Note discharge of all batteries of nematocysts. (Copyright ©*Medical Journal of Australia* 1980; 1; 15–20).

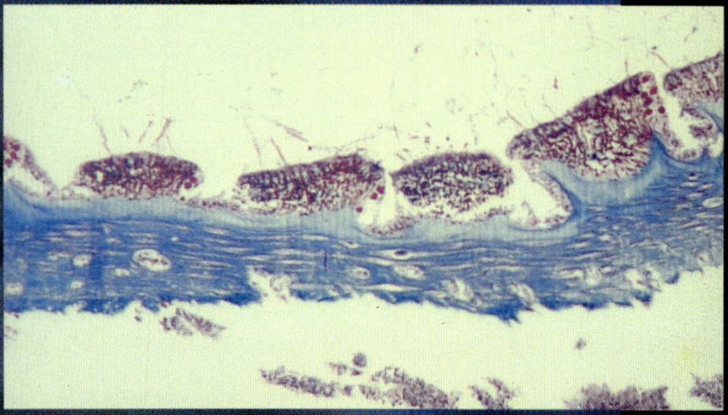

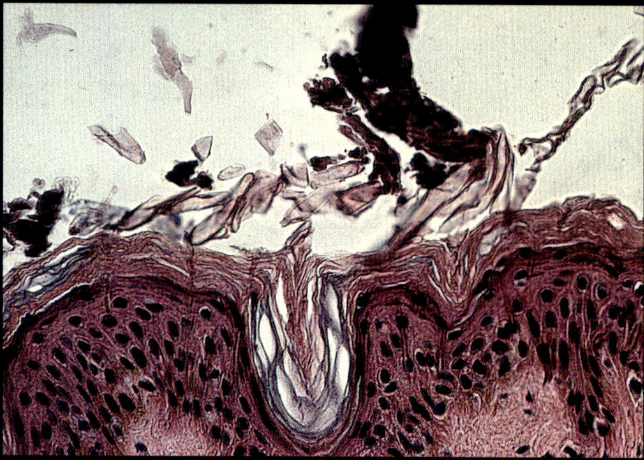

PLATE 9.69
A post-mortem skin biopsy from a fatal *Chironex fleckeri sting* in northern Australia, showing partially deformed, empty mastigophore nematocyst capsules firmly anchored to the keratin by the attached penetrating threads that have transfixed the epidermal layers during discharge. (Compare with Figure 5.7.)

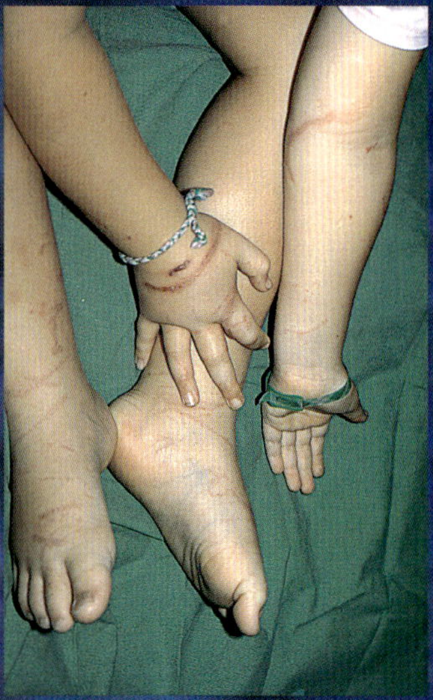

PLATE 9.71
A five-year-old female photographed two days after a serious *Chironex fleckeri* sting, during April in the Northern Territory of Australia. The patient developed marked sinus arrhythmia, but no circulatory compromise, during the acute envenomation. Note the tentacle marks on the swollen hand, both wrists and an upper forearm, as well as extensive leg stings.

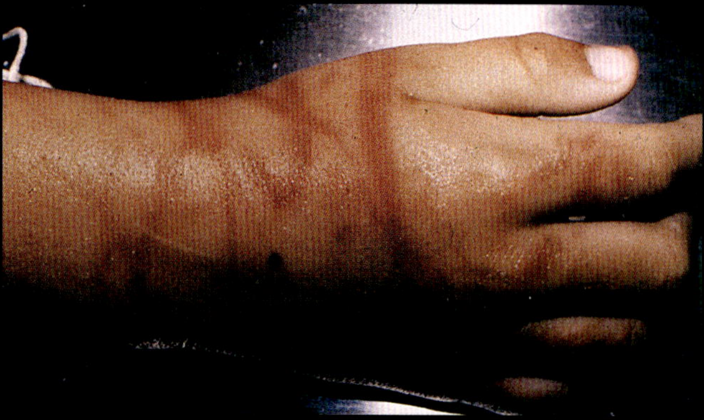

PLATE 9.70B
Post-mortem photograph of the child shown in Plate 9.70A. The involvement of the hands is typical of children in the water struggling with painful adherent tentacles. Note also the modest inflammatory response due to the rapid circulatory collapse (see also Plate 7.2).

PLATE 9.70A
Post-mortem photograph of a 5-year-old child following a fatal *Chironex fleckeri* envenomation in northern Australia in 1987. Only 4 metres of tentacle contact occurred, but massive envenomation resulted in circulatory arrest within minutes, which was unresponsive to resuscitation efforts.

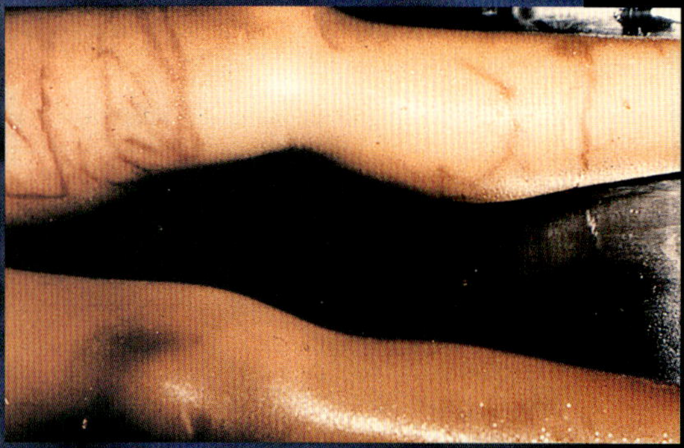

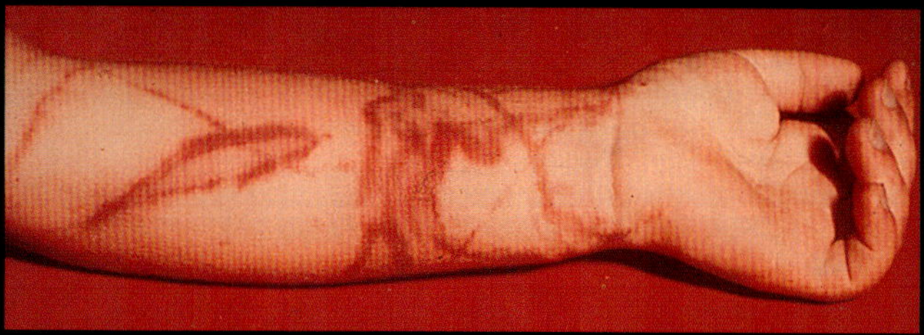

PLATE 9.72
The extensively stung forearm of a small child who sustained a massive *Chironex fleckeri* sting in north Queensland. (Courtesy of Dr Jack Barnes, Cairns)

PLATE 9.73
A portion of a serious *Chironex fleckeri* sting on the legs of a young adult female, seven hours after envenomation in Townsville, Australia. Note the acute inflammatory response of the wide tentacle marks (large animal). The inflammation has been partially controlled by the early administration of specific *Chironex fleckeri* antivenom.

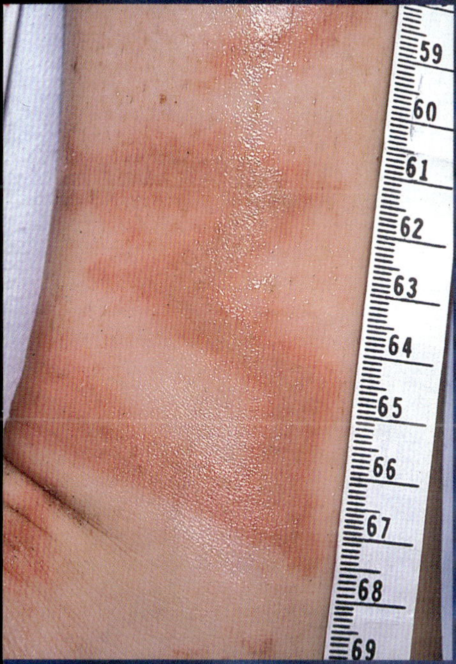

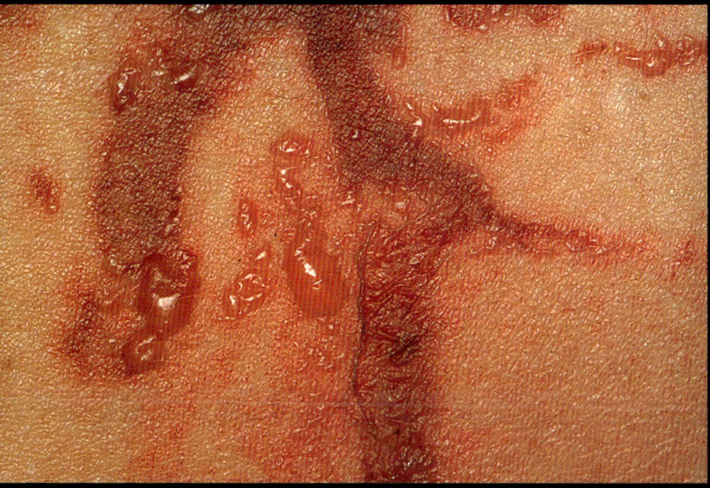

PLATE 9.74
Further progression of the acute inflammatory response in a life-threatening *Chironex fleckeri* sting in another young adult female in Townsville. The photograph taken 24 hours post-envenomation shows vesiculation and precarious skin survival in the wide tentacle marks. (Compare with Plates 1.2, 9.80A and 9.80B.)

PLATE 9.75
A young boy showing the characteristic sudden reflex backwards movement in shallow water at the moment of a chirodropid stinging on the submerged legs. This pattern of response is watched for by north Queensland lifesavers on patrol, and will initiate an immediate rescue sequence.

PLATE 9.76
Portion of a 30-minute-old, methylated spirits-doused, *Chironex fleckeri* sting showing withered but still adherent tentacle material. Methylated spirits should not be used to treat chirodropid stinging (HARTWICK *et al* 1980 — see text).

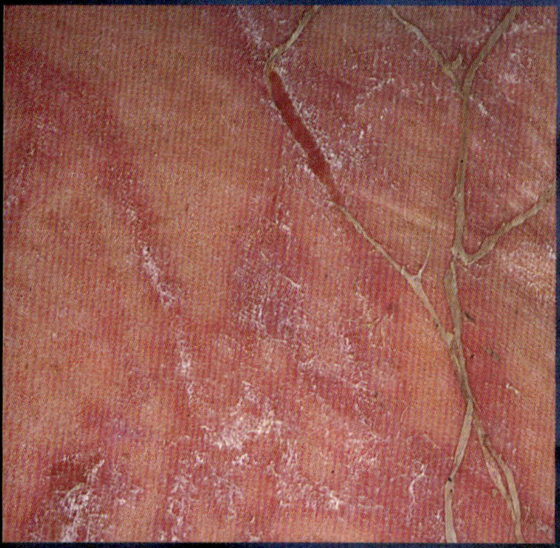

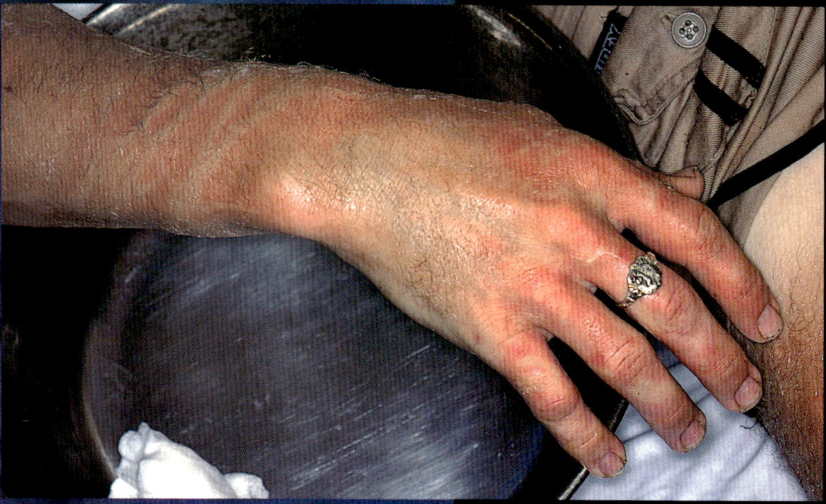

PLATE 9.77
The forearm of the companion of the massive *Chironex fleckeri* sting victim shown in Plates 1.2 and 9.74, who was stung simultaneously by the same large animal in knee-deep water in Townsville. Note the large welts on the dorsum of the wrist, but the much lesser skin damage due to the protection from the forearm hair. (See also Plate 1.3.)

PLATE 9.80A
A massive *Chironex fleckeri* sting (in which *Chironex fleckeri* antivenom was administered early), 20 hours after envenomation.

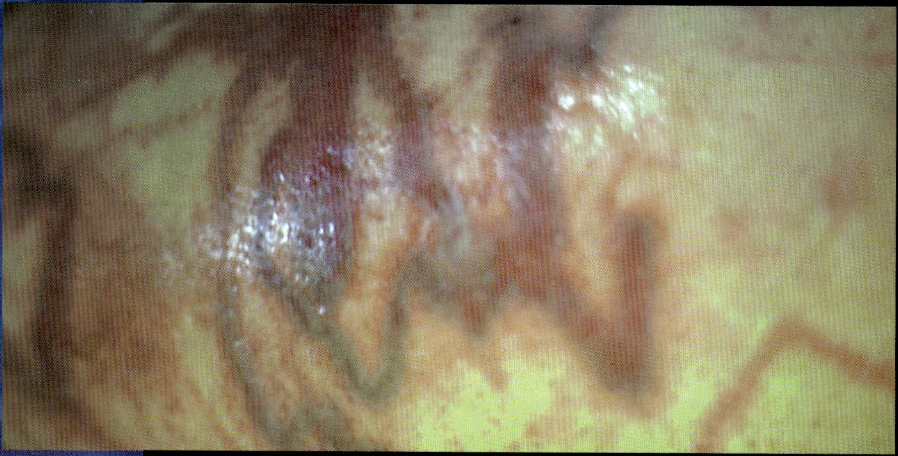

PLATE 9.78
Portion of the 30-minute-old, life-threatening *Chironex fleckeri* sting shown in Figure 1.2. Note the erythema, the 'frosted' appearance and the classical 'cross-hatching' in the large tentacle marks on the hairless skin of a young female. The cross-hatching has by chance been made more apparent by the earlier application of a spray of aluminium sulphate suspension, intended for acute therapy (see text).

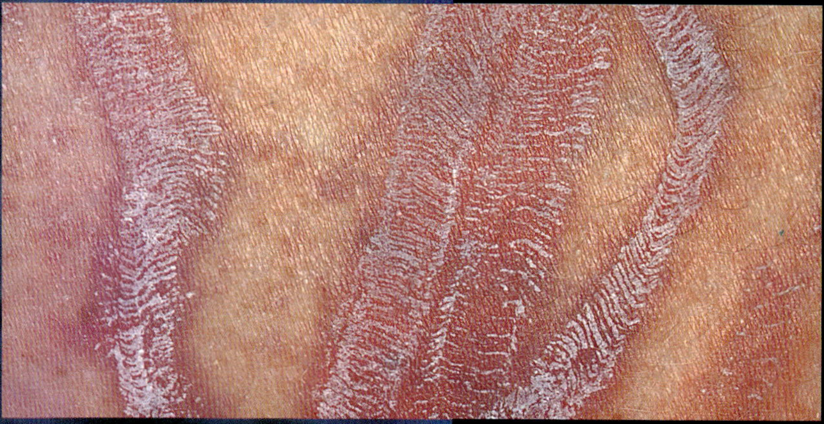

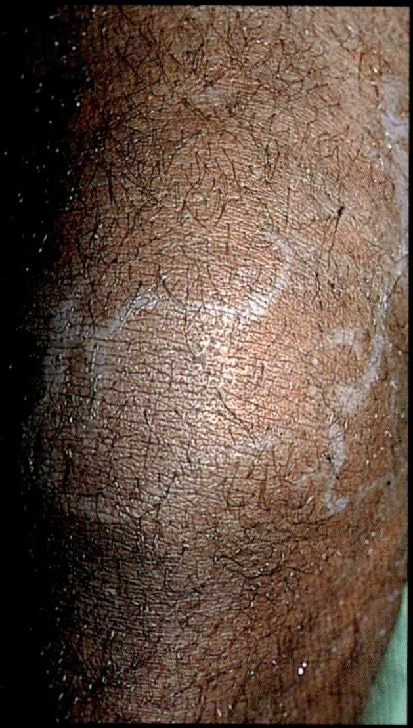

PLATE 9.79
A fresh (<20 hours) *Chironex fleckeri* sting on the knee of an Australian Aboriginal child in the Northern Territory. Note the white marks of the tentacle contact and the barely visible surrounding acute inflammatory response.

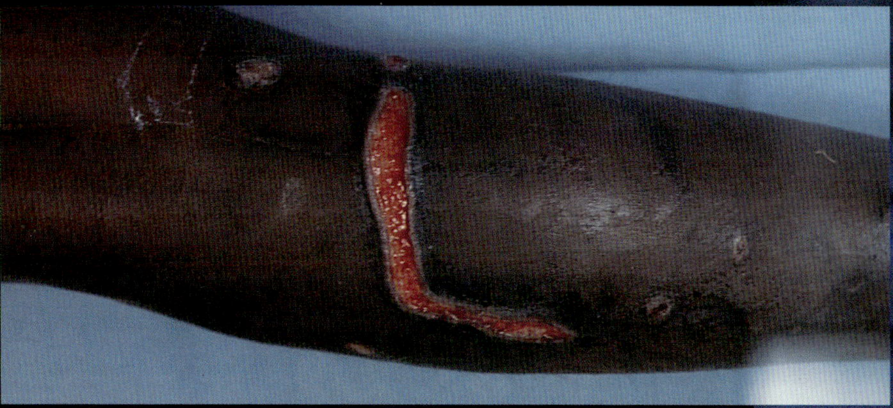

PLATE 9.81
The results of a *Chironex fleckeri* sting on the leg of an Aboriginal girl in north Queensland, photographed three weeks after the untreated sting, and showing total skin destruction in the line of tentacle contact. (Courtesy of Dr Ian Audley, Cairns)

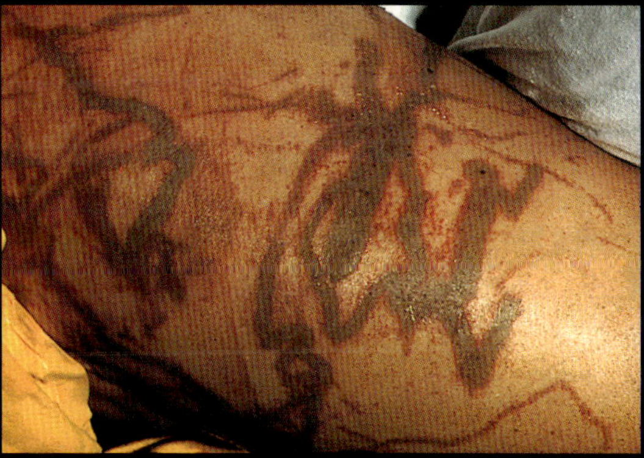

PLATE 9.80B
The *Chironex fleckeri* sting shown in Plate 9.80A, photographed 48 hours after envenomation. Note the rapid darkening as skin dies in the line of tentacle contact.

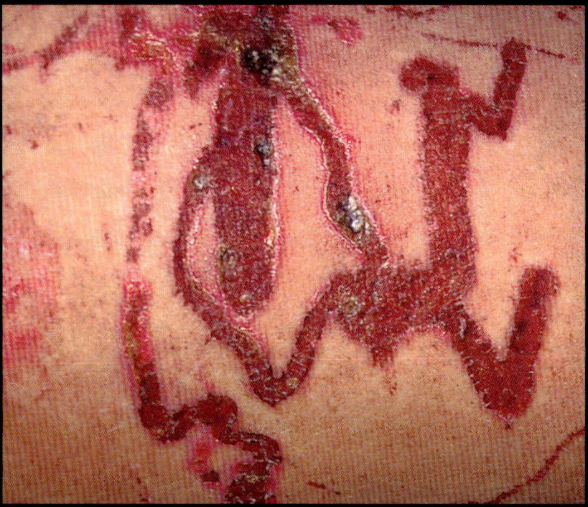

PLATE 9.80C
Eleven days after envenomation, scab formation is complete. These appearances should be compared with Plates 5.19.

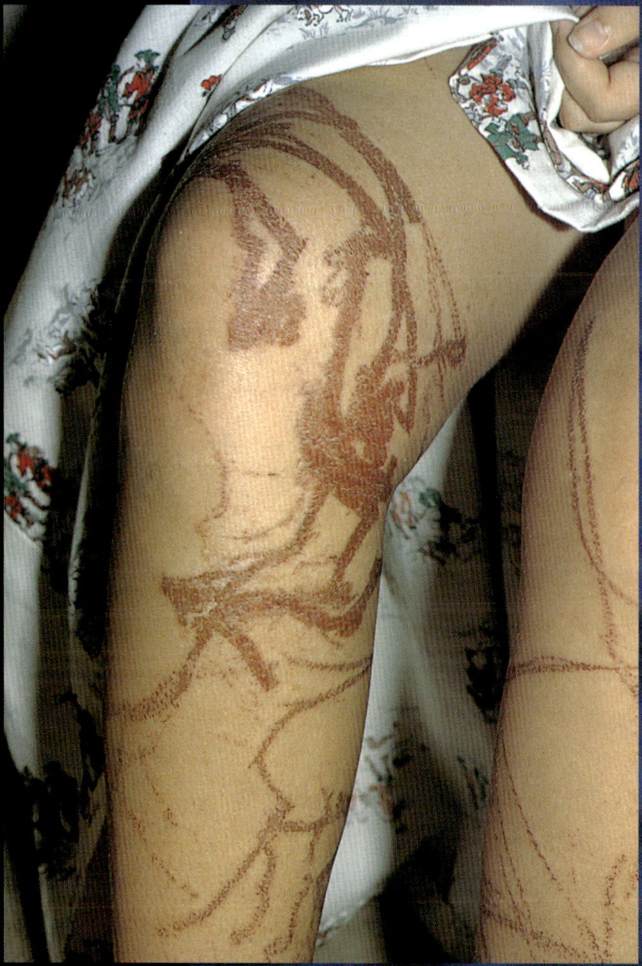

PLATE 9.82
A massive *Chironex fleckeri* sting on the legs of an Australian adult female, photographed two weeks later, and showing healthy scab formation free of infection. The patient's clinical state was never serious and did not correlate with the size of this sting (see text). Very little long-term scarring resulted (cf. Plates 5.20). (Courtesy of Dr Jack Barnes, Cairns)

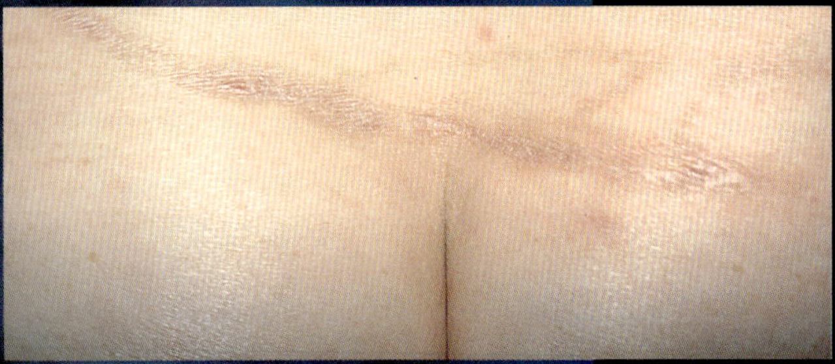

PLATE 9.83
Light scarring in a tentacle mark across the lower back of a north Queensland youth, photographed four months after the sting (cf. Plates 9.47 and 9.48)

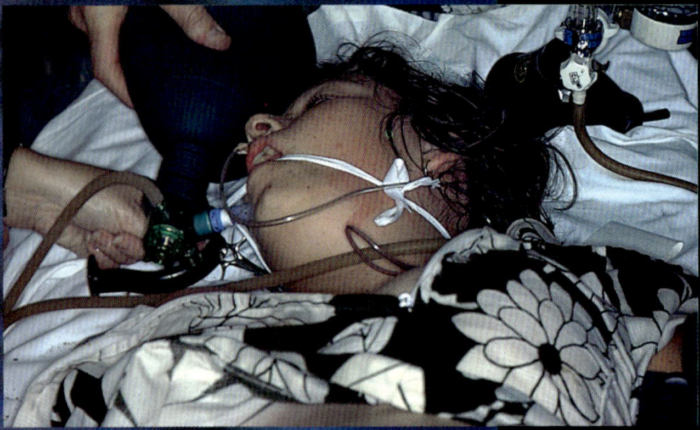

PLATE 9.84
A female patient with a life-threatening *Chironex fleckeri* envenomation, who arrived at the Emergency Department of a major hospital 30 minutes after envenomation. She was semi-conscious with acute pulmonary oedema upon arrival; controlled ventilation via an endotracheal tube is in progress.

PLATE 9.87
One ampoule (approximately 2.2 ml containing 20, 000 units) of specific box jellyfish (*Chironex fleckeri*) antivenom, prepared from hyperimmunised sheep by Commonwealth Serum Laboratories Melbourne.

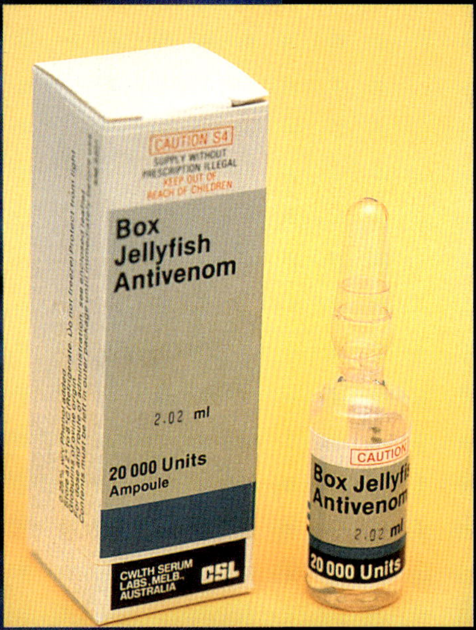

PLATE 9.85
A 22-year-old female with a history of asthma and hay fever, who had complete resolution of the sting from a *Chironex fleckeri* jellyfish, and 10 days after the sting developed this cutaneous papular urticarial reaction occured at the original sting site on the right leg.

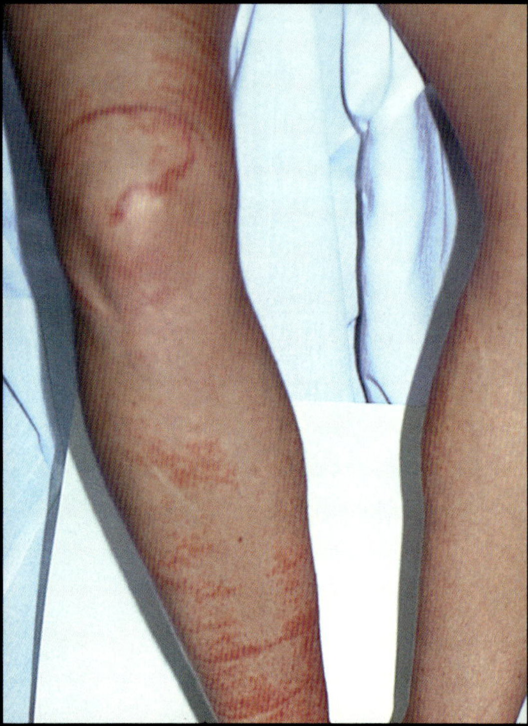

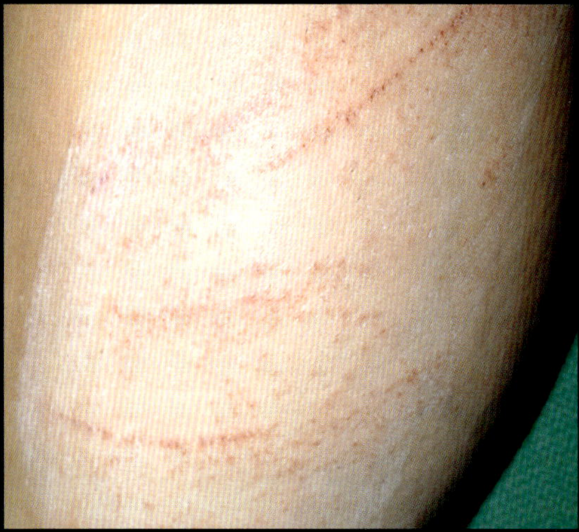

PLATE 9.86
A delayed papular urticaria to a *Chironex fleckeri* sting sustained 10 days earlier. *Chironex fleckeri* antivenom was administered at the time of the sting, with only moderate relief of symptoms, but subsequent complete resolution of the sting marks on the calf of the leg, prior to their reappearance shown here. The patient had no known allergies.

PLATES 9.88A–D
Surf Life Saving Queensland's Stinger Drill for the rescue and first aid resuscitation of a serious chirodropid envenomation in Australia. (Photographs by Mike Shapter and Peter Fenner, Townsville and Mackay) (see also Plates 21.1 and 21.2, and page 107)

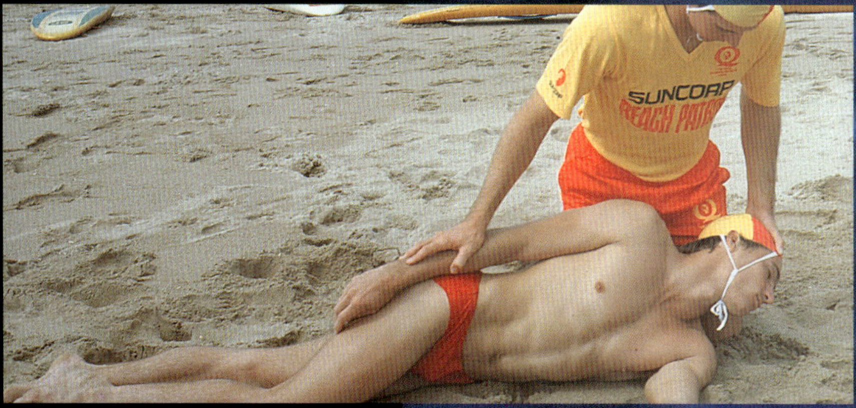

PLATES 9.88B
Adopt lateral position of victim unconscious.

PLATE 9.88A
An excellent simulation of a child with masses of adherent (simulated) *Chironex fleckeri* tentacles, being both restrained (arms held to prevent rubbing the adherent tentacles and sting area) and doused with vinegar, by North Queensland patrol lifesavers. (Reproduced courtesy of the Editors, *Journal of Wilderness Medicine* 1992; 3:422–431)

PLATE 9.88C
Patient unconscious, breathing absent; commence resuscitation. Apply pressure bandages, but do not interrupt resuscitation. (See page 110 with Figure 5.3.)

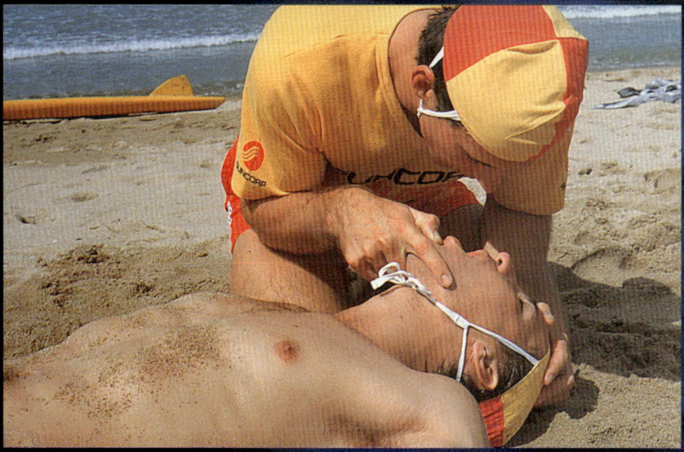

PLATE 9.88D
Expired air resuscitation commencing on a non-breathing victim.

PLATE 10.1
The Mediterranean 'moon jellyfish' *Aurelia aurita*. {Compare with Figures 6.3, 6.4 and 8.13}

PLATE 10.2
A recent (2 hours old) sting from a Mediterranean *Carybdea marsupialis*, showing erythema and small vesicles.

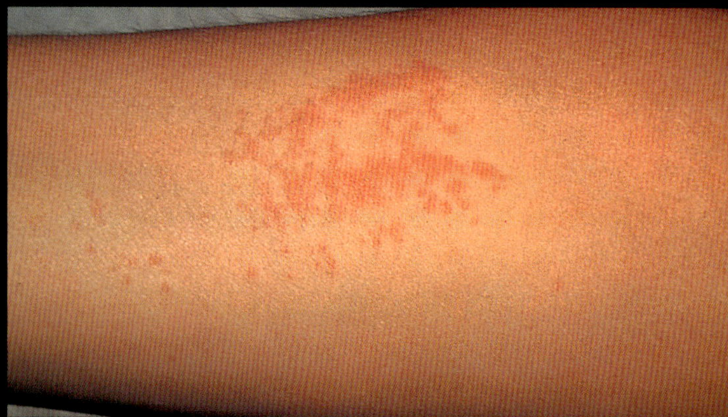

PLATE 10.3
A Mediterranean *Chrysaora hysoscella*, showing the typical radial dark brown markings on the yellowish umbrella or bell. (Photograph courtesy of Mr Ennio Vio, Dipartimento di Biologia, Universita Degli Studi di Trieste). (See also Plates 6.6 and 8.81)

PLATE 10.4
A Mediterranean *Chrysaora hysoscella* sting, showing a linear lesion on the ventral aspect of the wrist.

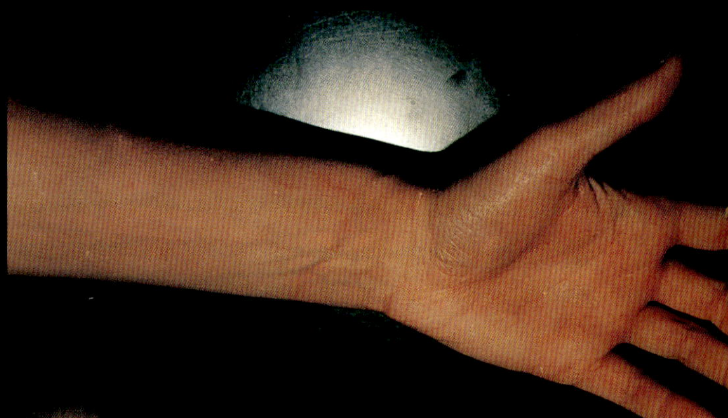

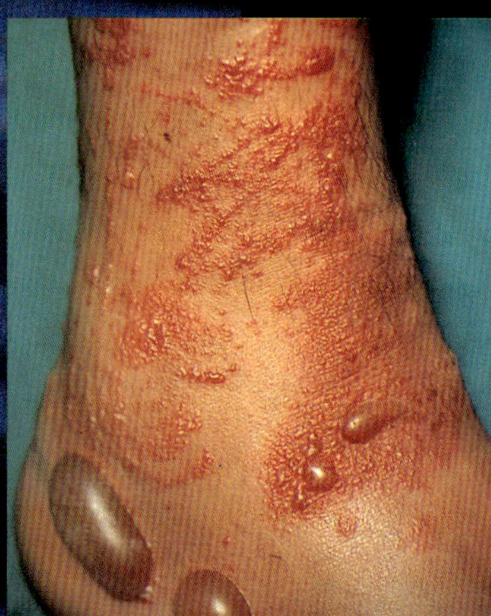

PLATE 10.6
The appearance of a three-day-old sting from the Mediterranean *Pelagia noctiluca*. Note the erythema, vesicles and large blisters of the acute dermatitis {see also Plates 8.85–8.87}.

PLATE 10.5
A fully submerged specimen of the Mediterranean *Cotylorhiza tuberculata* (Agassiz), photographed close to the surface in the Aegean Sea, near Phellasonika, Greece, during early December 1963. (Courtesy of Mrs Marie Bone and Dr Ron Southcott of Adelaide)

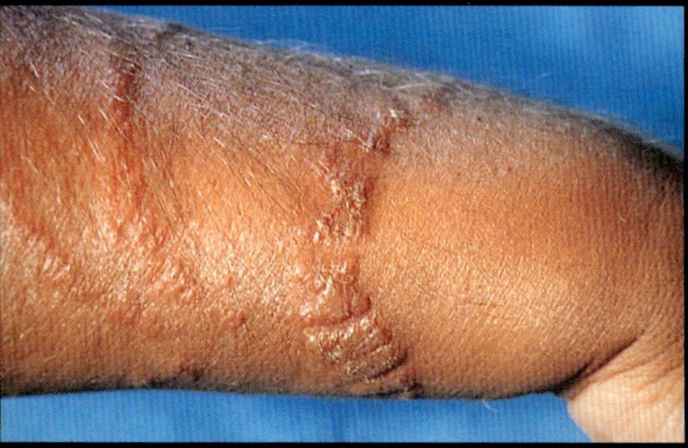

PLATE 10.7
Hypertrophic scar formation in a one-year-old sting from the Mediterranean *Pelagia noctiluca*.

PLATE 10.8
Rhopilema nomadica Galil, Spanier and Fergusson, 1990, in the eastern Mediterranean waters of the Israeli coast.

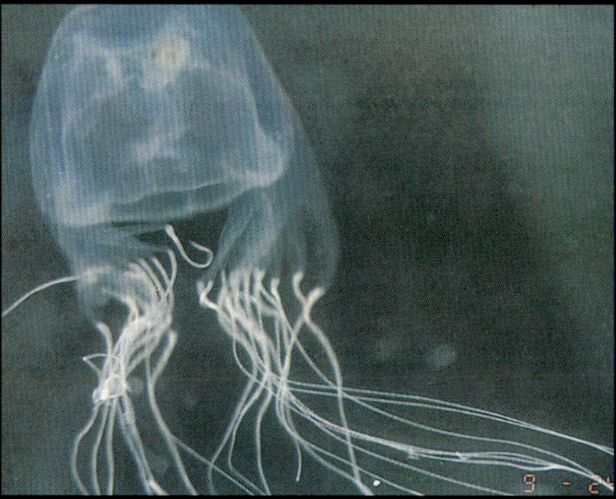

PLATE 10.9
'*Chiropsalmus quadrigatus* HAECKEL' from Okinawa, Japan. Note the similarity to the Australian *Chiropsalmus* sp. {see Plate 1.10}. (Photograph courtesy of Dr Yasuhiro Tomihara and colleagues, Okinawa)

PLATE 10.10
Photomicrograph of a discharged *Chiropsalmus quadrigatus* HAECKEL microbasic mastigophore from Okinawa, Japan, viewed unstained under low power (magnification unknown). Note the characteristic cigar-shape, and the butt spines on the thread, close to the empty capsule. Compare with Figures 6.10–6.12. (Photograph courtesy of Dr Yasutetsu Araki, Okinawa Prefectural Institute of Health, Ozato, Okinawa)

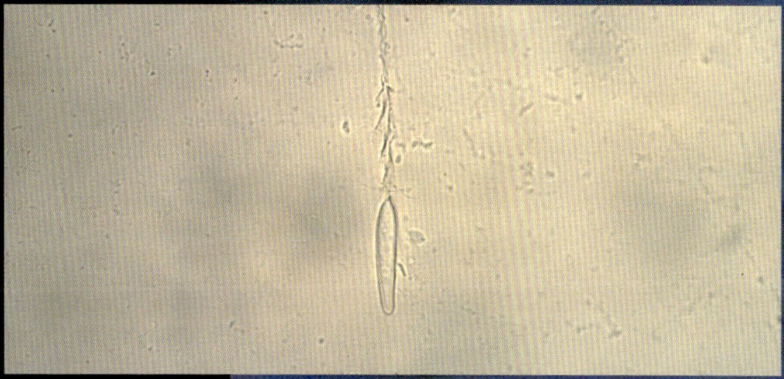

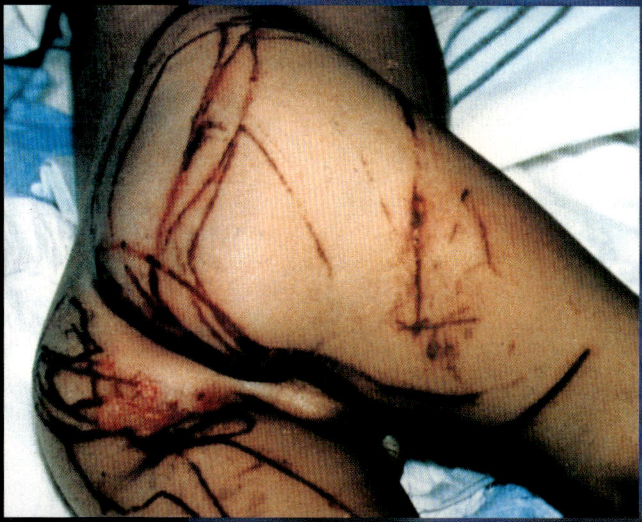

PLATE 10.11
A life-threatening envenomation in a two-year-old Japanese boy in Okinawa. He received expired air resuscitation on the beach from his father following rapid unconsciousness and apnoea, and was discharged from hospital, well but scarred, one week later. (Photograph courtesy of Dr Masakazu Nakamoto, Okinawa)

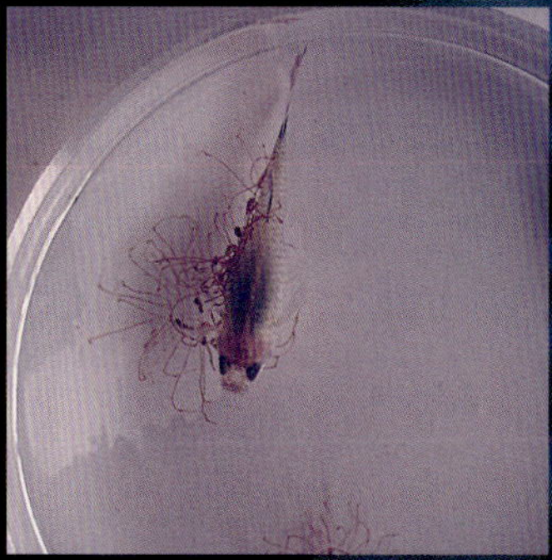

PLATE 10.12
The tiny Japanese medusa, *Gonionemus oshoro* UCHIDA, 1929. There are three medusae in the petri dish; the largest specimen is partly covered by a small fish. These medusae measure some 5–15 mm in diameter. Note the 'cross' pattern visible on the bell. Compare with Figure 8.52. (Photograph courtesy of Dr Ohtsuru, University of Ryukyus, Japan)

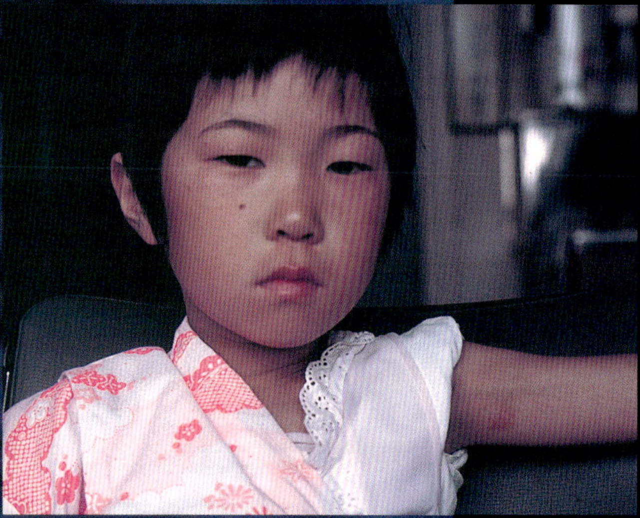

PLATE 10.13
Sting of *Gonionemus oshoro* UCHIDA, 1929 faintly visible on the upper left arm of a young Japanese girl. Often very difficult to see (cf. Irukandji sting Plate 9.23).

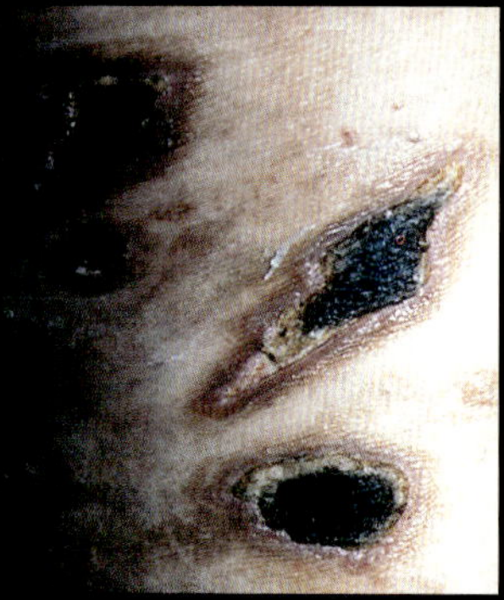

PLATE 10.14
Healing scars from a jellyfish sting on the upper posterior thigh of a Thai fisherman, presenting at the Siriraj Hospital, Bangkok. (Photograph courtesy of Dr Patchari Soontornpalin, Division of Dermatology, Mahidol University, Bangkok)

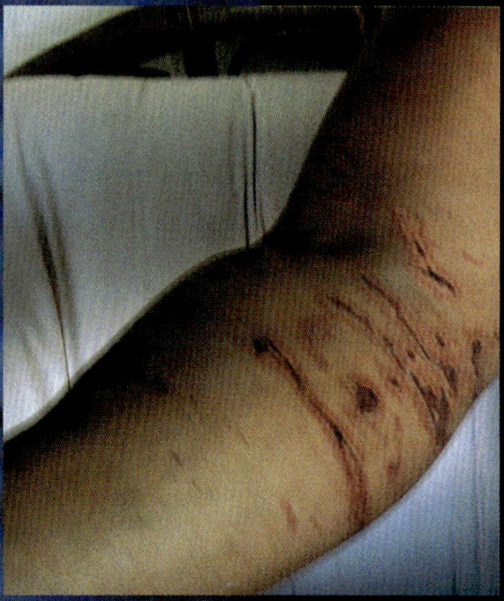

PLATE 10.15
The right forearm and lower arm of a young man stung by an unknown large jellyfish off Pattaya, Thailand. The 30 cm of wide (>1 cm) strap-like tentacle marks are shown on day 7 after the envenomation. Skin necrosis, slough formation and some eschar are present. Note the persisting oedema of the visible portions of the right arm. (Photograph courtesy of Dr Carl Edmonds, Sydney)

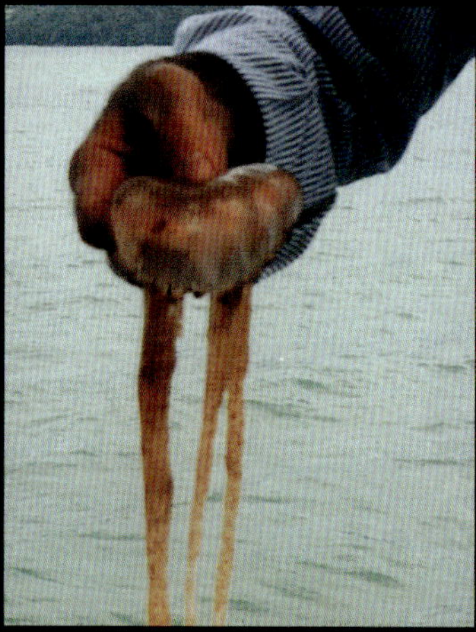

PLATE 10.16
A jellyfish with yellowish tentacles caught off Penang in the Malaccan Straits in October 1991, and identified from nematocysts by Dr Dale Calder in 1994 as possibly a *Chrysaora* sp.

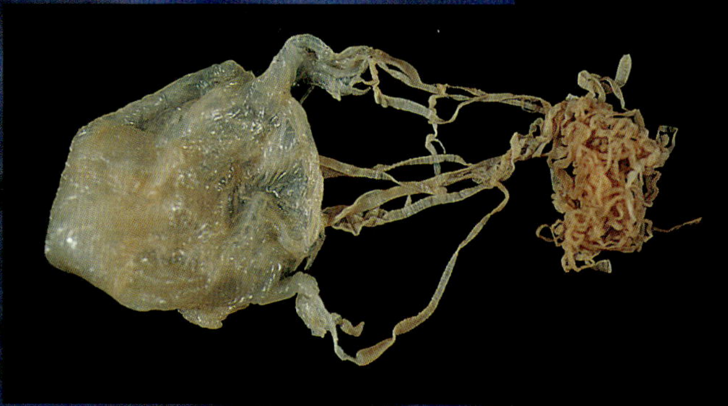

PLATE 10.17
A preserved specimen of *Chironex fleckeri* captured in Borneo.

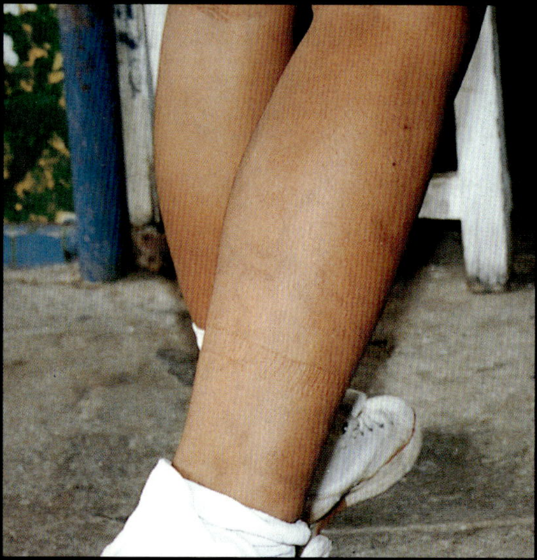

PLATE 10.18
The legs of an 11-year-old northern Labuan child photographed two years after a serious chirodropid sting which rendered him 'unconscious for 24 hours'. (Photograph courtesy of Associate Professor Nor Azila and Dr Iekhsan Othman, Universiti Malaya).

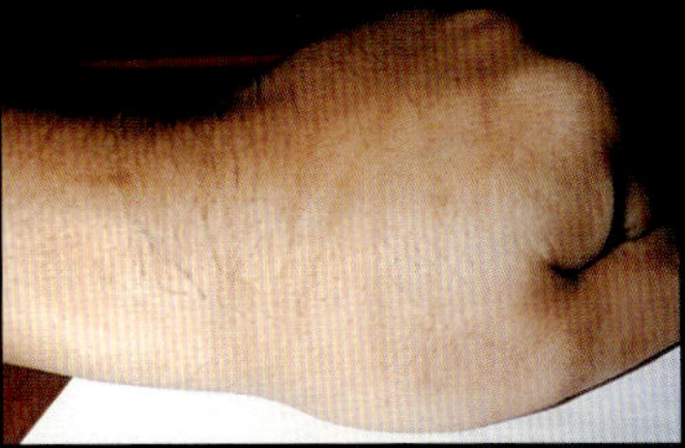

PLATE 10.19
Part of a sting on the hand of a 32-year-old male who was stung by a chirodropid jellyfish off the south coast of Sabah; he was 'unconscious for hours'. Note that scars are still visible. No antivenom was used. (Photograph courtesy of Associate Professor Nor Azila and Dr Iekhsan Othman, Universiti Malaya).

PLATE 10.20
The Puerto Rican *Carybdea alata*, photographed by Professor 'Chuck' Cutress, the original describer of this animal's life cycle (ARNESON & CUTRESS 1976) and Mrs Bertha Cutress, Puerto Rico. Compare with Plate 9.8.

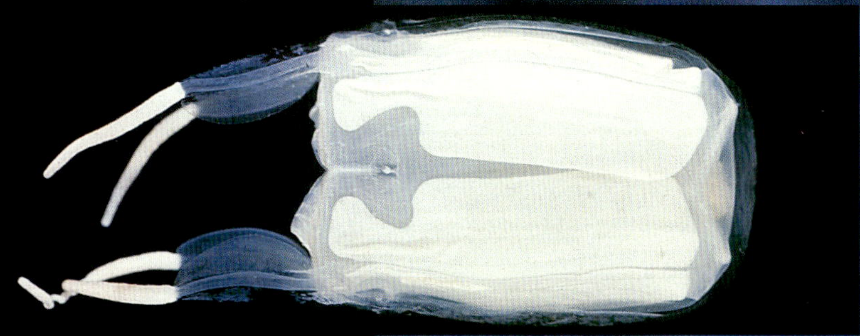

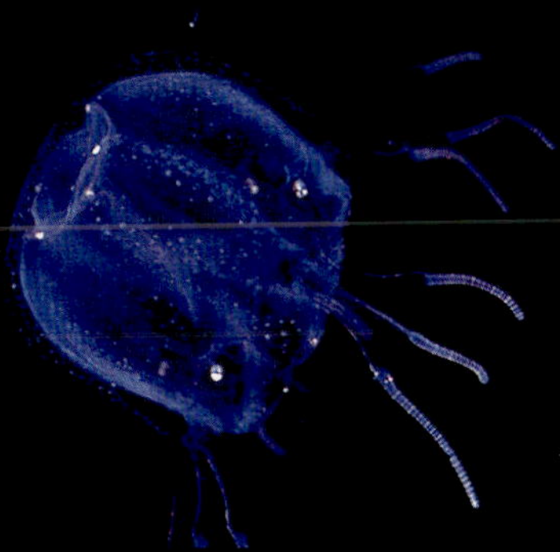

PLATE 10.21
The carybdeid cubozoan, *Tripedalia cystophora* in Puerto Rican waters. (Courtesy of Professor 'Chuck' Cutress and Mrs Bertha Cutress, Puerto Rico)

PLATE 10.22
The Atlantic *Physalia physalis* photographed in a Puerto Rican laboratory. See also Plate 8.43. (Courtesy of Professor 'Chuck' Cutress and Mrs Bertha Cutress, Puerto Rico)

PLATE 10.23
A preserved specimen of a 'reddish' cnidarian, suspiciously like a cubozoan, caught in the Gulf of Oman. The cause of this reddish colouration is not known, but it could tie with the description given by observers in COOPER's case report from Oman in 1985 {see Chapter 18, page 406}. A similar description was given by the victim in Case Report No. 2 from a catastrophic sting in Mangalore, India (see Figures 10.25A and 10.25B below)

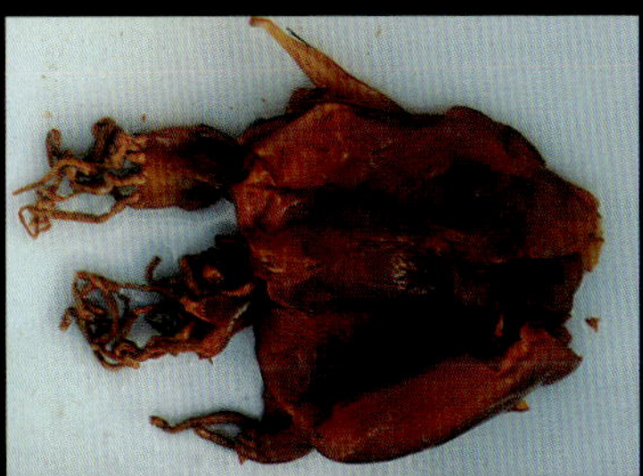

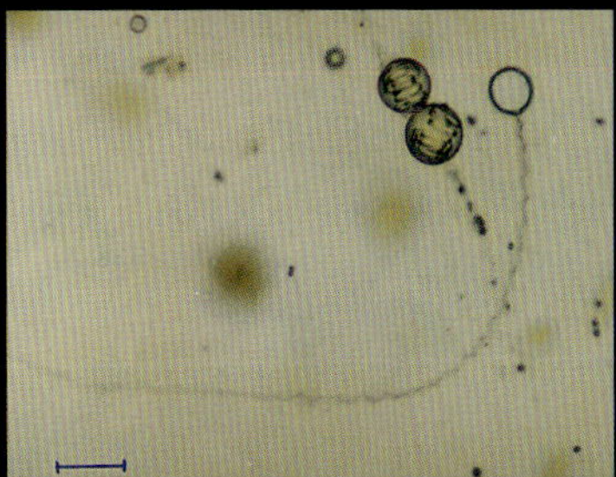

PLATE 10.24
A photomicrograph (x350) showing three spherical nematocysts from one of the tentacles removed from the *Physalia* sp. pictured in Figure 10.9. Note the two undischarged nematocysts, and the single 'empty' discharged one with its discharged threads still attached. The bar = 13.80 µm. Compare with Plate 8.41, page 26. (Prepared by and reproduced with the courtesy of Junaid M. Alam, Department of Pathology, Basic Medical Sciences Institute, Jinnah Postgraduate Medical Centre, Karachi)

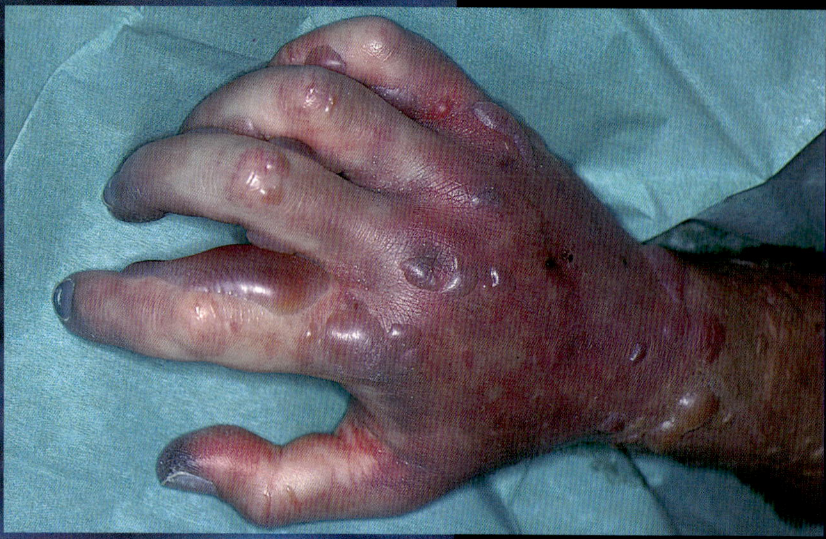

PLATE 10.25A
The catastrophic effects of a major cnidarian envenomation of a healthy 31-year-old female in the waters off Mangalore, India, in 1986. Note the cyanosis, vesiculation, oedema, devitalisation and fingertip necrosis of the patient's ischaemic right arm and hand. The left upper limb was similarly affected. Photographs taken six days after envenomation (Courtesy of Dr Viola Hach-Wunderle, Department of Internal Medicine, University of Frankfurt, Germany) (See also Plate 8.45.)

PLATE 10.25B
A closer view of the hand shown in Figure 10.36a. Photographs taken six days after envenomation (Courtesy of Dr Viola Hach-Wunderle, Department of Internal Medicine, University of Frankfurt, Germany) (See also Plate 8.46.)

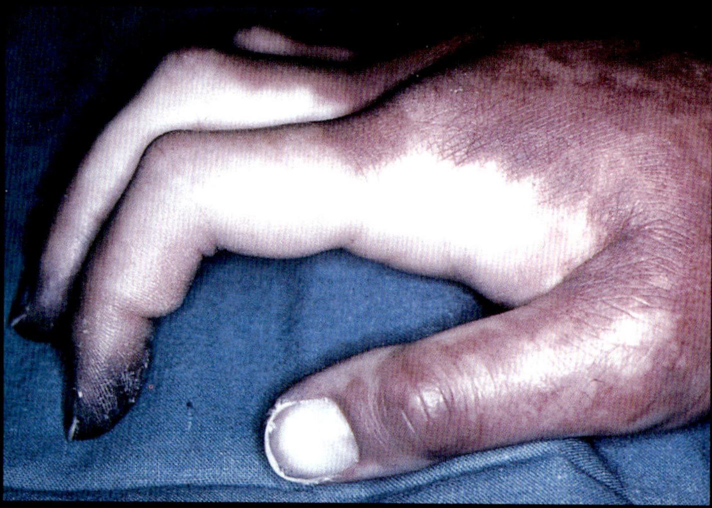

PLATE 10.26
A Miami aquarium specimen of a large *Physalia physalis*, illustrating the 'reddish' appearance of both the crest of the float and the proximal tentacle mass. (Photograph from HALSTEAD 1965, courtesy of the United States Office of the Assistant Secretary of Defense, Washington, DC)

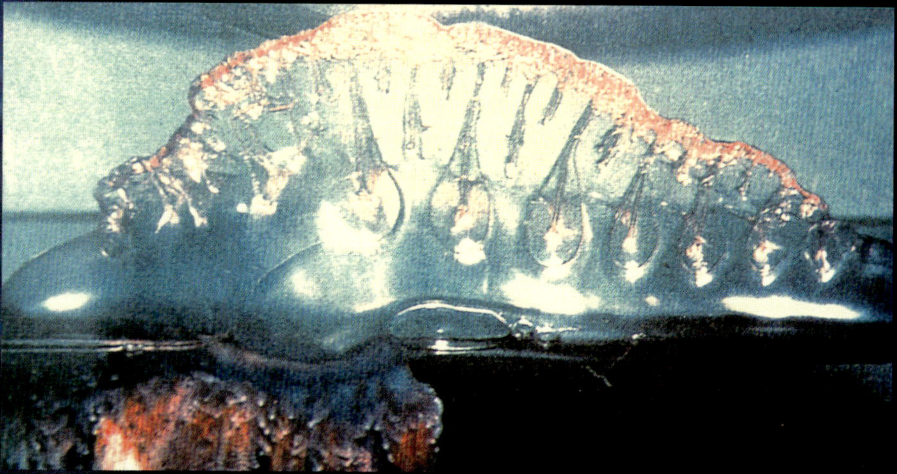

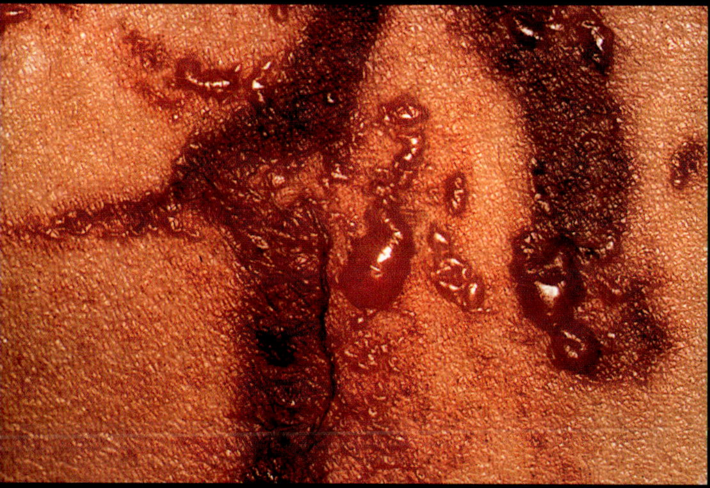

PLATE 11.1
Portion of a major *Chironex fleckeri* sting from northern Australia, 24 hours post-envenomation, showing vesiculation and threatening full-thickness skin death.

PLATE 11.2
Granuloma annulare lesions on the ankle of a caucasian female 10 weeks after a sting from a Pacific *Physalia utriculus*, on the island of Ohau, Hawaii. (Courtesy of Dr R.M. Mandojana and the publishers, *Journal of Wilderness Medicine* 1990; 1: 220–224)

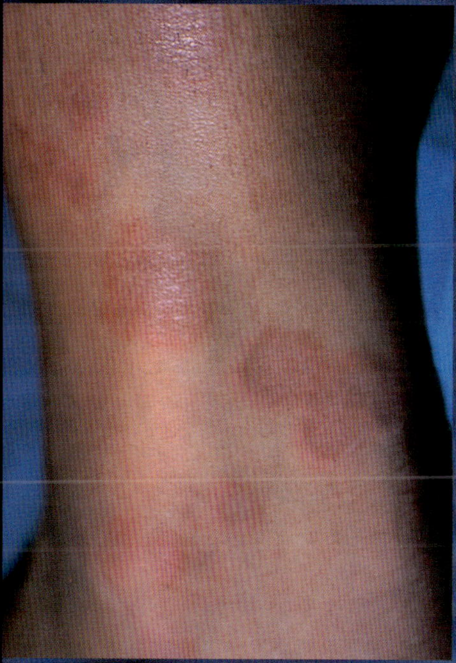

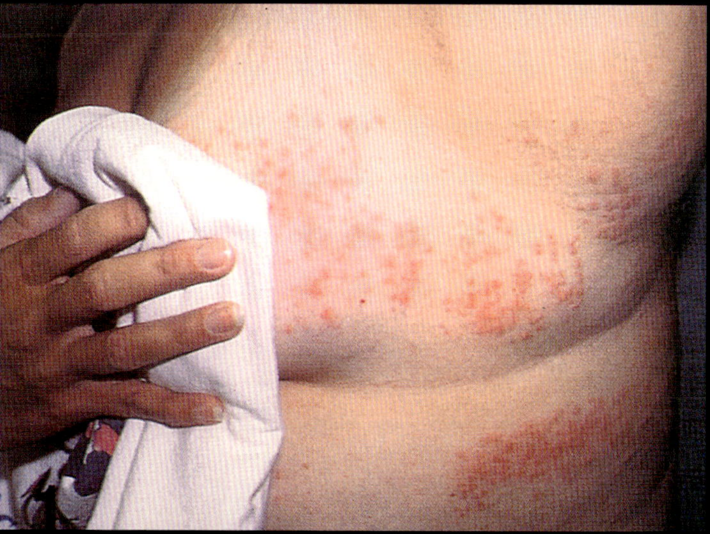

PLATE 11.4
Submammary and left axillary 'seabather's eruption' occurring under the bathing suit in an American female swimmer. (Reproduced from *J Am Acad Dermatol* 1994; 30: 399—406, courtesy of the publisher and the authors).

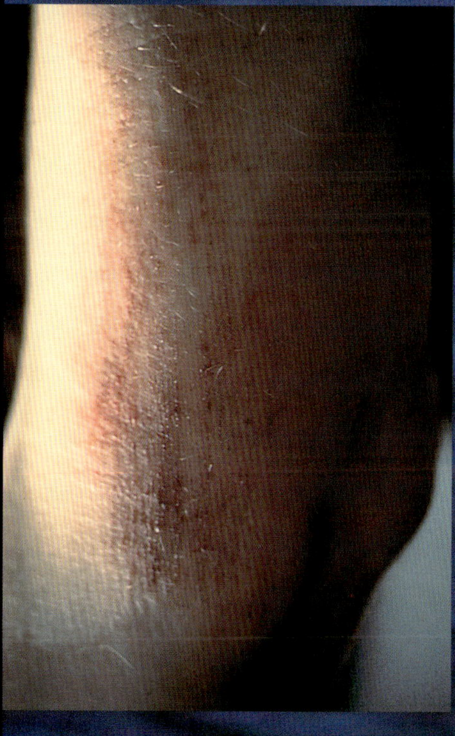

PLATE 11.3
Manta ray dermatitis on the knee of a patient.

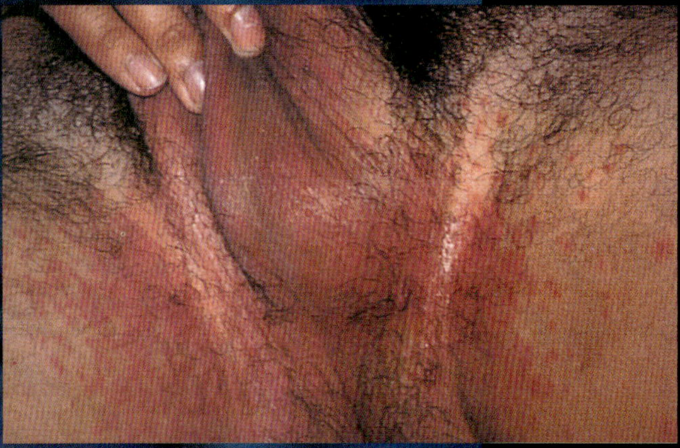

PLATE 11.5
Perineal 'seabather's eruption' in an adult male, showing inflammatory papules persisting one week after exposure. (Reproduced from *J Am Acad Dermatol* 1994; 30: 399–406, courtesy of the publisher and the authors).

PLATE 11.6
The nudibranch *Glaucus atlanticus* (*marginata*?). Nudibranchs are mollusc species lacking a protective shell. They feed on small nematocyst-bearing animals (e.g. *Physalia*, *Porpita*, *Velella*), and the ingested nematocysts pass via *Glaucus*' stomach into the ends of its feather-like appendages, where they are used to sting and capture its prey.

PLATE 11.7
The short-tailed non-stinging nudibranch *Ceratosoma brevicaudatum*, feeding on a toxic sponge, in 4 m depth of water at Point Turton jetty, South Australia. The nudibranch is itself highly poisonous as a result of absorption of the sponge's toxins. (Photograph courtesy of Karen Gowlett-Holmes, Collection Manager, Marine Invertebrates, South Australian Museum, Adelaide)

PLATE 12.1
A crown-of-thorns sea star clasped around (and feeding from) living coral. (Photograph courtesy of Dr V Marriott, Townsville).

PLATE 12.3
Close-up of the abactinal spines that protrude from the upper (aboral, dorsal) surface of the crown-of-thorns sea star. Toxins are thought to be found under the skin (integument) that encases the spines.

PLATE 12.2
Two specimens of the crown-of-thorns sea star *Acanthaster planci*, on a tabulate coral of the genus *Acropora*. The white skeleton of the coral is all that remains after the sea star has digested the thin cover of living polyps.

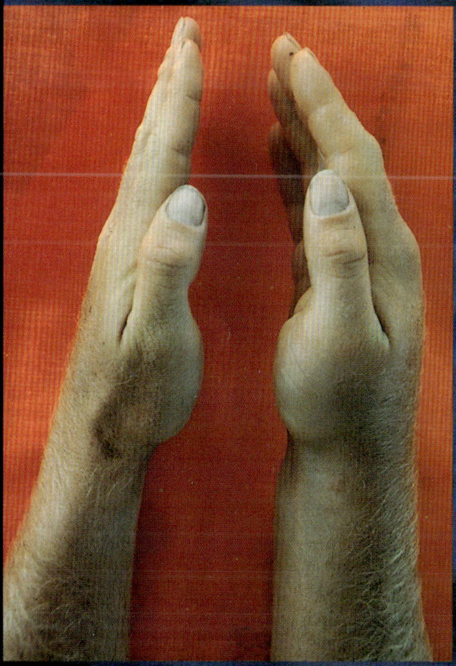

PLATE 12.5
Gross regional oedema of the right hand and wrist of a marine biologist scuba diver, 12 hours after injury to the right wrist and hand by three or four spines of a crown-of-thorns sea star. By this time there was little pain. (Courtesy of B. Miller-Smith, Townsville)

PLATE 12.4
Closer view of the spines of *Acanthaster planci*. Note the orange tips to the spines. (Photograph by Dr. V. Marriot, Townesville.)

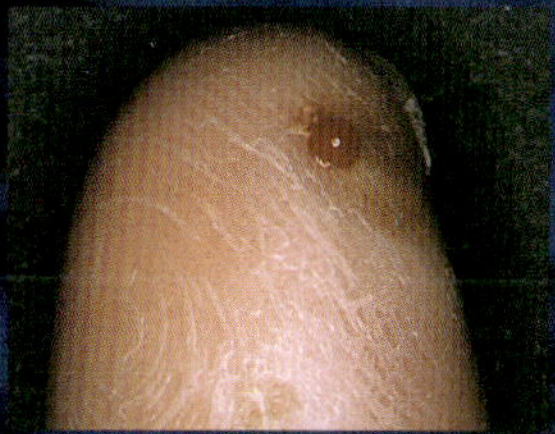

PLATE 12.6
The index finger of the author one hour after a single spine injury from a crown-of-thorns sea star on Queensland's Great Barrier Reef. Note the bleeding/oozing of the wound.

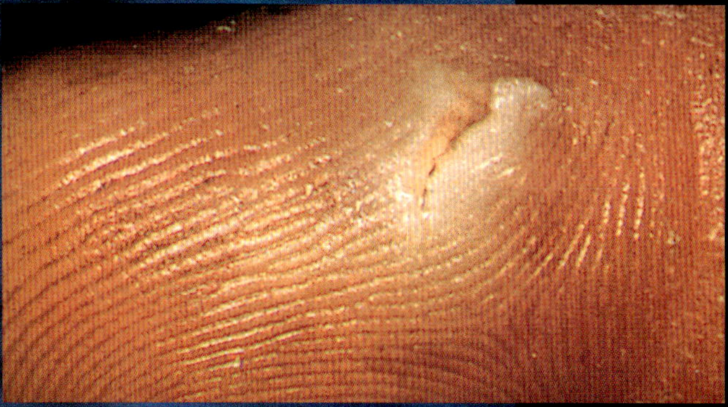

PLATE 12.7
Early infection in a crown-of-thorns sea star wound to an index finger.

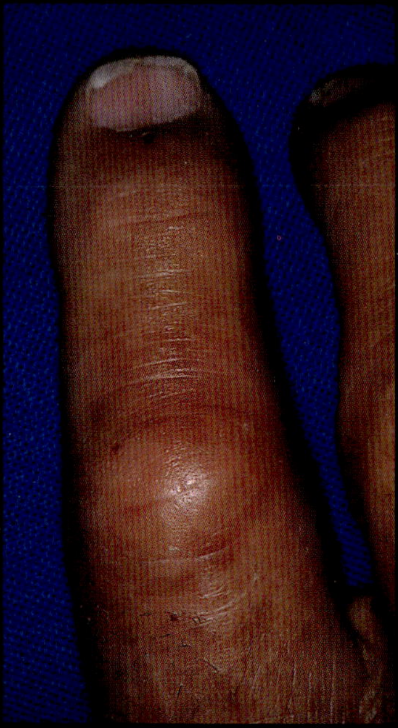

PLATE 12.8
The appearance of the proximal interphalangeal joint of a fisherman's right index finger, three days after penetration by a crown-of-thorns sea star spine. The joint is swollen, painful and stiff. No foreign body was seen on screening.

PLATE 12.9
The author's index finger several hours after a crown-of-thorns spine injury. Note the ecchymosis (blue discolouration of the tissues) around the spine wound (shown in Plate 12.6).

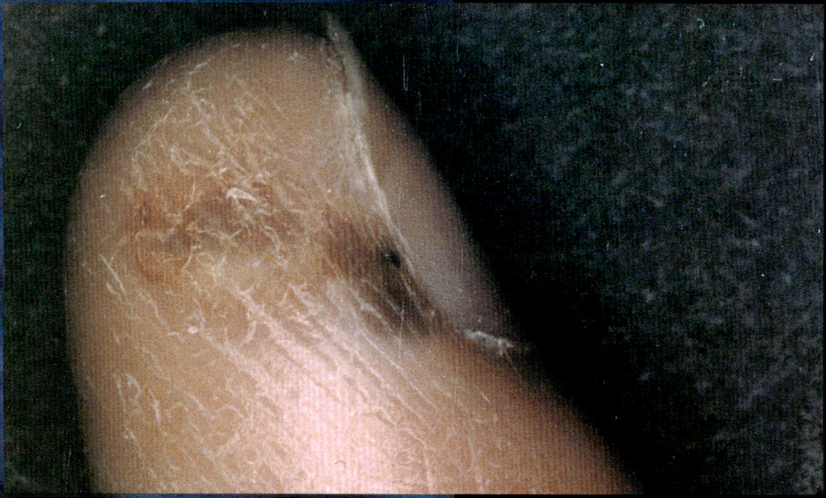

PLATE 12.10
The Caribbean sea urchin *Diadema setosum*. These animals abound in this region.

PLATE 12.11
The sea urchin *Toxopneustes pileolus* in Milne Bay, Papua New Guinea. (Photograph courtesy of Bob and Dinah Halstead, Cairns)

PLATE 12.12
A spectacular specimen of the sea urchin *Echinothrix calamaris*, photographed in 3–5 m depth of water in Bali, Indonesia. Note the impressive array of long spines. (Courtesy of Karen Gowlett-Holmes, Collection Manager, Marine Invertebrates, South Australian Museum, Adelaide)

PLATE 12.13
The severely-stinging sea urchin *Asthenosoma* sp., from Seragaki Beach, Okinawa Island, Japan. (Photograph courtesy of Dr Yasuhiro Tomihara, Okinawa)

PLATE 12.14
Species of sea urchin lying on an Australian shore. They are a hazard for bare-footed beachgoers.

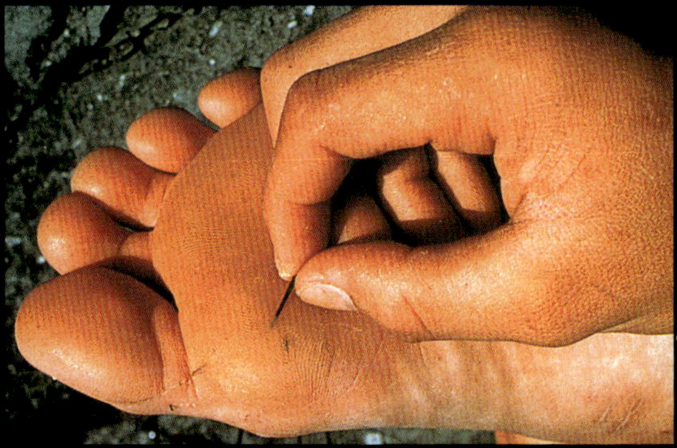

PLATE 12.15
A sea-urchin spine embedded in the sole of a beachgoer's foot. The tip of an embedded second spine is visible in the flesh adjacent to the large spine being removed.

PLATE 12.16
The common and toxic sea cucumber *Bohadschia argus*, shown with the toxic Cuvierian tubules extruded. Contact with these white tubular strands will result in intense skin irritation.

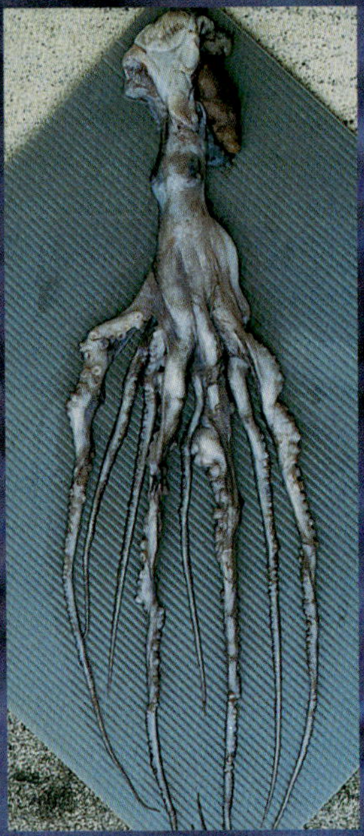

PLATE 13.1
A specimen of an Okinawan (Japan) octopus, illustrating the eight arms or tentacles, and the less apparent 'streamlined' shape (see Figure 13.4).

PLATE 13.2
The blue-ringed octopus *Hapalochlaena maculosa*, photographed in the non-swimming mode at Portsea, Victoria, Australia. This species is regarded as more common on southern Australian coastlines than in the north (Courtesy of Neville Coleman, Brisbane)

PLATE 13.3
The darker, somewhat larger northern Australian species of blue-ringed octopus *Hapalochlaena lunulata*, photographed at Dampier Archipelago, Western Australia (latitude 21°S). Note the prettily iridescing blue rings. Doubt exists about the taxonomic naming of the 'northern' Australian species (see text). (Courtesy of Neville Coleman, Brisbane)

PLATE 13.4
A blue-ringed octopus *Hapalochlaena maculosa* in 'swimming mode' with tentacles aligned behind the body, at Edithburgh Jetty, Yorke Peninsula, South Australia. Note the 'flashing' blue rings indicating that the animal is irritated. (Photograph by Karen Gowlett-Holmes, Collection Manager, Marine Invertebrates, South Australian Museum, Adelaide)

PLATE 13.5
The venomous cone shell *Conus textile*. Note the siphon projecting from the pointed end. Two white tentacles are seen projecting below the siphon, the closer one bearing a black eye-spot. The tip of the brown and white proboscis is also visible lying upon the white tentacles. (Photograph courtesy of Karen Gowlett-Holmes, Collection Manager, Marine Invertebrates, South Australian Museum, Adelaide)

PLATE 13.6
Conus striatus. (Note the shell slit running the full length of the shell).

PLATE 13.7
Conus striatus, showing a radula tooth held in the mouth. (Photograph courtesy of Clayton Bryce, West Australian Museum.)

PLATE 13.8
A specimen of *Conus geographus*. Note that the shell aperture extends the full length of the shell.

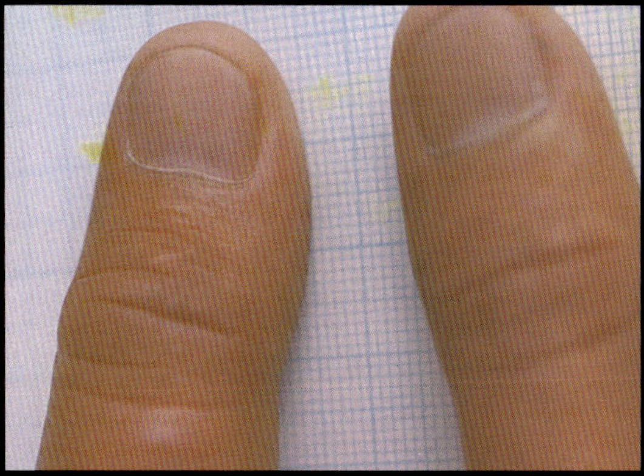

PLATE 13.9
The *Conus geographus*-envenomated (left) index finger of a Japanese patient in Okinawa. The patient survived. Note the minimal local signs; there may be faint erythema of the envenomated finger. (Photograph courtesy of Dr Yasuhiro Tomihara and colleagues, Okinawa)

PLATE 14.1
A toxic sponge of the genus *Neofibularia* (*N. irata*). Hook Reef, Australia. (Photographed by Dr J.N.A.Hooper, Queensland Museum)

PLATE 15.2
A blotched javelin, *Pomadasys maculatus*, caught in Platypus Bay (the western shore of Fraser Island, Queensland). Individuals of this species captured in Platypus Bay have recently caused several outbreaks of ciguatera. The lower lip of the mouth is capable of downward deflection, indicating specialisation for bottom foraging.

PLATE 15.1
A large coral trout, *Plectropoma maculatum*, from Australia's Great Barrier Reef. The flesh of this fish caused the death of a family cat, which was fed first! (See text.)

PLATE 16.2
The reef stonefish *Synanceia verrucosa*. Stonefish produce one of the world's most painful envenomations. This animal was found at a depth of 30 m. (Photograph by Neville Coleman)

PLATE 16.1
The scorpionfish (family Scorpaenidae) *Maxillicosta scabriceps*, in the sea at Yorke Peninsula, South Australia. (Photograph courtesy of Karen Gowlett-Holmes, Collection Manager, Marine Invertebrates, South Australian Museum, Adelaide)

PLATE 16.3
The forktailed catfish *Arius graffei*. (Photograph by Bill Stablum)

PLATE 16.4
Australian 'fantail' stingrays, *Dasyatis sephen*. (Photograph courtesy of Neville Coleman)

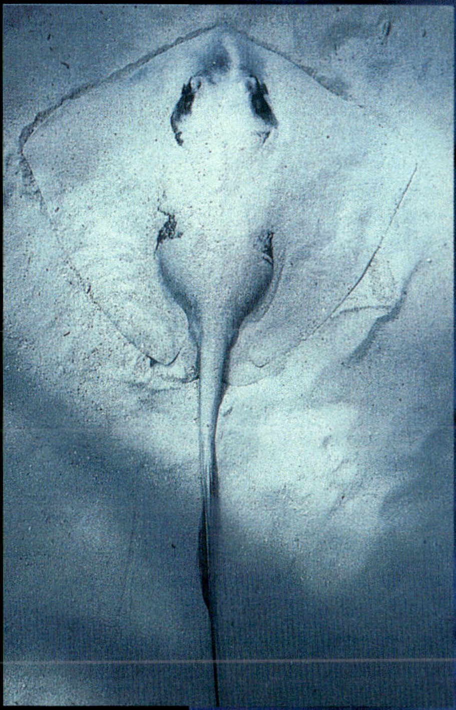

PLATE 16.5
A stingray in classical resting pose, lying partially buried, still and camouflaged, on the bottom of the sea. In shallow water they are easily stepped upon, resulting in lower leg injury (see text). (Photograph © National Geographic Society, by David Doubilet.)

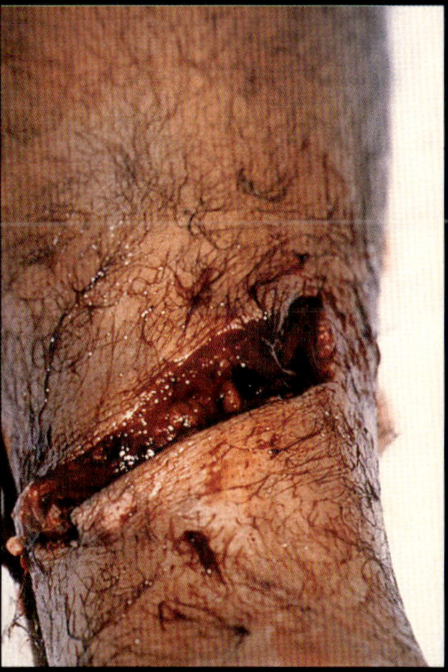

PLATE 16.6
A fresh shin laceration from a stingray spine, sustained in the manner illustrated in Figure 16.16. (Photograph courtesy of Dr Chris Acorr)

PLATE 16.8
A juvenile *Plotosus lineatus*. (Photograph by Bill Stablum)

PLATE 16.7A
Never spear a ray or bring it into a boat.

FIGURE 16.7B
An illustration of the powerful action of a stingray tail.

PLATE 16.9
The eel-tailed catfish *Tandanus tandanus*. (Photograph by Bill Stablum)

PLATE 16.10
A stonefish, *Synanceia horrida*, at Amity Point, southern Queensland. (Photograph by Bill Stablum)

PLATE 16.11
An algae-covered reef stonefish *Synanceia verrucosa*. (Photograph by Bill Stablum)

PLATE 16.12
The spines of the stonefish. The integumentary sheath has been removed, exposing some spines with intact pairs of venom glands; others with the venom glands absent (ruptured). In one case the venom gland from one side of the spine only remains intact. (Photograph by Neville Coleman, Brisbane.)

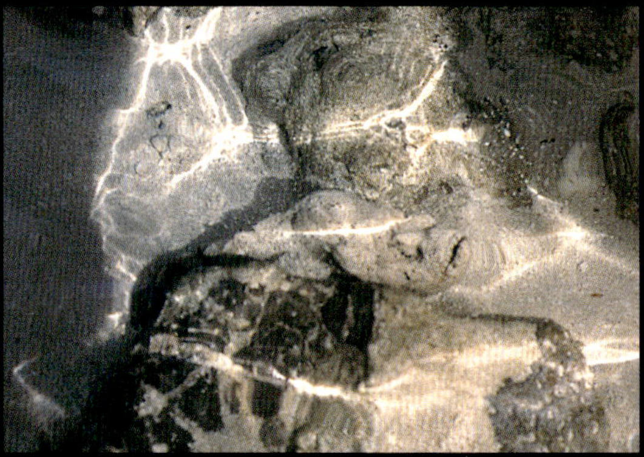

PLATE 16.13
The superb natural camouflage of the stonefish. (Photograph courtesy of the Queensland Museum)

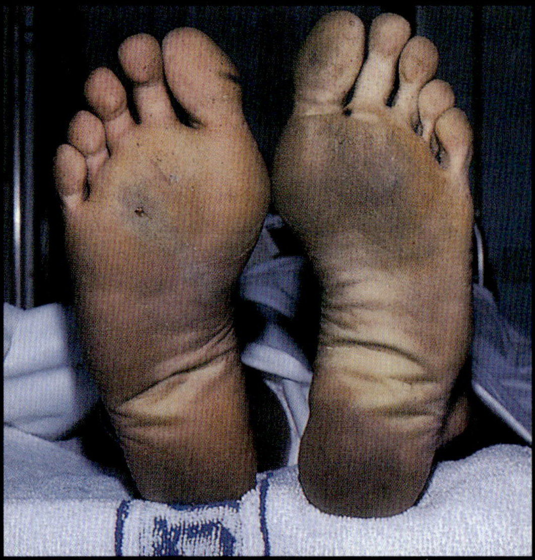

PLATE 16.14
The sole of a patient's (right) foot stung by a stonefish dorsal spine, showing the typical bluish discolouration surrounding the puncture site. The patient had an ankle block with local anaesthetic, permitting him to lie still for the photo!

Plate 16.15
The butterfly cod *Pterois volitans* (Photograph by Keith Gillet)

Plate 16.16
The scorpionfish *Notesthes robusta* (bullrout or kroki), a medically important scorpionfish from the east coast of Australia. (Photograph courtesy of Tom Gorringe)

Plate 16.17
Surgeonfish of the genus *Paracanthurus*. Note caudal peduncle spines retracted. (Photograph by Bill Stablum)

Plate 16.18
The scatophagid fish *Selenotoca multifasciata* (striped butterfish, scat, northern butterfish, poison bone). (Photograph by Bill Stablum)

Plate 16.19
A common Australian stargazer, *Kathetostoma laeve*, lying buried awaiting its prey at Portsea, Victoria, Australia. (Photograph by Karen Gowlett-Holmes, Collection Manager, Marine Invertebrates, South Australian Museum, Adelaide)

Plate 16.20
The spiny gurnard *Lepidotrigla papilio* in Portsea, Victoria, Australia. (Photograph by Karen Gowlett-Holmes, Collection Manager, Marine Invertebrates, South Australian Museum, Adelaide)

PLATE 17.1
The sea snake *Aipysurus laevis*, swimming in 10 m depth of water near Exmouth Gulf (latitude 22°S), Western Australia. Note the graceful appearance and the paddle-shaped swimming tail. (Photograph courtesy of Neville Coleman, Brisbane)

PLATE 17.4
Paired fang marks from a recent sea snake bite, on the volar aspect of a patient's forearm. Note the lack of local tissue reaction.

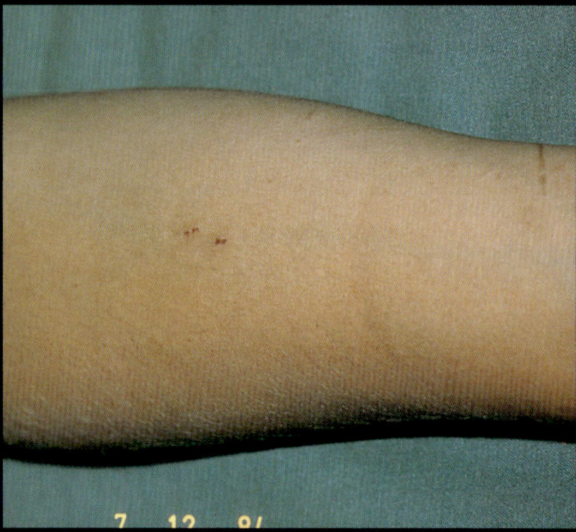

PLATE 17.3
A fresh bite by the sea snake *Aipysurus laevis* on a diver's finger. (Photograph courtesy of Dr Harold Heatwole, Armidale, NSW)

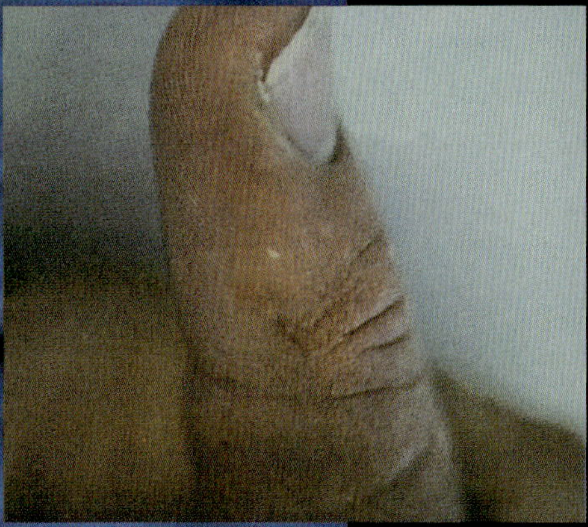

PLATE 17.2
The head of a submerged sea snake. Sea snakes are air breathers. Note the absence of visible nostril apertures, due to their sealing over by opercula during submersion.

PLATE 17.6
One ampoule of Commonwealth Serum Laboratories' sea snake (*Enhydrina schistosa*) antivenom. This is a refined preparation (pepsin digested IgG) of hyperimmunised horse plasma. Each ampoule contains 1000 units.

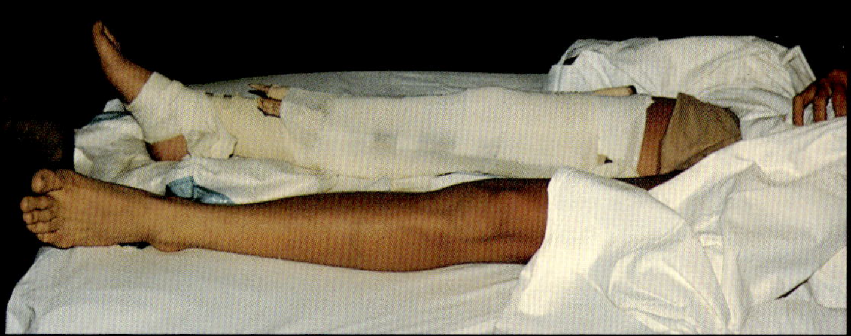

PLATE 17.5
A hospitalised patient receiving treatment for sea snake envenomation. The patient was bitten low on the right leg. Note the splint and compression-bandage still *in situ* on the right leg. (Photograph courtesy of Rockhampton Base Hospital, Queensland)

PLATE 18.1A
An Oman jellyfish sting on the back and arm of a 12-year-old boy, photographed two days after the sting.

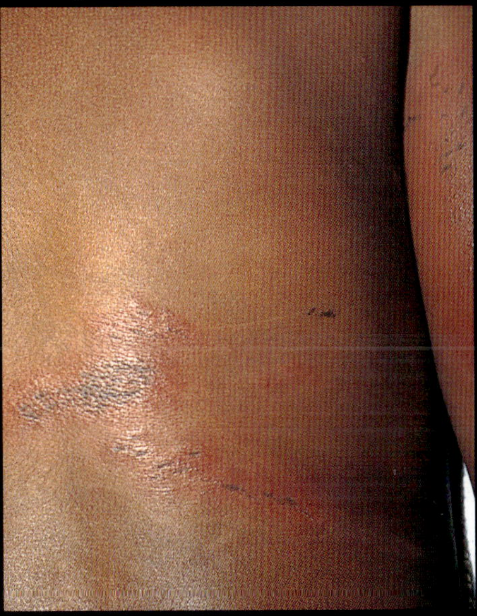

PLATE 18.1B
Closer photograph of the back of the patient shown in Plate 18.1A. Note the erythema, light scab formation, and the 'peau d'orange' appearance of the oedematous left (medial) extremity of the sting.

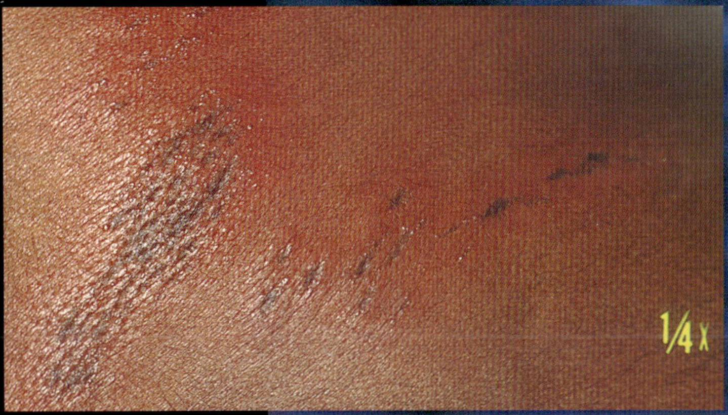

PLATE 18.3
Preserved specimens of Oman *Physalia* sp., captured in January 1987. Species identification is uncertain.

PLATE 18.2A
A large carybdeid captured in Oman waters and identified as a *Tamoya* sp.

PLATE 18.2B
Living specimens of a large carybdeid, *Tamoya* sp., captured and photographed by Professor Reginald Victor after direct encounters in the waters of a cove at Bandar Jissab, Gulf of Oman, on 31 December 1994.

PLATE 18.2C
As in Plate 18.2B, a living carybdeid photographed by Professor Reginald Victor on 31 December 1994. Sand is visible within the bell of the animal. Bell heights were approximately 38 cm. Two people nearby on this coastline were seriously stung during the same period.

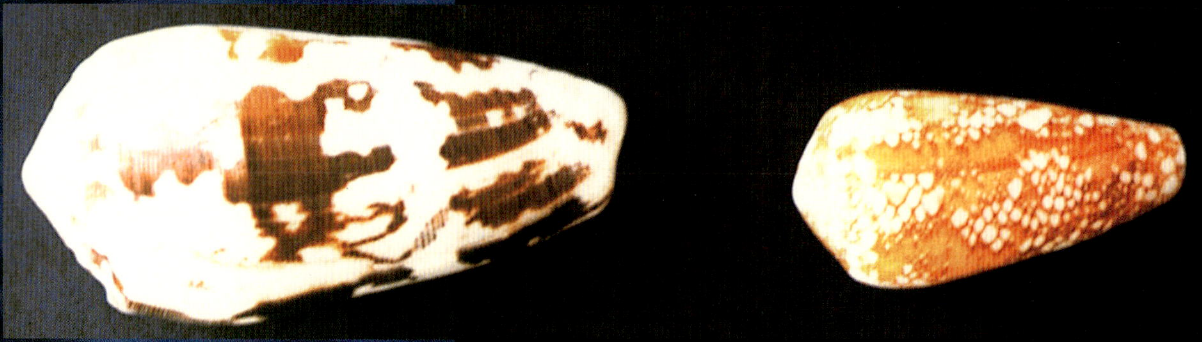

PLATE 18.4
Conus striatus (on the left) and *Conus textile* (on the right) from Oman waters. Compare with Plates 13.5 and 13.6.

PLATE 18.5
The sea urchin *Diadema setosum* from Oman.

PLATE 18.7
Three specimens of the catfish *Plotosus lineatus* from Oman.

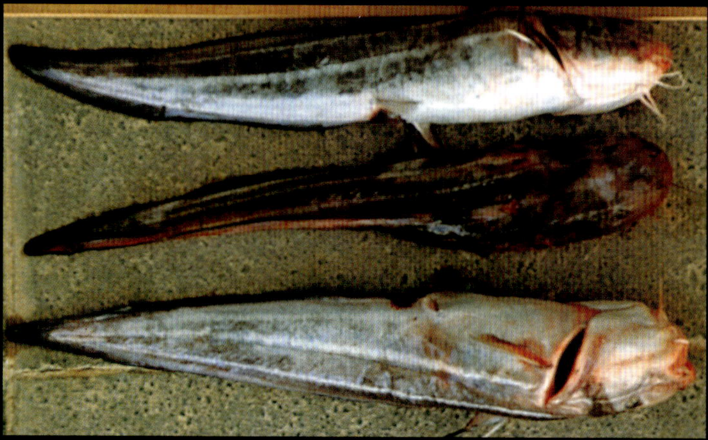

PLATE 18.6
The stinging *Urolophus* species from Oman.

Plate 18.8
An Oman catfish, *Arius tenuispinus*, implicated in the envenomation of a 19-year-old Omani male shown in Plate 18.9.

Plate 18.10
The venomous *Hydrophis cyanocinctus*, the 'annulated' sea snake from Oman. Note the flat paddle-shaped tail.

Plate 18.9
The healed wound on the hand of the 19-year-old Omani youth, envenomated by the specimen of *Arius tenuispinus* shown in Plate 18.8. Photographed six weeks after the injury.

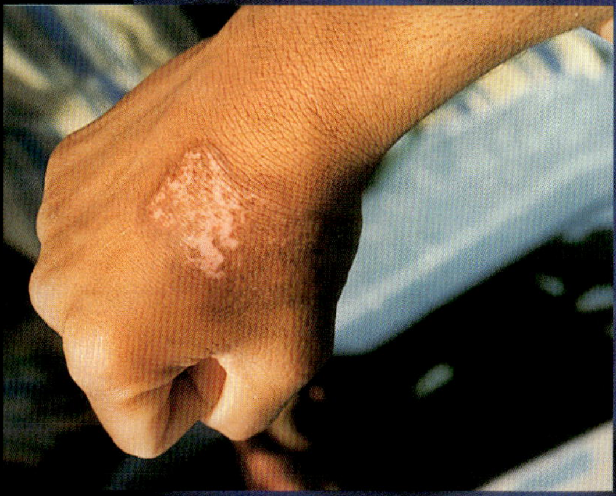

Plate 18.11
A juvenile specimen of an Oman sea snake of the genus *Hydrophis*. Note the contrast in appearance to the animal in Plate 18.10, and the difficulties of precise identification that may present.

PLATE 20.1
A Japanese (Okinawa) diver's method of carrying an underwater light strapped to one wrist. Needlefish are attracted to lights. (Photograph courtesy of Dr Yasuhiro Tomihara and colleagues, Okinawa)

PLATE 20.2
An Okinawan 'needlefish', *Tylosurus crocodilus*, of the kind responsible for injuries to Japanese divers. Note the long, thin, but powerful jaws, capable of delivering a stab wound. (Photograph courtesy of Dr Yasuhiro Tomihara and colleagues, Okinawa)

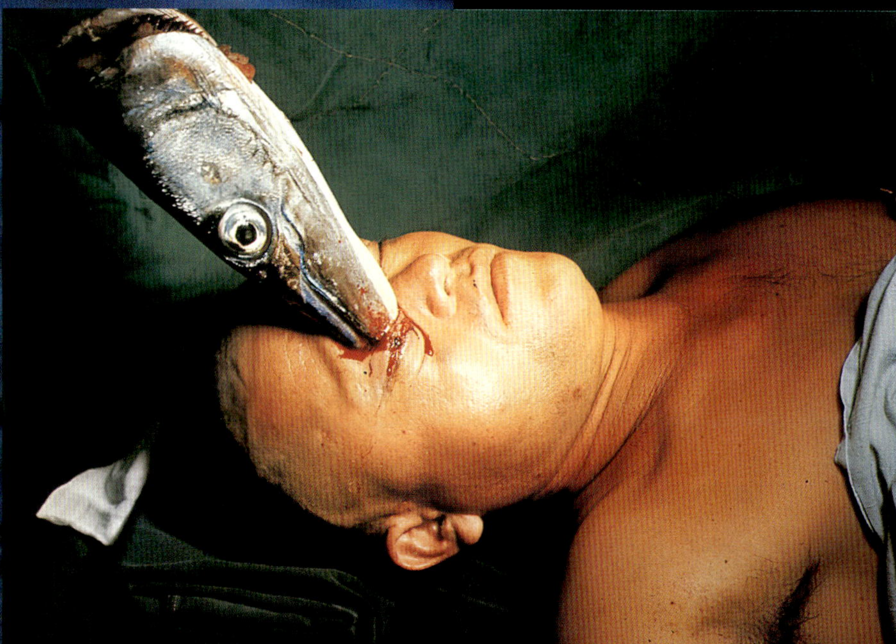

PLATE 20.3
A needlefish penetration of the orbit and entering the sinuses and soft palate — causing loss of the eye, but not proving fatal — in a Japanese diver. (Photograph courtesy of Dr Maeshiro, Okinawa, Japan)

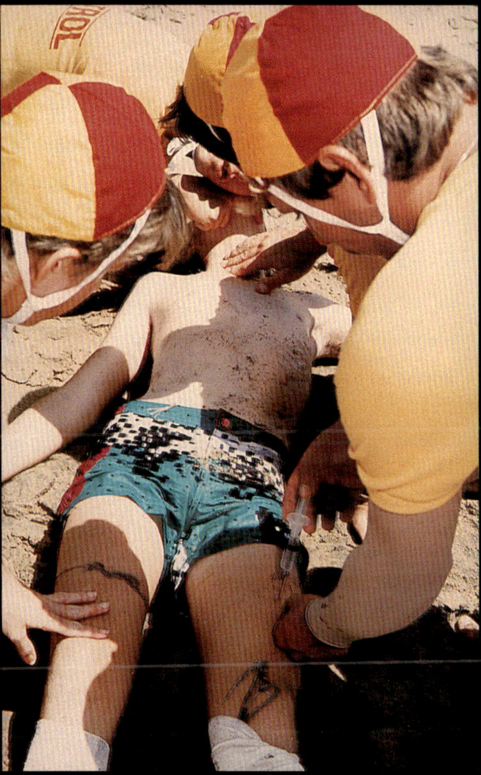

PLATE 21.1
North Queensland surf life savers practising a full-scale, on-beach rescue and resuscitation of a patient with simulated serious *Chironex fleckeri* envenomation. Note the (simulated) injection of specific box jellyfish antivenom (3 ampoules/60, 000 units, ≈6 ml) into the lateral aspect of the subject's left thigh; the (simulated) adherent, but vinegar-treated tentacle material still on the subject's legs and abdomen; the protective clothing worn by the rescuers; and resuscitation proceeding without interruption, irrespective of all other activities.

PLATE 21.2
A North Queensland Ambulance Service team arriving at the on-beach site of a major *Chironex fleckeri* envenomation. The officer in the foreground is carrying the on-beach box jellyfish sting rescue kit, which includes specific box jellyfish (*Chironex fleckeri*) antivenom. See also Plates 9.88.

Plate 22.1
The interface between humans and crocodiles is beloved of the media!

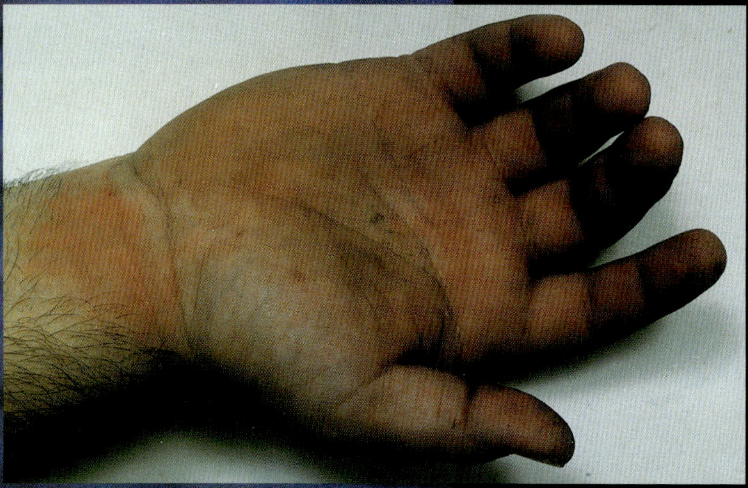

Plate 22.2a
A hydroid sting on the wrist of a Northern territory scuba diver approximately 18 hours after the sting. Note the intense local erythema, and slight oedema of the proximal palm. (Photograph courtesy Phil Alderslade and Dr Low Choy, Darwin.)

Plate 22.2b
As in Plate 22.2a, showing small vesicular lesions with early scabbing, scattered within the erythematous area. The sting pain was described as short, sharp, not lasting long, but quite intense. By day 6, the area showed a visible rash of raised purplish spots. (Photograph courtesy Phil Alderslade and Dr Low Choy, Darwin.)

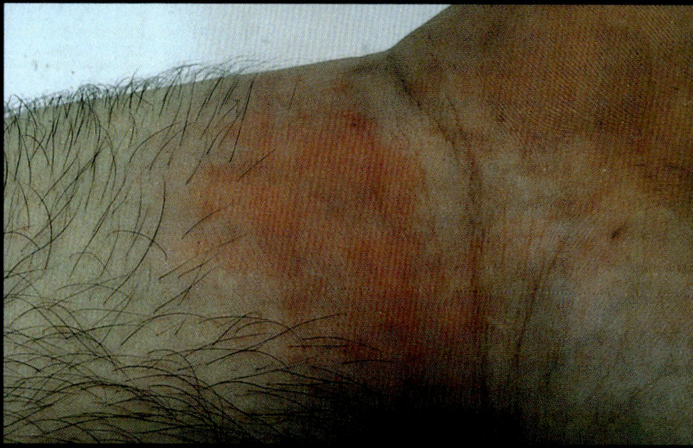

Plate 23.1
One of the editors (JWB) working independently and with initiative; on a Florida beach during a Christmas period!

Plate 25.1
The platypus *Ornithorhynchus anatinus*. Note the duck-like bill, the paddle-shaped tail, and the claws on the webbed feet for digging. (Copyright © *Medical Journal of Australia* 1992; 157: 829)

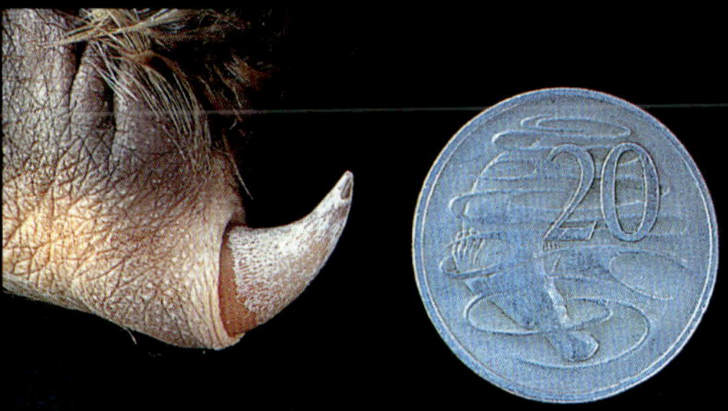

Plate 25.2
A harmless, juvenile spur on the hind leg of a young male platypus *Ornithorhynchus anatinus*. The Australian 20 cent coin is for size comparison. Only adult male spurs can envenomate. (Copyright ©*Medical Journal of Australia* 1992; 157: 829)

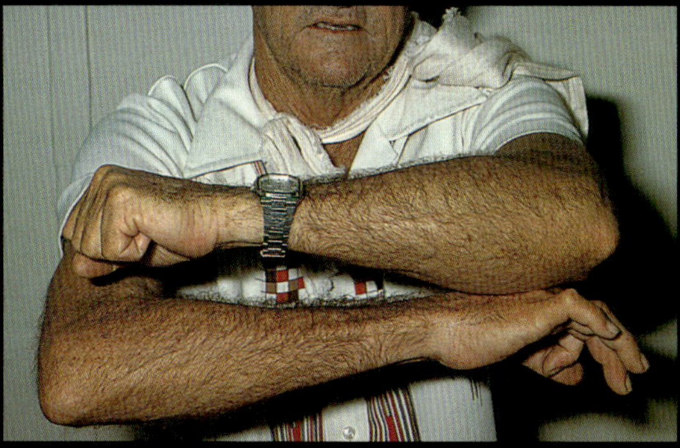

Plate 25.3
The patient photographed one month after platypus spur envenomation of his right hand. Note the muscle wasting in the right forearm, and the inability of the patient to make a fist with his right hand, because of persistent swelling, pain and weakness.

Plate 26.1
The newly declared 'Marine Jurisdictional Zones of Australia', according to the (United Nations) Law of the Sea Convention, 1994 (courtesy Dr Joe Baker PhD, and the Australian Geological Survey Organisation).

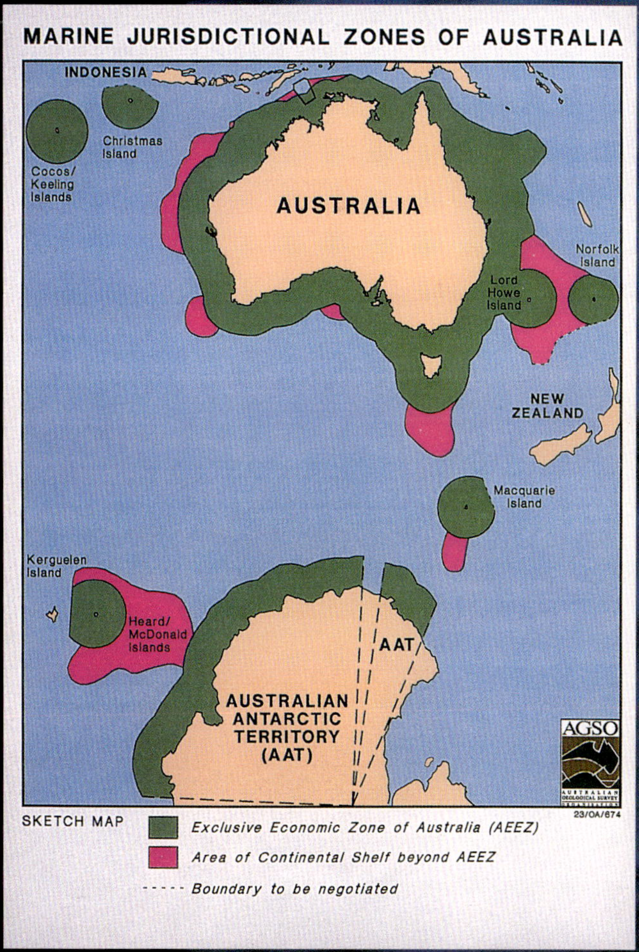

PLATE 26.2
The appearance of the left index finger of a 40- year-old male, 5 weeks after envenomation by a Portuguese man-o'-war (*Physalia physalis*) in Brazilian coastal waters in 1994. The ulcerated finger was subjected to surgical debridement 7 days following the sting, and exhibits restriction of extension due to cicatrisation.

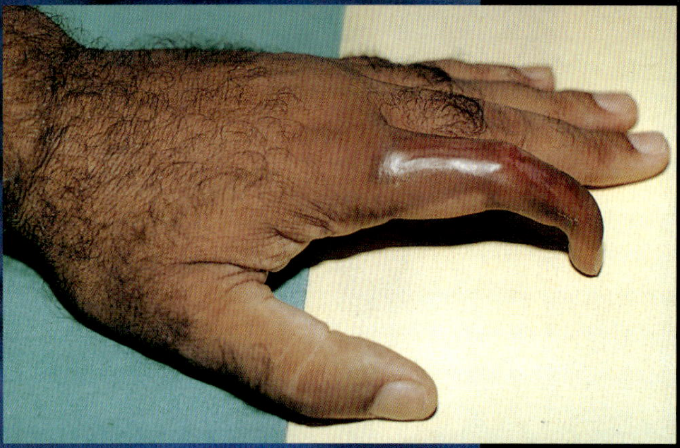

PLATE 26.3A
One of the pontoon colony of sea anemones in Hokimai Bay, Kawau Island, New Zealand, identified by marine biologists in New Zealand as *Actinothoe albacincta*, and considered most likely responsible for the significant envenomation of two male swimmers. (Maurice Miles, Auckland Health Care Services Ltd.)

PLATE 26.3B
The appearance of the chest rash on a 49-year-old male, 11 days after envenomation by numerous small sea anemones of the species shown by Figure 26.2A, in Hokimai Bay, Kawau Island, New Zealand, in November 1994. The lesions had 10 days of topical steroid and cold compress application, prior to the photograph. (Christopher Dickie and Auckland Health Care Services Ltd.)

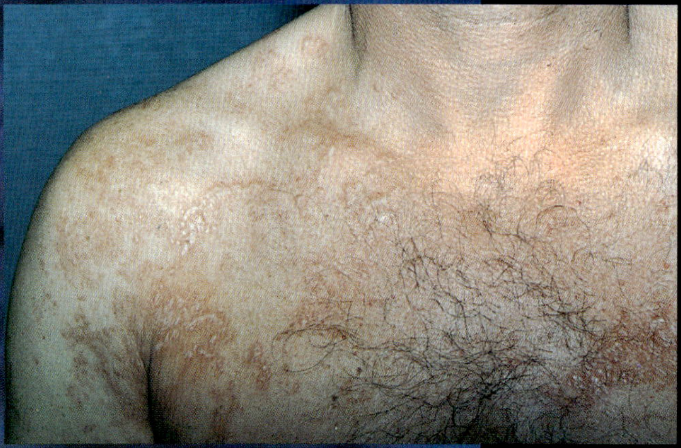

PLATE 26.4
A sea star, possibly a species of *Nardoa* (family Ophidiasteridae), observed by a scuba diver at 55m depth in Milne Bay, Papua New Guinea. A painful hand sting occurred, developing some minutes after handling, and taking 'a few' days to subside Wolfgang Zeidler, Senior Curator of Marine Invertebrates, Division of Natural Science, South Australian Museum, kindly examined the photograph. (Courtesy Mr Bob Halstead, Cairns)

PLATE 26.5
A two day old *Chrysaora quinquecirrha* sting on the left thigh of a young woman in Chesapeake Bay, eastern United States of America. There were no systemic effects.

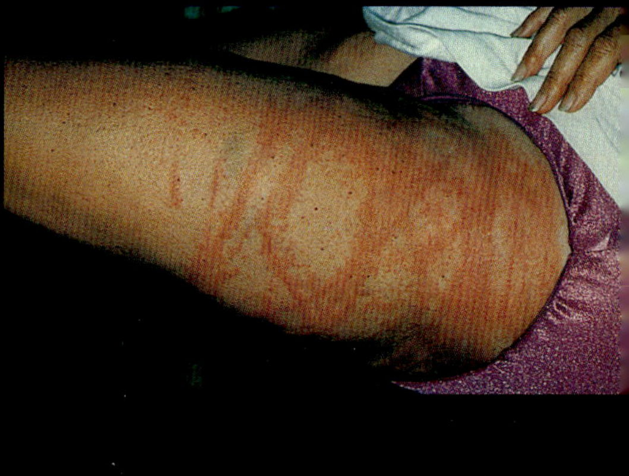

CLASS SCYPHOZOA

{SEE CHAPTER 6, PAGE 121}

Like all cnidarians, scyphozoans are physical variations of two body forms — a polyp and a medusa. As with the cubozoans {see Chapter 9, page 236}, the polyp form in the scyphozoans is small. The medusa is the dominant form of the life cycle; it can vary between species from a few millimetres in diameter to over 2 m with tentacles of enormous length (MAYER 1910; CLELAND & SOUTHCOTT 1965; HALSTEAD 1988). Many have attractive colouring (e.g. pinks or mauves), and their ocean and shore movement and distribution tends to be dominated by currents and tides, despite locomotion from their generally rhythmic, pulsating bell action.

Species of some genera, such as *Cyanea* (Plates 8.56 and 8.57), *Catostylus* (Plate 8.61), and *Pelagia* (Plates 8.84A and 8.84B) {see Chapter 10, page 284} may occur in large groups or 'swarms', but they can also occur as solitary individuals (Plate 8.57).

Members of this class are distributed worldwide, and some of them have medical significance. The scyphozoan orders Rhizostomeae, Coronatae and Semaeostomeae are considered here.

ORDER RHIZOSTOMEAE

BIOLOGICAL DETAILS — JACQUIE RIFKIN
MEDICAL DETAILS — PETER FENNER

This populous, worldwide order of scyphozoans includes the interesting genera *Cassiopea*, *Stomolophus* and *Catostylus*. The Mediterranean stinger *Rhizostoma pulmo* is discussed in Chapter 10, page 287. General anatomical and geographical distribution aspects of this order are described in Chapter 6, commencing page 144.

GENERAL DISTRIBUTION OF THE RHIZOSTOMES

Worldwide, although less common in the Arctic and Antarctic oceans.

The known species distribution in the Northern Territory is shown in Table 8.2.

TABLE 8.2
RHIZOSTOMES FOUND IN THE NORTHERN TERRITORY

Acromitoides spp.	*Phyllorhiza punctata*
Cassiopea andromeda	*Pseudorhiza haeckeli*
Catostylus mosaicus	*Rhopilema hispidum*
Mastigietta palmipes	

Courtesy of Phil Alderslade, Darwin (1994 unpublished observations).

GENERAL APPEARANCE

Unlike medusae in the order Semaeostomeae {see page 223}, which represent one kind of jellyfish appearance with a bell, single mouth and tentacles, rhizostomes have no tentacles or mouth as such. Instead they have eight nematocyst laden 'arms' {Plates 6.2A and 6.2B} which paralyse their prey, and which then supply multiple small 'mouths'.*

Despite this common feature of having arms rather than tentacles, the rhizostomes show a vast variation in size and appearance within the order itself. The bell diameter varies from a few centimetres in smaller species, to one or two metres in larger species. The bell may be opaque or almost transparent, may be white or contain various colours and/or patterns, again depending on species. The reader is referred to a specialist book such as KRAMP (1961) for further differentiation.

In keeping with the tremendous morphological diversity of rhizostomes worldwide, their stings can range from potentially fatal (*Stomolophus nomurai*), through troublesome (*Rhopilema hispidum*) to inconsequential (*Catostylus mosaicus*). Fortunately for medical and first aid personnel, the sting effects and management principles are similar for most species, and are considered at the end of the text on rhizostomes {page 221}. However, where actual data relating to the sting (and/or its management) of particular species of the order exist, these are described in association with the respective species.

* Some authors call the elongated structures surrounding the small mouths 'tentacles', and this terminology is retained here.

ENVENOMATION BY RHIZOSTOMES

Summary of evidence and general features

These animals usually induce a non-specific local reaction with mild skin irritation (FENNER 1993). However, severe localised pain with local redness, urticaria and swelling has been reported (CLELAND & SOUTHCOTT 1965).

HALSTEAD (1988) describes muscular cramps, severe whealing, mucoid expectoration, intense coughing, extreme mental depression and 'respiratory embarrassment'. Severe constitutional symptoms have been described as occurring after envenomation by a probable rhizostome medusa in the English Channel. These included breathlessness and anxiety, followed by extreme lethargy and depression. Another case is described in north Wales, when painful muscular spasms developed in the abdomen and neck, together with low back pain (cf. Irukandji syndrome, Chapter 9, page 236) (CLELAND & SOUTHCOTT 1965). MAYER (1910) suggested that another rhizostome medusa (*Lobonema smithii*) in the Philippines (Plate 8.69) inflicted a 'very severe sting'. *Rhopilema hispidum* {Plate 8.59} caused a serious sting on the arms of a biologist in Darwin Harbour in 1994 (see below) which later became necrotic (CORNELIUS 1994, unpublished observations). EVANS (1943) described 'a case of anaphylaxis with alarming symptoms' (unspecified) occurring immediately after a second sting, three weeks after an initial mild sting. He also described respiratory symptoms including sore throats, rhinorrhoea and bronchospasm in girls at Lowestoft, England, who inhaled dried jellyfish substance while beating fishing nets to clean them. Both LIGHT (1921) and BARNES (CLELAND & SOUTHCOTT 1965) describe the phenomenon of 'hot water' sensation with rhizostome jellyfish when skin tingling or stinging may be felt around the sea water area (or in the container water) where specimens may have been handled. This effect is believed to be due to the release of thousands of nematocysts into the surrounding water.

Blindness for two and a half days has been described by MARSH and colleagues (1986) after a severe *Phyllorhiza punctata* (Plate 8.70) sting in the eye where it is reported that 'over 50 nematocysts' were removed at operation from the cornea. No reference is supplied. [We are curious about how these structures were removed at surgery — Eds. (GLASSER *et al* 1992 and 1993).]

FATALITIES

Stomolophus nomurai stings in China have produced fatalities (MINGLIANG ZHANG 1987 & 1988; QIN SHEDE & MINGLIANG ZHANG 1990; MINGLIANG ZHANG & QIN SHEDE 1990 & 1991) {see page 214}. Other rhizostome medusae, despite evidence of some severe stings, are not known to have caused human death.

MEDICALLY IMPORTANT REPRESENTATIVES OF THE ORDER RHIZOSTOMEAE

The following rhizostomes are not necessarily placed in order of medical importance, nor is the list intended to be complete. On-going field studies and future research will significantly expand and modify it.

GENUS *RHIZOSTOMA*

Type species: *Rhizostoma pulmo*. Less commonly, another species, *R. luteum*, is described.

DISTRIBUTION

Rhizostoma pulmo (Figures 10.4 and 10.5, pages 288 and 299) is found in the Mediterranean Sea (AXIAK 1983) and the Black Sea {see Chapter 10, page 287}, on the west African Atlantic coast, in the English Channel and in the North Sea northwards to Norway.

APPEARANCE

Rhizostoma pulmo is a typical rhizostome with eight arms (described above) but also having 'terminal appendages' resembling tentacles extending from the mouth arms (Figure 10.4, page 288). These are usually thicker distally than proximally. The bell diameter may vary from 20 to 40 mm in smaller species but up to 50–60 cm in larger species. The bell may be colourless or an opaque white.

THE STING AND MANAGEMENT

This is discussed on page 221.

GENUS *RHOPILEMA* HAECKEL, 1880

DISTRIBUTION

The genus *Rhopilema* is composed of four species: *Rhopilema hispidum* (Plate 8.59), *R. verillii* and *R. esculenta* (= *R. asamushi*) (MAYER 1910) are found in the Indian Ocean, in South-East Asian waters as far as Hong Kong, in Japan and China, and along parts of the North American east (Atlantic) coastline. The fourth distinct species, *Rhopilema nomadica* is described in the Eastern Mediterranean {see Chapter 10, page 284 and Plate 10.8}.

Rhopilema hispidum is found all around the Indo-Pacific and as far north as Hong Kong, east to the Philippines, west to the Red Sea (MAYER 1910) and the Arabian Gulf (Plate 8.59). It was not described in Australia until 1994 (Darwin, Northern Australia [CORNELIUS & ALDERSLADE 1994, unpublished observations]). There are now increasing numbers, even swarms, of *R. hispidum* occurring in the Mediterranean Sea, as there appear to be no natural predators (ATTENBOROUGH 1994, unpublished observations). This new distribution has occurred since the opening of the Suez Canal (1869), linking the two previously unconnected seas. The taxonomic relationship between *Rhopilema hispidum* (Plate 8.59) and *R. nomadica* {see Plate 10.8} remains to be resolved.

Rhopilema verillii appears in late summer and autumn, from North Carolina to New England, including the Chesapeake Bay region (CARGO 1971).

APPEARANCE

The Indo-Pacific *Rhopilema hispidum* has a bell diameter of between 25 and 35 cm, and may be blue, white or dark red. Filaments that are thicker distally (furthest from the bell) hang below the 'arms', which may be brown, white, transparent or brown, whilst the gonads have a yellow tint. The specimen collected in Darwin Harbour was described as being large, with a 20 cm diameter bell covered with brown spots — a feature not reported in most of the literature.

Rhopilema in the Americas is a rare jellyfish, whose bell is approximately 2–4.5 cm wide by 1.5–3.5 cm tall. This structure is usually dome-shaped but may be hemispherical (Plate 8.58). Eight oral arms hang into the water to a depth greater than the bell diameter. The upper one-third of each arm is coalesced into a smooth furred mass. The lower portion of these stout, triangular or pyramidal arms has numerous frills and filaments. The oral and aboral surfaces of the bell {see Chapter 6, page 144} may be smooth or granular.

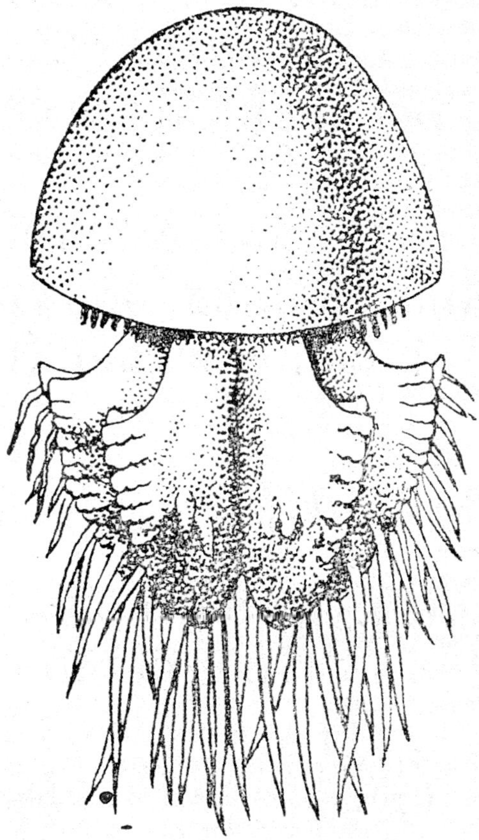

Рис. 54. *Rhopilema asamushi* Uchida (по Учида).

FIGURE 8.12
Rhopilema esculenta (= *R. asamushi*) Uchida. (After a published illustration in a Russian text, supplied by Dr Paul Cornelius.)

The variable colour of this medusa is similar to the Indo-Pacific species, with the bell being blue, white or dark red. The mouthparts are similarly brown, white or transparent, and the gonads usually have a yellow tint.

THE STING

Mild stings are the usual effects of this genus, although a more severe reaction was reported recently.

CASE 1

A small sting occurred on the finger of a 50-year-old marine biologist in Darwin in 1994 and caused moderate pain. A small pimple developed, which opened into a necrotic wound that took six weeks to heal despite being non-infected.

CASE 2

A probable sting, from a specimen later identified as *Rhopilema hispidum* occurred on the wrist of a 20-years-old male diver off Channel Island, Darwin, Australia, in 1990. Immediate mild to moderate pain was accompanied by the appearance of light patchy erythema of the sting area, which was protected by some hair. A couple of small areas of more marked erythema developed (Figure 8.73). Irritating discomfort continued for some hours, and the marks were still visible after six days (LOW CHOY and ALDERSLADE 1990, unpublished observations).

Human consumption

Rhopilema esculenta (Figure 8.12, page 211) is used as a food in Japan and China, where it is soaked in fresh water, flavoured with condiments and preserved in a mixture of alum, salt and steamed leaves (KISHINOUYE 1899, in MAYER 1910 [also in Spengel 1988], KIMURA *et al* 1983). It is also caught in Malaysia and preserved for export to Japan {see Chapter 10, page 296)}.

Rhopilema hispidum is on sale in some restaurants in the West End of London as a 'delicacy'. The taste was described recently by CORNELIUS (1994, unpublished observations) as 'dreadful'.

GENUS *CASSIOPEA* ('UPSIDE-DOWN JELLYFISH')

(Note that two different generic spellings exist: '*Cassiopeia*' and '*Cassiopea*'. The latter spelling is used here.)

Limited information indicates this animal is a significant stinger in certain regions, despite the scanty documentation. KRAMP (1961) describes up to 17 species of *Cassiopea*. Recognition certainly includes the following species:

Cassiopea xamachana R.P. Bigelow, 1892
Cassiopea andromeda Forskål, 1775
Cassiopea ornata Haeckel, 1880
Cassiopea ndrosia Agassiz & Mayer

DISTRIBUTION

Species in this genus have been sighted in all three major oceans, and particularly in the Indo-Pacific (SOUTHCOTT 1987b) and around Hawaii (see below). Documentation exists for the West Indies, Caribbean and the Netherlands Antilles (SOUTHCOTT 1959, 1987; CLELAND & SOUTHCOTT 1965), northern, eastern and southern Australia (T. STEPHENSON 1931; W. STEPHENSON 1962; SOUTHCOTT 1987b), Malaysia (CLELAND & SOUTHCOTT 1965), Fiji and New Caledonia (SOUTHCOTT 1987b).

APPEARANCE

A flattish circular but deeply scalloped bell is present, resulting in a ragged-edge appearance (SOUTHCOTT 1987b) (Plate 8.60 and Figure 6.24a, page 146). The colour is usually variable; it is generally a greenish grey-blue; and the exumbrella is patterned with a circle and radial stripe. The bell may attain 40 cm in diameter and has distinct concavity present in the middle. Five short lappets are found at the bell margin and the sense organs are arranged between them. Four pairs of adradial mouth arms arise from the mouth arm disc, each giving rise to primary branches, which themselves give rise to secondary branches. The stomach is cruciform in shape. The animal can lie on the bottom in an upside-down position, arms upwards, collecting planktonic food as debris drifts down. Further distinguishing features of *Cassiopea xamachana* R.P. Bigelow, 1892 are supplied by Jacquie Rifkin in Chapter 6, page 146.

THE STING AND MANAGEMENT

{See also page 221 below.}
The sting may be troublesome (SOUTHCOTT 1959). *C. xamachana* in the Caribbean has produced pain and local wheal formation on contact, followed by vomiting and joint pains (CLELAND & SOUTHCOTT 1965). In other areas (Girls Bank, Harbour Island, Eleuthra, Bahamas) this animal is harmless to humans. This same species has elsewhere produced generalised urticaria and facial oedema within an hour of the sting, suggesting a systemic allergic reaction to the venom (HOLZBERGER 1958 quoted in CLELAND & SOUTHCOTT 1965). In Hawaii it has produced a mild sting (FAUTIN 1986, unpublished observations).

GENUS *CATOSTYLUS* ('BLUBBER JELLYFISH')

The common 'blubber' is perhaps the most easily recognised jellyfish in the regions of its occurrence (Plate 8.61), partly because of its size and ubiquity, and partly because of its frequent occurrence in coastal shoals with their uninviting appearance to swimmers!

DISTRIBUTION

Catostylus spp. are described mainly in the Indo-Pacific region (AZILA 1990), with two or three species on the Atlantic coasts of Africa and Europe. None have been described on the Atlantic coasts of the Americas, or in either of the polar oceans (MAYER 1910). The type-species *Catostylus mosaicus* (QUOY & GAIMARD, 1824) was first described in Australia (QUOY & GAIMARD 1824), where it is very common in temperate and tropical waters, particularly on the eastern coastline (CLELAND & SOUTHCOTT 1965).

APPEARANCE

Up to 25–35cm across the bell, commonly coloured blue in Queensland, and yellow or brown in New South Wales (CLELAND & SOUTHCOTT 1961) because of the presence of symbiotic algae (zooxanthellae) {see also *Phyllorhiza pacifica*, page 218}. *Catostylus mosaicus* does not have tentacles as such, but there are eight long 'mouth arms' which contain the nematocysts. *Catostylus* are also reported in Northern Territory waters.

THE STING

Although there is a measure of contradiction in the literature, it is the editors' experience that the Australian specimens, at least, are almost devoid of an appreciable sting, a position supported by HALSTEAD (1988). There may be a mild, non-irritating skin reaction (FENNER 1993; FENNER 1988-94, Annual General Report, Surf Life Saving Queensland [in-house publication]). CLELAND and SOUTHCOTT (1965), however, describe varying stings from mild skin irritation to acute skin pain taking several hours to settle, and POPE (1947) records it at certain times of the year as giving a 'terrible sting causing nauseating pain'. LIGHT (1921) (of Philippine *Chiropsalmus quadrigatus* fame) {see Chapter 9, page 236}, initially had the opinion that the 'common black medusa' *Catostylus purpurus*, of Manila Bay was harmless; however, he revised his opinion, and describes a self-administered sting from the species, which caused hours of discomfort followed by an unpleasant Irukandji-like syndrome, whose symptoms were 'much alleviated' by '10 grains of aspirin'. No other international documentation of even moderately serious stings were found by the editors.

CASE REPORT

Contrary to the common Australian experience, a reliably witnessed sting from a confirmed *Catostylus* sp. sting on the Queensland Gold Coast, near Brisbane, in 1969, produced serious and progressive local effects. The young adult male patient, surfing naked (morning swim), was stung on the scrotum and rapidly developed swelling and vesiculation, accompanied by pain sufficient to prevent him donning his shorts! Later that day he was unable to walk and sought hospital treatment, although apparently otherwise recovered. No further details are available (M. READY 1994, unpublished recollections).

THE VENOM

There has been little study of *Catostylus* spp. venom, although preliminary work shows some effect on erythrocytes (AZILA 1990).

FIRST AID AND MEDICAL MANAGEMENT

{See page 221 below.}

GENUS *ACROMITOIDES* STIASNY, 1921

This genus has been described in the Philippines (LIGHT 1914) and Malaysia (KRAMP 1961), and recently in the Northern Territory of Australia (CORNELIUS 1994, unpublished observations). The bell diameter is about 100 mm and has three wide mouth arms (see *Pseudorhiza haeckeli*, Plate 8.72 and page 220). The bell may have brown spots, or the whole jellyfish may be a brown colour. The literature appears to make reference only to *Acromitoides purpurus* (SOUTHCOTT 1987). CLELAND and SOUTHCOTT (1965) record that *Acromitoides purpurus* (Mayer, 1910) referred to by LIGHT was identical to *Catostylus purpurus* Mayer {see above}.

GENUS *ACROMITUS* LIGHT, 1921

Acromitus spp. have been described from the Indian and Pacific oceans (KRAMP 1961). Species have bell diameters from 44 mm to 200 mm. Human stings have not been commonly described in the literature, but the venom of one species was investigated by GHOSH and colleagues (1990a; 1990b; 1993) {see below}.

ACROMITUS RABANCHATU

Acromitus rabanchatu is one of the most common venomous jellyfish found along the eastern coastline of the Bay of Bengal. It produces repeated and sometimes severe envenomations among fishermen, sea bathers and naval divers (GHOSH *et al* 1993).

The venom

Acromitus rabanchatu venom has a molecular weight of 182 000 daltons, which can induce a direct paralysing effect on animal muscle membranes (GHOSH *et al* 1993). Another component toxin with a molecular weight less than 20 000 daltons is heat labile and can cause hypoglycaemia in rabbits (GHOSH *et al* 1990b). The crude venom extract can produce hypotension and cardiac arrest, which is not blocked by acetylcholine, histamine or 5-hydroxytryptamine. Its activity on the contractility of guinea pig ileum was antagonised by atropine and cyproheptadine. In addition the crude venom is haemolytic and mildly dermatonecrotic (GHOSH *et al* 1990a).

GENUS *STOMOLOPHUS*

Two species are considered below:

1. *Stomolophus nomurai*
2. *Stomolophus meleagris* (cabbage-head jellyfish)

STOMOLOPHUS NOMURAI IN CHINA

DR MINGLIANG ZHANG AND PROFESSOR QIN SHEDE,
QINGDAO MEDICAL COLLEGE,
PEOPLE'S REPUBLIC OF CHINA

Stomolophus nomurai (Plate 8.62) is found in the temperate waters of Japan, Korea and northern China (the north-west Pacific). In China it is present in the North China Sea (the Yellow Sea and Bohai) and the East China Sea. Occurrence is seasonal, starting from the warmer southern waters moving northwards from June to November, with more frequent appearances in August and September.

It is often found floating in large numbers in shallow water, particularly during or after rain and onshore wind. As one of the most venomous jellyfish in the China seas, it causes severe stings and occasional deaths among swimmers and fishermen (MINGLIANG ZHANG 1988, 1992; MINGLIANG ZHANG & QIN SHEDE 1990, 1991).*

* The umbrella (bell) of the rhizostomes *Stomolophus nomurai* and *Rhopilema esculenta* are eaten following curing, as an hors d'oeuvre in Japan and China. The major edible component is the connective tissue protein, collagen (KIMURA *et al* 1983).

APPEARANCE

The body is dome-shaped with many long tentacles hanging underneath (Plates 8.62 and 8.63 [top right]). Many mature jellyfish specimens have bodies more than 1 m in diameter and may weigh about 100 kg.

Stomolophus is easily recognised by its numerous, non-stinging sand-like dots on the outer surface of the large bell (the Chinese name is 'sand jellyfish'). The body is usually translucent or milky white, but the oral arms and the tentacle bases are brown. Numerous mouth arms some 0.5–1 m in length hang under the bell (Plate 8.63 [top right]); some may reach more than 1 m in length (MINGLIANG ZHANG & QIN SHEDE 1991).

THE NEMATOCYSTS

Both the tentacles and the oral arms have large numbers of nematocysts, which can be divided into four types {see Chapter 6, page 168}:

1. atrichous isorhizas {Figure 6.43[VI]}
2. homotrichous anisorhizas {Figure 6.43[XI]}
3. homotrichous microbasic euryteles {Figure 6.43[XXVII]}
4. heterotrichous microbasic euryteles {Figure 6.43[XXVIII]}

Homotrichous anisorhizas appear to be responsible for the sting. Nematocysts are not present on the outer surface of the bell.

THE STING

Swimmers are usually stung on the arms and legs, and the tentacles produce multiple, linear, whip-like skin lesions (Plate 8.64). The forearms are easily stung, but as with most other jellyfish stings the palms are often not affected (at least, not immediately {see below}) because of the relative skin (keratin) thickness. Severity varies considerably with the size of the sting and other factors {see Chapter 9, page 271–2}. Chinese fishermen who have often been stung by the jellyfish usually show only mild reactions — comparable with Australian fishermen who show a similar acquired 'resistance' to the more severe chirodropid stings sustained from fishing net catches; the reason for this is unexplained, but may in part be due to keratinisation of the skin during exposure to the sun and salt water during the fishing season (FENNER 1993, unpublished observations).

SYMPTOMS AND SIGNS

Contact with the tentacles causes immediate severe pain or an intense burning sensation of the skin. The stung area quickly becomes erythematous and urticarial and after 2–4 minutes the skin develops red papules before vesiculating and forming petechiae and ecchymosis. The lesions of mild stings become dry and this may cause crusting of the skin before disappearing after five days. In severe stings, skin blistering and bullae occur with severe pain. Subsequent severe itching commonly develops over the next three to seven days and disappears without scars after about two weeks (MINGLIANG ZHANG 1987; MINGLIANG ZHANG & QIN SHEDE 1991; 1992).

The majority of stings have no severe systemic symptoms. Some patients with massive stings may develop severe muscle pain after one to four hours, which may be accompanied by fever (38–39°C), nausea, vomiting, profuse sweating, anxiety, central cyanosis and even occasional breathing difficulty {cf. Irukandji syndrome, Chapter 9, page 246}. Some cases may show a sinus tachycardia, haemolysis of red blood cells {see Chapter 7, page 214} and acute renal failure.

HUMAN DEATHS

All patients with fatal systemic symptoms develop acute pulmonary oedema one to four hours after the sting. In the summers of 1984, 1986, 1987 and 1988 about 3000 people were stung in Beidaihe beach (a city on the coast of Bohai in China) by *Stomolophus nomurai*, leaving eight victims dead from acute pulmonary oedema and cardiac arrest (MINGLIANG ZHANG 1988; MINGLIANG ZHANG & QIN SHEDE 1990 & 1992). The following details of the fatalities were verified:

1984	1 fatality at Quingdao, Heibei province
1986	2 fatalities at Beidaihe, Hebei province
1987	3 fatalities at Beidaihe, Hebei province 1 fatality at Haiyang county, Shandong province
1988	1 fatality at Longkou, Shandong province.

CASE REPORT

(courtesy of Dr Mingliang Zhang, MD)
At 4 pm on 23 August while bathing on a popular beach some 300 km east of Beijing, a healthy 32-years-old male was stung on both thighs, one arm and chest. There was a strong on-shore wind and numerous *Stomolophus nomurai* were present. At hospital 15 minutes later he had eruptions and severe whip-like lesions visible, particularly on his legs (Figure 8.80). He was in severe pain but his observations were satisfactory. Intravenous dexamethasone 10 mg and an unknown antihistamine were administered, and an application of '5% soda and fluocinolone acetonide' was applied to the sting. There was no relief.

Two hours following envenomation he became dyspnoeic and dizzy. He had 30 respirations per minute, a pulse rate of 150/minute and blood pressure of 90/60 mmHg. He had severe sweating, central cyanosis, and the sting area was purple with vesicles appearing. Numerous coarse moist breath sounds could be auscultated in his chest. Chest X-ray confirmed acute pulmonary oedema and ECG showed an acute right heart strain pattern. Blood and urine studies at this time showed intravascular haemolysis, haematuria with haemaglobinuria and a leucocytosis of 15×10^9 cells/l. Following further dexamethasone, scopolamine, antihistamines and other unknown injections he was intubated and ventilated artificially. A total of 1000 ml of fluid drained from his lungs via the endotracheal tube! A cardiac arrest occurred 3 hours post-envenomation; effective ventricular action responded episodically to cardio-pulmonary resuscitation, but with uncontrollable pulmonary oedema the patient died 24 hours following the sting.

FIRST AID AND MEDICAL MANAGEMENT

The following measures are currently recommended in China {see Chapter 5, page 106}:

1. As soon as possible, remove tentacles from the stung area to prevent further envenomation.
2. Do not rub the stung area with fresh water or any other substance {see Chapter 5, page 108}.
3. Keep the patient still, thus decreasing the absorption of the venom into the bloodstream.
4. Morphine is effective in patients with severe pain. Parenteral steroids or antihistamines, and the local application of cortisone creams has proved ineffective.

The patient must be observed continuously for at least four hours, as systemic or fatal reactions may occur one to four hours after the original sting. As in all serious envenomation cases, if serious systemic effects occur, care of airway, breathing and circulation take precedence.

Acute pulmonary oedema is treated in the usual manner {see Chapter 9, page 254}. However, large dose injections of scopolamine seem to be effective in patients with acute pulmonary oedema from *Stomolophus* envenomations, and the application of tincture of traditional Chinese medicine has been shown to alleviate pain and itching in all cases (MINGLIANG ZHANG & QIN SHEDE 1991).

NOTE: Vinegar has been shown to cause nematocyst discharge in *Stomolophus nomurai* and also had little effect on the patient's symptoms (MINGLIANG ZHANG 1993, unpublished observations).

STOMOLOPHUS MELEAGRIS

(CABBAGE-HEAD JELLYFISH, CANNONBALL JELLYFISH)

This animal (and to some extent, consequently, all *Stomolophus* spp.) had been regarded in Western countries as harmless to humans (TOOM & CHAN 1972), prior to the reports of Mingliang Zhang and Qin Shede. However, pioneer techniques for the collection of specific nematocyst populations were developed using *Stomolophus meleagris*, by TOOM and CHAN (1972); its cnidom contains only one organelle type, a heterotrichous microbasic eurytele.

DISTRIBUTION

Mainly the tropical and sub tropical eastern seaboard of the Americas.

Stomolophus meleagris occurs in abundance from time to time, along the US east coast from Chesapeake Bay to Texas, and can clog the nets of shrimp fishermen. As a possible food resource, specimens have been processed by brining and their composition studied (HUANG 1988). They consist of 68 per cent moisture, 25 per cent salt and 5.5 per cent protein.

APPEARANCE

(See Plates 8.66 and 8.67.)

Bell up to 18 cm wide, half egg-shaped, or almost globular and milky-blue or yellowish in colour; exumbrella pigmented brown, denser near the margin where numerous white or yellowish spots may also be present; thick and rigid gelatinous substance; bell margin pulsates incessantly with strength and rapidity; variable number of lappets, usually about 14 in each octant, with short grooves between them; eight marginal sense organs, 4 radial and 4 interradial; bifurcate ends of the mouth arms flare outwards; no marginal tentacles; central mouth with numerous slit-like mouths situated on the upper half of the manubrium, each free mouth edge being lined by small, incessantly-waving tentacles, which are also present on the mouth grooves at the end of the manubrium; mouth frills brownish-pink.

THE STING

The sole report of a significant human envenomation by *Stomolophus meleagris* occurred on a North Carolina beach in August 1984 and involved a healthy 32-year-old female in knee-deep water. The report described a primary, mild cutaneous lesion with a spontaneous recurrent eruption at the same site one week later accompanied by localised stiffness and oedema (BURNETT & CALTON 1985) (Recurrent eruptions had been previously reported following jellyfish stingings [MÅNSSON *et al* 1985]) {see Chapter 7, page 177}. This report clearly identified IgG antibodies to *Stomolophus meleagris* venom in the patient's serum, with a mononuclear perivascular cell infiltrate in the recurring skin lesion (Plate 8.68) {see also Plate 7.5}.

MANAGEMENT OF A STING

In the absence of specific data, this envenomation should be managed according to general first principles described in Chapter 5 {page 106}.

THE TOXIN

Animal studies with the toxin of *Stomolophus meleagris* showed cardiotoxicity (partially reversible with atropine [LARSEN & BILLINGS 1986]) and interference with sodium transport (TOOM & PHILLIPS 1975; TOOM *et al* 1975). The latter action was confirmed and extended to the sarcolemma of skeletal muscle (LARSEN *et al* 1987).

GENUS *LOBONEMA*

As with many other members of the order Rhizostomeae, precise information concerning the medical significance of this genus is slim, and what there is seems to be contradictory. The genus *Lobonema* is of historical interest, in that it is probable that the early recognition of cnidarian deaths in the Philippines (H.M. Smith in MAYER 1910; LIGHT 1914) were attributed to a variety of jellyfish, including *Lobonema mayeri* Light 1914 (which CLELAND and SOUTHCOTT [1965] consider was probably a synonym for *Lobonema smithii* Mayer, 1910 [Plate 8.69]).

LOBONEMA SMITHII MAYER, 1910

DISTRIBUTION

The Philippines (Plate 8.69) and 'Indochina' (Vietnam and Cambodia) (KRAMP 1961). It was first identified in Australia by CORNELIUS (1994, unpublished observations) from a specimen caught in Darwin Harbour by ALDERSLADE (1992, unpublished observations); recently large numbers have been caught in Darwin Harbour, usually in the summer (CORNELIUS & ALDERSLADE 1994, unpublished observations). Even more recently a specimen was photographed at Seringatapam Reef, Western Australia (ALLEN & STEEN 1994).

APPEARANCE

(See Plate 8.69) Thick, tough and rigid bell up to 24 cm in width; erect papillae on exumbrella, largest and most abundant at centre. They may be up to 4 cm in length, pointed, and full of nematocysts; 4 velar lappets in each quadrant; mouth arms 15 cm long, distal 3-winged portion 1.5 times as long as proximal portion; numerous long, spindle-shaped appendages; 16 radial canals (after KRAMP 1961).

THE STING

There is little documentation of stings. LIGHT (1921) considered that *Lobonema* in the Philippines (called *lanterna* by the Filipinos) could inflict a 'very unpleasant' sting but was 'not dangerous' (CLELAND & SOUTHCOTT 1965). It produced no immediate stinging sensation, but once symptoms did start they were of less intensity and duration than a sting from *Catostylus purpurus*, (the latter persisted for several hours {see page 213 above}).

MANAGEMENT OF THE STING

In the absence of specific data, it is recommended that the basic principles described in Chapter 5 {page 106} and at the end of this Chapter {page 221} be applied.

GENUS PHYLLORHIZA L. AGASSIZ, 1862

Phyllorhiza punctata von Lendenfeld, 1884 (This is the type species.) (Plate 8.70)
Phyllorhiza pacifica (Light, 1921) (Plates 8.71a and 8.71b)

PHYLLORHIZA PUNCTATA

(Plate 8.70)

DISTRIBUTION

Philippines and Japan (KRAMP 1961). Distribution southwards (towards Australia) is probable but unproven. In Australia it is common in the Northern Territory (ALDERSLADE 1994, unpublished observations), north Queensland (FENNER 1992, unpublished observations), West Australia and South Australia (MARSH & SLACK-SMITH 1986).

APPEARANCE

(See Plate 8.70) Bell size may attain 50 cm diameter (KRAMP 1961); filaments or appendages (looking like tentacles) may be present, hanging under the usual 'arms'. As its name implies, white spots cover the usually brown bell and 'arms'. It usually causes a mild sting or none at all, but stings to the eyes may be severe (MARSH & SLACK-SMITH 1986) {see Plate 5.10 and Chapter 11, page 310}.

PHYLLORHIZA PACIFICA (LIGHT, 1921)

PAUL CORNELIUS

(See Plates 8.71A and 8.71B) It has two synonyms: *Cotylorhizoides pacificus* (LIGHT, 1921) and *Phyllorhiza peronlesueuri* (GOY, 1990).

DISTRIBUTION

Previously described only in the Philippines (KRAMP 1961), its true distribution is hard to determine, because of limited reports, but may be widespread from north-west and north Australian waters north to at least the Philippines and other parts of the Indo-Pacific. It has also been reported from Pearl Harbour, Hawaii, by Cutress (in MANSUETTI 1963).

Australian distribution: Confirmed on three occasions in Shark Bay, Western Australia (Hamelin Pool, eight specimens, October 1988 [GOY 1990] as *P. peronlesueuri*; and more recently one in 1994 (CORNELIUS AND MARSH, unpublished observations). A specimen photographed by Loisette Marsh off the north-western coast of Western Australia might have been this species, as might one reported by Jack Barnes from Cairns, north Queensland on 20 December 1958 (KRAMP 1961). STIASNY (1924) described some specimens collected during January and February 1923 near Port Jackson (Sydney Harbour) as *P. punctata*. With them appear to have been some young *P. pacifica* which Stiasny described, but took to be a juvenile of the other species. The paucity of even dubious records from the Australian coast suggests that the species rarely occurs along most of it, at least with present current patterns, but it may be more common in tropical northern and western regions.

APPEARANCE

(The following description by CORNELIUS [1994, unpublished observations] is based on observations of a live population at Hamelin Pool, Shark Bay, WA, 25 January 1994.)

Large, up to 40 cm in diameter (stated as 2–3 cm [LIGHT 1921]). Umbrella slightly flatter than a hemisphere, larger specimens flatter than smaller; jelly moderately thick. Upper surface having many colourless mounds about 3 mm in diameter. Margin with several broad lappets per octant, comprising (from one rhopalium to the next) a small rhopalial, two singles about twice the size of the rhopalial, two doubles twice the width of the single and slightly longer, two further singles, and a rhopalial.

Gastrovascular system comprising eight straight radial canals passing from stomach to peripheral canal, in some specimens evidently four 'main' ones slightly wider than those between; 10–15 minor radial canals in each quadrant anastomosing over the whole region between stomach and peripheral canal; with an intermediate circular canal (mid-canal) midway between the stomach and periphery. Stomach rounded-Maltese-Cross shaped, its edge often coloured by the gonads surrounding it, so clearly visible from above through the jelly.

Mouth arms 4, each divided into two, distal regions Y-shaped in section, ends up-turned with up to 3 'windows' in the axil. Mouth arm length about 1.25 times the disc radius; when floating, projecting noticeably all round when seen from above. Mouths numerous, along the outward- and inward-directed branches of the Y; undivided basal trunk massive, translucent, visible from the side. Up to about 100 translucent clubs amongst the mouth-arms on each side of the 8 branches, and on the mouth-arm disc hundreds of fine filaments about as long as its diameter (in life, at rest).

With a few dozen long, thick, translucent, mostly parallel-sided clubs among the mouth arms, the ends tapered then slightly bulbous, somewhat three-sided, or at the tips three-rayed in section, in side-view shaped like the spade symbol in cards but without the pointed apex (termed [inverted] heart-shaped by GOY [1990]); these tips white in population studied, blue in Barnes' Cairns material possibly of this species {see page 218 above}; when contracted the white tips are conspicuous against the brown of the mouth-arms. Occasionally a club might extend as long as the total height of the bell plus mouth-arms. The three-rayed shape may be lost and regained from minute to minute through muscle contraction and relaxation.

Gonads white to pale yellow or blue, edging stomach and in many specimens visible as a rounded Maltese Cross of varied colour showing through the transparent jelly. Several dozen bulbous sacs of gonadial tissue hang down into lumen of sub-umbrella space within confines of mouth arm disc; white or creamy in some, ranging from dark to bright blue in others.

Colour (of Shark Bay population) somewhat complex and greatly varied. Ectodermal tissues of umbrella include patches of brown zooxanthellae (minute symbiotic algae) to a varied degree {see also *Catostylus*, page 213}, possibly absent in some specimens and commonly absent from central region. Zooxanthellae may also be present all over the bell or confined to the region outside the mid-canal; may be pale, so that the umbrella is a pale translucent grey, or as dark as mid-brown. Typically, but not in all specimens, upper surface and skirt are spotted, speckled or streaked with white pigment. In some specimens the white occurs in regular to irregular white dots or small patches but more typically is found in tortuous narrow streaks up to 3 cm long. In most specimens the streaks do not join and give an overall reticulate appearance. The streaks may spread uniformly over the whole umbrella or may be limited to the 'skirt'. They do not extend onto the lappets, nor onto a band immediately above them, so that the base of each lappet lacks the markings. A pale to deep blue pigment in the ectoderm and some other tissues of many, but not all, specimens, its distribution greatly varied within the population studied. This pigment may occur as a pale suffusion across the bell, over or edging the stomach alone (so that a solid or outlined blue cross shows through the jelly), or just over the branches of the outer region of the gastrovascular system, or finally as a pale suffusion to a deep blue over the entire bell. (Deep blue colour all over was not seen in the population studied, but a photograph of one showing that pattern from the north-west coast of Australia has been taken [MARSH 1993, unpublished observations]). Muscle fields in most specimens show narrow gaps in the colour along the 8 rhopalial canals. Mouth-arms are translucent, frilled brown (or blue in some specimens).

DIFFERENTIATION FROM *PHYLLORHIZA PUNCTATA*

By its bell shape (flat not hemispherical) (Plates 8.70, 8.71A and 8.71B); by presence of 'windows' in the mouth arms of mature specimens; by bell being irregularly streaked, not spotted in many specimens; by predominant colouration not being uniform and dark brown throughout a population but often buffish to bluish, in some bright blue; in many with the edges of the Maltese-Cross-shaped stomach visible from above; by the 8 radial canals being flanked by a narrow non-pigmented band as seen from above; by there often being a contrasting but indistinctly-edged colouration between the inner and outer regions of the bell; and, at least in life, by the flattened-ace-of-spades shaped ends to the mouth arm clubs (though muscular relaxation influences this).

THE STING AND VENOM

No sting. No information on any venom that may be present.

GENUS *PSEUDORHIZA* VON LENDENFELD, 1882

Pseudorhiza haeckeli is common in the south of West Australia to Victoria, but also recorded in Darwin Harbour (Northern Territory) and Mackay (north Queensland).

The bell diameter may be 20–25 cm (usually only 10 cm), with a height of up to 5–10 cm. Three wide, frilly appendages (similar to the mouth arms of *Chrysaora*) hang beneath the bell (Plate 8.82) and are brightly coloured blue, tipped with salmon pink (MARSH & SLACK-SMITH 1986). The 'arms' are much less obvious than in other rhizostome medusae. The animal produces a mild sting (CLELAND & SOUTHCOTT 1965), or no sting at all. We have no details of the venom.

GENUS *LYCHNORHIZA* HAECKEL, 1880

(Note that the generic term '*Lychnorhiza*' appears in the literature with two spellings: *Lynchnorhiza* and *Lychnorhiza*. The latter spelling [KRAMP 1961] is adopted here.)

At least two species of this genus are known: *Lychnorhiza lucerna* and *Lychnorhiza bartschi*. *Lychnorhiza* spp. have been described in South America, south India, Malaysia and Indonesia (Java).

LYCHNORHIZA LUCERNA

JOSEPH W. BURNETT

DISTRIBUTION

Lychnorhiza lucerna is an oceanic jellyfish species found worldwide but particularly off the east coast of South America. The related species, *Lychnorhiza bartschi*, is present across the Pacific Ocean.

APPEARANCE

Lychnorhiza lucerna has a flat thin gelatinous disc whose umbrella is approximately 1.5 cm wide (MAYER 1910). There are four separate septate cavities with ostia which allow the gonads to protrude downward. The bell is not quite hemispherical and contains granules and sharp-pointed projections on its surface. Sixteen sharply-pointed triangular lappets extend into the water, and eight thick mouth arms fall from the subumbrella. The mouth arms are twice as long as the radius of the umbrella. The outer surface of the upper half of the mouth arms is smooth, but the inferior half contains several wing-like membranes separated by vertical indentations or grooves which end just before the inferior margin. Minute clawed 'tentacles' hang from the structures on these membranes.

In addition long filaments, exceeding the length of the arms themselves, extend inferiorly from the middle third of the folded membranes.

THE STING

Lychnorhiza species are capable of delivering a significant cutaneous sting to humans (HALSTEAD 1988).

VENOM

No data available.

TREATMENT

Systemic analgesic measures should be initiated. Maintenance of the patient's respiratory and cardiovascular systems is essential. Ultrapotent topical corticosteroids can be tried to counteract the cutaneous sting.

FIRST AID FOR RHIZOSTOME STINGS

- There are usually no tentacles to remove, but any visible pseudotentacles and filaments should be removed (preferably with gloved fingers, or forceps if available) prior to any application, such as ice packs for pain relief. Deactivation of remaining nematocysts must await further definitive research. Scyphozoan research on nematocyst inhibition suggests so far that vinegar as well as methylated spirits should be avoided for this purpose (RIFKIN 1995, unpublished observations).
- If skin pain is present, cold packs are applied to the skin (see Chapter 5, First Aid, page 107) for 5–15 minutes, then reapplied if necessary.
- Refer to medical aid if further symptoms develop.

MEDICAL MANAGEMENT OF RHIZOSTOME STINGS

Where resuscitation is necessary this always takes absolute priority. The management of significant systemic symptoms — including severe muscle pain or cramps, breathlessness and anxiety — are dealt with in Chapter 5, page 99 (First Aid).

ORDER CORONATAE

GENUS *LINUCHE* ESCHSCHOLTZ, 1829

LINUCHE UNGUICULATA (SCHWARTZ, 1788)

MAYER (1910) referred to the Pacific specimens of *Linuche* as *L. aquila*. Both BIGELOW (1938) and KRAMP (1961) regard *L. aquila* and *L. unguiculata* (Figure 6.19, page 141) as the same animal.

DISTRIBUTION

Described in the tropical Atlantic (STRONG 1942) and Pacific oceans, reaching the Indo-Pacific region including Torres Straits and north Queensland (CLELAND & SOUTHCOTT 1965). MAYER (1910) describes *Linuche unguiculata* as far westward as the African coast but KRAMP (1961) states it is not apparent in the Indian Ocean proper. *Linuche unguiculata* swarms massively in the Caribbean Sea and the Indo-Pacific Ocean during various times of the year (ATTENBOROUGH 1994, unpublished observations). Outbreaks from this animal sting have been reported in Aruba, Cuba, Bahamas and regularly in the south-eastern Florida Atlantic coast during the spring (STRONG 1942; FRANKEL 1942; OSMENT 1976; WAISMAN 1989). The Caribbean swarms have caused nuisance stings.

APPEARANCE

This species is known as the 'thimble jelly' or 'button jellyfish' (Plates 8.73 and 8.74 {see also Figure 6.19, page 141}. It is slightly brown (Plate 8.75), measuring 1–10 cm long overall; maximum bell size 2 cm diameter, with the bell width slightly greater than its depth; bell apex is flat with vertical sides; there are eight stumpy tentacles which hang from pedalia. Juvenile forms are visible to the naked eye but virtually invisible in the water (MEBANE 1994). Larvae (Plate 8.76) are approximately 0.5 mm long and light green to brown. Some of the major distinguishing features of *Linuche unguiculata* are:

The bell is thinble-shaped, and between 13 and 16 mm in width. There are eight rhopalia situated on special lappets, and 8 stubby marginal tentacles are present hanging from pedalia at the bell margin. There are eight gonads which form four cleft crescents. The subumbrella has three circles of hollow protuberances arising from the radial stomach pouches.

THE STING

The clinical syndrome for the sting of this animal's planula larvae is known as 'seabather's eruption' (Plate 8.78) (SAMS 1949; STRAUS 1956; MEBANE 1994) {see Chapter 11, page 309}. Four to 24 hours after exposure to the animal, patients notice a prickling sensation which may or may not be accompanied by urticarial hive-like lesions (Plate 8.79). Occasionally some patients have a delay period of three or four days before a similar disorder unfolds. Rarely the patients may present with fine macular eruptions limited to the covered part of the body. The duration of these skin lesions may be seven to ten days, but recurrence may follow with secondary eruptions after intervals of three to ten days. In another group of patients urticarial lesions may be the predominant symptom and these may occur early after the sting or be delayed up to three or four weeks. Vesicular lesions may also accompany the exanthem. Any combinations of these eruptions can appear in individual patients. The eruptions are worse in, but not confined to, areas covered by the swimsuit (WONG *et al* 1994) (Plate 8.79).

Recurrent eruptions composed of urticarial or vesicular lesions may persist for different intervals up to a year (TOMCHICK *et al* 1993). One case of prolonged blurred vision accompanied by nausea, lethargy, fatigue, tightness of breath, weakness and abnormal taste sensation has been reported after the sting of this animal's medusa (BURNETT & BURNETT 1990).

Other complications of seabather's eruption include sore throat, cough, abdominal pain and diarrhoea. Constitutional symptoms such as fever, chills, headache, nausea, malaise and vomiting are rare but have been reported particularly in children (MOSCHELLA 1951). Conjunctivitis and urethritis may also accompany this disorder.

Recent studies have implicated *Linuche unguiculata* in seabather's eruption {see Chapter 11 , page 309} with resultant specific IgG antibodies in affected patients, against this medusa's antigen (BURNETT *et al* 1995).

THE VENOM

No data available.

PREVENTION AND TREATMENT

Simple precautions can be taken to minimise the extent of the disorder in endemic areas. These measures include wearing protective clothing (see the advice about protection against tropical stings from jellyfish on page 104), avoiding wearing large bathing suits or shirts and being sure to change the swimsuit as soon as possible after leaving the water since more sting is administered as the water within the suit evaporates. Swimsuits worn during an epidemic should be thoroughly laundered or discarded. If wetsuits are worn, they should have restrictive cuffs (TOMCHICK *et al* 1993).

High-potency topical corticosteroids (Class I) are of limited value but should be applied twice daily (WONG *et al* 1994). The use of topical or systemic antihistamines is ineffective, but anti-pruritic agents such as mentholated emollients may offer temporary relief. Systemic corticosteroids are not warranted.

NAUSITHOË SPECIES

JOSEPH BURNETT

DISTRIBUTION

World wide (KRAMP 1961). The genus *Nausithoë* (Figure 6.20, page 142), comprising seven species, can be found in any ocean (MAYER 1910). The commonest documentations to date concern the species *Nausithoë punctata* (Plate 8.80).

APPEARANCE

The scyphistoma larva or polyp form of *Nausithoë* is a hydroid that lives embedded within several different species of sponges and upon other marine objects (HYMAN 1940). This is an irregularly branched colony, with a basal disc, with or without a single whorl of internal teeth (JARMES 1991). This form of the species *Nausithoë punctata* Kölliker, 1853 (also known previously as *Stephanoscyphus racemosus* Komai, 1935 [KOMAI 1935; KRAMP 1961]) has tentacles projecting out of the sponge oscula and may be covered with an irregularly annulated chitinous perisarc.

The medusa of *Nausithoë punctata* (Plate 8.80) has a flat, thick discoid or bowlshaped bell ranging in size up to 1.5 cm in diameter; medusae up to 4 cm diameter occur rarely; the central part of the upper superior surface is thick and raised with a finely dotted surface comprising approximately 50 per cent of the diameter of the medusa; it is separated from its periphery by an annular radial furrow; 16 radial thickenings protrude from an umbrella; there are also 16 marginal lappets and eight adradial tentacles between the lappets; the colour of the medusae is variable ranging from clear to red-brown; eight large adradial gonads in the bell are coloured red-brown to bright yellow; small red crystalline spots may be visible on the ectoderm of the exumbrella.

Some of the major distinguishing features of *Nausithoë punctata* Kölliker, 1853 {see Figure 6.20, page 142} are detailed by Jacquie Rifkin on page 142.

Envenomation

Significant cutaneous eruptions can be produced by contact with the sponge spaces containing the polyp forms of this jellyfish (HALSTEAD 1980). Symptoms may vary from a mild skin prickling to an intense burning, throbbing or shooting pain. This has been well documented as the Japanese 'stinging algae', by KOMAI (1935) especially from the Seto region of Honshu. Significant fever has been reported for several days to weeks following a significant sting in certain individuals (CLELAND & SOUTHCOTT 1965).

The venom

No data available.

First aid

- Deactivation of remaining nematocysts is usually unnecessary.
- If skin pain is present, apply cold packs to the skin {see Chapter 5, page 98 and Plates 5.16–5.18} for 5–15 minutes, then reapply if necessary.
- Refer to medical aid if further symptoms develop.

Medical management

Unrelieved pain should be managed with systemic analgesics. Ultrapotent topical corticosteroids may be tried to counteract the cutaneous eruption.

Systemic symptoms are treated as described in Chapter 5, page 98 (First Aid).

Order Semaeostomeae

Genus *Aurelia* (Moon Jellyfish, Saucer Jelly)

(Unfortunately the term 'saucer jellyfish' is also used to describe all members of the order Semaeostomeae, and not just *Aurelia* spp., hence the value of scientific names.)

Two species of this genus attain 'mention' levels in medically oriented literature: *Aurelia aurita* (Linnaeus) (Figure 8.13, page 224) {see also Plate 10.1} and *Aurelia labiata* Chamisso and Eysenhardt, 1821. Both species have repeatedly been described as harmless (LIGHT 1914; CLELAND & SOUTHCOTT 1965; MARSH & SLACK-SMITH 1986) but recent data may modify this description {see below}.

Appearance

{See Plate 10.1} *Aurelia aurita* has been the most closely studied species, because of its apparent ubiquity. It has a flattened, saucer-like, transparent and smooth bell, with a fringe of numerous and very short tentacles attached around the smooth margin of the umbrella; the margin has eight small indentations; there are four long and thick oral arms bearing minute projections; the arms (and the gonads) may be violet or pink in colour (AXIAK 1983); the bell has a bluish-white hue in the water, through which the coloured gonads are visible; the bell may attain a diameter of 30–40 cm (AXIAK 1983). It has been reported in large numbers {see Chapter 10, page 285}.

Distribution

Aurelia spp. have been reported in Australia (PACY 1957; CLELAND & SOUTHCOTT 1965), Philippines (LIGHT 1914), Great Britain (HICKson 1906), the Adriatic Sea (Maretić 1975), the Mediterranean Sea {see Chapter 10, page 285}, Germany and the Netherlands (Pacy 1957), Norway (Russell 1970) and the Atlantic and Gulf Coasts of America (Burnett et al 1988). It is described by Cleland and Southcott (1965), and Marsh and Slack-Smith (1986) as 'world wide' in distribution. The few recent specific reports found by the editors may reflect the genus's currently believed relative medical insignificance. It is an uncommon sight, at least since 1971, in Australia's eastern tropics.

The sting

It has long been regarded as non-toxic (Mackerras 1958; Russell 1970) apart from an Australian report when *Aurelia* was blamed for a New South Wales episode of 'surfer's or

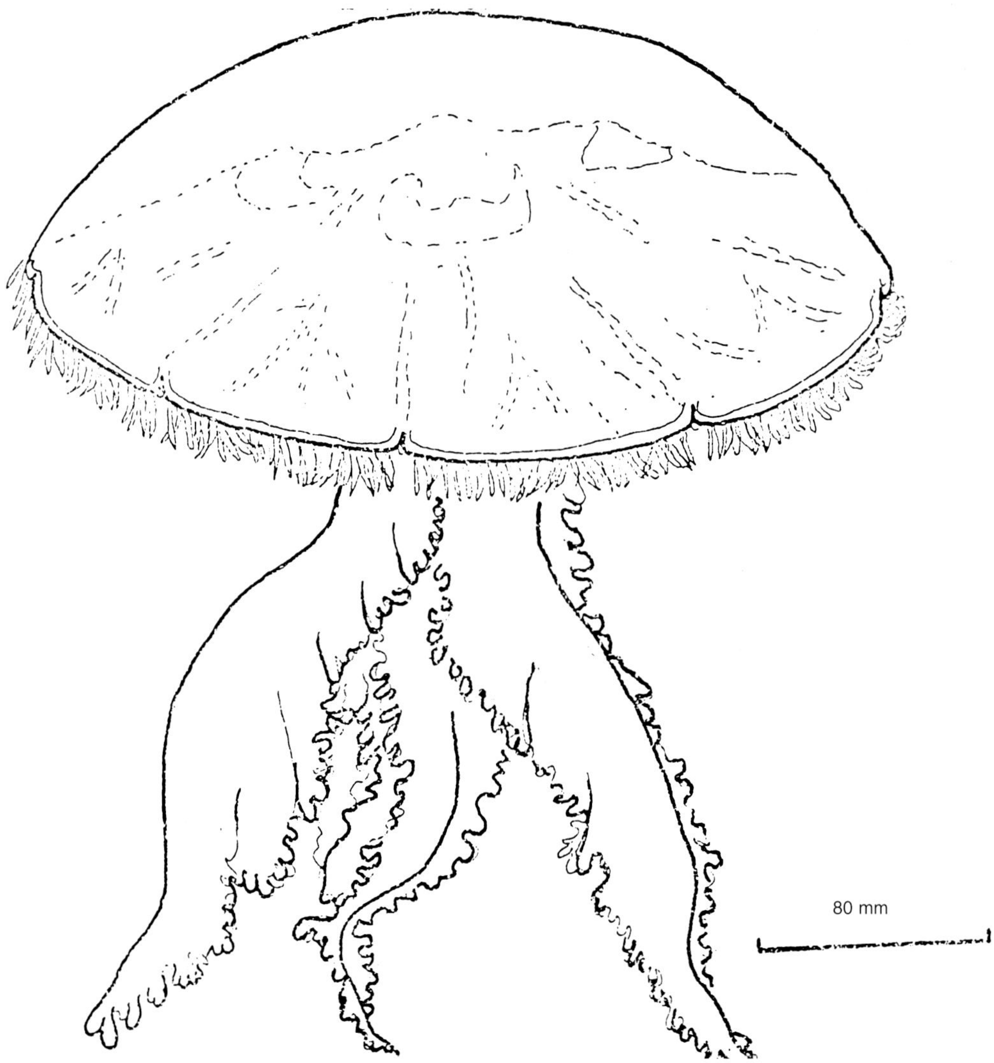

Figure 8.13
Drawing of the Mediterranean *Aurelia aurita* Linné, 1758. The bell may attain 30 cm in diameter. (Courtesy of Professor Victor Axiak, University of Malta, with the additional permission of the United Nations Environment Program.)

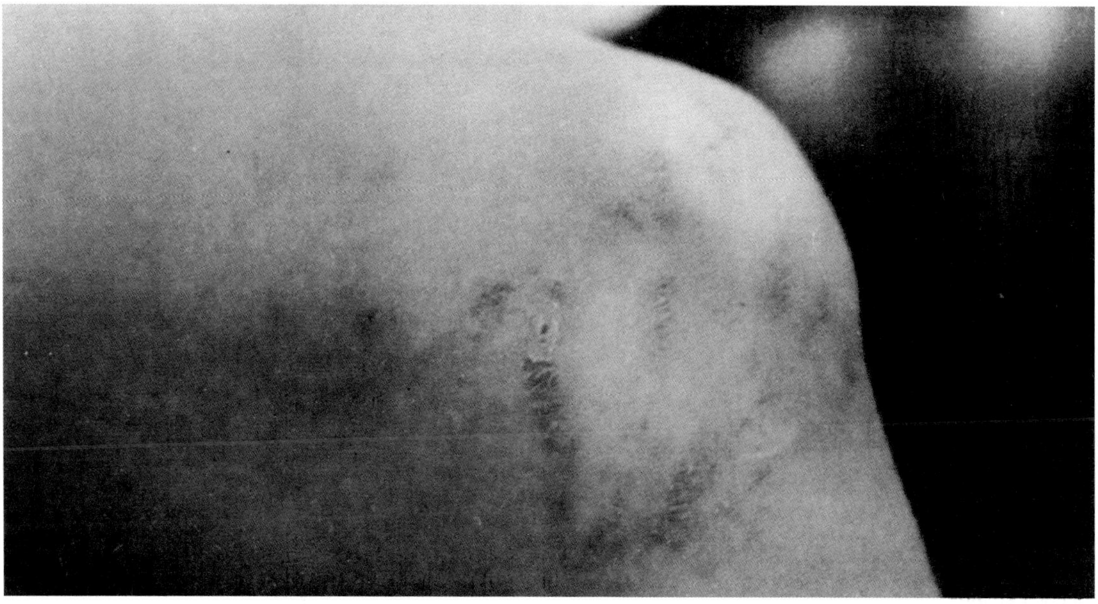

FIGURE 8.14
Erythematous and hyperpigmented linear lesions following a suspected sting from *Aurelia aurita* (at least in part—see text), photographed three days after envenomation. (Reprinted from *Toxicon* 26(2); 212, Burnett, Calton and Larsen, Significant envenomation by *Aurelia aurita*, the moon jellyfish. Copyright (1988), with kind permission from Elsevier Science Ltd, The Boulevard, Langford Lane, Kidlington OX5 1GB, UK.)

bather's itch' — subsequently shown to be a schistosome dermatitis (CLELAND & SOUTHCOTT 1965) {see Chapter 11, page 308}.

The literature includes the following descriptions of *Aurelia* stings: '*Aurelia aurita* is almost harmless...' (HICKSON 1906) and '*Aurelia aurita* ... produces only a weak burn.' (THIEL 1936).

In 1988, BURNETT and colleagues reported a significant envenomation from *Aurelia aurita* in Mississippi Sound, Gulf of Mexico. With better understanding of jellyfish venom toxicity and antigenicity {see Chapter 7, page 175}, and as these authors point out: '...animals thought to be harmless can be pathogenic to humans.' Their patient, stung on the right arm, developed local sting pain and piloerection of short duration, urticaria and ulceration. On day 4 post-sting, right axillary pain occurred and lasted 24 hours. The lesions healed in ten days and were followed by hyperpigmentation visible at the site two weeks later (Figure 8.14, this page). The swimmer, a marine biologist, was standing in a swarm of large *Aurelia* jellyfish. However, immunospecific serum antibodies were subsequently detected at equal titres with *Aurelia* and *Chiropsalmus* antigens (BURNETT *et al* 1988). Even allowing for any specificity limitations of the titrations, it is possible that this sting may not have involved only *Aurelia* (BURNETT 1992, unpublished observations).

MANAGEMENT

Non-specific, symptomatic management of *Aurelia* stings is appropriate with present understanding {see Chapter 5, page 99, First Aid}. Nothing is yet known about nematocyst inhibition in *Aurelia* spp.

GENUS *CHRYSAORA* ('SEA NETTLES')

This is a very widely dispersed member of the order Semaeostomeae (the scalloped medusae) {see Figure 6.21, page 143}. *Chrysaora hysoscella* is a familiar sight in the Mediterranean and Ireland (see Chapter 10, page 287) (Plate 8.81) {see also Figures 6.21, page 143 and 10.1, page 286 and 10.4, page 288}, and *Chrysaora lactea* swarms in Uruguay, South America, during the summer (MIANZAN *et al* 1988;

OLAGÜE *et al* 1990). However, the best known world species is *Chrysaora quinquecirrha* (DESOR, 1848) (Plate 8.82).

Note that *Chrysaora hysoscella* appears with the species name in three different spellings: '*hysoscella*' (KRAMP 1961), '*hyoscella*' and '*hysoscela*', {see Chapter 10, page 284}. The first-mentioned spelling appears to be more prevalent in Europe.

CHRYSAORA QUINQUECIRRHA (NORTH AMERICAN SEA NETTLE)

JOSEPH BURNETT

Chrysaora quinquecirrha (DESOR, 1848) occupies an important place in the evolution of understanding in cnidarian epidemiology, toxinology and immunology. It was originally known among marine scientists as *Dactylometra quinquecirrha*, prior to the separation by MAYER (1910) of the two genera *Dactylometra* AGASSIZ, 1862 and *Chrysaora* PÉRON & LESUEUR, 1809 (reported as 'Péron & Lesueur, 1807' [PÉRON *et al* 1807] in CLELAND & SOUTHCOTT 1965). *Chrysaora quinquecirrha* from the Philippines was associated with the initial disciplined observations by LIGHT (1914) and colleagues that led to the important recognition of chirodropid envenomation in those islands {see page 262 below}.

More recently, its large and seasonally recurring numbers on the extensive Maryland coast of the United States has provided the raw material for steady and important studies of its clinical stinging abilities (BURNETT *et al* 1983; BURNETT 1991; 1992) and consequent advances in cnidarian toxinology and immunology (BURNETT *et al* 1968; 1988; BURNETT 1992a; LONG-ROWE & BURNETT 1994b) {see Chapter 7, page 174}.

DISTRIBUTION

Chrysaora quinquecirrha (sea nettle) is found worldwide but its greatest known abundance is in the Chesapeake Bay of North America (BURNETT *et al* 1968). As well as the Atlantic coast of the United States, considerable numbers are found on the coasts of Argentina, southern Brazil, the Philippines (LIGHT 1914) the northern island of Japan (KUSUNOKIS. 1967. *The clinical study of poisonous jellybox stingings in the Hokkaido [Lake of Saroma]*, private monograph publication, 29 pages), and elsewhere in South-East Asia. The presence of the genus *Chrysaora*, but not the species, is confirmed in parts of Australia (CLELAND & SOUTHCOTT 1965; MARSH & SLACK-SMITH 1986).

APPEARANCE

(Plate 8.82)
{see also Figure 6.21, page 143}

CHRYSAORA QUINQUECIRRHA FROM CHESAPEAKE BAY

The bell of this animal ranges in size from 2 to 20 cm, with four central thick oral 'arms' {see Figure 6.21, page 143 and Plate 8.82}, and 16 peripherally mounted, thread-like ('fishing') tentacles; tentacles range in length from 2 to 250 cm; four lobulate gonads can be pink, light grey, clear or yellow; the bell is most commonly clear but may contain radial rusty red lines. These 'red-phase' medusae may have their red coloration extend down the tentacles. The cnidom of this jellyfish includes microbasic euryteles, atrichous isorhizas and holotrichous isorhizas (Plate 8.83) (BURNETT *et al* 1968).

Some important distinguishing features of the genus *Chrysaora*

PÉRON and LESUEUR, 1809 {see also Figure 6.21, page 143}

The genus *Chrysaora* varies greatly between species with respect to its anatomical feaures. The description above refers specifically to the *Chrysaora quinquecirrha* of Maryland. Other more general data appears on page 144 (details supplied by Jacquie Rifkin).

The general anatomy of a *Chrysaora* specimen is detailed in Figure 6.21, page 143. The bell margin is scalloped into lobe-like lappets (from 32 to 48), and the tentacles (24 or more) are situated between these structures. The marginal sense organs or rhopalia (3–9) are also arranged between the tentacles. The stomach is central and it gives rise to 16 separated, unbranched pouches. Ring canals are absent. Nematocyst warts are numerous on the bell.

THE STING

Instant cutaneous pain is usually perceived after contact with tentacles or arms.* The tentacles adhere transiently to the victim and usually wash off in the water. In most cases no subsequent skin eruption appears, although in more severe stings, linear urticarial erythematous lesions result. The lesion may last for several hours but the pain reaches its peak in a few minutes following the sting, only to disappear spontaneously within two hours.

In spite of the millions of local stings that occur annually around the world, few severe systemic symptoms are actually reported — despite repeated non-specific documentation. One prominent systemic reaction, when it occurs, seems to be a short-lived, sometimes delayed-onset, excessive upper respiratory tract mucous production with coughing, rhinorrhoea and lacrimation (WADE 1928; MARSH & SLACK-SMITH 1986). More severe claimed effects include diaphoresis, mental depression, subjective chest constriction, muscular cramps, weakness without prostration (WADE 1928) and even loss of consciousness (HALSTEAD 1988). CLELAND and SOUTHCOTT (1965) consider the lack of vomiting and abdominal pain in *Chrysaora quinquecirrha* stingings to be a differential diagnostic point from the Irukandji syndrome {see page 251 below}.

Corneal stings are uncommon but painful and may result in lacrimation, photophobia and delayed (1–2 days) increased intraocular pressure due to acute iridocyclitis, resulting in decreased visual acuity (GLASSER *et al* 1992) {see Plate 5.10, and Chapter 11, page 309}. Keloid formation in the skin does not occur, and post-inflammatory hyperpigmentation is rare. Reactivation of herpes simplex at the sting site has been seen (BURNETT & CALTON 1987).

*One of the editors (PJF) experienced little if any pain when he immersed his foot in a bucket full of fresh specimens of Chesapeake Bay *C. quinquecirrha*! {Plate 5.13}. It is more difficult to be stung by experimentally manipulating the medusa than it is to contact the animal in free water.

Experimental injections in animals can produce liver and kidney damage (MUHVICH *et al* 1991). Large local reactions to the sting have rarely appeared. No cases of recurrent eruptions from the jellyfish sting have yet been reported. (Note: One possible patient with acute arterial occlusion has been noted [WILLIAMSON *et al* 1988]; however, this patient had near-equivalent antibody titres also to *Chironex fleckeri* and *Catostylus*, so conclusive identification of the offending jellyfish is lacking.)

No fatalities have resulted from *Chrysaora* envenomations, but one patient stung by the related animal *Pelagia* had a positive laboratory test (basophil histamine release) against *Chrysaora* venom (TOGIAS *et al* 1985), indicating the strong possibility that anaphylaxis — albeit rarely — could occur. This is supported by the syndrome of upper respiratory tract mucous production with coughing, rhinorrhoea and lacrimation that seems to occur uncommonly (see above). A case of mononeuritis with hyperimmune serum to *Chrysaora* following a *C. quinquecirrha* sting is reported in Chapter 26, page 446.

THE VENOM

The isolation of *Chrysaora quinquecirrha* venom has been complicated by the combined problems of thermal instability, aggregation of heterologous proteins and peptides, and the loss of lethal potency during purification due to the presence of non-nematocyst contaminating proteins and proteases (LONG-ROWE & BURNETT 1994b).

Chrysaora venom is dermatonecrotic (skin-killing) and lethal to small animals. The venom is also weakly haemolytic in experimental animals (LONG & BURNETT 1989); this haemolysin is a short-lived bioactive peptide with a separate activity from the lethal factor(s), whose stability with freezing and thawing decreases with increased purification (LONG-ROWE & BURNETT 1994a). The venom contains many polypeptides ranging in molecular weight from 20 000 to 190 000 daltons; it appears that the clinically more important lethal proteins in the fishing tentacles are in the 100 000 to 200 000 dalton range (COBBS *et al* 1983). Cross-reactivity between *Chrysaora* and other jellyfish venoms is common (BURNETT *et al* 1988) and occurs with both humoral and cell-mediated immunity.

FIRST AID AND MEDICAL TREATMENT

Currently the best nematocyst discharge inhibitor for fresh *Chrysaora quinquecirrha* stings appears to be a topical application of a slurry of baking soda (50% $NaHCO_3$:50% water) (BURNETT *et al* 1983). This application makes no difference to the pain (BURNETT, FENNER & WILLIAMSON 1991 unpublished observations). Local first aid measures to counteract the pain (analgesia) do not appear to be dramatically more effective than counter-irritation, but uncontrolled field trials in the Chesapeake Bay area suggest that the topical application of cold packs {see Chapter 5, page 111} to the more painful *Chrysaora quinquecirrha* stings may offer some temporary relief (BURNETT, FENNER & WILLIAMSON 1991, unpublished observations). Systemic analgesics appear to be the best form of pain relief, as topical anaesthetics and corticosteroids, or systemic corticosteroids are ineffective. Corneal stings are best treated with topical anaesthetic drops and specialist medical referral, which should include measures to counteract any elevation in intraocular pressure (GLASSER *et al* 1992).

GENUS *PELAGIA*

PAUL CORNELIUS AND JOSEPH BURNETT

The scientific name of the animal implies its open ocean surface habitat and the fact that it luminesces in the water at night when struck by wave action or mechanical objects.

There seem to be at least two distinct stinging species: *Pelagia flaveola* Agassiz & Mayer (MAYER 1910) (Plates 8.84A and 8.84B), and *Pelagia noctiluca* (Forskål 1775) (Plate 8.86A). However, KRAMP (1961) lists from the literature some 15 different species names worldwide, commenting that STIASNY (1934) 'agrees that only one species exists'. Since KRAMP (1961), all *Pelagia* spp. (excepting *P. colorata* and *P. flaveola* recognised by RUSSELL 1964) have been merged into one. The status of *P. panopyra*, formally recorded from Australian waters and also

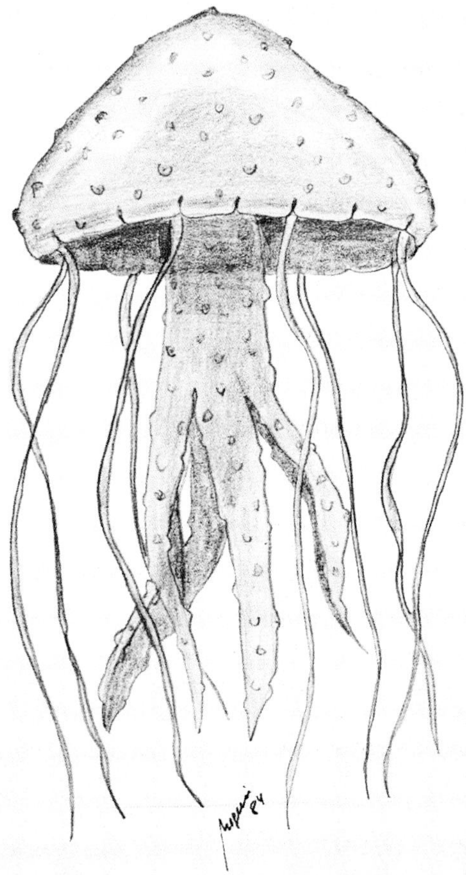

FIGURE 8.15
Drawing of the morphology of *Pelagia* sp. Note the exumbrella dotted with nematocyst batteries.

from the Philippines, Japan and Malaysia across to California, arguably remains open. It has been separated on its proportionately steeper bell shape, flatly rounded on top bending sharply downwards around the upper 'corners' as seen in side view, roughly 1.6 times wider than high (twice as wide in *P. noctiluca*), proportionately longer mouth arms, and small size (typically up to c. 5 cm). In the older literature *P. noctiluca* was considered an Atlantic form but in recent decades has been considered worldwide. Little is known about the various species of *Pelagia*, their distributions, and their various stinging powers.

The enormous pan-oceanic distribution of *P. flaveola* recorded in the older literature, if confirmed, would suggest that the lifecycle occurs entirely within the plankton as in *P. noctiluca*, whereas in *P. colorata* RUSSELL, 1964, a benthic stage has been demonstrated. This oceanic distribution, if independent of shallow coastal regions for the vegetative reproductive stage, suggests that *P. flaveola* has a greater geographical range than any other venomous organism except *Physalia*; yet little is known of the species.

APPEARANCE

The mauve stinger, *Pelagia noctiluca* is similar in appearance to *Chrysaora*, though generally smaller, reaching about 10 cm in diameter. Important distinguishing features of *Pelagia* (Figures 6.22, page 144 and Plate 8.87A) are supplied by Jacquie Rifkin, on page 143.

Bell colour is varied and ranges from brownish-yellow to pink or purple. Nematocyst warts of varied size and number are present. The tentacles and arms may be either red or purple. There are four oral arms with eight thin marginal tentacles, each of which alternates with eight sense organs or rhopalia. The central stomach gives rise to sixteen gastric pouches, which enter the sixteen lappets via canals.

PELAGIA NOCTILUCA (LITTLE MAUVE STINGER OR PHOSPHORESCENT FLOATER)

JOSEPH BURNETT

There appears to be general agreement that *Pelagia noctiluca* Forskål 1775 is the type-species of the genus (KRAMP 1961). It is certainly an important stinging jellyfish.

DISTRIBUTION

Although this animal has worldwide distribution in warm and temperate waters, large swarms have appeared in the Mediterranean, Adriatic and North Seas as well as the eastern Atlantic Ocean and south Queensland (BLOOMFIELD 1959 — possible misidentification [see page 228 above, and *P. flaveola* below, page 230]). Such swarms have been found to measure up to 2 kilometres in diameter and up to 15 metres in depth (AXIAK & CIVILI 1987). Its appearance and significance in the Adriatic and Mediterranean region are discussed by Dr Franco Kokelj, MD, in Chapter 10, page 285. In July 1987 *Pelagia noctiluca* was identified off the Abrolhos Islands (off the east coast of Brazil).

Pelagia sp. (identified as *P. noctiluca*) has been described around the Australian continental coastline, including inside major harbours, sometimes in very large numbers (BLOOMFIELD 1959; CLELAND & SOUTHCOTT 1965; MARSH & SLACK-SMITH 1986; MARETIĆ *et al* 1980). It is generally less abundant in colder oceans. Curiously, it has been seen much less frequently in Australian waters in the past 30 years.

APPEARANCE

The medusa of *Pelagia noctiluca* is mushroom-shaped measuring approximately 4–13cm in bell diameter with its four thick oral arms and eight thin peripheral tentacles hanging down as far as 3 metres (CLELAND & SOUTHCOTT 1965; HALSTEAD 1988; MARETIĆ 1987) (Figure 8.15, page 228). Nematocysts of this animal are located in batteries on all tentacles and on the external bell surface.

The cnidom of *Pelagia noctiluca* is composed of atrichous isorhizas, holotrichous isorhizas and heterotrichous microbasic euryteles (HALSTEAD 1965).

The medusa colours vary from light rose-red or violet to mauve-brown, pink or purple. The gonads are purple, and the nematocyst-bearing warts on the bell are violet — a real 'punk-rock' jellyfish!

THE STING

{See also Plates 10.6 and 10.7} Most stings produced by *Pelagia* are limited in time and extent. Instant pain and blanching followed by an urticarial lesion with oedema follows contact with the tentacle (Plate 8.85). Clinical lesions can be accompanied by localised burning, vesicle and bullae formation or crusting (Plates 8.86 and 8.87A) {see also Plate 10.6}. Later pruritus, persisting oedema and prolonged localised pain can appear (MARETIĆ *et al* 1987). The lesion may produce post-inflammatory hypo- or hyper-pigmentation; tatooing or keloid formation have also occurred (REED *et al* 1984) (Plate 10.7). Recurrent lesions at the same site following single envenomations have been reported (BURNETT *et al* 1987) {see Chapter 10, page 285}.

Rarely systemic complaints at the time of envenomation have included diaphoresis, fever, possible hypotension (MARETIĆ 1983) and 'collapse' (HALSTEAD 1988). Large local oedema reactions and one confirmed life-threatening anaphylaxis have been recorded secondary to (Mediterranean) *Pelagia noctiluca* envenomation (TOGIAS *et al* 1985).

THE VENOM

The proteinaceous venom of *Pelagia noctiluca* is antigenic in animals and humans, and haemolytic, dermatonecrotic as well as lethal to laboratory mice (CLELAND & SOUTHCOTT 1965). Clinical cross-reactivity to *Chrysaora quinquecirrha* venom and serological cross-reactivity to the Atlantic *Physalia physalis* venom has been demonstrated (OLSEN *et al* 1985). Multiple proteins with molecular weights ranging from 14 500 to 320 000 daltons have been detected (OLSEN *et al* 1985; MARETIĆ 1987).

TREATMENT

{see Chapter 5, page 103}

While general first aid and necessary medical management should follow the basic principles described on page 106, there are some special management features that apply to stings from *Pelagia noctiluca*.

Nematocyst discharge inhibition

Ionic calcium or magnesium have been found to be effective arresting agents against *Pelagia noctiluca* nematocyst discharge *in vitro* (SALLEO *et al* 1984). Maretić and colleagues (1987) advocated the use of a topical corticosteroid-anaesthetic preparation (but in uncontrolled studies). KOKELJ and BURNETT (1990) have reported the successful use of a topical hydroquinone bleach in cases of post-inflammatory hyper-pigmentation secondary to these stings. So far, clinical evidence indicates that special care should be taken with patients envenomated by *Pelagia noctiluca* if they have a medical history of allergy (TOGIAS *et al* 1985).

Recent studies suggest gadolinium [Gd^{3+}] is a powerful blocker of the calcium-dependent activation of *Pelagia noctiluca* nematocytes discharge [SALLEO *et al* 1994].)

PELAGIA FLAVEOLA ESCHSCHOLTZ, 1829 (LITTLE BROWN STINGER OR LITTLE YELLOW STINGER)

PAUL CORNELIUS

DISTRIBUTION

Old literature records suggest the species can be found across the Indian Ocean within a belt encompassing roughly the Arabian Gulf to Madagascar, over the entire northern Indian Ocean, to Malaysia, the Philippines, Papua New Guinea and across approximately to Mexico, perhaps occurring along part of the north-western coast of South America (ESCHSCHOLTZ 1829; AGASSIZ & MAYER 1902). It probably also occurs in Indonesia, and perhaps northern Australia. Probably only occasional in many places, it has been reported by local people as 'frequent' in northern Papua New Guinea. It is likely to prove open-ocean in distribution, hence arriving with currents and perhaps subject to long-term changes in occurrence as the current-patterns change.

APPEARANCE

(Plates 8.84A and 8.84B) Bell rounded, typically 3–5 cm in diameter, pulsating actively; flatter than a hemisphere, width being up to twice the height; upper surface warty, the warts being colourless except near the tip where they are white; the warts are rounded to somewhat pointed with the height of those in the centre being four times the width, becoming progressively less towards the margin; the edge of the bell has eight tentacles, having four deeply-cleft, rounded lappets (or two double lappets) between each tentacle; the tentacles are thick and smooth-sided, when partially contracted reaching to about the end of the mouth arms but when fully extended they are very fine and reach several times as far; there are four mouth arms with a length up to 1.5 times the bell height, sparsely warty on the outer side, the inner sides and ends wavy.

Gonads light brown, giving the predominant colour to the bell; tentacles and mouth arms light brown; jelly colourless and transparent. AGASSIZ & MAYER (1902) stated: 'disk of a dull amber-yellow colour and tentacles of a darker colour; exumbrellar warts and mouth-arms orange-red, gonads dull ochre'. MAYER (1910) stated: '[general appearance] light yellowish, tentacles citron yellow, gonads brownish-yellow, exumbrellar warts tipped with orange', suggesting that colour variation occurs.

THE STING

'Painful' (Papua New Guineans, CORNELIUS 1994, unpublished observations). Past confusion with other *Pelagia* makes comment pointless.

THE VENOM

No information is available.

TREATMENT

Treat as *P. noctiluca* {see above, and Chapter 5, page 103}.

GENUS *SANDERIA*

The single species in this genus could be of greater medical (i.e. stinging) significance than is currently appreciated.

SANDERIA MALAYENSIS GOETTE, 1886

PAUL CORNELIUS AND JOSEPH BURNETT

DISTRIBUTION

This medusa was first reported off Singapore (GOETTE 1886) and later confirmed elsewhere in the Indo-Pacific. It ranges from East Africa, the Gulf of Aden (CLELAND & SOUTHCOTT 1965), the Suez Canal, the Red Sea, the Arabian Gulf and Oman (KRAMP 1966) {see Chapter 18, page 404}, India, Malaysia (BROWNE 1926), the Philippines (MAYER 1910), and Japan (KISHINOUYE 1910; HIRASE *et al* 1927). Sightings have been reported at all times of year. Curiously the species has not been reported in Indonesia, Papua New Guinea or Australia.

Recent first-hand recorded information concerning this animal is lacking, as are recent illustrations of its appearance (VANHÖFFEN 1902; MAYER 1910) (Plate 8.88). There is a suspicion that the animal may be a cause of unidentified envenomations over a wider Indo-Pacific region than has been appreciated. We await further field studies.

APPEARANCE

This animal has a strikingly flat-topped bell typically 3–8 cm in diameter. A recent specimen delivered to the Natural History Museum, London (AL-HASSAN 1985, unpublished observations), had a diameter of 13 cm (CORNELIUS 1993, unpublished observations). The bell edge falls perpendicularly into a short 'skirt' approximately one sixth the bell diameter (MAYER 1910). Sixteen marginal tentacles hang from the skirt edge, and in the centre are the four frilled mouth-arms, which are slightly longer than the bell diameter. The bell, with a low, flat conical elevation in many specimens, is colourless to yellow. It may have groups of reddish spots extending radially over the bell surface and moutharms. The four gonads, located in the bell, vary from clear to brown or purple.

THE STING

A Japanese report described the tentacles of this jellyfish as 'violently poisonous' (HIRASE *et al* 1927). A description from the north-west Arabian Gulf (Kuwait) in 1985 described severe stings with local skin necrosis from *Sanderia*. Credence was lent to this report by the capture at that time elsewhere in the Gulf of several specimens of *Sanderia malayensis* (one with a bell diameter of 13 cm), now preserved in the Natural History Museum in London (AL-HASSAN 1985, unpublished observations).

This species may be responsible for the peripheral vasospasm and peripheral tissue necrosis that has been reported in two papers (ABU-NEMA *et al* 1988; WILLIAMSON *et al* 1988). Other sources also regard this jellyfish as possessing a venom of moderate to severe potency (AL-HASSAM 1985, unpublished observations; HALSTEAD 1988).

THE VENOM

No firm data are available.

FIRST AID AND ADVANCED MEDICAL TREATMENT

No definitive treatment studies of stings from *Sanderia* are available. Once the patient is stabilised, tentacular material should be carefully removed from the skin. STIASNY (1937) described *Sanderia* tentacles as 'exceedingly glutinous...if touched with a pencil or needle one cannot easily get rid of them'.* No objective data currently exist concerning nematocyst discharge inhibition in this species {see Chapter 5, page 101}.

Care should be taken to help preserve local limb blood supply by keeping the whole patient warm. (Present evidence suggests the greater concern is compromise of perfusion in the stung area rather than that from systemic absorption of venom.) Peripheral pulses over the involved extremity should be monitored by direct physical examination and/or Doppler measurements. Any circulatory compromise should be aggressively treated with local measures such as intra-arterial urokinase (50 000 i.u.) infused over a period of 30 minutes and repeated if necessary. Systemic heparin, dextran, corticosteroids, intra-arterial reserpine and intravenous pentoxifylline have been ineffective with patients. In one patient a cervical dorsal sympathectomy produced no improvement. Tolazoline intra-arterially was also of little apparent benefit. Early rehydration and maintenance of good fluid balance, aggressive analgesia and the administration of oxygen (preferably 100%) are essential. Judicious anticoagulation and intravenous infusion of lignocaine, vasodilators (such as papaverine), prostaglandin$_E$ or antiplatelet aggregation agents (i.e. prostacyclin) should be employed, with careful monitoring. Surgical intervention and/or hyperbaric oxygen therapy should be considered early for compartment syndromes.

GENUS *CYANEA* PÉRON AND LESUEUR, 1810

PETER FENNER AND JOHN WILLIAMSON

Common names

Australia: hair jelly(fish), hairy stinger, snottie

England: lion's mane

Denmark: den røde, vandmand (HALSTEAD 1988)

(Plates 6.1A and 6.1B and Figure 6.23 page 144, and Plates 8.89 and 8.91) This jellyfish genus has been made unjustly (in)famous by the fictional Sherlock Holmes short story called 'The Adventure of the Lion's Mane' by Conan Doyle, where a body found 'scourged' on a beach had supposedly been stung to death by a large 'lion's mane' ('*Cyanea capillata*') jellyfish (CLELAND & SOUTHCOTT 1965), referred to in that story variously as vile creature, miscreant, and fearful stinger! {see the Footnote on page 233}. In reality no deaths anywhere in the world are known from the sting of this jellyfish.

TAXONOMIC PUZZLES

MAYER (1910) stated that there were only two species of *Cyanea* — *Cyanea capillata* in the northern hemisphere — *Cyanea annaskala* in the southern hemisphere, and that they were present in the Polar oceans but not present in the tropics. However, KRAMP (1961) recognises at least seven species, distributed worldwide. HALSTEAD (1988) suggests that *C. capillata*, *C. arctica*, and *C. annaskala* LENDENFELD, 1882, are one and the same species; EDMONDS (1989) lists '*C. capillata*, *C. annaskala*, *C. arctica*, *C. nozakii* and others' as 'dangerous species', but no documentation is provided. There was even a differentiation suggested previously between *Cyanea* spp. in Port Phillip Bay in Victoria, Australia (*C. annaskala* variety *purpura*), and Port Jackson, New South Wales, Australia (*C. annaskala* variety *marginata*) (LENDENFELD 1887). Whatever the true taxonomic situation, there has undoubtedly been confusion of naming of various *Cyanea* species by different chroniclers over the decades (e.g. Editorial: '*Cyanea capillata*', *West Virginia Medical Journal*, January 1952; 48[1]: 16).

Current opinion in the marine envenomation literature favours reference to one principal stinging species worldwide: *Cyanea capillata* (CLELAND & SOUTHCOTT 1965; SUTHERLAND 1983; HALSTEAD 1988). MARSH and SLACK-SMITH (1986) make reference to a species of *Cyanea*, *C. mjobergi*, which is known in Western Australia; they do not state whether it specifically is responsible for stings. *Cyanea nozakii* (HEEGER 1994, unpublished observations) and *C. mjobergi* (COR-

* Cnidarian tentacles that both sting and demonstrate dense adhesion to human skin suggest a high population of discharging nematocysts, and possibly a larger venom dose injected. However, adhesion alone does not necessarily correlate with pain (e.g. the relatively painless but strongly adhesive [to skin] tentacles of the Moreton Bay anemone, *Stichodactyla haddoni*). (RIFKIN 1995, unpublished observations.)

NELIUS 1994, unpublished observations) have been identified on the east coast of Australia.

DISTRIBUTION

Cyanea spp. are distributed worldwide, from the Arctic to the Antarctic regions. However, they are more plentiful (and apparently smaller) in tropical and warm temperate oceans. They appear to be open-water jellyfish (although MAYER, 1910 describes them as occurring only in 'shallow waters along the shores, not animals of the high seas'). They have been commonly seen out at sea (e.g. by pilots of light aircraft) usually as lone, largish specimens (BURNETT & CALTON 1987c) {Plates 8.89 and 8.90}. There were no records found of sightings of the bell of the animal at depth; all reports are of specimens seen on the surface.

Documentation of sightings and stingings include: from Margate, England, believed to be the earliest account (WOOD 1874), Sweden (KRISTENSON 1949; UVNÄS 1960), Denmark (HEEGER *et al* 1992), south-west Africa and all oceans of the world including the Arctic and Antarctic (MAYER 1910), the western Pacific (FISH & COBB 1954), Hong Kong (WALKER 1977), Chesapeake Bay, Maryland (RICE & POWELL 1972), and Delaware Bay, Delaware (LONG & BURNETT 1989), USA. There is also copious documentation from Australia (BARNES 1960; CLELAND & SOUTHCOTT 1965; FENNER & FITZPATRICK 1986; MARSH & SLACK-SMITH 1986).

Cyanea spp. have been seen by the editors PJF and JAW in the waters of a number of tropical countries including Malaysia, Singapore, Indonesia and Thailand. *Cyanea* spp. are present all around Australia, where they may appear as single specimens or in large swarms. They are common in north Queensland tropical waters (swarms have been so big that they have actually blocked the engine cooler intake pipes on a Burmese naval vessel, preventing it from leaving Cairns Harbour [BARNES 1960]). Curiously they are seldom reported from the geographically adjacent Northern Territory.

APPEARANCE

The largest specimens occur in colder waters. MAYER (1910) recorded a specimen of *Cyanea capillata* variety *arctica* with a bell diameter of 2.3 m, and ocean-going *Cyanea* spp. probably attain the largest dimensions of all jellyfish. Australian specimens can grow to 1 m width across the bell (MARSH & SLACK-SMITH 1986) but the commonly-seen tropical size bell is just 10–25 cm in diameter. The bell is largely flattened and varies in colour from opaque white with yellow or brown colouration (*C. capillata*) (Plates 8.89 and 8.90) to blue (*C. lamarcki*) or violet with reddish mouth arms (*C. purpurea*).

The upper surface of the bell is saucer-shaped, with the edge scalloped into 16 lappets (BARNES 1960), but underneath are mucus-secreting mouth arms and more than 1000 long thin tentacles grouped in eight U-shaped areas (MARSH & SLACK-SMITH 1986), although these are V-shaped in other species (Plate 8.91) {see also Figure 6.23, page 144}. The name 'snottie' graphically describes the thick, slimy mucus with a strong 'fishy' smell that may be exuded by the mouth arms. The tentacles are each more than 1.5 m long when relaxed (BARNES 1960), are filamentous and invisible to swimmers and divers in the water when extended (Plates 6.1A, 6.1B and 8.93) and may attain extraordinary lengths (e.g. 12 metres?* [WOOD 1874]). Detached tentacles can still sting (BARNES 1960; SUTHERLAND 1983).

The description by BARNES (1960) of '...general resemblance to a mop hiding under a dinner-plate.' is often repeated.

Some other important distinguishing features of *Cyanea capillata* (LINNÉ 1758) {see Chapter 6, page 144 for details including a labelled diagram by Jacquie Rifkin}:

Highly developed circular and radial muscles, divided into 16 parts by radial septa are found in the subumbrella. The central stomach gives rise to gastric pouches which in turn lead to the branching canals in the lappets. The gonads are interradial.

* *Cyanea* tentacles described by MAYER (1910, p. 598) may be up to 25 times the bell diameter (up to 57m; HALSTEAD [1988] claims 36 m) when extended. In sheer dimension this would make in-water *Cyanea* comparable with the blue whale!

THE STING

Reports vary widely geographically and in severity. Chesapeake Bay and Australian coastal *Cyanea* spp. do not seem to be capable of producing painful stings (RICE & POWELL 1972; FENNER & FITZPATRICK 1986) (Plates 8.90, 8.92 and 8.94). Severe stinging effects are described (WOOD 1874; POPE 1953, 1963; NATVIG quoted in CLELAND & SOUTHCOTT 1960; divers in Torres Strait quoted by BARNES 1960). As few people swim unclothed in the Arctic, envenomation by the largest specimens are not described in the literature, to the editors' knowledge! With regard to tropical and temperate latitude *Cyanea* stingings generally, the editors agree with BARNES' north Queensland assessment (1960) that the appearance of the sting is often worse than the symptoms in the patient (Plates 8.90, 8.92 and 8.94).

NOTE: In the earlier part of this century in Norway, a 'jellyfish-cure' for 'neuralgias' consisted of the direct application of living *Cyanea capillata* tentacles on the 'diseased limb' of the patient, with the production of local stinging effects accompanied apparently by generalised autonomic, sensory and motor effects. This therapy was ceased when glottic oedema resulted and 'since individual dosage could not be controlled'! (GLAX 1924, quoting NATVIG in CLELAND & SOUTHCOTT 1960).

LOCAL STING EFFECTS

Most stings initially cause a sticky local sensation and burning pain, which may become more severe. Raised white wheals quickly develop with an impressive surrounding red flare (Plate 8.93). The wheals have a characteristic 'saw-tooth' pattern (Plate 8.94), seemingly caused by local contraction of the skin or the arrangement of nematocysts on the tentacle. These usually fade after 15 minutes or so to leave bright red, often zig-zag streaks, which BARNES (1960) considered a valuable diagnostic aid. BARNES has pointed out that despite the multitude of *Cyanea* tentacles, a single linear skin injury is not unusual (possibly because of a detached tentacle?). Rarely severe stings may cause local haemorrhage and/or blistering, particularly if many tentacles sting the same area. Large stings (Plates 8.90, 8.92 and 8.94) may cause initially troublesome skin pain but seldom systemic symptoms in Australian cases in the experience of the authors (PJF & JAW).

SYSTEMIC SYMPTOMS

Although Barnes and others have described systemic symptoms (mainly in children, and including nausea, backache and slight abdominal pain), and these tend to be repeated uncritically from text to text, evidence from CLELAND & SOUTHCOTT (1965) is the most convincing. They describe a case of a severe sting where the victim lost consciousness for 16 hours, although the wheals from the tentacles were comparatively slight. MARSH and SLACK-SMITH (1986) describe difficult breathing and palpitations (the heart 'skipping a beat') after some severe stings, but no first-hand documentation is identified.

Possible allergic reaction to *Cyanea* spp. venom is suggested by descriptions of 'glottic oedema', loss of consciousness despite slight sting marks, asthma-like symptoms from the inhalation of dust from dried *Cyanea* on fish nets, the presence of a 'histamine-releasing factor' in the venom (UVNÄS 1960) and the possible development of immunity to the stings (CLELAND & SOUTHCOTT 1965). No deaths have been reported in the scientific literature.

Following the early Australian description of corneal injuries from jellyfish stings (HERCUS 1944), and in particular from a *Cyanea* sting (MITCHELL 1962), such eye injuries are well documented (CLELAND & SOUTHCOTT 1965; GLASSER et al 1992).

CYANEA NEMATOCYSTS

BARNES (1960) first accurately described the differential micro-structural features of *Cyanea* nematocysts. Three types of nematocysts, atrichous (or basitrichous [HEEGER *et al* 1992]) isorhizas, holotrichous isorhizas and heterotrichous microbasic euryteles {see Figure 6.32, page 168} exist in the cnidom of the Chesapeake Bay *Cyanea capillata* (BURNETT 1971; RICE & POWELL 1972). However, HEEGER and colleagues

(1992) could not demonstrate a terminal opening in the tubule tip of *Cyanea capillata* nematocysts. Nematocysts from oceanic *Cyanea* specimens are many times larger than those from Chesapeake Bay animals. All nematocyst types in the *Cyanea capillata* cnidom appear to penetrate human skin (HEEGER *et al* 1992). As the tubules from *Cyanea* discharged nematocysts enter the skin, they follow a sinuous rather than a straight trajectory (BURNETT & CALTON 1987).

THE VENOM

Cyanea capillata venom is a toxic (possibly single) protein with associated non-toxic lipid (RICE & POWELL 1972) and an LD_{50} of 0.3 mg/kg in mice (WALKER 1977). It is experimentally dermatonecrotic (BURNETT & CALTON 1987), haemolytic in experimental animals (LONG & BURNETT 1989), and severely disturbs cardiac, smooth and skeletal muscle cell contraction in mammals (WALKER 1977; BURNETT & CALTON 1987). The correlation between experimental animal and clinical human effects appears poor.

PREVENTION

Interesting experiments are in progress on live human skin with *Cyanea capillata* tentacles, to evaluate the protective action of both sunscreen preparations and 'Parafilm' (an inert material) against actual nematocyst discharge (HEEGER *et al* 1992). No results are known to the editors. However, as with any cnidarian sting, protective clothing of any sort (e.g. a stinger suit, as shown in Plate 1.22) is completely effective.

FIRST AID

If there are remaining adherent tentacles, to prevent further envenomation they are picked off with the fingers. If in doubt, adherent material should be picked off with the fingers.

Vinegar should not be used. Although the sting or enevenomation syndrome is not severe, vinegar appears to cause mass discharge of the remaining nematocysts and could theoretically make the envenomation worse. Methylated spirits did not induce any discharge in one study (FENNER & FITZPATRICK 1986c), but produced massive discharge in another study (RIFKIN 1988a, 1996).

Cold packs or ice are applied to the stung area for 5–15 minutes (Plates 5.16 and 5.17). They may be re-applied if further pain recurs. Send for medical aid if pain is unrelieved or further symptoms develop {see Chapter 5, page 111}.

MEDICAL MANAGEMENT

Medical management is not commonly needed {refer to Chapter 5, page 113 (First Aid)}. *Cyanea* stings to the eyes (MITCHELL 1962) require immediate referral to an ophthalmologist for slit-lamp examination. Removal of nematocyst material from the cornea and treatment of corneal oedema is a specialist task {Plate 5.10, and Chapter 11, page 310}.

CHAPTER 9

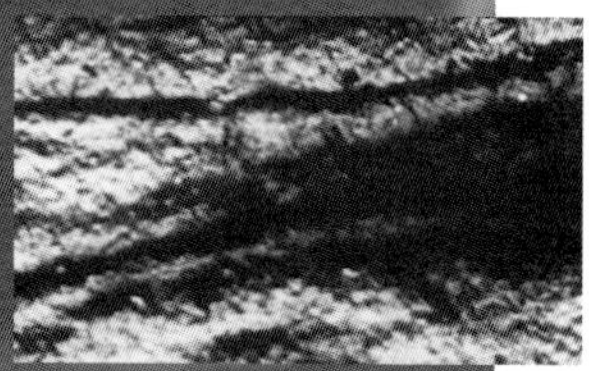

CUBOZOANS ('BOX JELLYFISH')

JOSEPH BURNETT, BART CURRIE, PETER FENNER, JACQUIE RIFKIN AND JOHN WILLIAMSON

The cubozoan jellyfish are of overriding international medical significance. Their suggested relationship among the cnidarians is shown in Table 6.3, page 147, with accompanying discussion.

TAXONOMIC OUTLINE OF THE CLASS CUBOZOA

The term 'box' jellyfish can be confusing, as there are many and differing species (Tables 9.1 and 9.2) referred to by that non-specific name. Box jellyfish all belong taxonomically to the class Cubozoa (WERNER 1975), being generally characterised by a roughly cube-shaped body or bell, with a fleshy pedalia (like a 'hand' or extension) in each lower corner of the bell, from which tentacles arise (Plate 9.1). They differ significantly from species to species in morphology (e.g. size and number of tentacles), stinging potential and clinical manifestations of stings.

The general public now tend to associate the term 'box jellyfish' with very serious stinging by chirodropids (WILLIAMSON *et al* 1980), particularly in the Indo-West Pacific region. It is preferable that cubozoan jellyfish should not be simply referred to non-specifically as 'box jellyfish' (or 'stingers'). If the correct scientific name is not known, then more specific 'popular' names are preferable, such as:

'*Chironex* box jellyfish'	for *Chironex fleckeri*
'Quaddie'	for *Chiropsalmus quadrigatus* (if this identity is suspected in north Queensland), or
'Cubo'	for *Chiropsalmus quadrigatus* in the Indo-West Pacific

'Irukandji'	for *Carukia barnesi* (and the like)
'Jimble'	for *Carybdea rastoni*

Considering the ever-present risk of misidentification with any other approach, these names may facilitate identification of the jellyfish species and therefore the application of correct treatment.

WORLD DISTRIBUTION

MAYER (1910) and KRAMP (1961) describe cubozoans in all oceans except the polar oceans (Table 9.1).

While previously recognised world localities are slowly expanding in extent, cubozoan jelly-

TABLE 9.1
WORLD-WIDE DISTRIBUTION OF CUBOZOAN JELLYFISH [SEE BELOW]

Genus	Ocean	Specific locations:
Carybdea	Atlantic	USA (east), Gulf of Mexico, Caribbean, South America, Mediterranean, Portugal, Africa (N, W & S), Sandwich Is
	Pacific	Hawaii, Borneo, Japan, Philippines, Malaysia, Sumatra, Papua New Guinea, Australia (S), Marquesas Is
	Indian	India (S), Sri Lanka
Tamoya	Atlantic	Brazil, Caribbean, USA (NE)
	Pacific	Samoa, Japan, Papua New Guinea, Australia (northern, & NSW)
	Indian	Malaya, Gulf of Iran, Africa (W & SW)
Tripedalia	Atlantic	Jamaica
	Pacific	Australia (N), Indonesia
Chirodropus	Atlantic	Africa, from Liberia on W coast to NE coast of South Africa
Chironex	Pacific	Australia (W, N & NE)
Chironex-like	Indian	Vietnam#, Philippines#, Brunei#, Sabah*
Chiropsalmus	Atlantic	USA (SE), Venezuela, French Guiana, Brazil, Argentina (N), Africa (W)
Chiropsalmus-like	Pacific	Australia, Singapore, Malaysia, Vietnam, Thailand, Philippines
Chiropsalmus	Indian	Maldive Is
Chiropsoides	Indian	Java, India (S), Sri Lanka

* *Chironex*-like; NORAZILA 1995, unpublished observation. # RIFKIN 1993, unpublished observations.
Note: Location names are as recorded in the publication source.

TABLE 9.2
COMMON MEMBERS OF CLASS CUBOZOA

Order Carybdeidae	Class Cubozoa	Order Chirodropidae
Carukia barnesi		*Chiropsalmus quadrigatus*†
Carybdea sivickisi		
Carybdea rastoni		*Chiropsalmus quadrumanus*
Carybdea marsupialis		
Carybdea alata		
Carybdea alata grandis		*Chiropsoides buitendijki*
Tamoya haplonema		*Chironex fleckeri*
Tripedalia spp.		*Chirodropus gorilla*
The 'Morbakka'	*Tamoya virulenta*††	
The 'Darwin carybdeid'		

Note: For distribution, see Chapter 9
† *Chiropsalmus* and *Chironex*-like specimens are recognised with reference to well-described Chirodropids within Australia.
†† See Chapter 6, page 148.

CARUKIA BARNESI (IRUKANJI)

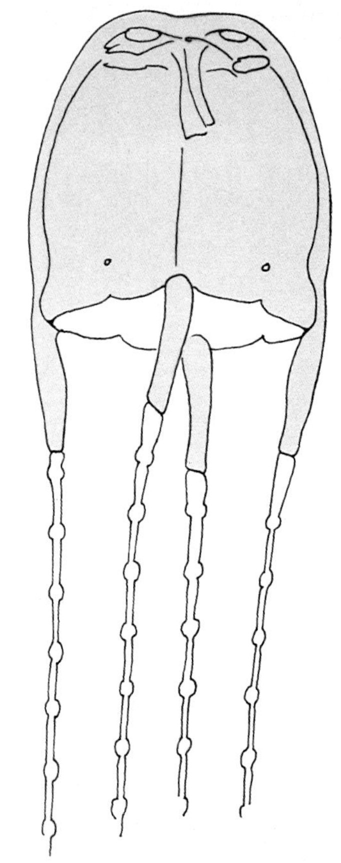

FIGURE 9.1
The typical gross features of a carybdeid medusa: an (often smallish) squarish bell, four pedalia and only one tentacle arising from each pedalium. (Drawn by Jacquie Rifkin.)

fish continue to be reported in the following locations:

- Indo-Pacific and Australia (CLELAND & SOUTHCOTT 1965; BARNES 1960; WILLIAMSON 1985; MARSH *et al* 1986; FENNER 1985- & RIFKIN 1991, unpublished observations)
- New Zealand (HOVERD 1985)
- Japan (Okinawa) (SHOKITA 1986; ARAKI 1994, unpublished observations) {see page 291}
- Gilbert and Ellice Islands (in territorial waters of present-day Kiribati and Tuvalu, respectively) (POPE 1951)
- Philippines (FENNER 1987; CORNELIUS, HEEGER 1995 personal comminication; RIFKIN 1995 this volume)
- Georgia (south-east Atlantic coast, USA) (e.g. BURNETT 1994, unpublished observations)
- Gulf of Mexico (PHILLIPS & BURKE 1970; BENGSTON *et al* 1991)
- Puerto Rico (ARNESON & CUTRESS 1976) {see page 300}
- Adriatic Sea (BOERO & MINELLI 1986) {see Chapter 10, page 285}
- West African coast (PAGÈS *et al* 1992) {see Chapter 10, page 301}
- Gulf of Oman {see Chapter 18, page 405}
- Sri Lanka (FENNER 1994, unpublished observations {see Plates 9.9B and 9.61}).

COMMON MEMBERS OF CLASS CUBOZOA

The class Cubozoa is divided into two main orders: Carybdeidae and Chirodropidae

Order Carybdeidae

Square or 'box-shaped' jellyfish usually having one tentacle arising from each pedalium at the lower corner of the bell, i.e. usually four tentacles only (Figure 9.1 and Table 9.2). These are commonly referred to as carybdeid medusae. A typical carybdeid medusa is *Carukia barnesi* (Plate 9.2).

Some less common carybdeids may have two pedalia (*Tripedalia binata*) (MOORE 1988) or three pedalia (*Tripedalia cystophora*) (CONANT 1897) in each corner. As each pedalium still bears a single tentacle (i.e. a carybdeid characteristic), the tentacles in each corner may in fact be considered to be paired (*Tripedalia binata*) or trebled (*Tripedalia cystophora*).

Order Chirodropidae

'Box-shaped' jellyfish with many tentacles (up to 15) arising from complex pedalia {see Figures 6.27, page 149 and 6.29, page 151}. They are commonly referred to as chirodropid medusae. A common member of this order is *Chironex fleckeri* (Plate 9.3) {see also Plate 1.4}. However, the order includes several other species; those that are currently recognised are listed in Table 9.2.

SIGNIFICANT CUBOZOAN FEATURES

SENSE ORGANS

All cubozoans studied to date have rhopalia (club-shaped structures which bear sense organs {see jellyfish anatomy Chapter 6, page 151}). There are four, located above the base of the bell, half-way between the pedalia. Some of the ocelli (eyes) are complex, and in at least one case have been shown to be image-forming structures {see Figures 6.7 and 6.8, page 127}; they may also detect vibration (HYMAN 1940). The complex ocelli were first described by CHUN (1897). They were later described as looking inwards (BERGER 1900) towards the stomach and its contents (STUDEBAKER 1972). HARTWICK (1991) later showed that in live specimens the 'eyes' are on stalks and can be rapidly rotated by musculature to look either inwards towards the stomach, or outwards to provide an image of the surrounding environment.

Early taxonomists (HAECKEL 1880; STIASNY 1922; MAYER 1910; BIGELOW 1912; KRAMP 1961; UCHIDA 1970) agreed that the niche in which the rhopalium is housed has characteristics that are of significant taxonomic importance (Figures 9.2 and 9.3, this page).

In *Tamoya* the outer rhopalium covering looks like a pair of lips, protecting the sensory niche (Figure 9.2). In many *Carybdea* spp. the lower 'lip' is split, giving the appearance of an upside-down 'hare-lip' (Figure 9.3).

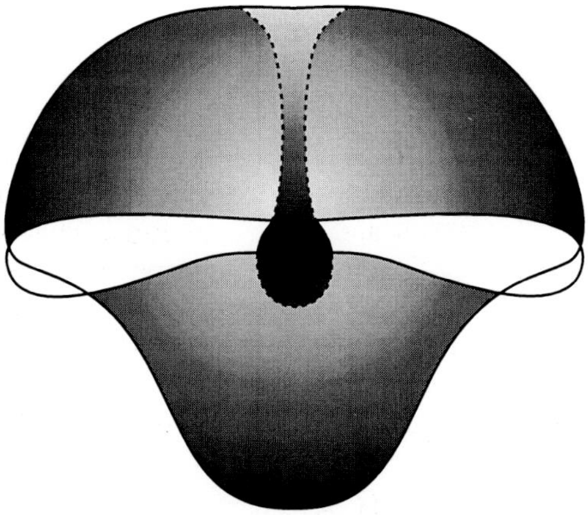

FIGURE 9.2
Opening of sensory niche in *Tamoya* species. The rhopalium covering looks like a pair of lips, one above and one below, protecting the sensory niche. (Drawn by Ronald Grendon, Grendon Design Services, Mackay, Queensland.)

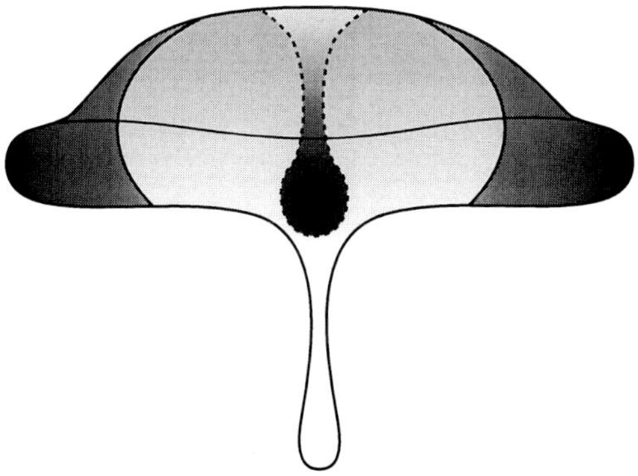

FIGURE 9.3
Opening of sensory niche in *Carybdea* species (except *Carybdea sivickisi*). The rhopalium covering has a split lower lip giving the appearance of an upside-down 'hare-lip'. (Drawn by Ronald Grendon, Grendon Design Services, Mackay, Queensland).

STING VARIATION

Stings vary clinically in appearance, severity and symptoms, in part because of qualitative venom differences. In addition, factors influencing the amount of injected venom during human stings include:

- The number, length and width of the tentacle contacts
- The extent of bell contact in those cubozoans with nematocysts on the bell (e.g. *Carukia barnesi*, Morbakka & Darwin carybdeid) {see Plate 9.16}
- The amount of body hair protecting the skin {see Plate 1.3}
- The skin (keratin) thickness at the sting site {see Plate 5.7}
- The length of time the skin is exposed to the nematocysts
- The number of nematocysts in the 'fire-ready' position {see Chapter 6, page 158}
- The length of time since the animal last fed (e.g. nematocysts previously discharged and not yet replaced)
- The extent of rubbing, handling, or inappropriate chemical treatment of the raw tentacles adherent to the victim's skin

It is also suspected that environmental factors affecting the animal, such as water salinity (BARNES 1966), the available food supply (HARTWICK 1987) and the physiological state of the animal at the time (ENDEAN 1988), could influence the amount and potency of venom actually injected. Further studies are warranted.

INHIBITION OF CUBOZOAN NEMATOCYST DISCHARGE

To date, tests on all species of the class Cubozoa show inhibition of undischarged nematocysts within about 30–40 seconds by 2–10% acetic acid in water. (This includes household vinegar which is 4–6% acetic acid in water) (Figures 9.4a, b and c) (HARTWICK *et al* 1980, FENNER *et al* 1985, 1986, 1987). This is also true in *Tamoya virulenta*, a large carybdeid (FENNER 1985; RIFKIN 1993-94, unpublished observations) {see page 148}. While the cubozoan nematocyst firing capability of alcohol (e.g. 'household methylated spirits', which is 96% ethanol plus 4% water and denaturants [HARTWICK *et al* 1980]) is undisputed (Figure 9.4B), Barnes maintained that the resultant firing was an aborted one, incapable of venom injection (KINSEY 1986). He also warned of the dangers of fresh water application to *Chironex* stings (KINSEY 1986) {see also Chapter 5, page 109}. Notwithstanding these caveats, current recommendation is that vinegar should be used in any cubozoan sting for nematocyst inhibition (BEADNELL *et al* 1992) {see Chapter 5, page 108}. If vinegar is unavailable, there may be slight benefit in using various beverages (CURRIE *et al* 1993); however, no reagent has yet been identified that compares with vinegar in inhibiting cubozoan nematocyst discharge (HARTWICK *et al* 1980).

If vinegar or the beverages referred to above are not available, it is preferable for the rescuer to pick the tentacles gently off the victim. Because of the thickness of keratin of the fingers, the rescuer is unlikely to be stung, or may feel harmless prickling. Although it is appreciated that further envenomation will occur, even if the tentacles are picked off gently, it is believed that this may be less than the resultant, continual discharge that occurs from tentacles left *in situ* on the skin, although this has not been proven. This treatment must not take precedence over any necessary resuscitation {see First Aid in Chapter 5, page 103).

FIGURE 9.4A
The undisturbed tentacle surface. (Copyright © *The Medical Journal of Australia* 1980; 1: 15–20. Reprinted with permission.)

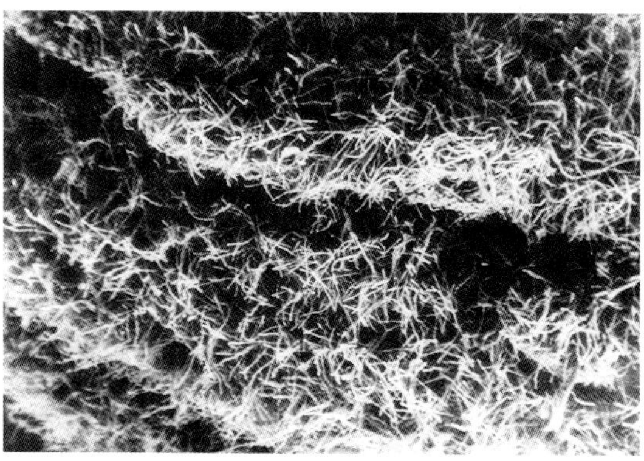

FIGURE 9.4B
Tentacle appearance after a 30-second immersion in methylated spirits. (Copyright © *The Medical Journal of Australia* 1980; 1: 15–20. Reprinted with permission).

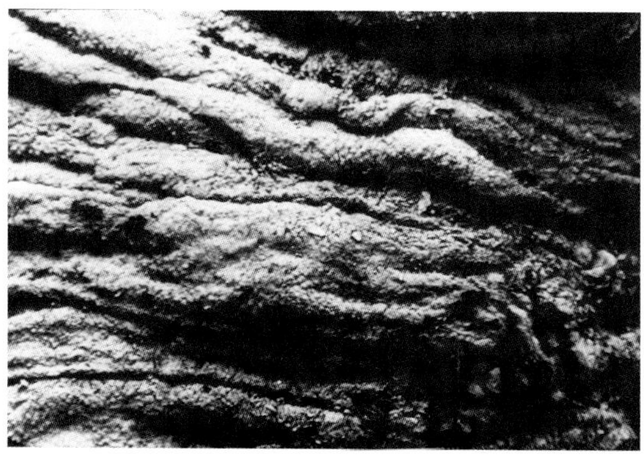

FIGURE 9.4C
Tentacle appearance after a 30-second immersion in 5% acetic acid in water prior to the same treatment as in b. (Copyright © *Medical Journal of Australia* 1980; 1: 15–20. Reprinted with permission.)

CARYBDEIDS

Carybdeids are 'box' jellyfish, in that the bell has the typical four sides with a single pedalium arising from the lower part of each corner, and each pedalium bearing a single tentacle.* Thus most described carybdeids usually have only four tentacles, one in each corner (Figure 9.1, page 238 and Plate 9.2).

Carybdeids are found worldwide except in the colder oceans. Species vary considerably in size with the largest having bell heights from 20 to 50 cm (Plates 9.4 to 9.6b) and the smallest only 6–8 mm (Plate 9.2). Most have transparent bells, which makes the larger ones difficult to see in their habitat and the smaller ones practically invisible. Tentacles vary in length and shape among different species. The 'large carybdeid' group are of current special interest medically and taxonomically (FENNER 1985; RIFKIN 1994 unpublished observations).

LARGE CARYBDEIDS

Correct identification of most of the larger carybdeid species remains difficult, although research is current in this area (BOERO & MINELLI 1986; FENNER *et al* 1985; FENNER 1991; RIFKIN 1994, unpublished observations). Many jellyfish taxonomists have described the various species characteristics, but the irregularities between these descriptions, even within a single species, make exact identification of specimens difficult.

A number of large carybdeids of varying shapes and sizes have been caught over the past decade, notably in Australian waters (FENNER *et al* 1985, MARSH & SLACK-SMITH 1986), the Mediterranean Sea (BOERO & MINELLI 1986) and Hawaiian waters (CURRIE & NOLAN 1993, unpublished observations, see below). Many have been damaged in trawler nets, some have been found in a decomposed state, and others have only been glimpsed in their natural habitat and inadequately described. Preliminary identification of some of these preserved specimens has included *Carybdea alata*, *Carybdea alata grandis*, *Carybdea marsupialis*, *Tamoya haplonema*, *Tamoya virulenta*, *Tamoya bursaria* and *Tamoya gargantua*. None of the captured Australian specimens completely matches the original description of any of these species, nor even their later descriptions by taxonomists (HAECKEL 1880; MAYER 1910; BIGELOW 1932; KRAMP 1961), but their identification is under investigation (FENNER 1985,1991; RIFKIN 1991–94, unpublished observations; CURRIE *et al* 1993, unpublished observations) {see page 148}.

This group of large carybdeids includes those with common or popular names such as 'Morbakka' (SOUTHCOTT 1985) 'fire jelly', and 'Moreton Bay stinger' (WILLIAMSON 1985). A suggested summary list of the large carybdeids is shown in Table 9.3.

TABLE 9.3
SOME LARGE CARYBDEIDS

Tamoya spp. including 'Morbakka' (see page 148)
Carybdea alata
Carybdea alata grandis
Carybdea marsupialis

DISTRIBUTION

The worldwide distribution of large carybdeids cannot yet be defined, but includes all tropical waters and possibly most oceans, the Mediterranean Sea and even New York harbour (MAYER 1910). POPE (1951) makes an interesting report of many specimens (described but not identified, as *Tamoya*) hanging on a line drying in the Gilbert and Ellice Islands (POPE 1951), prior to being consumed as a delicacy — also a common practice in Japan (KISHINOUYE 1889).

On Australia's Pacific coastline

On the east coast of Australia, large carybdeid specimens have occasionally been found as far south as Sydney (latitude 33° south) {see page 243} (FENNER 1994, unpublished observations), although the usual southern boundary appears to be Moreton Bay, near Brisbane (latitude

* *Manokia stiasnyi* is a rare carybdeid, originally called *Carybdea stiasnyi* (BIGELOW 1938) after STIASNY (1937), who first captured it in New Guinea. It was described by SOUTHCOTT (1967) as having an 'unusual branched state of the tentacles', leading him to rename it, creating a new genus.

27° south), where many Morbakka specimens are caught and stings occur. The jellyfish is locally referred to as the 'Moreton Bay stinger' or 'fire jelly'. Large carybdeid specimens are found in northern Queensland waters (Plate 9.7) (FENNER 1991).

Western Australia

Several large carybdeids have been caught off Port Hedland (latitude 20° south) in the north-west of Western Australia and identified as *Carybdea spp*. One very large specimen was identified by MARSH, of the Western Australian Museum, as *Carybdea alata grandis* (1991, unpublished observations). Further stings and specimens have been reported from Ningaloo Reef and Exmouth Gulf (latitude 21–22° south).

A specimen about 50 cm in height was photographed off Australian's North West Cape peninsula by Mr Pat Baker (Plates 9.6a and 9.6b). It was covered with orange nematocyst warts, and when touched produced a painful sting with systemic symptoms {see Case History 1, below}. As the specimen was not captured, it has not been identified.

CASE HISTORY 1

Whilst diving near Black Rock, off Ningaloo Reef, North West Cape in Western Australia, a diver saw a number of jellyfish, some of which were being eaten by a turtle {see Plate 9.54}. Touching the bell of one with his right hand, he found it to be 'knobbly'. Within a minute his hand began to tingle, and then it became more and more painful as if 'all the nerve ends were crying out'. The pain in the hand seemed to come on in 'waves'. Shortly after he felt a dull, localised pain under his left arm and his breathing became 'tight' and he felt 'unduly anxious'. The chest tightness passed off within an hour and the anxiety settled shortly after. The axillary pain, too, settled within a couple of hours, but the pain in the hand lasted some eight hours before finally abating. There were no further sequelae after these pains had settled.

CASE HISTORY 2

A healthy male scuba diver was stung on the left knee by a large carybdeid (*'Tamoya gargantua'*) through an intact lycra stinger suit, at 11 am in April 1994, on Ningaloo Reef, Western Australia. He was certain he had touched only the bell, and there was an immediate stinging pain at the contact site. He subsequently captured the specimen in a bucket (during which a fish which appeared to inhabit the subumbrellar region escaped). Despite only mild local erythema and no systemic symptoms, intense local pain developed within the hour, extending to involve his groin. After two hours the pain diminished, but a burning itch of the sting region persisted for 24–36 hours. The captured specimen measured 22 cm in bell depth, with tentacles of 8 cm. Brown nematocyst batteries were visible on the bell.

Northern Territory

A few large carybdeids have been found, but most have been the smaller 'Darwin carybdeid' type (RIFKIN & ALDERSLADE 1993, unpublished observations).

The presence of large carybdeids in other Australian waters remains uncertain.

CARYBDEA ALATA REYNAUD, 1830 (VAR. *GRANDIS*)

In Hawaii

CASE HISTORY 3

Report of sting from *Carybdea alata* (Bart Currie, Darwin, and Mike Nolan, Lahaina, Maui, Hawaii)

On 4 June, 1993, a 21-year-old male was scuba diving in 8 metres of water, 200 metres offshore from the Hawaiian island of Lanai. While at a depth of 3 metres he was stung on his right shoulder by a jellyfish. There was immediate local moderate pain. On return to the dive boat, he was treated with topical alcohol and a meat tenderiser containing papaya. There were no individual tentacle marks visible but the area over his right shoulder showed red blotches. The pain over the shoulder began to increase and after 15 minutes there was the onset of abdominal cramps and hyperventilation. The abdominal cramps and severe pain persisted until 45 minutes after the sting when they began to diminish. The abdominal cramps had fully resolved by 90 minutes, but the victim had general malaise with vague abdominal pain for the next two days. The shoulder erythema subsided and had resolved after 24 hours.

Around the time of the sting a number of beaches in Hawaii had been temporarily closed to swimmers because of the presence of box jellyfish. This is an occasional occurrence in Hawaii, and *Carybdea alata* has been identified in the past in these waters (MAYER 1910; CLARK & SIMS 1989). Subsequent to the sting the dive boat master photographed a carybdeid jellyfish of the species thought responsible for both the beach closures and the sting reported (Plate 9.8); with a bell diameter of up to 5 cm and tentacles up to 0.5 m long, the description of this carybdeid being consistent with *Carybdea alata*.

It should be noted that the use of alcohol is presently contraindicated in box jellyfish stings such as this, and vinegar is the treatment of choice to inactivate unfired nematocysts in cubozoans (HARTWICK *et al* 1980). Direct application to jellyfish stings of green papaya or papaya leaves is described as an old Hawaiian remedy (CLARK & SIMS 1989) but is scientifically untested.

In the Caribbean

The life cycle of *Carybdea alata* was first described in Puerto Rico here in 1973 (ARNESON & CUTRESS 1976). In this region *Carybdea alata* is principally an oceanic (deep water) animal; but in July 1973 and August 1974 it was found inshore near Aguadilla in large swarming numbers, under a night light and in Guantanamo Bay, Cuba in 1992 and 1994. *Tamoya haplonema* is also occasionally sighted in Puerto Rico (B CUTRESS 1992, unpublished observations).

APPEARANCE

Tamoya haplonema (MÜLLER, 1859)

(Plate 9.7) The bell is up to 9 cm in height, with a width up to 5.5 cm. 'White wart-like clusters of nematocysts' (MAYER 1910) are present on the exumbrella (cf. 'pink' nematocyst warts on 'Morbakka')

Tamoya gargantua (HAECKEL, 1880)

Bell up to 22 cm high and 13.5 cm wide, described by KRAMP (1961), but not by MAYER (1910). Nematocyst warts were *not* described by KRAMP (1961). May be a variation of *T. haplonema*.

Carybdea alata (REYNAUD, 1830)

(Plate 9.8) The bell is more a pyramid (a tall rocket-shape) with a height up to 8 cm and width up to 5 cm (KRAMP 1961). KRAMP (1961) states that nematocyst warts on the exumbrella are absent, but MAYER (1910) does not mention them.

Carybdea alata grandis

(Plate 9.10) Variety of *C. alata*. Bell height up to 23 cm. Nematocyst warts are not mentioned by MAYER (1910) nor KRAMP (1961). (A specimen 50 cm high with a width of 15 cm caught in Western Australia has been identified as *C. alata grandis* by Loisette Marsh of the Western Australian Museum (MARSH 1994 unpublished observations).)

Carybdea marsupialis (LINNÉ, 1758)

An elongated, exceptionally transparent bell some 5–15 cm in width (BOERO & MINELLI 1986), with a single tentacle at each of the four corners {Figure 6.6d, page 126; see also Chapter 10, page 285}. Numerous 'dull yellow' nematocyst warts (cf. 'Morbakka') mentioned by MAYER (1910) and KRAMP (1961).

CARYBDEA MARSUPIALIS (LINNÉ, 1758)

In the Adriatic Sea

The carybdeid jellyfish *Carybdea marsupialis* is described in the Adriatic Sea {see Chapter 10, page 285} but its distribution includes the whole of the Mediterranean Sea, as well as Portugal, the west African coast and the Caribbean Sea (KRAMP 1961). Although it is referred to by CLELAND and SOUTHCOTT (1965) and SUTHERLAND (1983) as being similar to *Carybdea rastoni* (Plates 9.26–9.29), the elongated cubic bell of an Adriatic Sea specimen can reach 15 cm in diameter (BOERO & MINELLI 1986) — much larger than *C. rastoni*. CLELAND & SOUTHCOTT (1965) attribute the distinction between *Carybdea marsupialis* and *Carybdea rastoni* to descriptions of the gastric cirri by HAACKE (1887) and BIGELOW (1938).

MORBAKKA TYPES *TAMOYA VIRULENTA*

A larger carybdeid with bell diameter from 6–18 cm, and a bell height of 6–14 cm (the bell height is usually greater than the diameter (Plate 9.7) {see also below}. Bell covered with numerous pinkish nematocyst warts in some cases (cf. above). Specimens are very similar in appearance to *Tamoya haplonema*, but lack the gastric cirri (see Chapter 6, pages 148–9).

THE STING

Because stings from the large carybdeids have many common features, the description of Morbakka stings serves as a general guide {see Plates 9.11 to 9.16}.

The nickname 'fire jelly' for Morbakka types is apt. Skin in the envenomated area looks (and feels) as if it is on fire (Plates 9.11 and 9.12). Initially the skin has raised white wheals (Plate 9.13) with a surrounding erythematous flare. The pain in this area is described as 'burning' and uncomfortable (FENNER *et al* 1985), but may be severe in some cases (see Case Histories 1 and 2 above).

Rarely, the 'bell' mark may be obvious (Plate 9.16)(nematocysts are present in the 'warts' on the bell — Plate 9.7). The sting marks may also show the characteristic length-wise tapering of the tentacle (Plate 9.13).

Occasionally 'ladder' markings, which are more characteristically associated with chirodropid (especially *Chironex*) envenomation (BARNES 1960) (see Plate 9.78), may be present from Morbakka stings. These match the transverse 'bands' clearly visible in the tentacles (Plate 9.14).

The wheals slowly change from pallor to redness, first becoming pink, then finally a dark red. The skin marks may persist for 24 hours, during which time the burning sensation fades, leaving the area itchy and tender. A papulovesicular lesion may last from three to ten days. Less commonly, actual skin necrosis may occur in a severe sting (Plate 9.15) leaving scars that take months to fade.

Large carybdeid stings and 'Irukandji-like' reactions

There are now many documented cases where systemic symptoms occurred after envenomation from these large carybdeids. These include the patient who sustained the potent sting shown in Plate 9.16 and the patients stung in Case Histories 1 and 3 and page 149. The symptoms in these patients (which were typical) consisted of the initial delay, then the start of 'boring' low back pain, muscle cramps, nausea, vomiting, headache and 'a feeling of doom'. These symptoms are very similar to, although less devastating than those following envenomation by *Carukia barnesi* (one of the Irukandji jellyfish — see below), and they may be described as an 'Irukandji-like' reaction. At present it is not known if all or just some of the morphological varieties of large carybdeid described above may produce this effect, or whether the response is simply dose-related.

Similar to these clinical observations, B. CUTRESS (1992, unpublished observations) of the Department of Marine Sciences, Mayagüez Campus, University of Puerto Rico, provided the following description of the reactions to local *Carybdea alata* stingings as recorded first-hand prior to 1970 by her late husband, the noted marine scientist and marine envenomation pioneer, Professor Charles ('Chuck') Cutress (Professor Cutress died on 17 January 1992):

> 'Two men, one a student and the other an aide in our Department, were stung when scuba diving off La Parguera. They were pulled into the boat by others...In a third case, by coinci-

dence my dentist, the sting occurred when he was scuba diving seaward of the Aguadilla pier...In all three cases the victims considered they would have drowned if people had not been there to pull them in as the stings virtually paralysed them...There are no known specific antidotes for cnidarian venoms. Treatments should be directed toward alleviating neurotoxic effects of the venom, relieving pain and controlling primary shock.'

Dermotoxicity lasting three days attributed to the sting of the Adriatic carybdeid *Carybdea marsupialis* is documented (KOKELJ *et al* 1992).

FIRST AID

{See Chapter 5, page 106}

Fortunately, accurate identification of individual 'large carybdeid' jellyfish is not of importance to the first-aider, but knowledge of the general type of sting, of the potential local and/or systemic effects and of basic treatment is important.

- Although adherent tentacles are rarely present in Morbakka stings, avulsed nematocyst material can be; discharge of any remaining, unfired nematocysts will be prevented by vinegar dousing.
- Apply a cold pack or ice wrapped in a cloth for 5–15 minutes to relieve local skin pain.
- If skin pain persists, or systemic symptoms develop, seek medical advice {see also First Aid Chapter 5, page 113}.

MEDICAL TREATMENT

Usually only required for severe stings. Potentially necrotic skin lesions should be covered with a sterile, non-stick dressing or the newer 'colloid-type' dressings which keep the skin lesion moist, thus allowing faster healing. Systemic antibiotics may be necessary when and if infection occurs {see Chapter 4, page 88}.

There are no documented cases of acute hypersensitivity to date. Itchy skin lesions may benefit from topical mentholated creams.

'Irukandji'-type reactions may need to be treated in hospital (FENNER *et al* 1986, 1988) (see below).

Muscle cramps respond to:

1. Pethidine (meperidine) 1 mg/kg body weight, up to 50 mg intravenously (IV) in adults, followed by an IV infusion of 30 mg hourly.
2. Morphine 0.1 mg/kg to 5 mg, as an IV dose, followed by 2.5 mg increments IV as necessary every five minutes, or as an infusion rate of 1–2 mg per hour after an initial IV dose of 5–10 mg.

   Note: In general, IV opiate usage should be in an environment where respiratory resuscitation skills and facilities are immediately available. Respiration may become shallow and ineffective, or even totally cease up to 30 minutes later in susceptible patients exposed to these drugs (OH 1990).

Alpha adrenergic blockade may be necessary for serious systemic hypertension. For management of severe hypertension by the occasional intensivist, the short-acting drug phentolamine 5 mg IV may be given initially, and 5-10mg IV repeated when necessary, with appropriate monitoring {see page 114}. More recently nifedipine has been suggested as the most appropriate treatment for the excess release of catecholamines with scorpion envenomation (GUERON *et al* 1993). In view of the similarities in systemic symptoms between the two syndromes, nifedipine might also prove to be the most appropriate treatment for the Irukandji syndrome.

SMALL CARYBDEIDS

DARWIN CARYBDEID

In 1992 two patients seen at the Royal Darwin Hospital were each fortuitously accompanied by the jellyfish responsible for their stings; the two specimens were newly-recognised species and were named the 'Darwin carybdeid' {Plate 5.4} (see Table 9.2) (CURRIE & ALDERSLADE 1992, unpublished observations). Nematocyst identification {Plate 5.5A} from skin scrapings of other Darwin cases has confirmed further stings from this species (CURRIE & RIFKIN 1993, unpublished observations).

APPEARANCE

These are small carybdeids up to the size of a matchbox, otherwise similar in appearance to the east-coast Morbakka. They have a transparent bell with white mammilations, and, similar to Morbakka, they lack gastric cirri (RIFKIN 1991, unpublished observations).

THE STING

CASE HISTORY 4

A 16-year-old local resident was swimming in 2 m of water from a boat anchored 50 m off a Darwin beach. It was a fine April day and the tide was outgoing. She swam into a jellyfish, which slipped beneath her one-piece bathing suit, stinging her on the abdomen, chest and arms. There was immediate pain at the sites of the stings but there was no nausea, vomiting, headache or altered conscious state. Her parents pulled her back into the boat and vinegar was immediately applied. She was subsequently brought to the Accident and Emergency Department at the Royal Darwin Hospital, where she arrived 50 minutes after the sting. At that stage she was generally well, although still complaining of moderately severe pain over the areas of the sting. On examination she had several areas of generalised erythema on her arms, and on her abdomen and chest there were linear welts which were pale centrally, with surrounding erythema. The total length of tentacle contact was measured at 1.5 m. The pain responded well to topical ice applications, electrocardiograph was normal and no further problems developed. She was discharged home 2 1/2 hours following envenomation. Fortuitously, the offending jellyfish was recovered from her swimming costume. It had a bell diameter of 3 cm with only one tentacle from each pedalium. Its identity was subsequently confirmed by Phil Alderslade of the Darwin Museum to be the new species Darwin carybdeid {see Plate 5.5B}.

From experiences so far recorded with this species it would appear that the sting of the Darwin carybdeid is:

- similar to a mild *Chironex fleckeri* sting;
- causes no systemic features of the Irukandji syndrome;
- has an excellent analgesic response to topical ice application.

The possibility is raised that some milder stings previously seen in the Darwin area attributed to *Chironex fleckeri*, may in fact have been from this four-tentacled jellyfish.

CARUKIA BARNESI ('IRUKANDJI')

This small, four-tentacled carybdeid was first captured by Dr Jack Barnes at Cairns in 1963 (BARNES 1964) and identified by him in a remarkable field experiment {see History Chapter 1, page 46} as capable of producing the Irukandji syndrome (KINSEY 1988) {see below}. Subsequently it was described as a new species (SOUTHCOTT 1967), and named *Carukia barnesi* in honour of its discoverer. Southcott's derivation of this name was from 'car' (carybdeid) and 'ukia', from 'Irukandji'. The term 'Irukandji syndrome' was actually coined in 1952 by Dr Hugo Flecker, a radiologist in Cairns with an interest in botany and envenomations (FENNER 1990) {see Chapter 1, page 43}. Flecker named this syndrome after a group of Aborigines who lived near Palm Cove, north of Cairns, where many stings were known to occur (FLECKER 1952). Despite its deceptively small size its sting may produce a delayed, but potentially devastating set of symptoms now known as the 'Irukandji syndrome'. The syndrome may totally prostrate a previously fit adult male {see below}.

Because *Carukia* is almost transparent and very small (about 2 cm bell diameter) (Plates 9.2 and 9.18), it is rarely seen or caught, and thus very little laboratory study has been undertaken on either venom toxicity or variations in morphology. (Capturing an Irukandji jellyfish is an unappealing prospect! The animal is too small to be easily seen, protective clothing is absolutely essential, and if a sting should occur, a wretched 12–24 hours not without risk is assured, even in expert medical hands! The editors recommend that only persons well versed in such capturing techniques, and equipped with the necessary knowledge, should attempt such field activities.)

Barnes himself always believed that more than one type of jellyfish was capable of causing the 'Irukandji syndrome', but he died in 1985 before confirming his theory (KINSEY 1988). It is now known that the 'Irukandji' syndrome can be caused by a number of jellyfish species (BARNES 1966; OTSURU *et al* 1973; KIZER 1984; FENNER *et al* 1986b, 1992, 1993; ZHANG MINGLIANG 1988; MIGAS & POPOVA 1988).

Jellyfish causing an 'Irukandji-like' syndrome

It is clear, as predicted by BARNES (in KINSEY 1988), that many jellyfish species may cause this syndrome. So far, these include species of *Carukia*, *Carybdea*, Morbakka {see page 148 and this chapter), *Physalia* {see Chapter 8, page 196), *Gonionemus* {see Chapter 8, page 203}, and *Stomolophus nomurai* {see Chapter 8, page 214}.

Distribution

Australia

The Irukandji syndrome has occurred in tropical Australian waters from Broome, Western Australia, across northern Australia, and southward along the Queensland coast to the beaches east of Childers — a distribution similar to that of *Chironex fleckeri* (FENNER 1993, unpublished observations). *Carukia barnesi* has also been caught and identified in the Northern Territory of Australia (RIFKIN 1992, unpublished observations). It thus remains speculative what proportion of Irukandji syndrome cases is due to stings from *Carukia barnesi* {see Plates 5.5a and 5.5b}; there may be regional differences (CURRIE *et al* 1992).

The southern-most recorded 'full' Irukandji syndrome to date occurred at latitude 26° south, in the waters east of Childers, central Queensland coast (FENNER 1994, unpublished observations).

Carukia is an ocean-going jellyfish, and stings have been identified from the Great Barrier Reef and open waters nearby. It has even been introduced into the swimming pools of cruise ships when they have been filled at sea (BARNES in KINSEY 1988). However, most stings occur in inshore waters. The current hypothesis is that although *Carukia* is mainly a deep-water ocean-going jellyfish, it can be swept close to shore by water currents or weather conditions. It may then surface in shallow water adjoining swimming beaches, or stratification by the specific gravity (e.g. salinity) of the water may force the jellyfish to the surface (BARNES 1964).

Such explanations may account for the 'periodicity' of seashore Irukandji stings. In some years in Australia very few stings are reported (1992–93, 1993–94 summers), whereas in others there may be hundreds (200 reported 1991–92 summer — INTERNATIONAL CONSORTIUM FOR JELLYFISH STINGS 1993). One beach may have a spate of stings, whereas another a short distance away may remain sting-free. On Christmas Day 1985, 36 people were stung within the old-style 'stinger nets' with the larger mesh (Plate 9.17) at Palm Cove, Cairns, north Queensland (FENNER *et al* 1986), and yet other nearby beaches recorded no stings. (This risk of within-stinger-enclosure stings has now been reduced by the use of smaller-mesh 'stinger-resistant' enclosures {see Plate 1.19}).

HARTWICK (1990, unpublished observations), while netting in open waters over the mid-continental shelf near Townsville, caught 72

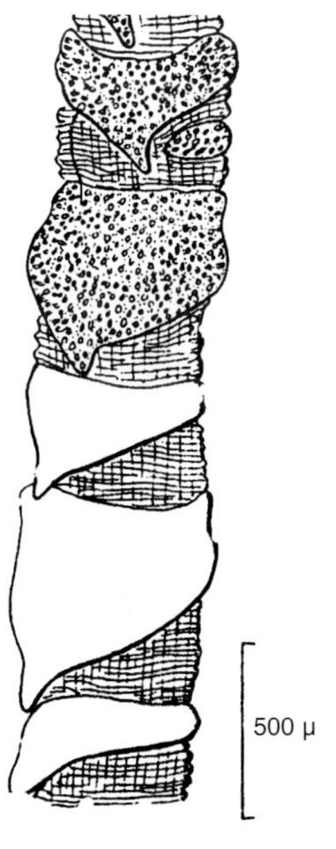

FIGURE 9.5
Diagram showing the 'pocket-handkerchief-like tails' of the encircling nematocyst-bearing rings on tentacles of *Carukia barnesi*. (After SOUTHCOTT 1967.) See also Plates 9.21 and 9.22

Irukandji between the years 1978 and 1983, during the tropical summer months of September to May (within this period the 'wet season' is usually between December and March). (See Table 9.4.)

TABLE 9.4
***CARUKIA* NETTINGS NEAR TOWNSVILLE, SUMMER 1978-1983 BY DR R HARTWICK, PHD**

September	1
October	3
November	1
December	12
January	23
February	13
March	12
April	4
May	3

Irukandji syndrome reported in other waters

Although *Carukia barnesi* has been found in Australian tropical waters, its accurate international distribution is currently unknown. However, there have been reports of marine stings causing an Irukandji-like syndrome {see below} from Papua New Guinea (CLELAND & SOUTHCOTT 1965), Hawaii (KIZER 1984), Japan (OTSURU *et al* 1973), and China (ZHANG MINGLIANG 1988) although these symptoms are probably caused by a number of other species {see below}.

APPEARANCE

Carukia barnesi SOUTHCOTT, 1967, is a small carybdeid characterised by the following morphological features:

- A pyramid-shaped bell with the height slightly greater than the width. The bell is usually about 2 cm in diameter (Plate 9.18).
- The bell has many small nematocyst warts or mammilations, more numerous near the apex. Each wart contains many nematocysts (Plate 5.2), which differ morphologically from the nematocysts of the tentacle (Figure 5.1, page 100). At present it is not known if the different types of nematocyst are responsible for different symptoms comprising the Irukandji syndrome.
- Eight large mammilations are present on the velarium, one in each octant {see Chapter 6, page 124}. These mammilations also contain nematocysts.
- Each pedalium has a canal that runs straight into the canal of the tentacle (unlike *Carybdea rastoni*, where a 'hook' is visible). {refer to Figure 9.7a}
- A single tentacle arises from each corner. It is usually 3–5 cm in length when contracted, but may extend to 35 cm. The tentacle gradually narrows from the origin to the tip.
  Nematocysts are carried on rings encircling the tentacle (Plates 9.19 and 9.20). They are close together near the bell, but become further apart near the tip. A number of these nematocyst rings may have small 'tails' and look like 'pocket handkerchiefs' tied around the tentacle (Figure 9.5, page 247 and Plates 9.21, 9.22 and 9.28).

NOTE 1: In specimens caught by PJF in the last few years these tailed rings have not been seen. Nor was there an appreciable difference in the tapering thickness of the tentacles (Plates 9.18 and 9.19). These observations may be due to maturity stages of the medusa, or it may be a different species as the pocket handkerchief structures would appear to be distinct and significant morphological structures.

NOTE 2: The anthomedusan, *Leuckartiara gardineri* Browne, 1916 ('the little red stinger') is a small jellyfish (bell 10-11 mm high) (CLELAND & SOUTHCOTT 1965) that has four long pink marginal tentacles (about 2.5 cm long [BARNES in KRAMP 1961]) and numerous minute thin tentacles (MARSH & SLACK-SMITH 1986); it delivers a 'surprising' sting, with no notable after-effects. It has been sighted in north Queensland and the Indian Ocean (MARSH & SLACK-SMITH 1986). It may be confused with *Carukia barnesi* as an Irukandji stinger. However, apart from the significant differences in the sting effects, *Carukia barnesi* also differs in that each of its four tentacles is borne on a pedalium (Plate 9.2 {See Chapter 6 page 121, Anatomy}.

THE STING

BARNES in KINSEY (1988) studied the epidemiology of Irukandji jellyfish stings in detail, and provided the first descriptions of many salient features (BARNES 1964). The International Consortium of Jellyfish Stings has data suggesting that Irukandji stings are more common on fine days, and in the afternoon (see Table 9.6). CURRIE and colleagues have recently compared patients' ages with the depths at which they were stung by Irukandji and *Chironex fleckeri* in Australia's Northern Territory during 1991. Their data suggests that *Chironex fleckeri* stings are more likely in shallow waters and Irukandji stings more likely in deeper and open sea waters (Table 9.5).

TABLE 9.5

A COMPARISON OF STINGS FROM *CHIRONEX FLECKERI* AND IRUKANDJI IN AUSTRALIA DURING 1991, WITH RESPECT TO THE WATER DEPTH THAT VICTIMS WERE IN

Stings in:	***Chironex fleckeri* (n = 34)**		**Irukandji (n = 20)**	
Shallow water (< 1 m depth)	76%	(26)	30%	(6)
Deep water (> 2 m depth)	9%	(3)	55%	(11)
Unknown depth	15%	(5)	15%	(3)
Scuba diving	6%	(2)	45%	(9)
Stung while entering water	14%		0%	

Based on data by CURRIE *et al* 1991 & 1992

The intensity of the initial sting pain on the skin of the victim is variable. It may be a sharp stinging pain, a vague underwater 'touch', or the victim may not even be aware of envenomation until systemic symptoms commence, commonly some 30 minutes later.

Unlike *Chironex* stings, which occur mainly on a victim's legs in shallow water {Plate 5.14}, most *Carukia* stings occur in deeper water (Table 9.5) and on the victim's body just below the water surface. Thus *Carukia* stings are usually on the trunk or upper arms of swimmers (FENNER *et al* 1986; BARNES in KINSEY 1988), or on the back (or back of the neck if a wetsuit is worn) in surfacing divers. The lower abdomen or legs are less often involved.

Sting marks left on the skin by this tiny animal's tentacles are usually difficult, if not impossible to see (Plate 9.23) (FENNER *et al* 1988). Any mark that does remain on the skin is usually the imprint of the tiny, nematocyst-bearing jellyfish bell (Plate 9.24). When visible this bell mark may be diagnostically helpful. (Nematocysts are plentiful at the apex of the bell where initial contact is likely.)

A couple of moments after the usual sting from the bell, the skin may develop a mild redness and a localised pilo-erection ('goose-pimple' effect) (Plate 9.23) which lasts for half an hour or so.

It can be seen from Table 9.6 that Irukandji stings are more common in the afternoons on still, fine days in Queensland. They are more common (74 per cent of total) on the upper extremities of swimmers. The neck is a common sting site on surfacing divers (even when they wear stinger suits, but without a protective 'hood') (refer to Chapter 22, page 423).

THE IRUKANDJI SYNDROME

This bizarre set of symptoms has been recognised for many years.*

It was first described as 'Type B stings' by Southcott in 1943 (SOUTHCOTT 1959) and then named the 'Irukandji syndrome' (FLECKER 1952). A number of investigators have since studied the syndrome (BARNES 1964; FENNER *et al* 1986 & 1988; MARTIN & AUDLEY 1990; CURRIE *et al* 1992).

After the initial sting there is a characteristic time delay before the onset of the severe sys-

* The first indication in the literature was probably by FLECKER (1945).

TABLE 9.6
IRUKANDJI STING RECORDS

	Number of stings reported	Percentage of stings
Total	301	
Morning	45	18%
Afternoon	213	82%
Males	185	63%
Females	110	37%
Fine weather	60	79%
Cloudy weather	16	21%
Children under 10 years of age	42	14%
Site of sting:	234*	
Trunk/abdo	92	39.3%
Legs	60	25.6%
Arms	58	24.8%
Head/neck	24	10.3%
State:		
Queensland	238	80%
Northern Territory	35	11%
Western Australia	28	9%
Wind:	162	
moderate or 15+ knots	30	18%

Taken from the sting report data base of the editors, 1987-93.
(Not all the sting reports contained information in each of the categories listed in this table.)
* 234 out of 301 cases contained sting site information.

temic symptoms that make up the Irukandji syndrome. The delay varies between 5 and 50 minutes, but is usually some 30 minutes after the initial envenomation. The cause of this delay is not yet known. Current hypotheses include the venom itself having a large molecular weight (BURNETT & CALTON 1987) with a consequent slow lymphatic absorption; the venom being a precursor causing a delayed release of an activator substance, and/or endogenous catecholamines; or a combination of these. The syndrome has three recognised clinical forms or patterns:

Acute effects:	1. Pain
	2. Catecholamine-like effects
Delayed complications:	3. Cardio-pulmonary decompensation

A victim may have any combination of the following signs and symptoms, but few patients, if any, escape the pain component.

1. Pain

(a) Low back pain is usually the presenting symptom and may be said to be a pathognomonic feature of the Irukandji syndrome. It is a severe 'boring' pain in the sacral area.

(b) Muscle pain or 'cramps' extend rapidly into all four limbs, then abdominal and chest wall muscles. (In the past this chest pain has been mistaken for heart attacks, and the abdominal pain for an intra-abdominal emergency (SOUTHCOTT 1959). The pain is described as severe, unbearable and coming on in 'waves' (similar to labour pains), although never fading completely.

(c) Chest pain or 'tightness' probably results from spasm of the intercostal muscles. So far, there has been no bronchospasm described with the Irukandji syndrome. A myocardial pain component has not yet been demonstrated, but may indeed present differential diagnostic difficulties (FENNER *et al* 1988).
(d) Severe headache (see below).

2. Catecholamine-like effects

Many of the signs and symptoms associated with the Irukandji syndrome resemble those of an adrenal medullary or catechdamine excess (e.g. as in phaeochromocytoma), a scorpion sting (FREIRE-MAIA & CAMPOS 1987), or even a funnel-web spider envenomation (SUTHERLAND 1983), all of which envenomations result in the excessive release of catecholamines into the bloodstream. This cause has not been established directly in *Carukia* envenomations, as captured venom has not been available in sufficient quantity to test adequately. A single, uncontrolled laboratory experiment using mice injected with trace amounts of Irukandji venom detected systemic hypertension and raised serum catecholamines associated with many of the characteristic signs (BURNETT 1989, unpublished observations). However, no biochemical evidence of elevated blood and urinary catecholamines has been consistently found in Australian cases of Irukandji syndrome to date (CURRIE 1993, unpublished observations).

(a) Sweating. This may be localised or generalised. In severe envenomation the sweating may be drenching. If localised it may be in the area of the envenomation, or it may affect a totally unrelated area of the body Plate 9.25).
(b) Pilo-erection (skin hairs 'standing on end'). This, too, may be generalised or localised, and related or unrelated to the sting area.
(c) Anxiety and 'wretchedness'. Victims are excessively anxious, much more so than would normally be expected after any other jellyfish sting. They look pallid, even grey, and feel absolutely dreadful with a feeling of 'impending doom' (a thought that may be shared by the attending first-aider!).
(d) Restlessness. Victims are restless and may move continuously, trying unsuccessfully to get comfortable. This is part of the central effects of the syndrome but is aggravated by the severe muscle pains.
(e) Headache. This is frontal or global and often of incapacitating severity ('the granddaddy of all headaches').
(f) Nausea, occasionally with intractable vomiting.
(g) Increased respiratory rate. Respiration is often 'sighing'.
(h) Tremor. A fine tremor, or fasciculation of the small muscles of the limbs may be present.
(i) Pallor, or peripheral cyanosis may occur, associated with an intense peripheral vasoconstriction.
(j) Oliguria. A reduced urine output may occur from a combination of hypovolaemia (sweating/vomiting) and vasoconstriction presumably affecting the renal vasculature.
(k) Tachycardia. The heart rate is often fast and may be irregular with ventricular extra-systoles.
(l) Hypertension. The systemic blood pressure may reach alarming levels in previously normotensive individuals, e.g. 280/150 mm Hg (FENNER *et al* 1986).
(m) Cerebral oedema. Cerebral oedema during an Irukandji syndrome has uncommonly-been observed (HEAZLEWOOD 1992, unpublished observations) (Case History 5).

CASE HISTORY 5

A 7-year-old male was stung in northern Australia by an Irukandji at 11 am on 20 December 1981. By 9 pm he was found to be confused, disorientated, tachycardic and tachypnoeic. He had several episodes of profound sweating, cyanosis and agitation. Chest auscultation revealed bilateral course crepitations, more marked on the left. Chest X-ray demonstrated increased markings consistent with pulmonary oedema. He had 4+ glycosuria and dextrostix of 130 mg/100 ml (N.R. 80–120 mg/100 ml). He had periodic verbal and visual hallucinations. On physical examination he was pale and in obvious respiratory distress needing 6 litres of oxygen/min via face mask to maintain his central colour. Examination of his fundi showed blurring of the disc margins.

Initial management consisted of intravenous dexamethasone and intermittent frusemide (Lasix). His symptoms persisted for the first 24 hours of admission, and on the second evening he became more disorientated and unresponsive to spoken words. He was given 100 ml of 20% mannitol intravenously. Two hours later he had a huge diuresis with a wet bed; three hours later was answering questions with a grunt, and four hours later was obeying commands and was easily roused. Ten hours later he was talking freely and answering questions and subsequently made a complete recovery with no neurological Fsequelae.

3. Cardio-pulmonary decompensation

Acute pulmonary oedema: Sudden breathlessness associated with acute pulmonary oedema may occur in victims some 15–18 hours post-envenomation (FENNER *et al* 1988; MARTIN & AUDLEY 1990), although it has occurred within two hours of some recent cases (see Case Histories 6–8, below). Such patients can (understandably) be misdiagnosed as acute myocardial infarction with ensuing pulmonary oedema.

FENNER and colleagues (1988) demonstrated (toxic) global cardiac dilatation, but could not exclude a pulmonary membranous toxicity with 'leaky lungs', whereas MARTIN and AUDLEY (1990) considered the oedema in their patient to be due to massive alpha-adrenergic stimulation.

Cardiac catheterisation to measure right atrial, pulmonary artery and pulmonary wedge pressures, together with comparison of pulmonary oedema fluid and plasma should provide further information. Echocardiography and radionuclear techniques investigating myocardial function may also assist diagnosis and thus clinical management.

CASE HISTORIES 6–8

Many of the severe acute symptoms described above developed in three cases (in March 1994). However, the 3 cases all developed varied clinical problems. The editors are indebted to Dr Ian Carney of Mackay for permission to use these as-yet unpublished case studies.

- A 32-year-old male, developed mild hypertension (maximum reading 220/116 mm Hg), which responded well to IV phentolamine (see below), severe pain which was controlled by morphine infusions, but with no other cardio-respiratory problems. An echocardiogram on the third post-envenomation day was normal.
- A 25-year-old female, had mild hypertension, which settled quickly with no medication, but remained tachycardic and then developed acute pulmonary oedema with an SpO_2 of 88% on air some nine hours after the original envenomation. Chest X-ray showed interstitial oedema with a normal heart size. The oedema failed to resolve with intravenous frusemide, morphine and a glyceryl trinitrate (GTN) infusion and a CPAP (continuous positive airway pressure) mask with PEEP of 5 cm and 100% oxygen was necessary. Cardiac enzymes at this time showed a CK of 596, CK:MB 74 (NR <8) and a ration of 12.4 (NR <1.6), AST 80, LDH 252 and a WCC of 19.1. Twenty-seven hours after envenomation, chest pain with ST depression in leads II, III and aVf occurred. The GTN drip which had been discontinued nine hours before was recommenced. A further nine hours later hypotension of 88/50 mm Hg persisted, and a dopamine infusion was started. The next day the CK had dropped to 305, the CK:MB to 20, the ratio to 6.6. An echocardiogram on the

third day showed a mildly dilated left ventricle with global hypokinesis, LVD 53 mm (NR 35–56), LVS 45 mm (NR 20–40), fractional shortening 15% (NR 25–45), dilated left atrium 27 cm^2 (NR <16), trivial mitral regurgitation and E point-septal separation 12 mm (NR <8 mm). Enalapril 2.5 mg was commenced and the patient discharged the next day. A German-speaking Swiss, she was unfortunately lost to follow-up.

- A 25-year-old female, was mildly hypertensive (110/110 mm Hg), tachycardic 120 bpm and was shaking severely; 80 mg propranolol, (a beta-blocker) was given, which resulted in a relative bradycardia of 80, and severe peripheral shutdown with femoral and carotid pulses only, and a BP of 60/25 mm Hg for which 500 ml of haemaccel (plasma expander) was given with moderate effect (BP 63/47 mm Hg). This hypotension caused impaired renal function with a serum potassium of 9.4 (NR <5.5), needing intravenous glucose and insulin, together with dopamine infusions (the patient later developed a BSL of just 1.8!). As soon as this patient's blood pressure returned to normal levels (120/70 mm Hg) the victim developed severe pulmonary oedema, needing IV frusemide, morphine and a GTN infusion. At this time the CK was 392, CK:MB 21 and the ratio 5.4. Both patients with life-threatening pulmonary oedema needed nitrate infusions and assisted ventilation via a CPAP mask for two days. An echocardiogram on the third day showed a mildly dilated left ventricle with mild global hypokinesis, LVD 54 mm (NR 35–56), LVS 41 mm (NR 20–40), fractional shortening 24% (NR 25–45), trivial mitral regurgitation and E point-septal separation 15 mm (NR <8mm). Enalapril 2.5 mg was commenced, and the patient discharged the next day. Also a German-speaking Swiss, she too was lost to follow-up.

Scuba diving and the Irukandji syndrome

The Irukandji syndrome occurring in divers is well known. As it resembles decompression illness, it may present a difficult and important differential diagnostic problem (WILLIAMSON 1985) {see Chapter 22, page 425}.

FIRST AID

{see also Chapter 5, page 106}

1. Immediately send others to tell remaining swimmers to clear the water, and call for medical aid.
2. Remain with the victim, give reassurance and encourage rest. Muscle activity increases the cardiac output and 'muscle-pump' absorption, with consequent increased systemic distribution of the venom.
3. Douse the area with vinegar, if a sting site is evident. Although, in previous cases, there have been no visible adherent tentacles, skin scrapings of the area have shown (as with all jellyfish stings {see Chapter 5, page 100}), there are still remaining unfired nematocysts (FENNER *et al* 1986b) {see Plate 5.2}. Note that vinegar prevents further discharge of nematocysts, but it does not stop pain or inactivate the injected venom (WILLIAMSON 1985).
4. Apply a vinegar-soaked pad, and bandage the area if possible with a compression bandage.

   Note: Compression immobilisation bandaging {Chapter 5, page 110} for chirodropid and Irukandji syndrome jellyfish stings currently lacks controlled experimental evaluation. The concept is an extrapolation from the known efficacy of the compression/immobilisation technique established by SUTHERLAND *et al* (1979) for the management of snake envenomation. In the small number of cases of Irukandji stings tried to date, there is little reported benefit (Dr J. HOLMES 1993, Cairns Base Hospital, Queensland, unpublished observations). Further studies are justified, particularly in view of the 'window of opportunity' that may be presented by the characteristic delay prior to onset of symptoms, following Irukandji stings. Additionally, recent simulation studies by HOWARTH and colleagues (1994) suggest that compression/immobilisation bandaging for any envenomation first aid should neither be too loose, nor too tight, to effectively restrict limb lymphatic flow of venom.

5. Transport the victim by ambulance to hospital. The ambulance may carry 'Entonox', a

mixture of 50% oxygen and 50% nitrous oxide, or 'Penthrane'; these agents may give some pain relief, but should be used only in conscious patients. More aggressive treatment should take place in hospital.

Note: Entonox should never be administered to a scuba or surface supply ('hookah') diver in pain after a dive. There is a risk of worsening or unmasking decompression illness (ACOTT & GORMAN 1992) {see Chapter 22, page 425}.

MEDICAL MANAGEMENT

There is no antivenom presently available. *Chironex fleckeri* antivenom has proven of no value for the Irukandji syndrome (FENNER *et al* 1986). Treatment is symptomatic, but effective when early and vigorous (FENNER *et al* 1986b, 1988):

1. Pain relief

(a) Pethidine (meperidine) 1 mg/kg, to a maximum total dose of 50 mg, as an immediate intravenous (IV) bolus dose and repeated when necessary, or as an infusion at the rate of 30 mg per hour in adults. BARNES (1964) expressed preference for pethidine (meperidine) as an analgesic for his Irukandji syndrome patients (KINSEY 1988).

Note: As respiratory depression, or even respiratory arrest may occur in susceptible patients (OH 1990) with the above regimes, they should only be carried out where respiratory resuscitation skills and facilities are immediately available.

(b) Morphine 0.1 mg/kg, to a maximum dose of 5 mg, as an IV dose, followed by 2.5 mg increments IV as necessary every five minutes, or as an infusion rate of 1–2mg per hour after an initial IV dose of 5–10 mg.

(c) BARNES (1964) advocated aspirin 600 mg orally every four hours for adults as the most effective analgesic, following initial control of the severe pain.

2. Reversal of catecholamine-like effects

For control of significant systemic hypertension, in the presence of adequate monitoring, phentolamine (a short-acting alpha-adrenergic receptor blocking drug) may be given intravenously (IV) as a loading dose of 5 mg, followed by 5–10 mg titrations IV. This may be carefully repeated if necessary until there is satisfactory control of elevated blood pressure (FENNER *et al* 1986b, 1988; MARTIN & AUDLEY 1990). Concomitant sweating, anxiety and tremor may also respond (FENNER *et al* 1986), even with phentolamine doses of 1–2 mg IV (MARTIN & AUDLEY 1990), but probably should not be regarded as indications for alpha-adrenergic blockade therapy in the absence of systemic hypertension {see below}. The use of longer lasting alpha-adrenergic-receptor-blocking hypotensive agents (e.g. phenoxybenzamine) requires continuous monitoring and an appropriate clinical environment and skills.

3. Treatment of pulmonary oedema

(a) Monitor haemoglobin oxygen saturation with pulse oximetry. Pulmonary oedema is treated with oxygen, and by diuretics, opiates, and nitrates by the intravenous route. Endotracheal intubation and intermittent positive-pressure ventilation may be necessary for developing or continuing arterial hypoxaemia (FENNER *et al* 1988). More recently continuous positive airway pressure (CPAP) masks with 5–10 cm positive end-expiratory pressure (PEEP) have proved effective (CARNEY & FENNER 1994, unpublished observations).

(b) Pulmonary oedema that fails to respond to the above therapy may require pharmacological inotropic support conducted in a critical care situation (FENNER *et al* 1986b, 1988; MARTIN & AUDLEY 1990) with controllable IV devices and continuous intra-arterial monitoring. For example,
(i) for a normotensive patient, dopamine in an IV infusion dose range 100–1500 µg/min (specific doses require careful individual titration);
(ii) for a patient with persisting systemic hypertension, glyceryl trinitrate (GTN) IV infusion (requires special polyethylene administration sets), in a dose range 5–100 µg/min (specific doses require careful individual titration).

Other medications have been suggested. Nifedipine has been used very successfully with the similar syndrome associated with scorpion envenomation (GUERON *et al* 1993). Others have been suggested, but their application for human cubozoan envenomations remain unproven in practice. They include:

- Calcium-channel blocking agents such as verapamil, which have enhanced the anti-lethal protection of *Chironex fleckeri* antivenom in experimental animals (BURNETT & CALTON 1983; BURNETT *et al* 1985; BURNETT *et al* 1990) (see also Chapter 9, page 281).
- Chlorpromazine and haloperidol as catecholamine antagonists, anxiolytics and analgesic potentiaters.
- Beta-adrenergic blocking agents, e.g. propranolol. These are not recommended in case of possible potentiation of the negative inotropic effects of Irukandji venom (see Case History 8 above).
- Clinical experience with the Irukandji syndrome has shown that vigorous symptomatic management under safely monitored medical conditions will not only provide acute amelioration of the symptoms, but may shorten the untreated course of the acute illness from 48 hours or more to some 12–18 hours. The protracted recovery from the envenomation in untreated patients can extend over 5–7 days (extending even into weeks in a minority of victims), and is characterised by muscle soreness, weakness and debilitation; acute management, as outlined above, may also shorten this recovery period by many days, and diminish its severity (WILLIAMSON 1985; 1989, unpublished observations).

It is emphasised that phentolamine-like drugs should not be used for the Irukandji syndrome in the absence of systemic hypertension, and then only in the presence of appropriate clinical skills, and blood pressure monitoring and resuscitation facilities. Such facilities should include intra-arterial pressure monitoring capability if required. This advice is based on the possible unexpected hypotension that may develop in susceptible patients exposed to hypotensive agents.

There is no evidence that repeated exposures to Irukandji stinging reduces the severity of the syndrome. However, the syndrome does present with widely variable clinical severity (BARNES 1964 & in KINSEY 1988). There is a clinical impression that syndrome severity may be 'venom-dose-dependent', in that larger or multiple Irukandji jellyfish stingings appear to result in more serious effects (FENNER *et al* 1988; WILLIAMSON 1989, unpublished observations).

CARYBDEA RASTONI HAACKE, 1886 (THE 'JIMBLE')

DISTRIBUTION

Carybdea rastoni (Plates 9.26 to 9.29) is found in most waters surrounding Australia but appears to be particularly common in the coastal waters of South Australia (where the animal was first described [HAACKE 1887; CLELAND & SOUTHCOTT 1965]), from Perth beaches southwards in West Australia (FENNER & WILLIAMSON 1987; INGRAM *et al* 1992), and in Broome waters (APPLEBY 1994, unpublished observations).

EDMONDS (1994, unpublished observations) states that in some summer months (usually December and January) small carybdeids (Plate 9.32) can be seen swimming in vast numbers out of Sydney Harbour; interestingly he has never observed them swimming into the harbour, away from the ocean. He believes he has observed this phenomenon over the past ten years or so since 1983, and that the specimens have increased in size during that time. Animals observed recently, with bell diameter of some 5 cm, are almost twice the size of those in his earlier sightings. To date, there appear to have been no reports of sightings, or stings, from surf lifesavers patrolling the Sydney beaches along the coastlines north and south of the heads of Sydney Harbour — although jellyfish specimens are rarely caught, and almost every sting is blamed on the 'Bluebottle' whether specimens are seen or not. Consequently some Sydney beach stings could be caused by carybdeids (I. MACKIE 1994, unpublished observations); Mackie's observations seem to tally with those of Edmonds. Morbakka-like specimens have been reported and identified in the Sydney area (RIFKIN 1993 and FENNER 1994, unpublished observations). It is possible that some of these larger carybdeids provisionally identified as *C. rastoni* (EDMONDS

1994, unpublished observations) may actually be Morbakka-type carybdeids (see page 148).

Carybdea rastoni has not yet been found in the Northern Territory. At certain times of the year, usually the warm summer months from December to the end of March, large numbers of stings can be troublesome, although not dangerous, in parts of Australia (CLELAND & SOUTHCOTT 1965).

On Perth beaches they may be encountered in large numbers swimming very close to the surface. They were more common in the mornings during 1987, but later in the afternoon in 1991; they caused many minor, but irritating stings during the National Life Saving Championships at Trigg Island, Western Australia, in March 1987 and 1991 (FENNER 1989, unpublished observations). Experienced pearl divers in Broome have long recognised the presence of *C.rastoni* (Jimble) stings in those waters (APPLEBY 1994, unpublished observations). Of interest is the observation by these divers of the increasing severity of the stings (producing a 'mini-Irukandji syndrome') with summer water temperatures in that region.

Carybdea rastoni are occasionally seen in large numbers by scuba divers, just offshore on South Australian beaches, on sandy bottoms at depths of about 5–8 metres. South Australian sea swimmers repeatedly report numerous summertime stings from multitudes of *C. rastoni* at dusk. The animals occur in surface swarms and are easily captured (SOUTH AUSTRALIAN STATE CENTRE, SURF LIFE SAVING AUSTRALIA, and WILLIAMSON 1991, unpublished observations).

International occurences

Carybdea rastoni have been reported from Japan (KISHINOUYE 1910), the Philippines (MAYER 1910), Malaya (Malaysia) (STIASNY 1937b; BURNETT 1990, unpublished observations), and the Marquesas Islands in the South Pacific (KRAMP 1961).

APPEARANCE

A small ('matchbox'-sized [INGRAM *et al* 1992]), transparent, box-shaped jellyfish, which is often very difficult to see under natural conditions. The bell is up to 3.5–5 cm high, 3 cm wide and has a single tentacle arising from each pedalium (Plates 9.26 to 9.29). The pedalium has a characteristic 'hook' shape in the central canal, which runs from the bell into the tentacle {see Chapter 6, page 148}.

The tentacles, gonads and gastric cirri (often clearly visible through the top of the transparent bell, clustered in each corner of the stomach) are usually distinct pinkish-brown, even after preservation in formalin. The tentacles (which may also be a yellowish colour) can be 6–8 cm long.

Morphological differences from *Carukia barnesi* (KRAMP 1961; SOUTHCOTT 1967)

It may be difficult for the untrained eye to differentiate *Carybdea rastoni* from *Carukia barnesi* (Table 9.7) {compare Plates 9.2 and 9.18 with Plates 9.27 and 9.28 however, the latter have only been found in tropical waters. Confusion exists in the literature regarding the morphology of these species. Nematocyst warts on the bell are not mentioned by MAYER (1910); KRAMP (1961) states they are present; SOUTHCOTT (1967) does not mention them when he differentiates *Carybdea rastoni* and *Carukia barnesi*. In the personal experience of one of the Editors (PJF), there were no warts on the bell of specimens studied in both Western Australia and South Australia (Plate 9.26), but a specimen of *Carybdea rastoni* photographed in South Australia by Karen Gowlett-Holmes shows obvious warty mamillations on the bell (Plate 9.27).

TABLE 9.7
FEATURES OF DIFFERENCE OF *CARYBDEA RASTONI* FROM *CARUKIA BARNESI*

1. Squatter, more cuboid bell shape
2. Pink gonads
3. 'Hook' in the pedalial canal
4. Obvious gastric cirri in corners of stomach {see Chapter 6, page 124} and Figure 6.6d, page 126
5. Causes local skin pain; no systemic symptoms
6. *Carybdea rastoni* is usually larger by some 50 per cent with bell diameter up to 4.5 cm

THE STING

The sting can vary in severity, but is usually minor, causing initial local pain or slight skin itching, which may not be noticed by the victim until leaving the water. Characteristically the skin sting mark consists of small, raised white weals surrounded by a bright red, erythematous flare (Plate 9.30). However, some stings may cause instant, severe local pain causing skin burning for several hours (Plate 9.31A). The centre of the wheal may sometimes become dark, necrotic and take several weeks to heal, leaving scars for up to several years (Plate 9.31B).

Few severe systemic reactions have been reliably documented to date (see above under 'Distribution', page 255), although allergy has occurred after recurrent, separate stings (INTERNATIONAL CONSORTIUM for JELLYFISH STINGS 1993). In the patient in Plate 9.31A large skin welts occurred, accompanied by erythema, nausea, lassitude and joint pains, which became progressively worse with successive stings. Serum showed characteristic raised immunoglobulin (Ig)G titres to *Carybdea rastoni* (BURNETT *et al* 1988) {see Chapter 7, page 174}.

First aid

{See Chapter 5, page 106}

1. Vinegar inhibits *Carybdea rastoni* nematocyst discharge (FENNER *et al* 1987) and should be used to douse the sting area, especially if the sting is severe.
2. Apply a cold pack {Plates 5.16 to 5.18}, or ice wrapped in a cloth, to the stung area for 5–10 minutes then reassess, and reapply the cold pack if necessary.
3. If skin pain persists, seek medical advice.

Medical management

Usually this is unnecessary. Systemic antihistamines and/or steroids will be required for any uncommon allergic response. Ultra-potent cortisone creams may help itchy, non-infected skin reactions. Manage severe reactions as for Irukandji syndrome.

Carybdea sivickisi Stiasny, 1926

The smallest cubozoan recorded to date (HARTWICK 1991a) was first described by STIASNY (1922 & 1926a). Only recently has knowledge of this small carybdeid been expanded (HOVERD 1985; HARTWICK 1991a).

Distribution

Recorded in the Philippines and Thailand (STIASNY 1922, 1926a & 1926b), in Japan (UCHIDA 1970) [mistakenly identified as *Tamoya alata* UCHIDA 1929], New Zealand (HOVERD 1985), and more recently in Australia (HARTWICK 1991a).

Appearance

(Figure 9.6) A small jellyfish with a 10–12mm bell diameter. The bell is covered with numerous 'warty mammilations' full of nematocysts. The four rhopalar niches (containing the sense organs) do not have the usual covering scales or 'lips' seen in others species of *Carybdea* (STIASNY 1926b) {see Figure 9.3, page 239}.

The four tentacles may be pale yellow to orange, and have rings of nematocysts, which may be separated by up to 1 mm in the relaxed state. The tips of the tentacles may have a bulb-like swelling (HOVERD 1985). The gonads and gastric region often have an orange appearance due to pigmentation (HARTWICK 1991a).

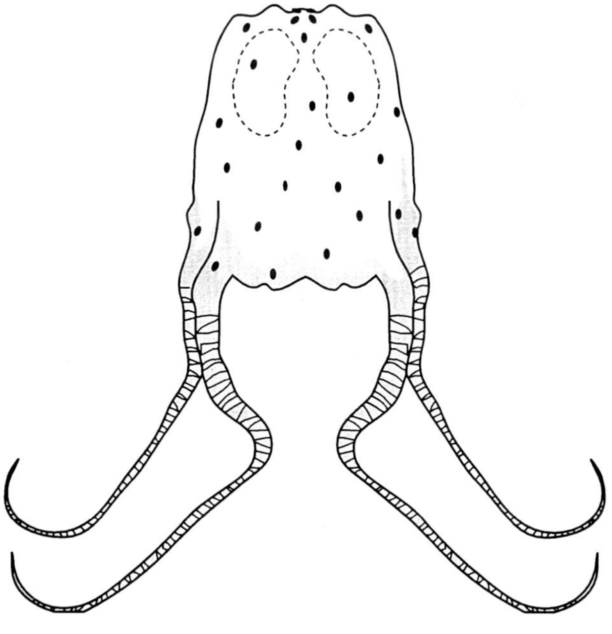

FIGURE 9.6
Carybdea sivickisi. (Drawn by Ronald Grendon, Ronald Grendon Design Services, Mackay, Queensland.)

HARTWICK (1991a) first reported two additional unusual features of *Carybdea sivickisi*:

- The production of an embryo mass, which then develops into hundreds of planulae; these later develop into polyps.
- Four areas on the bell, previously described, but not explained (STIASNY 1926b, HOVERD 1985), were actually adhesive pads. They contained nematocysts (oval microbasic eurytele) {see Figure 6.43, page 169} whose discharged tubules actually stuck *Carybdea sivickisi* firmly to any surface. Although this adhesion could not be easily broken by shaking, the animal could free itself at will.*

THE STING

Immediately painful skin stings, which often develop erythema and blistering that may last for hours.

- Multiple stings have not caused systemic symptoms to date.
- Stings have occurred on divers at night; the medusae appeared to be attracted towards the diver's light (HARTWICK 1991a).

FIRST AID

First aid measures for this cubozoan have not been previously described in the literature. Cold packs or ice should relieve the skin pain of the envenomation (see Chapter 5, page 111).

TRIPEDALIA CONANT, 1897

The genus *Tripedalia* is one of the four known genera that comprise the order *Carybdeidae* {see Tables 9.1 and 9.2 and Figure 6.25, page 148 and Plate 10.21}. *Tripedalia* spp. appear to be of minor, if somewhat uncertain, medical significance.

DISTRIBUTION

Tripedalia cystophora CONANT, 1897 occurs in the Caribbean Sea (Jamaica [CONANT 1897], Puerto Rico, Bahamas and the Philippines (STIASNY 1926b). The relatively recently described *Tripedalia binata* MOORE, 1988, has only been caught in the Northern Territory, Australia. MOORE (1988) has suggested a redefinition of the genus *Tripedalia* as a result of his original description of *Tripedalia binata*.

APPEARANCE

The cuboid bell is up to 10 mm tall and 14 mm in width, and has warty mammilations containing nematocysts. Depending on whether the species considered is *Tripedalia binata* or *Tripedalia cystophora*, there are either two or three pedalia present at each interradial corner of the bell. Each pedalium bears only a single tentacle, a feature diagnostic of all carybdeids. Nematocysts of *Tripedalia binata* are stenoteles and atrichous isorhizas in a ratio of about 10 to 1 (MOORE 1988) {see Figure 6.43, pages 168–173}.

THE STING

Tripedalia binata did not sting the inner aspect of a human wrist (ALDERSLADE 1988, unpublished observations), and *Tripedalia cystophora* caused no discomfort during collection, although *T. cystophora* is capable of capturing and digesting 'disproportionately large' fish (CONANT 1898).

Little useful documentation could be found about human stings from *Tripedalia* spp.

CHIRODROPIDS (MULTI-TENTACLED BOX JELLYFISH)

Chirodropids (the 'common' term for members of the order Chirodropidae, in the class Cubozoa) are the most venomous jellyfish known, and they rank among the most venomous creatures on earth (Plates 9.34–9.37). Currently three firm genera are recognised:

Chiropsalmus L. AGASSIZ, 1862
Chironex SOUTHCOTT, 1956
Chirodropus HAECKEL, 1880

A fourth, *Chiropsoides* SOUTHCOTT, 1956 = *Chiropsalmus* L. AGASSIZ, 1862 was suggested {see below, page 260} and has been adopted by the authors.

* Compare these comments with those relating to *Chironex* nematocyst discharge on page 271.

The medical significance of members of the genus *Chirodropus* and genus *Chiropsoides* remains to be determined.

WORLD DISTRIBUTION OF CHIRODROPIDS

JOSEPH BURNETT, BART CURRIE, PETER FENNER, JACQUIE RIFKIN, JOHN WILLIAMSON

The chirodropids will be considered in descending order of their medical significance, as this is presently understood:

1. Genus *Chironex*
   C. fleckeri
2. Genus *Chiropsalmus*
   C. quadrigatus
   C. quadrumanus
3. Genus *Chiropsoides*
   C. buitendijki
4. Genus *Chirodropus*
   C. gorilla

Chirodropid jellyfish inhabit warm tropical oceans (Plate 9.37) and have been caught near most of the large landmasses in such waters, (including close offshore islands) (KRAMP 1961; HARTWICK 1991) {see Tables 9.1 and 9.8}.

GENUS *CHIRONEX* SOUTHCOTT, 1956

Australia (Plates 9.1 and 9.3)

These jellyfish occur in predominantly tropical, north Australian waters extending from Roebourne (18° south latitude on the West Australian coast, near Dampier) in Western Australia, around the Northern Territory coastline (and offshore Australian islands) and down to 24° south latitude on the east Australian coast (Bustard Bay in Queensland, 5 kilometres north of Agnes Water, 90 kilometres due south of the Tropic of Capricorn). They are not open-ocean animals and are not encountered great distances from the shoreline (HARTWICK 1987, 1991b).

Remainder of the tropical Indo-West Pacific region

The full distribution is not yet known, but specimens resembling *Chironex fleckeri* have been identified from the following Indo-Pacific regions: Brunei (RIFKIN 1993, unpublished observations); Philippines (RIFKIN 1993, unpublished observations, from a specimen caught by MAYER [1910] and identified by RIFKIN 1995); and Vietnam (RIFKIN 1993, unpublished observations).

GENUS *CHIROPSALMUS* AGASSIZ, 1862

AGASSIZ (1862) erected the genus *Chiropsalmus*, its characteristics being the presence of side-wall (leaf-like) gonads and smooth finger-like 'gastric diverticula' projecting into the bell cavity. Descriptions and references in the literature may include several species {see pages 149–150}.

Australia

e.g. *Chiropsalmus quadrigatus* (up to the present known as *C. quadrigatus* HAECKEL, 1880 *sensu* BARNES 1965) {see Chapter 6, pages 150–154 for a discussion of the taxonomic difficulties} (Plate 1.10 and Figure 6.29, page 151)

Hartwick (1990, unpublished observations) suggested that in Australia *Chiropsalmus* sp. did not occur in creeks and rivers, although it is only encountered close to shore. This agrees with the observations made by ARAKI, TOMIHARA and KOHAMA (1994, unpublished observations) on the *Chiropsalmus quadrigatus* caught in the Okinawa region of Japan.* Barnes considered this animal's distribution in Australia to be much more limited than that of *Chironex fleckeri*, i.e. north Queensland only, in which area he believed 'mixed populations' of *Chironex fleckeri* and *Chiropsalmus* sp. Australia occurred 'not infrequently' but 'only under fairly specific conditions' (BARNES 1966); however, he predicted that this distribution was likely to prove to be much wider with further studies (BARNES 1966). Until recently, reliably identified specimens have been caught only in the Cairns to Townsville area. However, in June 1991 specimens were caught in Gove, Arnhem Land, in the Northern Territory, and preliminarily identified

* The editors are aware of Dr David Cargo's suspicion that the juveniles of these medusae may be found up in the creeks.

as *Chiropsalmus* sp. Australia (CURRIE 1992, 1993). This was curious, as June is dry season there, and chirodropids are generally regarded as occurring predominantly in the wet season months (HARTWICK 1987). Subsequently, numerous specimens of these *C. quadrigatus* have been captured in Gove each dry season (winter), but none has yet been identified in the wet season (summer) (CURRIE, ALDERSLADE, SCHRAVEN 1993, unpublished observations).

Remainder of the tropical Indo-Pacific

CHIROPSALMUS QUADRIGATUS HAECKEL, 1880

{See page 153 for a discussion of the taxonomic difficulties} (Figure 6.29, page 151 and Plates 9.35 and 9.38)

The first description of this species was of an immature specimen captured in Rangoon, Burma (HAECKEL 1880), which was inadequately described (see page 150). Later descriptions in the literature are probably of many different cubozoans, but reported as '*Chiropsalmus quadrigatus*' in the following locations: the Philippines (MAYER 1910 and LIGHT 1914), and more recently by FENNER (1987) HEEGER (1992) and CORNELIUS (1994, unpublished observations); the Maldive Islands in the Indian Ocean (STIASNY 1937); east Papua New Guinea (THIEL 1928); Malaysia (SEARLE 1957); the Gulf of Thailand (Indo-China) (KRAMP 1961); the southern tip of India (MENON 1936); and more recently, in the islands of southern Japan (Okinawa 26°N (3° north of the Tropic of Cancer), and Amami Islands, 28°N (5°N of the Tropic of Cancer (Plates 9.38A and 9.38B). Many stings occur each summer in these Japanese regions (SHOKITA 1986) (Plate 9.39 and Table 9.8, page 261). Their in-shore habitat is becoming understood (FENNER 1987; TOMIHARA et al 1994, unpublished observations) {see Chapter 10, page 292}, but in contrast to *Chironex fleckeri*, is not believed to be mangrove swamps (SHIMOJI 1994, unpublished observations).

NOTE: An animal originally named *Chiropsalmus zygonema* HAECKEL 1880 is believed in fact to be '*Chiropsalmus quadrigatus*' (THIEL 1928).

CHIROPSALMUS QUADRUMANUS (MÜLLER, 1859):

MÜLLER (1859) originally created the Atlantic chirodropid species *Tamoya quadrumanus*. Soon after, L. AGASSIZ (1862) erected the genus *Chiropsalmus*, and *Tamoya* was transferred to become *C. quadrumanus* (MÜLLER, 1859). The species was first reported in Brazil and from French Guiana northwards to Venezuela (AGASSIZ 1862). MAYER (1910) reported it in North Carolina, USA, and it has been reported in the Gulf of Mexico (GUEST 1959; PHILLIPS & BURKE 1970), in Puerto Rico (B. CUTRESS 1992, unpublished observations) (Plate 9.36) and on the coastline of South Carolina (CALDER & PETERS 1975). Its distribution thus appears to be right along the eastern coast of the Americas between the two Tropics (i.e. tropical waters).

GENUS *CHIROPSOIDES*

Chiropsoides buitendijki (HORST, 1907) (Plates 9.9A and B and 9.60) = *Chiropsalmus buitendijki* HORST, 1907

In describing the chirodropid *Chiropsalmus buitendijki* from Java, Horst stated that the tentacles were arranged in a linear series on the pedalium — a totally different pattern from other chirodropids (Plate 9.61). Because of this unique feature, both KRUMBACH (1925) and SOUTHCOTT (1956) believed that this species should be placed in a separate genus. Krumbach's chosen name *Drepanochirus* was unacceptable as it was previously catalogued for a beetle; Southcott proposed the new genus *Chiropsoides* (SOUTHCOTT 1956), although so far this has gained little recognition. In this book the editors have chosen to use the new name {see Chapter 6, page 150}.

Chiropsoides buitendijki was first reported in the former Java Sea (HORST 1907) and subsequently from Malaysia (then 'Malaya') (STIASNY 1919), Burma, Indo-China (RANSON 1945), and the southern tip of India (MENON 1936; RAJAGOPAL *et al* 1989). Further specimens have recently been caught by Dr Malik Fernando in Sri Lanka, and identified by one of the editors (PJF) {see Plates 9.9A, 9.9B, 9.60 and 9.61}.

Possible misidentifications

The following sightings and identifications have been reported outside geographical areas delineated above. Distribution patterns must continuously submit to scientific scrutiny

Chiropsalmus (now *Chiropsoides*) *buitendijki* has twice been reported in Australia (STIASNY 1926; POPE 1953), but the former identification has been questioned (BARNES 1966). *Chiropsalmus quadrumanus* was reported in Sumatra (KRUMBACH 1925), Australia (STIASNY 1926) and the Indian Ocean (RAO 1931). *Chiropsalmus quadrigatus* (*sensu* MAYER 1910) was reported in Australia (MCNEILL & POPE 1943) prior to the description of *Chironex fleckeri* by SOUTHCOTT 1956) (see Rifkin, Chapter 6 of this volume, Taxonomy of chirodropids, page 150).

Barnes repeatedly stated that for practical Australian public health purposes, *Chiropsalmus buitendijki* and *Chiropsalmus quadrumanus* (*Chiropsalmus*=*Chiropsoides*) could be disregarded (BARNES 1965 & 1966).

Further field and clinical studies are required, but at present *Chironex fleckeri* {Plates 1.4, 9.3 and 9.34} and the Australian *Chiropsalmus quadrigatus* {Plate 1.10} are the only currently-recognised chirodropids in tropical Australian waters — although another recently captured Australian chirodropid is under investigation (RIFKIN and ALDERSLADE 1993, unpublished observations).

GENUS *CHIRODROPUS* HAECKEL, 1880.

This genus includes two possible species:

- *Chirodropus gorilla* (Figure 9.9, page 268), reported on the West African coastline from

TABLE 9.8 (SEE ALSO TABLE 10.3, PAGE 305)
SUMMARY OF DOCUMENTED CHIRODROPID SIGHTINGS, BY COUNTRY AND REGION

Chironex fleckeri	***Chiropsoides buitendijki***
Australia	Indonesia — Java
***Chironex*-like chirodropids**	Malayan Archipelago
Borneo (island)	India — south
Papua New Guinea	Sri Lanka
Philippines	(Australia)
Vietnam	***Chirodropus gorilla***
Brunei	Liberia
Chiropsalmus quadrigatus	South Africa — west
(Australia) ie sensu Barnes	West Africa
***Chiropsalmus*-like chirodropids**	***Chiropsalmus quadrumanus***
Burma	USA — North Carolina, Texas
India (south)	(Sumatra)
Philippines	(Australia)
Maldive Islands	(Indian Ocean)
Vietnam	French Guiana
Gulf of Thailand	Brazil
Singapore	Amazon R mouth
Papua New Guinea	Venezuela
Brunei/Sarawak	(West Africa)
Japan	

* *Chironex*-like and *Chiropsalmus*-like categories are made with reference to certain well described chirodropids in Australia (see page 150)
Countries in brackets are possible misidentifications (but where a potentially lethal chirodropid has been found).
See also Chapter 10, page 284, International information on jellyfish stings.

Liberia (just north of the Equator) to as far south as the north-west coast of South Africa. More recently this distribution of *C. gorilla* has been reconfirmed in the Benguela Current, off the south-west African coast by PAGÈS *et al* (1992) {see this chapter, page 268}.

- *Chirodropus palmatus*, reported at St Helena Island in the South Atlantic Ocean.

FATALITIES

{SEE ALSO CHAPTER 3, PAGES 65 AND 70}

CHIRONEX FLECKERI

Australia

It is very likely that *Chironex fleckeri* caused deaths among the Aboriginal population of tropical Australia long before European colonisation (MOUNTFORD 1956; CLELAND & SOUTHCOTT 1965). Since the first firm documentation of deaths (both of them children [CLELAND & SOUTHCOTT 1965; WILLIAMSON 1983]) in Townsville, in the summer of 1884–1885 there have been some 63 documented human deaths from *Chironex fleckeri* envenomation in Australia: 36 on Queensland beaches, from Gladstone, northwards (FLECKER 1952; BARNES 1960; WILLIAMSON 1985; LUMLEY *et al* 1988) and 27 in the Northern Territory (CURRIE *et al* 1992) {see Chapter 3, page 70}.

Remainder of the tropical Indo-Pacific

Although *Chironex fleckeri*-like specimens have been identified from the Indo-West Pacific region north of Australia (see above), it is not yet possible to correlate actual deaths in this region with *the particular specimens captured.*

CHIROPSALMUS QUADRIGATUS CHIRODROPIDS TO 1995

Australia

No deaths are known with certainty from the sting of the Australian *Chiropsalmus* sp. {see Chapter 3, page 70}.

Remainder of the Indo-West Pacific

Chirodropid fatalities in the Indo-West Pacific are numerous, firmly documented (WADE 1927 & 1928; CLELAND & SOUTHCOTT 1965) and probably occurred for generations prior to their first formal recording. The earliest report was by US Naval Surgeon C.P. Kindelberger at Olongapo in the Philippines, in 1907 (OLD 1908). Recently FENNER (1987) reported chirodropids as responsible for some 20–50 deaths every year in the Philippines {see below and Chapter 3, page 65} (FENNER 1987; CORNELIUS 1994, unpublished observations). One death was reported in August 1961 in the north of Okinawa Island, Japan (ARAKI 1992, unpublished observations).

CHIROPSALMUS QUADRUMANUS

A human death in Texas, USA, in June 1990 is documented (BENGSTON *et al* 1991) {see Chapter 3, page 65}.

CHIROPSOIDES BUITENDIJKI

No deaths have been reported to date (see 'Stop Press' Chapter 26, page 442).

CHIRODROPUS SPP.

No deaths are documented, but they have probably occurred. The geographical distribution of *Chirodropus* is along a coastline where records of envenomations may be difficult to access.

CHIRONEX FLECKERI SOUTHCOTT, 1956

Common names: the northern Australian box jellyfish, *Chironex* box-jellyfish, sea wasp, big stinger, sea stinger.

DISTRIBUTION

Australia

Few other jellyfish have been subjected to such intense scrutiny with regard to their regional distribution (CLELAND & SOUTHCOTT 1965; HARTWICK 1987; HARTWICK 1991), reflecting the growing reputation of this impressive creature {Plates 1.4, 5.8, 9.3 and 9.34} as 'the world's most venomous animal' (CROPP & CROPP 1984; ENDEAN 1988; FENNER 1990; CASWELL 1992) — now a media catch-phrase! This species, presently the only species in the genus *Chironex*, was originally identified, on the basis of SOUTHCOTT'S (1956) description in northern Australia, where its presence is now legendary (BARNES 1960, 1965, 1966; CLELAND & SOUTHCOTT 1965; WILLIAMSON *et al* 1980; HARTWICK *et al* 1980; CURRIE 1994).

Chironex fleckeri medusa numbers, and sting numbers, are greatest in the hot summer months. Further north, on the Australian coastline at least, the medusae are present for longer periods in the year. The 'official stinger season' in the Northern Territory is from 1 October to 30 May (Figure 9.7, this page), but reports of *Chironex* stings have occurred in every month of the year except July (INTERNATIONAL CONSORTIUM for JELLYFISH STINGS 1993). The first sting recorded in Australia in the 1993–94 summer season was on 1 August 1993 at Gunn Point, Darwin, in the Northern Territory. Fatal stings have been documented in the Northern Territory in May and September (CLELAND & SOUTHCOTT 1965; CURRIE 1992 unpublished observations).*

In far north Queensland the season extends from 1 October to early June. HARTWICK (1990, unpublished observations) stated that the first *Chironex* in Townsville (latitude 19° south) usually occurred in November/December, and they were potentially lethal by December. He also stated that the temperature of the sea water was not important.

Calm summer sea conditions with onshore breezes (Plate 9.51) and the presence of young fish and shrimp close to shore provide favourable feeding conditions for the medusae (HARTWICK 1991). The free-swimming medusae thus favour protected sandy bays, free of snags and coral, and close to shore, when they are 'patrolling' for their food with tentacles extended. Unfortunately these are the very beach conditions that attract unwary swimmers, such as tourists visiting the popular northern Australian areas. Larger *Chironex* tend to swim against the current keeping tentacles stretched while feeding; smaller ones may swim in either direction (MULCAHY 1990, unpublished observations).

Chironex fleckeri have been sighted, caught and involved in human stingings over rocky shorelines in the Darwin area (ALDERSLADE & CURRIE 1992, unpublished observations; INTERNATIONAL CONSORTIUM for JELLYFISH STINGS 1993). Summer population densities of the larger full-grown adult *Chironex fleckeri* medusae can be intense in the coastal seas around Arnhem Land in the Northern Territory. Similar summer population densities are also likely to occur along the shores of the Gulf of Carpentaria.

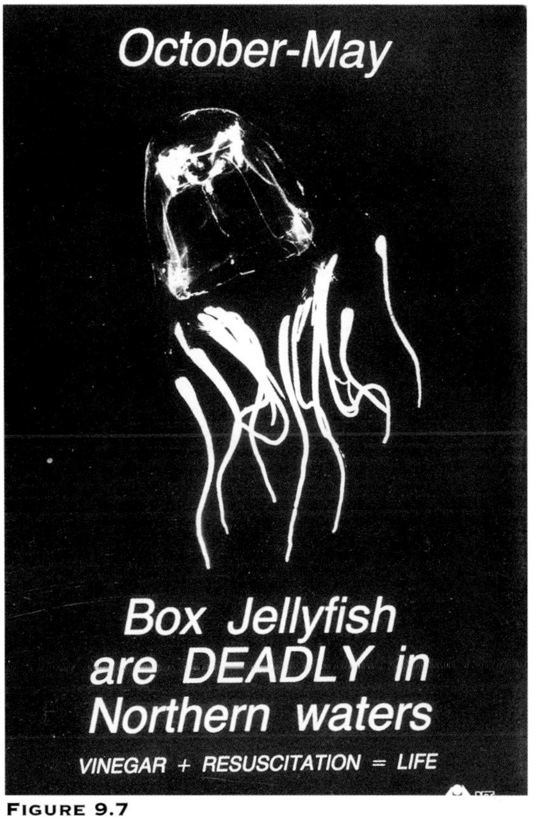

FIGURE 9.7
The Northern Territory Department of Health's striking warning poster against chirodropid stings. (Courtesy of NT Department of Health.)

Chironex are concentrated in shallow water but have been found 100–300 m out to sea, although rarely on the surface, and mainly in calm water (HARTWICK 1990, unpublished observations). However, there was a severe sting on the legs of a surf lifesaver in deep, rough water near Cairns in 1991 (FENNER 1991, unpublished observations) when a *Chironex* was washed on to the deck of the victim's surf ski while he was paddling some 200–300 m offshore. They have occasionally been trawled off the bottom by trawlers a couple of kilometres offshore (MULCAHY, 1988 unpublished observations). They can swim several kilometres in a day but this high-energy activity requires an abundance of food.

* Chirodropid fatalities throughout the Indo-West Pacific have been documented in every month of the year (CLELAND & SOUTHCOTT 1965).

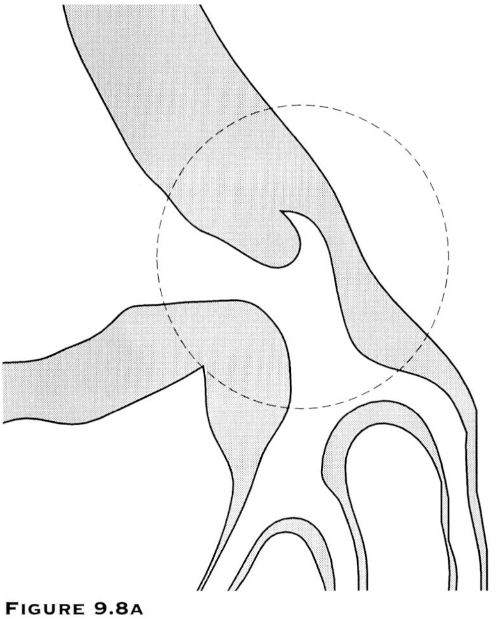

FIGURE 9.8A
'Hook-shaped' pedalial canal of *Chironex fleckeri* SOUTHCOTT, 1956. (Drawn by Ronald Grendon, Grendon Design Services, Mackay, Queensland.)

Documented human stings extend from Lagrange Bay, south-west of Broome in Western Australia (east Indian Ocean, latitude 18° south) (MARSH & SLACK-SMITH 1986), to Gladstone, Queensland, on the eastern Australian coastline (west Pacific Ocean, latitude 24° south) (LUMLEY *et al* 1988). Unconfirmed sightings extend from Exmouth in Western Australia (latitude 22° south) (MARSH 1991, unpublished observations) to Bustard Bay, south of Gladstone, Queensland, in the east (latitude 24° south) (MACKENZIE 1991, unpublished observations).

Remainder of the tropical Indo-Pacific

Chironex fleckeri is also now recorded in the other waters of the tropical Indo-Pacific (see RIFKIN, above).

LIFE CYCLE IN AUSTRALIA

BROWN (1973) was the first to suggest that *Chironex* may breed close inshore, after he caught 24 tiny *Chironex* juveniles only 6–9 mm diameter in shallow water in Horseshoe Bay, Magnetic Island, near Townsville, north Queensland, in the summer of 1972. His investigations of that island were prompted by the island residents', denial of any *Chironex* sightings during a year of particularly high prevalence of *Chironex* in north Queensland generally. His documentation (BROWN 1973) contains other valuable observations, including the then 'silent opposition' of tourist organisations

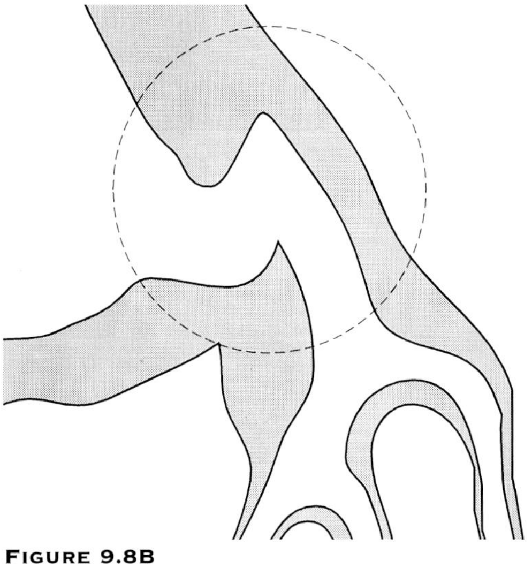

FIGURE 9.8B
'Volcano-shaped' pedalial canal of one of the chirodropids found in the Indo-Pacific region. (Drawn by Ronald Grendon, Grendon Design Services, Mackay, Queensland.)

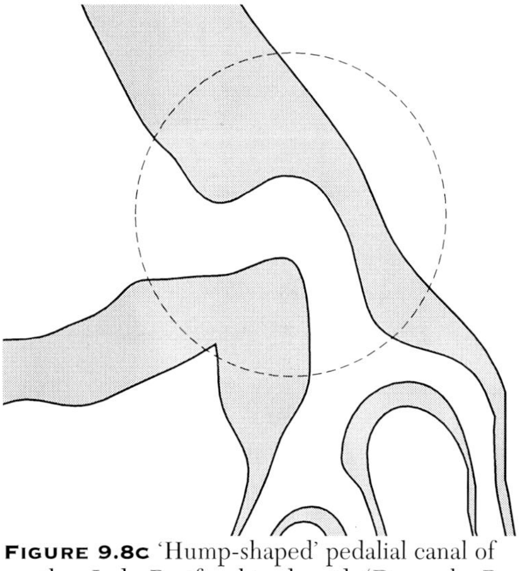

FIGURE 9.8C 'Hump-shaped' pedalial canal of another Indo-Pacific chirodropid. (Drawn by Ronald Grendon, Grendon Design Services, Mackay, Queensland.)

FIGURE 9.9
A diagramatic representation of a cluster of *Chironex fleckeri* tentacles, illustrating their 'banded' surface. Compare with Plate 9.43. (Drawn by Ronald Grendon, Grendon Design Services, Mackay, Queensland.)

to publicising the dangers of chirodropids in north Queensland waters and the need for protective clothing {see Plates 1.7, 1.8 and 1.22} against both sunburn and chirodropid stings.

Confirmation of Brown's suggestion, together with further understanding of the life cycle, has been brilliantly confirmed by Hartwick and colleagues and is reported in detail elsewhere (YAMAGUCHI & HARTWICK 1980; HARTWICK 1987, 1991a).

Although much remains to be learned and directly observed, the adult medusae at season's end probably swim up coastal estuarine saltwater creeks to habitats of the polyp forms (Plate 9.52) where the male and female adults reproduce sexually, probably by semelparous spawning (HARTWICK 1991a). However, it may be that they also reproduce asexually, as planula larvae that subsequently developed into polyps have been found live in the bell cavity of a mature *Chironex* (RIFKIN 1992, unpublished observations and Chapter 6, page 121). The surviving planulae develop into creeping ('roving') polyps that look for suitable habitats before attaching to the rocks and debris of the creek bottom, where they develop into sessile (and therefore fixed) polyps (Plate 9.52). On the way to adult maturation, these fixed polyps fall prey to various predator creatures, as a normal part of the biological food chain. Nevertheless they are surprisingly hardy and capable of asexual reproduction by budding, under favourable environmental conditions. The polyp forms require little food during the food-scarce winter period, but the growth rate of polyps (and medusae) varies with the food source and not the weather — although the food source is related to weather to some extent — thus there may be a range of sizes of jellyfish in the one area (HARTWICK 1990, unpublished observations). Larger numbers of polyps develop if the food supply (planktonic animals) is good and there is no overgrowth of other polyp species (HARTWICK 1990, unpublished observations). Eventually (typically during September in northern Queensland) the surviving polyps detach to become small free-swimming juvenile medusae (Plate 9.53), the survivors among which become the adult *Chironex* medusae (HARTWICK 1987). Contrary to common belief, this stage in the life cycle is not related to the coming of the rains, although January rains will kill off any remaining polyps in creeks.

The emergence of part-grown *Chironex* medusae in large numbers from selected saltwater estuarine creek mouths, in response to favourable environmental conditions, has been directly observed on occasions in far north Queensland (CROPP & CROPP 1984).

As with other life-cycle stages (planulae and polyps), the *Chironex* medusae fall prey to natural predators throughout their maturation. These predators include hawksbill turtles (Plate 9.54), butterfish (Plate 9.55), batfish and spinefoot fish (CROPP & CROPP 1984).

APPEARANCE

The bell is almost transparent, and often very difficult to see, even in very shallow water — as many victims will testify. This is because of its milky white colour, contrasted in larger specimens only by a deep blue, brownish, or yellow colour on the tentacle bases close to their attachment to the pedalia on the bell (BARNES 1965)

{Plates 5.8 and 9.56}. The bell size may attain the size of a human head, up to 22 cm in diameter or even more (HARTWICK 1991) {Figure 6.27 page 149} in north Queensland specimens, but it is generally smaller in the Northern Territory specimens (see this chapter).

On each pedalium are as many as 15 thick, robust, ribbon-shaped tentacles (Plate 9.43); each tentacle is capable of impressive retraction and extension to 3 metres or more in length. The tentacles are loaded with nematocyst battery rings with an estimated 4000 million nematocysts in the mature adult stage, according to Hartwick (WILLIAMSON 1985).

Among jellyfish, *Chironex fleckeri* are the élite — the most evolved. The adult animal possesses four rhopalia, the club-shaped structures in which sense organs are embedded. These lie just above the margin of the bell, equidistant between the four pedalia {see Chapter 6, page 128}. These rhopalia have both visual organs, which can differentiate light and dark, and balance organs, which determine the jellyfish position with respect to the surface and bottom of the water, and which also detect vibration, suggesting impending danger such as rough water from storms (or from humans!). Visually, a *Chironex* recognises the presence of obstacles in its path (e.g. humans) by shades of light and dark (shadows), and given time and opportunity will swim around them and avoid contact. Media statements of 'vicious attacks' by these jellyfish are absurdities, for the animal has no malice of forethought and will always avoid contact, given time and opportunity. Indeed most major stings produce extensive damage to the involved jellyfish, with avulsion of many tentacles and its subsequent death.

Chironex also possess an 'integrated' nervous system that coordinates an elegant swimming action by rhythmic and powerful pulsations of the bell, which can propel it, bell-first and tentacles trailing {see Chapter 6, page 128, jellyfish swimming behaviour}, up to the speed of a walking human (4–5 knots/h) for short bursts. In some cases the nervous system may also influence nematocyst discharge, possibly holding the nematocysts in the 'fire-ready' state or the relaxed state (RIFKIN & ENDEAN 1988; ENDEAN 1988a) {see Chapter 6, page 159}.

THE TENTACLE ARMAMENT

Chironex fleckeri has up to 60 highly retractile and extendable tentacles {Plate 9.43}, armed with millions of nematocysts {Plate 6.12}, attached to the four corners of its squarish bell {Plates 1.4 and 9.1}. The 'microbasic mastigophores' are probably those containing the lethal venom, although the other types of nematocysts also contain venom but are likely used for a different purpose (ENDEAN & RIFKIN 1975) (Plate 9.57) {see also Figure 6.43, page 168}.

CLINICAL ASPECTS OF *CHIRONEX FLECKERI* ENVENOMATION AND ITS MANAGEMENT

Clinical aspects are considered in detail later in this subsection, under 'Chirodropid stings', and sting management is presented in summary form at the end of the subsection {page 280}.

CHIRONEX STING RECORDS

It can be seen from Table 9.9 that *Chironex* stings are more common in the afternoons (although so, too, are swimmers) on still, fine days, especially in the Northern Territory. Stings can still occur on rough, windy days (>15 km/h winds) but are less common. Stings are most common (77 per cent) on the lower extremities of people walking or standing in the water, although stings may also involve other areas of the body (Plate 9.58). Stings of the neck and face are rare — fewer than 2 per cent. Few victims actually wore protective clothing.

CHIRODROPIDS NAMED *CHIROPSALMUS QUADRIGATUS*

DISTRIBUTION

These specimens have been caught all over the Indo-West Pacific from southern Japan (SHOKITA 1986) Plate 9.38 (A and B), the Philippines to the Maldive Islands in the Indian Ocean (RANSOM 1945), and Borneo and Sarawak

TABLE 9.9
***CHIRONEX* STING RECORDS**

	Number of stings	Percentage of stings reported
Total	270	
Morning	77	31%
Afternoon	174	69%
Males	177	64%
Females	90	36%
Fine weather	77	63%
Cloudy weather	46	37%
Children under 10 years of age	70	26%
Site of sting described:248		
trunk/abdomen	25	10%
legs	190	77%
arms	28	11%
head/neck	5	2%
State identified:	270	
Queensland	93	34%
Northern Territory	172	64%
Western Australia	5	2%
Wind details recorded: 164		
moderate or 15+ knots	16	10%

From the sting report database of the editors 1987–93

(FENNER *et al* 1995) {Plates 9.40A, 9.40B and 9.42}. However, the true distribution is probably even more extensive than presently appreciated (e.g. little is currently known about the Indonesian region). The identification of the jellyfish responsible for a fatal envenomation in a 10-year-old boy in Labuan, Malaysia (*Malaysian Straits Times*, 22nd July, 1992, page 1; AZILA & OTHMAN 1993) and a life-threatening envenomation of a child in Brunei (FENNER & WILLIAMSON 1996) cannot be made. However, examination of some specimens of chirodropids subsequently caught in this area have been identified as *Chironex*-like (RIFKIN 1994, unpublished observation) (Plate 9.59). Other chirodropid species are probably also present in this area.

APPEARANCE

In the simplest sense, these species have a box-shaped bell with groups of tentacles in each corner (Plate 9.59) {see also Plates 1.10 and 9.40A and B} — similar in appearance to a mature *Chironex fleckeri*. However, in view of the work reported by RIFKIN {see page 150}, it appears that investigators have been examining several different species under the one name. Morphological details are provided on page 151–3.

CLINICAL ASPECTS OF ENVENOMATION AND ITS MANAGEMENT

See page 269–270, 'Chirodropid Stings', and pages 280–3, 'Summary of Management of Chirodropid stings'. *Chiropsalmus* sp. Australia nematocysts may require expertise to distinguish them microscopically from those of *Chironex fleckeri* {compare Plate 10.10 with Plates 9.69 and 5.3}.

FATALITIES IN THE REMAINDER OF THE TROPICAL INDO-WEST PACIFIC

Human fatalities are known in the Philippines (OLD 1907; LIGHT 1912a). FENNER (1987) reported approximately one Philippine death each alternate year in adjacent villages, in the north of Luzon around the Bay of Sual, due to chirodropids. This would give between 20 and 50 deaths per year in the Philippines, a high estimate, but one corroborated by CORNELIUS (1994, unpublished observations). As no death certificate is currently required in the Philippines, accurate numbers are difficult to calculate and impossible to verify. Other deaths have been reported in the Indo-West Pacific, and one from Japan {see Chapter 10, page 291}.

CHIROPSOIDES BUITENDIJKI (HORST, 1907)

DISTRIBUTION

A specimen was first caught and described from Java (HORST 1907), with further sightings from the Malay Archipelago (STIASNY 1919), southern India (MENON 1936; NAIR 1951; RAJAGOPAL *et al* 1989), Indochina (RANSON 1945), and more recently in Sri Lanka (FENNER 1994, unpublished observations). The latter specimens were caught in deep water by Dr Malik Fernando while scuba diving (Plates 9.9 A and B); to date, they have not been reported as being caught off a beach.

APPEARANCE

A box-shaped bell up to 11 cm bell diameter, with groups of up to 8–9 tentacles in each corner, coming off in an unusual linear fashion on the outside of the main canal (Plate 9.61). There is also a very characteristic extension of the pedalial canal to the top of the pedalium (Plate 9.61).

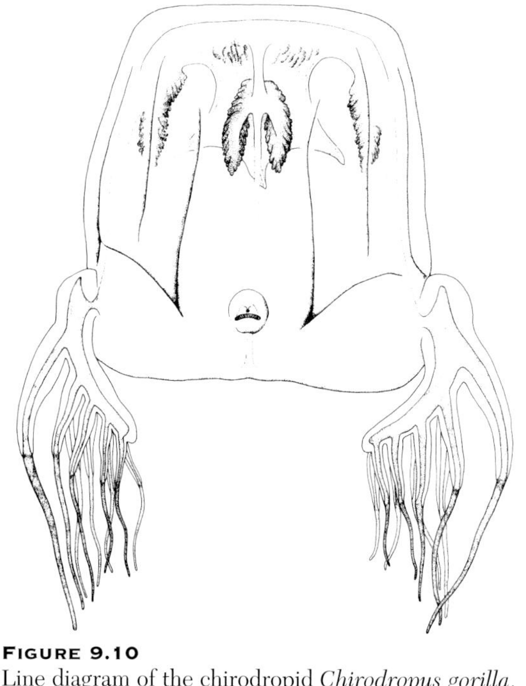

FIGURE 9.10
Line diagram of the chirodropid *Chirodropus gorilla*. (Courtesy of Professor Josep-Maria Gili, Barcelona.)

Eight long, superior, perradial gonads hang down to the level of the sensory organs, being of similar diameters through the majority of their length in the freshly-preserved specimen (although the gonads on some specimens come to a point [Plate 9.60]). These may shrink in preserved specimens (RIFKIN 1994, unpublished observations).

GENUS *CHIRODROPUS* HAECKEL, 1880

DISTRIBUTION

Chirodropus gorilla HAECKEL 1880 is reported from Liberia, 7° north of the Equator (KRAMP 1955), to South Africa, some 30° south (STIASNY 1931). It is under-reported in the northern distribution, as most chirodropid species appear to extend some 20° either side of the Equator.

An expedition from Spain recently caught several specimens in the Benguela Current off Angola, West Africa, 23° south of the Equator (PAGÈS *et al* 1992). A specimen from the expedition was kindly donated by one of its members, Professor J.-M. Gili of the Marine Sciences Centre in Barcelona, Spain, and is currently being investigated (RIFKIN & FENNER 1994, unpublished observations).

Chirodropus palmatus has been reported near St Helena, off the West African coast (HAECKEL 1880), but this may only be an immature, smaller specimen of *Chirodropus gorilla* (MAYER 1910).

APPEARANCE

Chirodropus spp. are essentially similar to the other chirodropids. Largest described 130 mm height, 110 mm diameter; maximum number of tentacles 9–11 per pedalium (36-44 total). MAYER (1910) describes gonads attached to the subumbrellar sides of the interradial septa and having grape-like clusters at their free edge; this was confirmed by KRAMP (1966). Mayer also describes a large manubrium hanging down into the bell cavity almost to the velarium, but qualifies his statement saying that 'a specimen had only been seen by HAECKEL'. PAGÈS *et al* (1992) describe the gonads as 'plumose' in appearance, each consisting of foliaceous plates bearing numerous transverse folds.

CHIRODROPUS ENVENOMATION

To the knowledge of the editors, no account of serious stings or deaths has been published. However, *Chirodopus* spp. are commonly encountered in geographical areas (including the west coast of Africa) from where there have been few reports of marine envenomation.

Although tests have not been carried out in this region, the authors suggest the first aid and medical treatment described below, under 'Summary of Management of Chirodropid Stings', page 280, would probably be useful for west African chirodropid stingings.

CHIRODROPID STINGS

Fortunately, and contrary to impressions created by many newspaper accounts, most stings from chirodropids are relatively minor or at least no threat to life. This was emphasised previously by BARNES (1966), but now bears restating. However, even small stings may be very painful (Plate 9.62), and require careful assessment and basic care to promote rapid healing, minimise scarring (Plates 9.47 and 9.64C) and avoid secondary infection (Plates 9.65 and 9.66) {see also Plates 5.22 A and B}.

Most data on chirodropid stings are based on studies of the sting and envenomation process of *Chironex fleckeri*. There is reason to believe that the similarities between a *Chironex fleckeri* envenomation in Australia and envenomations resulting from the (until recently) poorly described chirodropids found in the Indo-West Pacific (including all chirodropid types) are close {Plate 9.46}. The sting from the Australian *Chiropsalmus* sp., while similar in action to *Chironex* (BARNES 1960), is appreciably less dangerous (BARNES 1960; BARNES 1965; KEEN 1971; WILLIAMSON 1985; WILLIAMSON *et al* 1987; HALSTEAD 1988) and less disfiguring {Figure 9.60}.

Most intact chirodropids have a tremendous 'kill power' {Plate 7.2}. This is not some fiendish design to 'attack' and kill humans, but rather to enable them to kill their food prey quickly, before the latter struggles and destroys the predator; jellyfish, like humans, need to eat to live and breed. If jellyfish lose their tentacles, then they lose this 'kill power', cannot feed, and consequently die.

The stings (including potentially and actual fatal stings) are usually sustained by a swimmer or wader in shallow water close to shore, e.g. less than 2 metres depth (MAGUIRE 1968; WILLIAMSON *et al* 1980; LUMLEY *et al* 1988; CURRIE 1992; BEADNELL *et al* 1992) (Plate 9.67) {see also Plates 1.2 and 5.14}. Many stings have occurred when people enter the water, and children paddling in the shallows are particularly at risk (WILLIAMSON 1983) {Plates 7.2 and 9.45}. In Darwin a jogger was stung after treading on a *Chironex* in a small beach puddle more than 50 metres from the receding tide!

As *Chironex* inhabit very shallow water (Table 9.5, page 249), most stings are on the lower half of the body — a feature of significance for first aid (FENNER *et al* 1989) {see Table 9.9, page 267}. Most stings occur in children and women wading in very shallow water (WILLIAMSON 1983 & 1985a) consequently warning signs should read: 'Do not enter the water' rather than 'Do not swim'.

In contrast, most carybdeids are deep-water jellyfish (Table 9.5, page 249), and stings occur mainly on swimmers, thus stings are often on the upper torso or arms (Table 9.6). The proportion of stings on children is also consequently lower (BARNES in KINSEY 1988) (see Tables 9.6 and 9.9, pages 250 and 267).

The risk of a sting is greater under still, hot weather and calm water conditions, although mild surface agitation from wind does not exclude the risk.

The risk to the swimmer of more serious envenomation is also enhanced by bare hairless skin (more intimate tentacle contact) {Plates 1.2 and 1.3}, small body size (higher tissue and blood concentration of venom from a given venom dose) and rapid entry into the water, such as running and duck-diving (entanglement in tentacles resulting in the injection of a larger venom dose from discharging nematocysts {see Chapter 6, page 162}). Consequently children and small women are at greatest risk of a dangerous envenomation (Plates 9.81 and 9.82), a fact borne out by the serious morbidity (MAGUIRE 1968; WILLIAMSON et al 1980; FENNER *et al* 1989) and mortality (CLELAND & SOUTHCOTT 1965; WILLIAMSON 1983, 1985; LUMLEY *et al* 1988; VIMPANI *et al* 1988; CURRIE *et al* 1992) data {see Chapter 3, pages 70–73}.

A large adult box jellyfish may be invisible to the untrained eye under natural water conditions (WILLIAMSON 1985; HARTWICK 1987), even in bright sunlit weather (a hazard long appreciated by investigators [LIGHT 1914]){see Plates 5.8 and 9.56}, and the rapidly moving wader or swimmer can become entangled in the animal's tentacles. Duck-diving is even more dangerous, as there is the risk of diving directly into the chirodropid or surfacing directly under it. Fortunately fewer than 2 per cent of the total reported stings occur on the head or face (see Table 9.9, page 267).

Under the most dangerous conditions of hot, still, tropical water in a sheltered, sandy-bottomed inlet, chirodropids will cruise close to shore with their massive tentacle armament extended in the 'fishing' mode, following large concentrations of their natural food of shrimps (e.g. *Acetes australis*) and juvenile fish, commonly encountered in such localities (HARTWICK 1991a). The adult *Chironex fleckeri* of northern Australia may have a curtain of up to 60 tentacles trailing 3 metres behind its slowly pulsating bell. As stated above, these tentacles will be totally invisible to the unwary swimmer (HARTWICK 1987), who can blunder into them. However, chirodropids will always attempt to avoid contact and, given time and opportunity, will swim around a human in the water [Plate 5.8} and head for deeper water.

CHIRODROPID TENTACLE ARMAMENT

Again, in terms of the number and size of nematocysts, total stinging tentacle length, potency of injectable venom and speed of venom absorption, chirodropids in general and *Chironex fleckeri* in particular have few peers among the earth's venomous creatures, so far as humans as victims are concerned.

Tentacles number up to 60 in *Chironex fleckeri* (up to 15 on each pedalium), and apparently up to 36 (nine on each pedalium) in both the Philippine (RIFKIN 1993, unpublished observations) and Australian species of *Chiropsalmus* spp. (MAYER 1910; LIGHT 1914; CLELAND & SOUTHCOTT 1965; BARNES 1965). Although the width and extensible length of the tentacles of *Chironex fleckeri* exceed those of all other known chirodropids, the cnidae (i.e. types of capsular organoids [which include nematocysts] found among cnidarians) of the chirodropids have basic similarities (RIFKIN 1995, unpublished observations). However, they differ diagnostically from the cnidae of scyphozoan jellyfish (CALDER & PETERS 1975; RIFKIN & ENDEAN 1983).

CHIRODROPID NEMATOCYSTS {see Chapter 6, page 155}

Among the chirodropids, the cnidom of *Chironex fleckeri* has so far received most attention, particularly by Australian-based scientists (HARTWICK *et al* 1980; RIFKIN & ENDEAN 1983, 1988; ENDEAN 1988). Some work has been done on the cnidom of *Chiropsalmus quadrumanus* (CALDER & PETERS 1975), but apart from preliminary studies on the species known at the time as *Chiropsalmus quadrigatus* HAECKEL, 1880 (KINGSTON & SOUTHCOTT 1960; KINSEY 1986), surprisingly little is known first-hand about cnidae of other chirodropids. There are now known to be at least six different types of nematocysts on the tentacles of *Chironex fleckeri;* all the same types have also been found on *Chiropsalmus* sp. (Australian) tentacles (RIFKIN 1992, unpublished information), and probably five or six on those of *Chiropsalmus quadrumanus* (CALDER & PETERS 1975).

Commencing with some international pioneer observations (WEILL 1934; CARLGREN 1940; RUSSELL 1953) and more specific Australian observations (BARNES 1960; CLELAND & SOUTHCOTT 1965; ENDEAN *et al* 1969), the description of the microscopic structure (HALSTEAD 1988) (including fine structure [RIFKIN & ENDEAN 1983 & 1988], see Chapter 6, page 121 and Figures 6.32–6.40, pages 155–159), functions (RIFKIN & ENDEAN 1983; ENDEAN 1988a & 1988b) and tentacular assembly (HARTWICK *et al* 1980; RIFKIN 1988, 1996) of chirodropid nematocysts {Plates 6.12 and 9.57} has facilitated an understanding of the human envenomation process from these impressive animals.

A stained longitudinal section of an undischarged *Chironex fleckeri* tentacle in the contracted state {Figure 6.55} bears a resemblance to pseudo-stratified human mucosa with its layers in which the nematocysts are in approximate banks of three. On contact with the integument (skin) of the victim, some nematocysts penetrate to the dermis, injecting their venom {Plate 6.12} whilst others act as 'grappling irons' and effectively pull the integument of the prey closer to the tentacle (see below). In tightly contracted tentacles the higher layers may completely overlap the layers underneath, leaving much of the tentacle full of potent nematocysts which may never make direct skin contact (RIFKIN 1990, unpublished observations). In tentacles that are not as tightly contracted, the 'grappling iron' effect is to bring further nematocysts into action, and they discharge. The net effect of this is that the longer the tentacle is in contact with the skin, the greater the potential for increased envenomation — a fact influencing the first aid treatment {see Chapter 5, page 108}. Extended tentacles (tentacles stretched out while feeding) have all their nematocysts exposed (RIFKIN 1988a). It is now believed that the larger 'cigar-shaped' nematocysts, known as 'microbasic mastigophores' {Figure 6.43, page 168 and Plate 9.57} contain the toxins of *Chironex fleckeri* venom which may play a principal role in the systemic morbidity and mortality in human envenomation (ENDEAN & RIFKIN 1975; ENDEAN 1987). These important nematocysts, together with the other tentacular nematocysts that make up the cnidom of *Chironex fleckeri*, and the remarkable mechanisms that control the placement of these envenomating nematocytes in the 'fire-ready' position in the tentacle mesogloea [RIFKIN & ENDEAN 1988], are described in Chapter 6, page 154.

The *Chironex fleckeri* cnidom thus includes microbasic mastigophores, two kinds of trirhopaloid (=heterotrichous microbasic euryteles) {see Chapter 6, pages 168–9}, holotrichous isorhizas and atrichous isorhizas {Figure 6.43, page 168}. Trirhopaloid nematocysts appear to be both penetrants and injectors but their role in the envenomation process is unclear. Atrichous isorhizas probably serve an adhesive and entangling function upon prey but have not been seen penetrating skin (RIFKIN & ENDEAN 1983). Also it has been long appreciated that the tissues of jellyfish generally, quite apart from the nematocysts, may contain toxic material of their own (LANE 1960). This applies to *Chironex fleckeri* at least, among the chirodropids (ENDEAN & NOBLE 1971; ENDEAN *et al* 1993), and thus may modify the interpretation of jellyfish toxinological studies that have used whole tentacle extract. However, this information remains debatable, because it is possible that some of the toxic material in tentacular tissues may be absorbed mastigophore toxin that escaped from ruptured nematocysts during the experimental separation of tentacles and nematocysts.

The 'pseudo-stratified', contracted state of the tentacle {Figure 6.42, page 162} imparts a 'cross-hatched' appearance to the surface of the ribbon-like tentacle resulting from the transverse grooves formed between the surface 'villi'. This is well shown microscopically in Figures 9.9, page 265 and macroscopically in Plate 9.43, and is of medical significance {see below}.

When the tentacle is fully extended (relaxed), as occurs when the animal is 'cruising' and 'fishing' for its prey, the full armament of nematocysts is potentially presented on the surface of the tentacles (Plate 9.68). Under these circumstances, entanglement in this curtain of extended tentacles can result in a massive human envenomation.

Barnes noted from his field observations that while the apparent venom dose received from *Chironex fleckeri* stings could usually be judged by the total length and width of sting marks on victims' skin (KINSEY 1986; FENNER *et al* 1989), these criteria did not invariably correlate with the gravity of the patient's clinical condition; Barnes recorded extensive tentacle contact (e.g. 6–9 metres) in some non-fatal *Chironex fleckeri* stings (BARNES 1966 & 1980, unpublished observations). While the size of the animal and the total length of tentacle contact, and the size of the victim (WILLIAMSON 1983) and the amount of hair on the bare skin (WILLIAMSON 1985) (Plate 9.77) are indeed crucial determinants of the dose of injected venom, other less well understood variables may apply. These include the maturity of the tentacular mastigophores involved in the stinging (RIFKIN 1988a), the number of nematocytes in the 'fire-ready' position at the time (RIFKIN & ENDEAN 1988) and the possible central 'neurological' control over nematocyst positioning and discharge that may exist in *Chironex fleckeri* at least, depending on the physiological state of the animal at the time (ENDEAN 1988a).

LENGTH OF TIME OF TENTACLE CONTACT WITH SKIN

It is currently believed (FENNER *et al* 1989a; BEADNELL *et al* 1992) that the longer the tentacle is in contact with the skin, the greater the envenomation. Thus the first principle of treating envenomation is the inhibition of nematocyst discharge; in the case of chirodropids, vinegar dousing for a minimum of 30 seconds will totally inhibit the discharge of any remaining nematocysts present (HARTWICK *et al* 1980) {Figures 9.8A, 9.8B and 9.8C, page 264}.

NEMATOCYST DISCHARGE

{see Chapter 6, page 159}

With all cnidarian stings, nematocyst discharge occurs at the moment of contact between the tentacle epidermis and the jellyfish's normal prey (Figure 6.37, page 157) or human skin (Plate 9.76) {see also Plate 5.7}. The mechanism of cnidarian nematocyst discharge {see Chapter 6, page 159} has been under intense scrutiny for many years (e.g. HOLSTEIN & TARDENT 1984) and continues to be a subject of close investigation (ENDEAN *et al* 1991; RIFKIN 1996) (see also Chapter 26, page 450).

Following contact of the trigger mechanism of the nematocyst (cnidocil) with the surface of the prey or skin, a series of discharge events occur {Figure 6.11, page 130, and Plate 9.69} with explosive speed and force {Figures 6.11 and 6.38, pages 130 and 158}, which result in the penetration of the epidermis (in the case of a human sting), frequently down to the vascular papillary dermis (STRUTTON & LUMLEY 1988) {Plates 5.7, 6.10, 6.11 and 6.12} by a spiralling spine-bearing 'shaft' or 'butt' (RIFKIN & ENDEAN 1983) and distal-tubule-bearing mechanism which is ejected from the capsule of the nematocyst, everting as it goes (WEILL 1934; CLELAND & SOUTHCOTT 1965; HALSTEAD 1988). This event is accompanied by a rapid increase in the intracapsular pressure of the undischarged venom-containing nematocyst, which results in the everting tubule releasing matrix from the tubule continuously as discharge occurs {Figure 6.36, page 157}, and subsequently spewing out droplets of capsular matrix from the tubule tip (ENDEAN & RIFKIN 1983; ENDEAN *et al* 1991; RIFKIN 1992) {see Chapter 6, page 157, and Figures 6.36, 6.37 and 6.39, pages 157–8}.

This explosive 'chemico-mechanical' event occurs within milliseconds, with the emerging basal portion or 'shaft' of the everting tubule exerting tremendous force, capable of impressive penetration (HOLSTEIN & TARDENT 1984), e.g. through the tough cuticle of a prawn, or the scales of a small fish, or the keratin of human skin {Figure 6.38, page 158}. The penetrating tubules can be readily demonstrated on skin biopsies from the stung animal (e.g. a mouse {Figure 6.40, page 159} and human skin (STRUTTON & LUMLEY 1987; HALSTEAD 1988; LUMLEY *et al* 1988) {Plates 5.7 and 9.69}.

VENOM INJECTION

{see Chapter 7, page 174}

Each discharging microbasic mastigophore (MM) that penetrates the skin possibly deposits a micro-dose of *Chironex fleckeri* venom both along the path of the discharging tubule and from the tubule tip where capsular matrix is deposited. The trajectory of the average *Chironex fleckeri* MM is about 0.7 mm, with the largest reaching about 0.9 mm (HARTWICK *et al* 1980). When presented at right angles to the epidermis, such tubules may penetrate to the human papillary dermis (STRUTTON & LUMLEY 1988). With an estimated nematocyst population on the tentacles of an adult *Chironex fleckeri* in the order of four thousand million (HARTWICK 1983, unpublished observations; WILLIAMSON 1985), and even allowing for only a 10 per cent penetration rate, the total venom dose may be large when multiple tentacle contact is made. Furthermore this 'dose' is divided into millions of micro-doses deposited over an area of tissue — in sharp distinction to the depot of venom deposited in bulk, mainly at one tissue site, by a biting venomous animal {Figure 7.1, page 178} such as a sea snake. This presents a huge surface area of venom for absorption through venom-damaged dermal microvasculature. Such absorption will be enhanced by vigorous skeletal muscle contraction in the stung limbs (the 'muscle-pump' effect), as would occur in an unrestrained victim distraught with the savage pain.

As manifest clinically in serious stings, high (sometimes lethal) blood levels of venom may develop within minutes, following a massive *Chironex fleckeri* sting (MAGUIRE 1968; WILLIAMSON *et al* 1980; LUMLEY *et al* 1988; CURRIE *et al* 1992) (and probably other chirodropid stings [BENGSTON *et al* 1991]), especially in children (WILLIAMSON 1983 & 1985). This clinical onset of systemic effects is faster than can be explained by the simple intradermal or subcutaneous absorption described above — even allowing for the vascular damage from the highly toxic *Chironex* venom (ENDEAN & RIFKIN 1983; BURNETT & CALTON 1986). Animal studies have revealed that the downward thrust of some of the MM tubules into the dermis is associated with actual penetration and transfixion of small blood vessels {Figure 6.11, page 130}, providing opportunity for direct intravascular 'injection' of matrix from the discharging tubule {see also Chapters 6, page 121 and 126, page 442}. This microscopic scenario is likely to be repeated in human stinging (Plate 5.7).

HUMAN TISSUE RESPONSES TO *CHIRONEX FLECKERI* VENOM

Along with many other jellyfish venoms (BURNETT & CALTON 1977; BURNETT *et al* 1983; BURNETT & CALTON 1987a; BURNETT *et al* 1988; and see Chapter 7, page 176), *Chironex fleckeri* venom exhibits both toxic and immunologic (including allergic {Plate 7.4} effects (FREEMAN 1974; CRONE 1976; WILLIAMSON 1985; BURNETT & CALTON 1987b; BURNETT 1991a & 1991b). The toxic actions are by far the more common and more important clinically.

CLINICAL EFFECTS

Lethal chirodropid stings

Any serious chirodropid sting is among the most dramatic events in medicine and constitutes one of the most rapid envenomation processes known. The rapid process is sadly associated with equally rapid demise in the worst cases, within minutes (but not seconds) {Plates 7.2 and 9.45}. Death of experimental animals within seconds under laboratory conditions with intravenous injection of venom, has long been known (ENDEAN *et al* 1969). Unfortunately this time scale has on occasions been erroneously extrapolated to the human situation. There is always time to institute appropriate resuscitation measures, which in life-threatening stings will usually be life saving (MAGUIRE 1968; WILLIAMSON *et al* 1980; WILLIAMSON 1988; FENNER *et al* 1989). Only two cases (both children) are known to the authors in which a fatal outcome followed reasonably early and/or appropriate resuscitation (LUMLEY *et al* 1988; CURRIE *et al* 1992) (Plates 9.70A and 9.70B). However, in remote areas of coastal tropical Australia and in many other Indo-Pacific locations, poor public education, thinly scattered health resources and long distances to transport unfortunately indicate that fatal stings will continue.

To date there is firm documentation of 36 human deaths from chirodropid stings on the Queensland coast between Gladstone and Cooktown, since 1884 (CLELAND & SOUTHCOTT 1965; WILLIAMSON 1985; INTERNATIONAL CONSORTIUM for JELLYFISH STINGS 1993), and 27 in the Northern Territory this century (CURRIE *et al* 1992). Many more have occurred (CLELAND & SOUTHCOTT 1965; FENNER 1987) and continue to occur (FENNER *et al* 1995) throughout the Indo-Pacific region. Many of these have occurred on remote coastlines. The possible mechanisms of death are considered below {see under 'Venoms of *Chironex fleckeri* and other chirodropids'}.

Non-lethal toxic effects

The victim will seldom sight the animal at the time of the sting {see above and Plates 5.8 and 9.56}. The immediate effect of tentacle contact is searing pain at the sting site, which may render the victim (especially a child) incoherent in moderate as well as severe stings (FENNER *et al* 1989; BEADNELL *et al* 1992). Because the victim is usually in shallow water {Table 9.5, page 249}, (in contrast to Irukandji stings) most *Chironex fleckeri* stings are on the lower half of the body. However, especially in children, the victim's hands are frequently involved in tentacle contact and stinging, resulting from the reflex reaction of reaching down to the site of the severe pain (WILLIAMSON 1983, 1985; LUMLEY *et al* 1988) (Plate 9.70B). Generally, only the backs of the hands will be stung; the keratin on the palms may be too thick to permit nematocyst tubule penetration to reach to the depth of epidermal sensory innervation. However, ventral wrists and hairless forearms (e.g. those of women and children) are especially vulnerable (Plates 9.71 and 9.72) and may contribute significantly to the total venom dose injected (HALSTEAD 1988; LUMLEY *et al* 1988).

Following the onset of immediate and savage pain, an intense acute inflammatory response commences at the sites of contact with tentacle material {Figure 1.2, page 45} (BARNES 1960; MAGUIRE 1968; WILLIAMSON *et al* 1980; WILLIAMSON *et al* 1984; HALSTEAD 1988; LUMLEY *et al* 1988; CURRIE *et al* 1989). This inflammatory response is cut short in rapid fatalities {Plates 7.2 and 9.45, but progresses rapidly in survivors and non-life-threatening stings (Plates 9.73 and 9.74) {see also (Plates 1.2, 5.14 and 9.44}. The ensuing moments are usually among the most terrible for the conscious victim. While plucking vainly at the area of the body under the water where the terrible pain of the sting is felt the victim will scream and move rapidly backwards (Plate 9.75) and out of the water (although occasionally a small child or a massively stung adult or child may not make purposeful attempts to exit the water). If rescuers are present nearby (e.g. on a tropical beach patrolled by members of Surf Life Saving Australia {Plate 5.9}), they will be attracted by the agitation in the water and the cries, and will assist the victim to leave the water {see below}.

It is common in *Chironex fleckeri* stings (and probably other major chirodropid stings [BENGSTON *et al* 1991]), and indeed a valuable diagnostic sign, that substantial tentacle material is avulsed from the animal by the struggling victim in the water. It will often cling firmly to the victim's skin, as a result of its anchoring by thousands of discharged nematocyst tubules, and thus be adherent in the sting area on emergence from the water (BARNES 1960; CLELAND & SOUTHCOTT 1965; WILLIAMSON *et al* 1980; WILLIAMSON *et al* 1983; LUMLEY *et al* 1988; FENNER *et al* 1989; BEADNELL *et al* 1992) (Plate 9.76). Although this may occur with other serious jellyfish stings (e.g. Atlantic *Physalia physalis* {see page 198} and the larger carybdeids {see page 242}), the phenomenon is most consistently associated with chirodropid stings, especially *Chironex fleckeri*. Considering the colossal nematocyst armament of chirodropid tentacles, this adhesion does not seem surprising to the medical editors.

Adherent tentacles are of importance for other reasons. They may contain many thousands of unfired nematocysts and present a real risk of increasing the envenomation of the victim if any inappropriate actions are taken, such as rubbing them and the fresh sting area with sand (LUMLEY *et al* 1988) or bare hands, or applying alcohol (HARTWICK *et al* 1980). These tentacles may also present a sting risk (usually minor) to the rescuer, especially in victims with copious adherent tentacles (WILLIAMSON *et al* 1980; BEADNELL *et al* 1992). Hair on the rescuer will afford some protection (Plate 9.77). Finally the tentacle material provides a rich source of nematocyst material for later microscopic examination and possible species identification of the offending animal (BARNES 1960; CLELAND & SOUTHCOTT 1965; LUMLEY *et al* 1988; WILLIAMSON 1988) {see Chapter 6, page 167}.

The sting is an angry whip wheal, tracing out in a vivid linear erythematous flare with a white ischaemic centre, when viewed within the first few minutes (Plates 9.77 and 9.78), wherever tentacle contact has occurred; such dramatic erythema may be masked by skin pigmentation (Plate 9.79). This will have the appearance of a tangled mesh pattern in larger stings {Plates 1.2 and 9.78} and will quickly develop surrounding oedema and darkening of the most seriously envenomated ischaemic white areas (Plates 5.19A and 9.80A). The latter may proceed to vesiculation (Plate 9.74), actual partial or full thickness skin death (Plates 9.80B and 9.81 {see also Plate 5.14}) and scab formation (Plate 9.80C and 9.82), with the probability of scarring (Plate 9.83) {see also Plates 5.19C and 9.47}. Scarring is much more likely to result if secondary infection supervenes {Plates 5.20C and 9.64C}.

Close inspection of the larger wheals in the acute lesion may reveal a typical 'cross-hatched' or 'frosted-ladder' pattern imprinted on the victim's skin by the corresponding architecture of chirodropid tentacles (Plates 1.2 and 9.78). If present, this 'frosted-ladder' pattern is of diagnostic value, so it should always be looked for (BARNES 1960; WILLIAMSON *et al* 1980; SUTHERLAND 1983; WILLIAMSON 1985). The size of the sting will usually (but not always — Plates 9.70A and 9.82) correlate with the clinical gravity of the victim's condition {see Chapter 8, page 239} and should be judged by the following visual (and measurement when possible) criteria (FENNER *et al* 1989):

- Total length of tentacle marks:
  When priorities (see below), time and circumstance permit, measure the length of tentacle marks of the skin with a tape or piece of string. A sting indicating a total tentacle length of 2.2 metres on a child has been lethal (CURRIE 1991, unpublished observations; INTERNATIONAL CONSORTIUM of JELLYFISH STINGERS 1993) and, on another child, of 4 metres (LUMLEY *et al* 1988) {Plate 9.83}. The smaller the victim, the greater the risk (WILLIAMSON 1983).
- Maximum width of tentacle marks:
  This indicates the size of the animal involved, and the likely total dose of venom. Tentacle width greater than 6 mm indicates a large adult animal {Plates 9.44 and 9.73} and is associated with several fatality reports (BARNES 1966; WILLIAMSON 1992, unpublished observations).
- Total sting area:
  A useful guide is to observe the proportion of limbs that are stung, or their equivalent in the case of lower abdomen stings; any sting occupying an area equal to or greater than the area of one-half of a limb, fully encircled, is a serious threat to the victim (FENNER *et al* 1989).

As has been mentioned previously, a variety of additional factors may influence the dose of venom injected and the clinical response of the victim to a chirodropid sting. As well as the sizes of the victim, the animal and the sting, these include factors such as the extent of tentacle contraction (HARTWICK *et al* 1980) {Plate 9.67}, the number of nematocysts in the 'fire-ready' position (RIFKIN & ENDEAN 1988) and less well understood factors such as the physiological 'state' of the animal at the time, along with any 'central control' that the animal may exert over its nematocyst response (ENDEAN 1988a). Thus on occasions a sting that may by dimensional criteria indicate a life-threatening situation can result in surprisingly little serious effect on a victim (BARNES 1966) (Plate 9.82). However, any sting that produces impairment of consciousness and/or disturbances of heart action, heart rate, breathing or circulation must be treated as life-threatening, irrespective of the apparent sting size (LUMLEY *et al* 1988; FENNER *et al* 1989) {Plates 9.70 and 9.72}.

It is important to realise the possible significance of a sting victim who was loudly distraught and physically active, suddenly becoming quiet and 'co-operative'. Such a change may be a manifestation of serious developing central organ (e.g. brain) dysfunction (e.g. clinical shock) from increasing systemic venom levels (WILLIAMSON *et al* 1980), which would have been enhanced by the prior muscular activity induced by the savage pain (the 'muscle pump' effect) (WILLIAMSON 1988). The need for resuscitation may be imminent, and nothing takes priority over this aspect of first aid or medical management {see below and Chapter 5, page 106}, even in hospital (Plate 9.84).

Clinical immunologic effects of chirodropid stings {see Chapter 7, page 176}

All jellyfish venoms so far studied consist of mixtures of toxic and/or antigenic polypeptides and enzymes pathogenic to humans (BURNETT & CALTON 1987a, 1987b). Being antigenically active they initiate a specific immunological response in human victims, which is both humoral and cellular, and capable of later serological detection (BURNETT *et al* 1988; WILLIAMSON *et al* 1988).

The immunological aspects of jellyfish envenomation are discussed in Chapter 7, page 176.

No life-threatening allergic reactions to a chirodropid sting have yet appeared in the literature. However, delayed local reactions to *Chironex fleckeri* stings are well documented (WILLIAMSON 1985; CURRIE & ANSTEY 1989). The reappearance of these lesions at the original site may take the form of anything from a mild 'papular urticaria' to a recrudescence of a red, itchy and weeping vesicular eruption, some 10–16 days after the initial sting, and in the total absence of further contact with the venom (Plates 9.85 and 9.86) {see also Plate 7.4}; some local oedema also reappears. This event appears to be more commonly associated with patients who have a history of some immunologically-based abnormal condition, such as atopy or asthma.

VENOMS OF *CHIRONEX FLECKERI* AND OTHER CHIRODROPIDS

Since the earliest appreciation of the remarkable potency of chirodropid venoms against humans (LIGHT 1914; FLECKER 1952), many distinguished toxinological investigators have contributed to the knowledge of the actions of these complex venoms.

However, despite important studies, the exact mechanisms of action of *Chironex fleckeri* venom remain relatively poorly characterised; there have been conflicting results from a number of the studies. Difficulties have been encountered with venom retrieval, nematocyst venom isolation, thermolability of venom components and their propensity to aggregate and disaggregate and adhere to equipment surfaces (OTHMAN & BURNETT 1990).

Based on animal experiments, there is general agreement that *Chironex fleckeri* venom has dermatonecrotic, haemolytic and lethal factor components (SOUTHCOTT & KINGSTON 1959; BAXTER & MARR 1969; FREEMAN & TURNER 1969; KEEN 1970). Blood cell haemolysis has not been clinically documented in human envenomation, making death and skin damage the two important clinical manifestations of envenomation. The dermatonecrotic action of *Chironex fleckeri* venom requires further characterisation, but may include release of vasoactive peptides in addition to direct toxic tissue damage (KEEN & CRONE 1969; BURNETT & CALTON 1977; WILLIAMSON *et al* 1984a).

Comparison of the venoms from Australian *Chironex fleckeri* and *Chiropsalmus* sp. specimens showed similar activity, although death in laboratory mice was slower with *Chiropsalmus* than with *Chironex fleckeri* (KEEN 1971). A massive sting to an infant from *Chiropsalmus* sp. should therefore be managed as for a life-threatening chirodropid envenomation (WILLIAMSON 1983).

CURRENT CONTROVERSIES

The nature of the venom's lethal actions has engendered numerous studies, without a definitive answer yet emerging. Clinical observations suggest that the potentially lethal events with severe envenomation can be either respiratory impairment (MAGUIRE 1968; BOYD 1984; WILLIAMSON *et al* 1984b) or cardiac dysfunction (BARNES 1966; LUMLEY *et al* 1988; BEADNELL *et al* 1992; CURRIE *et al* 1992), or conceivably both. Respiratory depression, possibly by a central mechanism of action, was suggested in some animal studies (SOUTHCOTT & KINGSTON 1959; FREEMAN & TURNER 1969; BURNETT 1984). Evidence for cardiotoxicity from animal experiments includes findings of impaired cardiac contraction {see MUSTAFA *et al* 1995 in Chapter 26, page 453}, arrhythmias including various conduction defects and ventricular tachycardia, hypertension and hypotension and decreased coronary flow (FREEMAN & TURNER 1969; ENDEAN *et al* 1969; FREEMAN & TURNER 1971).

The clinical importance of both broad mechanisms of lethality — respiratory failure and cardiotoxicity — is reflected in case reports. There are two well-documented cases (MAGUIRE 1968; WILLIAMSON *et al* 1980) and one anecdotal report (SUTHERLAND 1983) of potentially fatal *Chironex fleckeri* envenomations where the victims survived after collapsing on the beach and receiving only expired air resuscitation. Conversely there have been two documented fatalities from cardiotoxicity despite respiratory support, which included endotracheal intubation and ventilation (LUMLEY *et al* 1988; CURRIE et al

1992). It is evident that both respiratory and cardiac support need emphasis in the clinical management of severe human envenomation.

As well as the uncertainty about the mechanisms of lethality, the actual nature of the relevant toxins and their actions remain contentious. This reflects the difficulties encountered when isolating and characterising the active venom components. It has been suggested that the methods of preparing the jellyfish material for venom studies may often have been unsatisfactory (ENDEAN *et al* 1993b). Using a method of isolating nematocysts from tentacles before undertaking venom studies, two myotoxins with molecular weights of approximately 600 000 and 150 000 have been identified (ENDEAN 1987). It has been suggested that these are the major cardiotoxins of *Chironex fleckeri* venom and that one or both of these may be absent or reduced in experiments using venom from the amnion-milking technique of BARNES (1967) or from homogenised whole tentacles (ENDEAN *et al* 1993b). This is a controversial area, especially as there are implications concerning the method of obtaining venom for antivenom production. In addition, it has been shown that crude extracts of tentacle material from which nematocysts have been removed contain neurotoxic and haemolytic fractions (ENDEAN *et al* 1993b). Despite the uncertainties raised by these recent findings,* animal studies including experiments using antivenom suggest that there is an appreciable overlap in venom components produced by the various methods (FREEMAN & TURNER 1969; BAXTER AND MARR 1974; BURNETT *et al* 1990).

CELLULAR MECHANISMS

The cellular mechanisms of the lethal and other actions of *Chironex fleckeri* toxins appear to result from abnormalities in ionic transport across membranes, leading to altered membrane permeability (BURNETT & CALTON 1977; MUSTAFA *et al* 1995, in Chapter 26, page 442). Calcium and sodium channels have been implicated, with animal experiments suggesting various tissues being involved, including smooth muscle, skeletal and cardiac muscle, cardiac conduction pathways and 'central' neurological pathways. In addition, the role of indirect venom activity through release of auto-pharmacological substances such as catecholamines requires elucidation. The definitive answers about the actions of *Chironex fleckeri* venom remain elusive.

In comparison to *Chironex fleckeri*, there has been little work done on the venoms of other chirodropids. *Chiropsalmus 'quadrigatus'* (i.e. *Chiropsalmus* sp.Australia) has venom with dermatonecrotic and lethal components similar to but less potent than *Chironex fleckeri* (KEEN 1971). Animal experiments with *Chiropsalmus* sp.Australia toxin extracted from tentacles suggested a cardiotoxic action resembling that of *Chironex fleckeri* (FREEMAN AND TURNER 1972).

NOTE: It can be speculated that the general (thermo)lability of *Chironex fleckeri* venom may result in a time-limited acute toxic action in the tissues of an envenomated human victim. Cases are documented where the initial, on-the-beach, acute life-threatening systemic toxic effects seem to recede after about 30–40 minutes of sustained resuscitation, even in the initial absence of antivenom, but always in the presence of early and effective resuscitation (MAGUIRE 1968; WILLIAMSON *et al* 1980). While on-going management problems may continue, such speculation lends support to the optimistic approach that should always be adopted by on-the-beach rescue teams who reach envenomated victims quickly {see flow chart, page 107}.

CHIRONEX FLECKERI ANTIVENOM

Chironex fleckeri antivenom (Plate 9.87) was first produced by the Commonwealth Serum Laboratories in 1970 (SUTHERLAND 1983) {see Chapter 5, page 114, and Chapter 21, page 420}.

The antivenom is made using hyperimmunised sheep serum, and rarely causes reactions, unlike snake antivenoms, which are horse antisera. In more than two decades of using *Chironex fleckeri* antivenom (at least 200 cases to date) there has only been one documented adverse reaction: a mild generalised rash 20 minutes after antivenom administration (SUTHERLAND 1990, 1992).

* Recent scrutiny of the tentacle preparation techniques by different workers may provide an explanation for this current enigma (J.W. BURNETT and K. LONG 1993, unpublished observations). Techniques for valid separation of nematocyst from tentacular toxins continue to be explored (LONG-ROWE & BURNETT 1994).

LABORATORY STUDIES OF *CHIRONEX FLECKERI* ANTIVENOM

Early experiments showed that this antivenom neutralised the lethal, haemolytic and dermatonecrotic effects of both 'milked' venom and whole-tentacle extracts of *C. fleckeri in vitro* (BAXTER & MARR 1974). Passive immunisation of mice with the antivenom reduced dermatonecrosis and protected against the lethal effect of the 'milked' venom (BAXTER & MARR 1974). However, in other experiments, *Chironex* antisera prepared from rabbits did not reduce dermatonecrosis (KEEN AND CRONE 1969). Subsequent experiments with mice showed antivenom could 'rescue' some but not all mice injected with *C. fleckeri* tentacle extract (BURNETT *et al* 1990).

Concern that the 'milked' venom used for antivenom production and used in some of the earlier animal studies may lack one of the significant lethal factors of *Chironex* venom is supported by studies showing the *in vitro* neutralising ability of the antivenom against crude venom to be markedly reduced when compared with the reported neutralising ability against 'milked' venom (ENDEAN & SIZEMORE 1988). In addition, the *in vivo* neutralising ability of the antivenom tested in both prophylactic and rescue experiments in mice using crude nematocyst venom was reduced approximately three-fold when compared with using 'milked' venom (ENDEAN & SIZEMORE 1988). The interpretation of some earlier studies is further weakened because successful use of antivenom in animal experiments has involved giving the antivenom either before injecting the venom or within a minute of venom injection (BAXTER & MARR 1974; ENDEAN & SIZEMORE 1988; BURNETT et al 1990). The implications of all these data is that higher doses of antivenom may be appropriate in life-threatening envenomations, and that further research is needed to ensure that animals used for antivenom preparation have been exposed to all the toxic fractions of the venom (CURRIE 1994).

Studies to establish the precise effectiveness of *Chironex* antivenom in human envenomations have not been possible, and survival (after both acute and protracted severe illness) is documented in massive envenomation with early and sustained resuscitation, before antivenom was immediately available (MAGUIRE 1968; WILLIAMSON *et al* 1980). While *Chironex* antivenom would be expected to neutralise circulating *Chironex fleckeri* venom, it remains uncertain if the 'milked' preparation of antivenom can completely neutralise venom already bound to tissues — in particular myocardial tissue (ENDEAN & SIZEMORE 1988; ENDEAN *et al* 1993c).

Despite all these considerations there is clear consensus among clinicians that antivenom plays a most important role in the management of severe envenomation from *Chironex fleckeri* (WILLIAMSON *et al* 1980, 1984a; FENNER *et al* 1989b; CURRIE & ANSTEY 1989).

INDICATIONS FOR ANTIVENOM USAGE

The indications for using antivenom in *Chironex* envenomation are:

1. A life-threatening situation with an unconscious and/or shocked patient or cardiac or respiratory complications
2. Severe, unrelieved pain from the skin lesions
3. Cosmetic concerns of skin scarring

There are clinical case reports to support all three indications.

Life-threatening stings

Life-threatening envenomations have been successfully treated with combined rescue protocols, which have included the use of *Chironex* antivenom (WILLIAMSON *et al* 1980; WILLIAMSON *et al* 1984; FENNER *et al* 1989; BEADNELL *et al* 1992). While the intravenous route of administration is preferable, the Queensland Ambulance Service has successfully used the antivenom intramuscularly on the beach (FENNER *et al* 1989, BEADNELL *et al* 1992). This is particularly important, given that death may occur within minutes of chirodropid jellyfish contact in serious stings (BARNES 1966), that the essential transport to hospital has inevitable delays and that risks are associated with the moving of any shocked patient. Although cause and effect is not established, cardiac arrhythmias have resolved following antivenom therapy in some cases (FENNER *et al* 1989; BEADNELL *et al*

1992; CURRIE 1992, unpublished observations). In other cases, cardiac arrhythmias have persisted despite the administration of antivenom, either transiently (CURRIE 1992, unpublished data) or till death (LUMLEY *et al* 1988).

A critical question is the appropriate dose of antivenom to use in life-threatening envenomation. Deaths have been reported from cardiotoxicity despite antivenom doses of 1 ampoule (20 000 units {see Chapter 21, page 420, and Plates 9.70A and B}) (LUMLEY *et al* 1988) and 2 ampoules (CURRIE *et al* 1992). The intramuscular dose contained in the kits used by the Queensland Ambulance Service is 3 ampoules (60 000 units) {Plates 21.1 and 21.2}, and when the antivenom is used intravenously up to 3 ampoules are recommended if the clinical response to a lesser amount is not sufficient (FENNER *et al* 1989).* Questions about the efficacy of the antivenom against specific components of *Chironex fleckeri* venom (ENDEAN & SIZEMORE 1988, LUMLEY *et al* 1988) justify consideration of even larger doses of antivenom in a critical situation, where survival is threatened despite full resuscitation and standard antivenom therapy. In one study it was calculated that about 12 ampoules of antivenom may be required to counter the effects of a theoretical envenomation containing twice the lethal dose of venom (ENDEAN & SIZEMORE 1988). There are obvious concerns about adverse reactions to such large doses of antivenom, despite the excellent safety record of this ovine product (SUTHERLAND 1990 & 1992). It is evident that more studies are necessary to clarify this matter and that emphasis must remain on preventing stings, early and sustained resuscitation, and early use of antivenom in life-threatening situations.

Pain relief

There has been documentation of impressive relief of pain using *Chironex* antivenom (WILLIAMSON et al 1984; BOYD 1984; HORNE 1988; CURRIE & ANSTEY 1989; KING 1991). However, complete resolution of pain does not always occur, and additional analgesic therapy such as narcotics may be required (FENNER *et al* 1989; BEADNELL *et al* 1992). In addition, antivenom is less likely to give pain relief in damaged skin when administered more than four to six hours after envenomation (WILLIAMSON *et al* 1984; CURRIE 1993, unpublished observations). Although previously believed to be ineffective for relieving the pain in severe envenomations (WILLIAMSON *et al* 1984), ice or cold-packs provide good pain relief in mild to moderate *Chironex fleckeri* stings (FENNER 1990, unpublished observations; CURRIE 1993, unpublished observations). Cold packs recently provided good pain relief in a lifesaver with *Chironex* stings to both feet (INTERNATIONAL CONSORTIUM for JELLYFISH STINGS 1993). At the Royal Darwin Hospital the majority of *Chironex* stings respond well to cold packs, with, in more severe stings, the addition of a single injection of a narcotic analgesic such as pethidine (meperidine) (CURRIE 1993, unpublished observations). Antivenom therapy is used if there is not adequate pain relief with this ice–narcotic combination.

Cosmetic benefits

In addition to pain relief, antivenom therapy has been reported to be associated with a significant reduction in the acute inflammatory response and possibly a better long-term cosmetic outcome (WILLIAMSON *et al* 1984; HORNE 1988; KING 1991). It is considered that for this cosmetic benefit, the antivenom needs to be given as early as possible and certainly within four to six hours of envenomation (BEADNELL *et al* 1992). This pure cosmetic role for antivenom remains speculative, but is supported by clinical experience with some followed-up cases (WILLIAMSON 1988, unpublished observations) {Plates 5.19A, B, C and Plate 9.73}. However, the acute inflammation of many *Chironex* stings resolves rapidly without any specific antivenom therapy, the majority of stings have little or no residual scarring, and (apart from secondary infection) scarring can still occur despite the use of antivenom (WILLIAMSON et al 1980; CURRIE 1993, unpublished observations). There have also been no systematic long-term follow-up studies of the acute scarring.

*The possible adjuvant therapeutic role of verapamil is discussed below.

CHIRONEX FLECKERI ANTIVENOM USE FOR STINGS FROM OTHER JELLYFISH

The efficacy of *Chironex fleckeri* antivenom against *Chiropsalmus* remains uncertain. In one study, *in vitro* neutralisation of the dermatonecrotic, haemolytic and lethal factors of the Australian *Chiropsalmus quadrigatus* venom occurred with the use of *Chironex* antivenom, while passive immunisation of mice with the antivenom reduced dermatonecrosis but did not protect against the lethal effects of *Chiropsalmus quadrigatus* venom (BAXTER & MARR 1974). Other studies using rabbit antisera prepared against *Chironex fleckeri* showed no *in vitro* cross-protection against the Australian *Chiropsalmus quadrigatus* venom (KEEN 1971). This uncertainty is of concern given the recent findings of large numbers of these *Chiropsalmus quadrigatus* during each 'safe' dry season in Arnhem Land waters of the Northern Territory (CURRIE 1992) and the many severe envenomations by chirodropids in the Philippines and other Indo-Pacific countries. At present the Commonwealth Serum Laboratories recommends the use of *Chironex* antivenom for '*Chiropsalmus*' sp. Australia envenomation, but it is not recommended for envenomation from any other non-chirodropid cubozoans, including the *Carukia* (Irukandji) envenomation (FENNER *et al* 1986). In view of the opinions expressed above, *Chironex fleckeri* antivenom could probably be used effectively for most, if not all, chirodropid stings in the Indo-Pacific.

SUMMARY OF MANAGEMENT OF *CHIRONEX FLECKERI* (AND OTHER CHIRODROPID) STINGS

Mindful of all of the foregoing, this summary is a compilation of the best available knowledge and is based on extensive field experience. As implied above, this sequence of management, with appropriate modification according to circumstance, serves as a guide for the management of any known chirodropid sting, anywhere in the world. Naturally, any of this procedure may be modified with advances in international knowledge and understanding.

FIRST AID

(See Plates 9.88 A–H, Stinger Drill sequence.)

- Retrieve the victim from the water and/or restrain if necessary.
- Send others for ambulance/medical help.
- Assess the conscious state and treat airway, breathing and circulation (ABC) as necessary {see Flow Chart, page 107}.
- Liberally pour vinegar over the stung area for a minimum of 30 seconds to inactivate remaining stinging cells on skin and on any adherent tentacles (Plate 9.88A).
- Apply compression bandages to major stings — those covering an area >50 per cent of one limb, or causing impairment of conscious state or impairment of breathing or circulation) {see also Figures 5.3A and 5.3B, page 110} — but only after at least 30 seconds of vinegar application. Do not waste time removing vinegar-treated tentacle material. Apply the bandages directly over the vinegar-treated tentacles (see Notes 3 and 4, below).
- Remain with the victim, treating with cardiopulmonary resuscitation if necessary, including during transport to health facility.
- If unconscious, or if there is evidence of life-threatening cardiac or respiratory decompensation 3 ampoules of antivenom (each containing 20 000 units [Plates 9.88F or H]) may be given intramuscularly by a trained health professional on the beach.
- In severe envenomation, use oxygen if available; Entonox (50% nitrous oxide 50% oxygen) can be administered for severe pain, but only in fully conscious and co-operative patients; its use should be discontinued if the patient's condition worsens.

MEDICAL MANAGEMENT

- Continue ABC (Airway/Breathing/Circulation) if necessary, and add oxygen to the breathing mixture; administer 100% inspired oxygen whenever possible.
- Monitor the oxygen haemoglobin saturation and electrocardiograph of patient.
- Secure intravenous access and administer a crystalloid solution.

- If unconscious, or if there is life-threatening cardiac or respiratory decompensation, or if a significant arrhythmia is present, administer a minimum of 1 ampoule of antivenom intravenously (each ampoule 20 000 units, diluted 1:10 with an isotonic crystalloid solution such as Hartmann's solution or isotonic saline, given over 5–10 minutes). In a life-threatening situation, up to 3 (or more) ampoules may be given intravenously consecutively if the clinical response remains inadequate.
- If after 3 ampoules of antivenom into a functioning blood circulation, there is persisting life-threatening cardiac decompensation or arrhythmia, then consider giving verapamil intravenously (0.1 mg/kg up to 5 mg adult dose) while continuing cardiopulmonary resuscitation. Verapamil can be slowly infused under ECG monitoring until the cardiac arrhythmia reverts or other unwanted effects occur. Inotropes such as adrenalin should be considered for persisting hypotension, but calcium should not be given.
- Cardiopulmonary resuscitation, employing oxygen-enriched air (preferably 100% oxygen) should be continued and not abandoned in the patient with ongoing circulatory arrest until after consideration of further therapy with even more antivenom (at least 6 ampoules total dose, if available) and consideration of more verapamil and inotropes. Intermittent positive pressure ventilation will assist in control of pulmonary oedema.
- In stings that are not life-threatening (no cardiac or respiratory decompensation, or impairment of conscious state) use cold packs for initial pain relief {see Plates 5.16 and 5.17}, together with intravenously administered analgesia if necessary (1 mg/kg of pethidine up to 50 mg adult dose initially). For pain not relieved by cold packs and narcotic analgesia, administer 1 ampoule of antivenom intravenously as above.
- Further intravenous narcotic analgesia may be administered when indicated, in conscious patients.
- Consider 1–2 ampoules of intravenous antivenom (administered as above) for potential cosmetic benefits in stings involving cosmetically sensitive areas (e.g. face or neck, especially in females).

NOTES:

1. The victim should be kept as still as possible, as movement may enhance venom absorption and compromise vital organ function in severe envenomation (clinical shock).
2. Do not rub the stung area either before or after the application of vinegar.
3. Once vinegar has been applied, it is unnecessary to waste time removing the tentacles in life-threatening envenomations, as they are rendered harmless by the vinegar after 30 seconds.
4. When vinegar is not available for a major sting, pick off the bulk of the adherent tentacles as quickly and as gently as possible without compromising resuscitation. Subsequently apply vinegar for at least 30 seconds at the earliest opportunity — if necessary, after compression bandages have been applied.

   The skin of the rescuer's fingers is too thick to permit significant penetration by nematocysts; although a faint prickling sensation may be felt on the fingers while handling tentacles that have not been treated with vinegar.
5. There is currently no substitute application as effective as vinegar in inactivating unfired chirodropid nematocysts. However, if vinegar is not available, beverages such as soft drinks may be of limited benefit as an alternative (CURRIE & ALDERSLADE 1993).
6. It should be remembered that vinegar cannot be expected to have any effect on the pain already inflicted by the initial sting.
7. It must always be remembered that the first priority remains care of Airway, Breathing and Circulation (ABC) {see Chapter 5, page 107}.

OTHER TREATMENTS

Cardiac drugs

Following resuscitation, respiratory and cardiac support and antivenom therapy, the roles of specific drugs in life-threatening *Chironex fleckeri* envenomation remain uncertain. Potential drug therapies relate to the cardiotoxicity of the venom. Because of the putative effect of the toxin on calcium influx release through voltage-dependent calcium channels, the slow channel calcium blocker verapamil has been studied in animals. Verapamil has been shown to delay death in envenomed mice (BURNETT & CALTON 1983), and to potentiate the action of *Chironex fleckeri* antivenom (BURNETT *et al* 1990). Despite these impressive studies there have been criticisms of the potential use of verapamil on theoretical grounds (FREEMAN 1984; CURRIE 1994; TIBBALLS 1994, unpublished observations), but further clinical experience with this drug for this envenomation is warranted before firm opinions are possible.

Life-threatening cardiotoxicity in *Chironex fleckeri*-envenomated humans is clinically manifest by poor cardiac output (low or unrecordable blood pressure), together with possible arrhythmias and pulmonary oedema (LUMLEY *et al* 1988, CURRIE *et al* 1992). Although under similar critical situations inotrope drugs such as adrenalin would be considered, it was suggested that these may conceivably exacerbate the pathologically contracted myocardial state caused by *Chironex fleckeri* venom (LUMLEY et al 1988). However, adrenalin was successfully used (later in the management) in one case of severe human envenomation before the availability of antivenom therapy (MAGUIRE 1968). While some animal studies suggested a beneficial effect of adrenalin and other inotropes (TURNER & FREEMAN 1969), other animal studies show no benefit from adrenalin (KEEN 1970). It has also been noted that excess sodium bicarbonate given during human resuscitation in general may depress myocardial contractility (CHAMBERLAIN 1989), and calcium supplements may also have adverse myocardial effects because of the postulated mechanisms of action of *C. fleckeri* venom (ENDEAN 1988b; LUMLEY et al 1988; MUSTAFA *et al* 1995 in Chapter 26). In addition to oxygen, therapy for pulmonary oedema may include the diuretic frusemide (BEADNELL *et al* 1992). Recent experience with pulmonary oedema from scorpion envenomation suggests that in severe pulmonary oedema, with or without hypertension, there may also be a role for angiotensin-converting enzyme inhibitors such as captopril and/or nitrates or prazosin (GUERON *et al* 1993). The use of calcium antagonists such as verapamil or nifedipine, angiotensin-converting enzyme inhibitors, nitrates and prazosin can all decrease blood pressure. Despite the theoretical concerns discussed above, inotropes such as adrenalin may still be required if hypotension persists despite use of these other drugs and measures.

Skin medications

Early animal experiments showed no benefit on skin necrosis of antihistamines, anti-5-tryptaminehydroxytryptamine compounds or cortisone acetate (KEEN & CRONE 1969). However, subsequent experiments suggested that the non-steroidal anti-inflammatory agent indomethacin and the anti-serotonin agent methysergide reduced the capillary leakage induced by *Chironex fleckeri* venom (BURNETT & CALTON 1986). At present there is no specific topical or oral therapy that can be recommended as routine for the stings of *C. fleckeri*. There is no clinical evidence supporting the use of topical steroid cream, and the stings should be treated in the same way as burns, including emphasis on avoiding secondary infection {Plates 5.20A–C and 9.66}. Antibiotics will sometimes be required if secondary skin infection develops. For the rare case of a delayed allergic reaction (Plates 9.85 and 9.86), topical ultra-potent steroid cream may be helpful {see Chapter 11, page 310}.

Pressure bandaging

First aid with the pressure-immobilisation technique using compression bandages {Figure 5.3A and B, page 110} has been advocated for major stings (WILLIAMSON *et al* 1984; LUMLEY *et al* 1988; FENNER *et al* 1989) {see also Plate 5.15}, but lacks controlled supporting data at present. This technique was originally developed to delay absorption of venom in snakebite (SUTHERLAND 1983; SUTHERLAND & KING 1991) and funnelweb spider bite (SUTHERLAND & DUNCAN 1980). The method would appear safe for chirodropid envenomation, provided it is used only after adequate application of vinegar, thus ensuring that enhancement of further venom release and its absorption does not occur with the bandage application. However, pressure-immobilisation first aid is most likely to be effective when venom is absorbed into lymphatics, and evidence suggests that *Chironex fleckeri* venom is also absorbed directly into the vascular system (ENDEAN *et al* 1969, KEEN 1970; ENDEAN & RIFKIN 1983; RIFKIN 1993). Other concerns with pressure-immobilisation are that in severe envenomation the area of stings to be covered may be extensive, making effective bandaging difficult and cumbersome during the time when the priorities are resuscitation, access to antivenom and pain relief. Application is easiest when a team of trained beach rescuers is available (Plates 9.88A–H). At present, pressure-immobilisation first aid is promoted by the editors (FENNER *et al* 1989; BEADNELL *et al* 1992) and is included in the protocols of the Queensland Ambulance Service, St John Ambulance (Australia) and Surf Life Saving Australia, but not in the protocols developed by the Northern Territory Department of Health and Community Services. The efficacy of this technique awaits controlled evaluation.

CHIRONEX FLECKERI VACCINE ATTEMPTS

In the 1970s the Commonwealth Serum Laboratories began studies of a potential vaccine for immunising life savers and others at risk of envenomation from *Chironex fleckeri*. A toxoid was developed that was thought to provoke an adequate immunity level in animals, but which caused skin nodules in monkeys (BAXTER & MARR 1975). Vaccination for people at risk is currently not considered a realistic option because of the rarity of severe stings and the fact the fact that regular booster injections would be required to maintain immunity. Most importantly, avoidance of jellyfish contact and therefore of envenomation should be the priority and is achievable with public education {see Chapter 5, page 103}.

CONCLUSION TO THE SECTION ON CUBOZOANS

These élite animals present the most important medical and taxonomic challenge of all venomous marine animals. Further studies concerning their world distribution, their envenomation mechanisms, the toxinology of their powerful venoms, and the development of therapeutic and prophylactic means of dealing with the phenomenal envenomation process of the chirodropids, are required. Researchers who work with chirodropids in their natural habitat cannot help but respect and admire them for their graceful elegance and envenoming power. The important ecological place they occupy in the marine environment should be recognised and respected.

CHAPTER 10

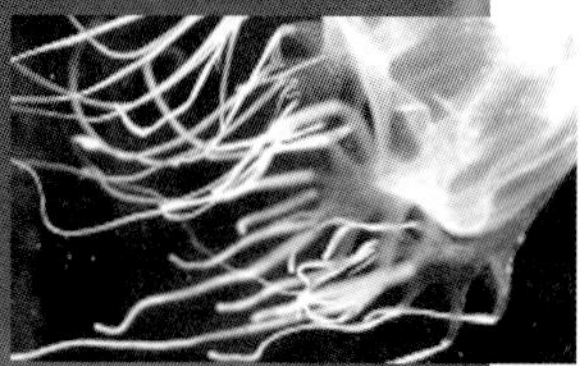

INTERNATIONAL INFORMATION ON JELLYFISH STINGS

JELLYFISH STINGINGS IN THE MEDITERRANEAN SEA

FRANCO KOKELJ

The presence of species of jellyfish in the Mediterranean Sea has been known for centuries (RICCI 1982). Aristotle (384–322 BC) knew of jellyfish stings, presumably in the Mediterranean region (STORER & USINGER 1957). In the Adriatic Sea one of the first known documented observations of jellyfish occurred in 1790 (GOY 1983), but systematic studies did not begin until the late nineteenth century. Although many reports have been published since then, the numerous jellyfish swarms and stings witnessed in the Mediterranean Sea in the last few years have stimulated a re-examination of the medical significance of these medusae, together with their histology, biology, distribution, and organisation.

THE ANIMALS

Eleven species of cubozoans have so far been identified in Mediterranean waters (Table 10.1).

Some of these animals occur very seldom (e.g. *Cotylorhiza tuberculata*). Others appear only occasionally and in small numbers (eg. *Paraphyllina intermedia*). *Aurelia aurita*, *Chrysaora hysoscella*, *Pelagia noctiluca* and *Rhizostoma pulmo* are the most common and appear cyclically in appreciable numbers in coastal waters. *Rhopilema nomadica* has recently assumed greater significance.

TABLE 10.1
SCYPHOMEDUSAE CUBOZOAN SPECIES KNOWN TO INHABIT THE MEDITERRANEAN SEA

Aurelia aurita
Carybdaea marsupialis
Chrysaora hysoscella
Cotylorhiza tuberculata
Discomedusa lobata
Drymonema dalmatinum
Nausithoë punctata
Paraphyllina intermedia
Pelagia noctiluca
Rhizostoma pulmo
Rhopilema nomadica

APPEARANCE OF THE ANIMALS, AND OF THEIR STINGS

Jellyfish species, endowed with nematocysts, can cause some damage after contact with human skin. The extent of this injury is both species and dose-dependent, and also varies with differing individual human susceptibility to the particular venom {see Chapters 3 and 8, pages 63 and 180, and pages 239–40}. Differences in our knowledge of these animals is due in part to their different habitats. For example, Mediterranean winter species such as *Drymonema dalmatinum* are less studied than those that appear in easily accessible coastal waters during the bathing season.

AURELIA AURITA (THE SAUCER OR MOON JELLYFISH)

Aurelia aurita has a milk-white or sky-blue, 20–40 cm diameter, saucer-shaped bell (RUSSELL 1970) (See Plate 10.1) {see also Figure 8.96}. It has long been known in small numbers in the Mediterranean Sea, but in 1962 and 1989 large numbers of *Aurelia* were reported in the Gulf of Trieste. This jellyfish, considered harmless by some, can sometimes cause significant skin lesions. RUSSELL (1970) reported two interesting clinical observations by colleagues: Ehremberg, who experienced no sensation when touching *Aurelia aurita* with his tongue and consequently judged this species to be inoffensive; and Weil, who wrote that *Aurelia aurita* was used successfully in the treatment of 'rheumatic pains and other neuralgia' at the baths of Sandfjord in Norway.*

Burnett and colleagues (BURNETT *et al* 1988a) reported an adult male patient stung by a Gulf of Mexico *Aurelia aurita* who developed painful urticarial skin lesions with post-inflammatory hyperpigmentation. This patient, a marine biologist, recognised an adult *Aurelia* next to his feet and developed a serological response to *Aurelia* antigens. Later analyses, however, showed that he also had titres to *Chiropsalmus* antigens. Although the cubozoan (box jellyfish) *Chiropsalmus* was not seen at the site, it is possible that this sting did not involve only *Aurelia* (BURNETT 1992, unpublished observations).

CARYBDEA MARSUPIALIS

Carybdea marsupialis (Class Cubozoa) {see Chapter 9, page 244} is an uncommon, small, Atlantic and Mediterranean box jellyfish (BOERO & MINELLI 1986). Small swarms can appear along the Riviera coast. Its precise world distribution remains uncertain. It has an elongated cubic bell of some 5–15 cm in diameter and of exceptional transparency. A 'true' carybdeid, it has a single tentacle at each of its four corners (CLELAND & SOUTHCOTT 1965) {see Figure 6.16, page 238}. Small swarms of this animal appeared along the Riviera and Adriatic coasts, during the summer of 1991, and in the summer of 1994 swarms of this jellyfish appearing along the same coasts caused significant stingings among bathers (KOKELJ 1994, unpublished observations).

* Editors' note: Our information from PACY (1957) and Pfleiderer (CLELAND & SOUTHCOTT 1965) is that this application of *Aurelia* occurred in the 'East Frisian Islands' (off the coast of the Netherlands) 'during the nineteenth century', and that it was *Cyanea capillata* that was used for 'jellyfish baths' in Norway. Furthermore, some serious reactions resulted in at least one patient. PACY (1957) also records skin eruptions to *Aurelia aurita* in five patients. On the present evidence it would seem imprudent to regard *Aurelia* spp. as harmless.

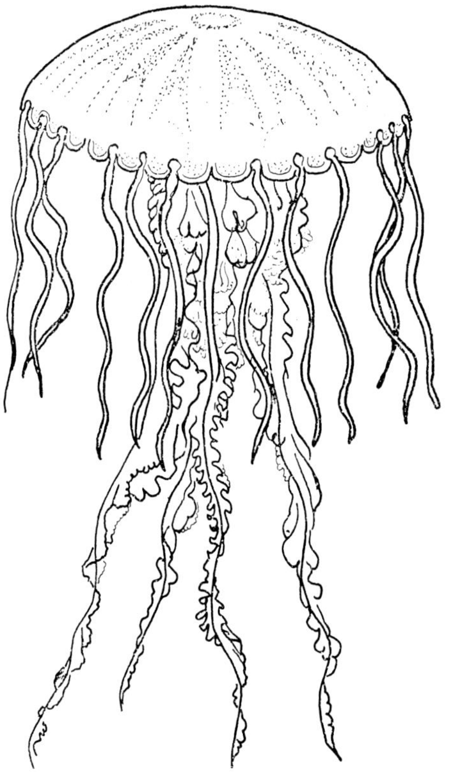

FIGURE 10.1
A schematic drawing of the Mediterranean *Chrysaora hysoscella* Linné, 1766. (Courtesy of Professor Victor Axiak, Department of Biology, University of Malta, and United Nations Environment Programme [UNEP].)

Stings caused by this animal result in burning pain lasting several hours, accompanied by a self-limiting (1–2 days) localised erythematous eruption. Small vesicles and urticaria can appear in the affected area (Plate 10.2) (KOKELJ *et al* 1992). On the basis of work with other cubozoans {see Chapter 5, page 108, and Chapter 9, page 240} these stings should be flooded with household vinegar, followed by the application of ice or cold packs to the stung skin for pain relief (EXTON *et al* 1989).

CHRYSAORA HYSOSCELLA

Chrysaora hysoscella (Plates 10.3 and 10.4), which is repeatedly observed in the Mediterranean Sea (although in small numbers), has also been described as a relatively innocuous medusa. The umbrella (bell) of variable colour has a flattened dome shape and can reach 20–30 cm in diameter. Its upper surface may exhibit radially dispersed markings in its central half, hence its other name (uncommonly used), 'compass jellyfish' (AXIAK 1983) {see also Figure 6.21, page 143 and Plate 8.81}. Goss (RUSSELL 1970) reported a case of a woman with dermatitis after a sting, and the author and his colleagues have recently observed a cutaneous eruption due to *Chrysaora hysoscella* in a marine biologist who handled this jellyfish without gloves (KOKELJ *et al* 1989); the dermatitis was characterised by a severely itching erythema of the hands (Plate 10.4).

COTYLORHIZA TUBERCULATA

Cotylorhiza tuberculata (Agassiz) is described by Axiak (1983) as a common Mediterranean jellyfish, but its stinging propensity attracts little mention (see below). It has a spectacular appearance (Figure 10.2 and Plate 10.5), a yellowish-brown sombrero-shaped umbrella with a central dome separated from a thinner peripheral margin by a shallow groove; this margin is divided into 16 lobes. There are eight oral arms with numerous purple-tipped projections. The umbrella may attain 15 cm in diameter.

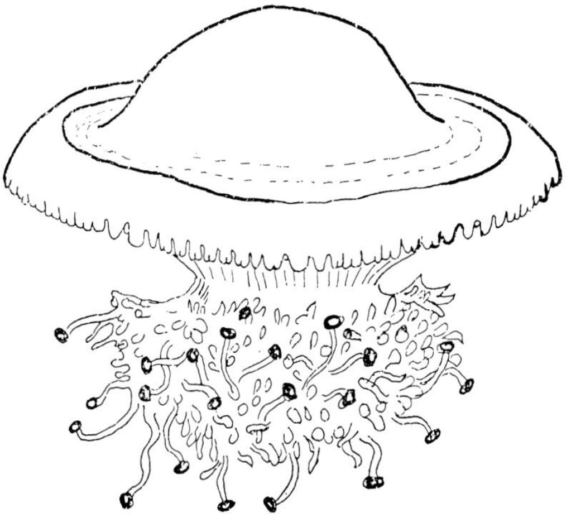

FIGURE 10.2
Drawing of the Mediterranean *Cotylorhiza tuberculata* (Agassiz). (Courtesy of Professor Victor Axiak, Department of Biology, University of Malta, and UNEP.)

PELAGIA NOCTILUCA (LITTLE MAUVE STINGER)

Pelagia noctiluca (Figure 10.3) {see also Figures 6.22 page 144 and 8.15 page 228, and Plate 8.86A} is found in many different oceans of the world and derives its name from its intermittent luminosity under natural conditions {see Chapter 8, page 228}. It is a small animal in the Mediterranean, usually not larger than 10 cm diameter (AXIAK 1983; VUCETIC 1985), and was considered rare there until recent years. Very large swarms, which were favoured by ecological and hydrological factors, occurred in the Gulf of Trieste during the summer of 1976 (AXIAK 1983 ROTTINI SANDRINI *et al* 1980), and again in the summers of 1977 to 1979 (MARETIĆ *et al* 1980), and on to 1986.

Pelagia noctiluca is certainly the most medically significant jellyfish of the Mediterranean Sea. Systemic reactions have been described, including the best documentation in the literature to date of anaphylaxis (TOGIAS *et al* 1985). A stung victim immediately feels a burning sensation at the site of contact; the only Mediterranean *Pelagia* sting adequately documented as having caused impaired consciousness was associated with an anaphylactic reaction (TOGIAS *et al* 1985). Following an initial phase of local skin ischaemia (blanching) {Plate 8.85}, an erythematous vesicular eruption, reproducing the imprint of tentacles, appears at the site of contact {see Plate 8.86B}. Wheals, papules and haemorrhagic blisters can appear; ulceration is rare but crusting with healing is not {Plate 8.86B}.

The rate and degree of healing depends on the size of the sting, the site of contact, the immunological reaction of the patient and the speed and efficacy of therapy. Sometimes lesions evolve into permanent scars, keloids (Plate 10.7) and hyper- and hypo-pigmented spots (KOKELJ & BURNETT 1990).

After multiple stings and/or in sensitised patients, vertigo, nausea, vomiting and possible hypotension are described (MARETIĆ 1983). Clinical and experimental observations support the hypothesis that *Pelagia noctiluca* venom has significant antigenic properties (TOGIAS *et al* 1985; BURNETT et al 1986a). Large local reactions,

FIGURE 10.3
Drawing of the Mediterranean *Pelagia noctiluca* Forskål, 1775. Compare with Plate 8.87A. (Courtesy of Professor Victor Axiak, Department of Biology, University of Malta, and UNEP)

delayed onset lesions and recurrent eruptions have been caused by *Pelagia* sp.

RHIZOSTOMA PULMO

Rhizostoma pulmo (Figure 10.4, page 288), a very common jellyfish in the Mediterranean Sea, has a large dome-shaped umbrella of different colours, which ranges from 15 to 60 cm in diameter (MAYER 1910; AXIAK 1983; HALSTEAD 1988). This animal usually appears in small swarms on the surface during the bathing season when the weather is fine and only occasionally causes significant stings in humans. It is considered essentially a coastal, rather than an offshore scyphomedusan and is a familiar sight along eastern Mediterranean shores (AXIAK 1983).

Eysenhardt reports that this jellyfish produces a burning sensation on the skin and particularly on the lips, while Von Zeyneck describes that sneezing and rhinorrhoea can follow handling of dried specimens of *Rhizostoma pulmo* as they are being prepared as hors d'oeuvres (RUSSELL 1970).

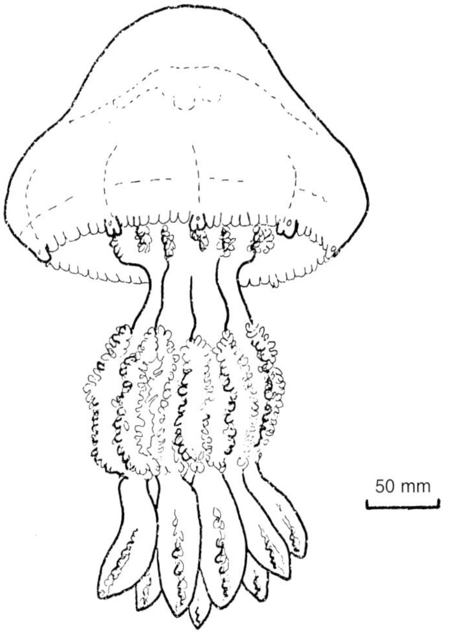

FIGURE 10.4
Drawing of *Rhizostoma pulmo* Macri, 1778, from the Mediterranean region. (Courtesy of Professor Victor Axiak, Department of Biology, University of Malta, and UNEP.)

Similarly Krumbach reported that sensitised people cannot stay in the room where *Rhizostoma pulmo* is being cut (RUSSELL 1970). Muir Evans has reported both urticaria and systemic symptoms such as dyspnoea and anxiety following *Rhizostoma pulmo* stingings (RUSSELL 1970).

Thus stinging effects by this animal appear variable, with two contrasting clinical phenomena having been observed. Children can play with and touch this jellyfish without apparent harm, whereas fishermen often report a burning sensation on their hands after tentacle contact. The rare dermatitis resulting from *Rhizostoma pulmo* is usually characterised by mild to moderate itching, with a light accompanying erythema that disappears spontaneously within a few hours.

RHOPILEMA NOMADICA

In 1990 another significant stinging species of jellyfish, *Rhopilema nomadica*, was identified along the Israeli coast by GALIL and colleagues (1990) (Plate 10.8). This jellyfish, whose hemispherical umbrella ranges from 10 to 85cm in diameter, has probably been present in the eastern area of the Mediterranean Sea for the past ten years, but its stings are likely to have been confused with those from *Pelagia noctiluca*.

Three types of nematocysts from *R. nomadica* have been described (AVIAN *et al* 1994). The venom of this jellyfish is dermatotoxic, as demonstrated both by skin tests (KOKELJ *et al* 1994) and from direct observation of stings sustained by bathers.

DIAGNOSIS OF MEDITERRANEAN JELLYFISH STINGS

As with any 'sting' it is essential to determine whether observed lesions are really caused by jellyfish {see Chapter 5, page 98}. If the stung person has not seen the animal (which is common), data on geographical area, weather conditions and season of the year are important. An exact medical history is (as always) fundamental. A physical examination (the earlier the better) should note the presence or absence of general symptoms, and the detailed characteristics and pattern of the lesions, including any regional lymphadenopathy (Table 10.2). These appearances may contain diagnostic clues (VUCETIC 1985) and may only become apparent in retrospect from high-quality close-up clinical photographs of the fresh skin lesions.

TABLE 10.2
IMPORTANT SYSTEMIC AND LOCAL REACTIONS RESULTING FROM ENVENOMATION BY KNOWN JELLYFISH SPECIES IN THE ADRIATIC SEA

Systemic reactions:	nausea
	vomiting
	shock (uncommon)
Local reactions:	erythema
	oedema (including angioedema)
	wheals
	blisters
	ulcers
Persistent (chronic) lesions:	scars
	keloids
	hyperpigmentation
	hypopigmentation

Next, an examination of specimens from the lesion(s) may be fruitful {see Chapter 5, page 98}. This includes superficial epidermal scrapings for adherent nematocysts, both discharged and undischarged (BARNES 1960; CLELAND & SOUTHCOTT 1965; SCHNADIG *et al* 1991) and selected skin biopsies, which are facilitated by expert interpretation. Then a sample of the patient's serum may permit the detection of specific immunoglobulin antibodies (especially IgG) to the offending jellyfish venom (BURNETT *et al* 1988b), although this evidence will often be 'after the fact', and such antibodies appear to offer no protection against the pain of envenomation {see Chapter 7, page 175}.

FIRST AID AND MEDICAL TREATMENT

First aid and medical management of stings due to Mediterranean jellyfish are the same in principle for all known species {see Chapter 5, page 106, and Chapter 8, page 231}, with the single exception of the possible use of vinegar for *Carybdea marsupialis* stings. Specific toxins within the different venoms of this group of animals have not yet been identified {see Chapter 7, page 175}. This is an important research goal.

SYSTEMIC REACTIONS

Systemic reactions to Mediterranean jellyfish envenomation are rare and may be due to a particular individual sensitivity of the patient (TOGIAS *et al* 1985) or a complication of the involved area due to large or multiple stings. In the former instance, basic resuscitation {see Chapter 5, page 106} and urgent medical attention is important, followed by hospitalisation. A similar response may occasionally be required for some regional and local complications of jellyfish stings (WILLIAMSON *et al* 1988). It is important to avoid rubbing the involved area and to limit the patient's muscular movements of the envenomated parts, to minimise the resultant enhanced absorption of toxins into the circulation by the 'muscular pump' (SUTHERLAND 1983; BURNETT *et al* 1987a).

Currently, the most likely serious systemic reaction to Mediterranean jellyfish stings is an allergic reaction (including anaphylaxis). Thus if pallor, hypotension and/or life-threatening dyspnoea and impaired consciousness rapidly develop it may be necessary, in addition to normal resuscitation measures {see the flow chart, Chapter 5, page 107}, to inject adrenalin (epinephrine) by the most effective route, as a matter of urgency and if necessary repeatedly (PIOLATTO & TARTAGLINO 1986). The recommended adult dose in an out-of-hospital anaphylactic or anaphylactoid emergency is 0.3–0.5 mg (i.e. 0.3–0.5 ml of 1:1000) adrenalin intramuscularly (IMI); for children use 0.01 mg/kg body weight IMI. These IMI doses may be repeated at five minute intervals. However, for circulatory failure or shock, or if there is no response to IMI adrenalin, intravenous (IV) adrenalin is desirable in a dose of 0.5 mg (adults) or 0.01 mg/kg body weight (children or small adults), given slowly and diluted to 1:10 000 (EXECUTIVE EDITORIAL BOARD 1989). Failing IV access, a double dose of adrenalin down the trachea can be tried, but this route is now recognised as being less reliable for drug administration (CHAMBERLAIN 1989). Cardiopulmonary resuscitation (preferably with oxygen) is essential to preserve brain oxygenation during the hypotensive/hypoxic episode, and to enable absorption of the injected adrenalin. In addition IV infusion, preferably with colloid solution, is urgently indicated in all serious cases. While the value of hydrocortisone and/or antihistamines as primary treatment for anaphylaxis is now discounted, they may have a place after the crisis is controlled (EXECUTIVE EDITORIAL BOARD 1989).

Patients particularly at risk of anaphylactic reactions include those with a history of previous stings. Patients whose envenomation produces systemic symptoms require close observation, because collapse may be delayed (BURNETT *et al* 1994a).

LOCAL SKIN REACTIONS

The aims of therapy for the local effects of jellyfish stings {see Chapter 7, page 178, and Chapter 9, page 282} are:

1. to avoid further discharge of unfired nematocysts in the damaged area
2. to remove safely tentacles still present on the skin
3. to administer therapy to minimise pain and arrest the local damage

4. to reduce the risk of secondary infection of the sting site, and to optimise cosmetic results.

Rationale

1. It is now apparent that the mechanism of jellyfish nematocyst inhibition by the local application of substances to the fresh sting site is a partially species specific and complex matter (HARTWICK *et al* 1980; BURNETT *et al* 1983; VUCETIC 1985; RIFKIN 1988a; FENNER *et al* 1993.). Several topical measures, including heat (PICCINETTI & PICCINETTI 1986), solar oils (sunscreen preparations), ethanol (SCARPA *et al* 1987; HALSTEAD 1988; HEEGER *et al* 1992) and aluminium sulphate suspensions (HENDERSON 1980; HENDERSON & EASTON 1980) have been espoused as generally effective, but controlled data are lacking.*

Further studies are necessary before specific recommendations can be made concerning nematocyst inhibition of the different jellyfish species of the Mediterranean area. However, the use of vinegar for all cubozoan stings so far tested is established {see above, and Chapter 5, page 108}. Similarly, suspicions that fresh water discharges nematocysts *in vivo* (VUCETIC 1985; BARNES 1966) requires further field study and interpretation (RIFKIN this volume).

2. Removal of adherent tentacle material should be performed as soon as reasonably possible, but this activity should not interrupt any resuscitation measures and ideally should be carried out only after local nematocyst inhibition. As already mentioned present levels of knowledge concerning nematocyst inhibition are inadequate, but fortunately many adherent jellyfish tentacles (with the exception of large chirodropid stings {Chapter 9, page 280} and certain carybdeid stings {Chapter 9, page 245}) can be removed without concern for patient or rescuer, even in the absence of prior nematocyst inhibition. Tentacles are to be picked off carefully with the fingers. The skin of the palms is usually too thick to sustain any significant pain or damage from handling the bare tentacles, but known or suspected sensitivity in the rescuer mandates the use of gloves. Do not rub or scrape off untreated tentacles, as this will initiate further nematocyst discharge. Washing off slippery fragments of fresh tentacle with sea water (not fresh water — see above) would appear safe and reasonably effective.

3. Again, despite the promulgation of a wide variety of topical and systemic substances as therapy for the sting; both as analgesia (HENDERSON 1980; ANGELINI & VENA 1991) and for the promotion of healing, few have been subjected to rigorous clinical trial and distinguished from a placebo effect. However, one satisfactory study showed topical corticosteroids or antihistamines relieved pain, burning and redness of jellyfish stings recorded in Greece faster than placebo (VLACHOS & KONTOS 1987). Proteases such as papain and bromolin can apparently digest jellyfish venom peptides, but applying such substances to a sting does not necessarily put them in contact with the injected venom. The local application of ice or cold packs has been shown to provide effective analgesia in some stings caused by other jellyfish (EXTON et al 1989). In the Trieste experience, oral analgesics should be given for the pain and potent topical steroids applied twice a day for three to five days for the eruptions caused by Adriatic jellyfish. We have used oral non-steroid anti-inflammatory drugs in an attempt to limit the local inflammatory reaction, but trials are lacking.

4. Secondary bacterial infection following a jellyfish sting is surprisingly uncommon but does occur {Plates 5.20A and 9.66} (WILLIAMSON *et al* 1984; BEADNELL *et al* 1992) {see Chapter 4, page 88}, and is usually preventable with attention to normal wound cleanliness.

Scars, keloids, hyper- and hypo-pigmentation changes at the sting site may present cosmetic problems, (Plate 10.7) particularly in young patients. Cryosurgery with liquid nitrogen and/or local steroid infiltration might be useful for keloids (BURNETT *et al* 1986), while topical hydroquinone has diminished the intensity of pigmented scars from jellyfish stings (KOKELJ & BURNETT 1990).

Acknowledgments: The author wishes to acknowledge the invaluable aid of Professor J.

* The editors have considerable reservations concerning some of these treatments, in particular 'heat application'. Some of these measures may later be found to be damaging.

W. Burnett of the University of Maryland and Professors C. Scarpa and L. Rottini Sandrini of the University of Trieste. Sincere thanks are also due to Dr John Williamson for assistance with revision of the manuscript, Mrs Marie Bone of Adelaide for her permission to use her photograph of the *Cotylorhiza tuberculata* and to Dr R. V. Southcott for provision of the same photo.

ADDITIONAL NOTES ON THE WESTERN MEDITERRANEAN REGION (SPANISH COAST)

JOSEP-MARIA GILI

The more common species causing human stings on Barcelona beaches include *Pelagia noctiluca* {Plate 8.87A} and *Chrysaora hysoscella* (Plate 10.3); these are more common during spring and early summertime, and may form dense swarms, resulting in many stings. The summer of 1994 (July–August) was a time of many stings on Barcelona beaches and the coastline of Greece. Stings have been reported from southern Spanish beaches (Malaga) and those of the Atlantic coastline, mainly by *Pelagia noctiluca*.

Of these two species, *Pelagia noctiluca* is the more troublesome. It becomes extremely abundant some 30–70 kilometres offshore, and onshore winds can bring them in onto the beaches.

Two other common but less troublesome species of cnidarians in the region are *Rhizostoma pulmo* (Figures 10.4, page 288) and *Cotylorhiza tuberculata* (Figure 10.2, page 286 and Plate 10.5). These rarely form dense swarms.

FURTHER INTERNATIONAL NOTES ON JELLYFISH STINGINGS

THE INDO-WEST PACIFIC REGION

JAPAN

YASUHIRO TOMIHARA, OKINAWA PREFECTURAL UNIVERSITY OF ARTS, AND YASUTETSU ARAKI, OKINAWA PREFECTURAL INSTITUTE OF HEALTH, DR MASIHIRO KOHAMA MD, AND THEIR SCIENTIST COLLEAGUES

A visit to Japan in 1992 by one of the editors (PJF) lead to contacts in Okinawa, an island in the very south of Japan (latitude 27° north). Chirodropid specimens are well known in this area.* They extend to Amami Islands in the north (28°) and have been tentatively identified as *Chiropsalmus quadrigatus* (SHOKITA 1986) (Plate 10.9) {see also Plates 9.38A, 9.38B}. The Japanese *Chiropsalmus quadrigatus*, like its Australian and Philippine chirodropid counterparts, is almost invisible in sea water under natural conditions {see Plates 5.8 and 9.56}. In Okinawa the colloquial name for these jellyfish is '*Habu-kurage*'. *Habu* means 'venomous snake' and *kurage* means 'jellyfish'. Nematocyst studies are in progress (Plate 10.10) {see Chapter 6, page 166}. Chirodropids cause up to 1000 stings annually in Okinawa (ARAKI 1994, unpublished observations). Children are common victims of stings.

Japanese engineers are currently planning 'stinger-resistant nets' similar to those in Australia (see Chapter 1, page 51) to help combat this problem (Note: three already *in situ* 1995). Unlike *Chironex fleckeri*, which is usually present on sandy beaches adjoining mangrove swamps, Japanese *Chiropsalmus* are often present on beaches with many rocky outcrops, making the use of stinger nets more difficult. Unlike in Australia, there are very few mangrove swamps, and local marine biologists believe that, unlike *Chironex*, the life cycle of this *Chiropsalmus* species does not include mangroves (ARAKI 1994, unpublished observations).

In Okinawa one fatal envenomation occurred in 1961, and two other severe envenomations (a two-year-old boy in August 1988 [Plate 10.11], and a seven-year-old girl in August 1989 [Plate 9.39]) required resuscitation on the beach within a few minutes of their stings.

* Editors' Note: Marine envenomation studies have long been pursued in Okinawa with vigour and discipline (SHIN KUSUNOKI 1970; OTSURU 1974).

CASE REPORTS

1. A fatal chirodropid envenomation
On 18 August 1961 a 14-year-old boy, while swimming at 10am at Shirahama, Shioya Bay, Okinawa Island, Japan (2° north of the Tropic of Cancer), was stung on both upper thighs in a water depth of 1 metre, only 2–3 metres from the shore. He died an hour later without receiving medical treatment. No other fatality is known (ARAKI, 1992 observations previously unpublished in English).

Serious envenomations

2 A two-year-old Japanese boy was stung on both thighs and hips by a *Chiropsalmus quadrigatus*, at 6.30 pm on August 1988, on an Okinawa Island seashore (Plate 10.11). Immediately after the sting he lost consciousness and stopped breathing. His father quickly commenced expired air resuscitation mouth-to-mouth, and his spontaneous breathing returned, however he remained cyanosed. He was carried to hospital where he developed haematuria, and blisters at the sting site. He was discharged a week later with remaining local scarring at the sting sites, but no systemic complications.

3 A Seven-year-old Japanese girl was stung on her left hand and right foot by a *Chiropsalmus quadrigatus*, at about 1pm in August 1989, on an Okinawa Island seashore. Five minutes after the sting she lost consciousness, and her father and a watching university student performed bare-handed cardiopulmonary resuscitation on the girl, on the beach, for the next 15 minutes, when spontaneous breathing returned. Twenty-five minutes after the sting an ambulance transported her to hospital. She was discharged well two weeks later with scars remaining on her right foot. {See also Plate 9.39.}

Seasonal occurrence of stings in Okinawa Island

Significant chirodropid stingings occur in Okinawa from June to September (the northern hemisphere summer) and usually in water of swimming depth (unlike Australia, where many chirodropid stings occur in wading depth shallow water). There were some 500 stings in Okinawa during the two-month period of July–August 1993. Victims are usually children under ten years of age (as in Australia [INTERNATIONAL CONSORTIUM for JELLYFISH STINGS 1993]), and scarring can result {Plate 9.39}.

Other cubozoans of no apparent clinical significance known to occur in Japan include the carybdeids *Tamoya virulenta*, *Tamoya gargantua* (KISHINOUYE 1910), *Tamoya bursaria (?gargantua)* (YAMASU & YOSHIDA 1976), and *Carybdea rastoni* (UCHIDA 1929; SATOH 1985; AZUMA *et al* 1986; OHTAKI *et al* 1990).

Multi-tentacled *Physalia* cause stings on the eastern coast of Honshu in the late summer (Japan Surf Life Saving 1992, unpublished data), but detailed information is not yet available.

Another significant clinical problem in Japan is a *Gonionemus* sp., a small hydrozoan which is discussed elsewhere in this text {Figure 6.5, page 125 and Plates 8.51 and 8.52}. Worldwide, *Gonionemus* has not been considered a particularly toxic medusa (HALSTEAD 1988), but recent information may well modify this opinion. *Gonionemus oshoro* UCHIDA, 1929 (Plates 8.52 and 10.12) causes numerous stings in Japanese swimmers on the island of Honshu (Plate 10.13). It occurs on the eastern rocky extremities of the Sea of Japan {Plate 8.51}, among seaweed that is harvested by female swimmers. This is on the opposite side of the Sea of Japan to the far eastern Russian coastline around Vladivostok, where the closely related (and equally toxic) *Gonionemus vertens* occurs {see Figure 8.8, page 202}. The latter animal, at least, can cause an Irukandji-like syndrome (AZNAURIAN 1958; OTSURU *et al* 1973) with apparent temporary psychiatric symptoms (Chapter 8, page 204).

Other Japanese jellyfish reports

In 1922, AOKI, writing in Japanese, reported via German literature the serious systemic effects of *Olindias formosa* {Figure 8.11, page 208} envenomation in Japan. HALSTEAD (1988) reports this species' distribution only from Japan. No first-hand details were obtained by the editors. Significant *Chrysaora quinquecirrha* stings have also been recorded in the Lake Saroma region of Hokkaido (KUSUNOKI, 1967). {See Chapter 20, page 414, for some details of other Japanese marine envenomation problems.}

PHILIPPINES

Chirodropids

On a personal visit in 1987, one of the editors (PJF) and Dr Robert Hartwick, established that chirodropid jellyfish were well known in the area and that stings occurred annually. Almost every village had a death apparently from this cause, every two to three years, usually a child. As there are some 50 small villages around the Bay of Sual (about 150km north of Manila) and many hundreds in tropical and subtropical areas of the Philippines, an estimate of an annual death rate from jellyfish stings in the Philippines of 20–50 was not thought excessive (FENNER 1987; CORNELIUS 1994, unpublished observations). As a death certificate is not required for burial in that region, verification of these estimates is impossible.

During this 1987 visit in late August, at the 'end' of the summer (jellyfish) season, several chirodropid specimens were caught in the Bay of Sual. They were tentatively identified by Dr Hartwick as *Chiropsalmus quadrigatus*, Haeckel 1880. These jellyfish were fairly abundant, and 11 were caught within the space of a hour in waist- to chest-deep water. All this time, three small naked children were playing in shallow water less than 100 metres away, totally unconcerned by the find.

Examination of these specimens showed that unlike *Chironex* the tentacles were very difficult to crush to permit the study of nematocysts. However, Dr Hartwick considered them to be similar to *Chironex*. Although the gonads were immature, Dr Hartwick considered probably one was a male with sperm, and another six were female with ova. Measuring the interradial diameters as accurately as possible with a simple plastic ruler, he determined the sizes of the 12 specimens were:

118 mm	(8 tentacles per pedalium
117 mm	(damaged pedalia)
113 mm	(two pedalia with 8 tentacles, one with 10 and the other damaged)
100 mm	(8 tentacles per pedalium [tpp])
95 mm	(8 and 9 tpp)
90 mm	(8 tpp)
63 mm	(7 tpp)
53 mm	(8 tpp)
52 mm	(8 tpp)
43 mm	(7 tpp)

The following clinical account was provided by a local doctor:

'We have many stings in the area and approximately 10 per cent of these cases die. The deaths usually occur within hours of the sting. The victim usually is in severe pain, sweats a lot, has trouble breathing and is usually very weak. As soon as possible, vinegar is used as it dissolves the tentacles. This has been used for as long as I can remember in this area' [Editors' note: LIGHT 1914b] 'and people often take it to the beach. I give victims an injection of "Antistine" or "Benadryl" [antihistamines] and usually a sedative and an analgesic' [Editors' note: not identified. Pethidine and morphine are banned in the Philippines]. 'If they have trouble with their breathing or have extreme tiredness I use 1 cc, 250 mg of intramuscular caffeine. Many cases get infected and I also treat them orally with antibiotics such as penicillin, tetracycline or chloramphenicol and then start them on "Decadron" [cortisone] and "Phenergan" [promethazine] 25 mg orally every four hours.'

'There have been three other deaths in the area in the past ten years. The first was a 35-year-old fisherman, thin but fit, who was found floating face down next to his banka with tentacles from a jellyfish all over his back. The next was a 10-year-old girl stung in Bolineo harbour five years ago. She walked back to the shore screaming and then collapsed on the beach and went a dark colour and died. She only had a few tentacle marks on the front of her stomach. The other death was two years ago when a seven-year-old boy was stung on the stomach and he died two hours later. None of these people saw a doctor.'

'This year many people have been stung but there have been no deaths. Two fisherman who were stung early this year caught the jellyfish, cut off the tentacle, soaked it in vinegar and then ate it! They say it toasted 'soapy' but good.'

'Other treatment is usually pouring gasoline over the tentacles, but sometimes vinegar and papaya [paw-paw] or lemon juice may be rubbed on the stung area.'

'It is interesting that it is fairly easy to distinguish between box jellyfish and other jellyfish when talking to many locals, because they have noticed the box shape and actually refer to them as "cubos"!'

Another clinical description sounded similar to an Irukandji syndrome {see Chapter 9, pages 249–52}. Although it is now believed that there are a number of jellyfish species capable of causing the Irukandji syndrome, no cases had yet been reported from the Philippines:

'I was diving in the south-east of Luzon in about 5 metres of water when a small jellyfish swam in front of me, I could see its tentacles behind, and one tentacle caught my lip. I couldn't see the jellyfish clearly as the water was murky and I just caught a glimpse of it, and a couple of tentacles in front of my mask. I carried on underwater for a couple of minutes but then started to feel I couldn't breathe properly and so I surfaced next to our boat. They told me I was unable to climb aboard and they pulled me in. I had pain in the centre of my chest and couldn't breathe, and they told me they had to do mouth-to-mouth resuscitation on me as I was breathing so shallowly. A wheal had appeared on my lip, and once I was breathing better (after five minutes) they gave me some coconut juice to drink that was supposed to help counteract the poison of the jellyfish. I then vomited for the next 10 minutes. They also put vinegar, Chinese herbs and hot-water compresses on my lip to help the pain that had started there where the tentacle had touched. They eased the pain but I could not tell which one specifically. The chest pain went after about 15 minutes, the breathing difficulty lasted about an hour, but the itching and burning pain from the sting on the lip lasted about six hours.'

Other jellyfish species

LIGHT (1914a) also reported significant stings in the Philippines from '*Dactylometra*', later renamed *Chrysaora quinquecirrha* {see Chapter 8, page 226} and distinguished them from chirodropid stings from the same area (CLELAND & SOUTHCOTT 1965). No evidence of further such stings was found during this 1987 visit. (Although numerous *Chrysaora* sp. stingings are reported in North America, most such stings are minor.)

THAILAND

One of the editors (JAW) visited Phuket, on the Andaman Sea, west Thailand, with Dr Robert Hartwick, in August 1985. Exploration of the local coastline, fishing villages, and excursions with local fishermen in their boats revealed *Catostylus* sp., and reports of nearby *Physalia* sp., but no chirodropids were seen. Local fisherfolk immediately recognised photos of chirodropids, and of human stings, and a village fisherman's wife proudly displayed leg scars from a sting three years previously which could have been a chirodropid. The fishermen knew the "summertime" chirodropids as *munga fi* {see Case Report below}, which translated approximately into 'fire jelly', and they associated hot, still weather and calm water with the risk of stings. Almost every fisherman admitted to *munga fi* stings at some time in his life. No deaths were known or admitted.

Later in Bangkok, at the Department of Dermatology, Siriraj Hospital, JAW learned that the presentation of patients with mainly chronic scarring from jellyfish stings sustained in the Bangkok area* was a regular occurrence (Plates 10.14 and 10.15). It appeared that keloid formation was a common reaction to jellyfish stings in Thais. A senior Bangkok scuba diving instructor, Lieutenant Colonel (Ret.) A.B.(Dusty) Rhodes, continues to report jellyfish stings and sightings in the Gulf of Siam and in the Andaman Sea, including one night time sighting of a chirodropid in the Andaman Sea {Plate 9.37}.

Studies in Thailand suggest that a locally occurring plant, '*phakbung-thale*', *Ipomoea pes-caprae* (L.) Roth, which is found in abundance along the warmer seashores of both hemi-

* Reference has been made elsewhere in the text to the occurrence of troublesome jellyfish stings in divers in the Gulf of Thailand {see Chapter 7, page 178}.

spheres, and which is used by Thai fishermen to prevent and treat jellyfish stings (WASUWAT 1970; APPLIED SCIENTIFIC RESEARCH INSTITUTE OF THAILAND 1977), may have a valid pharmacological basis (PONGPRAYOON *et al* 1991). Further studies are indicated.

CASE REPORT

(courtesy Dr Carl Edmonds, Sydney): A young male was stung on his right upper forearm and arm by an unidentified jellyfish, when he fell from his jet ski into the waters of Jom Teim Beach, Pattaya, Gulf of Thailand, at about 5pm in October 1994. The jellyfish was described as having a bell some 20–25 cm across, dome-shaped and with about 15 very thick tentacles, estimated as 1 cm thick and 50 cm long. The bell appeared to have some colour within it ("slight reddish" or "bluish"). Locals referred to the jellyfish as *magnopum* and described such jellyfish as tending to rise to the surface on cloudy days (it had been raining that day). The patient was aware of many such jellyfish present in the ocean at the time he was jet skiing.

The sting produced immediate burning pain and skin wheals at the contact site, followed within five minutes by early vesicle formation. Kerosene was applied without effect. Nematocysts on the skin were not sought. Extreme local pain persisted along with now-obvious vesiculation. One hour later he received an unidentified intramuscular injection, an unknown local application, and was commenced on antibiotics [type?]. (He remained on these and on oral analgesics and a non-steroidal anti-inflammatory agent for the next week.) The next day the pain had reduced, but generalised arm swelling developed on days 2 and 3 following the envenomation. The entire upper limb, including down to the distal interphalangeal finger joints, became balloon-like, more so with dependency, and painful on movement; frusemide had no apparent effect on the swelling. The patient terminated his holiday, and when seen in his home country on day 7, his arm swelling had reduced to involve just the upper arm and proximal one-third of the right forearm (Plate 10.15). There was about 30 cm of tentacle marks (in total) distributed in six major segments, each more than 1 cm wide, with necrosis, slough formation and scarring evident. The 4+ pitting oedema of the arm at this stage was not accompanied by venous or arterial impairment or signs of infection. Local skin itching was present.

He was maintained on ceftriaxone, and commenced on systemic prednisone 10 mg three times a day for three days, and local steroid ointment to the wound with elastic bandaging and elevation of the limb. Dramatic improvement followed within 24 hours. At 14 days post-envenomation, one week following this therapy, he had slight pitting oedema, and significant local pruritis controlled by fluocortolone cream, but had resumed full activities. The lesions remained strap-like with associated scarring.

SINGAPORE

One of the editors (JAW) visited Singapore in 1985, just prior to the death of a child on a Singapore beach from a blue ringed octopus envenomation (WILLIAMSON 1987) {see Chapter 3, page 69 and Chapter 13, pages 330–5}. No jellyfish were seen, and no stings heard of during this visit. Nevertheless, chirodropid jellyfish previously had been captured off Singapore (SEARLE 1957).

During a trip in 1987, another of the editors (PJF) observed that most of the island had a rock retaining wall built around to prevent erosion, and most of the original coastline vegetation had been destroyed and the edge of the sea altered by reclamation work.

There have been no recorded jellyfish deaths in Singapore, where all unusual deaths come to the notice of the government pathologist, hospitals keep accurate records, and certificates are issued for all deaths.

MALAYSIA

PETER FENNER, NOR AZILA AND IEKHSAN OTHMAN

Peninsular Malaysia (including Penang)

In 1985 one of the editors (JAW) visited Penang, where serious stings prior to 1965 are documented, including one fatality, (CLELAND & SOUTHCOTT 1965), and serious envenomations continue to occur (WILLIAMSON *et al* 1988; PEEL & KANDLER 1990; BURNETT *et al* 1994). In 1985 stings there were doused with a 'solution of ammonia', whose precise composition was not disclosed and whose efficacy appeared unproven {see Chapter 5, page 109} (WILLIAMSON *et al* 1988).* Some stingings on Batu Ferringhi beach involved jellyfish which were described as 'quite large and yellow' (see Plate 10.16) but which did not fit the description of chirodropids. No specimens were caught and identification was not possible. Of interest, a stinger enclosure (STARK & MOSS 1986 approx.; WILLIAMSON *et al* 1987) has been installed on Batu Ferringhi beach (BURNETT *et al* 1994).

In 1989 another of the editors (PJF) visited Kukup, a small fishing village in south-west Malaysia, where 'jellyfishermen' went to sea in large boats specifically to catch jellyfish. After return to port the catch of jellyfish was then dehydrated under compression by packing them in large jars with aluminium and magnesium sulphate between layers.† The dehydrated jellyfish were sold for human consumption, mainly to Japanese markets {see Chapter 8, page 212, and note on page 214}.

When shown photographs of chirodropids and their stings, local fishermen appeared to recognise them readily. On occasions they caught 'similar' jellyfish and also reported painful stings, although nothing more serious, as they took great care during their capture. They also displayed their 'ancient Chinese remedy' to put on such stings; it proved {interestingly — see Chapter 5, page 108} to be household vinegar (4–6% acetic acid in water [HARTWICK *et al* 1980]).

In October 1991 a third editor (JWB) visited Penang with a biochemist colleague, Dr I. Othman, PhD, and encountered a report of a 35-year-old stung on the neck and chest at Batu Ferringhi beach by an unidentified jellyfish and who became 'unconscious'. No reports of deaths were uncovered (OTHMAN and BURNETT 1991, unpublished observations). In 1989 a triathalon in Batu Ferringhi was cancelled because of multiple sea stings. Additionally, serious jellyfish (local Malay name *'ubur-ubur'*) envenomations by unidentified animals in Penang waters are documented (WILLIAMSON *et al* 1988; PEEL & KANDLER 1990; BURNETT *et al* 1994b). Field studies and enquiries by JWB in 1991 revealed the presence in Penang waters of the following:

- *Rhizostoma* spp. (see below)
- *Carybdea* spp.
- *Chrysaora* spp. (Plate 10.16)
- *Catostylus* spp., with several subspecies
- *Rhopilema* spp..
- *Lobonema* spp.

These same two workers subsequently claim to have identified *Chironex fleckeri* in Malaysian waters and demonstrated cytolytic properties of its venom, as well as of the venom from the Malaysian *Carybdea rastoni* (OTHMAN & BURNETT 1994).

However, *Cyanea* and *Aurelia* spp. were rarely encountered, and no evidence was found at that time for the presence of chirodropids or *Physalia* spp. (BURNETT & OTHMAN 1991, unpublished observations).

Known to occur in the Malaysian region are the rhizostomes *Thysanostoma flagellum*, *T. loriferum* and *T. thysanura*. Rhizostomes in the Indo-West Pacific are well known for their morphological diversity, where some 80 different species are currently recognised (CALDER 1991, unpublished observations). Furthermore, there appear to be considerable morphological differ-

* SEARLE (1957) has recorded the use by his colleague Dr J.R.Hendrickson of the juice of a slashed green pawpaw (papaya) on fresh jellyfish stings in the then Malaya, 'with beneficial effect' (CLELAND & SOUTHCOTT 1965).

† Aluminium sulphate is the principal component of a commercially available 'sting remedy'. See Chapter 5, page 109, for further discussion.

ences between Indo-West Pacific and American rhizostomes {see also Chapter 8, page 209}. Little objective detail is available concerning their stinging propensity, but it may be significant.

Sarawak

Correspondence since 1990 (REYNOLDS 1990–92, unpublished observations) with the Malaysian Department of Parks and Wildlife suggested there was a significant problem with chirodropid stings in Sarawak. The common name there for the jellyfish (see Plates 9.40A and B) is *malong*, and they are regarded locally as 'very dangerous', although no proof of deaths has been so far obtained. With the advent of tourism and beach activities, the development of a sting prevention and management programme is underway.

Labuan Island

Serious chirodropid envenomations and fatalities are documented (CLELAND & SOUTHCOTT 1965) and continue to occur (AZILA & OTHMAN 1993) (Plates 10.18 and 10.19) {see also Plates 9.46 and 9.47}. Many cubozoan stings affecting the people of Labuan are treated at home. The local people accept such stings as a normal part of life, and most victims neither go to hospital nor seek medical aid. Stings are often treated by rubbing the area with sand (cf. Australian Aboriginal treatment [CLELAND & SOUTHCOTT 1965]) and pouring on lime juice. Even the raw jellyfish bell may be eaten there with lime juice! The local jellyfish name is *balong api* (roughly translated means 'evil spirit of fire').

The jellyfish is usually present in April and May but may be present through to September. Locals seem to be aware of fatal stings, and fishermen wear long-sleeved shirts and long trousers — consequently suffering few stings.

LABUAN ENVENOMATIONS

(courtesy of Dr Nor Azila and Dr Iekhsan Othman, Universiti Malaya)

Case 1 A 12-year-old boy was stung on the legs while fishing with his father at night, off the north Labuan coast in 1991. His conscious state was impaired for some 24 hours in hospital. Plate 10.15 shows the appearance of his legs two years later.

Case 2 An 11-year-old girl was stung on the thighs in the same Labuan locality, at 12 noon, while swimming, in 1992. Her brother carried her home screaming with pain and with tentacles (long, segmented, glassy clear with a pinkish hue) still attached to the skin of her thighs. She became unconscious while the father was removing tentacles from her right thigh. She was transported to hospital with tentacles still attached to her right thigh, and the thigh was bandaged with the tentacles still attached. The patient developed locked jaw within 30 minutes and remained unconscious for 12 hours. Plate 9.47 shows the appearance of the scars 12 months later.

Sabah

The fishermen, wives and schoolchildren of Gaya Island, off Kota Kinabalu on the west coast of Sabah, are also well aware of the box jellyfish as they can be seen in the water around their dwellings from April to June. Deaths occur among children younger than nine years of age (AZILA 1993, unpublished observations), as do serious (Plate 10.19) and life-threatening stings (ANGKIE 1992).

Just south of Semporna, at Tawau, eastern Sabah, a large semi-dried unidentified species of chirodropid was seen for sale for human consumption in a market. It was described as being caught off Singapore. In Semporna there are many factories involved in processing non-toxic jellyfish (usually rhizostomes). Also in this area two small chirodropid specimens (6 cm bell height, seven tentacles arising from each corner) were caught and tentatively identified as *Chironex fleckeri* by MARSH (Western Australian Museum 1992, unpublished observations), although they were too immature for positive identification (RIFKIN 1993, unpublished observations). Other specimens have recently been collected from this region and sent to Australia {see Chapter 6, page 153 and Chapter 9, page 259}.

Current field and biological research into Malaysian cnidarians is being conducted by Dr Nor Azila, PhD, and Dr I.B. Othman, PhD, of Universiti Malaya in Kuala Lumpur (see Chapter 26, page 449).

INDONESIA

Indonesia sprawls across a huge Indo-Pacific equatorial area. The only chirodropids reported in this area have been *Chiropsalmus quadrumanus* (KRUMBACH 1925) {Plate 9.36 and Figure 10.7, page 301} and *Chiropsoides buitendijki* (HORST 1907) (=*Chiropsalmus buitendijki* {see Chapter 9, page 260}). The former identification is puzzling, as it is an atypical distribution for this species. It appears likely that *Chiropsalmus quadrigatus* would occur in the Indonesian area, but the nearest documented occurrences of chirodropids are in the Gulf of Thailand (RANSON 1945).

During a recent trip to a couple of the thousands of islands in Indonesia by one of the editors (PJF), the following observations were made:

Java

Java is the most heavily populated island in Indonesia, with a consequent great strain on its arable land. Most areas that are easily accessible presumably were once mangrove swamps and have been cleared for cultivation (similarly to Singapore) or made into fish and lobster farms. This mass destruction of chirodropid breeding grounds can be expected to have led to a reduction of the local chirodropids population.

Bali and Lombok

North Bali and west Bali have sandy beaches and are well cultivated with little sign of mangrove swamps, except in the north-west corner of Bali. East Bali is volcanic and had no sign of cultivation near the beaches, and no sign of mangroves — possibly destroyed in the devastating 1963 volcanic eruption.

South Bali appears to be a 'promising' area. Around Benoa Port, a large protected bay, there are many mangroves. However, a receding tide revealed a coral sea bed and not mangrove mud. This terrain is generally inhospitable to chirodropids.*

THE GREATER ATLANTIC REGION

THE AZORES (PORTUGUESE)

The editors have a recent case history from the Azores (1200 km west of Portugal, in the Atlantic Ocean) of a mild jellyfish sting in an adult male with a previous medical history of dysphasia, which produced a painful red mark on the underside of a foot, noticed the next day. This was followed by a three day period of 'hallucinatory fits'. The foot developed a 'very angry, red blistery rash' with localised sweating. Attempts to follow up this case report have not been successful (QUINN and FENNER 1993, unpublished observations).

THE BRITISH ISLES

DATA FROM SIMON MOORE

Distribution

The following species are recorded in the English Channel (usual maximum temperature 17°C in August): *Chrysaora hysoscella* {Plates 8.81 and 10.3}, *Cyanea lamarcki*, *Cyanea capillata* {Plate 8.90}, *Aurelia aurita* {Plate 10.1}, *Rhizostoma pulmo* {Figure 10.4, page 288} and *Pelagia noctiluca* {Plate 8.86A} (rare) (CAMPBELL 1989).

The following sightings, strandings and sting reports are documented in the British Isles dur-

*Editors' Note: North of the Equator chirodropids are usually present in the hot, dry season; south of the Equator they are usually present in the rainy season. This trip occurred at the end of the dry season. The close proximity of the Equator to the north of Bali (closer than to Darwin, which has chirodropids year-round) and yet the apparent complete lack of awareness of chirodropids or of stinging risk among the local Javanese questioned, would suggest that chirodropids are not present in South Bali — although in the previous experience of the editors, such evidence may be misleading. None were detected on this occasion. Further investigations are in progress.

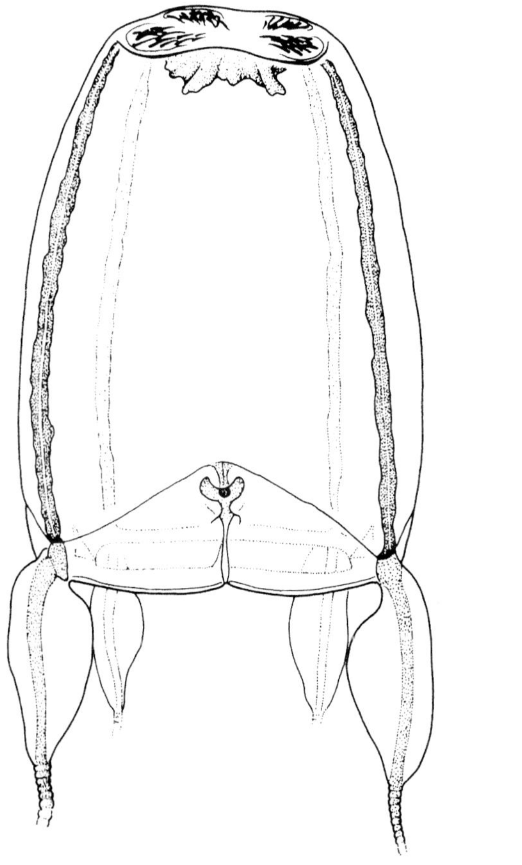

FIGURE 10.5
Diagram of *Carybdea alata*. Compare with Plate 9.8. (Courtesy of Professor Josep-Maria Gili, Barcelona.)

ing the summers of 1982 to 1988 (MOORE, 1992 unpublished observations*):

Rhizostoma octopus
: During June to October, 1983-85, numerous reports came from the British Isles.

Chrysaora hy(s)oscella (compass jellyfish)
: The Wash, North Sea and English Channel
: Ramsgate, July 1983
: Margate, August 1983
: Plymouth, June 1983
: Penzance, July 1984
: East Anglia, August 1984
: Southern Coast of Ireland (JWB, July 1990)

Aurelia aurita
: South Wales, Kent and English Channel (including one specimen 50 cm in diameter)

Physalia physalis
: Kent, July 1983
: South Cornwall, July 1983
: Dorset, June 1984

Cyanea capillata
: Dublin, summer 1994

Cyanea lamarcki †
: Devon and Cornwall, July and August 1984

Physophora hydrostatica
: Isle of Wight, June-July 1984

Velella (non-stinging) {see Plate 8.36 and comments in Chapter 18, page 406}
: South Wales, June-July 1983

Envenomation

Most stings in the British Isles are mild, and frequently not reported. However, severe constitutional symptoms have been described as occurring after envenomation by a probable rhizostome jellyfish in the English Channel. These

* During these same summer periods the following stingings/sightings were also recorded: *Pelagia noctiluca* in Malta, April-May 1982, and in Ibiza, Spain, in July-August 1983; *Cotylorhiza tuberculata* {Plate 10.5} in Ibiza, Spain, in September 1983; *Velella* {Plate 8.36} (sightings only) in St Tropez, France, in May-June 1983, in Sardinia in May 1983, and in Cork, south Ireland, in January 1988.

† Disciples of Sherlock Holmes will recall 'The Adventure of the Lion's Mane', in which the fatality on a Sussex (UK) beach was described thus: 'His [the body's] back was covered with dark red lines as though he had been terribly flogged by a thin wire scourge. The instrument with which this punishment had been inflicted was clearly flexible, for the long, angry weals curved round his shoulders and ribs. There was blood dripping down his chin, for he had bitten through his lower lip in the paroxysm of agony. His drawn and distorted face told how terrible that agony had been.' This makes for riveting mystery reading, and Conan Doyle's clinical description of the skin marks shows his research of the subject {see Chapter 8, page 234 and Plates 8.92, 8.93 and 8.94}. However, the symptomatic description is largely fantasy; a sting from *Cyanea* spp. (despite the dramatic account of WOOD [1874], which is thought to have been Conan Doyle's inspiration for the story [CLELAND & SOUTHCOTT 1965]) seldom produces more than annoyance.

included breathlessness and anxiety, followed by extreme lethargy and depression. Another case is described in North Wales, when painful muscular spasms developed in the abdomen and neck, together with low back pain (CLELAND & SOUTHCOTT 1965). EVANS (1943) describes 'a case of anaphylaxis with alarming symptoms' (unspecified) occurring immediately after a second sting, three weeks after an initial mild sting from a rhizostome jellyfish. He also describes respiratory symptoms including sore throat, rhinorrhoea and bronchospasm in girls at Lowestoft, England, who inhaled dried jellyfish substance while beating fishing nets to clean them. A recent confirmed *Cyanea capillata* stinging has occurred in Dublin (MOORE 1994, unpublished observations).

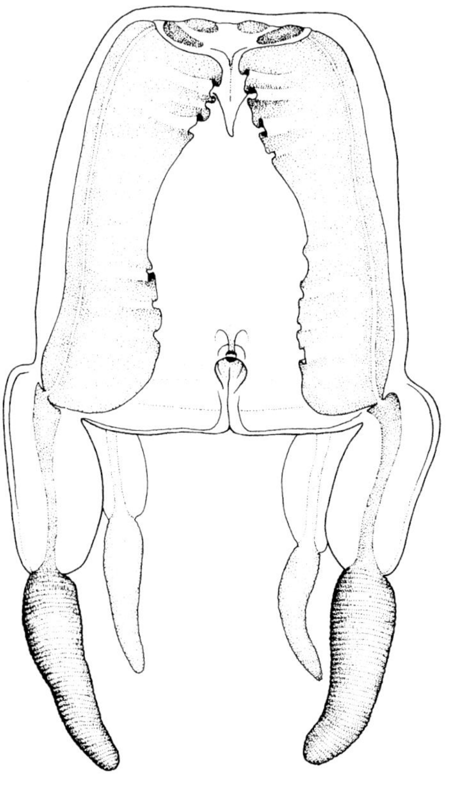

FIGURE 10.6
Diagram of *Tamoya haplonema*. Compare with Plate 9.7. (Courtesy Professor Josep-Maria Gili, Barcelona.)

THE CARIBBEAN AND ADJOINING REGIONS

BERTHA CUTRESS

Puerto Rico

The life cycle of *Carybdea alata* (Figure 10.5, page 299, and Plate 10.20 {cf. Plate 9.8} was first described here in 1973 (ARNESON & CUTRESS 1976). *Carybdea alata* stings can be quite troublesome in this area (CUTRESS 1992, unpublished observations) {see Chapter 9, page 244}. Also present in the waters are the other carybdeids, *Carybdea marsupialis*, *Tamoya haplonema* (Figure 10.6) {see also Plate 9.7}, *Tripedalia cystophora* (Plate 10.21), and the chirodropid *Chiropsalmus quadrumanus* (Figure 10.7, page 301) {see also Plate 9.36}, although no deaths have been reported from this region (however, a death from *Chiropsalmus quadrumanus* has been recorded in the nearby Gulf of Mexico's Galveston Island [BENGSTON *et al* 1991]). Not surprisingly, the Atlantic *Physalia physalis* (Plate 10.22) {also Plate 8.33} is also found in Puerto Rican waters.

Cuba

Carybdea alata sighting and stings have also occurred off-shore in Guantánamo Bay by the US Navy (BURNETT 1993, unpublished observations).

Netherlands (Lesser) Antilles

EDMONDS (1989) records multiple stings on a group of diving doctors from numerous small broken coelenterate tentacles (sea lice) {see Chapter 11, page 309}, in Bonaire in 1984. Identification of the animal is not recorded.*

The Gulf of Mexico

This region and the adjoining southern USA Atlantic coastline is beginning to increasingly feature prominently in documentation on venomous cnidarians, as well as serious and fatal marine envenomations (HEDGPETH 1954; PARRISH 1959; CLELAND & SOUTHCOTT 1965; BURNETT & CALTON 1985; BURNETT *et al* 1988a; STEIN *et al* 1989;

* At least one of us [JWB] suspects, on the basis of the evidence from WONG *et al* 1994, that this was caused by a *Linuche* sp. {see Chapter 11, page 309}.

BURNETT & GABLE 1989; BENGSTON *et al* 1991; BURNETT *et al* 1994), including some unexplained serious marine events (MEYER 1993). That chirodropids (Figure 10.7) {see also Plate 9.36} occur there has been known for over 20 years (PHILLIPS & BURKE 1970), and the whole Gulf and adjoining coastlines now warrant sustained field research concerning marine envenomation risks.

THE SOUTH ATLANTIC

HERMES MIANZAN AND FERNANDO RAMÍREZ

Brazil / Argentina

Olindias sambaquiensis {Plate 8.53} is a problem on the east coast of South America {see Figure 8.10, page 207}. *Physalia* spp. are also a problem to bathers {see Chapter 8, page 194}, and are carried far to the south to the more temperate seas and coastlines of southern Brazil and Uruguay {Figure 8.9 page 205} and Argentina (MIANZAN and colleagues 1993, unpublished observations), by the warm Brazil Current arm of the Atlantic south equatorial current. One such patient was found to have a definite contact-delayed hypersensitivity to the nematocysts of *Olindias sambaquiensis* (KOKELJ *et al* 1995).

Chiropsalmus quadrumanus {Plate 9.36} has been reported in French Guiana and the mouth of the Amazon (RANSON 1945), San Paulo, Brazil (VANNUCCI 1951), and Venezuela (KRAMP 1959). To the knowledge of the editors, there have been no reports of serious stings or deaths here.

Lychnorhiza lucerna Haeckel (also known as *Cramborhiza flagellata*) (Figure 10.8, page 302) is documented on the coast of Brazil and French Guiana (MACHADO 1943; HALSTEAD 1988). Presumably referring to the paper in Portuguese by MACHADO (1943), HALSTEAD (1988), describes significant local and systemic effects to the sting from this animal, but specific details are not provided.

West Africa

No account of serious stings or deaths has been found, but *Chirodropus gorilla* HAECKEL 1880 is reported from Liberia (latitude 7° north) (KRAMP 1955) southwards to South Africa (STIASNY 1931). An expedition from Spain recently caught several specimens in the Benguela current, off Angola, west Africa. A specimen {Figure 9.9, page 268} was kindly donated by one of the expedition members, Dr Josep-Maria Gili, Consajo Superior de Investigaciones Cientificas, Instituto de Ciencias del Mar, Barcelona, España; this specimen is currently being studied.

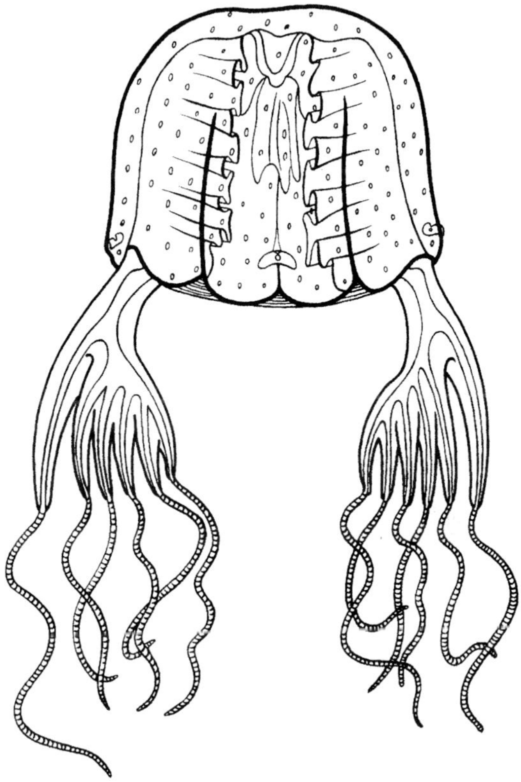

FIGURE 10.7
Line diagram of *Chiropsalmus quadrumanus*. Note the missing gonads. (Drawing based on R. Kreuzinger's from MAYER 1910.)

The mysterious *Chirodropus palmatus* has been reported near St Helena Island, in the Atlantic Ocean (latitude 15° south) (HAECKEL 1880). Its identification is based solely on this single report {see Chapter 9, page 262}.

SOUTH AFRICA

Physalia spp., both single- and multi-tentacled {see Chapter 8, page 195}, are an on-going problem to bathers along South Africa's coasts, the western (SHANNON & CHAPMAN 1983), southern and eastern (FENNER 1991, unpublished observations). SHANNON and CHAPMAN (1983) described a multi-tentacled species during a heavy infestation of *Physalia* on the south-western coast of South Africa in 1983.

The victim of a recent *Physalia* sp. sting in South Africa had a severe reaction with predominantly mental symptoms. Mental symptoms in the absence of circulatory compromise, and which appear to be directly toxin-induced from jellyfish stings, are notably associated with *Gonionemus vertens* {see Chapter 8, page 201} in the Sea of Japan, and the single case report from the Azores described above.

SUEZ CANAL AND THE RED SEA

Recent accessible published information on jellyfish stings from this area has been sparse. *Sanderia malayensis* Goette, 1886 {Plate 8.88} (BROWNE 1926, unpublished observations) and the sea anemone *Triactis producta* Klunzinger [LEVY *et al* 1970] have been reported in the region.

FIGURE 10.8
Lychnorhiza lucerna Haeckel (*Cramborhiza flagellata*). The bell may attain 12 cm diameter. (Diagram by R. Kreuzinger. (After MAYER 1910) from HALSTEAD 1965, courtesy of the United States Office of the Assistant Secretary of Defense, Washington DC.)

INDIAN OCEAN

EAST AFRICA

There are no references to chirodropids in East Africa to the knowledge of the editors. *Chirodropus gorilla* {Figure 9.9, page 268} is present on the west coast, and *Chiropsalmus quadrigatus* {Figures 6.29 and 6.30, pages 151–3} has been reported in the Maldive Islands (STIASNY 1937) — see below.

Sanderia malayensis {Plate 8.88} is also recorded in this region (MAYER 1937), north to Somalia (STIASNY 1937).

THE GULF OF OMAN AND THE ARABIAN (IRANIAN) GULF

A serious problem from a range of venomous marine animals exists in this region (COOPER 1985, 1991) (Plate 10.23). It is the subject of a specific chapter (Chapter 18) in this text by Professors Pratap Chand and Reginald Victor, and the reader is referred to page 404.

Work by Dr Paul Cornelius (1993, unpublished observations) {see above, and Chapter 8, page 231}, preceded by the efforts of others (STIASNY 1937; KRAMP 1956; MICHEL *et al* 1986; Dr Jassam AL-HASSAN, Biochemistry Department, University of Kuwait 1985, unpublished observations) implicates *Sanderia malayensis* Goette, 1886 {Plate 8.88} as a significant marine stinger in the Arabian Gulf.

PAKISTAN AND THE ARABIAN SEA

JUNAID ALAM, PHD, AND COLLEAGUES

Studies performed at the Department of Pathology, Basic Medical Sciences Institute, Jinnah Postgraduate Medical Centre, Karachi, Pakistan, and at the Department of Biochemistry, University of Karachi.

Physalia sp. (see Figures 8.5, page 195 and 10.9} are found on the northern shores of the Indian Ocean (Arabian Sea) in Pakistan, where they cause hundreds of stings each year (ALAM *et al* 1992). The stings of these hydrozoans are numerous on Karachi beaches (where the animals have occurred in enormous aggregations, e.g. Clifton Beach, Figure 10.10, page 305) and can be serious, causing severe skin pain and systemic effects which include muscle pain, sweating and anxiety. *Aurelia* sp. {Figure 8.13, page 224 and Plate 10.1} are also abundant in the Pakistani coastal environment (J.M. ALAM, S.M. ALAM, R. QASIM 1992, unpublished observations).*

Photographs (Figures 8.5, page 195 and 10.9) and sting reports suggest that the multi-tentacled *Physalia* species at least is present in Pakistan waters, and the single-tentacled *Physalia* species is present in nearby Omani waters.

The stinging *Sanderia malayensis* {Plate 8.88} is recorded off Karachi (GUARDAFUI, 1933, unpublished observations) {see above, and Chapter 8, page 231}.

SRI LANKA

Sea snake envenomation has occurred in Sri Lankan seas (FERNANDO & GOONERATNE 1983) and also cubozoan (chirodropid?) stings (FERNANDO 1991, unpublished observations) {see Plates 9.60 and 9.61}. Efforts to collect Sri Lankan specimens and study the incidence of stings there are currently underway, spearheaded by Dr Malik Fernando.

INDIA

The general and sometimes serious problem of jellyfish stings in swimmers is well recognised in India (ADIGA 1984; WILLIAMSON *et al* 1988; GHOSH *et al* 1993; Dr M.R.RAGHUNATH 1991, Central Institute of Fisheries Technology, Matsyapuri, Kochi-682 029, unpublished observations).

FIGURE 10.9
An apparent multi-tentacled *Physalia* sp. captured on a Karachi beach in Pakistan. Four long tentacles have been cut off for nematocyst studies. (Prepared by and reproduced with the courtesy of Junaid M. Alam, Department of Pathology, Basic Medical Sciences Institute, Jinnah Postgraduate Medical Centre, Karachi.) Compare with Figure 8.5, page 195.

The Indian Ocean constitutes one of the world's main reservoirs of venomous marine animals (CLELAND & SOUTHCOTT 1965; HALSTEAD 1988). Venomous marine cnidarian identification in seas immediately proximate to India include *Physalia* spp. (ADIGA 1984), *Sanderia malayensis* {see Chapter 8, page 231}, *Acromitus rabanchatu* (GHOSH *et al* 1993) and several cubozoan species, which are listed below. In view of the copious evidence of other venomous cnidarians to the immediate east, west and south of the Indian subcontinent, it is expected that future publications will reveal other resident jellyfish of significance.

* A survey and research effort is underway by Junaid M. Alam and colleagues (ALAM *et al* 1992 & 1993); this work includes the distribution of *Physalia* along the coast (Figure 10.10, page 305), the incidence and severity of stings, toxinological studies and an examination of local treatments. The last mentioned includes possible antiinflammatory properties of onion plants (*Allium cepa*). Although the main focus is presently on the Pakistani *Physalia* spp., associated studies are also involving the toxins of shellfish (*Perna viridis*), some *Conus* spp. and sea snakes. (J.M. Alam, S.M. Alam, R. Qasim 1992, unpublished observations.) (See Chapter 26, page 452.)

CASE REPORTS:

1. Brachial artery spasm (just above the elbow) occurred following a second *Physalia* sting two weeks after the first, at Mangalore in the Indian Ocean in the summer of 1983. The cause was attributed to hypersensitivity (ADIGA 1984; WILLIAMSON *et al* 1988). Success followed surgical decompression with local injection of reserpine 1 mg. The brachial artery started pulsating weakly, and in 36 hours the radial pulse returned. No residuae were present at discharge three weeks later.

2. In December 1986, while swimming off Goa in the Indian Ocean, a healthy 31-year-old woman was stung on the upper body and both arms by an unidentified jellyfish. She was flown, semi-conscious, and with gross limb tissue damage (Plates 10.25A and 10.25B) home to Frankfurt. After extensive but relatively unsuccessful medical treatment she underwent bilateral hand and lower arm fasciotomy (HACH WUNDERLE *et al* 1987). She made a fair recovery but remains partially incapacitated in both hands and arms. The patient described a "large reddish" mass in the water around her at the time of the sting {see *Sanderia malayensis* in Chapter 8, page 180; *Cyanea* sp. in Oman in Chapter 18, page 404; and comments with Plate 10.23}. Later serological examination for antibodies suggested (but did not establish with certainty) *Physalia* as the possible culprit (WILLIAMSON *et al* 1988). (Reddish-brown specimens of the normally- blue-coloured *Physalia* have also been observed [Plate 10.26].) Other antibodies detected included those of *Chironex fleckeri* and of *Chrysaora quinquecirrha*. Cross-reactions to jellyfish venom antibodies are known (BURNETT *et al* 1988b). *Sanderia malayensis* antigen was not available for testing.

As mentioned previously, chirodropids are present in Arabian/Indian seas. Chirodropids reported there and in adjacent ocean areas include *Chiropsalmus quadrigatus* HAECKEL 1880* {see Plates 6.8, 6.9 and 10.9} in Rangoon (HAECKEL 1880, STIASNY 1922), South India (MENON 1936) and Maldive Islands (STIASNY 1937); and *Chiropsalmus buitendijki* HAECKEL 1880 'India' (MENON 1936),* on the Trivandrum coast, southern India (NAIR 1951). Carybdeids in the same regions include *Carybdea alata* (Figure 10.5, page 299 and Plate 10.20) in the 'Indian Ocean off Madras' (MENON 1930) and the Red Sea and Indian Ocean (STIASNY 1940).

Such a reddish jellyfish as described above in Case 2, in Goa, might also have been *Sanderia malayensis* {see Chapter 8, page 231}. A case description of envenomation by this animal from Kuwait (JASSAM AL-HASSAN 1985, unpublished observations) contains some similarities.

CHINA SEAS

CONTRIBUTIONS BY MINGLIANG ZHANG AND QIN SHEDE

The only other non-cuboid jellyfish besides the Atlantic Siphonophore *Physalia* presently known to cause human death is *Stomolophus nomurai* in Chinese waters {see Chapter 8, page 214 and Plates 8.62–8.65}. Reports of at least five deaths have been published (MINGLIANG 1987, 1988, 1992), and the research of Professor Qin Shede, Dr Mingliang Zhang and colleagues, of the Department of Dermatology, Affiliated Hospital of Qingdao Medical College, is shedding further light on this problem.

Further field and laboratory work by Professor QIN SHEDE and Dr MINGLIANG ZHANG (1994, unpublished abstract for the 4th China-Japan Joint Meeting of Dermatology) has revealed some 32 species of cnidarians along the coast of China (13 not previously reported); the most venomous detected included *Aglaophenia whiteleggei* {see Plates 8.24 and 8.30}, *Physalia utriculus* {see Plates 8.31 and 8.32}, *Stomolophus nomurai* {see Plate 8.62}, *Cyanea nozakii*

* The chirodropid taxonomic names used here are as reported in the original literature. However, in the light of the new work presented by Rifkin in Chapter 6 {see page 150}, these namings should now be changed.

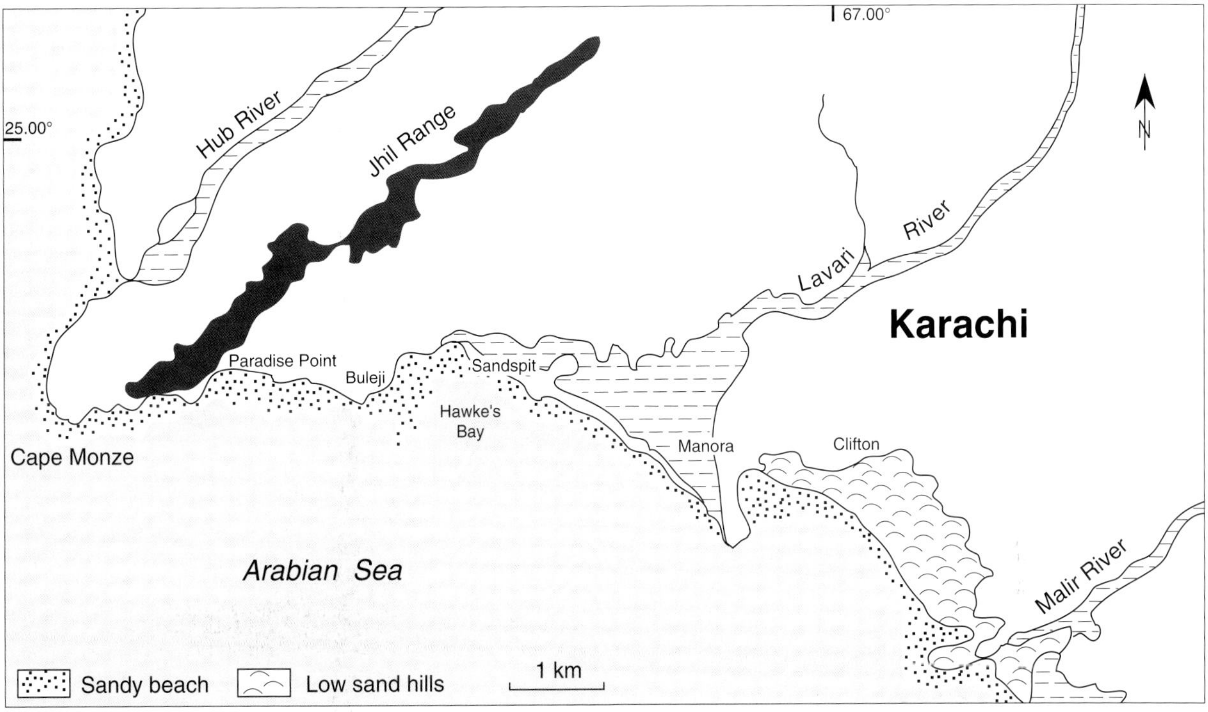

FIGURE 10.10
The coastal areas and major beaches of Karachi, Pakistan, where jellyfish sting surveys are presently being conducted. 1 Cape Monze. 2 Paradise Point. 3 Buleji. 4 Hawksbay. 5 Sandspit. 6 Manora. 7 Clifton. (Courtesy of Junaid M. Alam, Department of Pathology, Basic Medical Sciences Institute, Jinnah Postgraduate Medical Centre, Karachi.)

KISHINOUYE, 1891 and the sea anemone, *Haliplanella luciae* (QIN SHEDE & ZHANG MINGLIANG 1994, unpublished observations) {Plates 8.5, 8.12 and 8.13}.

Preserved and dried jellyfish for human consumption appear in Beijing markets at all seasons of the year; these would include *Stomolophus nomurai* (KIMURA *et al* 1983). These products are known as 'hydjoe' (BURNETT 1994, unpublished observations).

TABLE 10.3
SUMMARY OF THE KNOWN OR SUSPECTED DISTRIBUTION OF CHIRODROPID JELLYFISH

Chironex fleckeri*	***Chiropsoides buitendijke***	***Chirodropids****	***Chiropsalmus quadrumanus***
Australia	India — south	India	USA
? Papua New Guinea	Sri Lanka	Maldive Islands	North Carolina
? Borneo	Malayan archipelago	Burma	Georgia
Chirodropus gorilla	Indonesia — Java)	Gulf of Thailand	Texas
Liberia	(Australia)	Singapore	French Guiana
West Africa		Brunei/Sarawak	Brazil
South Africa (west)		Vietnam	Amazon mouth
		Philippines	Venezuela
		Japan	(West Africa)
		Papua New Guinea	(Indian Ocean)
			(Sumatra)
			(Australia)

'?' indicates a chirodropid detection reported but not verified.
Brackets round a country name indicate confirmation of the presence of a chirodropid, but not species identification.
*The reader is referred to the taxonomic considerations relating to Indo-Pacific chirodropids, described by Dr J. Rifkin in Chapter 9, page 150.

CNIDARIAN ENVENOMATION SYNDROMES

JOSEPH W. BURNETT

It should be emphasised that cnidarian envenomation syndromes may be quite variable because patients may be medicated for other diseases or afflicted with concurrent disorders. Jellyfish stings usually can be associated with cutaneous symptoms because of the pain they inflict. In two documented instances, however, pain-free stings were experienced: in the first, the patient was receiving prednisone, and in the latter, the patient had ingested significant amounts of alcohol (BURNETT *et al* 1987 [case 2]; FENNER *et al* 1993). Whether the role of concomitant administration of medication further affects the evolution of the clinical syndrome has not been established.

Usual development of the reaction to stings has been noted in patients who were pregnant (WILLIAMSON *et al* 1980) or those receiving ragweed rhinitis desensitisation (BURNETT 1995, unpublished observations). The interplay of other conditions on sting syndromes is not known. The effect of solar exposure on those disorders has not yet been evaluated. It has been shown experimentally that ultraviolet light exposure can affect an animal's cellular immune response against the venom (MIURA *et al* 1993). Indeed the venom itself can depress human cellular immunity (WACHSMAN *et al* 1991). These two negative actions on the immune system (venom and light-induced) are closely related and are probably complementary rather than additive. The level at which either or both must function before permitting clinical symptoms or other immune abnormalities to appear is speculative. The clinical counterpart of this observation in humans has not been recognised but remains a theoretical possibility.

The currently accepted list of cnidarian envenomation syndromes is as follows.

CLINICAL SYNDROMES PRODUCED BY JELLYFISH ENVENOMATION

Local reactions

Toxin-induced inflammation to skin, mucosa, and cornea
Exaggerated local reaction (angioedema)
Recurrent reactions up to four episodes
Delayed persistent reactions up to several months
Distant site reactions
Local lymphadenopathy
Seabathers' dermatitis

Long-term reactions

Keloids
Pigmentation
Fat atrophy
Contractions
Ulceration
Gangrene
Vascular spasm
Mononeuritis
Autonomic nerve paralysis
Ataxia
Increased ocular pressure
Mondor's disease (thrombophlebitis)
Deep venous thrombosis ('blueing')

Post-episode dermatitis

Herpes simplex
Granuloma annulare

Reactions from jellyfish ingestion

Gastrointestinal symptoms
Urticaria

Systemic reactions

Toxin-induced
Respiratory acidosis, pulmonary oedema
Blurred vision
Monarticular arthralgia and reactive arthritis
Pronounced vomiting
Psychosis, convulsions, coma, stupor
Fever
Muscle spasm

Fatal reactions

Toxin-induced
Immediate cardiac arrest
Rapid respiratory arrest
Delayed renal failure
Liver destruction
Allergy-induced
Anaphylaxis

CHAPTER 11

The Reaction of Exposed Body Surfaces to Cnidarian (and Related) Stings

Dermatological aspects of jellyfish (and related) envenomations ('Aquatic dermatology')

Joseph W. Burnett

Jellyfish-toxin-induced injuries to the skin, mucosa and the cornea of the eye (see below) result in instant pain or burning. In the case of the skin this is followed by local erythema and oedema within seconds; the cutaneous eruption may fade within hours or progress to vesiculation, bulla formation and possible subsequent necrosis and gangrene* (Plate 11.1). Exaggerated local reactions in the form of angioedema have been reported, as have recurrent episodes of the eruption for up to four cycles (BURNETT 1992a).

Recurrent eruptions are usually pruritic rather than painful, but their clinical appearance may match that of the initial episode. The duration of each recurrence shortens, and the interval free period (5–21 days) between cycles lengthens with successive exacerbation. Delayed persistent reactions in the form of granulomatous, nodular lesions have been reported (BURNETT 1992a). These reactions may scar and are reduced with intralesional or systemic corticosteroids. Inflammatory cutaneous reactions have been reported at sites distant to

*KOKELJ and colleagues (1993) recommend the comparative value of skin prick, skin patch and patch-scratch tests for the proper evaluation of cases suspected of having an allergic component in the eruption.

tentacle contact in *Linuche* stings (TOMCHICK *et al* 1993) and *Physalia* stings (MATUSOW 1980); in the latter instance the lesions occurred intraorally. Rarely, sting-induced cutaneous eruptions are accompanied by local lymphadenopathy, and although hypo- or hyper-pigmentation may follow the inflammation, scarring or keloid formation is rare. Fat atrophy, contractions, gangrene, local cutaneous thrombophlebitis (Mondor's disease) have been reported (INGRAM *et al* 1992). Instances of post-episodal dermatitis such as reactivated *Herpes simplex* or granuloma annulare are known (Plate 11.2) (BURNETT & CALTON 1987; MANDOJANA 1990). Aquatic dermatology is receiving increased attention (CLINICS IN DERMATOLOGY 1987; BURNETT 1987a; WONG *et al* 1994); dermatitis can result from contact with a variety of marine animals other than jellyfish, for example 'manta ray dermatitis' (Plate 11.3), and sea star dermatitis

SEA LICE

'Sea lice' is a colloquial, somewhat all-inclusive but descriptive title for things in the sea water that cause mostly minor, often multiple, irritating, itchy/stinging sensations on the skin. The symptoms may begin in the water, or later after the person leaves the water. Occasionally symptoms developing in the water may be sufficiently irritating for the swimmer to leave the water; a skin rash can appear (Plates 11.4 and 11.5) (BEARUP 1954; BEST & SABLIN 1964; CLINICS IN DERMATOLOGY 1987; TOMCHIK & RUSSELL 1993). The symptoms and signs at times develop underneath the bathing costume (BURNETT & CARGO 1979; WAISMAN 1989; BURNETT 1992b). Other terms used to describe such syndromes include:

'sea louse dermatitis'
'seabather's eruption (SBE)'
'swimmer's or bather's itch'
'Terrigal itch'
'pelican itch'
'seaweed dermatitis'
'*el Caribe*' {see below}
'Pearl Harbour itch'
'clam digger's eruption' (Atlantic coast of the USA)

Many of these terms convey different aetiologies to different investigators. For example, a distinction has recently been drawn between 'cercarial' (larval forms of trematodes) 'dermatitis' {see below} and 'seabather's eruption' (PARRISH 1993). The existence of the latter syndrome as a distinct entity has been questioned (MANDOJANA 1987). Furthermore, according to MEBANE (1994) the term 'sea lice' is a misnomer, applying only to fish parasites that have no effect on humans. He prefers the term 'sea bather's eruption'. While appreciating the academic precision of this point, the editors acknowledge the eloquently descriptive (from the sufferer's point of view!) term 'sea lice'. A recent review of seabather's eruption is by TOMCHIK and RUSSELL (1993).

DISTRIBUTION

The 'sea lice' phenomenon (skin irritation in swimmers, surfers or divers) has been widely documented, both as 'epidemics' and as isolated reports. Associated as it is with the activities of humans in the sea, it has been described literally all over the world (BAIRD & WEAR 1987). Specific marine regions include the south-east coastline of the USA (WAISMAN 1989; BURNETT 1992b), Long Island, New York (FREUDENTHAL 1991; FREUDENTHAL & JOSEPH 1993), the Caribbean islands (STRAUSS 1956), southern California (BEST & SABLAN 1964), Hawaii (GRAUER & ARNOLD 1961; IZUMI & MOORE 1987), Queensland (CLELAND & SOUTHCOTT 1965; WILLIAMSON 1940–49, unpublished observations; FENNER 1984–94, unpublished observations), New South Wales (BEARUP 1954; POPE 1972), and the Swan River estuary of Western Australia (MARSH & SLACK-SMITH 1986). Its incidence appears higher in warmer oceans.

A form of swimmer's itch occurs in Cuba during the winter months and is called '*caribe*' (WAISMAN 1989). (Curiously, *caribe* is also the name used for the infamous Amazon River piranha fish [family Characidae].)

AETIOLOGY

Almost certainly a variety of aetiological agents are involved; there is probably no single cause. The following sea creatures have been objectively implicated in separate 'sea lice' reports and/or 'outbreaks':

Minor cnidarians:

- *Olindias singularis* in Gulf St Vincent, South Australia (SOUTHCOTT in CLELAND & SOUTHCOTT 1965)

- Planula larval stage of the sea anemone *Edwardsiella lineata* on Long Island (New York) beaches in the 1980 and 1990 summers (FREUDENTHAL 1991 & 1993; BURNETT 1992b)
- Evidence implicating planulae of *Linuche unguiculata* (TOMCHICK *et al* 1993; WONG *et al* 1994), and unidentified planulae (larval cnidarians) (OWRE 1956), e.g. larval sea anemone (FREUDENTHAL 1987, unpublished observations, in MANDOJANA 1987).

Dispersed 'small broken coelenterate tentacles' in the sea apparently caused 'sea lice' symptoms in Bonaire (Caribbean region) in 1984 (EDMONDS 1989). How these fragments were so identified is not stated. In other claims for this cause in the literature objective identification data is lacking. (This should not be confused with *Gonionemus oshoro* envenomations, which occur among seaweed collectors in Japan {see Chapter 8, page 201}.) The sea comb jellyfish (ctenophore) is also commonly held directly responsible for 'sea lice', but firm evidence is lacking. However, the sea comb *Mnemiopsis leidyi* of the eastern United States coast, harbours a parasitic anemome larva, *Edwardsiella lineata*, which has been implicated in epidemics of Long Island 'sea lice' (FREUDENTHAL 1991; BURNETT 1992; FREUDENTHAL & JOSEPH 1993). One editor (JWB) suspects that *Linuche* species tentacles may have sometimes been a cause of this syndrome in the Caribbean region.

Marine algae:

- *Lyngbya majuscula* in Hawaii (GRAUER & ARNOLD 1961; IZUMI & MOORE 1987)
- *Microcoleus lyngbyaceus* in Ryukyu Islands and Hawaii (MARSH & SLACK-SMITH 1986)

Crab larvae:

- *Rhithropanopeus harrissii* in Maryland, USA (BURNETT & CARGO 1979)
- *Callinectes sapidus* in Maryland, USA (TRUITT 1939)

Small isopod crustaceans (Cymothoidism)

- *Cirolana woodjonesi* in Gulf St Vincent, South Australia and *Excirolana orientalis* in the tropics of Australia (HALE 1929)
- *Excirolana chiltoni* in California (BEST & SABLAN 1964)
- *Cirolana harfordi* in Fremantle, Western Australia (PRITCHARD 1991)*
- *Actaecia pallida* (McNEILL 1948). This infestation occurred from dry sand on a northern Sydney beach, affecting sunbathers!

Schistosome dermatitis (class Trematoda, the flukes):

- *Austrolilharzia terrigalensis* larvae (fork-tailed cercariae) in Narrabeen Lagoon, New South Wales. This irritating skin parasite uses seagulls, snails and certain siphon shells as hosts (BEARUP 1956; EWERS 1961). A similar episode has probably occurred in northern Tasmania (CLELAND & SOUTHCOTT 1965)
- *Microbilharzia variglandis* saltwater cercariae ('clam digger's eruption' along the Atlantic coast of the USA) (DELAWARE 1992)

Other 'stinging creatures' for which presumptive evidence exists to implicate them in the production of 'sea lice' symptoms are small slugs known as nudibranchs. Australian lifeguards recently reported many 'sea lice' stings in the water at Noosa (records kept at Queensland Surf Life Saving headquarters, Brisbane). They netted the area and caught many small slug nudibranchs.

Glaucus

Glaucus is a tiny (7 mm) marine mollusc, which floats on the surface of the sea (Plate 11.6). *Glaucus* is blue on top, and white underneath, so it has an excellent camouflage against natural predators, the sea birds.

Glaucus is an unusual creature that feeds on jellyfish and their tentacles. Other nudibranch species feed on other cnidarians. Yet by some unknown mechanism they do not stimulate nematocysts to discharge while they are 'eating' them. Swallowed nematocysts are then extruded from the stomach and pass into outpouchings ('cnidosacs') of these nudibranchs. Here they are discharged normally on contact with the integument (skin) of the nudibranch's prey, causing minor stings to humans ('sea lice'). Other non-stinging nudibranchs, such as the short-tailed nudibranch *Ceratosoma brevicaudatum*, inhabit the surface of, and feed

* *Cirolana harfordi* are about 1 cm long and are imported, carnivorous 'fish lice'. They were reported as having eaten through a diver's skin in less than one hour, causing bleeding that was difficult to stop, although the bites were not painful (PRITCHARD 1991).

off, toxic sponges (Plate 11.7), absorbing the sponge's toxins and becoming poisonous themselves, an effective deterrant to predators.

Carybdea sivickisi {see Chapter 9, page 257}

Carybdea sivickisi is a very small box jellyfish (cubozoan) that causes skin stings which may be termed 'sea lice' stings (Figure 9.6, page 257).

Not surprisingly, many skin irritations and skin rashes that accompany and/or follow swimming have had no cause identified.

Note the 'Sea Pest' — see Chapter 26, page 450.

MANAGEMENT

The treatment of sea lice irritations is symptomatic and should follow the basic principles of removal of the offending cause, topical application of ice packs for pain, and hydrocortisone or antihistamine creams for itch. Systemic pain and itch relief, or more serious inflammatory or allergic responses, require medical guidance {see Chapter 5, page 111}. High-potency topical corticosteroids are partially effective in seabather's eruption caused by *Linuche* sp.

Unrelieved scratching may lead to secondary infection, which will require medical selection of the appropriate antibiotic (AUERBACH 1987; AUERBACH *et al* 1987) {see Chapter 4, page 96}. Heavy skin contamination may dictate the need for tetanus prophylaxis.

Dr David Parrish of USA is examining the efficacy of dimethylphthalate repellents as protection against the skin effects of avian schistome induced 'clam digger's eruption' (BURNETT 1995).

OPHTHALMOLOGICAL ASPECTS OF JELLYFISH STINGS

Ocular stings have only relatively recently begun to be closely documented, with Australia taking a lead (PARADICE 1924; HERCUS 1944; MITCHELL 1962). The occurrence of such stings up to 1965 has been well documented by CLELAND and SOUTHCOTT (1965) and more recently by MARSH & SLACK-SMITH (1986). They have accounts of eye injuries from coral nematocysts, *Cyanea* sp., rhizostome jellyfish, sea anemone, *Physalia* sp. and *Catostylus* sp.

Ocular stings produce intense burning corneal pain, tearing and photophobia, and can result acutely in punctate epithelial keratitis, conjunctival and limbal injection, corneal oedema and anterior chamber flare {see Plate 5.10}. Subsequent iridocyclitis, increased intraocular pressure and decreased visual acuity are documented (GLASSER *et al* 1992). Such signs and symptoms can appear over 24 hours then resolve over 48 hours, including the decreased visual acuity.

More longstanding effects include mydriasis, decreased accommodation and peripheral anterior synechiae.

An unusual case of blurred vision lasting eight days, following a sting on the knee from a jellyfish is documented (BURNETT & BURNETT 1990).

DIAGNOSIS

In the presence of the history and the above clinical features, a slit-lamp examination of the cornea may reveal the presence of an adherent nematocyst on the cornea (GLASSER *et al* 1993), superficial punctate keratitis of linear abrasions of the cornea, and/or parallel puncture tracks from the nematocyst threads extending into the corneal stroma {Plate 5.10, upper left} (MITCHELL 1962; GLASSER *et al* 1992; GLASSER *et al* 1993).

MANAGEMENT

Advised therapy is directed towards pain relief, and control of the intraocular pressure, and of the inflammatory response. Adherent nematocysts will usually slough without active attempts at removal (with the attendant risks) (GLASSER *et al* 1992). Ocular injuries from jellyfish stings require immediate specialist referral.

SECTION

THREE

ECHINODERMS, MOLLUSCS AND SPONGES

CHAPTER 12

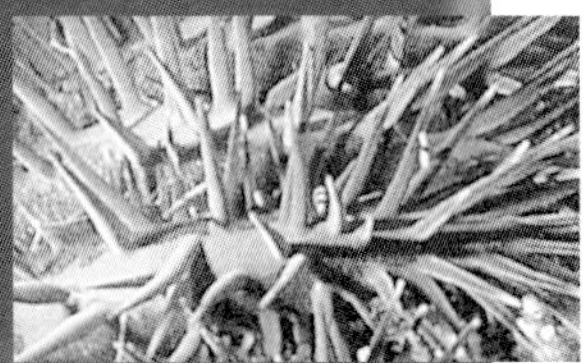

PHYLUM ECHINODERMATA

DAVID EXTON, P.J. MORAN,

JACQUIE RIFKIN AND JOHN WILLIAMSON

The phylum Echinodermata is a diverse group of organisms which are strictly marine. They have the distinctive characteristic of a fundamentally pentamerous (composed of five parts) radial symmetry which is derived from a bilateral ancestral form. Consequently these animals are not comparable with other phyla with radial symmetry, such as cnidarians, e.g. jellyfish.

The body wall contains a skeleton of calcareous spicules and spines, hence the name echinoderms, 'spiny skin'. The phylum contains only two classes that include venomous species: Asteroidea (sea stars) and Echinoidea (sea urchins). Other echinoderm classes, including members of the class Holothuroidea (sea cucumbers), contain poisons but no venom apparatus. The body toxins of echinoderms include steroid glycoside forms (see below) and terpenes (KREBS 1986). In general, spine or skin injuries from echinoderms do not constitute a major public health problem.

Note: Class Ophiuroidea

Although some brittle stars do possess toxins, these are used in the capture of organisms as part of the feeding process. MARSH and SLACK-SMITH (1986) state that *Ophiomastix annulosa*, which causes paralysis and death in small animals, should be handled with care.

CLASS ASTEROIDEA (SEA STARS)

The crown-of-thorns starfish *Acanthaster planci* has long been known as the only venomous sea star. However, some members of the genus *Echinaster* possess thorny spines and small pits from which poisons are secreted. The related *Plectaster decanus*, found in Australia, although it has not caused injuries, has produced contact dermatitis (POPE 1968). In addition, FISHER (1978) reports dermatitis from the northern hemisphere sun or rose star *Solaster papposus*.

THE CROWN-OF-THORNS SEA STAR: REACTIONS AND TREATMENT

P.J. MORAN

The crown-of-thorns sea star *Acanthaster planci* (Plates 12.1–12.4) is a well-known inhabitant of coral reefs in the Indo-Pacific region. Normally it occurs in low densities but from time to time the sea star forms large populations, termed 'outbreaks', which have caused extensive damage to corals (BIRKELAND & LUCAS 1990). Outbreaks of this sea star were first recorded in the Ryukyu Islands, Japan, in the late 1950s. Since that time they have occurred in places such as Australia's Great Barrier Reef (GBR), Guam, Fiji, Tahiti, American Samoa, and the Maldives. Outbreaks have been recorded on two occasions on the Great Barrier Reef: 1962–1974 and from 1979 to the present. Those that have occurred since 1979 have affected more than 50 per cent of reefs in the central third of the system (MORAN *et al* 1988).

The crown-of-thorns sea star has been the focus of intense public and scientific debate in Australia since outbreaks were first observed at Green Island in the early 1960s. At that time very little was known about the sea star, and some proposed that humans may have been responsible. Debates surrounding the cause of the outbreaks together with suggested solutions have continued for more than 30 years and have involved not only scientists but people from all sections of the community. Since 1985 the Federal Government has spent almost $10 million on research towards understanding why these outbreaks of sea star have occurred (ZANN & MORAN 1988; LASSIG & KELLEHER 1991).

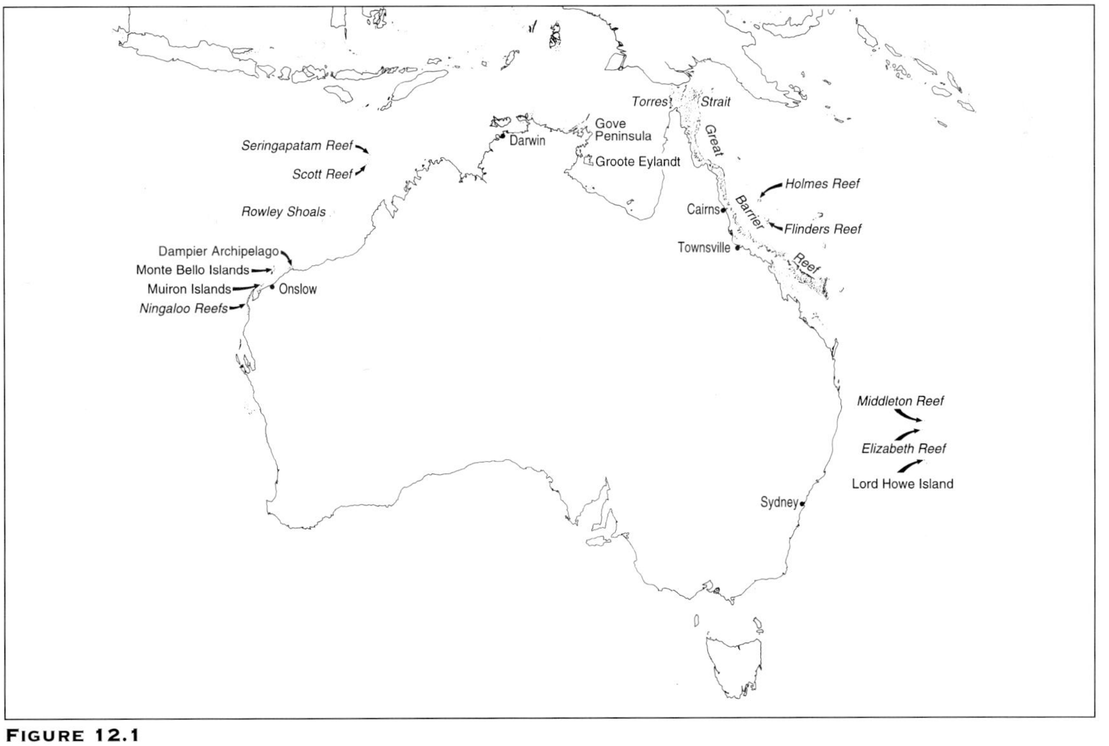

FIGURE 12.1
Distribution of crown-of-thorns sea stars in Australian waters. The map shows the main localities where the sea stars have been recorded in the past.

DISTRIBUTION

Crown-of-thorns sea stars have been reported from most reefs in the Indo-Pacific region from east Africa to central America (MORAN 1986). They have not been found on reefs in the Caribbean. HALSTEAD (1988) records the distribution as 'Indo-Pacific, Polynesia to Red Sea'. In Australia (Figure 12.1, page 313), outbreaking populations of this sea star have been reported not only on the Great Barrier Reef but also on reefs in the Torres Strait (WATSON 1990, unpublished observations), off the Gove peninsula (CHRISTIE 1991, unpublished observations) and on Elizabeth and Middleton Reefs (approximately 500 km east of Coffs Harbour, NSW) (HUTCHINGS 1989). Crown-of-thorns sea stars have been observed in low to moderate population densities on reefs around Lord Howe Island and the Solitary Islands (DeVANTIER & DEACON 1990) and on reefs located off the northwest coast of Western Australia (e.g. Ningaloo reefs, Dampier Archipelago, Muiron Islands, Monte Bello Islands) (SIMPSON & GREY 1989).

The sea star is known to be distributed widely over much of Australia's Great Barrier Reef, having been reported on reefs between Cape York and the Capricorn–Bunker complex (off Gladstone, Qld). Over the past 30 years outbreaking populations have occurred on reefs mainly between Lizard Island and those off the Whitsunday Islands, although some small outbreaks have been observed in the Swain Complex at the far southern end of the Great Barrier Reef (MORAN *et al* 1988).

Contrary to their formidable appearance (Plates 12.3 and 12.4), adult sea stars are relatively fragile and prone to damage by rough seas. Consequently, they prefer to live in sheltered habitats such as the lagoons and leeward slopes of reefs; they have also been found in very large numbers along windward slopes in deeper water (i.e. >3 m). Generally, they do not like to live in shallow water unless it is protected or the weather is reasonably calm for long periods. In such situations they pose a hazard to people walking on the reef. When crown-of-thorns sea stars occur in outbreaking populations (which may comprise as many as 2 million individuals) they also are a hazard to divers and trawl fishermen. The fingers and hands of the latter are at risk (Plates 12.5 to 12.9) as crown-of-thorns sea stars have been trawled (i.e. netted) from the sea bed between reefs.

APPEARANCE

Adult sea stars normally have a large disc with 14–18 arms, which they are able to regrow if the arms are eaten by predators. The normal size of an adult sea star is about 25–35 cm in diameter (arm tip to arm tip), although individuals as large as 70 cm have been reported (MORAN 1986).

Crown-of-thorns sea stars may display several colour variations. On the Great Barrier Reef individuals range in colour from reddish-grey with orange-tipped spines (Plate 12.4) to green with yellow-tipped spines. In other places, such as Thailand and the Maldives, the sea star may be a purplish-blue with dark red spines. As suggested by their name, crown-of-thorns sea stars are covered by a large number of long, sharp spines, which are toxic to a variety of animals, including humans.

There are four different types of spines on the upper (aboral) and lower (oral) surfaces of the sea star (BIRKELAND & LUCAS 1990). Each spine consists of a single crystal of magnesium calcite (WEBER 1969) covered by a 'skin' or integument which has three layers; an outer cuticle, an epidermis and a dermis. Epidermal cells are interspersed with neurosensory cells, pigment cells, and two types of glandular cells — acidophilic and basophilic cells. The acidophilic cells are thought to be responsible for producing toxic substances (HALSTEAD 1988).

The spines projecting from the upper surface of the sea star, termed abactinal spines (Plate 12.3), are the largest of the four types, each being up to 50–60 mm in length (Plates 12.3 and 12.4). Those projecting from the lower surface tend to be short and blunt. Each spine is articulated at the base so that it can be raised and lowered. The abactinal spines are extremely sharp and have three raised edges near their tips which aid in cutting through objects. As a consequence they are able to penetrate through wet suits, heavy gloves and most other underwater protective clothing (including stinger suits) with ease. Needless to say, extreme care should be taken when handling or in close proximity to crown-of-thorns sea stars, especially in the presence of surge or current.

THE VENOMS — BIOACTIVE SUBSTANCES AND THEIR ACTIVITIES

The bodies of crown-of-thorns sea stars contain compounds called saponins, which are sulphated steroidal glycosides (BIRKELAND & LUCAS 1990). Saponins are present in sea stars and sea cucumbers but not sea urchins, brittle stars or feather stars. They are thought to be important primarily in protecting crown-of-thorns sea stars from predators (LUCAS *et al* 1979). Saponins produce an unpleasant taste and have been demonstrated to be present in all phases of the life cycle of the crown-of-thorns sea star: eggs, larvae, juveniles and adults. Largest amounts of these compounds are found in the sea star's pyloric caeca and stomach, and constitute approximately 1 per cent of the total dry weight of the soft tissues of the sea star (BARNETT *et al* 1988).

Various studies have sought to isolate and identify the saponins present in crown-of-thorns sea star, largely because of their toxic nature. In all, nine saponins have been characterised from the sea star (ITAKURA & KOMORI 1988). Their activities are various, including immobilisation of spermatozoa leading to abnormalities and death in sea urchin and asteroid embryos, haemolysis of rat and mice erythrocytes, and antiviral activity (BIRKELAND & LUCAS 1990). Saponins also have surfactant properties (i.e. they can raise or lower the surface tension of a liquid).

A range of other bioactive substances have also been obtained from the spine venom of crown-of-thorns sea star: these have included a neurotoxin, some 5-hydroxytryptamine(5HT)-like substances, histamine-like substances, and a protein toxin (BIRKELAND & LUCAS 1990). They were isolated from crude extracts of spine tissue and have been demonstrated to have both haemolytic and lethal effects in experimental animals (TAIRA *et al* 1975). Injection of crude extracts into rodents (rats, mice and guinea pigs) caused writhing, salivation, lacrimation, diarrhoea, decreased activity, slowing of respiratory movements, hind limb paralysis and haemolysis of red blood cells (EVERITT & JUREVICS 1973) [Editors' Note: All such experiments are now subject to more stringent bioethical standards — see comment in Preface, page 23], as well as tissue swelling (oedema) (Plate 12.5), increased vascular permeability, myonecrotic activity, and protease and phospholipase activities (HEISKANEN *et al* 1973; SHIOMI *et al* 1985; KARASUDANI *et al* 1991).

The myotoxicity seen in experimental animals is said to be due to phospholipase A_2 activity of the venom (SHIOMI *et al* 1985; MEBS 1991). Recently, the extracts have been reported to produce hypotensive reactions (i.e. a decline in aortic pressure, and coronary, renal and aortic blood flows and an increase in heart rate and peripheral resistance) in dogs and rats (KINJO *et al* 1991; YARA *et al* 1992; SHIROMA *et al* 1994). Pretreatment with indomethacin greatly reduced these symptoms in the dogs but not in the rats, suggesting that the component toxin's mode of action may be species dependent. YARA and colleagues (1992) found that the hypotension in rats could be significantly inhibited through use of a platelet activating factor (PAF) antagonist. Recently SHIROMA and colleagues (1994) have demonstrated indomethacin-suppressible hypotension, together with a spontaneously reversible thrombocytopaenia and leukopaenia in dogs, in *Acanthaster planci* spine venom experiments. The thrombocytopaenia and leukopaenia were not affected by indomethacin.

SHIOMI and colleagues (1988b) isolated a lethal factor from crude spine extracts that had a LD_{50} (i.e. dose that kills 50 per cent of experimental animals in a given time) of 0.43 mg/kg in mice. The purified factor was a glycoprotein of 20 000 to 50 000 molecular weight which caused flaccidity, numbness and occasional convulsions in mice. Subsequently, SHIOMI and colleagues (1990, 1991) found that the factor produced degenerative changes in mouse liver, suggesting hepatotoxicity in the active substance.

Some of the effects from extracts of crown-of-thorns sea star spines, such as inflammation and increased capillary permeability, may be mediated by the presence of histamine-like compounds (EVERITT & JUREVICS 1973) and kinins (BIRKELAND & LUCAS 1990). However, as these substances are present in low quantities in the spine venom it has been concluded that the toxins probably affect victims by causing the release of mediators and/or the endogenous production of kinins. SHIOMI and colleagues (1989) have suggested that the venom contains a substance that acts on mast cells, leading to the release of histamines and a subsequent increase in capillary permeability and local tissue oedema (Plate 12.5). SHIROMA and his colleagues (1994) postulate that the hypotensive effects in dogs are due to the endogenous production of prostaglandins, including prostacyclin.

THE INJURY

There is no evidence to suggest that a venom is actively injected into a victim when punctured by a crown-of-thorns sea star spine (FLEMING *et al* 1972). However, a variety of substances, including dermal tissue, mucus, bacteria (terrestrial or marine) {see Table 4.2, page 91} and venom, are likely to be passively deposited into the wound. The immediate pain and bleeding (Plate 12.6) associated with such an injury may be due to the presence of saponins in the spine dermal tissue of the sea star. Other spine toxins apart from saponins would appear to be implicated because the range of symptoms experienced by victims is far greater than would be expected from saponins alone (MORAN & WILLIAMSON 1986). Given that quite small wounds bleed freely for some time it is possible that the venom also has anticoagulant properties (ODOM & FISCHELMANN 1972), which could be related to the presence of surfactant substance(s).

Generally most injuries from crown-of-thorns sea stars have been reported on the extremities, particularly the fingers, hands and feet. Victims vary quite widely in their reaction to an injury. Symptoms may include rapid onset of intense local pain, throbbing or burning sensation, continued bleeding, redness (erythema) and oedema of the injured area, itchiness of swollen tissues, numbness and stiffness of injured extremities, leaking of serous fluid from the wound (Plate 12.6), secondary infection of the wound (Plate 12.7) and enlargement of regional lymph glands (BARNES & ENDEAN 1964; POPE 1964; CLELAND & SOUTHCOTT 1965; WILLIAMSON 1968; WEBER 1969; MORAN & WILLIAMSON 1986). Systemic (general) effects seldom follow single puncture wounds. Local allergic reactions and general sensitisation are documented (BARNES & ENDEAN 1964; CLELAND & SOUTHCOTT 1965; WILLIAMSON 1968, MORAN & WILLIAMSON 1986; BIRKELAND & LUCAS 1990). No deaths are known, although a leg amputation has been reported (Chairman of the Ministry of Social Services, Cook Islands, unpublished observation) following injury by a crown-of-thorns sea star (ANDREWS 1986). EDMONDS (1989) records a protracted acute illness following multiple thigh punctures in a breath-hold diver in Fiji.

CLINICAL FEATURES

Usually, but not invariably, severe burning pain is experienced immediately after the injury. Care should be taken particularly by divers because the pain can be so intense as to cause loss of control, disorientation, panic and unsafe ascent {see Chapter 22, page 424}. Wetsuits offer a measure of protection but the spines have remarkable penetrating capabilities. Many authors have commented that the pain (which can throb) is often far greater than expected given the extent of the wound (BIRKELAND & LUCAS 1990). Reports indicate that the pain may last for several hours. Punctures into finger joints (Plate 12.8) or other sensitive areas (e.g. under nails) can lead to pain, discomfort, and limitation of joint movement for several weeks (MORAN & WILLIAMSON 1986).

Injuries involving multiple punctures from spines are more likely to result in severe pain as well as swelling of regional lymph glands, fever, and protracted nausea and vomiting according to EDMONDS (1989). However, such documentation is sparse. Edmonds states that such vomiting and retching 'may last for several days'. In such cases, secondary bacterial infection could occur in the absence of early medical management {see Chapter 4, page 96}. Claims of 'paralysis' are difficult to substantiate but could perhaps be lay descriptions of severe limb pain with multiple punctures.

Following the initial bleeding, tissues immediately around the puncture wound may later turn a purplish-blue colour (Plate 12.9). This ecchymotic discolouration may extend and eventually become surrounded by a region of erythema (ODOM & FISCHELMANN 1972) which can persist for several days. The area may also swell considerably; multiple punctures (e.g. on a hand) can cause the entire hand and forearm to swell (Plate 12.5). This condition may persist for up to a week, during which time the wounds may leak serous fluid, a local numbness and stiffness may be felt, and swelling of the regional lymph glands may occur; such features would require that secondary infection be excluded.

Case reports (including the author's own) indicate that the entire area can become 'itchy' whilst the oedema persists. Abnormal swelling and itchiness could indicate that certain susceptible victims may be allergic to the toxins produced by the crown-of-thorns sea star. Thus there is the theoretical possibility that some individuals could suffer anaphylactic reactions (ODOM & FISCHELMANN 1972) although this has not been documented to date. [Editors' Note: Anaphylactic reaction to other venoms is now well documented (TOGIAS *et al* 1985; RYAN & CARAVATI 1994; REISMAN 1994).] It is likely that immunologic sensitivity to these toxins can be built up over a period of time as a consequence of repeated exposure to such injuries.

Victims will react differently to injury from crown-of-thorns sea star spines, depending on the site and extent of the injury, the sensitivity of the victim to the toxins, the size (maturity?) of the sea star, and how rapidly appropriate first aid and medical treatment can be applied.

FIRST AID

Any visible, protruding (but only easily removable) spine fragments should be pulled out immediately. If this is followed by bleeding, simple direct pressure will suffice. However, in the absence of a serious general allergic reaction (anaphylaxis) (which is rare — see below and Chapter 8, page 230) compression/immobilisation bandaging {Figure 17.2, page 400} should never be used for crown-of-thorns envenomation. Buried or fixed fragments should not be sought by first aiders; this is a matter for early medical assessment.

The immediate and often increasing acute local pain of the spine wound(s) can receive some relief by the immersion of the envenomated part (usually an extremity) into hot (not boiling!) water {see Chapter 5, page 112}. (The temperature of the water should be tested by the first aider, not by the victim.) It is the experience of the author that very cold water also provides some analgesia (MORAN 1992, unpublished observations). Whatever temperature is chosen for the water, repeated immersion of the affected area may be necessary as the symptoms often reappear upon removal of the envenomated part from the water.

MEDICAL MANAGEMENT

All but the most trivial crown-of-thorns spine injuries require early medical assessment. The currently recommended management (similar to that for stonefish envenomations) {see Chapter 5, page 114, and Chapter 16, page 378} has been outlined previously by MORAN & WILLIAMSON (1986) and is listed below:

1. Wash the wound carefully as traces of mucus and dermal tissue may still be present.
2. Remove any superficial fragments of spine that are still embedded in the wound. Because the spines are quite brittle, deeper fragments often remain; radiological examination may be appropriate in selected cases (Figure 12.2, page 318), but the decision to remove deeply retained spine fragments should be taken only by experienced clinicians.
3. Under skilled medical supervision, inject local anaesthetic around the wound or administer strong analgesics in order to relieve on-going pain.
4. Dress and bandage the wound, but never apply a tourniquet. Pressure bandages are also

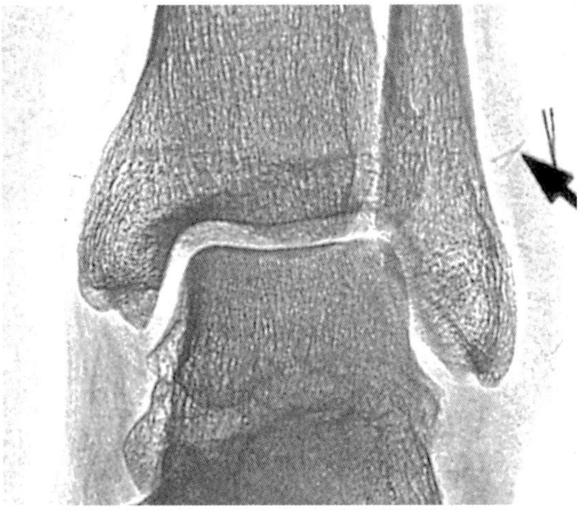

FIGURE 12.2
Soft-tissue X-ray of a diver's ankle, showing a fragment of a crown-of-thorns sea star spine embedded in the tissue (arrow tip). Adjacent to the arrow are two metal pins placed on the patient's skin to mark the point of skin entry by the spine. (Courtesy of Dr Denis Campbell and Marty Thysson, formerly of Townsville.)

contraindicated unless a general allergic reaction (anaphylaxis) — which is rare — develops.
5. Administer a tetanus toxoid injection, and/or verify the patient's tetanus immunity.
6. Manage prophylaxis or treatment for secondary infection.
7. Arrange bed rest, elevate the injured area (where possible). Reassure the patient.

Any patient showing unrelieved local allergy (rapid swelling, itchiness and skin rash), irrespective of the apparent size of the wound, should be referred for prompt medical attention.

Divers can phone the Divers Emergency Service for advice (within Australia: 1-800-088200. From outside Australia: 61-8-83735312) {see Chapter 22, page 426}. The topical application of antihistamine creams and local anaesthetic jelly has proved disappointing in this situation.

In the rare event of developing generalised allergic reactions (rapid local swelling, rapid weak pulse, extensive itchiness and rashes, breathing difficulty or clinical shock) immediately apply a compression/immobilisation bandage to the injured part {see Figure 17.2 page 400, and Chapter 5, page 110}, administer 100% oxygen and summon urgent medical assistance. This complication is more likely in victims with a personal history of other allergies.

Fragments of spine that are deeply embedded in tissue and detected radiologically (Figure 12.2) may indicate anaesthesia and surgery for removal. Such a decision is a skilled medical one only.

Failure to remove pieces of spine can uncommonly lead to complications (Plates 12.7 and 12.8). BIRKELAND and LUCAS (1990) reported a case in which the right index finger of a victim became purulent within three days. The bacterium *Staphylococcus aureus* was isolated from the wound. An aneurysm of a superficial artery (as a result of the bacterial infection) was subsequently diagnosed and rectified with surgery.

People who are known to be allergic to the sea star should wash their hands thoroughly after having touched the sea star or any material that may have been in contact with the animal. Allergic responses can be initiated from just touching. Toxins can be transferred via the hands to open cuts or the eyes. The author has experienced swelling and itchiness around the eyes after touching the face with fingers that had been used to handle sea stars.

CURRENT RESEARCH NEEDS

There is still much to be learned about the bioactive substances released by crown-of-thorns sea stars. In particular we need to know:
1. All the toxins that are present in the sea star.
2. Their chemical structures.
3. Their individual and cumulative effects on human tissues.
4. Where they are produced in the animal.
5. Where they are stored in the animal.
6. How they are released.
7. Their metabolism within human tissue.
8. How they are denatured in the human system.
9. Their mode of action, particularly in relation to producing serious local and general responses (e.g. severe allergies).
10. Whether their effects may be overcome, perhaps through use of antivenoms or anti-inflammatory agents.

FURTHER READING

Further scientific information on the crown-of-thorns sea star can be obtained from MORAN (1986) and BIRKELAND and LUCAS (1990). More general explanations of the crown-of-thorns phenomenon and the research that has been conducted can be found in ZANN and EAGER (1987), MORAN (1988), and ENGLEHARDT and LASSIG (1992). Results of the latest research have been presented in a comprehensive video (16 minutes duration) compiled by the Australian Institute of Marine Science and the Great Barrier Reef Marine Park Authority, Townsville, Australia. A list of all important articles pertaining to the sea stars is given by MORAN and colleagues (1991).

ACKNOWLEDGMENTS

Dr J.T. Baker, PhD, provided many useful comments on an earlier draft of this manuscript. Dr J. Williamson's editorial and medical advice was invaluable.

CLASS ECHINOIDEA (SEA URCHINS)

DAVID EXTON AND JACQUIE RIFKIN

The sea urchins are circular, ovoid or flattened echinoderms without arms. Although most sea urchins can be handled safely, some cause injury to humans by virtue of their spines, their jaw-like pedicellariae, or their poisonous flesh. This group contains the majority of venomous echinoderms.

DISTRIBUTION

Sea urchins (also known as sea eggs) are found in all oceans, particularly on rocky foreshores and reefs. The most venomous species tend to be concentrated in tropical and subtropical marine regions. In the Indo-Pacific region (to which most of this section refers) there are several species of venomous sea urchin (see Table 12.1). These include species of needle-spined sea urchin, *Diadema setosum* Leske, 1778 (Plate 12.10), and *Echinothrix* and the 'flower sea urchin', *Toxopneustes pileolus* (Lamarck 1816) (Plate 12.11). These genera are widely distributed throughout the Indo-Pacific area and northwards to Japan. *Diadema setosum* is also found on the east Africa coast (HALSTEAD 1988), Oman {see Plate 18.5}, and *D. antillarum* in the Caribbean (BURNETT 1987b). In Australia, the southen range of *Toxopneustes pileolus* extends to Sydney (New South Wales) and Shark Bay (Western Australia) in Australia (MARSH & SLACK-SMITH 1986).

APPEARANCE

Sea urchins are globular, cushion-shaped animals with numerous spines, of many shapes and colours. These may be long and visible (Plates 12.10 and 12.12) or short and hidden (Plate 12.13), depending on the species. They have a roughly spherical internal skeleton or 'test', composed of closely fitting calcareous plates covered with a thin epithelium as are the spines which project from this test. The spines are moved by fine muscles and each is attached to a tubercle on the test by a ball-and-socket joint. Stalked, movable, jaw-like pedicellariae also extend from the test and these aid in keeping the surface clean and free of foreign growths.

Although the majority of sea urchins can inflict a traumatic injury on the unwary surfboard rider, fisherman, diver or beach fossicker (FISH and COBB [1954] listed ten harmful species from the central and western Pacific region alone), some sea urchins are especially dangerous because of the venom associated with their spines or their pedicellariae.

MARSH and SLACK-SMITH (1986), to whose documentation of sea urchins the reader is referred, have identified a number of sea urchin species on the Australian coastline and grouped them according to possession of venomous long spines, or venomous short spines, or venomous pedicellariae (Table 12.1). Earlier records have been made by CLELAND and SOUTHCOTT (1965).

TABLE 12.1
CATEGORIES OF VENOMOUS SEA URCHIN

Family Diadematidae
With venomous long spines:
Diadema setosum (Leske, 1788)
Diadema savignyi (Michelin, 1845)
Echinothrix calamaris (Pallas, 1774)
Echinothrix diadema (Linnaeus, 1758)
With long spines and venomous pedicellariae:
Centrostephanus tenuispinus H.L. Clark, 1914
Centrostephanus rodgersii (Agassiz, 1863)

Family Echinothuriidae
With venomous short spines:
Araeosoma owstoni Mortensen, 1904
Araeosoma tesselatum
Araeosoma thetidis (H.L.Clark, 1909)
Asthenosoma varium Grube, 1868
Asthenosoma ijimai Yoshiwara, 1897
Asthenosoma intermedium Clark, 1938
Family Toxopneustidae
With venomous pedicellariae:
Toxopneustes pileolus (Lamarck, 1816)
Tripneustes gratilla (Linnaeus, 1758)
Family Echinometridae
Heterocentrotus mammillatus (Linnaeus, 1758)
Family Temnopleuridae
Salmacis sphaeroides (Linnaeus, 1758)

MARSH 1995 personal communication

MECHANICAL INJURIES FROM NON-VENOMOUS SEA URCHINS

Skin punctures by the solid-spined sea urchins (these are the majority) produce localised pain, with the spines frequently breaking off in the wound (SMITH 1977; MARSH & SLACK-SMITH 1986; EXTON 1987), but there appears to be no toxic substance present (SMITH 1977).

ENVENOMATION BY SEA URCHINS

Both spines and pedicellariae of sea urchins are capable of producing an envenomation but usually only one or the other is venomous.

SPINES

(The reader is also referred to Chapter 18, page 408, for an elegant description of sea urchin spines.) Doubt formerly existed as to whether an actual venom was injected by sea urchin spines (SMITH 1977; HALSTEAD 1988). However, the symptoms that result from many recorded sea urchin spine injuries, and the local signs of redness and paraesthesiae, have strongly suggested envenomation. While most sea urchins have solid spines of calcium carbonate in the form of calcite intermingled with organic material, there are a few species possessing hollow ones. It is the latter that produce the most troublesome envenomations.

Sea urchins with venomous long spines (see Table 12.1)

The needle-spined diadematid sea urchins, *Diadema setosum* (Plate 12.10) and *D. savignyi*, have exceedingly sharp, elongate primary spines (Figure 12.3), the secondary spines being shorter. These brittle spines are hollow and have whorls of pointed scales, which make removal extremely difficult. The spines are easily capable of inflicting a deep penetrating injury. The venom in diadematids is probably present in the thin skin surrounding the spines (MARSH and SLACK-SMITH 1986) or in the lumen of the spines (see Chapter 18, page 408).

Only the hollow, secondary spines on the sea urchin *Echinothrix calamaris* (Plate 12.12) inject venom. The well-developed retroussé barbs on these sharp spines (Figure 12.3) make the spines impossible to pull out once embedded. Venom

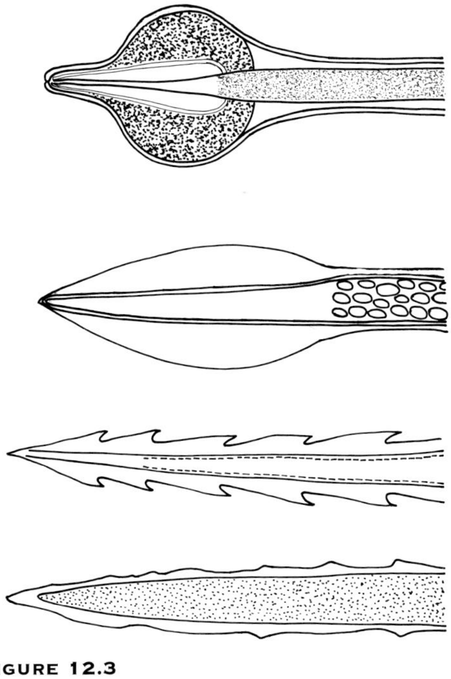

FIGURE 12.3
Tips of a selection of sea urchin spines, showing barbing and venom sacs (glands): (a) *Asthenosoma* sp. (b) *Araeosoma* sp. (c) *Echinothrix* sp. (d) *Diadema* sp. (Drawing by Jan Hynes, after Marsh & Slack-Smith 1986.)

glands are present in the skin surrounding the spine tips. Wounds from *Echinothrix* are far more painful than those of *Diadema*, and spine fragments from the former species are not absorbed and may be seen in the tissues many months after the original injury (MARSH and SLACK-SMITH 1986).

Sea urchins with venomous short spines

The 'tam-o'-shanter' sea urchin *Asthenosoma* sp. is well known for its bright colours and short, sharp spines each encased in skin with inflated venom glands at the tip and (in *A. varium*) with swellings along their length (Figure 12.3A) ((MARSH and SLACK-SMITH 1986). Found in the Indo-Pacific region and in Japan (Okinawa) (Plate 12.3) it can deliver a severe sting at the slightest touch. The venom glands of *Araeosoma* spp. have similar characteristics (Figure 12.3B). In echinothurid sea urchins, downward pressure on the sheath surrounding the spines forces the venom into the victim's tissues, as has been described for the echinothurid *Phormosa bursarium* (MORTENSEN 1943; HALSTEAD 1988).

PEDICELLARIAE (Figure 12.5)

Located between the spines and scattered over the test of most sea urchins are small, stalked, mobile grasping organs called pedicellariae, each of which is equipped with a set of three movable jaws or pincers. Contact with the sensitive hairs on the pedicellariae causes rapid closure of the pincers. Globiferous pedicellariae found in some sea urchins have venom glands associated with the piercing jaws. As mentioned above, pedicellariae may be venomous or non-venomous.

Sea urchins with venomous pedicellariae

The flower sea urchin *Toxopneustes pileolus* (Plate 12.11) is recognised by its very short, white-tipped spines hidden in a plethora of flower-like pedicellariae, and is reputed to be the most venomous of all sea urchins (FISH & COBB 1954; HALSTEAD 1971). FUJIWARA (1935) stated that the pedicellariae are capable of injecting venom, though it appears that specimens are sometimes handled with impunity by the collector (UNDERHILL 1987; EXTON 1993; unpublished observations, MARSH 1995, personal communication). COLEMAN (1994 personal communication to JFR) stated that this species grows to an unusually large size off Lord Howe Island, off the New South Wales coast, and that the large pedicellariae on the oral side of the test are quite capable of penetrating thick keratin especially if collected from the underside with bare hands (Figure 12.4 and Plate 12.11).

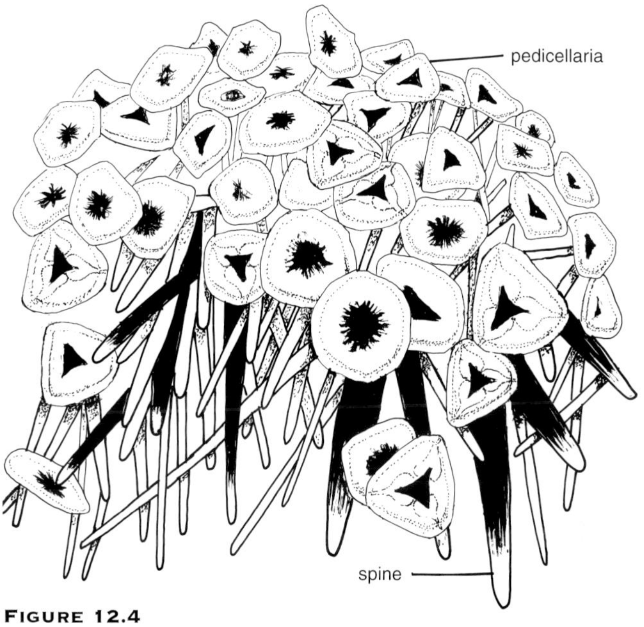

FIGURE 12.4
Drawing illustrating the relative arrangement of pedicellariae and spines on a *Toxopneustes* sea urchin. The disc-shaped pedicellariae are on long thin stalks, interspersed with the thicker, randomly pointing spines (shown shaded). Only the pedicellariae of *Toxopneustes* are venomous; the spines are non-venomous. (Drawing by Jan Hynes.)

Pedicellariae of *Tripneustes* are much smaller but are still capable of envenomating if they penetrate tender skin. The pedicellariae will continue envenomating once they are dislodged from the skeleton (CLELAND & SOUTHCOTT 1965; MARSH & SLACK-SMITH 1986).

The members of the genus *Centrostephanus* are interesting sea urchins because in addition to long needle-like venomous spines they have pedicellariae equipped with venom glands — and yet there has so far been no report of an Australian injury from one of these species (MARSH & SLACK-SMITH 1986).

Poisonings: ingestion of sea urchin organs

Sea urchin fisheries exist in several countries, and the roe is eaten as a delicacy/aphrodisiac in Japan (EDMONDS 1989); however, sea urchin ovaries (the main part eaten) may be poisonous at certain times. Non-fatal poisonings are well documented (CLELAND & SOUTHCOTT 1965; HALSTEAD 1988), but external injury and envenomation from the spines and/or pedicellariae are more common and important problems.

Sea urchin fatalities

Human deaths are referred to in the literature (FREYVOGEL 1972; MARSH & SLACK-SMITH 1986; EDMONDS 1989), but no precise documentation of a death could be identified. For example, it is difficult to obtain details of deaths from *Toxopneustes pileolus* in Japan (SMITH 1977) (this species has venomous pedicellariae but non-venomous spines, see Figures 12.4, page 321, and 12.5). One such reported fatality was that of a female Japanese pearl diver who apparently drowned following loss of consciousness in the water as a result of contact with a *Toxopneustes* species (HASHIMOTO 1979).

Sea urchin venom pharmacology

Most research into the toxinology of sea urchin venom has been conducted on pedicellariae venom (HALSTEAD 1988). Venom from the pedicellariae of *Tripneustes gratilla*, a common Indo-West Pacific sea urchin, is thought to be a thermo-labile protein (ALENDER *et al* 1965; FLEMING & HOWDEN 1974). BURNETT (1987) reports that *Diadema* venom is composed of proteins,

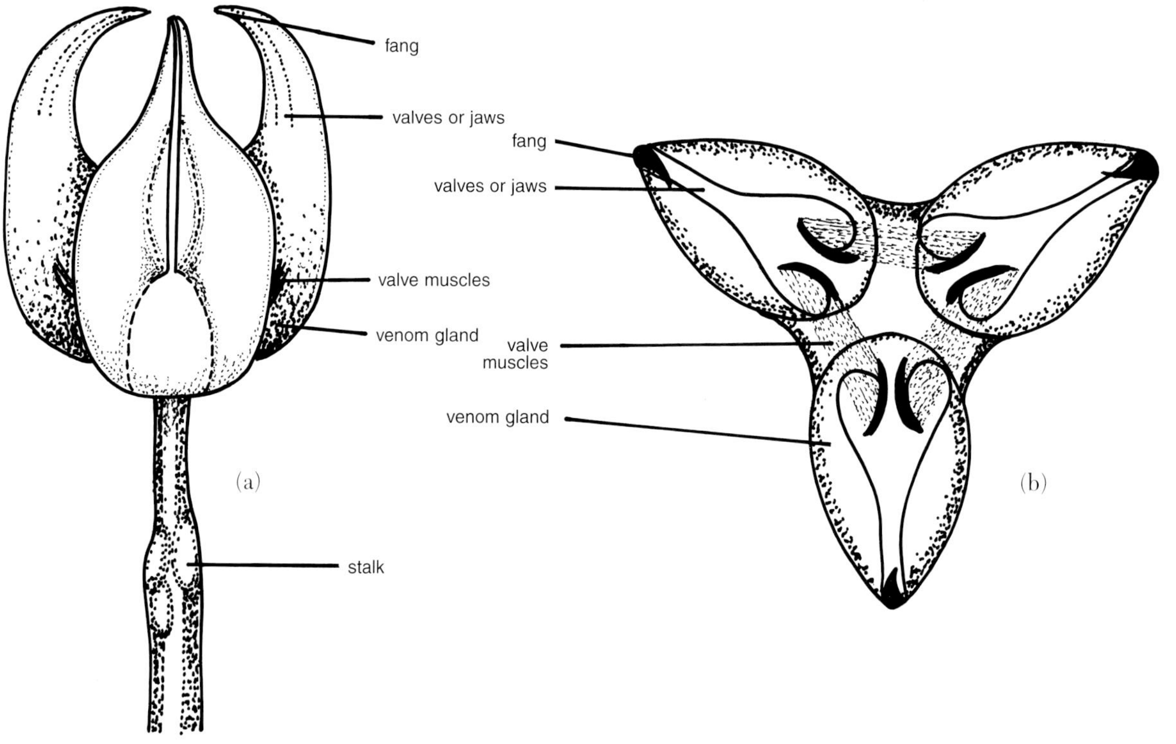

Figure 12.5
Diagram of a venomous pedicellaria from the sea urchin *Toxopneustes*, showing (a) in elevation and (b) in plan, the triple jaws each with a fang and a venom gland. (Drawing by Jan Hynes, after Marsh & Slack-Smith 1986.)

steroid glycosides and various mediators of inflammation. It seems likely that both acute toxic and acute allergic (e.g. migraine [HALSTEAD 1988]) reactions are possible.

SYMPTOMS AND SIGNS

ENVENOMATION BY SPINES

Local effects

Pain from *Diadema* or *Echinothrix* stings is immediate and intense, and wounds develop redness, swelling (oedema), and often bleed profusely (BADEN 1987). The pain far exceeds that produced by injury alone, suggesting the introduction of a venom (SMITH 1977) {see above}. The pain subsides after several hours, reappearing as direct pressure is applied (e.g. walking on a spined foot) (Plate 12.15). A wound from black sea urchin spines subsequently becomes discoloured with a violet-black hue (BADEN 1987), presumably because the pigments from the dark spines leech out into the tissues; the discolouration may remain up to four days.

If a spine enters into a joint, it may induce troublesome synovitis (CRACCHIOLO & GOLDBERG 1977). Certain Panamanian sea urchins produce problematic lesions (LESSIOS 1983).

Secondary infection of sea urchin spine puncture wounds is a common occurrence {see below and Chapter 4, page 91}, and as with some other marine wounds, indolent ulceration may also occur (AUERBACH 1988).

Systemic effects

If several venomous spines have penetrated the skin, especially if they are deeply embedded, the victim may less commonly develop systemic symptoms, including nausea, local numbness, muscular weakness and respiratory distress (LINAWEAVER 1967) (see Case Report below). The pain of multiple stings may be sufficient to cause delirium (AUERBACH 1988).

The epithelial covering of the venom glands at the tips of the secondary spines of *Echinothrix calamaris* and *Echinothrix diadema* contain a vaso-pressor tentatively identified as noradrenaline (MARSH & SLACK-SMITH 1986). PUGH (1913) documented non-specific 'cardiac irregularities' in a case of sea urchin poisoning, but a clear link between such observations in sea urchin envenomation could not be found in the literature.

CASE REPORT

Recently KOERNER (1993) described the development of a Guillain-Barré syndrome in a 47-year-old male following an injury to his right knee from 30 sea urchin spines (thought to be a *Diadema* sp.). Six days following the removal of most of the spines aching of his right leg and flank commenced; three further days later dysphagia and dysarthria progressed to bilateral facial paresis. Steady deterioration over the next two weeks led to respiratory support with tracheostomy and jejunostomy. He made a partial eventual recovery over the ensuing three months in hospital, and had mild residual facial paresis and dysarthria at ten months post-envenomation.

Delayed reactions

Secondary infection, granuloma formation weeks after the injury (BURNETT 1987; EDMONDS 1989) and wound tattooing (BURNETT 1991) are all described. Foreign body granulomas from the inorganic substances in the spine (BADEN 1987; BURNETT 1991) are the most common, occurring three or four months after spine penetration. The granulomas are commonly 3–4 mm in diameter, pink to purple in colour, painless, with a keratic surface and rarely suppurate; some are oedematous and cystic (BURNETT *et al* 1986b). The more classical vesicular form of delayed hypersensitivity has also been reported but is relatively uncommon (VALENTINE 1982).

MANAGEMENT OF SEA URCHIN SPINE INJURIES

First aid for envenomation by spines

Basic resuscitation, if indicated (rarely), as always takes absolute priority {see Chapter 5, page 107}.

Manual removal of an obviously protruding spine from a peripheral area (e.g. foot, see Plate 12.15) appears sensible. However, it commonly breaks off, leaving an embedded buried piece in the tissues. Do not 'dig' for this without appropriate medical supervision or advice.

Immersion of the envenomated part in hot (not boiling) water will usually ease the immediate pain within a few minutes, but not eliminate it, in the editors' experience {refer to Chapter 5, page 112}. Sustained hot water immersion may be necessary to maintain the pain relief, pending medical intervention. However, some authorities deny the benefit of this measure (MEBS 1994), and further practical research in different world regions is required. The oft-quoted advice to crush the retained spine into fragments in the skin is not recommended; this will not necessarily lead to the fragments 'dissolving', may cause further pain, envenomation or infection, and will certainly make the retained pieces difficult to find (AUERBACH 1987). Do not apply a compression/immobilisation bandage unless unmistakable signs of serious systemic allergy (anaphylaxis) are developing acutely {see Chapter 5, page 110}.

MEDICAL MANAGEMENT OF SPINE INJURIES

The fragility of the spines makes them very difficult to remove. A xeroradiograph with a skin tag marker should be taken to identify retained spine fragments. Embedded spines may become absorbed, but it is more common for them to be extruded through the skin (BADEN 1987) — hence the rationale for not recommending routine surgical removal of all spine fragments. The spines of *Echinothrix* spp., however, are not absorbed. Under experienced surgical supervision, deeply embedded spines, particularly those producing troublesome granulomas, may require excision (BADEN 1987).

Principles of surgical removal of the spine fragments (if indicated) under appropriate anaesthesia, pain relief, prevention of secondary infection, tetanus prophylaxis and the management of delayed reactions are described previously {see Chapter 5, page 113}. If the spine has penetrated an interphalangeal joint, the finger can be splinted acutely until spine extraction is performed, to limit fragmentation and further penetration. It may also help reduce the fusiform finger swelling commonly noted following a puncture (O'NEAL *et al* 1964)

The risk of secondary infection from sea urchin spines is appreciable, and may become serious {see Chapter 4, page 93}. A recent patient of one of the editors (PJF) with a multiple spine injury from sea urchin became severely infected, with a systemic toxaemia leading to hospitalisation and intravenous antibiotic infusions (FENNER 1993, unpublished observations).

In the absence of secondary infection, delayed allergic effects and retained non-absorbable spine fragments, the affected site will usually resolve over one or two weeks.

Corticosteroids have no place in the non-allergic lesion. Their efficacy in an allergic reaction is variable. Tetanus prophylaxis should always be considered for a marine wound, especially in the non-immunised patient.

ENVENOMATION BY PEDICELLARIAE

Local and systemic effects

Envenomation by pedicellariae (e.g. of *Toxopneustes pileolus*) can be severe. Intense radiating pain, numbness, respiratory difficulties, and 'paralysis' lasting some six hours, are all stated (FISH & COBB 1954; ENDEAN 1961). The pain (but not necessarily the paralysis) can be of short duration — about 60 minutes. WALKER (1988) stated that as few as seven pedicellariae from (unspecified) sea urchins are enough to give a variety of symptoms including immediate pain with central radiation, haemorrhage, numbness, 'paralysis', aphonia, bronchospasm and hypotension. These symptoms generally resolve within a few hours. (Editors note: These events must be uncommon.) MORTENSEN (1943) quotes a report by Fujiwara 1935, of giddiness, dyspnoea, and facial paralysis of six hours duration following a *Toxopneustes* spp. envenomation.

MANAGEMENT OF PEDICELLARIAL ENVENOMATION

Hot water immersion should also be applied for the pain of pedicellariae envenomation (see above). Pedicellariae detached from the test and attached to the skin must be removed mechanically or envenomation will continue.

ROSCO (1977) suggests applying shaving foam and gently scraping with a razor. Otherwise the principles for the care of any envenomated marine wound apply (see above for the management of sea urchin spine injuries — excluding the aspects related to penetrating and retained spine fragments). Pain relief, wound toilet and prevention of secondary infection (including tetanus), and the management of any systemic toxic or allergic response are all indicated {see Chapter 5, page 113}.

Acknowledgement: The authors thank Loisette Marsh of the Western Australian Museum for the information contained within Table 12.1 and helpful advice concerning the biology of this section.

CLASS HOLOTHURIOIDEA (SEA CUCUMBERS)

DAVID EXTON AND JOHN WILLIAMSON

These sausage-shaped animals are better known in their prepared food form as 'trepang' (Malay) or 'bêche-de-mer' (MARSH & SLACK-SMITH 1986).

DISTRIBUTION

Holothurians are encountered principally in tropical seas, in a variety of habitats ranging from hidden under rocks and corals, among seaweeds, or simply lying exposed on flat sandy sea bottoms. As bêche-de-mer or trepang, they are a prized delicacy for Chinese and Malaysian gourmets, and for some Pacific Island communities (e.g. Fiji). However, more than 40 such species so far examined — including all shallow water species used for food — have been shown to be poisonous unless correctly prepared (MARSH & SLACK-SMITH 1986; HALSTEAD 1988).

APPEARANCE

Sea cucumbers are elongate sausage-like echinoderms without free arms but with a cluster of tentacles around the mouth at the anterior end of the body. They have no spines or pedicellariae. Like other echinoderms they move about on tube feet which extend through the body wall. Some species possess internal thread-like organs called 'Cuvierian tubules', which are ejected through the anus when the animal is disturbed or irritated (Plate 12.16).

SEA CUCUMBER CHEMISTRY

The chemistry of echinoderm toxins focuses upon saponins, common glycosides among the plant world but uncommon in animals (HALSTEAD 1988). Saponins are found in many sea stars and are known as asterosaponins {see *Acanthaster planci* above}, and in sea cucumbers — from whence the first animal saponin was identified (holothurin) (YAMANOUCHI 1943). The saponin chemistry of echinoderms has been reviewed by BURNELL and APSIMOV (1983). The distribution of saponins among holothurians was studied by HASHIMOTO (1979).

SEA CUCUMBER TOXINS

Sea cucumber flesh mixture contains 'holothurin' toxins. Holothurins are elaborated in the body wall, especially in the skin. Intoxications from sea cucumbers occur either from ingestion of the flesh (poisoning) or from direct contact with the toxic Cuvierian organs (envenomation).

The Cuvierian tubules of most sea cucumbers are highly toxic, and skin contact triggers pain and a vigorous inflammatory reaction (e.g. *Bohadschia argus*, Plate 12.16 (MARSH & SLACK-SMITH 1986). The toxic material is also highly irritant to the eyes if transferred there on fingers; claims of 'blindness' made in the literature remain unsubstantiated, and are doubtful (UNDERHILL 1987). The toxin exhibits neural blockade in animals (ENDEAN 1982).

Sea cucumber toxin may be released into the sea water continuously, or only when the animal is molested. This material is usually absorbed or ingested by predators.

Holothurin is water soluble but at least partially heat resistant. It is hydrolysed by (gastric) acid into non-toxic products, which may be one of the reasons for the lack of correlation between toxicity studies and human poisoning from sea cucumbers. Also 'edible' species of sea cucumbers probably contain less toxin than other species (YAMANOUCHI 1955). In fact a temperate species, low in toxicity, is eaten in Japan (MARSH

1995, personal communication). Holothurins are potent cardiac glycosides, capable of causing severe illness or fatality upon ingestion. Deaths from sea cucumber poisoning have been claimed (SAVILLE-KENT 1893). Anti-tumour activity has been detected in holothurins.

PREVENTION OF HOLOTHURIN POISONING

Dried sea cucumber products have little toxicity, due to lengthy contact with water during preparation, which involves gutting (eliminates the Cuvierian organs), boiling for 1.5 hours, and washing and drying. Consequently most of the poison is leached out of the body wall (MARSH & SLACK-SMITH 1986).

Do not touch the white Cuvierian tubules extruded from the anus of a sea cucumber. Divers should not clear their masks after handling holothurians; eyes may become inflamed with tubule fragments or toxin.

MANAGEMENT OF SEA CUCUMBER INTOXICATION

For systemic effects, animal studies suggest that anticholinesterase therapy may be effective (FRIESS 1963). In any case, first principles should be applied {see Chapter 5, page 106}.

AUERBACH (1988) suggests that management of holothurin-induced contact dermatitis is supportive. Topical or systemic corticosteroids may be necessary to manage a severe irritant dermatitis. If an eye is involved, it should be anaesthetised with 1–2 drops of 0.5% proparacaine HCl and then irrigated with 250–500 ml of normal saline to remove any residual foreign matter. The cornea should then be stained with fluorescein to identify corneal defects. Prompt referral to an ophthalmologist is essential. A proper slit lamp examination is optimal to determine whether inflammation extends into the anterior chamber or involves the iris. If there is no signs of infection, a moderate approach to the inflammatory keratitis includes regular instillation of cycloplegic, mydriatic, and corticosteroid ophthalmic solutions.

CHAPTER 13

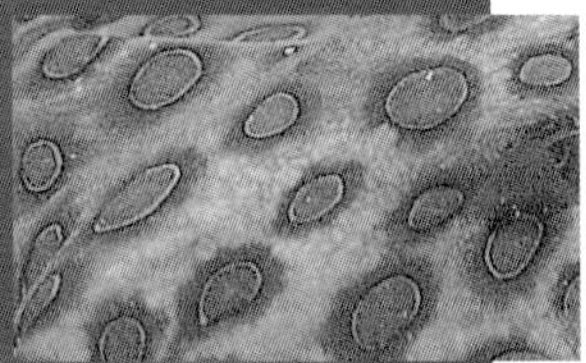

PHYLUM MOLLUSCA

JOHN WILLIAMSON AND

JACQUIE RIFKIN

INTRODUCTION

HALSTEAD (1988) emphasised the dominating global importance of molluscs (mollusks) as a source of marine foods. Most species of mollusc are harmless and of no direct toxicological significance to humans. On the other hand, because many molluscs consumed by humans act as accumulators of marine toxins thought to originate in dinoflagellates (HUMPAGE *et al* 1993), and many also possess intrinsic venoms and poisons of medical importance, molluscs make up the largest group of marine invertebrates toxic to humans {see Chapter 3, page 84}. The knowledge of toxinology of marine molluscs is increasing but already includes recognition of neurotoxins of great potency and of importance in biological research. Quite apart from the venoms and poisons contained within the softer body organs, the external skeletons of oysters, and other bivalves have caused considerable mechanical injury to humans.

CLASSIFICATION OF PHYLUM MOLLUSCA

Class Aplacophora — gastroverms
Class Polyplacophora — chitons
Class Monoplacophora — monoplacophorans (HALSTEAD 1988)
Class Gastropoda — gastropods, e.g. snails and slugs, nudibranchs
Class Bivalvia — bivalves, such as oysters, scallops, clams

Class Scaphopoda — tusk shells
Class Cephalopoda — celphalopods, e.g. squid, octopus, cuttlefish, nautilus

HUMAN MORTALITY FROM MOLLUSC TOXINS

There has been significant human mortality {see Chapter 3, page 67}, principally from paralytic shellfish poisoning (MEYER 1953; McFARREN *et al* 1957; HALSTEAD 1988; SMART 1994), with cases reported in Europe, North America and Japan, as well as the Indo-Pacific {see Chapter 3, page 67}. Deaths from blue-ringed octopus (FLECKER & COTTON 1955; SUTHERLAND & LANE 1969; WILLIAMSON 1987) and from cone shells (FLECKER 1936; HALSTEAD 1988) are documented but are relatively infrequent (see Chapter 3, page 73).*

The true incidence of morbidity from mollusc intoxications (poisoning predominantly, MEYER *et al* 1928; MYER 1953; HUMPAGE *et al* 1993) can only be estimated, and remains a significant global public health problem. It may be increasing (HALLEGRAEFF 1993).

BIVALVE MOLLUSCS

DISTRIBUTION

Bivalve molluscs are found worldwide. Cases of paralytic shellfish poisoning (PSP) after eating bivalves have been recorded from Europe, North America, Japan (MEYER 1953), Thailand, Philippines, Brunei, Sabah, Papua New Guinea, New Britain, Vanuatu, Chile (HALSTEAD 1988), and South Africa (CLELAND & SOUTHCOTT 1965). SCHANTZ (1971) states that PSP is largely restricted to tropical seas 30° north or south of the Equator. In regions further from the Equator, however, the toxin in the Californian mussel, in the Alaskan butter clam and in the dinoflagellate *Gonyaulax catenella* appear identical chemically, pharmacologically and biologically (SCHANTZ *et al* 1960). Paralytic shellfish poisoning has spread internationally through ships releasing contaminated ballast water in the harbours and off the coasts of other nations. *Gonyaulax* sp. (which produces a true 'red tide'), has led to contamination of indigenous Australian bivalves (JONES 1991).

THE TOXINS

Saxitoxin (STX) is a water-soluble toxin produced by dinoflagellates (marine algae) (SMART 1994), believed to have the formula $C_{10}H_{17}N_7O_4$ and a molecular weight of 299. Its action is similar to tetrodotoxin (KAO 1966), and the relationship between ciguatera and paralytic shellfish poisoning is currently being researched (SCHUETT & RAPOPORT 1962), reviewed by HALSTEAD (1982) {see also tetrodotoxin, page 331 below}.

Poisoning occurs after eating shellfish which have ingested and accumulated STX from poisonous dinoflagellates such as *Gonyaulax* spp. This occurs most commonly when the latter appear in huge numbers as 'red tides' or 'water bloom', usually in the warmer months (March to October in the northern hemisphere, September to March in the southern hemisphere). The 'dangerous' season, when STX is present, may be indicated by the deaths of significant numbers of fish and sea birds. Seawater temperature and solar radiation appear to correlate strongly with outbreaks in Sabah, Brunei and Chile. Previously, several days were required to test shellfish toxicity, long enough for shellfish beds to increase significantly in toxicity (HALSTEAD 1988), but YENTSCH (1981) has developed a reasonably rapid assay method for STX in dinoflagellates. The concentration of STX in bivalves may persist in them for several weeks (SUTHERLAND 1983).

POISONING

Systems of poisoning may occur in humans within 20 minutes of ingestion and begin with paraesthesia of the hands and mouth, weakness of the limbs and a 'floating' feeling. Ataxia (staggering gait), vomiting and headache may occur, and bulbar and respiratory paralysis may super-

*Recent direct communication with colleagues in Okinawa, indicates that both blue-ringed octopus and cone shell fatalities occur there {see Chapter 3, pages 68, 69 and Chapter 20, page 414}. Numbers are presently being investigated (ARAKI and TOMIHARA 1994, and KOHAMA 1996, unpublished observations).

vene in some 8 per cent of cases (WILLIAMSON 1992a) within 12 hours of ingestion, causing death (LAMBERT 1973). The symptoms of PSP (Paralytic Shellfish Poisoning) poisoning resemble those of horseshoe crab intoxications (FUSETANI *et al* 1982). PSP has been identified in some Okinawan coral reef crabs (YASUMOTO *et al* 1983), and animal tests have identified a similar toxin in Australian coral reef crabs (LLEWELLYN & ENDEAN 1988, 1991).

Eye irritation may occur in swimmers when red tides occur at swimming beaches (FENNER 1992, unpublished observations). However, this may not be a STX effect. Upper respiratory and eye symptoms have occurred in people living near beaches when dinoflagellates are beached and dry out (LAMBERT 1973).

HUMAN DEATHS

{See also Chapter 3, page 67.} Fatality rates have previously been estimated at about 8 per cent (MEYER 1953; HALSTEAD 1982). In a careful recent review of 2334 worldwide cases of PSP, SMART (1994) documents a fatality rate of 5.9 per cent; he emphasises the relatively high mortality among children with PSP. Human autopsy records exist (LINSTOW 1894; MEYER *et al* 1928; TAKAHASHI *et al* 1955). (Minimata disease, caused by eating mercury-impregnated contaminated shellfish, is referred to on page 67.)

MEDICAL MANAGEMENT

The clinical similarity to tetrodotoxin poisoning (TTX) dictates that first aid and medical management for PSP should be conducted in an identical manner. There is presently no antiserum for either PSP or TTX poisoning.

OCTOPUS (CLASS CEPHALOPODA)

Cephalopods (see Figure 13.1) have well-evolved brains and nervous systems, and are quick reactors and learners, active predators and swift swimmers. All cephalopods are marine, and

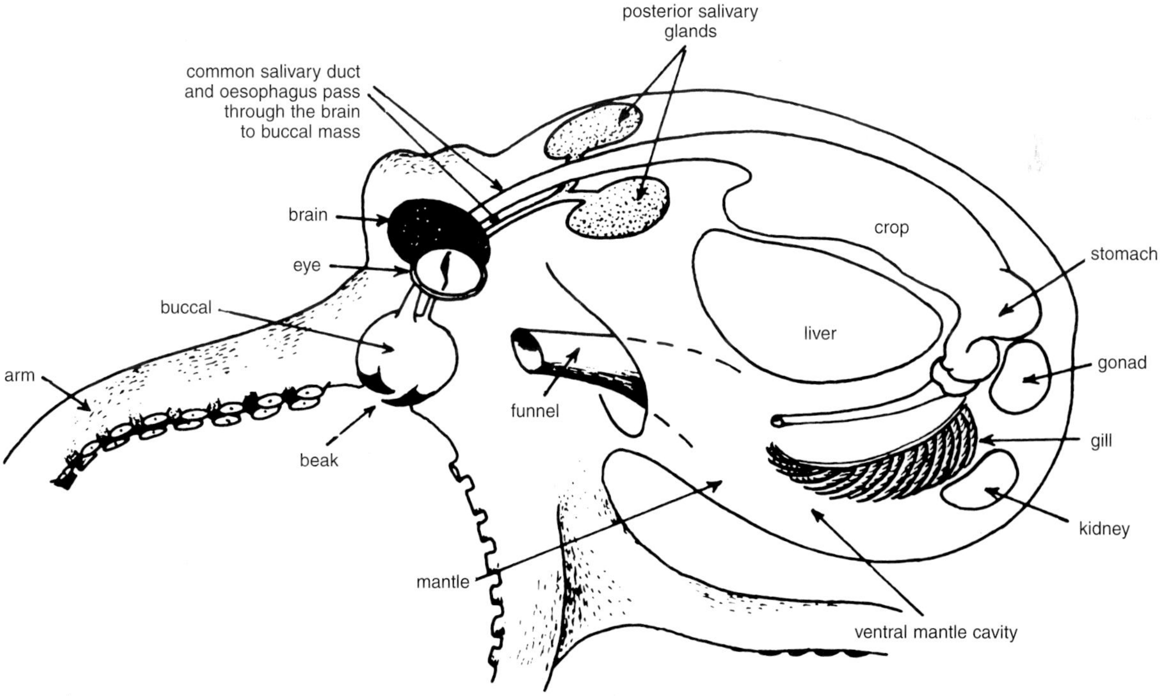

FIGURE 13.1
A schematic drawing of the blue-ringed octopus *Hapalochlaena maculosa*, showing the large, venom-containing (paired posterior) salivary glands and connecting duct. (Drawing reproduced from *Australian Animal Toxins*, courtesy of Professor Struan Sutherland and Oxford University Press, Melbourne.)

intolerant of salinity extremes (HALSTEAD 1988). Their sensory organs include efficient eyes. The mouth of an octopus,* as well as of squid and cuttlefish, contains a pair of chitinous jaws shaped like a parrot's beak, plus 'a many-toothed radular ribbon' for macerating food (MARSH & SLACK-SMITH 1986). The mouth is surrounded by eight tentacular arms equipped with suckers. Octopus possess two sets of salivary glands. The larger set behind the eyes of the blue-ringed octopus secretes tetrodotoxin-like substances (see Figure 13.1, page 329).

The endurance of octopus for high-speed swimming (by jet propulsion, forcing water out of the mantle cavity under pressure) is short (see Plate 13.4); they usually creep over a surface, changing colour to match their surroundings. Colour changes also occur when the animal feels excited or threatened. Most (not all) species of octopus have an ink gland, the contents of which can be squirted out suddenly as a 'smoke screen' when the animal is escaping (HALSTEAD 1988).

HISTORICAL ASPECTS

POWELL (1946) reported one of the only documentations of an octopus actually attacking a human. The writer himself was a victim of an unprovoked attack in New Zealand, from an octopus with arms each about one metre in length. The creature apparently slid its whole body out of the water and fastened two of its arms around his leg. Powell used an open pocket knife to slash at the arms, but the animal retreated only after he had severed three of them. The lurid tales of battles between seafarers and giant octopus belong to the realms of fiction (WILLIAMSON 1987). However, large (but non-lethal) cephalopods undoubtedly exist (SMITH 1977). As far as is known, octopus belonging to the genus *Hapalochlaena* are the only cephalopods that present a significant envenomation risk to humans, though AUERBACH (1988) reports that *Octopus joubini* of the Caribbean can be harmful but to a lesser degree.

* It is an on-going puzzlement to the editors which form of the word 'octopus' is plural. In this text 'octopus' is used as the plural form; the words 'octopi' or 'octopod' are not used. General advice on taxonomy in this section was provided by Dr Richard Willan, Curator of Molluscs, Museums and Art Galleries of the Northern Territory, Darwin.

BLUE-RINGED OCTOPUS (*HAPALOCHLAENA* SPP.)

There is now extensive documentation of these animals and their venom (principally tetrodotoxin [TTX]-like). While the uncommon poisoning (including some mortality) from ingestion of the flesh of certain cephalopods (squid and octopus) is documented from Japan (KAWABATA *et al* 1957), this intoxication will not be considered further here.

DISTRIBUTION

Known in the central part of the Indo-West Pacific Region, including Southern Japan, the Indo-Pacific and New Zealand (WILLIAMSON 1987; HALSTEAD 1988; EDMONDS 1989), islands off eastern Papua New Guinea (MARSH & SLACK SMITH 1986), the Indian Ocean (HALSTEAD 1988) and right around Australia (FLECKER & COTTON 1955; CLELAND & SOUTHCOTT 1965; SUTHERLAND 1983; MARSH & SLACK-SMITH 1986).

The animals frequent shallow rocky reefs and can be encountered by beach and coastal fossickers and waders. They are not usually found in surf conditions. They may inhabit the empty shells of dead gastropods. In urbanised coastal areas they may be found in tins, bottles and human-produced crevices {see the editors' comments in the Preface, page 23}.

APPEARANCE

Easy recognition of blue-ringed octopus is usually possible (Plates 13.2 and 13.3). Like other octopus they have impressive camouflage, with their colour merging with that of the substrate of rock or weed. Bites are possible as a consequence of a victim disturbing an animal before it is seen.

The body and arms are striped with darker brown patches, which tend to encircle the arms and impart a mottled appearance (Plate 13.2), enhancing camouflage. The normally almost invisible blue rings superimposed upon and contained within the darker brown patches can be displayed in a startling and attractive way when an animal is agitated (Plates 13.3 and 13.4). This may increase their visual appeal to the unwary, especially children.

When swimming they align the tentacles in a streamlined manner behind the body (Plate 13.4). The small brownish octopus measures some 12–20 cm from arm tip to arm tip; such small creatures could easily fit into the palm of a curious child's hand!

The uncertainty concerning the taxonomy of the species in the genus *Hapalochlaena* continues (CLELAND & SOUTHCOTT 1965; MARSH & SLACK-SMITH 1986; HALSTEAD 1988) but this is currently being addressed by T. Shanks in Melbourne, Australia.

THE ENVENOMATION

It is stressed that the blue-ringed octopus is generally non-aggressive to humans, and the majority of bites so far recorded have occurred only out of the water upon people handling or interfering with an animal (FLECKER & COTTON 1955; SUTHERLAND & LANE 1969; SUTHERLAND 1983).*

While some doubt has been expressed about the mechanism of venom administration by blue-ringed octopus to their natural prey (e.g. crabs) (MARSH & SLACK-SMITH 1986), there seems little doubt that an actual bite is necessary for human envenomation. The bite is so slight that it may not be noticed by the victim (FLECKER & COTTON 1955; SUTHERLAND & LANE 1969; WALKER 1983; SUTHERLAND 1983) — an important clinical observation, as this may falsely reassure those involved. The bite may, however, produce a small bleeding spot on the skin (FLECKER & COTTON 1955). This puncture may apparently vary from minute to a small laceration 5 mm deep to a blood-filled cyst (SUTHERLAND & LANE 1969; EDMONDS 1989). Not every bite results in symptoms of envenomation, and similarly all grades of clinical envenomation effects probably occur, the milder of which will not necessarily come to medical notice (WILLIAMSON 1990, unpublished observations). Near-deaths, as well as deaths {see above}, are well documented (SUTHERLAND 1983; WALKER 1983; WILLIAMSON 1987) or recognised (WILLIAMSON 1991, unpublished observations) (See also Chapter 3, page 73).

*There is an anecdotal Australian report of significant envenomation occurring in a diver who unwittingly leaned on an unseen blue-ringed octopus while photographing underwater. A faint skin prick was felt, the diver looked and saw the animal under his elbow with its blue rings flashing, and the diver wisely left the water immediately. Some 90 minutes later, collapse from muscle paresis with respiratory embarrassment resulted. Fortunately, recovery occurred without active medical intervention (SOUTHCOTT 1980, unpublished observations).

THE VENOM

This is a salivary toxin of great biological interest. The posterior salivary glands in the blue-ringed optopus have been the source of the salivary toxin for studies. They are situated above the 'brain', in the 'neck' region and joined by ducts (one from each gland) which pass through the brain to form a common duct opening at the base of the tongue (radula) (Figure 13.1, page 329) (SUTHERLAND & LANE 1969).

Composition of the salivary toxin from the posterior salivary glands

A summary of the present understanding of the components of this mixture in Australian species (WILLIAMSON 1987), based on the work of SUTHERLAND and LANE (1969), HOWDEN and WILLIAMS (1974), SAVAGE and HOWDEN (1977) and SHEUMACK *et al* (1978), is as follows:

- Tetrodotoxin-like substances (maculotoxins) (see below)
- Hapalotoxin
- Antigenic, non-toxic components, mw 30 000 and 600 000
- Small quantities of histamine, tyramine, serotonin, and possible other phenolic amines
- A potent hyaluronidase. MARSH and SLACK-SMITH [1986] suggest that 'a spreading agent' may also be a component of the saliva from the smaller paired anterior salivary glands.

'Maculotoxin', 'tetrodotoxin', 'hapalotoxin', and 'saxitoxin'

Some confusion remains concerning the precise molecular and toxinological distinction between

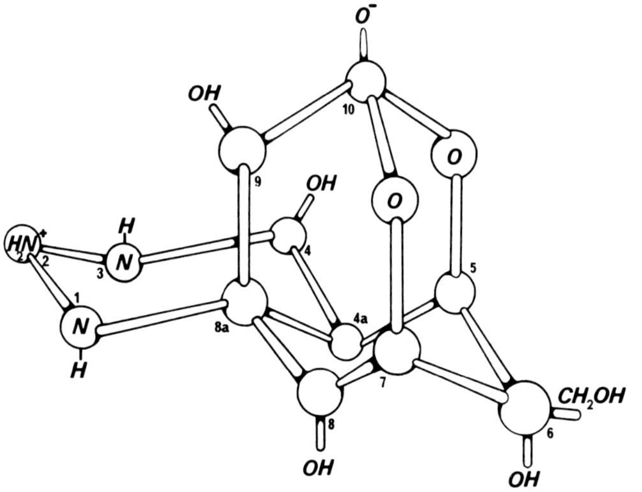

FIGURE 13.2
Molecular structure of tetrodotoxin (TTX), $C_{11}H_{17}O_8N_3$. (Reproduced with permission, from J. Harris, in (J. Pearn, ed.) *Animal Toxins and Man*, Queensland Health Department, Brisbane, 1981, p25.)

these toxins and terms. Owing to this confusion it is presently customary among some workers to call tetrodotoxin from pufferfish 'TTX'. Tetrodotoxin from other organisms is considered to be 'TTX-like'. FREEMAN and TURNER (1970) (who coined the term 'maculotoxin' for a toxin they extracted from the posterior salivary glands of *H. maculosa*), JARVIS *et al* (1975) and subsequently FREEMAN again (1976) agreed with SHEUMACK *et al* (1978) that 'maculotoxin' and 'tetrodotoxin' (TTX) are so similar pharmacologically and biochemically as to be indistinguishable and are probably identical (FREEMAN 1980, unpublished observations). However, CROFT and HOWDEN (1972) and DULHUNTY and GAGE (1971) suggested that maculotoxin and tetrodotoxin were not identical. CROFT and HOWDEN (1972) and later SAVAGE and HOWDEN (1977) identified two toxins in the secretion of this posterior gland saliva. One was tetrodotoxin, and Savage and Howden named the other 'more polar toxin' (HALSTEAD 1988) 'hapalotoxin'. This 'minor toxin' had an intraperitoneal toxicity in mice of one-fifth of that of the 'main toxin'. In contrast, saxitoxin (STX) (paralytic shellfish toxin), $C_{10}H_{17}N_7O_4$, mw 299, is a water-soluble substituted tetrahydropurine base (at least in part), LD_{50} in mice 2.4 µg/kg intravenously (SUTHERLAND 1983). Saxitoxin and tetrodotoxin are closely similar pharmacologically (KAO 1966); experimental work links their actions {see below}.

A single adult blue-ringed octopus weighing only 25 g, with posterior salivary glands weighing 300 mg, theoretically possesses enough venom to fatally paralyse 10 humans, each weighing 75 kg (SUTHERLAND 1983). This equates approximately with the LD_{50} range of the more dangerous Australian snake venoms (SUTHERLAND 1978).

NOTE: TTX-like substances were first recorded as associated with blue-ringed octopus venom in Australia by SHEUMACK *et al* 1978 .

Tetrodotoxin (TTX) and TTX-like toxins

Tetrodotoxin $C_{11}H_{17}O_8N_3$, molecular weight (mw) 319.3 (Figure 13.2) is found in certain tissues of marine fish (phylum Chordata) and its ingestion causes poisoning in humans (TIBBALLS 1988; SUN *et al* 1994) {see Chapter 3, page 86}. Its isolation and purification as part of TTX fish poisoning research, and the final elucidation of its chemical structure in 1964 (Sankyo Company literature), was a Japanese and American research triumph.

Apart from the presence in blue-ringed octopus, TTX-like substances are found in freshwater puffer fish from Thailand (SAITANU *et al* 1991), in the skin of central American frogs (genus *Atelopus*); in the cutaneous glands of a Californian newt (genus *Taricha*); in a species of Pacific goby (marine fish with spiny fins, and ventral fins modified into a suction disc); in certain shellfish, such as the ivory shell *Babylonia japonica* and trumpet shell *Charonia lampas*; in the sea star *Astropecten polyacanthus* and in flatworms (see Chapter 3, page 65) (AUERBACH 1988).

Toxinology of TTX and TTX-like substances

A TTX-like substance is, if not the only {see above}, then certainly the principal toxic component of *Hapalochlaena* spp. venom. TTX-like toxins are important, occupying a wide distribution in nature {see below} and an important role in both experimental (KAO 1966; BLANKENSHIP 1976) and clinical toxinology. They are regarded as non-antigenic (SUTHERLAND *et al* 1970) {see below}. They are very potent in their effect on mammals; the intravenous or subcutaneous LD_{50} in mice is 9 µg/kg (NARAHASHI *et al* 1967; SUTHERLAND 1983).

TTX (similarly to shellfish toxin saxitoxin [STX]), blocks the movement of sodium ions (Na^+) across excitable membranes by impeding exclusively this ion's movement through Na^+ channels (HILLE 1975; BLANKENSHIP 1976). It has been suggested that this is achieved by attaching to an adjacent receptor to the Na^+ channel so that the planar guanidinium group (common to both TTX and STX molecules) projects over the Na^+ channel as a loose-fitting cap. This would prevent Na^+ flux without affecting the Na^+ channel's gating action (KAO 1983). Such properties have naturally attracted the close attention of research neurophysiologists and biologists (BLANKENSHIP 1976; HARRIS 1981; AZUMA *et al* 1986; STYS *et al* 1993).

Thus TTX and TTX-like substances cause particularly (but not only) mammalian somatic nerve conduction blockade, leading to a dose-dependent, flaccid voluntary muscle paresis or paralysis, with respiratory insufficiency or failure, and consequent cerebral hypoxia (WALKER 1983; SUTHERLAND 1983). TTX also causes potent vascular smooth muscle depression with hypotension in experimental animals. Human autonomic nerve dysfunction and systemic hypotension, although less evident than expected from animal studies, does occur {see below}.

There is no direct action of TTX upon either the myocardium, neuromuscular junctions {see Figure 3.2, page 83} or the brain. The dominating threat to life is cerebral hypoxia from respiratory failure. This initial effect of brain-sparing and certain autonomic effects in human victims have important clinical connotations. It is of interest that TTX has been trialled as a long-acting local anaesthetic, spinally and epidurally, in dogs, with equivocal results (TREETRAKAN *et al* 1979).

CLINICAL FEATURES OF BLUE-RINGED OCTOPUS ENVENOMATION

Clinical features of envenomation will develop in a dose-dependent manner, but are evidenced earlier and more seriously in the weak or people of small body mass, and especially children. People actually in the water appear to be at no risk of a bite unless they actively interfere with the animal. (NOTE: One accidental in-water envenomation involving a scuba diver is recorded [see above and Chapter 22, page 423.)

As described above, the bite may pass unnoticed, with minimal local skin signs. Blood on the skin (FLECKER & COTTON 1955), local oedema and a local stinging sensation lasting six hours, have been reported (EDMONDS 1989). Circumoral paraesthesiae may occur (AUERBACH 1988) and may herald more serious effects.

The first sign of serious envenomation (most commonly following a known handling) may be the rapid onset (within 10 minutes) of progressive muscle weakness, with speech and respiratory difficulty, dysphagia and visual disturbance (FLECKER & COTTON 1955; SUTHERLAND & LANE 1969). Nausea and vomiting may occur, with the risk of aspiration of stomach contents into the trachea. Collapse from muscle weakness soon occurs, followed by hypoxic and possibly hypotensive unconsciousness in the absence of effective resuscitation. The autonomic effects of the toxin may contribute to the apparent pulselessness and fixed dilated pupils, suggesting actual death.

OTHER CLINICAL FEATURES OF TTX ENVENOMATION OR POISONING

The absence of direct brain effects of TTX and TTX-like substances can result in an apparently dead patient who is receiving skilled resuscitation listening helplessly in a paralysed state to everything that is said and done. This frightening experience is well documented in a survivor of serious envenomation from a blue-ringed octopus bite (HOPKINS 1964), and rescuers should be mindful of such a possibility. Both TTX and TTX-like substances are readily absorbed either by injection or ingestion, appearing in the urine from either of these routes of entry.

Although TTX would appear to be non-antigenic {see below}, one case report exists of envenomation in a female with previous shellfish allergy in whom there developed generalised urticaria, pruritis and joint effusions, with a haemorrhagic vesicle at the bite site; this patient also manifested simultaneous nerve conduction blockade effects (EDMONDS 1969).

CLINICAL FEATURES OF TTX POISONING

A brief history of TTX intoxication from puffer fish poisoning is given by SUN *et al* [1994] {see Chapter 3, page 66}.

AUERBACH (1988) describes a disseminated-intravascular-coagulation-like syndrome following TTX intoxication after pufferfish ingestion, heralded by petechial skin haemorrhages, which may progress to bullous desquamation along with haematemesis, and other stigmata of prolonged coagulation. Of all such victims he claims 60 per cent will die, most within the first six hours. Survival past 24 hours is a good prognostic sign.

TANI (1945), working with Japanese puffer-fish, suggested the lethal human dose of TTX by ingestion is about 2 mg (10 000 mouse units). An estimated mortality rate of 51 per cent {see Chapter 3, page 66} was reported when at least 3000 people were poisoned by pufferfish in Japan between 1955 and 1975 (SEUNAGA and KOTOKU (1980).

FIRST AID

PUFFERFISH POISONING

First aid may be of life-saving importance (HOPKINS 1964; WALKER 1983; SUTHERLAND 1983). Early expired air resuscitation (EAR) takes absolute priority in cases of respiratory failure {Chapter 5, page 106}. Prevention of hypoxia by intermittent positive pressure (IPPV) or as EAR {see Chapter 5, page 107}, together with airway protection will result in functional neurological recovery after two to eight hours, or even longer. (TTX intoxication ['tetrodotoxication'] from puffer fish poisoning requiring 13 hours' successful ventilatory support is documented [SUN *et al* 1994].) However, following blue-ringed octopus envenomation, diminishing weakness may persist for some days.

BLUE-RINGED OCTOPUS BITE

As soon as possible after recognition of a blue-ringed octopus bite, apply a compression/immobilisation bandage {see Chapter 5, page 110, Figures 5.3A and B, page 110, and Figure 17.2, page 400} over the bite site and then the entire limb. Bites on the trunk should have sustained compression applied with a hand-held pad. Immobilise the victim. Do not release the compression bandage until the patient is under direct and skilled medical supervision. Monitor the patient's airway and pulse. Call for immediate medical or skilled paramedical aid. Do not leave the victim. Commence EAR as soon as breathing becomes weak or central cyanosis (blue lips and tongue) develops. Commence cardio-pulmonary resuscitation (CPR) if the carotid pulse is lost {see Chapter 5, page 98, and flow chart, page 107}. Do not cease resuscitation until medical supervision is available. Appearances of death in this envenomation can be totally misleading.

MEDICAL MANAGEMENT

(APPLIES ALSO TO PUFFERFISH POISONING)

Currently treatment for tetrodotoxication (blue-ringed octopus envenomation and puffer fish poisoning) is supportive. Gain control of the airway with endotracheal intubation, administer 100% oxygen, and correct hypotension with intravenous fluids and, if indicated, inotropic support (SUN *et al* 1994). Monitor urine output as a guide to adequate volume replacement and maintenance. The case for anticholinesterases (e.g. edrophonium, neostigmine) in this envenomation (SUTHERLAND 1983), although advanced for puffer fish poisoning (TORDA *et al* 1973), remains unestablished (TIBBALLS 1988), as does that for steroids and antihistamines (LYN 1985).

Prevent or treat pulmonary aspiration of gastric contents as appropriate. In the absence of hypoxic damage, full recovery can be anticipated. (SUN and colleagues [1994] point out that a patient with tetrodotoxication who presents unconscious may already have suffered hypoxic damage.) Be alert to the possibility of awareness in the apparently unconscious, paralysed patient (HOPKINS

1964). There is currently no specific antivenom for blue-ringed octopus envenomation {see below}. Consider tetanus prophylaxis if appropriate.

Further suggestions for therapy for tetrodotoxication, based only on experimental evidence, include (SUN *et al* 1994):

- opiate antagonists (e.g. naloxone), based on the similarity in molecular configurations of opiates and tetrodotoxin (which itself has opiate activity)
- monoclonal neutralising antibodies (SIMS & OSTMAN 1986) {see below}
- cysteine, based on two possibilities:
  - (i) cysteine opens sodium channels, likely through its excitation of glutamanergic and GABA-ergic neurones, thus opposing the nerve-conduction-blocking actions of TTX (MINEGOLOMB *et al* 1989)
  - (ii) TTX decomposes quickly in the presence of cysteine (SIMS & OSTMAN 1986)

PROSPECTS OF ANTIVENOM DEVELOPMENT

Although no antivenom is presently available for tetrodotoxin (TTX) envenomation (blue ringed octopus) or poisoning (e.g. pufferfish), or poisoning from shellfish, crabs and other organisms (STX) (SUTHERLAND 1983; THEAKSTON & WARRELL 1991), recent research is encouraging. Tetrodonic acid, a relatively non-toxic derivative of TTX, has been used successfully to develop a mouse spleen cell produced monoclonal antibody that reacts partially *in vitro* with TTX (WATABE *et al* 1989). Subsequently, passive protection of mice *in vivo* challenged with either TTX or saxitoxin (shellfish poison) has been possible using a rabbit antiserum that was developed against a tetrodotoxin-formaldehyde-keyhole limpet haemocyanin conjugate (KAUFMAN *et al* 1991).

CONE SHELLS

TAXONOMY

Cone shells (Plates 13.5–13.8) belong to class Gastropoda, family Conidae (TAYLOR *et al* 1993). Two of the major conoidean families are the Conidae (cone shells) and the Terebridae (auger shells). In fact all members of the superfamily Conoidea possess a venom apparatus, but only members of family Conidae, subfamily Coninae have been known to envenomate humans (HALSTEAD 1988).

HUMAN ENVENOMATION

According to Kohn (1958) and KIZER (1990) there are at least 18 species of cone shell that have been implicated in human envenomations, some of them occasionally causing deaths. A list of seven of the more commonly known of these cones is presented below. The species considered to be dangerous to humans are confined to tropical and subtropical waters (SUTHERLAND 1983).

Conus geographus	geographer cone
Conus gloriamaris	glory of the sea
Conus omaria	feathered cone
Conus striatus	striated cone
Conus textile	textile cone
Conus aulicus	princely cone
Conus tulipa	tulip cone

ABBOTT (1973) suggests that variations in feeding habits and venom production account for variations in venom toxicity. There are three categories of cone based on diet: fish eaters, mollusc eaters and worm eaters. Worm eaters are far more abundant than mollusc eaters, which are in turn more abundant than fish eaters. The cones that feed on fish tend to be more venomous to humans than most mollusc-eaters except *Conus textile* and all worm-eaters. OLIVERA *et al* (1988) suggest that 10 per cent of all cones are fish-eaters. Most human fatalities have been caused by cones that feed on fish, although one species claimed to be associated with human fatality, *Conus textile* (Plate 13.5) (HALSTEAD 1988), is a mollusc eater (see ABBOTT 1973, above).

HUMAN DEATHS

Some details are considered in Chapter 3, page 68. It is the opinion of the editors that perhaps too much emphasis has been placed on the cone shell as a public health problem. The editors suspect that this may in part have been because of the early effective research upon the animal and its toxin, stimulated by the relative ease of capture of specimens and the unquestioned elegance of its envenomating apparatus

(SUTHERLAND 1983) (Plate 13.7, and Figure 13.3, page 337).

Nevertheless deaths have occurred (RUMPHIUS 1705; FLECKER 1936; TOMIHARA and colleagues 1994, unpublished observations), although relatively rarely {see Chapter 3, page 68}. The claimed incidence of human fatalities worldwide is uncertain, ranging from seven to 16, with precise verifiable documentation often lacking (SUTHERLAND 1983). HALSTEAD (1988) states 'more than eight deaths' in 'more than 28 cases' 'reported in the literature', but no details are provided. STANISIC (1987) provides references totalling 11 human deaths (FLECKER 1936; KOHN 1963; ENDEAN & RUDKIN 1963). These accounts almost certainly omit some recent data from Tomihara {see Chapter 3, page 68, and Chapter 20, page 414}, who reports at least four human deaths from cone shell envenomation in Okinawa, Japan, since 1927, all attributed to *Conus geographus* (Plate 13.8). The only fatality reported in Australian literature was caused by *Conus geographus* (FLECKER 1936).

Historical aspects of cone shell envenomation are provided by CLELAND and SOUTHCOTT (1965), and a good recent account is by MARSH and SLACK-SMITH (1986).

DISTRIBUTION

There are about 300 species of cone shells worldwide (WALLS 1980), with the majority occurring in the tropical waters of the Indo-Pacific, and some species found in temperate and subtropical waters worldwide (e.g. New Zealand and Okinawa Island, Japan). Approximately 80 species occur in Australian waters (WALLS 1980), mainly on coral reefs such as the Great Barrier Reef (or reefs off Western Australia) where rubble and sand provide a suitable habitat.

Cone shells are usually found buried in rock crevices or under sand or coral rubble in shallow waters, although some species are found in deep water. They commonly remain buried during the day and emerge to feed at night.

Stingings from marine cone shells have been reported from:

- Australia (FLECKER 1936; KOHN 1958; J WEINER 1993, unpublished observations). Most books dealing with Australia's Great Barrier Reef record episodes of cone shell stings.
- Papua New Guinea and New Britain (HINDE 1885; CLELAND & SOUTHCOTT 1965)
- Moluccas (Spice Islands) (ADAMS 1848)
- Okinawa, Japan (TOMIHARA 1994, unpublished observations)
- Kurile Islands (north of Japan), various western Pacific islands, Hawaii (HALSTEAD 1988)

APPEARANCE

Cone shells are usually easily identified by their cone-like shape with a long, narrow aperture running almost the full length of the shell (Plates 13.5–13.8). Some specimens of *Conus leopardus* may be as long as 23 cm. The largest Australian species, *Conus aulicus*, reaches just 15 cm. *Conus geographus*, one of the species most commonly implicated in serious human envenomation, reaches 13 cm in length. However, the majority of cone shells are rarely longer than 10 cm (SUTHERLAND 1983).

Although under natural conditions cone shells are usually well disguised by a thick horny covering forming the outer layer of the shell, underneath they have a vast array of colours and patterns. This causes problems in identification of species, but at the same time makes them very attractive to shell collectors. HALSTEAD (1988) points out that many stings have occurred while the collector was scraping encrusted organic debris from the shell of the animal to reveal the colours, and postulates that such scraping could stimulate the living cone to strike. In fact, this occurs with great regularity in the laboratory where even light pressure applied to the shell of *C. geographus* causes forceful expulsion of a dart from the proboscis (JP BINGHAM 1995, personal communication to JR)

As described above, the aperture runs almost the full length of the shell. The venomous proboscis which holds the envenomating harpoon can emerge from this slit to reach almost anywhere on the length of the shell, making the recommended handling of even the larger, blunt end of the shell somewhat risky. Also arising in this slit, at the narrow end of the shell, is the siphon, which is often confused with the proboscis (Plate 13.5). The cone shell draws in water through this siphon and passes it over the gill for respiration and over a sensory organ to detect the taste or smell of its prey in the water.

PREY CAPTURE BY CONE SHELLS

Cones may approach their victims slowly. Suddenly the proboscis emerges, usually from the narrow end of the shell, and fires a small harpoon (radula tooth) (Plates 13.5 and 13.7) into the victim. Fish-eating cones lie submerged in the sand and track nearby fish with the siphon. When the fish is near, the cone shell will rapidly emerge from the sand and plunge its harpoon into the hapless victim. The fish, once stunned, is enveloped by the proboscis and devoured. Although many cones approach the victim and appear to stab it, *Conus geographus* approaches the prey and scoops it up in its greatly enlarged, funnel-like rostrum (part of the proboscis). Envenomation occurs inside the rostrum (JOHNSON & STABLUM 1971). Swallowing then occurs.

ENVENOMATION APPARATUS

As stated above, mollusc-eating cones other than *Conus textile*, are not dangerous to humans (FREEMAN *et al* 1974), but the fish-eating cones are usually larger and have a different venom that has caused human deaths.

The venom is produced in a special convoluted, duct-like venom gland (Figure 13.3). At the distal end of this tube is a muscular bulb which can contract, forcing venom out of the gland to the tip of the proboscis, where a radular tooth is located, in readiness for prey capture. This harpoon-like structure has a barbed tip and usually has teeth along the length of the shaft. It is one of several (up to 20) which are formed and are stored within the radular sac (Figure 13.3) in various stages of development. A mature radular tooth in a fish-eating cone is hard, consisting of a rolled-up sheet of chitin which may reach 20 mm in length (STANISIC 1987). It is hollow to allow passage of venom (FREEMAN *et al* 1974). When required, a tooth is pushed to the tip of the proboscis, still attached by a ligament to the dart sac. After envenomation occurs a tooth is not reused (Plate 13.7).

CONE SHELL VENOMS — POTENTIAL SOURCE OF DRUGS

The venoms from cone shells consist of peptide toxins called conotoxins. Cone shells have a unique cocktail of compounds known to be biologically active. Some of these are attracted to receptors in the human brain (KURIGAN 1995).

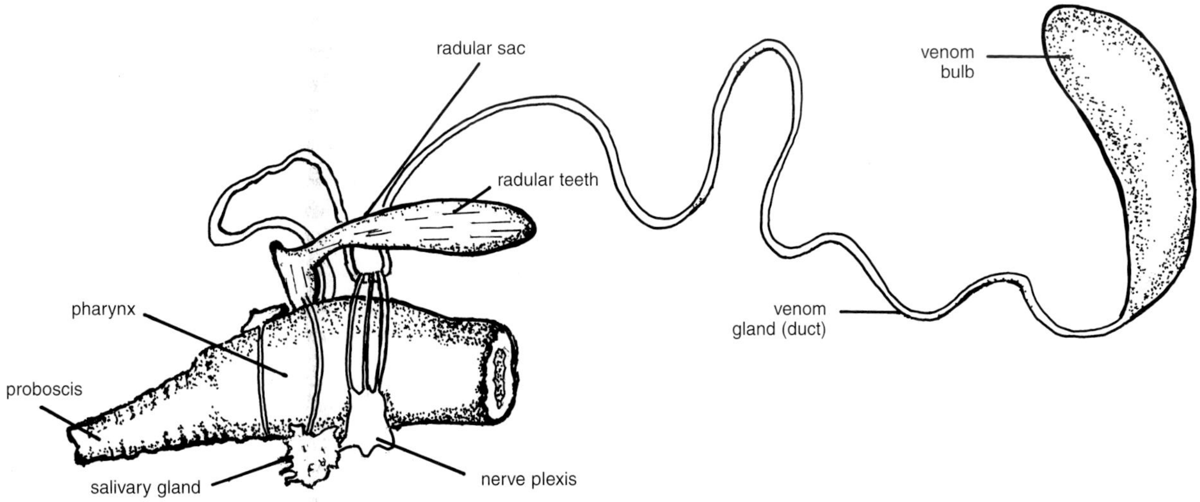

FIGURE 13.3
The venom apparatus of *Conus* species, after Endean and Duchemin (Toxicon 4: 1966). (Drawing by Jan Hynes.)

The most extensively studied coneshell toxin is from the fish eater *Conus geographus* (e.g. HERMITTE 1946; KOHN *et al* 1960; WHYTE & ENDEAN 1962; ENDEAN & RUDKIN 1963; FREEMAN & TURNER 1972; TURNER, FREEMAN 1974; GRAY *et al* 1977, 1981). Conotoxin peptides range in length from 13 to 30 amino acids.

WHYTE and ENDEAN (1962) showed that venom from *Conus geographus* paralysed the skeletal muscle of rats and guinea pigs, and ENDEAN and colleagues (1974 & 1977) showed that although the venom caused flaccid paralysis of vertebrate skeletal muscle (including respiratory muscles), its effects were reversible and it did not affect smooth or cardiac muscle. However, venoms of other fish-eating cones such as *C. magus* contain components that are also active in both cardiac and smooth muscle (ENDEAN & IZATT 1961, 1965, ENDEAN *et al* 1974), and envenomations of humans by this and other fish-eating species are more dangerous.

*There are at least four major families of conotoxins (KURIGAN 1995):

- α-conotoxins — compete with the neurotransmitter acetylcholine (ACH) for binding to the ACH receptor, and cause flaccid paralysis of skeletal muscle;
- μ-conotoxins — close voltage-dependent sodium channels in skeletal muscle causing flaccid paralysis of skeletal muscle;
- ∞-conotoxins — close voltage-gated calcium channels causing flaccid paralysis of skeletal muscle. This was the first family of conotoxins tested for medical use;
- ∂-conotoxins — cause skeletal muscle rigidity and paralysis by maintaining voltage-sensitive sodium channels in their open state. They include the 'lockjaw' peptide (responsible for muscle rigidity) and the so-called 'King Kong' peptide. The latter, identified in *C. textile* only, appears to allow the test animals to assume dominant behavioural patterns in competitive situations! (KURIGAN 1995). Some of these toxins target specific phyla. Other major toxin classes indentified in *C. geographus* venom include a peptide that induces sleep in experimental animals and a vasopressin analogue.

While every documented account of cone shell envenomation concerns the direct toxic effects of the venom, *Conus* venom would certainly appear to be antigenic (KOHN *et al* 1960). Furthermore, the antigenic activity of the venom demonstrating itself clinically as a serious allergic reaction would seem theoretically possible, and may already have occurred (J. WEINER, M.D. Melbourne, 1993, unpublished observations). Attempts to develop an antivenom have taken place (SUTHERLAND 1983)

SYMPTOMS AND SIGNS OF ENVENOMATION

A sharp pain is usually felt at the site of penetration of the radular tooth. Occasionally it is mild, occasionally severe and even excruciating. The site of the sting may show little reaction (Plate 13.9), or it may blanch, or develop a bluish tinge, and this is followed by numbness and local swelling. Severe stings may induce symptoms that require two to three weeks to resolve (AUERBACH 1988).

In serious envenomations incoordination and muscular weakness may develop rapidly, and swallowing, speech, vision and even hearing may be affected. Nausea may occur and generalised pruritus may be distressing; respiratory paralysis may develop (SUTHERLAND 1983). A sting from *C. geographus* is claimed to have the greatest malignant potential, with a reaction that may rapidly progress to respiratory arrest, coma and cerebral oedema, disseminated intravascular cagulopathy

*Research on coneshell venoms is currently being carried out at the University of Queensland's Centre for Drug Design and Development as well as at Neurex Inc. in San Diego. It appears that the small peptide molecules of cone shell venoms are particularly valuable in the field of peptide mimetics and have particular advantages for the production of drugs with fewer side effects. Studies include molecules that would specifically target and control the release of cellular toxins and so limit ischaemic damage following strokes. These experiments have already been carried out using rats in the laboratory. Other work uses toxins from *Conus geographus* to target receptor-mediated responses thought to be involved in Alzheimer's disease, and still other work uses worm-eating cone toxins for internal parasite control in animals without producing the side effects typical of anthelmintics on the market today (personal communication JP Bingham 1995 to JFR, and see also KURIGAN 1995).

and cardiac failure (RICE & HALSTEAD 1969), though Queensland workers studying C. *magus* consider its sting to be more dangerous (ENDEAN *et al* 1974). In the Asia-Pacific region, particularly off Singapore, stings are accompanied by severe pain, and spreading numbness, nausea and hypersalivation. In more serious stings, pain is excruciating. The venom causes blurred vision, lack of muscular coordination and respiratory paralysis, and death (MING *et al* 1990).

DEATHS

Recorded deaths appear to have been the result of hypoxia by paralysis of the respiratory muscles (ENDEAN & RUDKIN 1963), which is probably reversible. (Note the similarity clinically to blue-ringed octopus envenomation described above.) There is no evidence that any of the documented fatalities received effective respiratory resuscitation. Clinical recovery has been documented after four hours of assisted ventilation (PETRAUSKAS 1955).

PREVENTION

Prevention is probably the most important aspect, as almost all fatal cases have occurred while the victim was collecting shells.

1. No cone shell should ever be handled carelessly as most have a venom apparatus capable of causing pain and discomfort. Even the larger worm-eaters can inject large amounts of venom (STANISIC 1987). If handling is necessary, thick heavy gloves should be worn and the shells should be held by the blunt posterior end.
2. Cone shells should never be placed in a pocket or a bag held close to the body as the darts can easily penetrate clothing.

FIRST AID

After envenomation, immediate pressure immobilisation over the area {see Figures 5.3A and B, page 110}, together with enforced rest for the patient, is important {see Chapter 5, page 110}. The pressure bandage should be left in place until resuscitation facilities are available. Expired air resuscitation may be required {See Chapter 5, page 107} and will be life saving when indicated.

MEDICAL MANAGEMENT

Assisted ventilation may be necessary for many hours. Antivenom is not available {see above}. Edrophonium, used with some effect by TORDA and colleagues for tetrodotoxin poisoning (1973), may be worth consideration for paralysis from *C. geographus*, as a slow intravenous dose of 10 mg (adult) (SUTHERLAND 1983); but its efficacy for use in *Conus* envenomation remains unproven. Antibiotics may be necessary {see Chapter 5, page 115}, and tetanus toxoid prophylaxis must be given if the patient's immunity is not current.

FUTURE RESEARCH

Scientists have barely tapped the surface of the rich store of toxins present in the cones, only one of which, *Conus geographus*, has been intensively studied (OLIVERA *et al* 1988).

Acknowledgments: The authors are grateful to Dr Richard Willan of the Museums and Art Galleries of the of the Northern Territory and to Shirley Slack-Smith of the Western Australian Museum for helpful suggestions for the manuscript.

CHAPTER 14

Phylum Porifera (sponges)

Jacquie Rifkin

Introductory biology

Sponges were formally accorded the status of animals by zoologists in 1835 (HALSTEAD 1988), although John Ellis had in fact delivered a lecture in which the original concept was proposed, at the London Royal Society in 1765. Early descriptions of sponges likened them to clones of protozoans, and they were placed with the unicellular organisms as a separate class by Lütken in 1876. They were then finally attributed formal phylum status in the early part of the 20th century (HOOPER and WIEDENMAYER 1994). Sponges are mainly marine, but there are other families of sponge, including the Spongillidae, which are found worldwide in fresh water.

There is a high frequency of bioactive compounds occurring in sponges and it is suggested (e.g. PROKSCH 1994) that these act as a chemical defence against various environmental stress factors such as predation, competition for space and overgrowth by fouling organisms.

Distribution

Most formal reports of envenomations from marine sponges have come from Australia and the western tropical coastline of the Atlantic (HARTMAN 1967; SOUTHCOTT & COULTER 1971; SOUTHCOTT 1987a; HOOPER *et al* 1991). However, toxic and stinging sponges are found worldwide. Documentation is poor at present.

Although most sponges live in shallow waters, some are found at considerable depths. Almost without exception sponges will attach themselves to some form of substrate, such as rock, seaweed or even hard-shelled animals. They are more diverse in warmer tropical waters but are equally abundant in tropical regions and polar seas (CLELAND & SOUTHCOTT 1965).

PROKSCH (1974) suggests that sponges which are growing in exposed waters are usually more toxic than those not so exposed, though HOOPER (1995, personal communication) suggests that other forms of toxicity may be prevalent in more cryptic species.

APPEARANCE

Adult sponges are sessile, plant-like individuals, with a variable body form which is influenced to some extent by the type of substrate and the amount of water movement. However, different groups have characteristic shapes. Species that consist of rounded or flattened clumps or flexible, whip-like growth forms often live in areas of strong water movements; massive or bushy forms may be found in calmer waters (Plate 14.1).

Colours may vary from white, grey, black or drab to brilliant hues that range through the colour spectrum. The more brilliant green, red yellow, orange, pink or mauve sponges tend to be found in shallower waters. These colours are often due to carotenoid pigments derived from their plankton diet and have a function thought to be related to UV protection (HOOPER 1995, personal communication).

CELL STRUCTURE

Individual sponges consist of a collection of totipotent cells, bounded by a continuous epithelium, which enclose a system of canals and chambers of varied complexity, and which are connected to the outside by small pores. Special flagellated cells that line the 'in-current' canals carry water currents, bearing food particles and oxygen, in through the pores. The water carrying waste materials passes out through exhalant apertures or oscula. This honeycomb structure, together with the sessile habit of sponges, provides suitable living quarters for a great range of other small sea creatures (including marine bacteria) and sea slugs; so much so that sponges have been likened to marine 'hotels' (Plate 11.7) (WILKINSON 1984).

Most sponges contain both male and female sex cells. Larvae leave the parent through the oscula and finally settle to the sea bottom to form a sponge. However, asexual reproduction is also common and in fact is thought to be the most prevalent means of dispersal and recruitment in shallow waters (HOOPER 1995, personal communication). Sponges have remarkable regenerative powers.

SKELETAL COMPONENTS

Sponge cells lie within a proteinaceous matrix supported by a skeletal framework of small transparent needle-like spicules, and some also have a fibrous (collagenous) skeleton called spongin — all of which tends to make them unattractive prey. Not all sponges have some form of skeleton (see below), but skeletal components (i.e. spicule types) provide us with a basis for taxonomic classification (HOOPER & WIEDENMAYER 1994). Sponge skeleton may be composed of:

- spongin fibres with or without silicious spicules (most members of Class Demospongiae)
- a basal calcareous and argonitic skeleton and free silicious spicules (some members of the class Demospongiae)
- no mineral skeleton (some members of the class Demospongiae)
- calcium carbonate spicules (class Calcarea)
- siliceous (glass) spicules (Class Hexactinellida)

NOTE: Most toxic sponge injuries result from abrasions by members of the class Demospongiae.

Marine sponges of commercial value have only spongin skeletons and the collection of these constituted a substantial industry prior to the advent of synthetics. (The finest commercial sponges came from the Mediterranean Sea [MOORE 1908]). It seems that all types of skeletons containing spicules are capable of harmful effects on human skin. It is unclear whether living, skeleton-free marine varieties offer a lesser risk. Nevertheless some sponge species are non-toxic (DE LAUBENFELS 1932).

Sponges also possess chemical defences, so that both mechanical and toxic chemical injury may be sustained by individuals who handle specimens (Plate 3.1).

TOXIC SUBSTANCES FOUND IN SPONGES

As stated above, sponges generally are rich in biologically active compounds, and this subject has been reviewed by SOUTHCOTT (1987), STONIK and ELYAKOV (1988) and by GARSON (1994). These include cytotoxic, cytostatic, antibiotic, muscle relaxant, hypotensive, arteriospastic, antispasmodic, and neuroactive (affecting presynaptic function) agents. It was earlier shown that certain sponges contain antibacterial substances (JAKOWSKA & NIGRELLI 1960; SHARMA & VIG 1972). STONIK and ELYAKOV (1988) stated that some polyether and macrolide compounds are accumulated by sponges from ingested microorganisms or produced by symbiotic algae and bacteria. It also appears that some sponges contain metabolites that act as strong anti-tumor toxins. The microorganisms, often present in symbiotic association, may be responsible for or contribute to the biosynthetic processes detected (GARSON 1994).

CLINICAL ASPECTS

Although many sponges are in fact abrasive to the skin, according to SOUTHCOTT (1987a), the chemicals within the sponge are the more important cause of skin irritation and dermatitis.

Commonly there is no particular sensation on initial skin contact (often both hands, but bare skin anywhere may be affected). A juice or slime may be observed by the victim coming from the sponge onto the skin during contact. Rapid removal of this material by washing may be to no avail in preventing subsequent symptoms (SOUTHCOTT & COULTER 1971). A variable time later (hours or days) a burning, stinging and/or itching sensation commences that may increase to unbearable intensity and persist for days or longer. Descriptions such as '...for days he was like a madman', and 'absolute hell' are documented (CLELAND & SOUTHCOTT 1965; SOUTHCOTT & COULTER 1971). The symptoms and signs are invariably confined to the skin and to the contact area or immediately adjacent areas. However, local joint stiffness may also occur. An acute inflammatory reaction with erythema and oedema (sometimes severe) develops simultaneously with the symptoms. The oedema can begin to compromise the local circulation (see Case Report below). Papules, vesicles or bullae, followed later by desquamation, are described. This syndrome is commonly referred to as 'stinging sponge dermatitis' (BURNETT *et al* 1987). Of particular concern may be iritis or corneal damage, which may follow eye contact with the sponge or, more likely, with fingers contaminated with toxic sponge material.

It is clear that the ever-increasing popularity of scuba diving {see Chapter 22, page 423} is leading to more injuries from marine sponges (SOUTHCOTT & COULTER 1971).It is not surprising that a single sponge may produce injuries in entire families or groups if passed around (CLELAND & SOUTHCOTT 1965). One case of erythema multiforme 10 days after sponge contact is reported (BURNETT *et al* 1987).

CASE REPORT

A healthy and athletic surgeon, with no history of allergy, was scuba diving in South Australian waters wearing a glove on one hand only, when he touched a sponge with both hands. No symptoms were experienced until some six to eight hours later at his home. Burning, itching and pain of increasing severity developed in the ungloved hand only, accompanied by rapid local swelling. Local applications and oral analgesics failed to provide relief or reduction of the oedema. On the second day all five fingertips had become dusky blue. He cancelled his operating lists. On day 4 he received a systemic steroid injection (Hydrocortisone 100 mg intramuscularly) followed by oral steroids for one week. At the same time the pain and swelling began to diminish, and the fingertips became pink over 24 hours after the commencement of this therapy {Plate 3.1}. His hand was completely recovered two weeks after the sponge contact. The sponge species was not identified.

DOCUMENTED SPONGE STINGINGS AND DERMATITIS

Despite repeated reference to 'poisonous sponges' in the literature, there is no record of human poisoning from the ingestion of marine sponges. Reports all refer to intoxication or injury from handling sponges. The following documentations were found of human injury from marine sponges, all of them primarily dermatological:

- *Tedania ignis* ('fire sponge') and *Neofibularia nolitangere* (DUCHASSAING and MICHELOTTI) (= *Fibularia nolitangere* De LAUBENFELS, 1936) in the West Indies (DUCHASSAING & MICHELOTTI 1864; VERRILL 1907; De LAUBENFELS 1936; HARTMAN 1967; YAFFEE 1970)
- *Tedania ignis* and *Neofibularia nolitangere* in the USA (BURNETT *et al* 1987d)
- *Tedania* sp. in New Zealand (BERGQUIST in SOUTHCOTT & COULTER 1971)
- *Microciona prolifera* ('the red sponge') in north-eastern USA (CORSON & PRATT 1943); this sponge contains antimicrobial substances
- *Neofibularia mordens* in southern Australia (SOUTHCOTT & COULTER 1971); this sponge causes a severe dermatitis and prickling skin sensation comparable to the 'fire sponges' in the West Indies; *N. mordens* is the most toxic sponge in the Australian region (SOUTHCOTT 1987a).
- *N. hartmani* in New Caledonia coastal waters (WILKINSON 1978; HOOPER & LÉVI 1993)
- *N. irata* on the Great Barrier Reef (Australia) — histamine like reaction with severe inflammation and swelling (WILKINSON 1978)
- *Lissodendoryx* sp. in southern Australia (SOUTHCOTT & COULTER 1971)
- *Haliclona* sp. of Moreton Bay, southern Queensland, Australia (McCAFFREY 1983)
- *Biemna saucia* in the Darwin region and on the North-West continental shelf of Western Australia; this sponge can cause a contact allergy (HOOPER *et al* 1991).

The Mediterranean sponge diver's disease (*maladie des plongeurs, maladie des pêcheurs d'éponges, Skevas-Zerfus disease*) an irritating skin rash, is caused not by the sponges themselves but by a sea anemone growing on the sponges (ZERVOS 1903) {see Chapter 8, page 186}.

FRESHWATER SPONGES

Although sponges are generally resistant to desiccation, some freshwater species are remarkably so (CLELAND & SOUTHCOTT 1965). Furthermore, some freshwater sponge species have been reported to have abrasive properties. Injected extracts from the European sponge *Spongilla lacustris* have caused major systemic disturbances and some fatalities in experimental animals (SOUTHCOTT 1987a).

MANAGEMENT OF MARINE SPONGE INJURIES

The literature reports a range of local applications for sponge injuries and sponge dermatitis, which have had no effect. One report of 'dramatic relief' from the local application of a proprietary antihistamine cream is described in an asthmatic nine-year-old boy (CLELAND & SOUTHCOTT 1965). Results from oral antihistamines and analgesics have also proved disappointing. The impression from the literature is that many sponge injuries in humans run their natural and painful course over weeks, despite various first aid and medical interventions.*

On the basis of inadequate existing knowledge, documentation and experience, and pending advances in understanding, the following nonspecific management guidelines are suggested.

FIRST AID

1. Remove spicules from skin with fine forceps if skin has been pierced.
2. Wash the contact area as soon as possible with soap and fresh water.
3. Try cold water soaks to reduce local symptoms (BURNETT *et al* 1987).
4. Apply ultra-potent local steroid cream to the contact area, even possibly before the development of local symptoms. (Steroid creams should never be used on infected lesions.)

* HOOPER *et al* (1991) described cases in which considerable relief was obtained when a proprietary topical burn cream, butesin picrate, containing analgesic and antiseptic properties, was applied to injuries from toxic sponges.

5. Offer oral analgesics (aspirin, paracetamol) for developing pain.
6. Seek early medical advice if other than trivial symptoms and signs begin to develop.

MEDICAL MANAGEMENT

1. Use parenteral narcotics for adequate pain relief; there is no current documentation concerning the value of non-steroidal antiinflammatory agents.
2. Manage systemic allergy in the usual way (EXECUTIVE EDITORIAL BOARD 1989) {see Chapter 5, page 98}.
3. Use short-term parenteral steroids for unrelieved local symptoms and/or compromised local tissue viability (excluding infection).

PREVENTION

As most sponges are toxic in one way or another, it is probably best not to touch any sponge either alive or dead without adequate protective skin covering including gloves.

Acknowledgements: The author thanks John Hooper, Curator of Lower Invertebrates at the Queensland Museum, for helpful advice with the manuscript.

HISTORY

While strong suspicions exist of historic awareness of ciguatera poisoning from the writing of Ch'en T'sang-chi during the Tang Dynasty of China, AD 618-907 (READ 1939), a firm early report of ciguatera intoxication was from the New Hebrides in 1606, when sailors with the Spanish explorer Fernandez de Quiros were affected by a fish called *pargos* (snapper) (de QUIROS 1606). This report was followed by many others from the Caribbean over the ensuing decades (RUSSELL & EGAN 1991). Early reports from the Pacific were by ANDERSON (1776) and FORSTER (1778).

Poisoning by the then eighteenth-century staple Caribbean seafood, the marine turban shell *Turbo pica* (*cigua* in Spanish), is thought to be the origin of the term 'ciguatera' (PARRA 1787). From this specific marine poisoning the term came to be applied to any marine fish poisoning (RUSSELL & EGAN 1991).

The late A.H. (Hank) Banner was a modern pioneer of ciguatera research at the University of Hawaii in the mid-1950s (SCHEUER 1994).

DISTRIBUTION AND INCIDENCE

Fish that cause ciguatera (ciguateric fish) are restricted to those inhabiting the warmer waters, virtually around the world, particularly coral reef waters (e.g. Plate 15.1). Outbreaks of ciguatera occur sporadically and unpredictably in both time and space; they can often be curiously localised.

In Australia most ciguateric fish are caught in Queensland and Northern Territory waters (Figure 15.1) (GILLESPIE *et al* 1986; LEWIS 1987; LEWIS *et al* 1988). Three Australian ciguateric 'areas' are Maryborough to Gladstone, especially the inside coast of Fraser Island (HOLMES *et al* 1994), Bowen to Port Douglas (Great Barrier Reef) and Gove (Nhulunbuy). Fish from northern New South Wales and Western Australia rarely, if ever, cause ciguatera. The risk of eating a toxic fish in Australia, while significant {see below}, is comparatively low.

Ciguatera is much more significant as a health concern to inhabitants of island countries in tropical and subtropical waters, especially the smaller island nations of the Pacific basin (LEWIS 1992a; KALY & JONES 1994), and the West Indies (VERNOUX & LEJEUNE 1994). Figure15.2, page 348, shows the incidence of ciguatera for Pacific island countries (see also RUFF & LEWIS 1994). Of these, it is the atoll island countries that are worst affected.

The incidence of ciguatera in Queensland (0.16 cases per 10 000 population) is similar to the incidence in Tonga. These figures are likely to represent only about 20 per cent of actual cases. Reasons for under-reporting of ciguatera include non-reporting of confirmed cases and misdiagnosis, with mild cases often mistaken by

FIGURE 15.1
Map of Queensland and part of the Northern Territory indicating known ciguatera-prone areas (hatched zones). Darker shading indicates areas of higher risk.

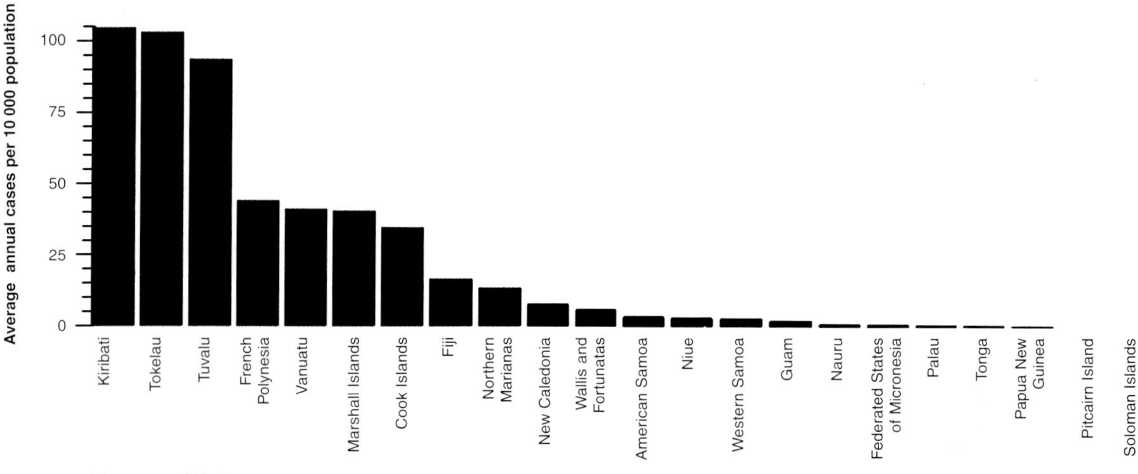

FIGURE 15.2
Incidence of ciguatera 1985–1990 in Pacific Island countries (average annual cases per 10 000 population) Asterisks indicate incomplete reporting by these countries to the South Pacific Epidemiological and Health Information Service (SPEHIS). For more recent data the reader is referred to RUFF & LEWIS 1994 and HABERMEHL *et al* 1994 in Chapter 26, page 446.

victims and their medical attendants for more common illnesses.

FISH IMPLICATED IN AUSTRALIA

Many warm-water fish may potentially carry ciguatoxins (DALZELL 1994). Fish that have caused ciguatera in Queensland include coral trout (Plate 15.1), narrow-barred Spanish mackerel, reef cod, barracuda, emperor *Lutjanus sebae*, grouper, trevally, queenfish and kingfish (GILLESPIE *et al* 1986).

The moray eel *Gymnothorax undulatus*, red bass *Lutjanus bohar*, chinaman *Lutjanus nematophorus* and paddletail *Lutjanus gibbus* are reputed, on anecdotal evidence, to be regular carriers of ciguatoxin and are not widely sold by Queensland marketing authorities. However, research is required to confirm that these species do indeed pose a greater risk than other reef species.

There is a ban on the capture of narrow-barred Spanish mackerel and barracuda in Platypus Bay (between Fraser Island and the Queensland mainland, Figure 15.1, page 347). These fish are frequently toxic, this area representing the only true 'hot spot' for ciguatera in Queensland (HOLMES *et al* 1994). Since the introduction of this ban, the incidence of ciguatera from Spanish mackerel has declined. In Platypus Bay a small bream-like fish, the blotched javelin fish (Plate 15.2), has recently caused a number of cases of ciguatera (LEWIS *et al* 1988; HOLMES *et al* 1994) and should also be avoided.

Similar fish to those causing ciguatera in Australia cause ciguatera in Pacific island countries (GLAZIOU *et al* 1995). Implicated fish include many species of cod, surgeonfish (Plate 16.17), parrotfish and the red bass.

BIOLOGICAL ORIGIN OF CIGUATERA

The precursors of the ciguatoxins, known as gambiertoxins, are produced only by certain strains of *Gambierdiscus toxicus*, a benthic dinoflagellate (a small bottom-living algae) (HOLMES *et al* 1991; HOLMES & LEWIS 1992, 1994). *G. toxicus* lives on a variety of macroscopic algae, eaten incidentally by herbivorous fish. Fish apparently bioconvert the injected gambiertoxins to ciguatoxins, which accumulate in the tissues of these fish. Large carnivorous fish then prey on these smaller fish and the toxins are concentrated. As the final link in this food chain, humans are then exposed to the ciguatoxins. At low levels these toxins appear to have little effect on fish, although fish may die from

CHAPTER 15

CIGUATERA (FISH POISONING)

RICHARD J LEWIS

WITH MEDICAL CONTRIBUTIONS

BY GEOFFREY K KING

Although it may be confused with a variety of other marine poisonings, ciguatera poisoning is a major public health problem in the tropical and subtropical Pacific region (WHO REPORT 1981; GILLESPIE 1987; LEWIS & SELLIN 1992; BOWEN & POTTER 1993; RUFF & LEWIS 1993; GLAZIOU *et al* 1995) and is the subject of intensive research (MEMOIRS of the QUEENSLAND MUSEUM, PART 3, 1994; this issue contains an 'In Memoriam' acknowledgment to the late Noel Charles Gillespie, a pioneer in ciguatera research and aquaculture.)* The true incidence of actual deaths is less clear (SUTHERLAND 1983) and may have been previously overestimated {see Chapter 3, page 67}.

Ciguatera is a disease caused by the consumption of fish* contaminated with lipid-soluble toxins called ciguatoxins (GILLESPIE *et al* 1986; LEWIS *et al* 1991; MEMOIRS of the QUEENSLAND MUSEUM, PART 3, 1994). In some ciguatoxic fish, water-soluble toxins may also be present (ENDEAN *et al* 1993a, 1993b). At present there is no simple and reliable means of detecting these toxins in fish (LEWIS 1994a, 1994b; VERNOUX 1994; PARK 1994) and there is no known culinary method that renders a contaminated fish safe to eat. The disease is characterised by distressing and often debilitating short-term gastrointestinal and longer-term neurological symptoms (BOWERMAN 1980; GILLESPIE *et al* 1986; RUFF & LEWIS 1994).

---

*See also Chapter 26, pages 445–447.

SECTION FOUR

TOXIC (POISONOUS AND VENOMOUS) MARINE VERTEBRATES

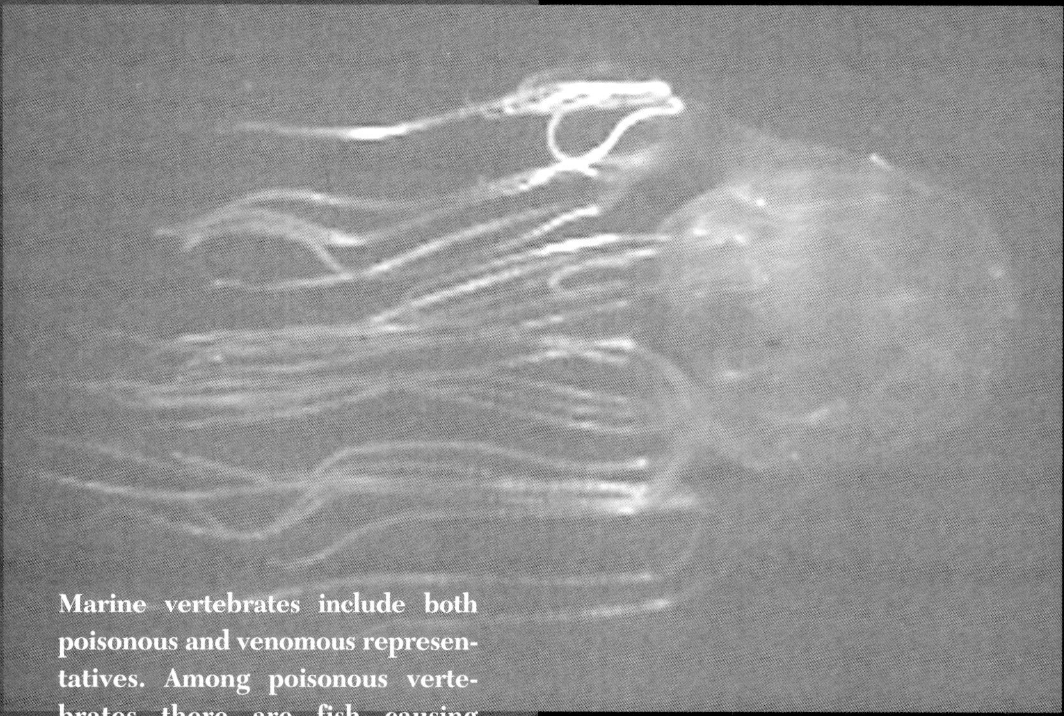

Marine vertebrates include both poisonous and venomous representatives. Among poisonous vertebrates there are fish causing ciguatera poisoning, and pufferfish, which causes *fugu* poisoning (tetrodotoxication) (see Chapter 3, page 86, and Chapter 13, page 334), as well as large mammals such as whales and seals (see page 87), which cause poisoning when their liver is eaten. Envenomation frequently occurs upon contact with an enormous range of fish species equipped with venemous spines, and upon being bitten by sea snakes. The resulting injuries and illness (notably fish spine injuries and ciguatera poisoning) are common, particularly among people whose occupations and/or recreational activities put them in contact with these marine animals.

exposure to high levels (LEWIS 1992b). This may explain why ciguatera is rarely fatal in humans.

Reef disturbance has been implicated as a factor that can increase the risk of ciguatera in an area (BAGNIS *et al* 1988; BAGNIS 1994; KALY & JONES 1994), though many instances of reef disturbance do not result in a flare-up of ciguatera. Research is required to elucidate the environmental factors that favour the growth and toxin production of *Gambierdiscus toxicus* before it will be possible to minimise the risk of ciguatera through controls on human activity. The origin and transfer of toxins in ciguatera has recently been reviewed (LEWIS & HOLMES 1993).

THE TOXINS

Structures for the toxins responsible for ciguatera have recently been determined (MURATA *et al* 1990; LEWIS *et al* 1991; LEWIS & BRERETON 1994). Three major ciguatoxins (CTX-1, -2 and -3) are present in the flesh and viscera of ciguateric fish (Figure 15.3) (LEWIS *et al* 1991; LEWIS & SELLIN 1992; HOLMES & LEWIS 1994). These toxins are closely related polyether compounds that arise through oxidative metabolism of ciguatoxin precursors (i.e. the gambiertoxins produced by *G. toxicus*, e.g possibly gambiertoxin-4b [see above]). Interestingly, this bioconversion amplifies toxicity by up to tenfold. The ciguatoxins bind with high affinity to voltage-dependent sodium channels (LEWIS *et al* 1991)(see also Chapter 26, page 446), causing the voltage-dependence of activation to be shifted in the hyperpolarising direction. Such an action results in the depolarisation of excitable cells, especially nerve cells* (GILLESPIE *et al* 1986). The high potency of the ciguatoxins means that levels as low as 0.1 part per billion in fish may result in mild ciguatera. Detection of these low levels poses a major obstacle to the development of a simple method for the detection of ciguateric fish.

*Intraperitoneal ciguatoxin induced slowing of mixed and motor nerve conduction velocities and amplitudes of adult rat ventral tail nerves. Also the duration and magnitude of supernormal electrophysiological response in the nerve was increased. Intraperitoneal lignocaine (lidocaine) modified or blocked all these disturbances (CAMERON *et al* 1993).

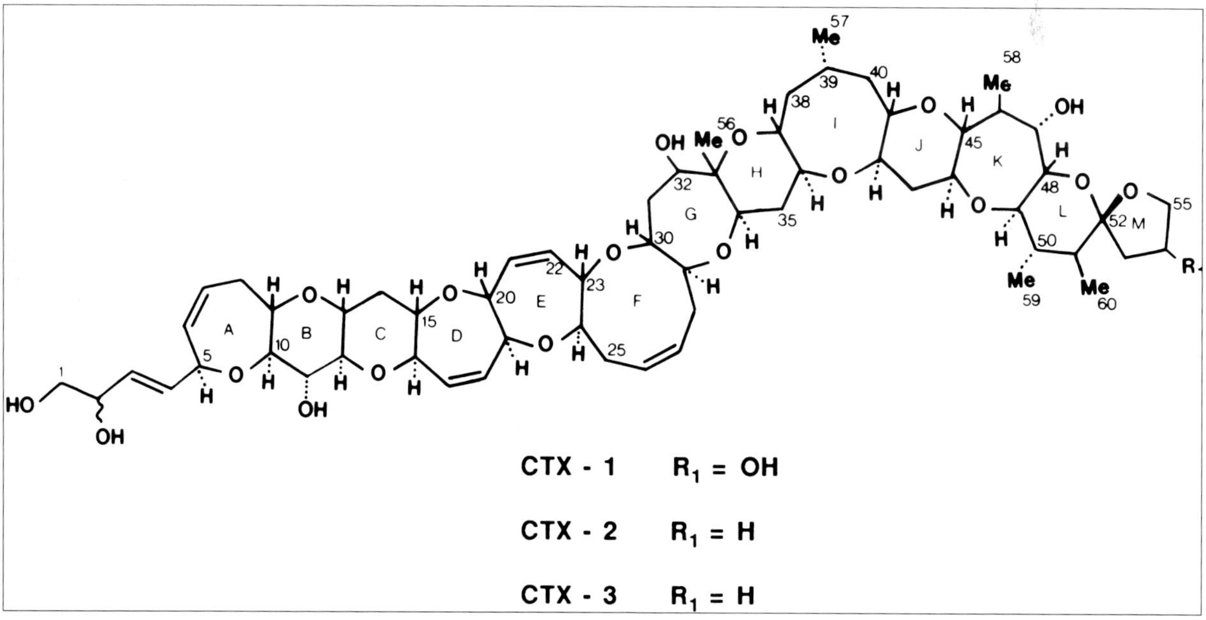

FIGURE 15.3
Structures of the three major ciguatoxins (CTX) found in the flesh and viscera of ciguatoxic fish. An altered stereochemical feature at carbon 52 is the only difference between CTX-2 and CTX-3.

SYMPTOMATOLOGY AND DIFFERENTIAL DIAGNOSIS

Ciguatera is still the most frequently reported seafood-related illness in Australia and the Pacific basin. Considerable under-reporting still occurs in Australia, which has implications for both investigation and control of outbreaks (KRAA *et al* 1994). Symptoms of ciguatera commence from less than 1 to more than 24 hours after the consumption of a toxic fish meal. The time to onset of symptoms varies with the toxicity of the fish, the amount of fish eaten and the susceptibility of the consumer. All patients exhibit symptoms by 36 hours (RUFF & LEWIS 1994). RUSSELL and EGAN (1991) suggest that, worldwide, ciguatera may be classified clinically into four 'patterns':

1. Gastrointestinal/neurological, with the former symptoms predominating
2. Neurological/gastrointestinal, with the former symptoms predominating
3. Purely neurological symptoms
4. Purely gastrointestinal.

The first and second are much more common than the third and fourth.

DALZELL (1994) has recently reported the ranking of ciguatera symptoms in the South Pacific (n=380).

Symptoms of ciguatera in Australia have been well documented (GILLESPIE *et al* 1986; LEWIS 1987; LEWIS *et al* 1988; RUFF & LEWIS 1994) and, in approximate order of frequency may include the following: (Percentages indicate the proportion of patients exhibiting that particular combination of symptoms, out of the total number only under that respective group heading, e.g. 'gastrointestinal', 'neurological').

1. Gastrointestinal:
- diarrhoea (>60 per cent of patients)
- abdominal pain, nausea(>50 per cent of patients)
- actual vomiting (>30 per cent of cases)

2. Neurological:
- myalgia (muscle aching) (>80 per cent of patients)
- burning sensation of skin on contact with cold water, pruritis (itching), arthralgia (joint aching) (>70 per cent of patients)
- paraesthesiae (tingling and numbness) of hands, feet, mouth and lips, headache (>60 per cent of patients)
- mood disorders (depression, irritability, anxiety) (50 per cent of patients)
- ataxia (unsteadiness), or vertigo (sensation of spinning), sweating, eye pain (>40 per cent of patients)
- dental pain, tremor (>30 per cent of patients)
- neck stiffness, demonstrable paresis (decreased power in specific muscle groups) (>20 per cent of patients)
- salivation (10 per cent of patients)

3. Non-Specific:
- fatigue and lassitude (90 per cent of patients)
- chills (40 per cent of patients)
- skin rash, shortness of breath (including exertional), pain on urination (20 per cent of patients)

Less commonly, blurred vision, bradycardia and low blood pressure occur. Cardiovascular manifestations are uncommon in Australia but are documented elsewhere (ADAMS 1993) (see also Chapter 26, page 442). However, in general, the above symptoms are similar to those reported for ciguatera in other countries and ocean localities (BAGNIS *et al* 1979; BOWERMAN 1980; RUSSELL & EGAN 1991; ADAMS 1993).

Patients may experience a few or a conglomerate of the signs and symptoms listed above. The more severe the case usually the greater the number of symptoms experienced. Gastrointestinal symptoms last for one or two days, whereas some neurological symptoms typically persist for several weeks. In some victims the neurological symptoms can persist for many months, or even years.

There have been two Australian cases of local symptoms transmitted sexually, one male to female and one female to male.

The variations in symptomatology (presentation, duration, severity) of ciguatera poisoning has led to attempts to develop a severity rating scale, to facilitate the defining of individual cases and the contrasting of different outbreaks (LANGE 1993).

EXACERBATIONS AND SENSITISATION

Symptoms of ciguatera may be exacerbated by alcohol. BROADBENT (1985) considers the rapid potentiation of skin itch by alcohol — particularly on the soles of the feet and the forearms — to be pathognomonic of ciguatera poisoning in suspected cases.

A small number of sufferers complain of an allergy-like syndrome that can last several years. In these cases, symptoms typical of ciguatera can be brought on by the consumption of non-toxic fishes, e.g. cold-water or estuarine fish species. Non-fish products such as chicken and pork may occasionally cause similar problems. It is recommended that people refrain from eating fish (including shellfish) for three to six months after confirmed ciguatera poisoning, to avoid a possible relapse that can follow consumption of fish.

DIFFERENTIAL DIAGNOSIS

The diagnosis of ciguatera intoxication is clinical, and requires awareness and a systematic approach (PEARN 1994). One set of diagnostic criteria requires a history of reef fish ingestion less than 24 hours previously accompanied by one neurological symptom and one other symptom.

Medical examination should exclude other conditions such as botulism, or shellfish or scombroid poisoning. Scombroid poisoning is due to the ingestion of fish flesh contaminated with bacteria (e.g. *Proteus morgani*), resulting in the development of a high content of histamine-like substances (e.g. histamine and cadaverine, probably arising from saurine), in the fish muscle (KNIGHT 1987; EDMONDS 1989). Decompression illness may require exclusion in affected divers (ADAMS 1993) {see below, and Chapter 22, page 425}.

TREATMENT

As always, support of any depressed vital functions {see Chapter 5, page 98} takes absolute priority in severely affected cases (exceedingly rare in Australia).

INTRAVENOUS MANNITOL

Until recently the treatment of the common forms of ciguatera has been only symptomatic and supportive. The recent anecdotal observations that intravenous mannitol solution can significantly reduce the severity and duration of ciguatera symptoms (PALAFOX *et al* 1988; PEARN *et al* 1989; WILLIAMSON 1990; BLYTHE *et al* 1994) may be a tentative forward step in the management of ciguatera. Mannitol is a crystalline alcohol obtained from glucose. It is used in medicine as an osmotic diuretic. However recent studies of the actions of mannitol upon ciguatoxin in mice suggest both a central neurological action of ciguatoxin in those animals, and an absence of any marked benefit of intravenous mannitol (LEWIS *et al* 1993). Mannitol did not reverse the positive inotropic action of ciguatoxin-1 upon human atrial trabeculae (LEWIS *et al* 1992). (Some other recent animal and *in vitro* studies [e.g. C. PURCELL 1994, unpublished observations] also do not lend clear support to these anecdotal clinical observations, and until controlled clinical trials are completed [PALAFOX 1994], caution is necessary.)

The treatment involves a 1 g/kg intravenous infusion of 10% or 20% mannitol solution over 30 minutes, given as soon as possible after ciguatera is diagnosed, and in the presence of appropriate resuscitative skills and facilities. Recent evidence suggests that the treatment is most effective if given early, ideally within two days of the onset of symptoms (PEARN 1994), before the patient begins to recover. Some patients report a dramatic improvement in the neurological symptoms, usually within several hours of the infusion. The reason(s) why some people do not respond (ADAMS 1993) requires further study, but may be related to delayed timing of the treatment or inadequate dosage. Some people report a recurrence of symptoms about 24 hours after a first administration of mannitol {see Case Report 1, below}. In these instances, a second infusion of mannitol is recommended.

CASE REPORT 1

In 1992 in Port Douglas, far north Queensland, a Sri Lankan family of three ate a 1 kg Spanish mackerel for lunch. That afternoon all three developed symptoms. The father (aged 31) had diarrhoea and vomiting and altered sensation to feet and legs, and in the arms to a lesser extent. The mother (aged 30) had nausea, abdominal pains and diarrhoea, and calf pain. The child (aged 3) had pains in the abdomen and feet. Upon referral to hospital all three received intravenous (IV) mannitol with dramatic immediate improvement. They were discharged from hospital asymptomatic the next morning. That afternoon all three represented with: (father) joint pains and peri-oral tingling, treated symptomatically; (mother) some joint pains requiring no further treatment; and (child) diarrhoea and vomiting which was settling, and required no further treatment. All three remained well thereafter.

Mannitol should not be administered until the patient is adequately rehydrated. So far the treatment has proved safe and free of the documented side-effects of mannitol (PEARN *et al* 1989; WILLIAMSON 1990). These include the very uncommon anaphylaxis to mannitol (four nonciguateric cases since 1967, identifiable beforehand by a history of atopy) (McNEILL 1985).

Research into the mechanism by which mannitol can reverse the symptoms of ciguatera has revealed that mannitol does not chelate ciguatoxin, nor does mannitol displace ciguatoxin from its site of binding (LEWIS *et al* 1992; LEWIS 1994, unpublished observations). PEARN and colleagues (1989) have suggested that mannitol may reverse ciguatoxin-induced oedema of axonal Schwann cells (ALLSOP *et al* 1986). To explain the lasting effectiveness of mannitol, it has been hypothesised that (1) mannitol prevents long-term nerve damage (e.g. Schwann cell lesions, hypoxic zones) which probably results from such oedema, and that (2) ciguatoxin remains bound in the body only for a few days, and the longer-term effects of ciguatera relate to the initial nerve damage. These hypotheses can explain why mannitol would be most effective when given early in the disease. Interestingly, it has been repeatedly observed in Australia that the diuresis that would normally follow an infusion of mannitol (an osmotic diuretic) does not appear when mannitol is used to treat ciguatera, even in patients where an intravenous infusion of rehydrating solution has been administered prior to mannitol. The explanation for this observation is not apparent, but it does suggest that mannitol's diuretic action is not responsible for its beneficial effect in ciguatera.

A blinded, controlled study of the use of mannitol in ciguatera poisoning is currently being undertaken in the Marshall Islands by Palafox and colleagues (PALAFOX 1994).

Traditional treatments using locally-growing plants are still widely used by islanders in the western Pacific (BOURDY *et al* 1992)

SYMPTOMATIC MEASURES

The symptomatic management of ciguatera with sedatives, vitamins B_6 and B_{12}, antihistamines and tricyclic antidepressants (BOWMAN 1984; SUTHERLAND 1983; GILLESPIE *et al* 1986; PEARN *et al* 1989; RUFF & LEWIS 1994) has met with variable success. Other relatively unsuccessful medications have included calcium, atropine, lignocaine, phentolamine, nifedipine, morphine, tocainide, pralidoxine and steroids. However, intravenous rehydration is always helpful in cases with diarrhoea and vomiting, and may be life-saving in seriously affected cases (SUTHERLAND 1983; HALSTEAD 1988). At present there is no compelling reason to use multiple-dose charcoal to treat ciguatera, especially if the potential for adverse side effects of this therapy are considered (TENENBEIN 1991).

PREVENTION

There is no certain way to avoid ciguatera, because the toxin cannot be simply detected at present. (There are current efforts to develop a test kit to enable detection of ciguatoxin in fish flesh; research is proceeding, and is encouraging [LEWIS & SELLIN 1993; LEWIS 1994b].) However, the following measures will reduce the risk of contracting ciguatera from warm-water fish.

- Fish species locally implicated in ciguatera should be avoided.
- Moray eel, red bass, chinaman fish and paddletail should not be eaten.
- Small portions (<50 g) from several fish should be consumed in a single meal, rather than a large portion (>200 g) from any one suspect fish.
- The liver, viscera or roe of potentially toxic fish should not be eaten.

Further meals of a toxic fish will result in a progression to more severe symptoms.

REPORTED METHODS OF PRIOR DETECTION OF AFFECTED FISH

South Pacific Islanders test their fish by feeding the liver, ovary or testes to cats and observing for signs of illness (see Case Report 2, below). (Flies on the fish at market are regarded as a reassuring sign!) BROADBENT (1985) describes a 'characteristic' ascending paralysis with bradycardia, excess salivation and skin itch, in poisoned cats and dogs. He considers increased nipple sensitivity in lactating animals is a diagnostic sign. However, it should be appreciated that the absence of clinical signs in a test animal may not exclude the presence of lower levels of ciguatoxin in the fish.

FUTURE RESEARCH

Future research will be directed at ways to improve the management of ciguatera. Areas of research may include: (1) determining the factors responsible for the appearance of ciguatoxins in fish; (2) development of an assay that reliably and cost-effectively distinguishes toxic from non-toxic fishes; and (3) development of improved treatment regimes, including a more simple orally effective therapy.

To facilitate ciguatera research in Queensland, samples known to be from a fish that has caused ciguatera can be saved, frozen and forwarded to the local health inspector. Cases of ciguatera are interviewed on a regular basis by the Queensland Health Department, and form the basis of a database (see LEWIS *et al* 1988) that is maintained by the Queensland Department of Primary Industries for research purposes only. The South Pacific Epidemiological and Health Information Service (SPEHIS) also compiles a valuable data base of fish poisoning (DALZELL 1994).

CASE REPORT 2

In Mossman, far north Queensland in 1991, two domestic cats were fed from the same Spanish mackerel, and a third cat shared the flesh of a cod with its owner couple (both of whom subsequently developed mild but characteristic ciguatera symptoms). Both cats that fed from the mackerel became unconscious and moribund. The third cat became unwell and off its food. A local veterinary surgeon who was consulted administered IV mannitol to all three cats. The unconscious cats woke up immediately, and the third cat made an immediate recovery. All three animals were released well the next morning. There were no apparent relapses.

CHAPTER 16

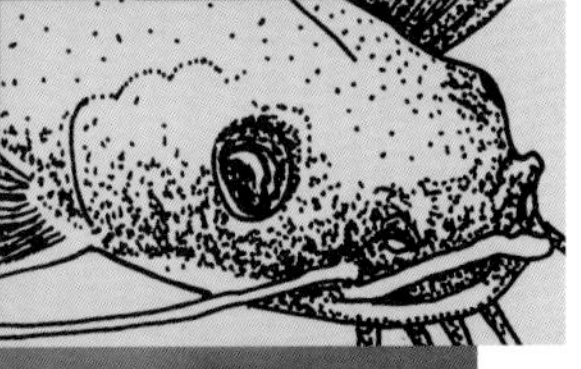

VENOMOUS FISH

JACQUIE RIFKIN AND JOHN WILLIAMSON

GENERAL FEATURES OF VENOMOUS FISH

JACQUIE RIFKIN AND JOHN WILLIAMSON

Many fish have well-developed defensive organs in the form of armour, teeth, spines, poison-containing skin and/or venom glands, or poisonous flesh; these traits are illustrated by the hypothetical venomous fish, 'Pandora' (Figure 16.1, page 355). In addition, many fish have evolved special defensive behaviours. Scorpionfish, for example (Figure 16.2, page 356 and Plate 16.1), actually erect their venom-laden spines as a means of defence against predation or injury.

Fish are generally regarded as venomous if they possess a traumatising apparatus, such as pungent spines, capable of both puncturing the skin and depositing a venom within the puncture wound. In fact relatively few fish are dangerous to humans. HALSTEAD (1970) stated that 200 of approximately 20 000 species of fish are regarded as venomous. However, the non-venomous wounds fish inflict — for example, puncture wounds from the needle-like jaws of the longtom or needlefish (family Belonidae){Plate 20.2} — can result in secondary infection (AUERBACH *et al* 1987) {see Chapter 4, page 88}.

Venomous fish that have been reliably reported to be hazardous to humans occupy several groups. Some of these include the stingrays, stonefish (SMITH 1951; WIENER 1965), catfish (both forked-tailed and eel-tailed) (BREAULT 1991; McKINSTRY 1993), scorpaenids, scatophagids and weeverfish (MARETIĆ 1988). In Australia, only the stingray is reported to have caused fatalities (WRIGHT-SMITH 1945; FENNER *et al* 1989b), although elsewhere deaths have also been attributed to stings from the stonefish (MARETIĆ 1988), lionfish (RUSSELL 1971), catfish (HALSTEAD *et al* 1953) and weever fish (MARETIĆ 1988) {see Chapter 3, page 69}. In Australia most of the above-named fish cause significant complaint (morbidity), excepting the Australian weeverfish [families Mugiloididae and Parapercidae], which are not venomous.

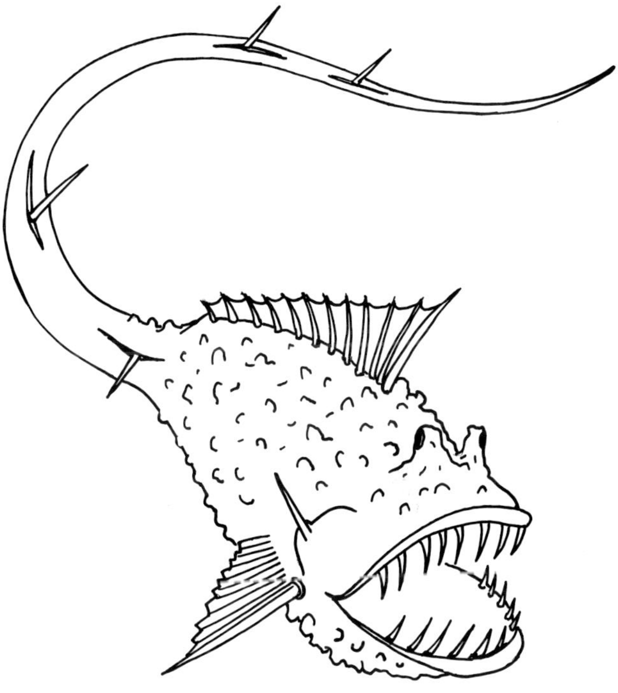

FIGURE 16.1
Pandora — a fish ready to fight. Note the needle-sharp teeth, raised dorsal fin with multiple spines, extended pectoral fin spines, opercular spines, scalpel-like spine on lateral, proximal portion of tail, erect dorsal spines on tail, and wart-like skin glands containing poison. (Drawing by Jan Hynes.)

THE TRAUMATISING APPARATUS

VENOMOUS SPINES

Spines in close association with cells containing poison capable of entering a wound when the spine sheath is ruptured are considered to be venomous. They may be present on all or some of the dorsal, ventral, anal or pectoral fins. Venomous spines may occupy the operculum or preoperculum, the 'shoulder' or cleithrum, the head, or the tail (see Figure 16.1). The spines may have a median dorsal or median ventral ridge, with venom glands arranged along grooves. Some venomous fish have serrated spines which produce severe lacerations in victims (Plate 16.6). Other spines are hollow to allow passage of the venom into the wound (Figure 16.50, page 393).

Venom glands

Fish venom glands are comprised of concentrations of unicellular glands (CAMERON and ENDEAN 1972 A and B, 1973), which may be located either in the integumentary sheath surrounding the spine or adhering closely to the spine itself (see Figures 16.4A and B, page 357). Both situations may occur on a single fish spine. The three types of unicellular gland found in association with fish spines (Figure 16.3, page 356):

(a) mucous glands
(b) clavate glands (club-shaped)
(c) sacciform glands

Mucous glands are spherical cells filled with a slimy secretion (Figure 16.3a, page 356). During normal growth of the skin individual gland cells work their way to the surface where they are ruptured, releasing secretions when the fish rubs against any surface. In the same way, mucus is released during handling or when the integument, including that surrounding the spine, is damaged (see below).

Clavate cells are irregularly shaped cells (Figure 16.3b, page 356) that contain proteinaceous poison; like mucous glands they need to be ruptured before their contents are released. Thus, when certain fish are handled, both mucus and poisons are released causing skin irritation (see page 371). The contents of the clavate cells may cause excruciating pain and significant generalised symptoms in humans when in association with sharp penetrating spines.

Sacciform glands (Figure 16.3c, page 356) have an apical pore through which venom is released without the gland itself rupturing.

Well-developed compact venom glands, which are usually closely apposed to a spine, are bound together internally by a tough sheath of

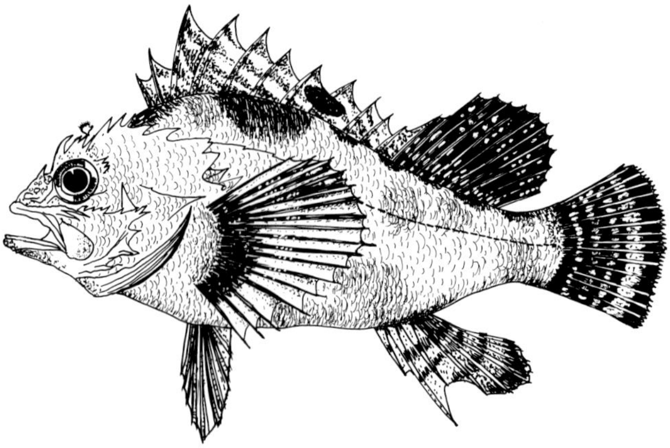

FIGURE 16.2
A scorpionfish, *Scorpaena ergastulorum* with venomous dorsal spines. It raises its dorsal spines as a form of defence. This fish is sometimes called 'poor man's lobster'. (Drawing by Jan Hynes.)

connective tissue (Figure 16.30, page 376). The integumentary sheath surrounding the whole spine is covered by an epithelium usually containing unicellular glands (see Figure 16.2 above, and Figure 16.4, page 357). When a victim is stabbed by a spine (Figure 16.31, page 377) the integumentary sheath is ruptured, and the glandular secretions of the epithelium are introduced into the wound along with the material contained within the compact internal venom glands. These glands have been shown to be capable of regeneration (CAMERON & ENDEAN 1966).

CAMERON and ENDEAN (1972b, 1973) suggest that in general there is a positive correlation between a reduction of scales and an increase of skin toxins (crinotoxicity). Thus most fish in which scales are absent or reduced have toxins associated with their skin.

VENOMOUS TEETH

A cosmopolitan reef fish capable of delivering a venomous bite is the sabre-toothed blenny. FISHELSON (1974) has shown that the canine teeth of the blenny fish *Meiacanthus nigrolineatus* are grooved and are associated with a venom gland. The venom gland is a compact organ, which is formed by aggregations of epithelial cells, and enveloped in tough connective tissue. A venomous secretion is pressed into the wound along the grooved fangs.

COLEMAN (1991) described the sabre-toothed blenny *Plagiotremus rhinorhynchos*, commonly called the lyretail blenny, a reef fish up to 8 cm in length, with large venomous fangs in the lower jaws. These fish will dart quickly out of the holes where they live and bite off a piece of fin from a passing fish. McKAY (1994, personal communication) stated that these fish are responsible for painful bites in divers.

There have been reports of a poison associated with the teeth of some Pacific Ocean moray eels

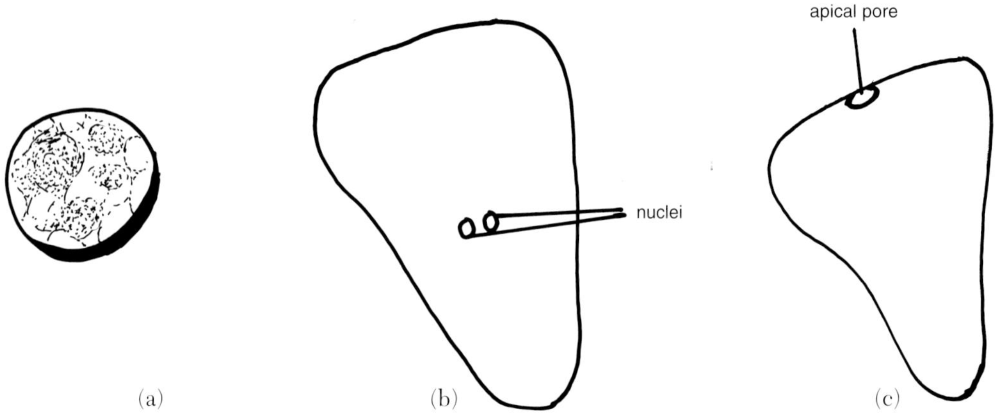

FIGURE 16.3
Three types of unicellular gland cell found in fish epidermis: (a) mucous gland; (b) clavate gland; (c) sacciform gland. (Drawing by Jan Hynes.)

(R. KUITER, personal communication, 1996, HALSTEAD 1988), but not for Atlantic specimens (MARETIĆ 1988).

FISH DEFENSIVE TOXINS VS A SEDENTARY LIFESTYLE

CAMERON and ENDEAN (1972b, 1973) suggested that venom glands, as well as skin toxins evolved in order to allow the fish to exploit a sedentary lifestyle. {For the distinction between a *toxin*, a *poison* and a *venom*, see the Glossary}.

In describing the habits of certain fish, Cameron (1976) stated that stonefish, scorpionfish, scorpaenids, batrachoidids and weeverfish are examples of venomous fish possessing such a sedentary lifestyle. Among cartilagenous fish, many rays also share this characteristic. Knowledge of the habits of various fish allows the wader, diver, or fisherman to be aware of the possibility of envenomation.

Stonefish (Plates 16.2, 16.11 and 16.13) habitually blend themselves to the sand and rubble, remaining in the same spot for periods of up to four months (STABLUM 1994, personal communication). These animals have developed the most venomous of fish spines to deter predators that would swallow them.

Scorpaenids, all of which possess a venom apparatus (McKAY 1994, personal communication), spend extended periods of time in one place, being 'cryptic' in colouration and blending in well with the background. Some zebrafish of the genera *Pterois* (Plate 16.15) and *Dendrochirus*, are notable exceptions, with their brilliant colouration, long fins, aggressive behaviour and many with habits of swimming out in the open. However, even these fish tend to be site-attached. (Figure 6.44). Batrachoidids live on the ocean bottom (i.e. are benthic) and have similar sedentary habits. In Australia batrachoidids are sometimes called 'bastard stonefish'.

In the North Sea, weeverfish are abundant and *Trachinus vipera*, the most venomous of the weeverfish species enter shallow water when spawning, burying themselves in sand. Like the stonefish and some stingrays, some European weevers may be stepped on in the shallow waters they inhabit. Other weeverfish species bury themselves in the substrate in deeper water; stingings occur when they are trawled up (MARETIĆ 1988).

As fish venom glands have evolved, selection for a more efficient traumatising apparatus combined with a more toxic venom appears to have occurred (CAMERON & ENDEAN 1972a). Many fish in which protective scales are reduced or absent have protein- and mucus-producing cells in the skin. The toxins from the skin arise from the proteinaceous cell line (see above). CAMERON and ENDEAN (1972b) stated that it is the outer epithelium that usually gives rise to the compact, paired venom glands associated with the spines. An example of the most primitive form of piscine

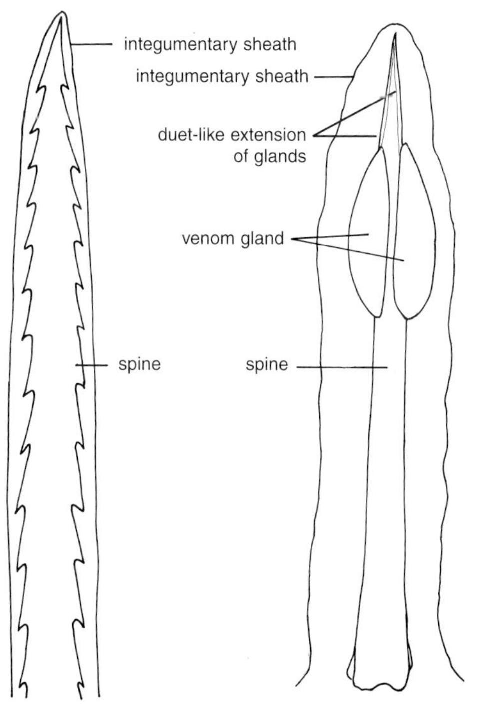

FIGURE 16.4A
The simple venom apparatus of *Arius graffei*, in which poison-containing epithelial cells (not to scale) overlie a traumatising spine. It is possible that the well-developed venom glands of the stonefish evolved from just such an arrangement. (Drawing by Jan Hynes.)

FIGURE 16.4B
Simplified drawing of the venom apparatus from the stonefish *Synanceia horrida*, with paired connective tissue bound glands and a duct-like distal portion. The whole spine is covered in a loose, integumentary sheath. (Drawing by Jan Hynes.)

venom apparatus occurs in the forktailed catfish *Arius graffei* (Plate 16.3 and Figure 16.4A, page 357). In these fish the proteinaceous and mucous glands of the skin are continuous with the thickened layers of integumentary sheath surrounding the spines (Figure 16.4A, page 357). Other catfish such as the eel-tailed catfish *Tandanus tandanus* (Plate 16.9) show a higher evolutionary development of the venom apparatus by the increase in the ratio of poison cells to mucous cells. Sections through the spines of the striped catfish *Plotosus lineatus* (Figure 16.23, page 372, and Plate 16.8) show a migration of unicellular poison glands from the integumentry sheath to sites surrounding the spine itself. The plaque formed internally increases the efficiency of the operation of the venom apparatus (CAMERON & ENDEAN 1972b).

In more complex venom glands, such as those found in juvenile acanthurids (surgeonfish, Plate 16.17 and Figure 16.43, page 389) and stingrays (Figure 16.12, page 362), large bundles of unicellular poison-containing cells are arranged in postero- or antero-laterally placed grooves of the spines (Figure 16.14, page 363).

In the scorpaenid (e.g. zebrafish) venom apparatus, the venom glands are placed in more intimate contact with grooves and are surrounded with a tough sheath of connective tissue (Plate 16.15, and Figure 16.35, page 385). This sheath directs the venom upwards into the wound and prevents it from leaking out of the ruptured integument.

The stonefish (Figures 16.28, 16.29, pages 374–375 and Plates 16.10, 16.11) possesses a very sophisticated venom apparatus (Figure 16.30, page 376). The gland is partially enclosed by the lateral grooves of the spines. A tough connective tissue sheath connects the glands to the spine and also surrounds the glands them-

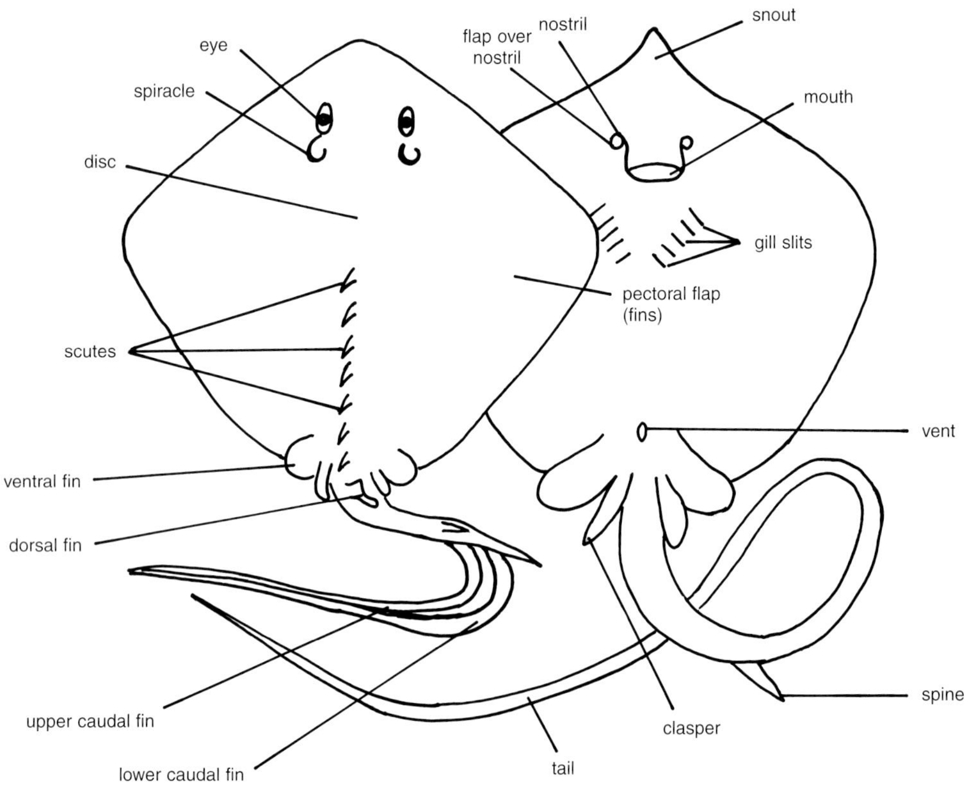

FIGURE 16.5
Stingray external anatomy (dorsal and ventral views) showing some morphological variations. (Redrawn by Jan Hynes, after COLEMAN 1974.)

selves (Figure 16.14B, page 363). In the proximal region of the spines (closer to the point of attachment), the groove surrounds only part of the gland, although in the distal region (remote from point of attachment), the duct-like portions of the glands are entirely surrounded by it. Venom is ejected in a single burst. The local toxicity of the venom in this sting is high.

In some toadfish of the genus *Thalassophryne* (Figure 16.49, page 392), the ductless venom glands are found at the base of hollow spines (Figure 16.50, page 393). When pressure is applied to the base of the spine, the venom passes through the spine and into the wound. Although this is a very highly developed venom apparatus, the venom is less potent than that of the stonefish.

Specific features of venomous fish stings are considered below. The general management of a venomous fish sting is dealt with in detail under that for 'Stonefish' on page 378.

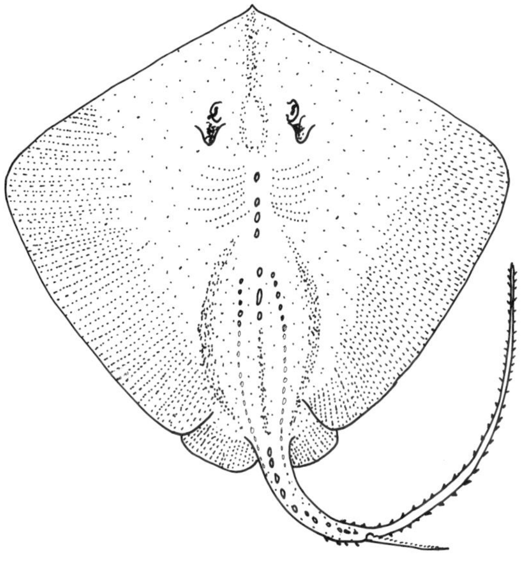

FIGURE 16.6
Dasyatis thetidis, commonly called the black stingray, smooth stingray, or thorntail stingray. (Redrawn by Jan Hynes, after SOUTHCOTT 1975.)

CLASS CHONDRICHTHYES (CARTILAGINOUS FISH)

Comprising the rays, sharks and chimaeras, the members of this class possess well-developed jaws and fins. Members of the groups mentioned below have been the cause of human fatalities.

SUBCLASS ELASMOBRANCHII, ORDER RAJIFORMES (RAYS)

These animals (Figure 16.5, page 358 and Plate 16.4) are dorso-ventrally flattened cartilaginous fish with gills on the ventral surface. The pectoral fins extend from the snout to the pelvic fins. Many species have a whip-like tail up to twice the body length. On the tail is at least one (and occasionally up to seven) sharp, retro-serrated (retroussé), barbs (Figure 16.14A, page 363). In some species the spines are located at the distal end of the tail, and in others the spines are located more proximally.

Rays vary in size from 25 cm to 2 m in width (most measure at least 1 m) with the larger species particularly common in the warmer tropical waters.

When stingrays are not gracefully swimming with slow beats of their pectoral fins, they tend to rest on the bottom, partially burying themselves under the sand and silt. A pair of dorsal nostrils and a pair of dorsal eyes are well placed for watching for food and danger while they lie there, partially buried on the bottom (Plate 16.5). They may be very difficult to see and may be easily stepped upon in the shallows.

Stingrays possess a good set of dental grinding plates ('teeth') with powerful jaws, which are used to crunch up their usual food of oysters, shellfish and other crustaceans. It is not the mouth parts that cause injuries to humans but 'the sting in the tail'. Members of the following families have stings on the tail.

FAMILY DASYATIDIDAE (STINGRAYS, WHIPRAYS)

(Figure 16.6) There are five or more genera in the family, with more than 60 species. These fish usually occur in coastal marine or estuarine conditions in shallow to deep waters of the Atlantic, Indian and Pacific oceans. Most species are demersal, bottom-dwelling, and inhabit continental shelves; some live mainly in the open sea.

These include some of the largest of the rays, discs ranging in shape from circular to rhomboidal. The head is slightly raised from the rest of the disc, but there is no separate head region. The tail is whip-like, and one or more spines are present near the base apposed to the tip of the tail in most species. In most cases, the tail is longer than the disc. Dorsal, anal and caudal fins are absent but membranes may be present. One species, *Dasyatis brevicaudata*, the 'smooth stingray', found off the New South Wales coast, the southern Australian states and Western Australia, is the largest recorded, growing to 4.2 m in length, 2 m wide and to more than 350 kg in weight. This species has a very large, sharp spine capable of inflicting severe, even fatal stings.

Family Gymnuridae (butterfly rays)

(Figure 16.7) Two genera comprise this group, *Gymnura* and *Aetoplatea*, with at least 12 species. They occur in warm temperate and tropical seas around the world. Most species live on the bottom on the upper continental shelf.

The disc is very broad, more than 1.5 times as broad as it is long, resembling a butterfly in shape. The head is not elevated above the fins and is not demarcated from the pectoral fins. The tail is short and slender; it may possess a dorsal fin and/or a stinging spine, but it lacks caudal and anal fins. Butterfly rays have a poorly developed sting which may be absent in some cases, but if present it originates from near the base of the tail.

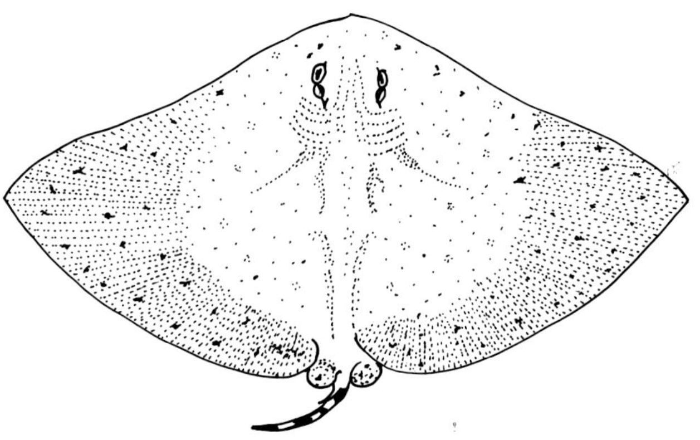

FIGURE 16.7
Gymnura australis, commonly called the butterfly ray or rattailed ray. In this species the sting is reduced or absent. (Redrawn by Jan Hynes, after SOUTHCOTT 1975.)

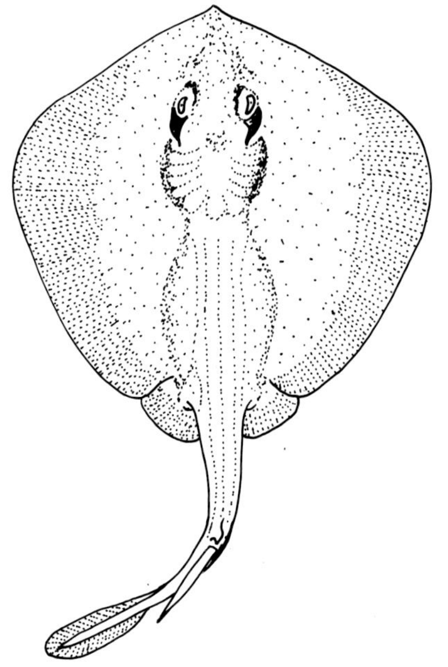

FIGURE 16.8
A stingaree, *Trygonoptera* sp. B = *Urolophus testaceus* (the common stingray). (Redrawn by Jan Hynes, after SOUTHCOTT 1975.)

Family Urolophidae (stingarees)

(Figure 16.8) Three genera are currently recognised in Australia: *Plesiobatis*, *Trygonoptera* and *Urolophus*. Worldwide there are about 40 species, 22 of which are present in the Australian region (LAST & STEVENS 1994).

The eastern shovelnose stingaree *Trygonoptera* sp. B(= *Urolophus testaceus*) is present in southern Australia. A similar species, *Trygonoptera testacea* (= *Urolophus testaceus*) is found from Cape Howe, NSW to Caloundra, southern Queensland. These rays, often called the 'common stingray', are found in Australia in sandy flats in shallow waters and to depths of about 150 m (LAST & STEVENS 1994). SOUTHCOTT (1975) believes that these species are responsible for many of the stings sustained by waders in shallow waters.

Like the stingrays, members of this group have an oval to circular shape. The head is not prominently raised above the pectoral fins. The tail is slender and is almost equal to the length of the disc. The caudal fin is lobe-like and well developed. They may have a small dorsal fin and usually one or more spines.

FAMILY MYLIOBATIDIDAE (EAGLE RAYS OR BAT RAYS)

(Figure 16.9) There are four genera, *Aetobatus*, *Aetomylaeus*, *Myliobatis* and *Pteromylaeus*, with 22 species (NELSON 1984). They occur in the Atlantic, Indian and Pacific oceans and are commonly found in reef waters. In Australia they are found in coastal waters of all states, to a depth of about 100 m. They are more pelagic in their habits than are most rays, and when not feeding or resting on the bottom they may be seen swimming near the surface and leaping clear of the water.

These rays are of medium to large size and are relatively inoffensive, some species growing to more than 2 m across the disc. SOUTHCOTT (1975) asserts that they are capable of upsetting small boats by their sheer weight. The disc is much wider than it is long, with prominent wing-like tips. The front of the head, which extends beyond the disc, is formed into a thick fleshy flap for digging up molluscs and other invertebrates. There is a small dorsal fin at the base of the tail; the caudal fin is absent. The tail is very long (longer than the disc) and whip-like, with a small dorsal fin and one or more spines. Stings, when present, are found at the base of the tail. The wings of these rays are commonly sold in fish markets as 'skate' (SOUTHCOTT 1975).

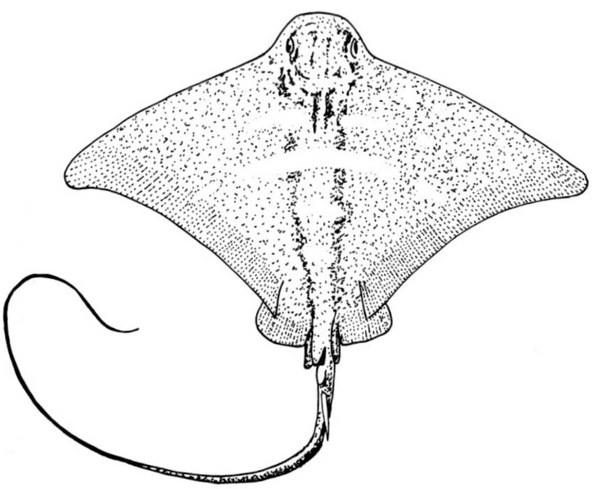

FIGURE 16.9
An eagle ray, *Myliobatis australis*. (Redrawn by Jan Hynes, after SOUTHCOTT 1975.)

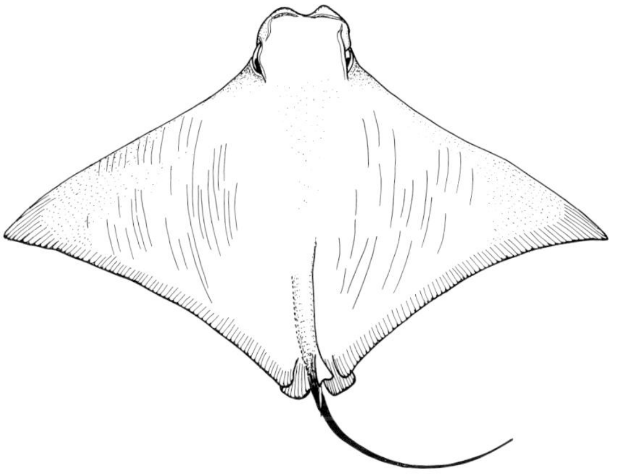

FIGURE 16.10
A cownosed ray, *Rhinoptera neglecta*. (Redrawn by Jan Hynes, after LAST and STEVENS 1994.)

FAMILY RHINOPTERIDAE (COWNOSE RAYS)

(Figure 6.10) There is one genus, *Rhinoptera*, with ten species. They are found in warm or temperate seas around the world.

Cownose rays are medium to large in size, with a disc much wider than it is long. The head has an indented snout with 2 fleshy lobes and an internasal flap, extending out from the disc. The tail is usually longer than the disc, the caudal fin is absent, and a dorsal fin is present. One or more spines are present near the base of the tail (LAST and STEVENS 1994).

The cow nose ray *Rhinoptera bonasus* (Mitchell) is found in abundance mating and nursing young in the summer at Chesapeake Bay, and wintering in Yucatan.

FAMILY MOBULIDAE (DEVIL RAYS)

(Figure 16.11)
Devil rays occur in pelagic waters of the Pacific, Atlantic and Indian oceans, mainly at the ocean surface.

They are the largest of all rays. The disc is broader than it is long. The head is demarcated

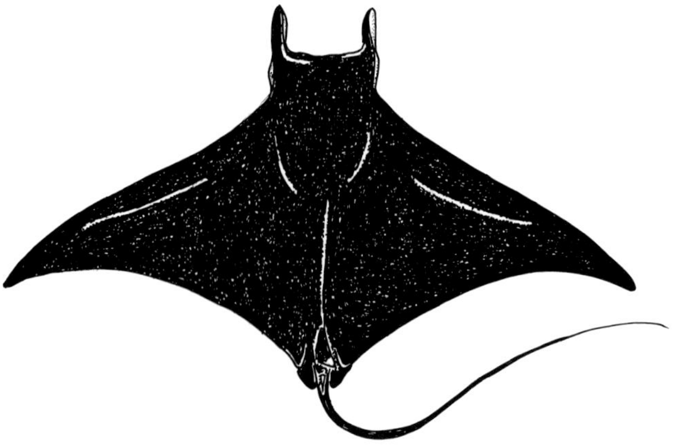

FIGURE 16.11
The Japanese devil ray *Mobula japanica*. (Redrawn by Jan Hynes, after LAST and STEVENS 1994.)

from the rest of the disc, and there are anterior extensions of the pectoral fins on either side of the head. Some devil rays, such as the Japanese devil ray *Mobula japanica*, do have caudal spines (LAST and STEVENS 1994).

Manta birostris, the harmless manta ray, which is also known as the devilfish, manta, munguna, and Prince Alfred Ray, is found mainly in tropical seas. Beloved of underwater photographers, it is a huge fish with a disc reported to 910 cm across, although specimens with discs of about 400 cm across are more common. Caudal spines are absent in this species (LAST and STEVENS 1994).

FAMILY HEXATRYGONIDAE (SIXGILL STINGRAYS)

(Figure 16.12) Family members are unusual, with six pairs of gill openings, long, pointed snouts, and without dorsal fins. There may be either one one or two stinging spines on the tail. The first specimen was collected in 1985 off Queensland. There is one genus, *Hexatrygon*, with 5 species (LAST and STEVENS 1994).

FAMILY POTAMOTRYGONIDAE

This family includes mainly the freshwater 'river stingrays' of South America (CASTEX & LOZA 1964).

ANATOMY OF THE STINGRAY VENOM APPARATUS

(Figure 16.13) According to HALSTEAD (1988) the stingrays' venom apparatus consists of:

- the caudal appendage or tail {see below}
- the barbed spine with enveloping integumentary sheath and its associated venom glands
- cuneiform (wedge-shaped) area of the integument, with which the spine is in contact when at rest. When the spines are lying flat against the dorsal surface of the fish they are bathed in mucus and venom (MARETIĆ 1988).

THE STINGRAY STING

The stingray 'sting' is a commonly misunderstood term (HALSTEAD 1988). It refers in fact to the spine, including its sheath and associated venom glands. The 'spine' refers only to the rigid surface of the spine which is made of vasodentin and covered by a layer of vitrodentin (MARETIĆ 1988). The barbs are the backwards-pointing (retroussé) teeth or serrations associated with the lateral aspects of the spine (see Figure 16.14, page 363)

One or more spines are present on the dorsal surface of the stingray tail. The barbs contribute to the tearing of the integumentary sheath and the broadening of the wound.

HALSTEAD and BUNKER (1953) stated that there are four possible types of caudal appendage on sting rays (Figure 16.15, page 364), and each has a different stinging ability:

(a) The gymnurid type is characterised by a small spine, up to 2.5 cm in length, usually situated on the middle or proximal third of the

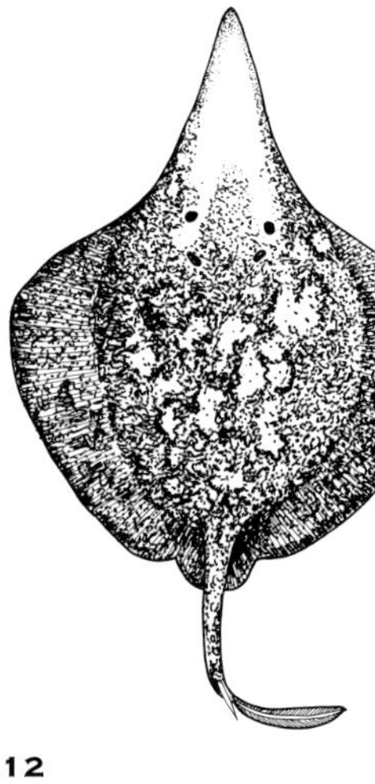

FIGURE 16.12
A sixgill stingray, *Hexatrygon* sp. (Redrawn by Jan Hynes, after LAST and STEVENS 1994.)

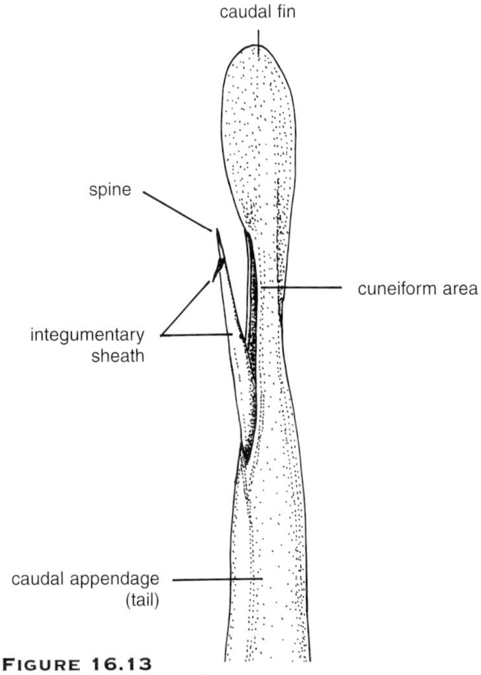

FIGURE 16.13
External anatomy of a stingray venom apparatus. (Redrawn by Jan Hynes, after HALSTEAD 1988.)

characteristically small tail. These apparently present the least danger to humans (HALSTEAD 1988). The butterfly rays (family Gymnuridae) possess this type.

(b) The myliobatid type has a moderate to large spine that reaches 12 cm in length and is situated at the root of the long whip-like tail. This type of caudal appendage is possessed by the eagle rays (family Myliobatididae), cownose rays (family Rhinopteridae) and devil rays (family Mobulidae). Although this type of appendage is capable of serious stinging, its striking ability according to HALSTEAD (1988) is comparatively limited.

(c) The dasyatid or stingray type has a sting measuring up to 37 cm in length, located at the base of a long whip-like tail, and usually closer to the 'whip' than in the Myliobatidae. The striking ability of this appendage is somewhat greater than that found in the myliobatid rays. The true stingrays (family Dasyatididae) have this type of caudal appendage.

(d) The urolophid type is characterised by a short muscular tail with a sting about 4 cm in length, providing a very well-developed striking ability. The round rays (family Urolophidae) and most freshwater stingrays (family Potamotrygonidae) possess this type of caudal appendage.

Of these four existing types of caudal appendage, those of the myliobatids (eagle rays) and the dasyatids (stingrays) are considered by HALSTEAD (1988) to be the most dangerous to humans. GRAINGER (1985) measured the sizes of spines in the myliobatid types and dasyatid types of rays with equal 'wingspans' and found that the sting from the stingray type was much larger than that of an eagle ray of equal size, even though the eagle ray possessed more than one spine along the length of the tail. Usually, the larger the ray, the more power behind the lash of the tail. ENDEAN (1982) suggested that the rays possessing distally-placed spines would have more power in the flick of the tail and cause more damage.

LOCATION OF VENOM-PRODUCING CELLS WITHIN THE STING

PEARSON (1967) and other workers have examined the stings of a number of Australian stingrays and found that venom-producing cells were present within the epidermis of the integumentary sheath and contained a venom with a

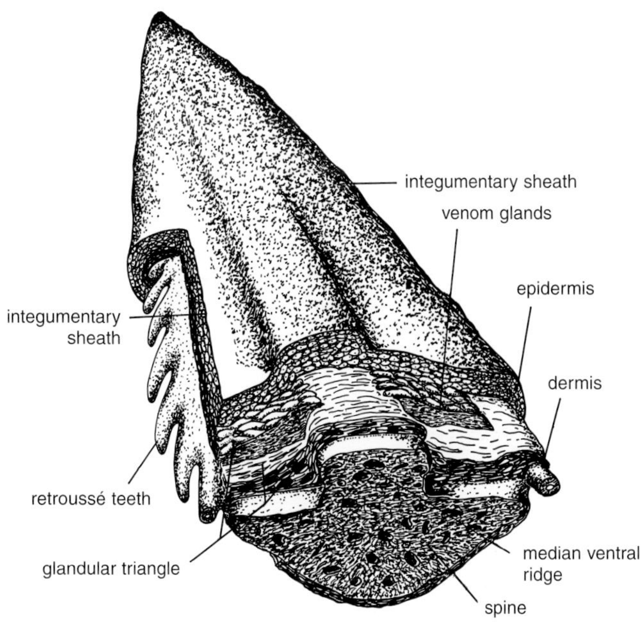

FIGURE 16.14
Internal anatomy of a stingray sting. (Redrawn by Jan Hynes, after HALSTEAD 1988.)

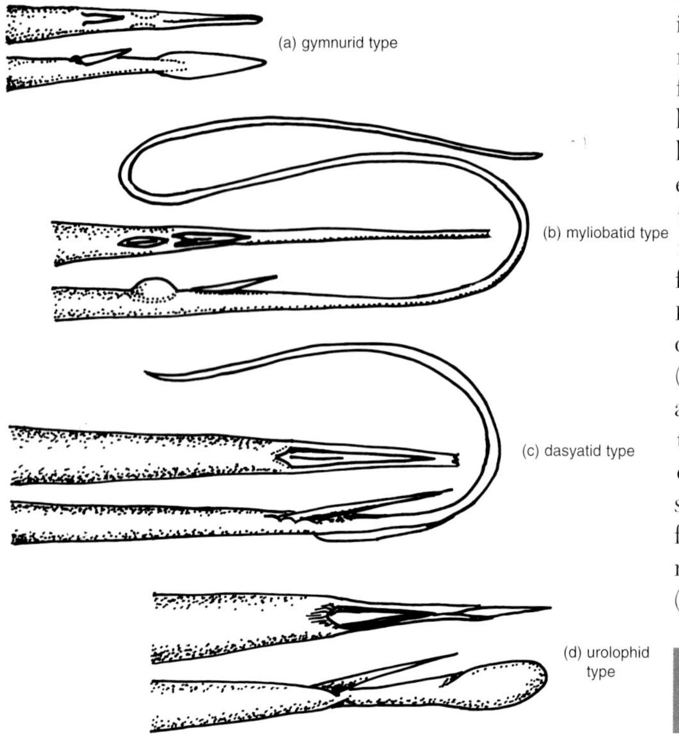

FIGURE 16.15
The four basic types of ray sting/caudal appendages. (Redrawn by Jan Hynes, after HALSTEAD 1988.)

microtubular component (SMITH *et al* 1974). Although most glands were found in the epidermis of the ventrolateral grooves of the spine (the glandular triangle) (see Figure 16.14, page 363), they could also be found in dorsolateral and ventrolateral epidermis surrounding the spines.

MODE OF STINGRAY INJURY AND/OR ENVENOMATION

The stingray is a non-aggressive and remarkably intelligent creature ('the pussycat of the sea'), but if frightened, cornered or captured it can react purposefully and with great speed. Rays are not capable of voluntarily raising and lowering their stings. The wound comes from a powerful arching forward flick of the tail (Figure 16.16 and Plate 16.6).

During an envenomation the tip of the spine and the barbs pierce the integumentary sheath and lacerate the skin of a victim. Venom simultaneously passes into the wound. Stingrays frequently envenomate persons who capture them and then unwisely attempt to bring them into a boat (Plates 16.7a and 16.7b). Just as commonly, victims accidentally tread on the flattened ray buried in the sand, while wading or launching a boat. The tail flicks up and the lower portion of the leg may be lacerated by the erect spine (Figure 16.16). It is also dangerous to swim directly over a stingray, as the spine may pierce the person's trunk, with possible fatal consequences (CADZOW 1960; CROSS 1976; FENNER *et al* 1989b). Stingray woundings also occur regularly in trawl fishing operations (JONES and DERBYSHIRE 1987), when fishermen are sorting commercial species from the rest of the catch. Increased tissue damage may be caused when a victim pulls away from the source of the pain and the spine is dragged forcibly out. The spine is relatively fragile and readily breaks off, embedded in the wound (Figure 16.7, 360).

SYMPTOMS AND SIGNS OF STINGRAY INJURY

As indicated above, the injury may have two aspects: direct trauma from the powerful stabbing action of the spine, and envenomation

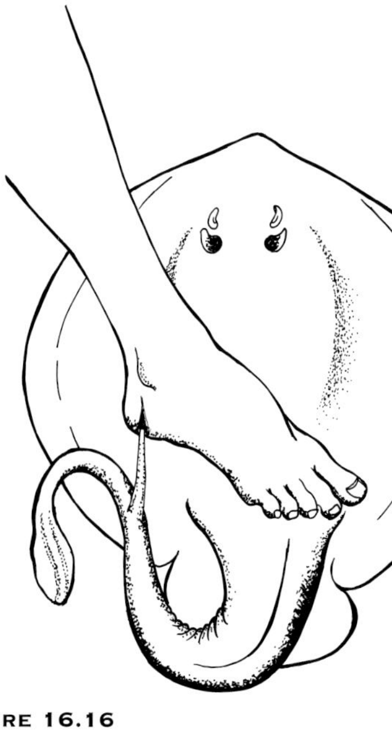

FIGURE 16.16
The mode of envenomation in stingrays. (Drawing by Jan Hynes.)

effects. These two effects may work dangerously in combination in certain circumstances (FENNER *et al* 1989b). It is best to consider all rays equipped with spines as hazardous though not all members of every ray family listed above have spines.

TRAUMA

The clinical features will be determined by the site of the injury. Many injuries are lacerations on the lower limb and they may be multiple and deep (RUSSELL *et al* 1958). Trauma to the lower shin or foot is common (Plate 16.6 and Figure 16.17) when people shuffle in the shallows launching boats or wading in. Stingrays will not necessarily evade approaching feet. Once trodden on, they reflexly arch the tail over and drive the spine into the offending foot, penetrating clothing and even rubber boots with ease. If a major blood vessel is penetrated, haemorrhage is an obvious acute hazard and, if high on a limb, can be fatal (e.g. KNIGHT 1989).

RUSSELL and colleagues (1958) quote an incidence of '5 per cent of stingray wounds' in which a broken-off fragment of the spine is retained MARETIĆ (1975) showed with his own guinea pig experiments that even after spine fragments were meticulously removed, some barbs always remained. It is the editors' experience that retention of spine fragments is a commonly observed event in Australian stingray wounds — especially if the wound is X-rayed.

NOTE: Penetration of any part of the body trunk (chest, abdomen, groin or back) (LIGGINS 1939; WRIGHT-SMITH 1945; KNIGHT 1989) is a grave medical emergency, partly because of the inexorable necrotising action of any venom deposited on penetrated viscera or organs (FENNER *et al* 1989b); all such wounds must receive immediate and expert medical attention.

ACTIONS OF THE VENOM

According to MARETIĆ (1988), who summarised the work on stingray venoms as outlined below, an adult stingray has more venom than either a weeverfish or a scorpionfish. Furthermore all stingray venoms are apparently very similar.

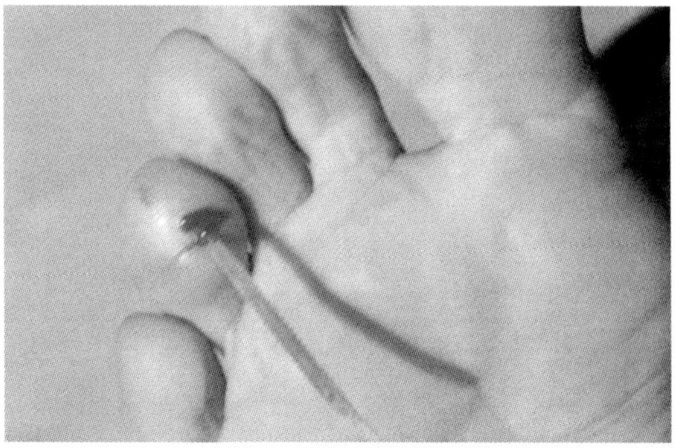

FIGURE 16.17
A broken-off piece of a stingray spine in a toe wound. (Photograph by Dr Peter Fenner.)

The venom of *Urobatis halleri* contains serotonin, 5-nucleotidase and phosphodiesterase, but has no protease or phospholipase activity (RUSSELL 1972). The venom of *Urobatis halleri* has a damaging effect upon the mammalian cardiovascular system. Low concentrations of the venom cause peripheral vasodilation or vasoconstriction and may induce atrioventricular blocks grades I–III in cat heart. Higher doses provoke changes of ST segment and T wave symptomatic of ischaemia and lesions of the cardiac muscle. The venom has a disturbing action when applied directly to cardiac muscle; it induces changes in heart rate, in the amplitude of systole, and induces ectopic rhythms which may lead to cardiac arrest. The venom also depresses respiration and may cause convulsions in experimental animals, but does not affect neuromuscular endings (RUSSELL 1972; RUSSELL & VAN HARREVELD 1954; RUSSELL *et al* 1958a, 1958b).

Local venom effects

The venom has an insidious and powerful tissue-necrosing action, which will continue unabated in untreated wounds (BARSS 1984; FENNER *et al* 1989b). Even minor untreated stings can thus simulate a bacterial cellulitis (FISHER 1978). Deposited venom is also powerfully pain-producing. Stingray injuries, even relatively trivial ones (Figure 16.17), are typically followed by immediate and persistent local pain. Paraesthesias may be present (MARETIĆ 1988). (When proven stingray wounds are not painful, it is probably because the integumentary sheath has previously been ruptured and venom is not released, or the integumentary sheath does not penetrate or rupture [RUSSELL *et al* 1958b]. Alternatively the spine could have been broken previously.)

Venom-mediated tissue necrosis or subsequent secondary infection {see Chapter 4, page 88} in a dependent area such as in the lower limb always delays healing. Such lesions may become chronically infected, causing persistent sinuses or ulcers for many months, defying subsequent standard treatments (FENNER *et al* 1989b).

Stings from freshwater stingrays (family Potamotrygonidae)

According to MARETIĆ (1988) these are frequent in some South American countries. For example, hospitals in Colombia register several thousand injuries annually (MARINKELLE 1966). Over a five-year period, in one small provincial hospital alone, there were eight fatalities, 23 amputations of lower limbs, and 114 patients incapable of working for up to eight months.

Systemic venom effects

Large peripheral envenomations have been described as causing disturbance of central nervous and heart action in humans (RUSSELL *et al* 1958, SUTHERLAND 1983), although clear cause and effect of the venom in such cases is not fully established. MARETIĆ (1988) reported symptoms following the sting of *Dasyatis thalassia*, which included nausea and profuse sweating with flattened or biphasic T waves in the ECG.*
However, he states that the most serious complications are secondary infections or injuries to vital organs.

Described systemic effects include anxiety, headache, tremors, maculopapular skin rash, nausea, vomiting, diarrhoea, diaphoreses, generalised pallor, restlessness, delirium, peripheral neuropathy or neuritis, limb paralysis, lymphangitis, abdominal pain, arthritis, fever, hypertension, difficult breathing, cardiac arrythmias, congestive heart failure, hypotension and syncope (WIENER 1959a). However, it is the editors' impressions from the literature, and experience, that stingray venom has minimal systemic effects. Based on experience with other marine venoms, allergic reactions could be expected; these are not yet documented.

DEATHS

Seventeen deaths have been attributed in the literature to stingray injury (LIGGINS 1939; WRIGHT-SMITH 1945; RUSSELL *et al* 1958; MARINKELLE 1966; RATHJEN & HALSTEAD 1969; FENNER *et al* 1989b). Three of these reports are closely documented, and another three include probable identification of the stingray species responsible (LIGGINS 1939; RUSSELL et al 1958; MARINKELLE 1966). One inadequately documented report claims (unconvincingly) the cause of death to be due to stingray envenomation; another death, reported in the same paper, resulted from tetanus six days after envenomation (RATHJEN & HALSTEAD 1969). A sixth, the first reported Australian death, was attributed at autopsy to stingray sting penetration of the heart, although the actual event was not witnessed (WRIGHT-SMITH 1945). In 1989 the editors reported a delayed fatality in a 12-year-old boy after a freak accident in a small boat; the sting of an airborne stingray caused direct penetration of the left lung and the heart, although this was not obvious at the time; symptoms compatible with a penetrating chest injury were present for only a very short period and disappeared. There were no clinical signs apparent. The venom caused a delayed, insidious myocardial necrosis resulting in right ventricular rupture and cardiac tamponade (FENNER *et al* 1989b). The most recent known documented death was in Fiji when a diver exsanguinated after a sting from a giant stingray punctured his femoral artery (KNIGHT 1989). Stings from freshwater stingrays (family Potamotrygonidae) in Colombia, South America, have been reported by MARINKELLE (1966) (see above) to account for eight fatalities over five years, in one small province alone.

PREVENTION

Stingrays that are speared, netted or caught on a line should never be brought on board a boat (FENNER *et al* 1989b) (Plates 16.7a and 16.7b). The most serious injuries have resulted from unin-

* The editors have yet to see direct electrocardiographic abnormalities in human victims from peripheral stingray envenomation in Australia.

formed persons grappling with a live, frightened stingray on the deck of a boat, under which circumstances the animal, not surprisingly, will actively try to envenomate. Stingrays sometimes mount unprovoked attacks on human beings (R. KUITER, personal communication, 1996).

Avoidance of the most common injury, a sting from a stationary stingray to the lower leg in the shallows, can be avoided by inspecting the bottom of the launching area in clear water first, or shuffling with well-clad feet and legs when launching or wading in murky shallows. If a stingray is seen while snorkelling or diving, leave it alone! If a stingray is caught from a boat, cut the line or release the net and allow it to escape. Never pull it on board.

Note: Captive stingrays in aquaria make the most fascinating and friendly inhabitants, when treated kindly, gently and with respect. Their intelligence far exceeds that of turtles and sharks, in the experience of one editor (JAW). However, their handling under even these circumstances should not be attempted except under expert supervision.

MANAGEMENT OF STINGRAY WOUNDS

Stingray injuries (mostly minor) are very common, but significant morbidity can result unless the wound is carefully cleaned as soon as possible {as for a coral abrasion, see Chapter 8, page 180} to remove any venom, remaining integumentary sheath or broken spine pieces (RUSSELL *et al* 1958, SUTHERLAND 1983, BARSS 1984; FENNER *et al* 1989b). These measures may require skilled medical intervention, and as a general principle all stingray wounds should receive qualified medical inspection at the earliest opportunity.

FIRST AID

- Control visible haemorrhage.
- Do not apply compression/immobilisation bandaging; stingray venom has a locally necrosing action but little apparent systemic envenoming property in humans, according to present clinical evidence.
- Immediately place the affected area in water as hot as possible (the first-aider to test it first!) {see Chapter 5, page 112}. The duration of immersion is determined by response, but withdrawal of the envenomated part will not uncommonly lead to a quick return of the pain.
- If heat is unavailable or fails to provide relief, then ice or ice packs are applied to the area for up to 30 minutes and repeated when necessary (MULLINS *et al* 1957; FENNER *et al* 1987b, and see FENNER 1996).
- When possible, perform early wound toilet using clean fresh water, a weak antiseptic/detergent and a soft bristle brush.
- Seek medical aid for all but the most trivial stingray injury (WILLIAMSON 1985).

MEDICAL MANAGEMENT

1. If necessary, inject local anaesthetic (1% plain lignocaine) (AUERBACH 1988) directly into the wound, and around the site of envenomation to ease the pain. It has been observed that the early use of (plain) lignocaine for analgesia in these wounds may actually contribute to tissue sparing from the necrosing action of the venom [FENNER *et al* 1989b]. A regional nerve block may be necessary, performed by an appropriately skilled person (FENNER *et al* 1989b).
2. X-ray the wound area for traces of retained foreign material.
3. Explore the wound carefully under appropriate anaesthesia, following the full length of any tract left by the penetrating barb. Remove all traces of broken barb or integumentary sheath and clean the wound using an appropriate non-irritant antiseptic. Experienced surgical decision is often required (BARSS 1984), especially concerning excision of the wound margins, and the use of an operating microscope may be helpful.
4. Leave the cleaned area open to granulate; do not suture tightly closed. Even if no surgery is required, the patient should be observed for three or four hours to exclude any systemic reaction, e.g. allergy (AUERBACH 1988).
5. Tetanus prophylaxis is given if indicated.

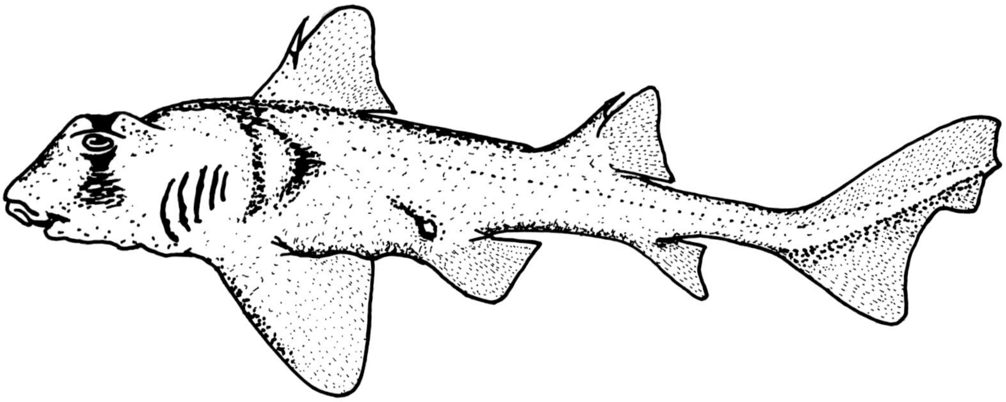

FIGURE 16.18
The Port Jackson shark *Heterodontus portus-jacksoni*. (Redrawn by Jan Hynes, after SOUTHCOTT 1975.)

6. Antibiotics should be considered in delayed or other selected circumstances {see Chapter 4, page 96}.
7. All stingray wounds thought to have penetrated the abdomen, groin, chest wall (front or back) or lumbar region should be referred as a matter of urgency to experienced medical centres for appropriate imaging and further specialised assessment (FENNER *et al* 1989b). This urgency is not lessened by the absence of symptoms or signs at the time.
8. Follow up all wounds until they are healed.

Notes:

- It has been suggested (COLQUHOUN 1990, unpublished observations) that a worthwhile investigation for a possible stingray spine puncture of the pericardium or myocardium is echocardiography. This would reveal a pericardial effusion, however small.
- Cardiac enzymes may be indicators as the creatinine kinase MB fraction may rise significantly within eight hours of necrosis. This may be a critically late detection (FENNER *et al* 1989b).
- Colour flow imaging of the myocardium is now available and may give an improved idea of tissue viability.
- FENNER (1995) recently found overriding medical benefits in packing stingray lacerations with Kaltostat (Bristol Myers/Squibb), an alginate dressing derived from seaweed.

VENOMOUS SHARKS

FAMILY HETERODONTIDAE (BULLHEAD, OYSTER CRUSHER, TABBIGAW, HORN OR PORT JACKSON SHARKS)

Members of the genus *Heterodontus* are distributed in tropical Indian and Pacific oceans: South Africa to Japan, Australia and New Zealand; and eastern Pacific from California to Galapagos Islands and Peru (NELSON 1984). The Port Jackson shark *Heterodontus portusjacksoni* (Meyer), also commonly known as 'doggie' or 'oyster crusher', is a bottom feeder but it breeds in the shallows in non-tropical waters. It has a large blunt head, marked supra-orbital crests, a stocky body (Figure 16.18), two fixed dorsal fin spines, one in the front of each dorsal fin; an anal fin, and dark harness-shaped markings on the body. The spines are considered to be venomous (HALSTEAD 1988). This species has been reported by SOUTHCOTT (1970) to have stung a victim on the wrist; the wound became very painful, a severe pain developed in the axilla, and there was local muscle weakness. The effects lasted about two hours. According to SOUTHCOTT (1975), a Californian shark species gives similar effects. For sting management, see page 378.

ORDER SQUALIFORMES (DOGFISH SHARKS)

The dogfish shark, *Squalus acanthias*, of the Pacific Ocean has been reported to be venomous (SOUTHCOTT 1975). For sting management, see page 378.

VENOMOUS CHIMAERAS

FAMILY CHIMAERIDAE (SHORT-NOSE CHIMAERAS, RATFISHES)

Found in the Atlantic and Pacific oceans, chimaeras have a wide geographical range from north temperate to south temperate waters.

These primitive fish are scavengers, with long, smooth, laterally compressed bodies, and small mouths, rodent-like teeth, and a single gill opening on either side leading into a bronchial chamber; the true gill clefts open into this. They possess a very narrow long tail (Figure 16.19). Two dorsal fins are present, each preceded by a spine.

VENOM APPARATUS

The first dorsal fin is triangular with a strong serrated, erectile spine capable of causing considerable damage. This is also the venom apparatus (HALSTEAD 1988). These fish are also capable of inflicting serious bites with their well-developed dental plates.

Where the spine rests against the dorsal fin it is apparently bathed in secretions. According to HALSTEAD (1988) glandular epithelium is situated in the interdentate depression, in the epidermal layer of the anterolateral grooves, and the connecting membrane between the spine and the first dorsal ray. Members of the genera *Chimaera* and *Hydrolagus* have caused injury to humans. MAGEROY and BAERHEIM (1991) reported a case in which a fisherman was stung on his right calf by a ratfish (*Chimaera monstrosa*). He experienced immediate burning pain which was relieved by hot water. A few days later the calf and back of the thigh became numb and he required crutches to walk for the next 3 weeks. Symptoms persisted for 9 weeks.

Manage as for a stonefish sting {see page 378}.

CLASS OSTEICHTHYES (BONY FISH)

This class includes all fish with true bony skeletons. They are cosmopolitan, exhibiting a wide diversity of size, colour and form, and are of great nutritional and medical significance to humans. As described earlier (see page 86), the toxins of poisonous and/or venomous bony fish may be encountered in their flesh (e.g. ciguatoxin, scombrotoxin), their blood (HALSTEAD 1988), their reproductive tissue, or in their glandular tissue. A traumatising organ (e.g. spines or teeth) may or may not be associated with the glands). The typical external anatomy of bony fish is shown by Figure 16.20, page 370.

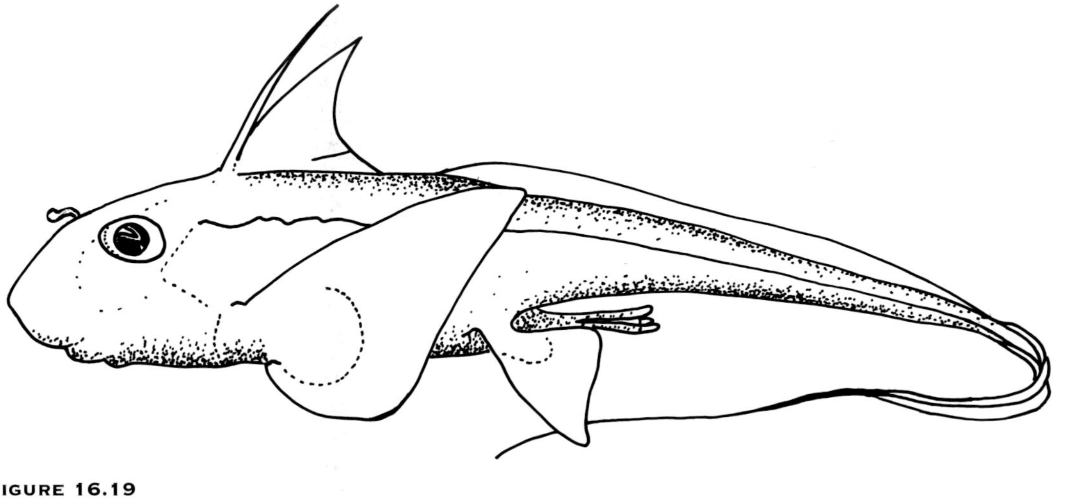

FIGURE 16.19
A shortnose Chimaera. (Redrawn by Jan Hynes, after NELSON 1984.)

ORDER SILURIFORMES (CATFISH)

Most members of this order are bottom dwellers, living in marine, estuarine or freshwater environments. Catfish are so-named for their harmless, fleshy, whisker-like sensory barbels which extend from the snout in most species. These are used for detecting food.

Scales are reduced or absent, and catfish skin is slimy and contains poisons. The secretions arise from unicellular mucous and poison glands which form the bulk of the epidermis of the fish. Fish handlers may experience severe pain from the toxic secretions merely by handling the fish by the tail. The epidermis of the integumentary sheath covering the spines is comprised of similar cell types.° Most catfish possess a sharp spine on the leading edge of the dorsal fin and on each pectoral fin, which can inflict painful wounds. These can be locked in the extended position, making handling dangerous. Each spine possesses retroussé barbs which cause increased tissue damage on penetration and make removal difficult.

Several species of catfish are known to be venomous (HALSTEAD 1988). Most species are passive stingers, and victims are often stung when removing the catfish from nets and fish hooks. However, some species such as *Heteropneustes fossilis* from India (Figure 16.21, page 371) will actively attack their victims and produce a painful sting. Stings from the eel-tailed catfish *Plotosus lineatus* (Figure 16.23, page 372, Plate 16.8) may (it is claimed) uncommonly result in fatalities (NELSON 1984).

ENVENOMATION

Catfish possess a mechanism that allows the dorsal and pectoral fins to spring out from three sides and remain locked in place. This mechanism is also employed while the fish is asleep so that any potential predator is repelled. When the integumentary sheath covering the spine and any compact venom glands present are ruptured, their contents are forced into the wound. In addition, toxins from the epidermis of the integumentary sheath (CAMERON and ENDEAN

° AL-HASSAN and colleagues (1991) working with the Arabian Gulf catfish *Arius bilineatus* Valenciennes has demonstrated accelerated wound healing properties of the epidermal secretions of this catfish, in test rodents.

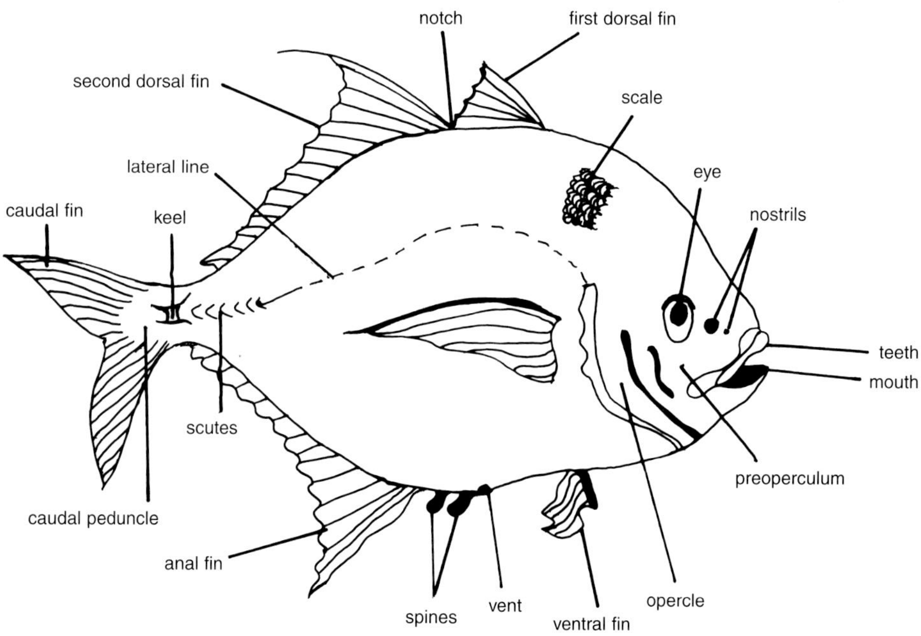

FIGURE 16.20
External anatomy of a bony fish, class Osteichthyes. (Redrawn by Rebekah Lewis, after COLEMAN 1974.)

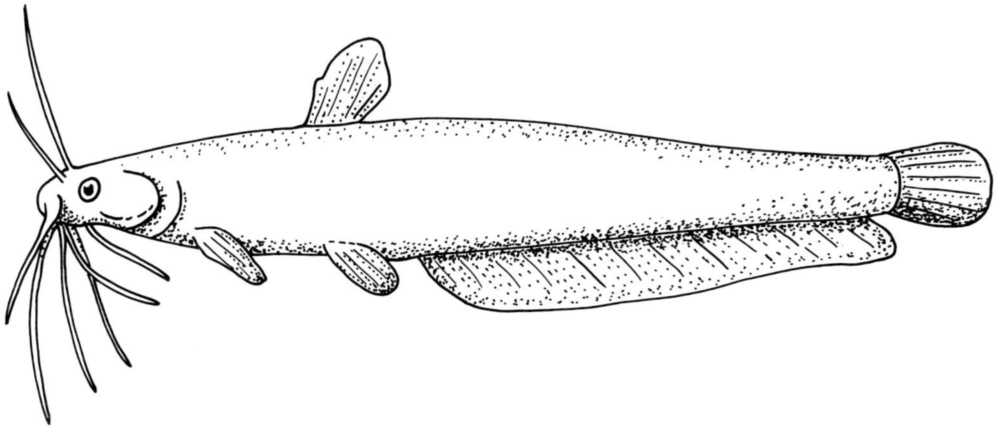

FIGURE 16.21
Heteropneustes fossilis, an aggressive catfish from India, which will actively attack its victims. (Redrawn by Jan Hynes, after NELSON 1984.)

1972, SHIOMI *et al* 1988, McKINSTRY 1993) may enter the wound along with the venom. (The proteinaceous skin secretion of an Arabian Gulf catfish *Arius thalassinus* Ruppell contains components that are vasoconstrictors *in vitro*; these vasoconstrictor actions are variously blocked by indomethacin and atropine [AL-HASSAN *et al* 1986]). It is dangerous to handle a wriggling catfish even when holding it by the tail.

CLINICAL FEATURES

MANN and WERNTZ (1991) described a case of catfish envenomation to the hand in which the fingers developed gangrene and required amputation. They stated that victims developed symptoms from haemolytic, dermatonecrotic, oedema-promoting, vasospastic and 'lethal' components of the venom. AUERBACH (1988) considered that the sting from a catfish resembled that of a stingray in terms of local and systemic effects.

The wound is immediately painful, described as stinging, throbbing or scalding, and the pain radiating. A mild envenomation may produce pain for 30–60 minutes. Large tropical catfish may apparently cause pain for up to 48 hours (FISHER 1978). The soft tissue surrounding the wound becomes ischaemic in appearance, with cyanosis, reactive hyperaemia and oedema (which may be marked). As with any marine wound, secondary infections may ensue.*

Systemic effects include local muscle spasm, sweating and fasciculations. Less frequently,

* The editors wonder if some of the "deaths" were from secondary infection and septicaemia.

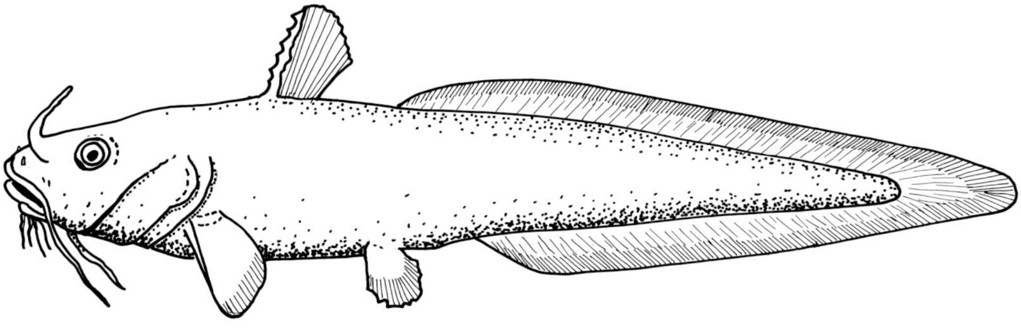

FIGURE 16.22
An unidentified species of the family Plotosidae, the eel-tailed catfishes. (Drawing by Jan Hynes, after NELSON 1984.)

peripheral neuropathy, lymphoedema, lymphadenopathy, lymphangitis, weakness, syncope, hypotension, respiratory distress (and death) are described (HALSTEAD *et al* 1953).

MANAGEMENT

Manage as for a stonefish sting {see page 378}.

MANN and WERNTZ (1991) used local or regional anaesthesia to relieve pain and vasospasm. They suggested that skin problems required irrigation and that spine fragments needed to be removed.

PREVENTION

Envenomation from catfish generally occurs as a result of careless handling. Care must be taken when releasing these fish from lines, nets or sorting trays.

CATFISH ASSOCIATED WITH FATALITIES

There have been no Australian fatalities documented and only one in the USA, which was secondary to septicaemia (McKINSTRY 1993); but accurate details concerning other claimed fatalities are lacking (HALSTEAD 1988).

FAMILY PLOTOSIDAE (EEL-TAILED CATFISH)

The striped catfish *Plotosus lineatus* (Figure 16.23 and Plate 18.7) lives in marine, estuarine and freshwater habitats in the Indian Ocean, Arabian Sea and the western Pacific from Japan to Australia.

In Australia species in the family Plotosidae are found in tropical Western Australia,

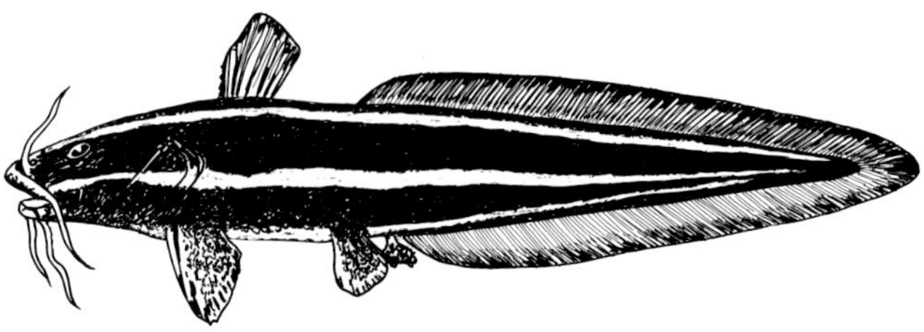

FIGURE 16.23
Plotosus lineatus, the striped catfish, is often found in shoals or pods as a juvenile, but is usually solitary as an adult. Fatalities are claimed from stings from this fish. (Redrawn by Jan Hynes, after SOUTHCOTT 1975.)

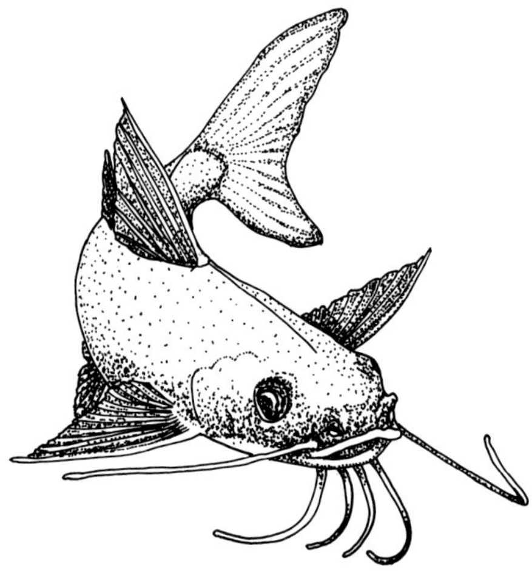

FIGURE 16.24
The forktailed catfish *Arius graffei* causes minor injuries in humans. There are three extended, lockable spines at the leading edges of dorsal and pectoral fins. (Drawing by Jan Hynes.)

Northern Territory, Queensland and northern New South Wales, in bays and inlets, and in muddy or sandy areas among eel grass or on reefs (COLEMAN 1974).

These catfish are generally a brownish-olive colour dorsally and pale ventrally. As juveniles they have three narrow yellowish longitudinal lines or stripes on each side (Plates 16.8 and 18.7). Young fish normally travel in circular pods or shoals, revolving slowly at the surface (GRANT 1975). As they grow, their stripes become less obvious. The shoal formations break up and the fish become solitary, moving into estuaries or into rivers.

In Queensland they rarely exceed a length of 30 cm, though they may reach 1 m in length in other parts of Australia and the world (HALSTEAD *et al* 1953).

Structure of the venom apparatus

The first dorsal spine and the two pectoral spines lock in place, easily causing lacerations. The structure of the barbed pectoral and dorsal fin spines is similar. There are three components to the catfish fin sting (Figures 16.4A and 16.25, pages 357 and 373):

- The epithelium of the outer integumentary sheath, containing the the unicellular glands, and in some cases a compact inner layer of

poison-containing cells, forming a plaque which closely adheres to and surrounds most of the spine (not shown)
- The middle bony portion of the sting
- The central canal

CATFISH CAUSING HUMAN MORBIDITY

Family Plotosidae (eel-tailed catfishes):
Cnidoglanis macrocephalus
Plotosus lineatus
(Figure 16.23 page 372, Plate 16.8)
Euristhmus lepturus
Tandanus tandanus (Plate 16.9)

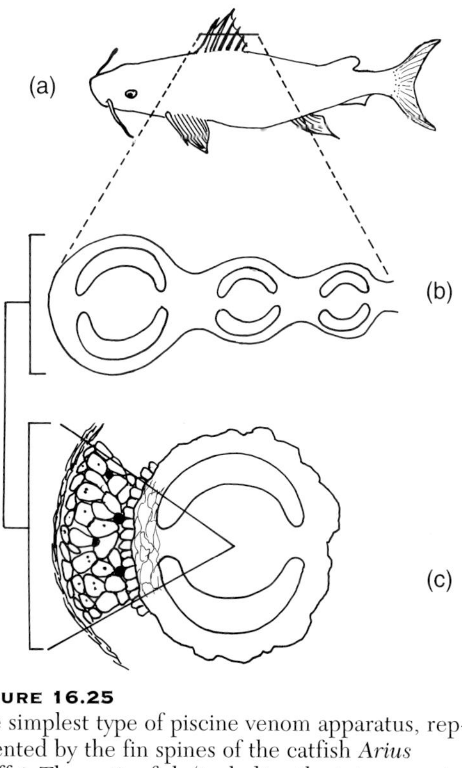

FIGURE 16.25
The simplest type of piscine venom apparatus, represented by the fin spines of the catfish *Arius graffei*. The entire fish (including the integumentary sheath of the first dorsal fin spine and the pectoral fins' spines), is covered by a glandular epithelium: (a) catfish *Arius* sp., with dorsal fin locked; (b) horizontal section of the first three dorsal spines, including part of the fin ray; (c) drawing of a section through the integumentary sheath of the dorsal fin spine. Note the large, poison-containing clavate cells, and small, round mucous cells (black). (Drawing by Jan Hynes.)

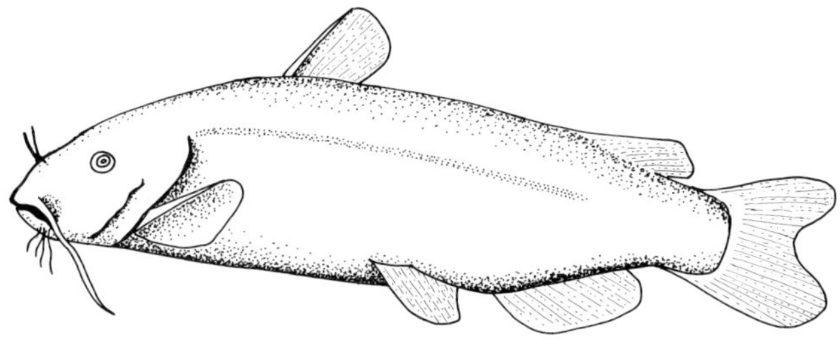

FIGURE 16.26
A catfish of the genus *Ictalurus*, responsible for many serious stings in the United States. (Redrawn by Jan Hynes, after NELSON 1984.)

Family Ariidae (forktailed catfishes)
Arius thalassinus
Arius graffei (Figure 16.24, page 372)
Family Ictaluridae (Figure 16.26)
Ictalurus nebulosus—brown bullhead
Ictalurus punctatus—channel catfish
Ictalurus furcatus—blue catfish
Ictalurus catus—white catfish
Family Trichomycteridae (*candirú*) (Figure 16.27, page 374)

CLINICAL ASPECTS

It is considered by some that marine catfish are the more venomous (AUERBACH 1988). Stings from catfish of the family Plotosidae, such as *Cnidoglanis macrocephalus*, small specimens of *Plotosus lineatus*, and *Euristhmus lepturus*, may produce pain levels requiring hospitalisation (CAMERON 1981). Members of the family Ariidae (forktailed catfishes), are mainly marine and are found in tropical and subtropical waters. *Arius thalassinus*, the Arabian Gulf catfish causes significant envenomations (AL-HASSAN *et al* 1986). *Arius graffei* produces localised pain of shorter duration (CAMERON 1981). In North America the ictalurid catfish (Figure 16.26) and 'the Carolina madtom' *Noturus furiosus* are most often responsible for serious catfish envenomations (SCOGGIN 1975; MANN & WERNTZ 1991).

The family Trichomycteridae contains the pencil or parasitic catfishes (Figure 16.27, page 374). These are found in fresh water, in Costa Rica, Panama and South America. Some members of this family pierce the skin and suck blood from their victims. The Amazonian pygidid catfish

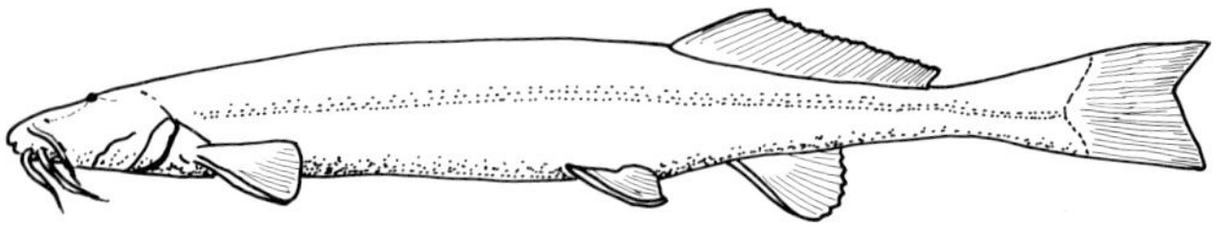

FIGURE 16.27
A representative of the family Trichomycteridae, a small (matchstick-sized) parasitic catfish capable of invading the human urethra and vagina, with potentially serious consequences. (Redrawn by Jan Hynes, after NELSON 1984.)

(*Urinophilus* sp. according to HALSTEAD [1988], but of the genus *Vandellia* according to BREAULT [1991]) (common names *candirú*, *canero*, *carnero*, or urethra fish) (Figure 16.27) can invade human urethral and vaginal openings, sometimes causing wounds that require surgical intervention (GUDGER 1929; BREAULT 1991).
Sting management: manage as for a stonefish sting {see page 378}.

ORDER SCORPAENIFORMES (MAIL-CHEEKED FISH)

The head and body are usually spiny or with bony plates; the pectoral fins are usually rounded (Figure 16.28); the membranes between the spines of the dorsal fin are usually deeply serrated. All members of this group are considered to be of potential medical importance because of their highly developed venom apparatus (see below).

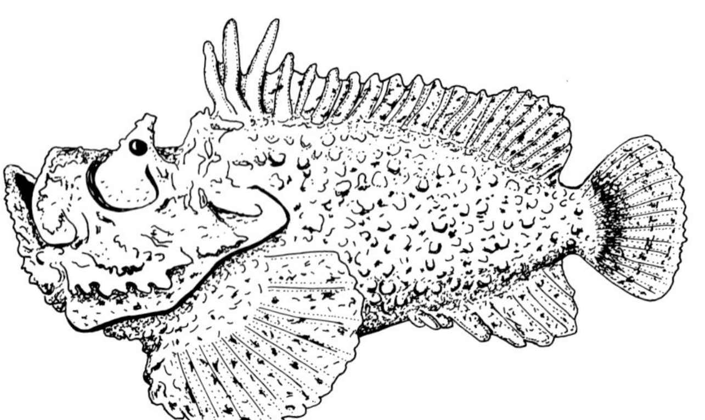

FIGURE 16.28
The estuarine stonefish *Synanceia horrida*. The third dorsal spine may reach 40 mm in length. (Redrawn by Jan Hynes, after McKAY 1987.)

SCORPAENIFORMES CAUSING CLAIMED FATALITIES

FAMILY SYNANCEIIDAE (STONEFISHES)

There are six genera with a total of 10 species (NELSON 1984) but only the genus *Synanceia* will be considered here. This genus has been extensively studied (e.g. ENDEAN 1961).

The estuarine stonefish *Synanceia horrida*= *S. trachynis* (Figure 16.28 and Plate 16.10) is the most venomous fish known. It is spread throughout the Indo-Pacific, Indian Ocean, China and Australia. In Australia it is found from the northwestern part of Western Australia, around to northern New South Wales. The reef stonefish *Synanceia verrucosa* (Figure 16.29, page 375 and Plate 16.11) inhabits coral reefs and muddy or sandy areas around India, Indonesia, China, the Philippines, the Red Sea and Australia (HALSTEAD & VINCI 1988).

Accurate knowledge of the prevalence of stonefish is hindered by their magnificent camouflage (Plate 16.13). These fish typically possess bony and fleshy facial protuberances, with algae-encrusted and warty skin* (CAMERON 1981). This makes them extremely difficult to see, when they lie partially buried in or between rocks, sand or coral rubble. They will not move when approached, and the dorsal spines are usually not raised unless the fish is touched.

It is not widely appreciated that stonefish are plentiful in the sea, but because of their superb camouflage they are not commonly sighted. (This may even apply to the animal that delivers a sting being stepped upon in shallow water.) Similarly, as records of stings show (SUTHERLAND 1983; INTERNATIONAL CONSORTIUM for JELLYFISH STINGS 1993), stonefish stings are themselves not rare.

The large pectoral fins used as scoops for digging out hiding places in the mud or sand make them clumsy swimmers; and yet they are capable of rapid movement over short distances. The

* Most of the warts on the skin of the stonefish and other 'stonefish types' are in fact skin tubercle glands which contain toxins (CAMERON 1981).

estuarine stonefish has small eyes which are elevated (Figure 16.28, page 374), while the reef stonefish has eyes that are not as elevated (Figure 16.29). Prey such as shrimps and small fish are sucked in very rapidly. Although they may be preyed upon by sharks and rays, their defensive toxins act to protect them.

Stonefish are not always covered in a coating of algae and other organisms, and particularly when found at depth may present a relatively clean, even modestly coloured appearance, while blending with their background (Plate 16.10). The toxins found within the tubercle glands (see above) may allow for the periodic shedding of the the fouling organisms and they have anti-protozoal actions.

Apparently stonefish are capable of surviving for more than 24 hours out of the water in a moist environment (MARETIĆ 1988).

Venom apparatus

Stonefish possess 13 dorsal fin spines each with paired venom glands (Figures 16.28 and 16.12). The three anal and two ventral fin spines are not functional. All spines of the stonefish are covered by a thick integumentary sheath. Dissection of the venom apparatus or rupture of the sheath exposes the spine and the paired venom glands (Plate 16.12). Although in Australia stonefish spines are commonly relatively short and stout with medially-located venom glands (Figure 16.33d, page 382), those of Singapore specimens are longer, especially the second and third dorsal spines, with more distally located venom glands, as shown in Figure 16.33, page 382) Compare with those of *Scorpaena* and *Pterois* types (Figures 16.33a and b, page 382). The third dorsal spine is usually straight and long, reaching up to 3 cm in Australia (ENDEAN 1961) and 4 cm in length in waters around Singapore (GOPALAKRISHNAKONE and GWEE 1993). The first three spines on the fin ray are angled forwards and upwards, while the others are angled slightly backwards. The spines partially enclose most of the spindle-shaped gland and fully enclose the tapered distal end of it. A tough fibrous sheath surrounds the glands (Figures 16.30, 16.31 and 16.33, pages 376, 377 and 382) (ENDEAN 1961; GOPALAKRISHNAKONE & GWEE 1993). Gland cells have an irregular membrane ranging from round to polygonal (hexagonal) in outline. Some of the cells contain granules, others contain a homogeneous matrix. Vacuoles may be present as well (ENDEAN 1961). Like other fish, stonefish gland cells disintegrate, releasing granules, and the glands are capable of regeneration (ENDEAN 1961).

The envenomation process

When the stonefish is stepped on, the integument is ruptured as the foot is punctured by the spine and the blockages at the distal end of the venom glands are cleared. As more pressure is applied by the stepping motion, the glands are compressed, conveying the venom granules and cellular material into the foot in an explosive burst (Figure 16.31, page 377). COLEMAN (1974) stated that a stream of venom from a venom gland could be shot to a height of 15–23 cm when the gland is under pressure. It is believed the spine(s) need to penetrate the victim's tissues only up to 0.6 cm in a juvenile fish for envenomation to occur, whereas for adult stonefish the penetration must reach 1 cm before venom delivery occurs (CAMERON 1988).

Clinical features of stonefish sting

The pain associated with stings by the stonefish is immediate and excruciating. Both juvenile as well as adult stonefish are capable of delivering serious stings. Most punctures are delivered when the stonefish are accidentally stepped upon. However, painful stings may be delivered to hands when stonefish are caught in nets or on fishing lines. The spines are tough and are capable of penetrating sandshoes or thongs.

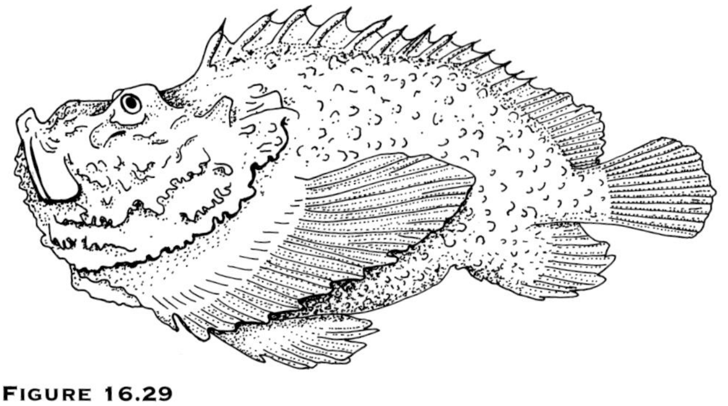

FIGURE 16.29
The reef stonefish *Synanceia verrucosa*. See also Plate 16.11. (Redrawn by Jan Hynes, after McKAY 1987.)

Although there remains some controversy about the frequency with which stonefish produce human fatalities (FLECKER 1956; CAMERON 1988) {see below, and Chapter 3, page 69}, there is none concerning the absolutely devastating agony of stonefish envenomation. It belongs to one of life's most dreadful experiences. An eloquent description at the turn of last century by an Australian Aboriginal, who spoke limited English, is quoted by SUTHERLAND (1983) from *Confessions of a Beachcomber* by E.J. BANFIELD (1908): 'Suppose that fella nail go along your foot, you sing out all a same bullocky all night. Leg belonga you swell up and jump about. Bingie [belly] belonga you, sore fella. Might you die.'

The pain is immediate, local, savage and totally incapacitating. Victims writhe about incoherently. The envenomated part displays at least one puncture wound, which develops a classic bluish discolouration around it (Plate 16.14). Movement of the limb will be resisted, and the pain may spread to involve the entire limb and regional lymph glands. Regional ischaemia around the puncture wound has not been observed by the editors.

Despite the massive local oedema, which can spread to involve an entire distal limb, and the unhealthy local tissue discolouration, tissue necrosis (in the absence of secondary infection) in stonefish envenomation seems conspicuous by its absence — this is in contrast to stingray envenomation {see page 367}. Local capillary refill is normally retained.

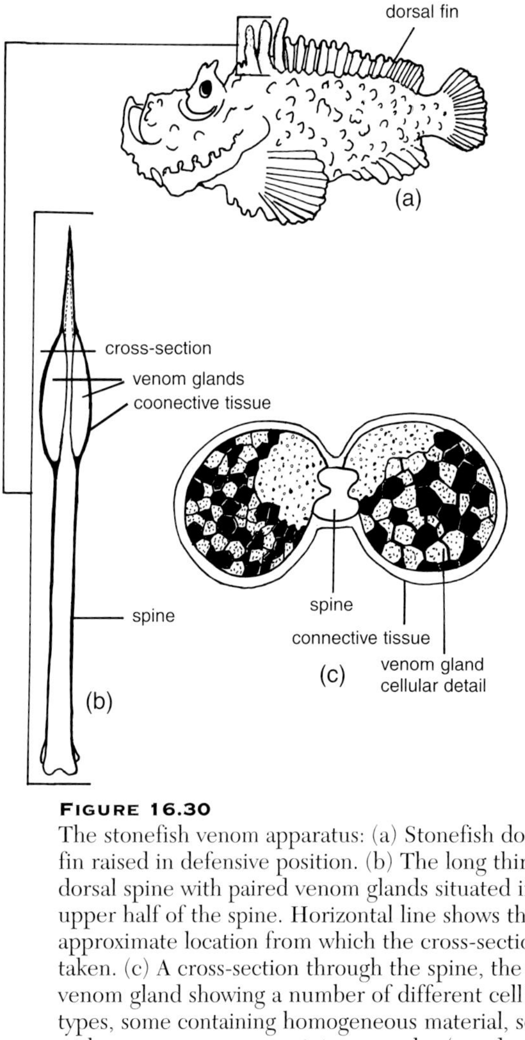

FIGURE 16.30
The stonefish venom apparatus: (a) Stonefish dorsal fin raised in defensive position. (b) The long third dorsal spine with paired venom glands situated in the upper half of the spine. Horizontal line shows the approximate location from which the cross-section is taken. (c) A cross-section through the spine, the venom gland showing a number of different cell types, some containing homogeneous material, some with compartments containing granules (not drawn to scale). The granules are released into intercellular spaces as part of holocrine secretory activity. Granules along with cell debris are released in a holocrine secretion when the glands are ruptured. (Drawn by Jan Hynes from specimens, sections and drawings.)

Systemic effects and possible fatalities

No Australian fatalities from stonefish stings have been confirmed to date. An anecdotal account of a Thursday Island death ('within a few days') of a prominent medical officer in 1915 attributed to a stonefish sting on the foot (*The Post-Graduate Bulletin*, March 1963, page 48), has been identified as inaccurate (FLECKER 1956). The international literature repeats descriptions of 'shock', 'cardiovascular collapse', 'cardiac arrhythmias', 'convulsions' and even 'haemorrhagic pulmonary oedema', but with occasional exception (SMITH 1957; SUTHERLAND 1983) specific details are never supplied. SMITH's (1957) description of human fatalities in the Seychelles and Mozambique included agonising pain following envenomation in the foot (both cases), followed by rapid 'collapse', central cyanosis, pulmonary oedema ('frothing at the mouth'), unconsciousness, and death 'well within the hour'. Direct intravascular deposition of the venom was postulated by Smith in the Mozambique case, as the victim's foot showed no local reaction. One may wonder if anaphylaxis was possible. Also, does the venom of the animals in Smith's report differ (qualitatively and/or quantitatively) from the Australian species?

PHOON and ALFRED (1965) referred to deaths in the Malaysian peninsula region. Records of deaths days or months following envenomation (Le JUGE 1871; TASCHENBERG 1909) must raise suspicions of non-specific complications such as infection. Secondary tetanus infection is certainly a theoretical possibility from a stonefish wound (SUTHERLAND 1983). Despite the superb description by an Aborigine, quoted above (a description compatible with delayed death from secondary infection), the editors have been unable to verify specific instances of other acute toxic deaths from stonefish. Also, considerable first-hand experience of two of the editors (JAW and PJF) in the management of Australian stonefish envenomations has failed to reveal a single case in Australia where life-threatening symptoms or signs developed. Vaso-vagal syncope has been observed in case of a 17-year-old male (WILLIAMSON 1974, unpublished observations). Although it seems certain that no Australian fatality has ever been recorded, the Australian media continue to describe Australian stonefish as 'lethal'. We agree with the earlier observations of FLECKER (1956), who stated 'Press agents are always eager to report the injury for its news value, not being very particular about the truth.'

Stonefish venom

Synanceia venom was the first scorpaenoid venom to be studied (BOTTARD 1889; DUHIG & JONES 1928), and Australian workers have since sustained a distinguished place in this research (WIENER 1959a; 1959b; ENDEAN 1961; AUSTIN *et al* 1965). Studies have been made of the venom from all species of *Synanceia* (SAUNDERS 1959; AUSTIN *et al* 1965; KREGER 1991). The venom obtained from members of the genus contains heat-labile and antigenic protein toxins (WIENER 1959a) with molecular weights up to 158 000, and which exhibit myotoxic, neurotoxic, vascular-leakage and myocardial effects in

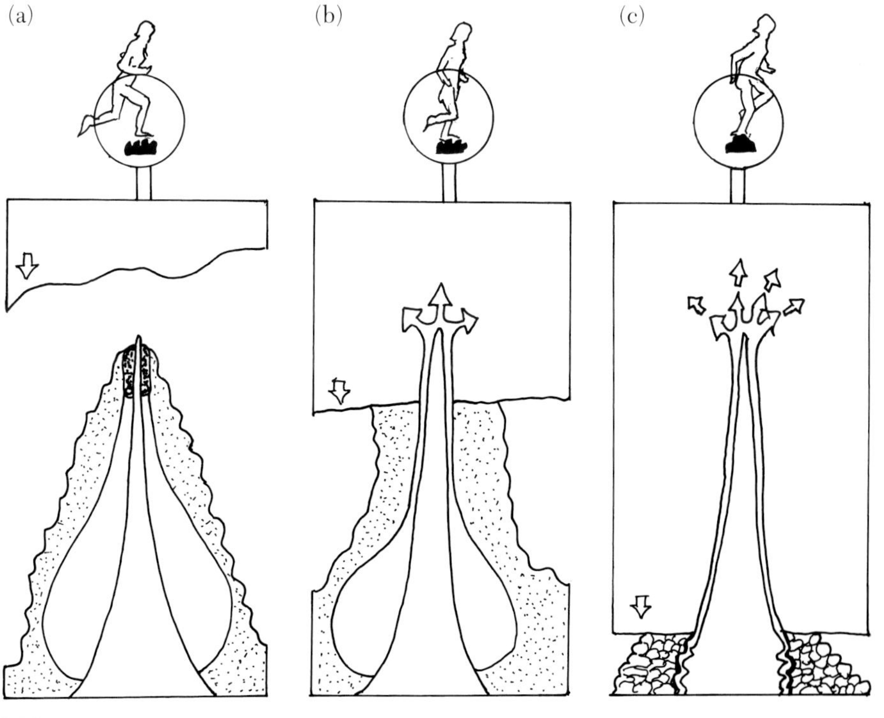

FIGURE 16.31
Mechanism of envenomation by stonefish: (a) Pairs of venom glands each with a blocked duct-like extension leading to the top of the spine; the whole spine is covered by a warty integumentary sheath. (b) The action of stepping on a stonefish strips the integumentary sheath from the spine; the blockage is cleared and the gland is compressed. (c) The venom and cellular material shoot into the wound with explosive force. (Redrawn by Rebekah Lewis, after UNDERHILL 1987.)

experimental animals, but no effects on blood coagulation (WIENER 1959a; SUTHERLAND 1983; AUSTIN *et al* 1965; KREGER 1991). The much-studied haemolytic effects are of no clinical significance (SUTHERLAND 1983). The haemolytic, lethal and vascular permeability-increasing effects are probably produced by the same molecule, which may be more accurately described as 'cell membrane damaging', or cytolytic (KREGER 1991). Venom in the venom glands retains its toxicity for at least several days after the fish has died (DUHIG & JONES 1928) — a point of practical significance to handlers. (The animal itself may remain alive for a day or more out of water.)

The intra-peritoneal LD_{50} in mice is 1.3–2.0 mg/kg (WIENER 1959b; KREGER 1991). Intravenous (IV) injections into mice were fatal within 30 minutes, but animals surviving beyond this time did so indefinitely (SAUNDERS 1959). In a guinea pig of 250 g body weight a subcutaneous injection of 4 mg causes muscular weakness, coma and then death within three hours (WIENER 1959a). Intravenous injections of either *Synanceia verrucosa* or *Synanceia horrida* venom into rabbits produced systemic hypotension, respiratory arrest, atrioventricular block and ventricular fibrillation (SAUNDERS 1959) (these two venoms are now considered to be identical [HALSTEAD 1988]). Life was not extended in these animals by ventilatory support. Hypotension and respiratory failure result from IV injections of *Synanceia horrida* venom into the dog and the chicken (WIENER 1959a). In mice, the postmortem showed haemorrhagic lungs of a gelatinous consistency (WIENER 1959a). The venoms probably possess myotoxic properties in animals, which seriously affect diaphragmatic musculature before that of the myocardium (AUSTIN *et al* 1961). Structural studies of the venom molecules have not been documented.

MANAGEMENT OF A STONEFISH STING AND OF VENOMOUS FISH STINGS IN GENERAL

Editors' note: The description that follows is full, and because the features of stonefish envenomation typify the worst scenario likely to be encountered in most venomous fish stings, this description should be taken as a model for any venomous fish sting management.

FIRST AID

TRADITIONAL REMEDIES

(FROM: ABORIGINAL COMMUNITIES OF THE NORTHERN TERRITORY OF AUSTRALIA 1988)

Australian Aboriginals were well aware of the effects of the stonefish sting, and the excruciating pain from an envenomation. The symptoms of the stings are vividly enacted in the traditional aboriginal dances or corroborees. Two treatments for the pain of marine stings such as stongray injuries include application of *Ipomea pes-caprae*, or goats foot, a succulent plant commonly found growing on sand. Containing a mild antihistamine, it apparently has antihistaminic, anti-inflammatory, haemostatic, antiseptic and anodyne properties. The leaves are prepared by heating on hot stones until soft and brownish. The preparation is applied and changed until the pain is gone. This treatment is still used around Darwin and the Gulf of Carpentaria, and it is widely used throughout tropical parts of the world including Mexico, Guam, India, New Guinea (PAIJMANS 1976), Vietnam, Thailand (APPLIED SCIENTIFIC REASEARCH CORPORATION OF THAILAND 1977) and Tanzania.

Another treatment for pain relief after stingray and stonefish stings is the chewing of and application of wood ash from the white mangrove *Avicennia marina*. A solution of the wood ash is alkaline, astringent and antipruritic. It is not known whether pain relief is achieved through denaturation of the venom or by local analgesia. It is used in Darwin and the Gulf of Carpentaria, Mozambique and Zaire. European fishermen have used this plant as primary treatment of stingray stings for pain relief, prior to hospital treatment. {See also ALAM 1995 in Chapter 26}

FIRST AID: CONVENTIONAL MEDICINE

Necessary resuscitation measures for any effects on consciousness or breathing {see Chapter 5} must always take priority, even though, as stated above, this need has not yet been encountered for Australian stonefish envenomation by any of the editors. However, first aid relief of the conscious patient's agony is of the greatest importance and should be regarded as a matter of urgency.

The first aid principles espoused here apply equally well to any venomous fish sting {see also Chapter 5} and are summarised thus:

1. Resuscitation as required.
2. Plunge the envenomated part into comfortably hot (not boiling!) water. Do not ask the patient to gauge the temperature of the water. The patient will be too distraught. Rescuers should do this themselves first. This measure will usually reduce (not eliminate) the pain but will be thankfully received by the suffering patient. Maintenance of immersion, and of the water temperature, is necessary to prevent full return of the pain.
3. If the skills and facilities are available on site (and in the absence of the immediate availability of specific antivenom [see below] in the case of a confirmed stonefish sting), local infiltration of the envenomated area with no more than 2 mg/kg body weight of plain lignocaine or bupivicaine, will also ease the pain. (Normal care must be taken not to administer an accidental intravascular injection.)
4. Seek skilled medical aid. All but the most trivial of venomous fish stings warrant early medical contact.

Further comments

- Never apply an arterial tourniquet to a venomous fish sting.
- Do not apply compression/immobilisation bandaging to wounds from any venomous fish stings (SUTHERLAND 1983). Retardation of venom at the site will only increase local pain and tissue damage.
- Elevation of the envenomated limb may not be possible if hot water immersion is in use. Immersion takes priority.
- With the exception of stingray injuries {page 367} and catfish spines {page 373} (McKINSTRY 1993), it is relatively unusual for a wound from a venomous fish to have a visible foreign body (e.g. a broken-off piece of spine) retained. (This is in contrast to spine injuries from sea urchins or the crown-of-thorns sea star.) However, such a fragment may be present, hidden in the depths of the wound, and will be revealed only by a competent medical examination and imaging {see below}.
- Do not incise the puncture wound, unless informed medical opinion has been sought. 'Amateur' incision is seldom of benefit and carries risks.
- Do not inject potassium permanganate or other irritant solutions into the wound. Although previously recommended (EVANS 1943), their efficacy is unclear and they may produce further tissue damage. Although emetine hydrochloride is on record as providing impressive pain relief in stonefish stings (WEINER 1958), it is now safer to use hot-water immersion and/or local anaesthetics.

MEDICAL MANAGEMENT

Again, the principles of clinical management described below, allowing for specific minor variations, apply equally well to any venomous fish sting.

1. Adequate pain relief is an urgent requirement. This is most adequately achieved by specific antivenom administration in the case of a stonefish sting only (SUTHERLAND 1992), or by a regional anaesthetic block (e.g. ankle block, wrist block). Correctly employed local anaesthesia can be dramatically effective (Plate 16.14) {see also platypus envenomation, Chapter 25, page 438}.
2. If the identity of the offending animal can be confirmed with certainty to be a stonefish (often difficult or impossible [FLECKER 1956; WILLIAMSON 1985]), and the patient has sig-

nificant pain and/or local effects, specific Commonwealth Serum Laboratories' stonefish antivenom (WIENER 1959b; SUTHERLAND 1983) should be administered (see below). This will also produce effective analgesia (WILLIAMSON 1985; SUTHERLAND 1992). The efficacy of this antivenom for other spine injuries from venomous fish is possible but unclear at present.

3. Contrast radiography is advised to detect or exclude any retained foreign body.
4. The decision to incise and explore the wound will vary according to the features of each individual case. Skilled surgical exploration under appropriate anaesthesia, performed early, with removal of any foreign material and the successful washout of deposited venom, may improve outcome. However, in the majority of cases this measure is not necessary, and is contraindicated for the unskilled.
5. Tetanus prophylaxis and wound toilet are advisable.
6. The management of delayed, severely contaminated or infected wounds requires surgical wound debridement and toilet, culture of infected material, careful tetanus prophylaxis, appropriate antibiotic therapy and follow-up care. Awareness of the special requirements for marine pathogens is important (AUERBACH *et al* 1987) {see Chapter 4, page 96}.
7. Bed rest and immobilisation of the injured part is helpful.

Further comments

- Attempts to control the acute pain of stonefish (and some other venomous fish) stings by the administration of systemic narcotic analgesics alone is a strictly medical matter, may be ineffective (WIENER 1958) and may produce respiratory depression and impaired consciousness.
- Suspect a retained foreign body or a marine pathogen {see Chapter 4, page 96} in any non-healing venomous fish wound being actively treated.
- Venomous fish wounds not given early medical treatment may remain painful and disabling for months (FLECKER 1956; SUTHERLAND 1983).

ANTIVENOM THERAPY

Australia's Commonwealth Serum Laboratories (CSL) in Melbourne produce a specific stonefish antivenom, which is a hyperimmunised horse antiserum prepared against the venom from the dorsal spines of *Synanceia horrida* (WIENER 1959b; SUTHERLAND 1983; 1992). The antivenom has been shown to destroy the venom's *in vitro* lethal, vascular permeability-increasing and haemolytic properties (KREGER 1991). One ampoule of 1.5–3.0 ml will neutralise *in vitro* 20 mg of venom (SUTHERLAND 1983). Its clinical efficacy for both analgesia and reduction of venom-induced tissue damage is now well established (WEINER 1965; WILLIAMSON 1985; SUTHERLAND 1992; FENNER *et al* 1993).

At present levels of understanding, this antivenom should not be routinely administered for other than stings from *Synanceia* species. However, there is circumstantial evidence that it may be effective for serious or non-responding stings from other scorpionfish (RUSSELL 1965; HALSTEAD 1988).

CSL stonefish antivenom is recommended only for intramuscular (IMI) use. Premedication is recommended (SUTHERLAND 1992), and specifics are discussed elsewhere in this text {see Chapter 5, page 114}. Patients with an allergy history require especially careful premedication considerations. Facilities for resuscitation and the management of anaphylaxis must be immediately to hand. The recommended stonefish antivenom doses (SUTHERLAND 1983) are:

One or two puncture wounds:
give 1 ampoule IMI (2000 units)
Three or four puncture wounds:
give 2 ampoules IMI (4000 units)
More than four puncture wounds:
give 3 ampoules IMI (6000 units)

There is evidence, especially following the use of a large volume of antivenom, that a short course of prophylactic steroids (e.g. 100 mg hydrocortisone or equivalent, daily for four days), commencing with the antivenom administration, may reduce the risk of delayed serum sickness (SUTHERLAND 1992).

PREVENTION OF VENOMOUS FISH STINGS

The principles of prevention of stonefish envenomation apply to any venomous fish sting.

1. Be aware that a stonefish will usually not be seen until a sting occurs, and even then maybe not! Their camouflage is total.
2. In the event that a stonefish is caught in a net while dragging for baitfish, never use bare hands to release the fish. The fish should be held down firmly with a solid stick while it is being untangled, then scooped up in a bucket or landing net. The authors suggest throwing the fish back into the water as it will swim away. (Common sense is required in shallow waters, heavily populated with stonefish, where people commonly wade).
3. If handled, stonefish should be lifted only by placing one hand underneath the body and pulling the tail towards you with the free hand (GRANT 1995, personal communication). Alternatively, the stonefish may be picked up by the tail (STABLUM 1995, unpublished observations). However, be forwarned that they will raise their spines in response. Since they are slimy and tend to 'flick' their heavy bodies when being held in this way, a dangerous situation could arise if the fish falls on someone. (STABLUM 1995, unpublished observations)
4. Always wear thick-soled shoes or heavy-duty boots for wading in the shallows, pushing boats or fossicking on sea shores or intertidal reefs. The spines can penetrate thin shoes.
5. Never reach for a submerged 'rock' with bare hands or feet. Touch only what you or your guide identify as safe.
6. Do not search by hand into hidden crevices in the sea.
7. Do not 'tease' marine life or handle animals unnecessarily.
8. Do not jump barefoot from a boat into the shallows unless the bottom is clearly visible and free of marine life.
9. After filleting or cleaning any fish, always dispose of the remains safely. Spines may be capable of injecting potent toxins for many hours, or even days after the spines are removed, depending upon prevailing conditions and temperatures. Fish spines themselves could take several months to break down. They should be considered to be as dangerous as broken glass or any other sharp object, capable of producing severe infection if the skin is pierced.
10. Wear a bracelet alerting your companions if you have a serious allergic condition and/or are on potent medications.
11. Take particular care during trawler sorting operations, wearing gloves when practicable.

HIGHLY VENOMOUS FISH RELATED TO STONEFISH

GENUS *INIMICUS* (BEARDED GHOUL, DEVIL FISH, DEMON STINGER)

SOUTHCOTT (1975) stated that the genus *Inimicus* contains a number of highly venomous species with venom effects similar to those from *Synanceia*. He also stated that this genus is of considerable importance in the Indo-Pacific region (e.g. *Inimicus japonicus* of Japanese waters). In Australia two species are commonly caught by trawlers, from Shark Bay (Western Australia) to Moreton Bay, southern Queensland. GRANT (1985) states that bearded ghouls prefer outcrops of rocks and reefs, and lie almost completely buried in the sand. In addition, he states that they lie motionless in the

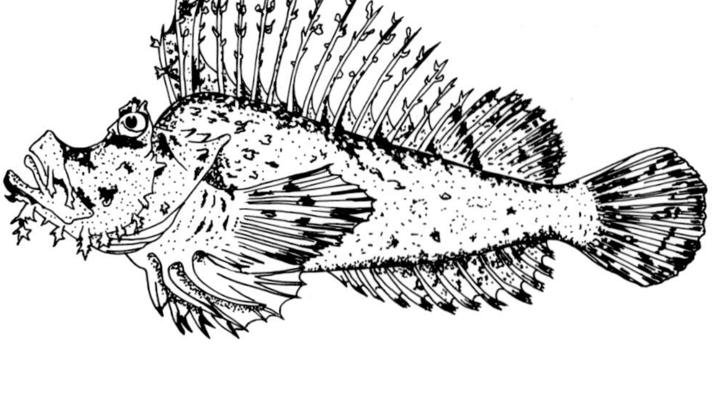

FIGURE 16.32
The bearded ghoul *Inimicus caledonicus*. (Drawing by Jan Hynes, after SOUTHCOTT 1975)

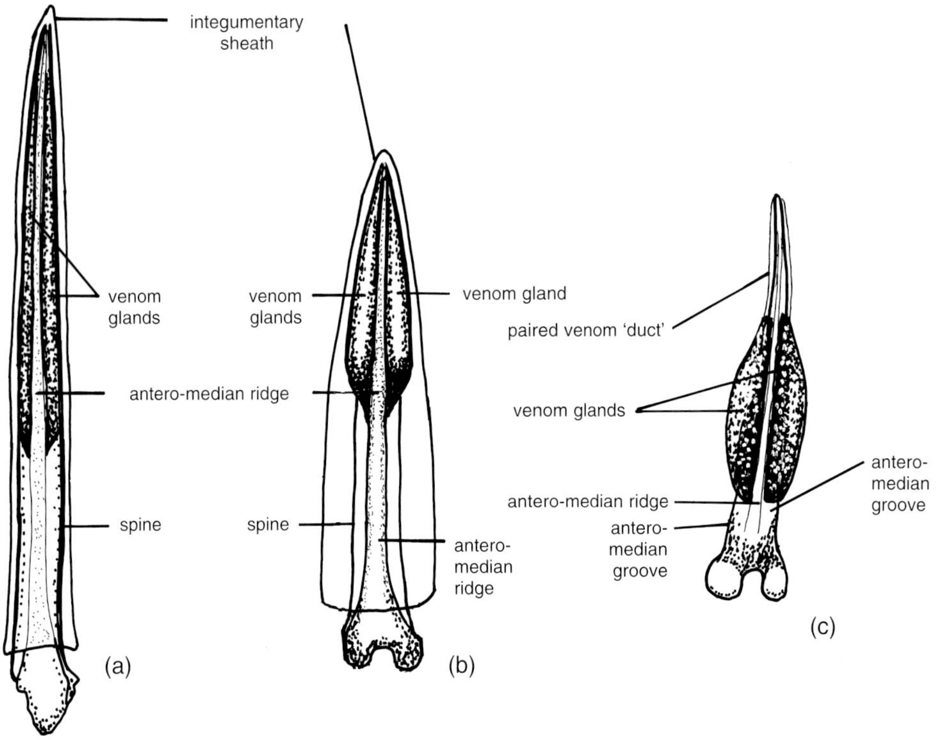

FIGURE 16.33
The three basic scorpaeroid spine types, represented by the genera *Pterois*, *Scorpaena* and *Synanceia*: a. *Pterois* type — spine long and thin, with thin integumentary sheath and thin well-developed venom glands without duct-like extensions. b. *Scorpaena* type — spine moderately long and heavy, with a moderately thick integumentary sheath, moderately sized, well-developed glands, without duct-like extensions. c. *Synanceia* type — spine usually relatively short and stout, with a moderately thick integumentary sheath, large, well-developed glands and well-developed duct-like extensions.(Redrawn by Jan Hynes, after HALSTEAD 1988.)

sand with dorsal spines erect when the area is approached by divers.

The body of *Inimicus* (Figure 16.32, page 381) is covered in warts or skin lumps, many skin tubercle glands similar to the stonefish. Short fleshy filaments cover the upper portions of the fins and these form a 'beard' in the lower part of the head and cheek area. There are from 12–18 dorsal spines and the anal fin has two spines. There are eight species (NELSON 1984). The genus *Choridactylus* is closely related.

FAMILY SCORPAENIDAE

(SCORPIONFISH AND ROCKFISH)

Scorpionfish are distributed in all oceans, from polar waters to the equator though mainly throughout tropical and temperate seas (HALSTEAD & VINCI 1988). According to RUSSELL (1971) at least 80 members of the family Scorpaenidae have been implicated in human envenomations.

All scorpionfish possess a bony ridge below the eye, with a heavily-armoured spiny head. They may take on the appearance of the rocks and plants around which they hide, and some even change their colouration to match their environment. Scorpionfish are encountered in different habitats. They may be generally slow-moving swimmers (e.g. the butterfly cod, Plate 16.15), or they may be more stationary (e.g. bullrouts, Figure 16.34, page 384). Some scorpionfish are considered to be good for human consumption. GRANT (1985), for example, considers that the red stinger or black-spot stingfish, *Scorpaena ergastulorum* (see below and Figure 16.2, page 356) caught off the Queensland coast, is an excellent, white-fleshed pan fish, regardless of its venomous spines.

VENOM APPARATUS

The dorsal spines, the anal spines and the ventral spines usually are involved in envenomations. In these spines the glands are usually situated beneath the epithelium, adjacent to the more distal region and mid-region of the spine, enclosed by connective tissue (Figure 16.33). There is no duct to the venom gland and the venom is released when the integumentary sheath is torn away. The family Scorpaenidae is generally thought to consist of two arbitrary groups, *Scorpaena* and *Pterois*. These groups are based on the structure of the fin spines, the thickness of the integumentary sheath, and the extent of development of the venom glands. The structure of the venom organs of these groups may be contrasted with the structure of those of members of the family Synanceiidae (Figure 16.33, page 382). In Australia, members of the family Synanceiidae usually possess short, stout fin spines apart from the third dorsal spine (ENDEAN 1961; GOPALAKRISHNAKONE & GWEE 1993). In Singapore, the second and third dorsal spines are the largest and straightest. The spines generally taper to a sharp tip. The integumentary sheath is very thick and warty, the venom glands are very large and highly developed, and distally, there is a very well-developed duct-like process which extends to the spine tip (Figures 16.30 and 16.33, pages 376 and 382).

SCORPAENIDAE CAUSING FATALITIES

Fatalities have been reported as a result of injury from spines of butterfly cod (genus *Pterois*) and related species.

FAMILY SCORPAENIDAE PTEROIS GROUP (BUTTERFLY COD, LIONFISH, TURKEYFISH, ZEBRAFISH)

Species of *Pterois* and *Dendrochirus* inhabit coral reefs of the Red Sea, the Indian Ocean, China, Japan, Africa, Australia, Melanesia, Micronesia and Polynesia (HALSTEAD & VINCI 1988).

These fish are beautifully coloured and patterned, with delicately frilled fins and spines, and are popular with underwater photographers (Plate 16.15). They are found around jetty piles and boat wrecks, where they slowly patrol their territory. They normally do not show fear of humans and these fish have been reported to drive their dorsal spines into the fingers of aquarium enthusiasts or divers when nets are not used to catch them. They are popular saltwater aquarium fish the world over.

Venom apparatus

The venom apparatus consists of 13 dorsal spines, three anal spines and two pelvic spines, all enclosed by colorful integumentary sheaths. Glandular tissue closely adheres to the anterolateral grooves of the spines. The dorsal spines are the longest of the spines, and the spine grooves extend their entire length. The glandular tissue of the dorsal spines is found in three-quarters of the spine length. The venom is released by mechanical pressure when the integumentary sheath is ruptured. Table 16.1 lists the comparative aspects of the venom organs of the *Pterois* group and the *Scorpaena* group.

TABLE 16.1
GENERAL CHARACTERISTICS OF THE VENOM ORGANS OF THE PTEROIS GROUP AND SCORPAENA GROUP OF THE FAMILY SCORPAENIDAE (SCORPIONFISH AND ROCKFISH)

***Pterois* group:**	
	Fin spines: elongate, slender, thin
	Integumentary sheath: thin
	Venom glands: thin and well developed without a well-developed extension leading to the spine tip
***Scorpaena* group:**	
	Fin spines: moderately long and heavy
	Integumentary sheath: moderately thick
	Venom glands: moderately sized and well developed without a well-developed extension leading to the spine tip

Adapted from HALSTEAD & MITCHELL 1963

Envenomation

Envenomation by lionfish may be severe. In fact, according to MARETIĆ (1988), the symptomatology is more severe than that from stingrays, weeverfish and stonefish [the editors doubt this]. Hardly surprisingly, aquarists are among the most commonly envenomated (KIZER *et al* 1985). The sting produces immediate and intense skin pain with severe lasting oedema and local discolouration, blisters and necrosis. Claimed generalised symptoms uncommonly include respiratory distress and 'shock'. Deaths from 'zebrafish' are claimed (RUSSELL 1971; MARETIĆ 1988) but must be rare. Philippine zebrafish sting deaths seem likely (HERRE 1952).

A five-year study of envenomations by scorpionfish in San Francisco showed that 45 out of 51 cases of scorpionfish envenomations were due to envenomation by lionfish. Victims experienced either complete or moderate relief from stings when hot water was applied to the site of the wound (KIZER *et al* 1985).

Pterois venom

Only limited study of *Pterois* venoms has been carried out, but in general they are believed, along with *Urolophus* (stingray), *Trachinus* (weeverfish) and *Scorpaena* venoms, to be broadly similar to stonefish venoms (RUSSELL 1965). Envenomations should be managed accordingly. A (polyvalent?) horse serum antivenom for scorpionfish stings (also used for weeverfish, see below), for intravenous use, was available from the Institute of Immunology, at Rockefellerova 2, Zagreb, (THEAKSTON & WARRELL 1991). but production of all fish antivenom from this institution ceased in 1994 (MEIER 1995).

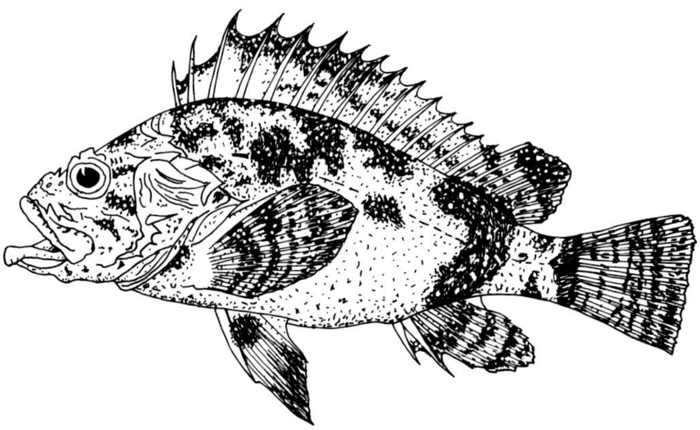

FIGURE 16.34
The scorpionfish *Notesthes robusta* (bullrout), common in estuarine and freshwater creeks. Like other scorpionfish it has venomous fin spines. When approached, it raises its 15 venomous dorsal spines, often resulting in deep multiple punctures of the foot. (Drawn by Jan Hynes, after SOUTHCOTT 1975.)

SOME OTHER SCORPAENIDS CAUSING MORBIDITY

FAMILY SCORPAENIDAE *SCORPAENA* GROUP

Scorpaenids belonging to this group include a large number of genera that are well adapted to living on the sea bottom. NELSON (1984) reports that there are about 60 genera with about 310 species.

MARETIĆ (1988) stated that in the Pacific *Scorpaena* stings occur during fishing; in the Mediterranean, it is mostly women who are stung by scorpionfish when they are scaling fish. RUSSELL (1965) considered the sting from *Scorpaena halleri* to be more severe than that of the weeverfish *Trachinus vipera* (see below).

Symptoms following *Scorpaena* spp. stings include radiating pain, nausea, chills, fever, increased heart rate, sweating, pallor, dyspnoea, restlessness, vomiting or fainting, migraine; envenomation symptoms differ from species to species. MARETIĆ (1975) noted wandering of the pacemaker, ST segment depression, and prolongation of the PQ interval, electrocardiographically, in animals.

Some Australian species which illustrate many typical features of the group, are included here. The venom apparatus of the bullrout is considered in more detail below. In Australia *Scorpaena ergastulorum* (Figure 16.2, page 356) commonly known as the red rock cod, cardinal fish, red stinger or black-spot stingfish, is a red fish, up to 38 cm in length, with a black spot on the dorsal fin membranes between spines 9 and 11. This fish is found in waters off south-eastern Australia. Fishermen may receive minor scratches from these fish, with soreness that may last up to a few days. Punctures, however, can cause severe stings.

WATKINS (1969) (in SOUTHCOTT [1975]), described immediate and almost 'unbearable pain'. If the wound is in the hand, the pain spreads up the arm to the chest over a half-hour

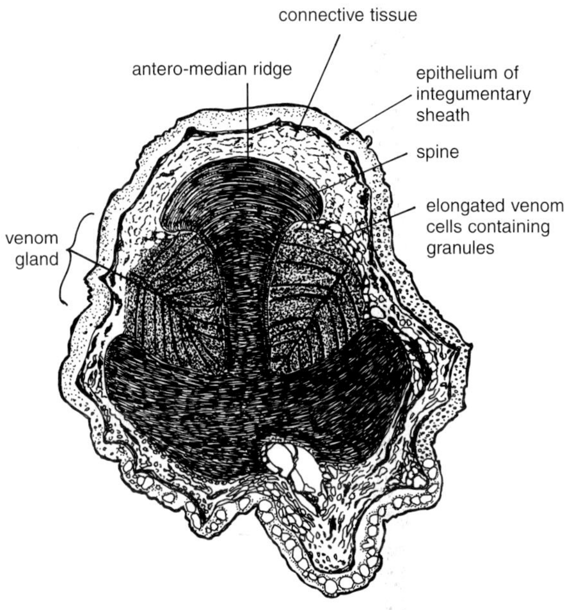

FIGURE 16.35
The transverse section through a spine from a '*Scorpaena*-type' venom apparatus. Note the glands containing elongated granulated cells present within the anterolateral grooves of the spine. The entire spine is shown in Figure 16.33b, page 382. (Drawing by Jan Hynes.)

time period. Swelling of the affected limb occurs within 10 minutes. General symptoms that may occur at the same time include pallor, writhing, vomiting and prostration. Swelling lasts up to two days. GRANT (1978) describes immediate severe pain in serious stings, hot and cold flushes, with apparent paralysis and swelling of the affected limb developing within 10 minutes. He has also recorded nausea, laboured breathing and chest pains. Grant stated that these symptoms usually ease within 2 hours, though swelling may not subside for 48 hours.

Notesthes robusta (bullrouts or kroki)

Bullrouts (*Notesthes robusta* Günther) (Figure 16.34, page 384, and Plate 16.16) are virtually confined to the east coast of Australia, extending from the central Queensland coast to the Clyde River in southern New South Wales (LAKE 1971; HARRIS & PEARN 1987). They are examples of scorpionfish that have adopted a stationary mode of existence and are commonly found sitting on the sea floor in estuarine waters or among freshwater grasses. They have been reported to be present in freshwater as well as estuarine habitats and are often found in large numbers below weirs and dams (HARRIS & PEARN 1987).

Bullrouts are compressed in shape, but are sedentary, cryptic and generally solitary. They are usually a dull yellowish colour, with darker grey or brown markings, and difficult to see under natural conditions, as they blend in so well with weeds or stones.

Venom apparatus

CAMERON and ENDEAN (1966) described the anatomy of the bullrout venom apparatus. The fish possesses 15 spines on its dorsal fin, three spines on the anal fin and a single spine on each of the ventral fins (Figure 16.33, page 382). Each of these spines is associated with a paired venom gland. HARRIS and PEARN (1987) also describe long stout venomous spines on the upper jaw and operculum. The paired venom glands are partially enclosed within grooves in the spines (Figure 16.35). The venom gland ranges from a third of the spine to half the spine length and differs from the stonefish venom gland (the venom of which is forced out in a single burst [ENDEAN 1961]). The venom is believed to be proteinaceous in nature.

The dorsal spines, which are erected when the fish is accidentally stepped on by an unwary or careless wader, usually inflict severe pain. When the spine penetrates the skin the integumentary sheath is pushed to one side and the spine and venom glands enter the puncture. The venom cannot pass out of the more proximal portion of the gland as it is partially enclosed in the spine grooves and surrounded by collagen; thus the venom must enter the wound via the distal end of the gland, since this end is easily ruptured. The gland regenerates, at least partially, within six days (CAMERON & ENDEAN 1966).

Symptoms and signs

Human stings usually involve deep penetrations, many of which are multiple, and like a stonefish sting the pain is immediate and excruciating. However, unlike stonefish stings the pain will seldom sustain its severe intensity for

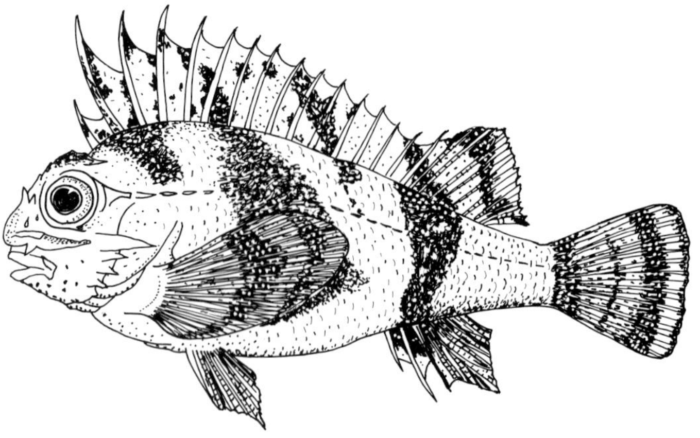

FIGURE 16.36
The fortescue *Centropogon australis*, which produces painful stings. (Redrawn by Jan Hynes, after SOUTHCOTT 1975.)

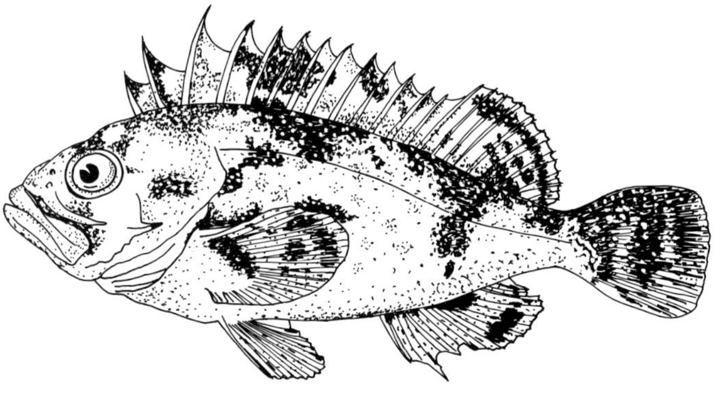

FIGURE 16.37
The cobbler *Gymnapistes mormoratus*. (Redrawn by Jan Hynes, after SOUTHCOTT 1975.)

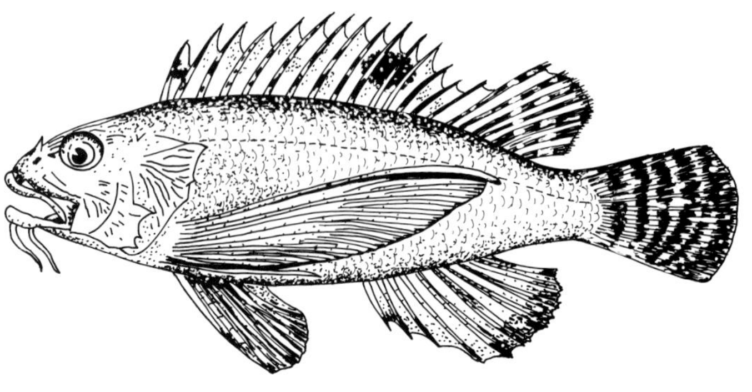

FIGURE 16.38
The waspfish *Hypodytes carinatus*. (Redrawn by Jan Hynes, after SOUTHCOTT 1975.)

longer than four to six hours or so. Similarly to a stonefish wound, the site around the bullrout puncture will become bluish.

Management of envenomation

As for stonefish stings, an earlier report claimed efficacy from the local infiltration of potassium permanganate into the wound from bullrout stings (SHORT 1957, in WIENER 1958), this is no longer recommended. To avoid bullrout envenomations, the wader should shuffle when walking so the fish can move away.

SOME OTHER HARMFUL AUSTRALIAN SCORPAENIDS

Centropogon australis (fortescue)

(Figure 16.36) This fish is found in coastal and estuarine waters, in Queensland and New South Wales. It is a small robust fish up to 15 cm in length with broad vertical bands coloured in shades of dark brown. The fin and head spines of both mature and immature specimens are responsible for causing painful stings (SOUTHCOTT 1975).

Gymnapistes mormoratus (cobbler)

(Figure 16.37) This fish occurs in shallow waters from Western Australia to Victoria and Tasmania. It is a small fish, up to 20 cm in length. The body is mottled, beige or yellow and dark brown. According to SOUTHCOTT (1975), these fish may be caught by seine or line fishing, and punctures of the skin by the spines of the dorsal and head regions can cause exquisite pain.

Hypodytes carinatus (ocellated wasp fish)

(Figure 16.38) This lilac-coloured fish, silvery-white below, has a large black spot ('ocellus') near the middle of its dorsal fin and a small dark spot at the front end of the same fin. It grows to about 17 cm in length and favours warmer waters of Queensland and Western Australia. Inhabiting deeper waters (30–70m) it is rarely captured except in a trawl net (SOUTHCOTT 1975). It has formidable dorsal, pectoral and anal fin spines.

Neosebastes thetidis (the thetis fish)

(Figure 16.39) This is a deep-water fish occurring widely around the southern half of

Australia, from Western Australia to Tasmania and New South Wales; it is usually encountered during trawling operations.

The genus *Neosebastes* contains a number of species that inflict painful stings. *Neosebastes thetidis* is robust, reddish brown with black mottling and grows to about 33 cm in length.

Neosebastes nigropunctatus (black spotted gurnard perch) is a slender species growing up to 36 cm in length. It is a reddish colour with black dots and has been recorded off Western Australia, Tasmania and South Australia. SOUTHCOTT (1975) reported that a fisherman was stung on the foot and ankle by five spines of one of these fish while it was flipping about in the bottom of the boat. The pain was severe and it spread from the ankle up to the knee. The pain lasted for six or seven hours; the ankle swelled in circumference by 1 cm. The tenderness and local oedema lasted for three months.

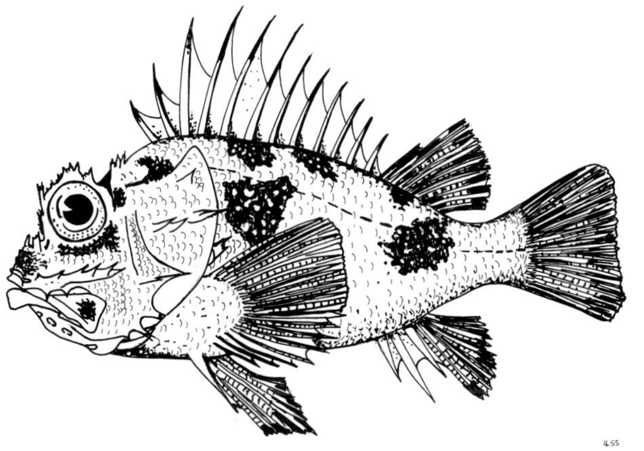

FIGURE 16.39
The thetis fish *Neosebastes thetidis*. (Redrawn by Jan Hynes, after SOUTHCOTT 1975.)

FAMILY GNATHANACANTHIDAE

Gnathanacanthus goetzeei (red velvet fish)

(Figure 16.40) This fish grows to about 28 cm in length. The colour of mature specimens is reddish with orange tints. Two large, separated dorsal fins and a large anal fin also characterise this species. It is found in deeper waters of the southern coast of Australia. SOUTHCOTT (1975) reported a painful sting of a fisherman in South Australian waters, from a dorsal spine of this species.

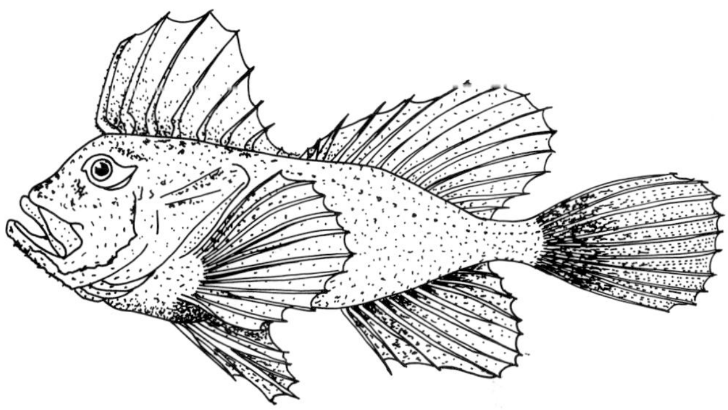

FIGURE 16.40
The red velvet fish *Gnathanacanthus goetzeei*. (Redrawn by Jan Hynes, after SOUTHCOTT 1975.)

ORDER PERCIFORMES (PERCHLIKE FISHES)

Fatalities as well as considerable morbidity have been reported from the European weeverfish, genus *Trachinus*.

FAMILY TRACHINIDAE (WEEVERFISH)

These are all marine. Their precise world distribution is still not clear, but they are most common in the Mediterranean Sea, Black Sea and in the Pacific off the coast of Chile. Weeverfish also exist along the eastern Atlantic

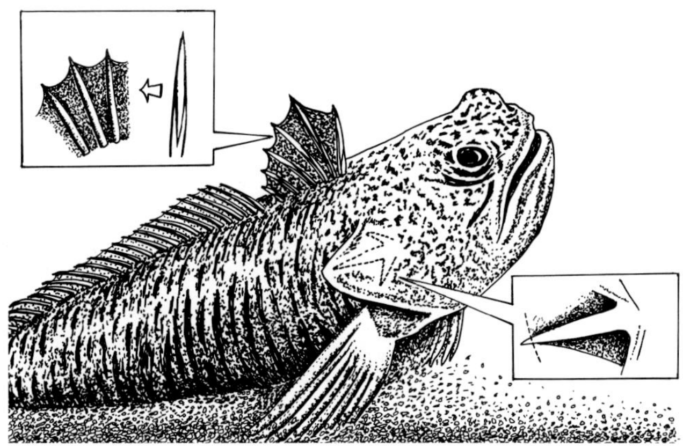

FIGURE 16.41
The weeverfish *Trachinus draco*, showing detail of dorsal and opercular spines. (Redrawn by Jan Hynes, after HALSTEAD 1988.)

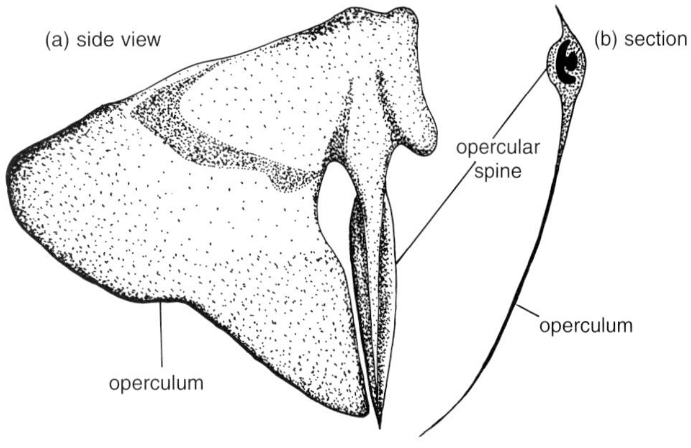

Figure 16.42
The opercular bone and associated dagger-like spine from the weeverfish. (Redrawn by Jan Hynes, after HALSTEAD 1988.)

coastline as far north as Denmark. They are usually inhabitants of flat, sandy or muddy bays.

Australian weevers are unrelated to the European species. One genus, *Parapercus* lives in tropical and subtropical waters of the Indo-Pacific (including Australia) and the Red Sea.

In Europe there are four species of weevers: the greater weever *Trachinus draco* Linné (Figure 16.41, page 387), the lesser weeverfish *Trachinus vipera* Cuvier, and the southern (European) weevers *Trachinus radiatus* Cuvier and *Trachinus araneus* Cuvier.

These are all small fish with an elongate body, a standard length being 21 cm. The largest species is *Trachinus draco*, which reaches up to 46 cm in length. They feed on shrimp, annelids (worms) and small fish, and commonly bury themselves in sand or mud for camouflage, from whence they dart out with impressive speed to spine their prey.

The anal and second dorsal fins are long. Venom glands are associated with the operculum or gill cover, and the first dorsal fin. These fish are sedentary animals but when disturbed may attack, raising their dorsal fins and expanding their opercula. (MARETIĆ (1988) has reported that these fish will attack without provocation and should not be disturbed.)

The toxin remains potent for hours in dead weeverfish, especially when they are refrigerated (SKEIE 1966). Weeverfish are regarded as among the most venomous fish found in temperate waters (HALSTEAD & MODGLIN 1958).

Venom apparatus

Weeverfish possess five readily visible dorsal spines (plus a couple less visible) and an opercular spine on each side of the head (Figure 16.41, page 387), all venomous. The five to eight dorsal spines and the two opercular spines (Figure 16.42) each have an investing integumentary sheath. The associated venom glands are surrounded in connective tissue, and connective tissue may also invade the gland, separating it into pockets. As in other venomous fish, each compact gland associated with a spine consists of aggregations of unicellular glands that discharge all of their secretions and need to be regenerated once they are ruptured. Also typical of scorpaenoid venom cell histology, these cells contain membrane-bound spaces filled with granules (see stonefish Figure 16.30, page 376) (PERRIÉRE & GOUDEY-PIERRIÉRE 1989). The entire sting is covered by an integumentary sheath (HALSTEAD & MODGLIN 1958).

The venom glands extend to the tips of the spines, the first three dorsal spines being most highly developed. Opercular spines (Figure 16.42) are razor sharp and dagger-like, and grooved on the free surfaces. The ductless triangular-shaped venom glands run along the grooves situated on the lateral surfaces of the opercular spines. The association of the operculum and the spine is shown in Figures 16.41 and 16.42 (pages 387 and 388). Specific muscles are involved in the raising and lowering of this impressive spine (HALSTEAD 1988).

Venom pharmacology

Following major European contributions during the early decades of the twentieth century, considerable toxicological, pharmacological and chemical research into *Trachinus* venom, using experimental animals, has occurred (RUSSELL & EMERY 1960; SKEIE 1962; CARLISLE 1962; MARETIĆ 1988). (This is a natural consequence with a venomous animal prevalent among European civilisations.)

The venom is a protein mixture, not all components of which are toxic in animals (SKEIE 1962). It is antigenic, labile with storage and handling, can produce intense local tissue destruction, and has no effect on neuromuscular transmission (RUSSELL & EMERY 1960). The venom produces respiratory depression, hypotension and cardiac ischaemic changes with arrythmias in cats (SKEIE 1962). *Trachinus vipera* venom appears to contain 5-hydroxytryptamine (CARLISLE 1962) and weeverfish venom has been observed to produce peripheral vascular disorders (LINARES DEL RIO *et al* 1989). Russell considered the systemic toxic effects of stingray, scorpionfish, weeverfish and stonefish to be similar (RUSSELL 1965).

Fatalities

Fatalities from weeverfish envenomation were recorded by ULMER (1865), GREIG (1930) and EVANS (1943), though RUSSELL (1965) stated that the three cases of fatalities that he studied were in fact due to secondary infection. The lesser weever *Trachinus vipera* is considered to be the most toxic of these fish because of the high concentrations of serotonin in the venom (MARETIĆ 1988). It is the smallest of the weevers and reaches only 14 cm in length.

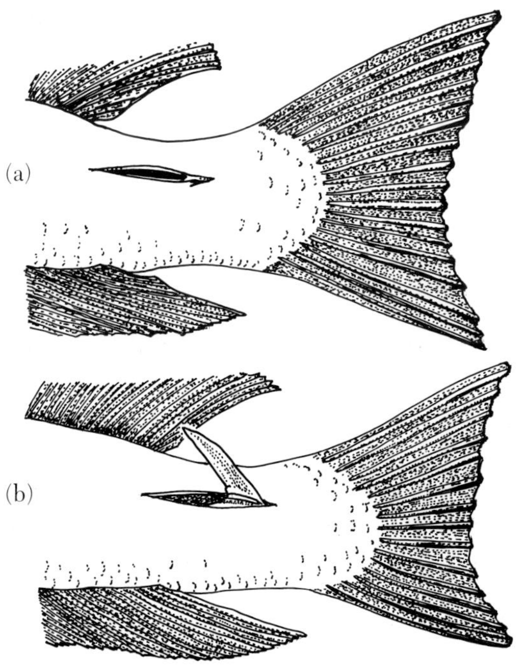

FIGURE 16.43
The surgeonfish venom apparatus from the caudal peduncle: (a) retracted position (b) extended position. (Redrawn by Jan Hynes, after HALSTEAD 1988.)

Clinical features of the sting

Weeverfish envenomations produce severe pain in professional fishermen as well as bathers in sandy coastal areas. The sting is capable of penetrating a leather shoe. The pain is described as 'more intense' than that of stingrays (RUSSELL 1965), and the similarity to stonefish stings, with the sometimes writhing prostration of the hapless victim from the pain, is notable. Usually the entire limb is affected; the wound bleeds profusely, the finger (which is most often envenomed) blanches from ischaemia and then turns red or blue. An important distinction between stonefish and weeverfish stings is the local tissue necrosing action of the latter's venom, which can lead to local gangrene (HALSTEAD 1988). Local tissue temperature increases, as does swelling. Lymphangitis, parasthesiae and weakness in the stung extremity may develop. Swelling has been known to last up to one year (MARETIĆ 1988). Reported complications include fibrosis of finger joint capsules, with loss of mobility (MARETIĆ 1975).

General symptoms, when they occur, include dry mouth, prostration, dizziness and collapse. Weeverfish have produced peripheral vascular disorders in several human victims (LINARES DEL RIO *et al* 1989).

First aid and medical management

The stings are treated with local or regional anaesthetic blockade {see Chapter 5}. The venom is heat-labile; in any case, dramatic relief is effected by immersion of the affected hand in hot water. For further details, refer to 'Management of a stonefish sting' on page 378.

Unlike stonefish sting management, selected cases of weeverfish envenomation may require local excision of necrotic tissue by skilled medical persons.

Antivenom therapy

There is a report of a horse antiserum for intravenous use against weeverfish envenomation, prepared at the Institute of Immunology in

Zagreb (THEAKSTON & WARRELL 1991), but production of all fish antivenoms from this source ceased in 1994 (MEIER 1995).

OTHER VENOMOUS FISH OF THE ORDER PERCIFORMES

Several perchlike fish have a venom apparatus, and injuries have resulted from some of these.

SUBORDER ACANTHUROIDEI FAMILY ACANTHURIDAE (SURGEONFISH)

Acanthurids are marine and are found in all tropical seas. They also occur in Japanese waters. Juvenile surgeonfish (adults have not been shown to have venom glands) have a venom apparatus associated with the fins, and some also have a knife-like spine on the tail.

Subfamily Acanthurinae

These fish have one or more spines on the caudal peduncle (tail stalk) (Plate 16.17 and Figure 16.43, page 389), which when extended can form a formidable weapon. There are two genera in this subfamily, *Acanthurus* and *Paracanthurus*. The latter surgeonfish have blue body markings and a yellowish tail (NELSON 1984). According to McKAY (1994, personal communication) the venom in these spines is not particularly toxic.

Venom apparatus

TANGE (1955) studied the venom apparatus of the acanthurid *Prionurus microlepidotus* and found that it consists of nine dorsal spines, three anal spines, two pelvic spines, and venom glands each surrounded by an integumentary sheath. The venom glands, consisting of large unicellular gland cells, are not completely surrounded by connective tissue as in scorpaenids, and the glands themselves are smaller than in other venomous fish. No glandular ducts are present. Some species of acanthurids also possess caudal spine venom organs that can be extended out from the body or lie flat against it. Although the integumentary sheath in some species is quite substantial, there is questionable histological evidence to support the presence of venom gland cells in association with this spine.

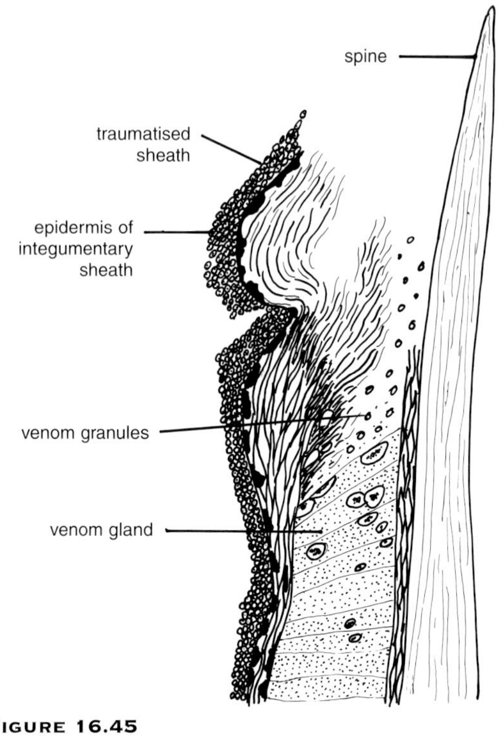

FIGURE 16.45
Siganid venom apparatus. Longitudinal section through a dorsal fin spine of a siganid fish, showing the manner of venom release once the spine tip is traumatised. (Redrawn by Jan Hynes, after HALSTEAD 1988.)

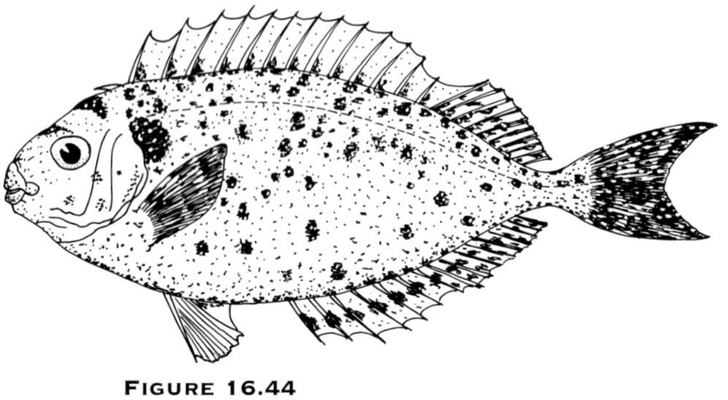

FIGURE 16.44
Siganus sp. (known as rabbitfish, mi-mi, happy moments). (Redrawn by Jan Hynes, after SOUTHCOTT 1975.)

FAMILY SIGANIDAE (RABBITFISH, SPINEFOOT, HAPPY MOMENTS, MI-MI)

Siganids generally live in reefs or around rocks in the Indo-Pacific area, Polynesia, east Africa, the Red Sea and eastern Mediterranean. In tropical waters some are schooling species and others live among coral (WOODLAND & ALLAN 1977).

They are fish of moderate size, with laterally compressed, oval-shaped bodies (Figure 16.44, page 390). All have slippery skin and very small scales. They possess sharp spines on their dorsal, ventral and anal fins. There are two genera: *Lo* and *Siganus* (= *Teuthis*).

Venom apparatus

The venom apparatus of siganid fish is formidable, consisting of 13 dorsal fin spines, seven anal fin spines, four pelvic fin spines, musculature, venom glands and the enveloping integumentary sheath. In all species the gland must be ruptured before venom is released (Figure 16.45, page 390).

The venom glands of *Siganus vulpinus* (HALSTEAD *et al* 1971) sit in anterolateral glandular grooves, which are situated on each side of the anteromedian ridge. In all siganids these grooves extend throughout the entire length of the spine. The venom glands of the spines occupy the distal third to the distal half of the spines. All spines are capable of erection and depression. These fish sleep with their spines in a locked position.

Medical aspects

The stings from these fish cause excruciating pain lasting several hours, and comparable to the stings produced by the Scorpaenidae. The pain may radiate up the affected limb (CAMERON 1981). MING and colleagues (1990) report that *Siganus canalicularis* and *S. javus* of Asia-Pacific waters produce a sudden intense pain at the site of the sting. This pain increases in intensity and spreads to adjacent areas. The pain lasts from a few hours to a few days, and the puncture wound is prone to infection. Clinical features include swollen, tender lymph nodes, nausea and vomiting and increased heart rate.

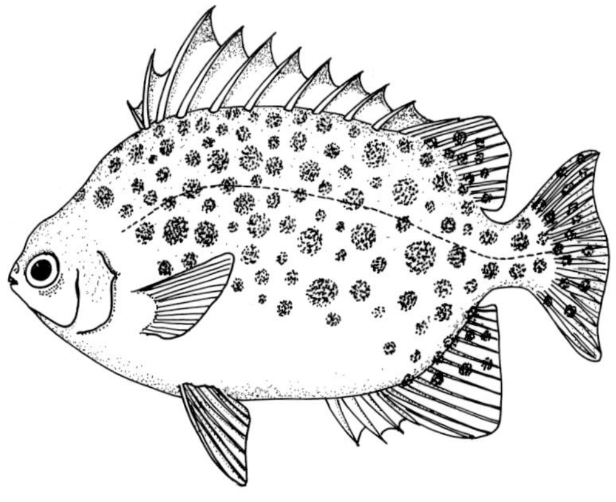

FIGURE 16.46
Scatophagus argus (scats). These fish feed off detritus, hence the name. (Redrawn by Jan Hynes, after SOUTHCOTT 1975.)

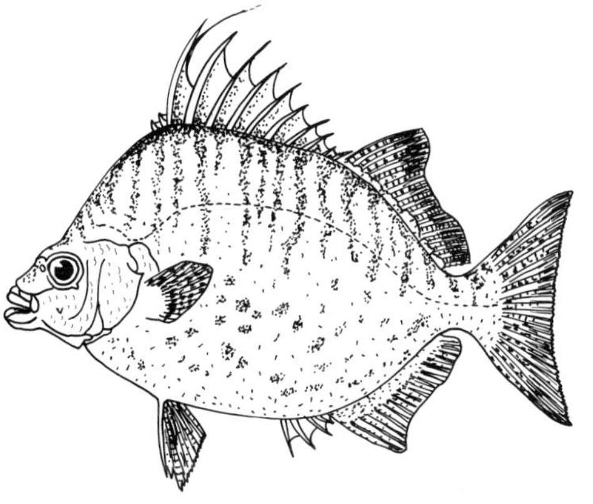

FIGURE 16.47
Selenotoca multifasciata. (Redrawn by Jan Hynes, after SOUTHCOTT 1975.)

Treatment is similar to that of scorpionfish and stonefish stings {see page 378}.

FAMILY SCATOPHAGIDAE (SCATS)

Scatophagus argus (Linné) (spotted butterfish, spadefish)

(Figure 16.46) This species is found in the Indo-west Pacific. It is deep bodied; grey-green above, silvery below and grows to a length of 32 cm in southern Queensland. It feeds off detritus (particles of debris) and dung in its marine or estuarine environment.

Selenotoca multifasciata (Richardson) (striped butterfish)

(Plate 16.18 and Figure 16.47, above) This species is found on the eastern, northern and western coasts of Australia and in Papua New Guinea (HALSTEAD 1988). *Selenotoca multifasciata* is silvery, with a greenish back, and dark vertical bars. It is typically small, usually attaining a length of less than 35 cm. The Scatophagidae feed off detritis, and especially dung, in their marine or estuarine environment.

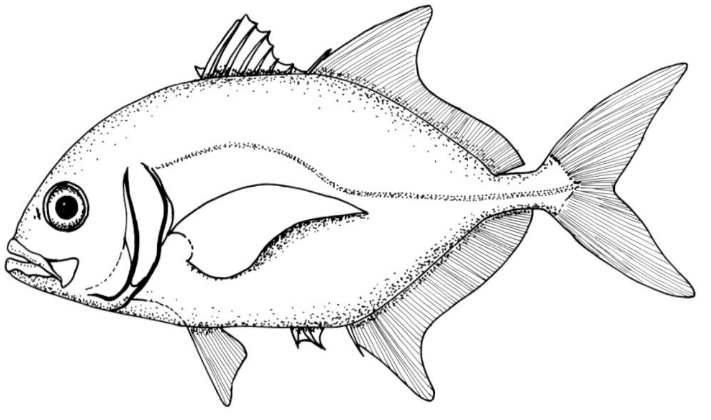

FIGURE 16.48
A member of the family Carangidae (jacks and pompanos). There is questionable evidence about the presence of a venom apparatus in these fish. (Redrawn by Jan Hynes, after NELSON 1984.)

Venom apparatus of Scatophagidae

Species in both genera have very sharp dorsal, anal, and ventral fin spines. *Scatophagus argus* possesses 11 dorsal spines, four anal spines and a pair of ventral spines, and *Selenotoca multifasciata* possesses 12 dorsal spines, four anal spines and a pair of ventral spines (CAMERON & ENDEAN 1970). According to HALSTEAD (1988), all spines bear venom glands. The venom glands are composed of a slender, elongate tissue, which lies within grooves on both sides of the spines. The glands are not enclosed by a connective tissue layer as in scorpaenids; and as in catfish of the family Ariidae, consist of thickened layers of glandular cells within an integumetary sheath (CAMERON & ENDEAN 1970). Before venom can pass into a wound, the toxin-containing cells within the epidermis must be ruptured, as would occur when the spine lacerates human skin in a fishing accident.

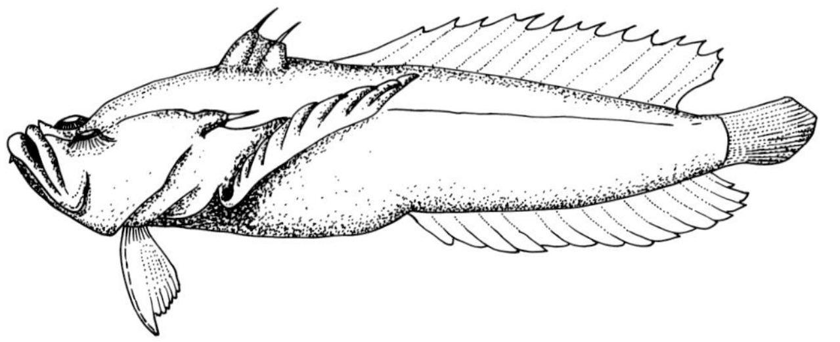

FIGURE 16.49
The venomous toadfish *Thalassophryne* sp. Note the two venomous dorsal spines. (Redrawn by Jan Hynes, after HALSTEAD 1988.)

These fish are not dangerous, but envenomations by the juvenile scatophagids (which are the only ones to have effective venom glands) cause temporary pain at the site of the wound. In the adult members of both species, scales grow up the integumentary sheath of the spines and venom production is impeded.

FAMILY CARANGIDAE (JACKS AND POMPANOS)

(Figure 16.48) Members of this family are found in coastal and oceanic waters in the Atlantic, Indian and Pacific oceans, and according to HALSTEAD (1988) they are found near coastal and oceanic coral reefs and islands. They are fast-moving, medium-sized to large, metallic-skinned fish. One of the most distinctive characteristics of this group is the presence of the two spines in front of the anal fins.

Venom apparatus

HALSTEAD and colleagues (1972) studied the venom apparatus of *Scomberoides sanctipetri*. It consists of seven dorsal spines, two anal spines (particularly venomous), their muscles, venom glands and integumentary sheaths. Unlike the scorpionfish, there are no well-developed antero-lateral grooves on any of the spines, although they have a locking mechanism, as in catfish. Mucous and poison-containing cells are found within the epidermis of the integumentary sheaths. As such, they resemble the histological situation of the fin of the catfish *Arius graffei*, described above.

The sting is instantly painful, 'bee-like', and the pain persists for some 30 minutes. Hawaiian fisherman are commonly stung by these fish. Hot-water immersion is partially effective for pain relief.

ORDER BATRACHOIDIFORMES (VENOMOUS TOADFISH, FROGFISH [AUSTRALIA])

FAMILY BATRACHOIDIDAE (TOADFISH AND FROGFISH)

These fish are unattractive, marine bottom-dwellers. They live mainly near the coast in benthic waters and rarely enter brackish waters, though some species live only in fresh water.

They are found in the Atlantic, Indian and Pacific oceans, including South America. These fish lie buried in bottom sand or mud, and are easily stepped upon. Their bodies are usually scaleless, drab-coloured, with a large head, and eyes more dorsal than lateral. The mouth is large (Figures 16.49 and 16.51). Some can produce sounds with their swim bladders, and are capable of living out of water for several hours.

Batrachoidids possess the most highly developed venom apparatus found among fishes, though the venom is not as powerful as that of the stonefish. No Indo-west Pacific batrachoidids, including *Halophryne,* are considered to be venomous (CAMERON & ENDEAN 1970), although SUTHERLAND (1983) states that these fish are considered to cause 'severe local pain' and occasional mild general symptoms.

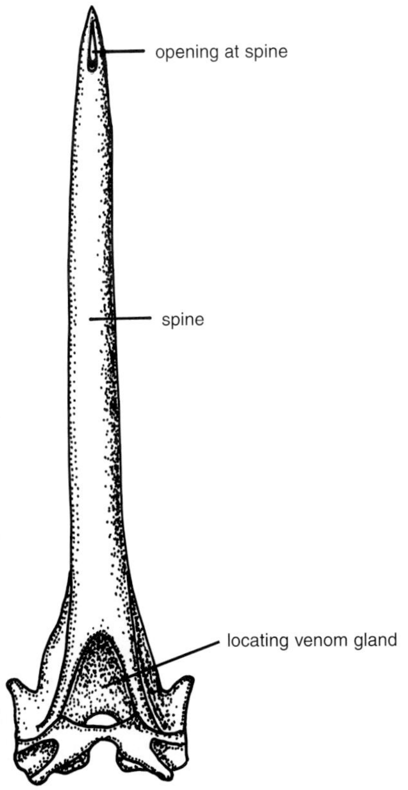

FIGURE 16.50
A dorsal spine of the venomous toadfish *Thalassophryne* sp. The hollow spine serves as a hypodermic syringe, transferring venom from the basally situated venom gland (not shown) into the victim, once pressure is applied to the spine. (Redrawn by Jan Hynes, after HALSTEAD 1988.)

They are classified in two subfamilies, each having a different spine type.

Subfamily Thalassophryninae

These are found in the eastern Pacific, the western Atlantic and the Caribbean Sea. Batrachoidids of the two genera *Thalassophryne* and *Daector* possess fin spines that are hollow to the tip, for the passage of venom (COLLETTE 1966) (Figure 16.50). *Thalassophryne* and *Daector* include 11 species.

Venom apparatus of toadfish

The toadfish venom apparatus consists of the outer integumentary sheath, a glandular zone, the spine and the central canal (Figure 16.50). The central canal is replete with large cells and an amorphous secretion.

A whitish venom gland surrounds the greater part of the spine, although not at the tip. There is a hollow at the base of the spine and one near the tip through which the venom passes. The two dorsal stings are similar in structure to the opercular stings, though the spines are longer and more slender and the two dorsal spines are enclosed in a single integumentary sheath. The spines of both the operculum and the dorsal fin are capable of movement controlled by muscles.

Subfamily Batrachoidinae

These are found off the coasts of the Americas, Africa, Europe, southern Asia and Australia. Compact venom glands are absent. Members of the genus *Halophryne* (frogfish) (Figure 16.51) do not possess hollow spines, and poison-containing cells are confined to the outer

FIGURE 16.51
The Australian banded frogfish *Halophryne diemensis* has been shown not to be venomous, although it has caused injuries. (Redrawn by Jan Hynes, after GRANT 1975.)

epithelium. *Halophryne diemensis* is present in Australia.

The toadfish sting

Toadfish stings produce intense pain, local swelling, and secondary infections are described. Treat as for scorpionfish or stonefish stings {page 378}.

FISH IN WHICH THE PRESENCE OF A VENOM APPARATUS HAS BEEN QUESTIONED

FAMILY URANOSCOPIDAE (STARGAZERS)

Stargazers are found in warmer waters in the Mediterranean Sea and the Atlantic, Pacific and Indian oceans. They are small, carnivorous bottom-dwellers.

The head is large and square and the body is naked or covered with small scales. The eyes are dorsally placed and look upwards (Figure 16.52 and Plate 16.19). On the cleithrum or shoulder, which is situated just above the pectoral fin and behind the operculum, there are two large double-grooved spines with a venom gland at each base. Fin spines are absent.

Venom apparatus

HALSTEAD (1988) believes that the venom apparatus consists of the two shoulder (cleithral) spines, gelatinous material (a venom gland?) associated with the cleithral bone and associated spines, and the integumentary sheaths that surround them. The spine protrudes from the integumentary sheath at its distal end. He suggests that this material is forced into the wound through the spine grooves once the integumentary sheath is ruptured.

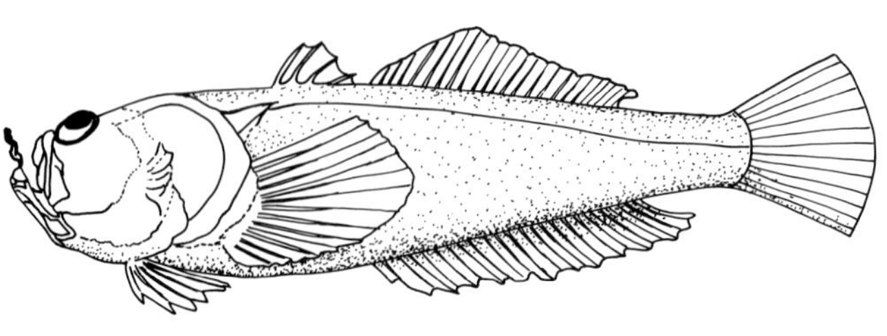

FIGURE 16.52
A representative of the family Uranoscopidae, (stargazers). These non-venomous fish inflict painful wounds. (Redrawn by Jan Hynes, after HALSTEAD 1988.)

Although HALSTEAD (1970) reports evidence of a spongy material in association with the bone and cleithral spine in stargazers, MARETIĆ (1975) (cited in MARETIĆ 1988), stated he was not able to find hard evidence of envenomations. Furthermore, when he removed glandular tissue associated with cleithral spines of *Uranoscopus scaber*, his 'stabbing' experiments showed no toxic effects. HALSTEAD (1988) {see Chapter 3} makes non-specific reference to Mediterranean fatalities, but confirmation is lacking. McKAY (1994, personal communication) states that the stargazer stings in Australia are not serious.

Sting management: hot-water immersion, and then as for a stonefish sting, according to severity of symptoms {page 378}.

OTHER FISH IN WHICH THE PRESENCE OF A VENOM GLAND IS QUESTIONED

(MARETIĆ 1988). The deep-sea scaly dragonfish (family Stomiatidae), squirrel-fish (family Holocentridae), horse mackerel, butterflyfish (family Chaetodontidae), old wives (family Enoplosidae), mojarras (family Gerridae), buglerfish (family Percidae), scats (family Scatophagidae), seabass (family Serranidae), lettered perch, snake mackerel (family Gempylidae), sculpins (family Cottidae), searobins (family Triglidae), dragonets (family Callionymidae), and goosefish (family Lophiidae). (See also Note 2, below).

In Australia, several other spine-armed fish are expected to possess venom glands associated with their spines. However, insufficient work has been carried out on them to determine whether this is the case (UNDERHILL 1989). These include the 'old wife' *Enoplus armatus*, and the flathead *Platycephalus fuscus* and *Thysanophrys cirronasus*.

Footnotes

1. Fish venoms are predominantly unstable large proteins (HALSTEAD 1988). As such mol-

ecules are dissociated with changes in pH and temperature, hot-water immersion might cause denaturation of the venom in the tissues. However, the return of pain on extraction of the part from hot water casts some doubt on this rationale. The analgesic efficacy of hot-water immersion for venomous fish injuries cannot be disputed and should always be adopted by first-aiders as a first measure for pain relief in venomous fish stings. But see Chapter 26, page 445.

2. McKAY (1987) reported that the stings sustained from catfish, bullrouts, cobblers, waspfish, and stingrays during trawling operations and net fishing, especially at night, are usually managed on board unless further attention is required. However, he stated that stonefish (Figures 16.28 and 16.29, pages 374–375; Plates 6.10 and 6.11), bearded ghoul (Figure 16.32, page 381), gurnard perch (Plate 16.20) and goblinfish *Glyptauchen* require medical attention at all times, as the pain may cause delirium and last for several days. In the view of the editors, such practice should be exercised with caution. In addition to the unrelenting pain, some such marine wounds (in particular stingray spine penetration [FENNER *et al* 1989b]) are prone to serious complications in the absence of early and skilled medical intervention.

CHAPTER 17

SEA SNAKES

CHRIS ACOTT AND JOHN WILLIAMSON

INTRODUCTION

Sea snakes (family Hydrophiidae) are principally marine, air breathing, front-fanged venomous reptiles. More than fifty species of sea snakes (subfamily Hydrophiidae) and sea kraits (subfamily Laticaudidae)[*] have been identified (SMITH 1926), at least 32 of these in Australian seas (LIMPUS 1987) {see also Chapter 18, page 409}. WHITE (1995) has given a recent comprehensive listing of known venomous species. Their obvious relationship to elapid (cobra) venomous terrestrial snakes has led to speculation that one group has evolved from the other (WHITE 1995).

Many sea snakes possess a potent venom, and a number of species have been responsible for human fatalities {see below and Chapter 3, page 67}.

SPECIES DOCUMENTED AS CAUSING HUMAN ENVENOMATION

No sea snake should be assumed to be harmless. Claimed species identification by self-styled 'experts' should be regarded with caution. Although

[*] Sea kraits (genus *Laticauda*) are more heavily built 'sea snakes' and are an amphibious genus, whose species tend to come onto dry land (SUTHERLAND 1983). Unlike other sea snakes, their nostrils are not situated on the upper surface of the snout (HALSTEAD 1988). However, they may usefully be considered as and with sea snakes generally. Two sea kraits — the banded sea snake *Laticauda colubrina* and the black-banded sea snake *Laticauda laticaudata* — are the most frequently recorded (SUTHERLAND 1983; LIMPUS 1987).

the identity of any sea snake that bites a human is often not established, even in well-documented reports (FULDE & SMITH 1984; DOBB 1986; PATTERSON & SWALLOW 1991), the following species have certainly been implicated in human envenomation. Here they are listed in approximate descending order of importance (i.e. frequency of recorded envenomations):

Enhydrina (Enhydra) schistosa, the beaked sea snake (REID & LIM 1957; REID 1975; HALSTEAD 1988), considered the commonest cause of sea snake bite (ZIMMERMAN & HEATWOLE 1987)
Astrotia stokesii (MERCER *et al* 1981)
Hydrophis species (REID & LIM 1957)
  H. cyanocinctus (HIGA *et al* 1990)
  H. spiralis
  H. klossi (LIMPUS 1978)
  H. elegans (probable)
Kerilia jerdoni (REID & LIM 1957)
Lapemis hardwickii, Hardwick's sea snake (REID & LIM 1957)
Microcephalophis gracilis (REID & LIM 1957)
Pelamis platurus, the yellow-bellied sea snake (SWAROOP & GRAB 1954).

The following well referenced additional species are listed by HALSTEAD (1988) as having caused human envenomation:

Hydrophis species
  H. obscurus (*H. chloris*)
  H. ornatus (*H. ocellatus*), reef sea snake
  H. nigra (*Thalassophina viperina*)
Thalassophis anomalus

FATALITIES AND MORBIDITY

Firm documentation of deaths from sea snake envenomation are from Malaysia (REID & LIM 1957), Indonesia (BOKMA 1942), Madras and Burma (HALSTEAD 1988). While numbers of deaths are difficult to gauge, between 1944 and 1955 in 17 Malayan north-west fishing villages (population 40 100), out of 144 known sea snake bites (presumably untreated) 41 were fatal (REID & LIM 1957). Informed estimates of fatality rates are about 3 per cent (REID 1956 & 1957). WHITE (1995) estimates by calculation an annual worldwide sea snake fatality rate of at least 150. Other locations of documented fatalities include Java (Indonesia), Madras and Burma (HALSTEAD 1988). A fatality in a Batinah coast (Oman) fisherman is reported in 1991 {see Chapter 18, page 409}. A fatality from the widely distributed reef sea snake, *Hydrophis ornatus*, is recorded in United States armed forces literature (NAVMED, 1965 approx.). Fatalities in Mozambique are claimed (SWAROOP & GRAB 1954). No confirmed Australian fatalities are yet known. Despite the formidable envenoming apparatus and reputation of the Australian and western Pacific olive-brown sea snake *Aipysurus laevis* (SUTHERLAND 1983) (Plates 17.1 and 17.3), no firm record of a fatality from this species was found by the editors.

Similarly, morbidity figures remain speculative worldwide, and reporting probably depends on public education in populations at risk, such as fishing communities (WHITE 1995). Peninsular Malaysia (then Malaya) was certainly such a community, in which it was estimated that 1500 cases occurred annually (REID 1975a).

DISTRIBUTION

There is extensive documentation of the world geographical distribution of sea snakes (SMITH 1926; BARME 1968; SUTHERLAND 1983; LIMPUS 1987; TU & FULDE 1987; HALSTEAD 1988; WHITE 1995). They are widely dispersed throughout the tropics and subtropics in the Indian and Pacific oceans. They are identified north to the Persian Gulf (BARME 1968) {see also Chapter 18, page 409}, to Japan (latitude 35° north) (HOMMA *et al* 1964) {Chapter 20, page 415}, west to the eastern African coast of the Indian Ocean (BARME 1968; SMITH 1977) and east to the Californian coast (HALSTEAD 1988), central America (BARME 1968; TU & FULDE 1987) and along Colombian, Equadorian and northern Peruvian coastlines (WHITE 1995). They have not been found in the Atlantic Ocean proper (SUTHERLAND 1983) {see comments in Chapter 24, page 435}, although there have been reports of them in the Caribbean (via the Panama Canal?) (ACOTT 1983). It should be noted that only the yellow-bellied sea snake *Pelamis platurus* is found in the deep open ocean. All other known sea snakes dwell close to shorelines or coral reefs.

Sea snakes are repeatedly found well up tidal rivers; freshwater 'sea snakes' are known in the Philippines and the Solomon Islands but not in

Australia (LIMPUS 1987). Some species have been found in estuaries or even freshwater lakes that have had previous connections to the sea (ACOTT 1983; WHITE 1995).*

The most widely distributed species is the ocean-going yellow-bellied sea snake *Pelamis platurus* (Linné) (BARME 1968; SUTHERLAND 1983; HALSTEAD 1988); and *Hydrophis ornatus* (Gray) (BARME 1968; SUTHERLAND 1983). It is notable that REID and LIM (1957) found a close correlation between the incidence of different Malayan (now Malaysian) sea snake species associated with fatal bites and their respective frequency of occurrence in Malayan (Malaysian) waters.

Within their habitat range, sea snakes are ubiquitous and agile; the increasing activities of scuba divers are shedding light upon the wide depth ranges of these remarkable animals. However, they do not tolerate extremes in temperature well and are rarely found where the average sea temperature is less than 20°C. They avoid the higher temperatures of hot tropical waters by swimming submerged. As mentioned, sea snakes are found close to shore or around coral reefs (SUTHERLAND 1983; LIMPUS 1987), although the widely distributed yellow-bellied sea snake *Pelamis platurus* is pelagic (ocean-going). The majority of sea snakes are bottom feeders, but *Pelamis platurus* is a surface feeder.

APPEARANCE

Their general appearance is that of a land snake, but they lack ventral scales (needed for land locomotion) and have a paddle-like tail (flattened in the vertical plane) to aid in swimming (BARME 1968; ACOTT 1983; LIMPUS 1987) (Plate 17.1). Recorded lengths have ranged from 30 cm to 2 m, with the *Hydrophis* spp. tending to be the longer (BARME 1968). They are wonderfully graceful and agile swimmers.

They are often mistaken for eels, but lack the eel's fins and gills. They also have scales, and nostrils (which are closed by an operculum during submersion) (Plate 17.2).

They are air breathers equipped with a large right lung at least (considered the sole lung by BARME [1968]) and can stay submerged for up to two hours at the extreme (HEATWOLE 1975a). They are well suited to the marine environment, their adaptive physiology being complex and interesting (HEATWOLE & SEYMOUR 1975; ACOTT 1983). They are closely related to elapid snakes, and some authorities have postulated that the sea snakes have evolved from the Indo-Australian elapids or vice versa (SUTHERLAND 1983; LIMPUS 1987).

All sea snakes have venom glands, with the venom duct opening near the tip of the front fangs (SUTHERLAND 1983; LIMPUS 1987 HALSTEAD 1988).

ENVENOMATION

Sea snake bites have been reported as painless (BARME 1968; SUTHERLAND 1983; LIMPUS 1987; EDMONDS 1989), although in a small Australian series of eight sea snake bite victims of one of the authors (CA), all the victims complained of a painful bite (ACOTT 1983). The appearance of typical fresh bites is shown in Plates 17.3 and 17.4.

Most bites are single, although repeated separate strikes have been recorded {see also Plate 7.1}. As with other snake bites (and all other venomous creatures), sea snakes may bite but fail to inject venom or enough venom to cause any significant clinical effect (REID & LIM 1957; REID 1975a; BARME 1968; FERNANDO & GOONERATNE 1983; SUTHERLAND 1983; WHITE 1995). Thus clinical illness from envenoming is uncommon. Overall, about 80 per cent of bites will result in no envenoming or a trivial systemic disturbance (REID 1975a). As with envenomation generally, children are particularly at risk of serious effects, a fact borne out by both Malaysian and Australian documentations.

The largest clinical series of envenoming was that of REID, in which he described sea snake envenoming in the Malayan fishing communities of the late 1950s and early 1960s. (The reader is referred to his pioneering work: REID 1956a, 1956b, 1957, 1959, 1961a, 1961b, 1962, 1975a, 1975b.) Bites are commonly directly associated with net fishing (DOBB 1986; JELINEK *et al* 1991; PATTERSON & SWALLOW 1991).

* Colubrid water snake envenomation of a human has apparently occurred (D'ABREU 1931).

Sea snakes are curious, generally non-aggressive creatures. However, interference with them during mating, during an ascent to the surface to breathe, or rough handling, may provoke an attack (HEATWOLE 1978; LIMPUS 1987; ZIMMERMAN & HEATWOLE 1987).

Their venom yield tends to be smaller than that of comparable size terrestrial elapids. Fresh venom is usually a clear colourless liquid which dries to a white solid. The venom has neurotoxic and myotoxic (but no coagulopathic) properties, and only one of these two properties, may dominate the clinical presentation. The neurotoxin produces post-synaptic acetylcholine inhibition (causing paralysis) (Figure 17.1), while the myotoxin causes muscle necrosis (MARSDEN & REID 1961; SUTHERLAND 1983; LIMPUS 1987). Reconstruction of damaged muscle commences one or two weeks after the bite. Myotoxins in sea snake venom appear to act similarly to the phospholipase A_2 myotoxins found in some elapid and viper snake venoms (WHITE 1995). In pharmacological preparations, sea snake whole venom has little or no effect on heart or smooth muscle (LIMPUS 1987).

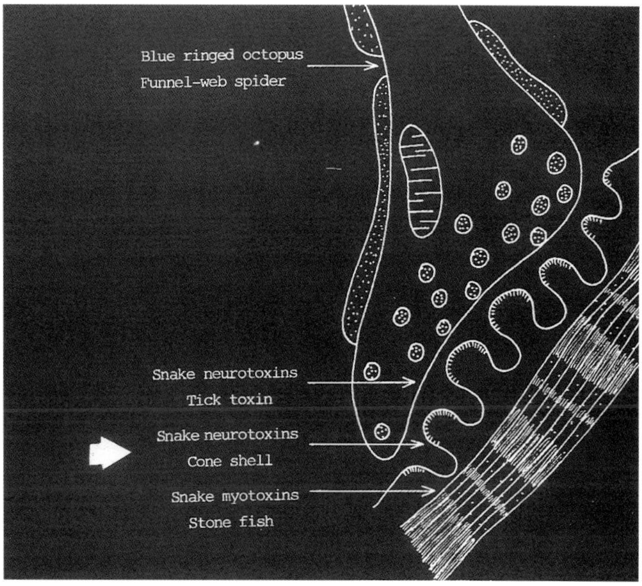

FIGURE 17.1
Schematic diagram of a human somatic neuromuscular junction, showing the post-synaptic site of action of sea snake neurotoxins. Myotoxins act directly upon the sarcolemma. (Photograph courtesy of Associate Professor Vic Callanan, Townsville.)

CLINICAL FEATURES OF ENVENOMATION

If no effective first aid measures have been employed (pressure immobilisation [Figure 17.2, page 400 and Plate 17.5]) then systemic signs of envenoming may occur between 30 minutes and 3.5 hours (REID 1961).

Generalised muscle aches and pains and non-specific weakness are usually the first symptoms. These may be associated with or preceded by signs and symptoms also associated with anxiety, e.g. nausea, vomiting, dizziness, dry throat, shallow breathing and rapid pulse. The painful muscles are usually those of the neck, face, trunk, arm or thigh. Often the muscle pain is described as stiffness which progresses to severe pain on movement. Jaw muscle pain leads to trismus (REID 1961; MARSDEN & REID 1961; SUTHERLAND 1983; ACOTT 1983). This movement pain produces a reluctance to move (the patient will remain motionless), which clinically may be mistaken for paralysis, suggesting neurotoxin involvement.

Myoglobinuria is proof positive of envenomation and may be evident three to six hours after envenoming if a myotoxin is active. Plasma creatine kinase (CPK) will be elevated (SUTHERLAND 1983), indicating muscle necrosis. Secondary renal damage may occur as a result of the myoglobinuria (ZIMMERMAN & HEATWOLE 1987), and hyperkalaemic deaths were reported by REID (1961, 1975; MARSDEN & REID 1961).

True paresis may be evident within minutes to some hours after the bite and is characterised by blurred vision, ptosis, drooling, inability to swallow, dysarthria, flaccid limb paralysis and respiratory distress (inability to cough, rapid shallow breathing, inability to maintain an airway and finally apnoea as the diaphragm is paralysed) (ACOTT 1983).

VENOM DETECTION IN SEA SNAKE BITE

Sea snake venom detection techniques are not used clinically at present, although there has been one report of cross-reactivity with the antivenom of the Australian king brown snake (FULDE & SMITH 1984).

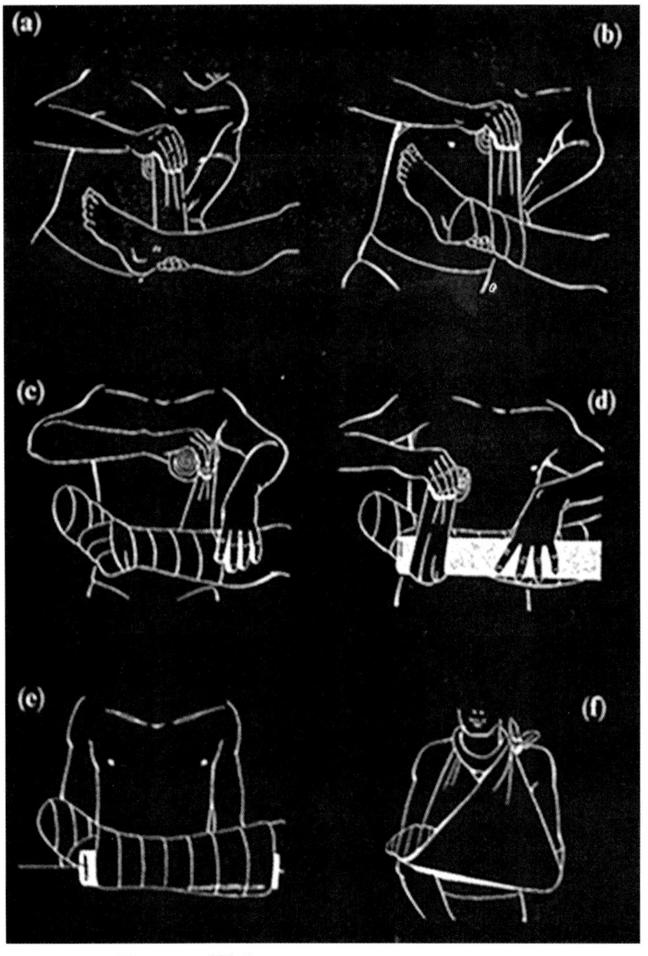

FIGURE 17.2
The compression/immobilisation bandaging technique of entrapping venom in the tissues. See also Plate 5.15.

BIOCHEMISTRY AND HAEMATOLOGY OF ENVENOMATION

Biochemically there is an elevation of the levels of CPK and liver enzymes, especially serum glutamic oxalacetic transaminase (SGOT), if myolysis is present. Renal impairment will be indicated by a rise in urea and creatinine. Potassium levels need to be monitored if there is any evidence of renal involvement. A non-specific leucocytosis is common (MERCER *et al* 1981).

Following sea snake envenoming there is no evidence of coagulopathy, thrombocytopaenia or haemolysis (REID 1956; SUTHERLAND 1983; ACOTT 1983).

MANAGEMENT

Patients who exhibit no objective evidence of either myglobinuria or clinical envenoming in the absence of first aid, for more than four hours after the bite are unlikely to develop serious envenomation effects. However, first aid should always be applied, in case of an exception to this pattern.

FIRST AID

The recommended first aid worldwide for sea snake bites is the application of the compression immobilisation technique (compression bandage and immobilisation of the limb) (SUTHERLAND 1983; WILLIAMSON *et al* 1987; SUTHERLAND 1994c), described in Chapter 5, page 110 of this text {see Plate 5.15, and Figure 17.2}. These should not be removed until the patient is in medical hands; and even then they may not be disturbed (Plate 17.5). If no compression bandage is available direct compression over the bite site with the hand should be used, along with early immobilisation (ACOTT 1983), perhaps combined with makeshift direct pressure over the bite area.

If the victim is a diver using a full wet suit, the suit may well act as a compressive bandage, particularly if the bite is on an extremity (SUTHERLAND 1983). It is unlikely that serious envenomation can occur through most wet suits (HEATWOLE 1975b; SUTHERLAND 1983) (see below). However, the wet suit top must not be allowed to interfere with resuscitation efforts {see also Chapter 22, page 426}.

As in all envenoming, attention to the airway, breathing and circulation is paramount. Expired air resuscitation (EAR) will be life saving for respiratory paralysis {see Chapter 5, page 107, and Plate 9.88}.

MEDICAL MANAGEMENT

All sea snake bites seen within six hours of a bite, even if symptom free, should be regarded as an emergency and assessed urgently. Venom detection kits (SUTHERLAND & KING 1991) are of no use clinically in the management of obvious sea snake bites {see below}. In any case the victim usually will have been bitten while in the water {cf. blue-ringed octopus bite, page 331}, so any surface venom will have been washed off. In most cases the diagnosis will be obvious from the history. However for those victims who have been bitten while on the beach or in a boat, a careful history on the mobility of the offending snake and the use of a venom detection kit will give a clue to the snake's identity (sea snakes other than sea kraits [see above] are incapable of land locomotion because of the lack of ventral scales [BARME 1968; ACOTT 1983]).

Cessation of first aid measures should occur only when a hospital has been reached and full resuscitative measures are immediately available (including ampoules of antivenom). Sea snake antivenom is only indicated when there is clear evidence of envenoming (REID 1962; SUTHERLAND 1983; ACOTT 1983). Detection of myoglobinuria three to six hours after the bite, CPK elevation and the development of any of the symptoms or signs previously described (muscle weakness and movement pain and impending paralysis) are indications for antivenom use. Any patient who is initially symptom free should be frequently reassessed clinically. Urine output should be carefully monitored and checked for myoglobin. Envenoming is a dynamic process and 'out of date' results may be only one hour old (SUTHERLAND 1983).

SEA SNAKE ANTIVENOM

Sea snake antivenom {see also Chapter 21, page 418} is made by Commonwealth Serum Laboratories (CSL) in Melbourne — now with the collaboration of the Australian Venom Research Unit within the Department of Pharmacology at the University of Melbourne — and is available throughout the Indo-Pacific Area (SUTHERLAND 1983; ACOTT 1983) (Plate 17.6). The anti-sea snake component of the antivenom is made by hyperimmunising horses with the beaked sea snake *Enhydrina schistosa* venom, and has been shown to be effective against the venoms of other sea snakes (Table 17.1) (BAXTER & GALLICHIO 1974; SUTHERLAND 1983; WILLIAMSON *et al* 1987). Workers in the Haffkine Institute in Bombay in 1980 also prepared an effective monovalent antivenom using *Enhydrina schistosa* venom (GAWADE *et al* 1980).

CSL tiger snake (*Notechis scutatus*) antivenom can also be used (SUTHERLAND 1983; LIMPUS 1987) but is only recommended if sea snake antivenom is not available. Several ampoules of tiger snake antivenom are required to be equivalent to an ampoule of sea snake antivenom (a ratio of 3:1 is often quoted) (BAXTER & GALLICHIO 1974; ACOTT 1983, 1986), with the commensurate increase in risk of delayed serum sickness with the larger volumes (SUTHERLAND 1992). The minimum dosage is 1 ampoule of sea snake antivenom, but multiple ampoules may be required depending on the severity of the envenoming (REID 1956; SUTHERLAND 1983). One of the recurring clinical problems with the use of any antivenom is that not enough antivenom is given.

As with all snake antivenoms, sea snake antivenom should be given intravenously and diluted 1:10 with a crystalloid solution (SUTHERLAND 1983 & 1992; ACOTT 1983) {see also Chapter 21, page 419}. However, the dilution (not the total dose) may depend on the size and medical history of the patient. The antivenom should be administered slowly initially and be completed within 30 minutes. (Before any antivenom is used, check its expiry date.)

TABLE 17.1
KNOWN SEA SNAKE VENOMS NEUTRALISED BY THE AUSTRALIAN SEA SNAKE ANTIVENOM, IN DESCENDING ORDER OF 'EFFECTIVENESS' (BASED ON THE DRIED WEIGHT OF RESPECTIVE VENOMS NEUTRALISED BY A SINGLE AMPOULE OF THE ANTIVENOM)

Hydrophis spiralis
Lapemis hardwickii
Hydrophis cyanocinctus
Hydrophis elegans
Astrotia stokesii
Laticauda semifasciata
Aipysurus laevis
Disteria major
Hydrophis stricticollis
Enhydrina (Enhydra) schistosa

After WHITE 1995.

Antivenom complications

Pre-treatment with subcutaneous adrenaline before antivenom use is somewhat controversial (SUTHERLAND 1992 & 1994) and is discussed elsewhere in this text {Chapter 5, page 111, and Chapter 21, page 421}; but it should be considered if there is a history of allergy or asthma. However, appropriate measures for the treatment of anaphylaxis (including importantly adrenaline) should be ready whenever antivenom is given (ACOTT 1983; EXECUTIVE EDITORIAL BOARD 1989). (It is the authors' clinical experience that anaphylactic reactions to any of CSL's antivenoms are rare [SUTHERLAND 1992], and antivenom should never be withheld if clinically indicated, simply on the grounds that the envenomed patient is thought to be allergic to it.)

As with other antivenoms, a delayed reaction (serum sickness) may occur. There is evidence that it may be prevented by steroid premedication (SUTHERLAND 1992). Treatment is with oral steroids if severe. The incidence of serum sickness is generally related directly to the volume of antivenom used (SUTHERLAND 1992). It is an immune complex disease and may become evident at any time up to 14 days after antivenom use. Symptoms are fever, generalised lymphadenopathy and a urticarial rash (SUTHERLAND 1983).

ON-GOING MEDICAL MANAGEMENT

Care of the airway and breathing is paramount (especially if glossopharyngeal [9th cranial nerve] function is impaired) and will require endotracheal intubation and ventilation. Respiratory paralysis is treated in the usual way (intubation and ventilation) (MERCER *et al* 1981). If compression bandaging and antivenom are used appropriately then the envenoming process will be halted and reversed in some cases. However, the role of antivenom in reducing the extent of myolysis is at present uncertain.

Other measures in the management of myolysis include good hydration and the maintenance of a good urinary output. Renal failure should be treated along standard lines (WHITE 1991).

Infection as a complication of a sea snake bite wound is rare. However, tetanus prophylaxis is recommended.

Any patient suspected of being envenomed by a sea snake should be observed for 24 hours following the cessation of the appropriate first aid measures.

PREVENTION OF SEA SNAKE BITE

As with any land snake, 'Leave them alone'. They may investigate swimmers and divers in the water, but the evidence is that they will not bite unless provoked. If hauled on board a vessel in a net or trawl, great care with their release is required. Ocean-going 'boils' of sea snakes should not be interfered with.

Divers have a special responsibility not to interfere with or provoke sea snakes {see Chapter 22, page 427}. Wet suits (but not stinger suits {see Chapter 22, page 425}) offer excellent protection against sea snake envenomation, despite the observation that the fangs of at least *Astrotia stokesii* and large specimens of *Aipysurus laevis* may be long enough just to penetrate a tropical (3 mm) wet suit (HEATWOLE 1975b).

NOTE: In 1986 a letter appeared in the medical literature describing the use of electric shock in the treatment of (land) snakebite. Not surprisingly this 'medical miracle' was taken up by the media, with the attendant dangers of displacing the central importance of compression/immobilisation and correct antivenom application, in the minds of the uninformed (RUSSELL 1987). There followed in 1990 an article claiming efficacy for such measures based on anecdotal cases of 'confirmed or suspected' spider bite, using kilovoltage currents! This dangerous and unfounded practice has been laid to rest (BUCKNALL 1991). Happily not even transient enthusiasm for this manoeuvre impacted upon the management of any marine envenomation, before it was discredited.

SECTION

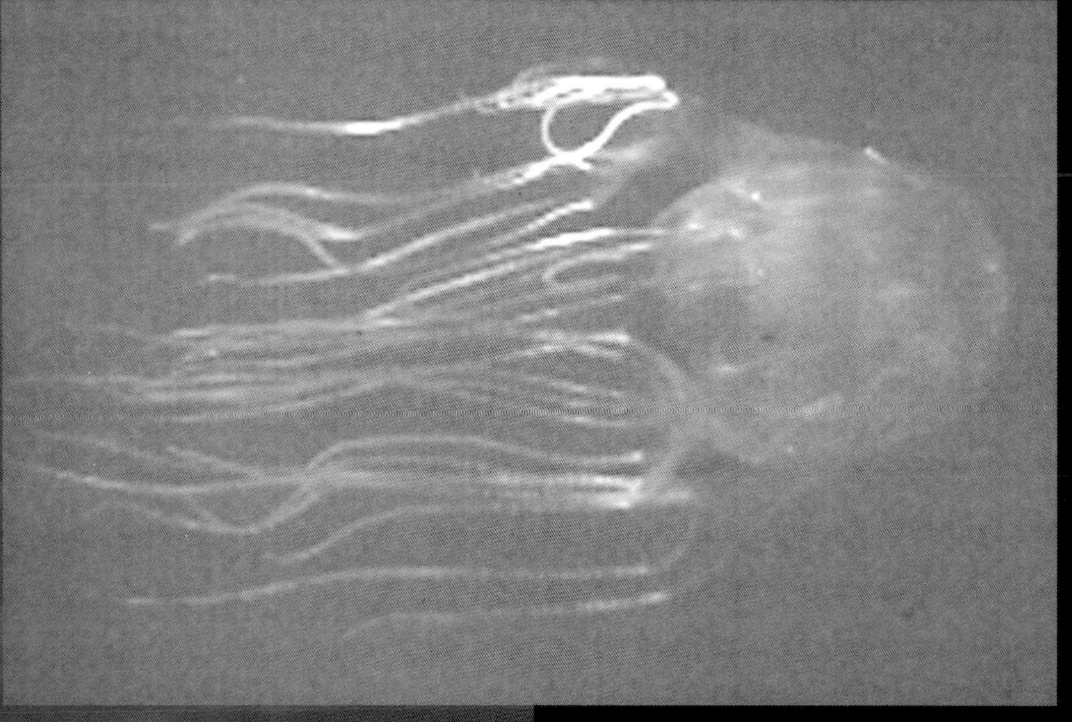

FIVE

INTERNATIONAL TOXIC (POISONOUS OR VENOMOUS) MARINE ANIMAL OCCURRENCES

CHAPTER 18

MARINE STINGS IN THE GULF OF OMAN

PRATAP CHAND AND REGINALD VICTOR

INTRODUCTION

The Sultanate of Oman, with 300 000 square kilometres of very varied terrain, occupies the south-eastern part of the Arabian peninsula. The coastline of Oman stretches over 1700 km, facing the Arabian Sea and the Gulf of Oman, up to the Straits of Hormuz in the north (Figure 18.1).

Fishing is an important occupation and industry for many Omanis, and a survey of the Sultanate in 1990 revealed 19 296 fisherman of whom 18 546 were involved in traditional fishing as their vocation (ALLISON 1991). These fishermen and their families are constantly exposed to marine stinging organisms. In addition, the beach and the sea are popular for recreation for not only the Omanis but also for the sizeable population of expatriates working in the country in the oil industry and other occupations. The hazards of marine envenomation are becoming increasingly appreciated. There is a need for a systematic study of the problem in the Sultanate of Oman.

THE PREVALENCE OF MARINE ENVENOMATIONS

Since 1991 the authors have embarked on a systematic study of the problem in the Sultanate of Oman, to assess the prevalence of both stinging marine species and actual stings in patients seen by medical professionals along the coastline from Musandam in the north to Salalah in the south. A nationwide questionnaire was sent to major hospitals and health

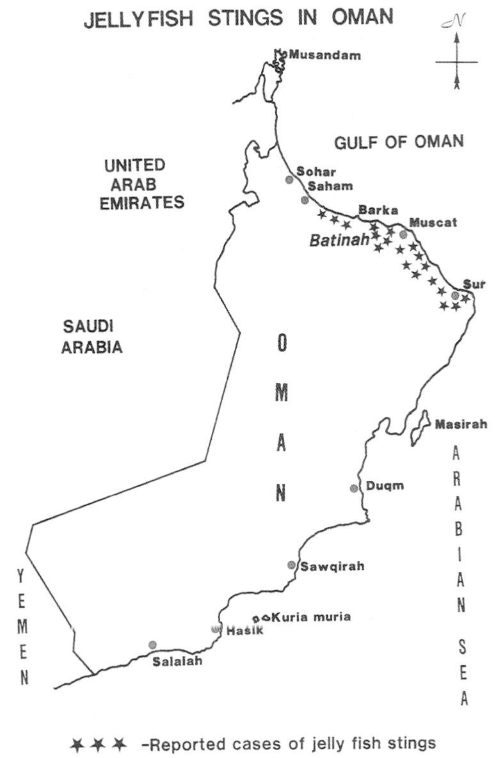

FIGURE 18.1
Map of the Sultanate of Oman, showing geographical distribution of jellyfish stings.

centres along the coastline of Oman enquiring about prevalence, seasonal variations and clinical features of stings. 'Stinger report forms' {see also Chapter 23, page 429} were left in all the hospitals and health centres to be completed and returned whenever a case of stinging occurs. In such cases the area would then be visited, the doctor, the patient and local people interviewed, and attempts made to collect specimens. Based on data collected from the survey up to August 1994 we describe the prevalence of marine stinging organisms and provides clinical profiles of their stings in the Sultanate of Oman.

CNIDARIA

The most important marine envenomations in Oman are caused by 'jellyfish', a term which probably covers several taxa (groups) {see Chapter 6, page 120}. No case of stinging reported so far identified the stinger conclusively; however, stings continue to occur (Plates 18.1A and 18.1B). The search for jellyfish has been independent, isolated and not intensive, often based on the number of stingings reported in Muscat and its suburbs, which comprise the capital area of Oman.

The cubozoans {Chapter 9, page 236} seem to be represented by two species:

- A small carybdeid reported between the last week of June and the third week of July 1992 (THANGARAJA 1992, unpublished observation). The specimens collected were not preserved well but probably represented a species of the genus *Carybdea* {Figure 9.1, page 238}.
- A large carybdeid identified as a *Tamoya* species (Plates 18.2A–C) {compare with Plate 9.10} reported in 1991 (COOPER). The Oman animal is characterised by relatively short, greenish-brown tentacles when the latter are retracted. It was prevalent in August 1992. A 15-year-old Omani swimmer who came in contact with a specimen (which was collected), experienced mild irritation. In December 1994 one of the authors (RV) encountered a swarm of hundreds of *Tamoya* (Plates 18.2A–C) some 20 metres offshore in the Gulf of Oman. Subsequently (January 1995), on nearby beaches, two bathers were badly stung. Captured specimens, difficult to see under natural conditions, were positively identified (Plate 18.2A–C) (VICTOR 1995, unpublished observations).

Speculations and anonymous claims (1994) exist in Oman about the presence of *Chironex* sp. {Plate 1.4} and *Chiropsalmus* spp. {Plate 9.35} (AL FAHAL 1984; ANDERLINI 1984; GOONEWARDENE 1986; COOPER 1991), but factual evidence is still lacking. The distributional range of *Chiropsalmus quadrigatus* {Plate 9.35} and *Chiropsalmus quadrumanus* {Plate 9.36} is stated to include Oman waters (HALSTEAD *et al* 1990), but specimens are yet to be seen {see also comments in Chapter 10, page 302}.

The presence of the ubiquitous hydrozoan *Physalia* in Oman is certain. This species was seen in Bandar Jissah, Qantab and Sidab in the capital area (Muscat) in small numbers between the last week of April and the end of May 1992. Records of *Physalia* in other months and years are available; the preserved specimens shown in Plate 18.3 were collected in January 1987. Large specimens typical of *Physalia physalis* are not available in Oman collections; the taxonomic identity of *Physalia physalis* in Oman distinguishing it from *Physalia utriculus* requires further study {see also Chapter 8, page 192}.

Another pelagic chondrophore, *Velella* sp. {see Plate 8.36}, has been implicated as one of the painful stingers along the Oman coast (ANDERLINI 1984). [Editors' note: Curiously, the species *Velella velella* is considered to be harmless in most other areas of the world, see Chapters 8 and 10.] This hydrozoan is a modified polyp with a concentric float or disc with a characteristic chitinous 'sail'. All around the edge of the disc are hollow stinging tentacles; the budding stalks surrounding the mouth also have nematocyst batteries. This taxon is supposedly rare in Oman, and there appears to be some confusion concerning its identity.

ANDERLINI (1984) reported the presence of two rhizostome taxa, referred to as 'brown mushroom' and 'white mushroom' in Oman waters. The former was about 10 cm in diameter and appeared to be a permanent resident in the area, while the latter with a bell diameter of about 10–15 cm, probably *Rhizostoma pulmo* {see Figure 10.4, page 288}, occurred along the Batinah coast when the water temperature was very warm. GOONEWARDENE (1986) reported the presence of *Cyanea capillata* in Oman, and several fishermen on the Batinah coast, on being shown photographs, identified *Rhizostoma* sp., *Cyanea* sp. and *Chrysaora* sp. as taxa found in Oman waters. (The *Rhizostoma* identification may be mistaken [CORNELIUS 1994, unpublished observations].) Other jellyfish such as *Olindias*, *Sanderia*, *Pelagia*, and the 'upside down jellyfish', *Cassiopea* have been recorded elsewhere in the Gulf of Oman (ANDERLINI 1984; DIPPER & WOODWARD 1989) [Editors' note: All these genera are described in Chapter 6, page 119, and Chapter 8, page 180.]

The coral-like milleporid hydrozoan, *Millepora* sp. {see Plate 8.25}, commonly known as stinging or fire coral is found in southern Oman (SALM 1992, unpublished observation); its occurrence appears to be rare. The distributional ranges of plumulariid hydroids such as *Aglaophenia* sp. {see Plates 8.24 and 8.26} and *Lytocarpus* sp. covers Oman waters (HALSTEAD *et al* 1990). These hydroids are known to colonise wrecks, and the stinging sea fir (stinging hydroid) *Lytocarpus* sp. has earlier been recorded on a wreck at Hamriya in the Arabian Gulf (DIPPER and WOODWARD 1989). Because old wrecks are favoured diving haunts in this region, snorkellers and divers should be aware of these hydroids {see Chapter 22, page 423}.

Published clinical reports of jellyfish and other marine stings from Oman have so far been limited to a few case reports. The most alarming cases were those of two European children seriously stung by jellyfish in the Ras al Hamra area of the capital (AL FAHAL 1984; COOPER 1985 & 1991). Both children needed intensive hospital care, and this prompted ANDERLINI (1984) to suggest that the species responsible was 'undoubtedly...*Chironex fleckeri* or a close relative'. However, one of the victims, a 12-year-old boy, recalled seeing 'a mass of reddish brown strands' (COOPER 1991).°

COOPER (1985 & 1991) described the above two cases in detail. One of the patients had hypertension and atrioventricular block, but recovered. The other child was thought to have initial anaphylactic shock [Editors' note: as opposed to endotoxic shock] {see Chapter 3, page 77}, and later myoglobinuria and renal failure which necessitated haemodialysis.

In 1986 GOONEWARDENE reported four stings, one by *Physalia* and three by an unidentified jellyfish, all from the Muscat region. Two of the patients developed pain, difficulty with breathing, nausea, vomiting and abdominal cramps.

° This description also fits with that of lion's mane jellyfish or red jellyfish, *Cyanea* sp., which has earlier been reported from the capital area (GOONEWARDENE 1986). It could also fit the colour description of a 'cubozoan-like' species known to inhabit Oman waters {see Plate 10.23}. Even a small specimen of *Cyanea* has been reported to raise large wheals on a victim of its sting (PEARSE *et al* 1987), though many stings from *Cyanea* spp. elsewhere in the world, although capable of producing dramatic-looking skin reactions {Plate 8.93}, apparently cause minor, even trivial effects {see Chapter 8, page 234}.

One patient developed clinical shock, was resuscitated and subsequently had myoglobinuria and renal impairment. The fourth patient had only minor injuries. All patients made full recoveries. Fatalities due to jellyfish stings have not been reported from Oman so far.

RESULTS OF SURVEY OF THE PREVALENCE OF MARINE ENVENOMATION

The preliminary answers from our survey indicate that there is a geographical concentration of reported jellyfish stings along the Batinah coast in the Gulf of Oman and also further east as far as Sur, a coastal town facing the Arabian Sea. So far no cases have been reported further south than Sur. A seasonal prevalence has also been reported in June, July and August, reaching a peak in July. These months are the hottest in the year, with air temperatures reaching 48°C, and also correspond to the upwelling season of the Arabian Sea during the Southwest Monsoon, with the maximum nutrient elevation in July and August (SHEPPARD *et al* 1992).

The maximum number of jellyfish stings in 1991–92 were seen in Khoula Hospital in Muscat. Review of their records from September 1991 to August 1992 revealed one case in March, three in May, 42 in July and 29 in August 1992. Of these 75 patients, 52 were males; 11 were children. Information about the time of stings was available only in 55 of the above patients; 73 per cent were stung in the evening, 18 per cent at noon and 9 per cent in the morning. Clusters of cases occurred on particular dates, attesting to the presence of jellyfish in shallow waters on those days.

Analysis of the symptoms revealed local pain and a characteristic rash in all patients (Plates 18.1A and 18.1B). Itching was present in six patients, chest pain in two and body ache in two. A rapid pulse rate (tachycardia) was observed in seven and high blood pressure (hypertension) {see Chapter 9, page 254–55} also in seven patients; two had extensive stings, were hospitalised and made subsequent complete recoveries. No fatality occurred. In none of these stingings was the causative organism caught. Descriptions were available from patients however, and in August 1992 many patients saw a large transparent jellyfish with a bell of about 20 cm diameter and tentacles about 1 m long. In August 1992 two of these specimens were caught from the sea and identified as *Tamoya* sp. Thus it is likely that the stings seen in July and August 1992 were therefore caused by jellyfish such as *Tamoya* sp. (Plates 18.2A–C). However, attempts are being made to collect more specimens from fishing trawlers, and this could in future add more information concerning the local prevalence of different species of jellyfish.

MOLLUSCA

The gastropod molluscs of the family Conidae {see Chapter 13, pages 335–39}, commonly known as cone shells, are represented by at least 26 species in Oman (BOSCH & BOSCH 1982). Opinions on species dangerous to humans are variable, and a list of 16 species could be compiled from worldwide literature (MARSH & SLACK SMITH 1986; EDMONDS 1989; HALSTEAD *et al* 1990). Two species included in such a list, *Conus textile* and *Conus striatus*, are found in Oman (Plate 18.4); an additional local species, *Conus pennaceus*, is also known to be dangerous (BOSCH & BOSCH 1982). No known case of cone shell injury or envenomation has been reported from Oman. If any hospital record exists, it is probably masked as 'fish bite' or 'sea animal bite'. Fisherman rarely encounter cone shells and are therefore indifferent to the dangers posed by them. On the basis of experience elsewhere {see Chapter 13, page 335}, swimmers, divers and those exploring shallow waters, especially children, may be vulnerable to cone shell stings.

POLYCHAETA

Interesting reports on probable marine bristle worm envenomations {see Plate 3.3} exist along the Batinah Coast, especially in the vicinity of Al-Musnah. This animal is referred to as '*Ghool*', which in Arabic means 'snake'. It supposedly lives in burrows at the edge of the water and delivers a painful 'bite' {see Plate 3.2} to anything that blocks the opening. This is often a human foot in reported cases. It is 'red, pink and long', and hence gets the name *Ghool*. The injury is characterised by 'twin' puncture marks, and the excruciating pain lasts about four hours; the affected foot becomes swollen and inflamed. In one case the entire leg was swollen (AFZAL & FAROOK 1992, unpublished observations). The description of this animal indicates that it could probably be an amphinomid polychaete of the genus *Eunice*, a widely distributed taxon in tropical areas (HALSTEAD *et al* 1990).

ECHINODERMATA

Dr A.C. CAMPBELL, now at Queen Mary and Westfield College, London, England, worked on the echinoids of Oman for several years and kindly supplied the information compiled here. The crown-of-thorns sea star *Acanthaster planci* {see Plate 12.1}, known for its notorious venomous spines, occurs on the coral reefs of Oman. Swimmers and divers should be warned about this species, although no known case of crown-of-thorns sting has so far been reported.

In Oman there are five dangerous species of sea urchins, belonging to three genera, *Diadema*, *Echinothrix* and *Toxopneustes* {see Plates 12.10, 12.11 and 12.14}. Two species, *Diadema setosum* (Plate 18.5) and *Diadema savignyi*, despite an inconclusive debate over their taxonomic status in the Arabian Sea, could be considered together for stinger considerations. The long primary spines and secondary spines distributed over the test are hollow and brittle. {see Chapter 12, page 319}. Venom is contained in the spine tissue or in the lumen. The spine tips have minute, backwards-pointing whorls of barbs which allow penetration of skin easily, but not removal {see Figure 12.3, page 320}; spines tend to break off easily and become lodged in tissue. These stings cause considerable discomfort; once in the flesh, spine fragments are difficult to remove, and the natural inflammatory response of tissue hardening and healing takes three to four weeks.

Echinothrix diadema and *Echinothrix calamaris* are relatives of *Diadema* spp. In these species only the secondary spines are sharp enough to penetrate the skin. In *Echinothrix diadema* the needle-sharp secondary spines are 2.5–3.5 cm long; stings are much more painful than those of *Diadema* spp. *Echinothrix calamaris* is smaller, with aposematic stripes on the hollow primary spines. None of these species should be handled when encountered. *Toxopneustes pileolus* {Plate 12.11} is covered with blunt spines and minute flower-like pedicellariae. The spines are harmless, but the pedicellariae bite! According to FUIJWARA (1935), the pedicellariae are capable of biting human tissue and injecting venom from the venom sac present in each of the jaws, causing eventual respiratory failure; subsequent verification of this effect is lacking. Other accounts of *Toxopneustes pileolus* injury are vague and uncertain. Although the real risk of this species is yet to be known, its importance as a venomous species cannot be ignored.

FISH

Envenomation from marine fish {see Chapter 16, page 354} is common in Oman. Most of the cases of marine stings reported to the village health centres along the coast are fish-related. The fish fauna of the Arabian Gulf and the Gulf of Oman are diverse, and cataloguing all species capable of inflicting venomous injuries is beyond the scope of this chapter. Only the most important species feared by the local fishermen and expatriate divers are presented here.

Although there are several stingray species representing at least five families in Oman waters, most human envenomations reported are probably caused by members of two families, Urolophidae and Dasyatidae. Despite a relatively smaller size, *Urolophus* sp. (Plate 18.6) can inflict painful injuries with its spined muscular tail {see Chapter 16, page 364}. The stingray illustrated by GOONEWARDENE (1985) may be a *Taeniura* species commonly found in shallow waters; two species, *Taeniura melanospila* and *Taeniura lymma* are known from this area (DIPPER & WOODWARD 1989; HALSTEAD *et al* 1990). These animals lie partly buried in the sandy shallows {see Plate 16.5} and whip out their tails if stepped on. Omani fishermen

on the Batinah coast receive most stingray injuries during launching or beaching fishing boats. A stingray of the small urolophid or dasyatid variety when retrieved from a net is either thrown back into the water or is 'relieved' of its offending tail.

Catfish stings affect Omani fishermen, but these less-enticing creatures pose a smaller risk to swimmers and divers. The venom apparatus of catfish involves the dorsal and pectoral spines (McKINSTRY 1993) {see Chapter 16, page 370}. Most injuries occur when retrieving fish from nets, and inadequate medical treatment subsequently aggravates the problem. The species *Plotosus lineatus*, commonly known as the oriental or striped catfish, was collected on the Batinah (Plate 18.7). These catfish move about in large schools and are considered one of the more dangerous venomous fish (HALSTEAD *et al* 1990) {see also Chapter 16, page 372}. Other common catfish in the region, of the genus *Arius*, consist of four well-defined species (AL-HASSAN *et al* 1986). The representative, *Arius tenuispinis* shown in (Plate 18.8) was implicated in the stinging of a 19-year-old Omani youth. The photograph of the healed wound, six weeks after the incident shows a pigmented scar (Plate 18.9).

The term 'scorpionfish' includes many genera and species of the family Scorpaenidae {see Chapter 16, page 383}. The grouping followed here on the basis of the structure of the venomous organs is that of HALSTEAD *et al* (1990). Only three genera, *Pterois, Scorpaenopsis* and *Synanceia*, are known from the Arabian Gulf and the Gulf of Oman. *Pterois* spp. are commonly known as zebrafish or lionfish. The scorpionfish described by GOONEWARDENE (1985) is a member of this genus. Four species, *Pterois antennata, P. russelli, P. volitans* and *P. radiata*, are likely to occur in the Omani region. These are the most beautiful of venomous fish, not aggressive, popular for aquaria and frequently photographed. Worldwide, injuries are common among aquarium handlers.

The stonefish *Synanceia* sp. {see Chapter 16, page 374} is not a common marine hazard in the Gulf of Oman. Often the scorpionfish *Scorpaenopsis* sp. is misidentified as the stonefish. The stonefish is considered deadly by some {see Chapter 3, page 69, and Chapter 16, page 374}, whereas the scorpionfish inflicts only a very painful sting. Only a single case of stonefish envenomation from Qatar is well documented, although several divers claim to have seen the stonefish (DIPPER & WOODWARD 1989). Both *Synanceia* sp. and *Scorpaenopsis* sp. are bottom dwellers in the shallow waters; they lie dormant in the intertidal zone, well concealed by mud, coral and rocks. Other records of injuries specifically caused by these two animals are so far not known in Oman.

SEA SNAKES

Earliest records of sea snakes in Oman date back to 1887. The taxonomy of sea snakes in the Arabian Gulf and the Gulf of Oman is well documented (GASPERETTI 1988); at least nine species are likely to occur in Oman waters (GALLAGHER 1990). Although *'Hanaysh il Bahr'* in Arabic refers to sea snakes, the blanket term *'Ghool'* (see above) covering all snakes is widely used. Sea snakes are often caught in fish nets and traps. Omani fishermen seem to handle them with ease, but they are well aware of the danger {see Chapter 17, page 397, and Chapter 20, page 415}. One fatality was reported in 1991 at a fishing settlement on the Batinah coast; the fisherman, accidentally bitten during a routine fishing trip, died on the way to the nearest health centre (VICTOR & AL RAWAHI 1992, unpublished observations). Sea snakes also become stranded on the beach and appear helpless, but knowledgable locals promptly kill these creatures.

The species known from Oman belong to six genera, *Hydrophis, Lapemis, Pelamis, Enhydrina, Microcephalophis* and *Praescutata*. Four species represent *Hydrophis* while all other genera are represented by one species each. The hook-nosed, beak-nosed or beaked sea snake *Enhydrina schistosa* {see Chapter 17, page 397} is referred to as the most dangerous species, and the annulated sea snake *Hydrophis cyanocinctus* is considered as the second most dangerous species in the region (GALLAGHER 1990).

Sea snakes are readily recognised by their flat, paddle-shaped tails (Plate 18.10) {see also Plate 17.1). Field identification of sea snakes using

colouration alone could be misleading in this region; the shades of green, grey, blue and yellow, with or without bands and spots, occur in more than one species; the varying colour patterns of juveniles, especially the hydrophids (Plate 18.11) are an added complication. Correct identification of sea snakes, if and when collected {see Chapter 17, page 397}, is a job for the specialist — just as it is for terrestrial snakes (SUTHERLAND 1983). The identification of representative species illustrated here (Plates 18.10 and 18.11) were made by GARDNER (1991 unpublished data).

CONCLUSIONS

Our survey and review of existing literature reveal the prevalence of a wide range of stinging marine organisms and actual stings in the waters of the Sultanate of Oman. Information about first aid and management of marine stings is being imparted to the coastal health centres and hospitals. A nationwide comprehensive documentation of stings through 'Stinger Report Forms', and also collection of implicated organisms, is underway and should provide more information in the future. Isolation and study of the chemistry and toxicology of venoms from local marine species and attempts to develop specific antivenoms {see Chapters 21 and 23} are likely future developments.

Acknowledgments: The authors, Associate Professor Pratap Chand and Professor Reginald Victor, thank Professor Nicholas J.Y. Woodhouse for initiating this project, and the Director General of Curative Medicine for his kind help in conducting the survey. They also thank Drs Thangaraja, Andrew Campbell, Rod Salm and Andrew Gardner for their permission to use their respective unpublished material. They thank Dr David Clayton for identifying the fish and Mr Sayeed Farook for some of the photography. Gratitude is expressed to CET, Sultan Qaboos University. Finally they thank Mrs Mariam Gentle for her invaluable secretarial help and Mrs Rose P. Natividad for typing the manuscript.

Uranoscopus scaber and the stingrays *Dasyatis pastinaca* and *Myliobatis aquila* (both called '*ratão*'), are caught up in fishermen's nets. Well known and feared by fishermen, they are all carefully handled, their spines are immediately cut away, and they seldom have the opportunity to wound anyone (GONZAGA 1985a & 1985b). If a sting does occur, it is extremely painful and is generally treated like the lesser weever sting.

CASE REPORT

A 34-year-old woman who was 6 weeks pregnant was stung on the foot by a weeverfish, in Portugal (GONZAGA 1985c). She was a primagravida with a clear menstrual history since menarche at the age of 14. After the sting she experienced immediate and local excruciating pain in her foot, followed by general muscular and colicky abdominal pains; the latter were similar to but more severe than the pains of menstruation. She developed a blood-stained vaginal discharge, although her foot responded to local treatments. After three weeks of vaginal loss she had metrorrhagia followed by abortion of her pregnancy. Although European weeverfish venom has been observed to produce peripheral vascular disorders (LINARES DEL RIO *et al* 1989), and respiratory depression, hypotension and cardiac ischaemic changes with arrhythmias in cats (SKEIE 1962), the editors could find no documentation of direct harmful actions on skeletal or uterine muscle.

CNIDARIA

In southern Portuguese waters, and more rarely in the north, some painful urticarial wheals following contact with 'jellyfish-like' organisms are probably cnidarian stings. On occasions these stings are accompanied by mild systemic reactions. The offending animal is rarely, if ever, captured.

The presence of the Portuguese man-o'-war *Physalia physalis* {see Chapter 8, page 192} (known locally in Portugal as '*fisália*'), of *Pelagia noctiluca* ('*alforreca*') {Chapter 8, page 228, and Chapter 10, page 286} and of some other cnidaria, is well documented along Portuguese seashores (GONZAGA 1985b). *Physalia physalis* are rare and may be less venomous than in other latitudes {see Chapter 3, page 65}. *Pelagia noctiluca* in the Portuguese region are small and seldom appear in significant numbers near the shore, but they can readily inflict painful stings {see Plates 10.6 and 10.7}.

Mild urticarial reactions are sometimes reported after handling opelets *Anémonia (Anemonia) sulcata* (known as '*anemona-do-mar*').

CURRENT CNIDARIAN STING MANAGEMENT IN PORTUGAL

Adherent tentacle fragments of the Portuguese man-o'-war are 'inactivated' by the application of 'rubbing alcohol', and then removed by shaving after drying with talcum powder. [Editors' note: The whole question of cnidarian nematocyst inhibitor substances for first aid use is complex (HARTWICK *et al* 1980; FENNER *et al* 1993; RIFKIN *et al* 1993) and is examined in detail elsewhere in this book {see Chapter 5, page 98, and Chapter 9, page 280}. A local corticoid ointment may be applied. A systemic analgesic and an antihistamine will be used in selected cases.

SEA URCHINS ('*OURIÇOS-DO-MAR*')

Painful punctures from the spines of sea urchins {see also Chapter 12, pages 313–25, and Chapter 18, page 418} occur when people step on a specimen of *Paracentrotus lividus*, an *Echinus acutus* or an *Echinus esculentus*, or handle a specimen of *Centrostephanus longispinus* caught in a fishing net. Management consists of soaking the wounded area with vinegar 'to dissolve the spines' (SALDANHA 1980; GONZAGA 1985). (Editors' note: This technique is unexplored in most other world regions; see Chapters 5 and 12).

CHAPTER 19

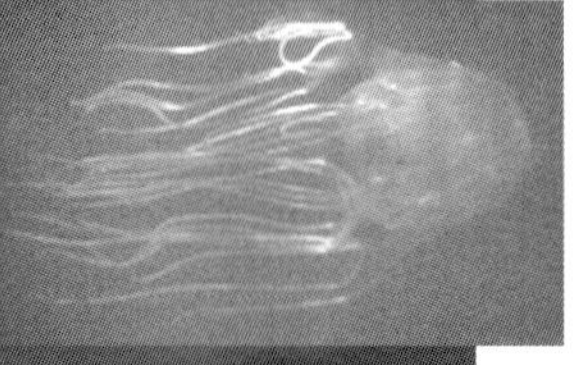

MARINE STINGS IN PORTUGAL

ROGÉRIO A F GONZAGA

INTRODUCTION

Portuguese seaside waters are close to the interface between the Mediterranean Sea and the Atlantic Ocean. Salinity is about 35 parts per 1000 (3.5 per cent), and the water temperature can easily reach 25°C in summer, at the southern province of Algarve. In the Atlantic Portuguese isles of Madeira and Açores, the situation is similar.

For climatic reasons, Portuguese venomous marine species are not very dangerous, and the majority of sting accidents are quite mild indeed. Consequently local health facilities treat most such cases on an empiric basis and do not make a great effort to identify the specific offending marine animal.

VENOMOUS FISH

The vast majority of stings are inflicted by weever fish {see Chapter 16, page 388} known locally as '*peixe-aranha*' — the lesser weever *Echiichthys* or *Trachinus vipera* being very common on all Portuguese shores (GONZAGA 1985a). Treatment consists mainly of local measures such as injection of local anaesthetic and/or the local application of heat to 'inactivate' the venom {see Chapter 5, page 112, and Chapter 16, pages 378–80}. Occasionally a systemic analgesic is administered for more severely affected patients.

Sometimes more dangerous species — the greater weever *Trachinus draco*, scorpionfishes *Scorpaena scrofa*, *S. porcus*, *S. maderensis*, and *S. notata* (all known by the local name '*rascasso*'), *Chimera monstrosa* ('*peixe-rato*'), the stargazer

CONE SHELLS

The small Mediterranean cone shell is *Conus ventricosus*. These animals are present in deeper waters of Portuguese shores. However, no known case of cone shell envenomation has yet been confirmed in Portugal (GONZAGA 1985b). {See also Chapters 13 (pages 335–39), 18 (page 407) and 20 (page 414)}

CHAPTER 20

JAPANESE MARINE ENVENOMATIONS, AND SOME DEATHS IN OKINAWA

The following data about envenomation and subsequest deaths in Okinawa have been provided by Dr Yasuhiro Tomihara and Yasutetsu Araki and their biologist and medical colleagues, and was collected from hospital records, the relatives of patients, and newspaper reports in 1990–92. Some illustrations have been provided by Dr Yasutetsu Araki, Okinawa Prefectural Institute of Public Health, 2085, Ozato, Okinawa 901–12.

CNIDARIANS

Japanese venomous cnidarians of significance are also discussed in Chapter 8, page 201, Chapter 9, page 260, and Chapter 10, page 291–93.

CUBOZOANS

The Okinawan chirodropid, probably *Chiropsalmus quadrigatus* HAECKEL, 1880, or a closely related species {Plates 9.38A and B and 10.9} has produced at least one confirmed death in Okinawa, in 1961 (ARAKI 1992, unpublished observations). Other deaths are suspected but remain unconfirmed. Stings from this animal are common {Chapter 10, page 291–94} and well recognised {Plates 9.39 and 10.11}.

Other cubozoans of little apparent clinical significance known to occur in Japan include the carybdeids *Tamoya virulenta*, *Tamoya gargantua* (KISHINOUYE 1910), *Carybdea rastoni* (UCHIDA 1929; SATOH 1985; AZUMA *et al* 1986; OHTAKI *et al* 1990), and *Tamoya bursaria* (?*gargantua*) (YAMASU & YOSHIDA 1976).*

---

* During the screening on television in Australia of a Canadian–Japanese program produced in 1991, dealing with

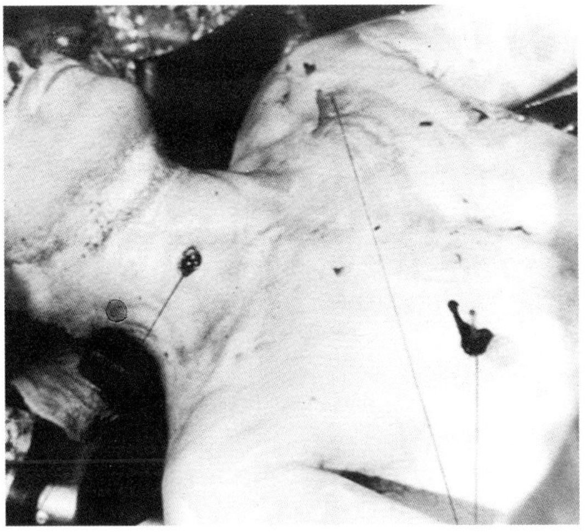

FIGURE 20.1
Fatal multiple stab wounds from needlefish, in a Japanese diver in Okinawa. The sites of penetration (neck, left shoulder and right chest) are indicated. (Photograph courtesy of Dr Yasuhiro Tomihara and colleagues, Okinawa.)

OTHER CNIDARIANS

Stings occur from multi-tentacled *Physalia* sp., *Gonionemus* spp. {Plates 8.52, 10.12 and 10.13}, and possibly *Olindioides formosa* (AOKI 1922).

CONE SHELLS

Conus geographus {Plate 13.8} is reported as causing four deaths in Okinawa (YASHIRO 1939; ARAKI 1994, unpublished observations) {see also Tables 3.3A and 3.3B, pages 68–69}:-

1927 a 5-year-old boy stung on the finger while clamming on a beach
1935 a 32-year-old male stung on the right thumb while clamming in the sea
1944 a 13-year-old male stung on the left finger while fishing in the sea
1954 an 11-year-old stung on the left hand while clamming on the beach

environmental rescue efforts of the hugely polluted Tokyo Bay (latitude 35–36° north), clear pictures of healthy carybdeid jellyfish were shown. The dimensions and the exact location of the animals were not indicated (WILLIAMSON 1994, unpublished observations).

BLUE-RINGED OCTOPUS

At least one species of *Hapalochlaena* {see Plates 13.2 and 13.3} is common, but no deaths have been reported so far.

SEA URCHINS

The brightly coloured and venomous 'tam o'shanter' sea urchin *Asthenosoma* sp. occurs on Okinawa beaches {Plate 12.14}. It stings severely when touched.

MARINE VERTABRATES

Marine vertebrates (primarily fish and sea snakes) are thought to have been responsible for 14 deaths in Okinawa to 1989, another in 1992 and one of a date unknown.

NEEDLEFISH

Five deaths recorded. These injuries usually result from night divers wearing torches (Plate 20.1), which attract the needlefish. These fish* have very long, sharp and strong needle-like jaws (Plate 20.2) capable of causing fatal penetration injuries (Figure 20.1). The fatal penetration injuries known are anterior intracranial via the orbit, via the breast, via the abdomen, and via the neck (Figure 20.1). The needlefish *Tylosurus crocodilus* (Plate 20.2) caused one death in 1992, but no details are available.

SEA SNAKES

Hydrophiidae species have caused four deaths.
Laticauda sp. two deaths (1989).
Hydrophis cyanocinctus, two deaths (1989).

* Other common names for various species of fish that have elongated narrow jaws include 'longhorn', 'gar', 'aiguillette', 'flying fish' and 'skipper'. They are capable of impaling humans in the water or above the water (i.e. flying garfish). Apart from Japan, serious and fatal stab injuries are recorded from Hawaii (MANDOJANA & SIMS 1987) and Papua New Guinea (BARSS 1982).

STONEFISH

Synanceia verrucosa produced one (indirect?) death in 1983. This was a male who was envenomed on the foot in 40 cm of sea water, when trying to put a stonefish in a bag. He had 4 puncture marks and 'fell down and was drowned' (ARAKI & TOMIHARA 1990-92, unpublished observations).

'ACCIDENTAL DEATHS'

ARAKI (1994, unpublished observations) reports that each year there are some 50 'accidental deaths' in the sea in Okinawa alone: 41 in 1987, 47 in 1988, 43 in 1989, 45 in 1990, 52 in 1991, and 46 in 1992. They all occur in the sea and occur while people are gathering sea shells and swimming, but definitive causes of these deaths have not been determined. {Refer also to Chapter 3, page 68.}

SECTION

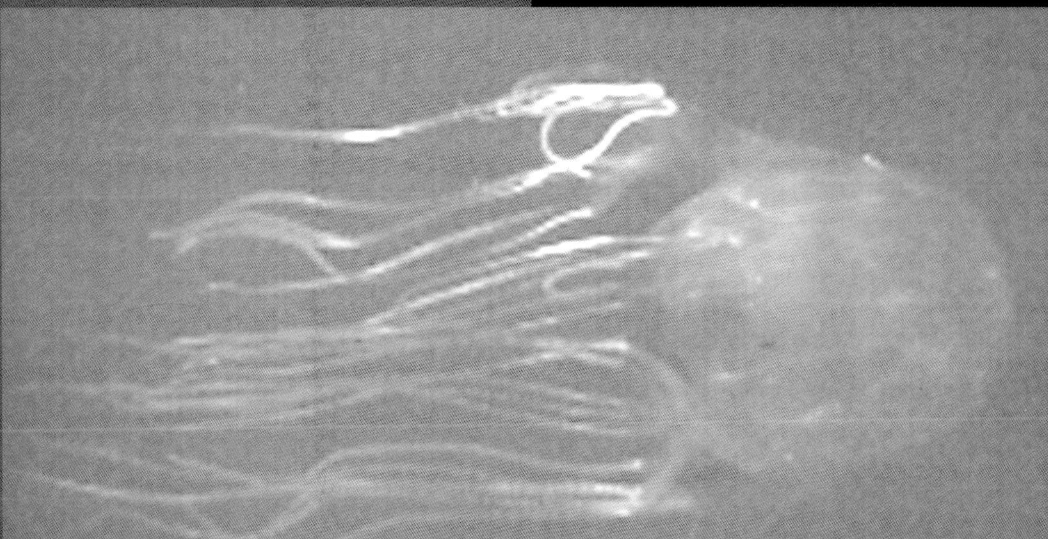

SIX

SPECIAL CONSIDERATIONS

CHAPTER 21

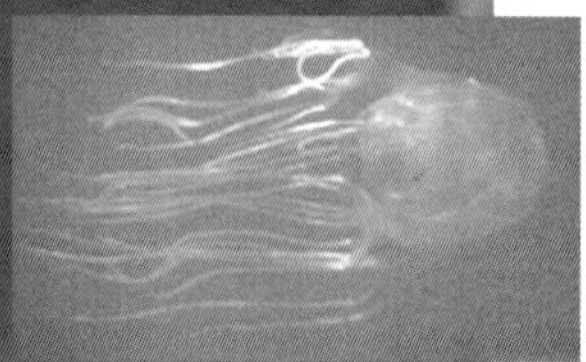

MARINE ANTIVENOMS

AVAILABILITY

Marine antivenoms (along with other hyperimmune sera) available commercially have been listed by THEAKSTON and WARRELL (1991) (using the format of CHIPPAUX and GOYFFON [1983]) and more recently by MEIER (1995). In comparison with antivenoms available for envenomating by terrestrial animals, such as snakes, scorpions and spiders, marine antivenoms remain few.

INTERNATIONAL MARINE ANTIVENOMS (EXCLUDING AUSTRALIAN)

SEA SNAKE ANTIVENOM

Japanese sea snake antivenom ('Dried Antivenin') is a rabbit anti-serum prepared against the venom of the sea snake *Hydrophis cyanocinctus.* (Interestingly, this antivenom is produced at the specific request of China.) It is a lypholised preparation with a 10 year shelf life (Japanese law), requiring reconstitution in sterile water prior to slow intravenous injection. It is ineffective against the venom of the sea krait *Laticauda semifasciate.* This antivenom costs 10 000 Japanese Yen per vial (approx. Aus$125.00), and is available from:

Japan Snake Institute
Dr Michihisa Toriba, PhD, and
Dr Yoshiharu Kawamura, PhD [Director])
Yabuzuka-honmachi, Nittagun
Gunma Prefecture, 379–23 Japan

Indian monovalent antivenom against *Enhydrina schistosa* venom (GAWADE *et al* 1980) — also effective against the venoms of *Hydrophis cyanocinctus*, *H. ornatus* and *H. spiralis* — was available from:

Haffkine Biopharmaceutical Co. Ltd
Parel, Bombay 400012, India*

FISH ANTIVENOMS (HORSE SERUM)

Monovalent and polyvalent, against weeverfish *Trachinus* spp., scorpionfish *Scorpaena* spp. and *Uranoscopus* spp. venoms, were at one time produced at:

Institute of Immunology,
Rockefellerova 2,
Zagreb, Yugoslavia.*

However all fish antivenom production at this institution ceased on 1 January 1994 (MEIER 1995).

AUSTRALIAN MARINE ANTIVENOMS

The story of the development of Australian antivenoms is a proud and enthralling one and has recently been chronicled by Professor Struan Sutherland (SUTHERLAND 1994b).

AUSTRALIAN MARINE ANTIVENOMS AVAILABLE (1995)

Beaked sea snake (*Enhydrina schistosa*)
Stonefish (*Synanceia horrida* = *Synanceia trachnys* and *Synanceia verrucosa*)
Box jellyfish (*Chironex fleckeri*)

All Australian marine antivenoms were previously made by and available from the Commonwealth Serum Laboratories, now officially known as CSL Ltd, at 45 Poplar Road, Parkville, Vic 3052, Australia. However, Australian venom research is now centred at:

The Australian Venom Research Unit,
Department of Pharmacology
University of Melbourne
Parkville, Vic 3052, Australia,
Phone: 61-3-93447753
Fax: 61-3-93482048

All enquiries should be directed to Associate Professor Struan Sutherland, Foundation Director of the Unit.

SEA SNAKE ANTIVENOM

This is a refined preparation of plasma from horses immunised with both beaked sea snake *Enhydrina schistosa* venom and Australian tiger snake *Notechis scutatus* venom (SUTHERLAND 1983). Each ampoule contains 1000 units in 15–35 ml, where 1 unit of antivenom neutralises 0.01 mg of *E. schistosa* venom in an *in vitro* assay (THEAKSTON & WARRELL 1991). It is an expensive antivenom but is the antivenom of choice for the treatment of envenomation by the following sea snakes (SUTHERLAND 1983):

E. schistosa, Aipysurus laevus,
Astritia stokesii, Disteira (Hydrophis) major,
Hydrophis cyanocinctus, H. elegans,
H. nigrocinctus, H. spiralis, H. stricticollis,
Lapemis hardwickii, Laticauda semifasciata,
Microencephalophis gracilis

Consequently (and in sharp contrast to envenomation by Australian land snakes, for example) identification of the species of sea snake responsible for the envenomation is not necessary to ensure usage of the correct (specific) antivenom. If sea snake antivenom is unavailable, the CSL monovalent antivenom against the Australian land tiger snake *Notechis scutatus* is effective. As a last resort for antivenom usage, if both these antivenoms are unavailable, CSL polyvalent snake antivenom (Australia–Papua New Guinea) can be used (SUTHERLAND 1983); although this will include the increased risk of delayed serum sickness associated with any large volume antivenom therapy (SUTHERLAND 1992) {see below}.

AUSTRALIAN STONEFISH ANTIVENOM

This antivenom, presented in ampoules containing approximately 1.5–3.0 ml, is a pure equine antiserum prepared against the venom of *Synanceia horrida* (= *S. trachynis*) (WIENER 1959b). It is also effective against the venom of *S. verrucosa* (SUTHERLAND 1983).

*It appears that currently marine antivenom production outside Australia is uncertain.

BOX JELLYFISH *CHIRONEX FLECKERI* ANTIVENOM

This is the world's only currently available cnidarian antivenom. Available since 1970 from CSL Limited in Melbourne, this antivenom is prepared from the serum of sheep hyperimmunised with the venom of the northern Australian box jellyfish, *Chironex fleckeri*. One ampoule of 1.5–4.0 ml contains 20 000 'units' (a 'unit' is the amount of the antivenom that neutralises one intravenous LD_{50} mouse dose) (SUTHERLAND 1983). It is effective against many effects of *Chironex fleckeri* venom *in vivo* in humans, particularly for pain relief and reduction of skin damage (WILLIAMSON *et al* 1984a; CURRIE & ANSTEY 1989; BEADNELL *et al* 1992).

THE DEVELOPMENT OF BOX JELLYFISH *CHIRONEX FLECKERI* ANTIVENOM

In 1967 Barnes developed his innovative venom collection technique, which led to production of the CSL's box jellyfish antivenom (BARNES 1967; SUTHERLAND 1983). Using a small vacuum flask with a human amniotic membrane (off a placenta 'borrowed' from the local hospital's maternity ward!) stretched across the neck {Plate 1.9A–C}, he applied gentle suction to the flask through a side arm, which pulled the membrane slightly downwards. Into this centre dip were placed fresh, live *Chironex fleckeri* tentacles. A small direct electric current was then passed across this membrane though electric terminals secured to the sides. In this way Barnes was able to induce the stinging cells (nematocysts) to discharge through the membrane, enabling pure venom to be collected beneath the membrane.

This venom was then frozen, sent to CSL in Melbourne, and there injected into sheep where antibodies were formed, enabling the first production in 1970 of the Commonwealth Serum Laboratories box jellyfish antivenom (BAXTER *et al* 1968; SUTHERLAND 1983 & 1994b). It was the first really effective treatment for severe *Chironex fleckeri* stings and remains a cornerstone of the management of any major *Chironex fleckeri* envenomation (WILLIAMSON *et al* 1984; FENNER *et al* 1989b; BEADNELL *et al* 1992). The venom is no longer collected in this way, and further research is suggesting some possible ways to increase further its undisputed clinical effectiveness (ENDEAN & SIZEMORE 1988; BURNETT *et al* 1990; ENDEAN *et al* 1993c).

STORAGE AND SHELF LIFE OF MARINE ANTIVENOMS

In general it is recommended by the manufacturers that antivenoms are stored away from light and kept cold (e.g. 2–8°C) but not frozen, until needed. The expiry date of the antivenom is printed on the label of most (including all Australian) antivenoms. No further details concerning shelf life were available (at the time of writing) for other international marine antivenoms.

SAFE CLINICAL USAGE OF MARINE ANTIVENOM PREPARATION

All antivenoms contain foreign (animal) protein and carry the risk of allergic reactions (MALASIT *et al* 1986) and serum sickness ('delayed reactions') in the recipient (CORRIGAN *et al* 1978; SUTHERLAND & LOVERING 1979; SUTHERLAND 1992). However, in the presence of clear clinical signs of envenomation, appropriate marine antivenom should not be witheld solely because of such considerations. SUTHERLAND (1992) considers one reason for the recently observed decline in the adverse reactions to Australian snake antivenoms at least, is the appropriate use of premedication. Attention has also been paid to the method of administration of antivenoms (SUTHERLAND 1977), including marine antivenoms (WILLIAMSON 1985).

PREMEDICATION

Premedication for land snake antivenoms has been controversial, centring on concern about the risk of a cerebrovascular accident resulting from premedicant adrenaline in the face of snake venom-induced coagulopathy (SUTHERLAND 1992, 1994); this matter is becoming resolved. Happily, neither sea snake venom (REID 1956; SUTHERLAND 1983; ACOTT 1983) {see Chapter 16, page 354} nor any other known marine venom interferes significantly with systemic coagulation *in vivo* in humans, so that this particular controversy does not apply to marine antivenoms.

A suitable premedication for the administration of any of the marine antivenoms listed above is:

1. Subcutaneous or intramuscular (IMI) adrenalin 0.25 mg (0.005mg/kg in children). The IMI route is preferred in the presence of clinical shock.
2. A parenteral antihistamine of choice (e.g. IMI promethazine 25 mg [0.5 mg/kg body weight in children]).
3. In patients known to be allergic to antivenom, in addition IMI corticosteroids (e.g. 100 mg hydrocortisone, or 2 mg/kg body weight in children) should be added.

Delayed serum sickness, yet to be reported in the literature following any marine antivenom administration, has also declined in frequency following Australian land snake antivenom usage associated with the prior administration of corticosteroids (SUTHERLAND 1983). This complication is associated with large volume antivenom administration (e.g. polyvalent ampoules, or many monovalent ampoules used) and would also appear to be more common with other world antivenoms (CORRIGAN *et al* 1978). On these grounds an argument can be presented for the inclusion of corticosteroids in the above premedication regimen for any marine antivenom.

ANTIVENOM DOSAGE

It is important for therapists to remember that the purpose of any antivenom is to neutralise **the dose of venom** *in vivo*, not the patient! Therefore, the dose is not reduced for small patients or children. Total dosage is a matter of careful titration against the clinical response of the closely monitored patient. In this context, the appropriate use (including the judicious release and/or re-application if necessary, under direct medical supervision) of compression bandaging (SUTHERLAND *et al* 1979) {see Chapter 5, page 98} can assist in the manageable rate of systemic release of venom during this clinical titration process. (It should be noted that compression immobilisation bandaging should not be used for stonefish or stingray envenomation, nor for venomous fish stings generally {see Chapter 16, page 354}.)

MARINE ANTIVENOM ADMINISTRATION

Ideally, administer the antivenom slowly intravenously (excepting stonefish antivenom), diluted in 500 ml (or at least 1 in 10 [SUTHERLAND 1992]) in isotonic intravenous fluid (e.g. 0.9% sodium chloride) (WILLIAMSON *et al* 1984a). An administration rate of 500 ml over 20 minutes is suitable for adults, the clinical state of the patient permitting. Have full resuscitation facilities together with measures to counteract anaphylaxis (EXECUTIVE EDITORIAL BOARD 1989) immediately to hand.

The second but important exception to this advice, as a matter of practical necessity, is associated with the emergency on-beach management, by trained paramedical personnel, of a major chirodropid jellyfish envenomation {see Chapter 5, page 98, and Chapter 8, page 180}. This consists of the immediate intramuscular injection (into the lateral thigh) of 3 ampoules (60 000 units) of specific CSL box jellyfish (*Chironex fleckeri*) antivenom (Plate 21.1) (FENNER *et al* 1989b). This initially controversial policy has been adopted by the Queensland Ambulance Service in north and central Queensland coastal areas (Plate 21.2), and has so far proved successful (FENNER *et al* 1989b; BEADNELL *et al* 1992). There are only two possible skin reactions (BOYD 1984; FENNER *et al* 1993) and no serious allergic reactions to CSL *Chironex fleckeri* antivenom have been reported despite more than 100 administrations since 1970

(SUTHERLAND 1983 & 1992; FENNER *et al* 1993). No premedication was used in most cases. The editors recommend the adoption of this Queensland training and practice policy on a wider national and international scale. Nevertheless, in their opinion, routine premedication along the above lines is recommended, when it is possible and practical.

FUTURE RESEARCH AND MARINE ANTIVENOMS

Marine antivenom technology and international availability are presently inferior, both qualitatively and quantitatively, to terrestrial antivenoms (THEAKSTON & WARRELL 1991; WILLIAMSON 1992; MEIER 1995). The balance of research effort now needs to be corrected somewhat in favour of marine antivenom development. Antiserum to the sea anemone toxin ATX II (BÉRESS & BÉRESS 1971) {see Chapter 8, page 180} has been prepared (MAIER & RATHMAYER 1982). Early efforts are being made for 'non-antigenic' tetrodotoxin and also saxitoxin (WATABE *et al* 1989; KAUFMAN *et al* 1991); the high molecular weight (FREEMAN *et al* 1974) polypeptide-containing (NAKAMURA *et al* 1983) toxins from several cone shell species offer prospects for antivenom development (KOHN *et al* 1960; SUTHERLAND 1983); however, none is currently available. There are also no existing antivenoms available for ciguatoxin poisoning, or for envenomation by stingrays, crown-of-thorns sea star or, importantly, Irukandji stings. Cloning techniques may offer future promise for marine antivenom development (NAGUIB *et al* 1988; WATABE *et al* 1989; COLLINS *et al* 1993).

IMMUNISATION AGAINST MARINE ENVENOMATION (VACCINES)

Immunisation would seem a practical proposition for occupationally or environmentally exposed population groups — Pacific islanders for ciguatera poisoning and shellfish poisoning, and fishermen for venomous fish stings and perhaps jellyfish stings. Earlier attempts were made to develop an immunising agent (toxoid) against *Chironex fleckeri* venom (BAXTER & MARR 1975), but a combination of clinical impracticality and certain laboratory difficulties led to the efforts being discontinued (SUTHERLAND 1983). No available vaccines against marine envenomation or poisoning, whose clinical efficacy is established, are presently known to the editors.

CHAPTER 22

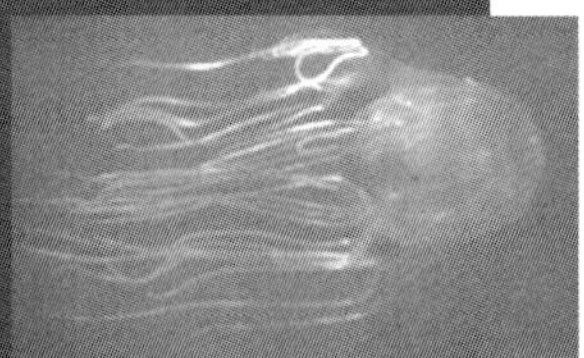

SCUBA DIVING PERSPECTIVE

JOHN WILLIAMSON

ENVENOMATION HAZARDS TO DIVERS

There has been a huge worldwide increase in self-contained underwater breathing apparatus (scuba) and surface supply breathing apparatus (ssba or 'hookah') diving in the past 20 years. The major component of this increase has been in recreational divers (WILKS 1993), who now number in the millions (LINAWEAVER & BOVE 1990). Hand in hand with this increase has been a steady increase in the number of injuries to divers from marine animals in all tropical and temperate oceans (EDMONDS 1989, 1992; WILLIAMSON *et al* 1993; BURNETT *et al* 1994a; AUERBACH 1995; WILLIAMSON 1995).

While divers are no more exposed to poisoning by the ingestion of marine fish than many other groups, by virtue of their underwater breathing capability divers may now invade marine animals' territories previously unassailable by humans. (Skilled breath-hold divers [snorkellers] are in a similar, if somewhat more restricted category.) Divers may approach venomous marine animals from new directions (e.g. ascent from below into the tentacles of a jellyfish on the surface [RUSSELL 1966; KIZER 1984b; HALSTEAD 1988; BURNETT & GABLE 1989; WILLIAMSON *et al* 1993; BURNETT *et al* 1994a]). Also, unlike fossickers, paddlers or aquarium handlers, they are exposed to envenomation and to animal injuries on any part of the human body by virtue of their total immersion (HARTWICK 1991a) {see Case Reports 1 and 2 in Chapter 8, page 187}.

There is an on-going need for diver education to protect divers, the animals and the underwater environment (WILLIAMSON 1988 & 1993).

The following cases have been documented (or can be verified by the editors), of fatalities or serious morbidity resulting from marine animal injuries, or fish flesh poisoning, in scuba, 'hookah' and breath-hold divers. For specific details of the animals mentioned below, and of the injuries they may produce see the relevant sections in this text.

Note: A world pioneer in the documentation of marine creatures that present a hazard to divers is Dr Carl Edmonds of Sydney (EDMONDS 1989, 1990, 1992, 1995).

Note that this book does not cover in any detail shark trauma (COPPLESON 1958; DAVIES 1964; AUSTRALIAN SHARK ATTACKS 1981; CAMPBELL & SMITH 1993), eel trauma (EDMONDS 1989) or crocodile trauma (Plate 22.1). Needlefish trauma fatalities in Japanese divers in Okinawa are described in Chapter 20, page 414. None of these events involves intoxication (at least by the animals!).

FATALITIES

Jellyfish	Atlantic *Physalia* sp. {Plates 8.33 and 10.22} envenomation (BURNETT & GABLE 1989)
Stingray	Breath-hold diver in Fiji (KNIGHT 1989) {Plates 16.7A and B}

SERIOUS MORBIDITY

Jellyfish	Atlantic *Physalia* sp. envenomation (RUSSELL 1966; HALSTEAD 1988; BURNETT *et al* 1994)
	'Irukandji Syndrome' {page 00} (WILLIAMSON 1985; WILLIAMSON *et al* 1991 & 1993; FENNER & WILLIAMSON 1994, unpublished observations)
	Jimble *Carybdea rastoni* {Plates 9.26 and 9.27} summertime stings among Broome (Western Australia) pearl divers (APPLEBY 1994, unpublished observations)
Sea anemone	*Cryptodendrum adhaesivum* envenomation on the buttocks (Plates 8.10 and 8.11).

> Editors' note: During November 1994 at least two snorkellers came into contact with a clump of sea anemones (species not identified) around the pylons of a marine jetty, in Waitemata district, New Zealand. They developed early contact pain, vesiculation, rashes {Plates 26.3A and B} and, in one case, subsequent protracted 'giddiness, inability to walk comfortably without being off balance, inability to concentrate, and extreme tiredness'. This patient's skin rash responded slowly to brief systemic corticosteroid therapy, but the central neurological effects continued and interfered with his occupation. Other neurological diagnoses have been excluded; serological investigations are continuing. The other patient developed no such neurological symptoms and made an uneventful, if somewhat protracted, recovery from his skin damage (M.A. MILES 1994 unpublished observations {see Chapter 26}).

Corallimorpharian *Discosoma* sp. (Ruppell and Leuckart 1828), producing a mononeuritis (BURNETT *et al* 1994) {Plates 8.21B}

Ciguatera poisoning from eating coral trout (*Cephalopholis miniatus*) speared on Marion Reef (ADAMS 1993)

Blue-ringed octopus envenomation occurred in a male diver while taking underwater photographs of something else (SOUTHCOTT 1980 unpublished observations) {see Chapter 13, page 331}

STINGINGS THAT SPOIL A DIVER'S DAY!

Stinging hydroid colonies {e.g. Plates 8.2 and 8.24} can produce smart stings in unwary or unprotected divers (Plates 22.2A and 22.2B), especially those without gloves and a wet suit or stinger suit, and those who may inadvertently trap fragments of hydroid within the mask.

Spine-bearing, sedentary marine creatures, such as the crown-of-thorns sea star {Plates 12.1–12.4}, and sea urchins {Plates 12.10–12.14} present well-known hazards to divers.

Divers who wish to touch or handle marine sponges would do well to wear gloves and not touch their bare skin with the gloves subsequently during the dive {see Chapter 14 and the Case Report on page 342}. The same hazards continue for the out-of-water handling of marine sponges after the dive. Earlier reference has been made to the Mediterranean sponge diver's disease (*maladie des plongeurs, maladie des pêcheurs d'éponges,* Skevas-Zerfus disease) {see Chapter 8, page 186}, an irritating skin rash on these divers, which is not caused by the sponges themselves but by a sea anemone growing on the sponges (ZERVOS 1903).

Less well known is the report by AUERBACH (1988) of a contact dermatitis (and presumably eye irritation) resulting in a diver who cleared his/her mask in close vicinity to, but not actually touching, the exposed Cuvierian tubules of a sea cucumber {Plate 12.16}; these contain the toxic saponin holothurin {see Chapter 12, page 325}).

Night diving puts divers at increased risk of contact with particularly nocturnally active venomous marine animals, such as groups of the tiny stinging carybdeid *Carybdea sivickisi* [HARTWICK 1991a] {Figure 9.6, page 257} and sea urchins {Chapter 12, page 319}).

One Australian authority on venomous fish (COLEMAN 1991) claims that a species of sabre-toothed blenny, *Plagiotremus rhinorhynchos*, commonly called the lyretail blenny, a reef fish up to 8 cm in length with large venomous fangs in the lower jaws, is responsible for painful bites in divers (McKAY 1994, unpublished observations).

SECONDARY INFECTION OF MARINE WOUNDS

Although this section makes no specific reference to the hazards to divers of non-venomous marine animals that inflict trauma, such as sharks, eels, and certain fish — the trigger fish (RANDALL & MILLINGTON 1990) and the Japanese needlefish {Plates 20.1–20.3}) — nor to coral cuts and abrasions {Plate 8.3}, both minor and major wounds are prone to secondary infection from marine pathogens (BUCK *et al* 1984; INTERNATIONAL SYMPOSIUM on DIVING in HAZARDOUS WATER 1991; KLONTZ *et al* 1993), which may be serious if untreated. See Chapter 4 {page 93} for specific management details of such complications, and also for references to possible inhaled infections from contaminated snorkels.

SPECIAL CONSIDERATIONS FOR DIVERS

DIFFERENTIATING MARINE ENVENOMATION FROM DECOMPRESSION SICKNESS

Pain, weakness, or other neurological signs or symptoms in any diver following a dive immediately raises the possibility of decompression illness (DCI), including cerebral arterial gas embolism (GORMAN 1991). Many marine envenomations in divers are obvious (EDMONDS 1989; BURNETT *et al* 1994). All cnidarian stings and most other marine envenomations are accompanied by immediate and localised pain. Exceptions to this, apart from the actual ingestion of a marine toxin (e.g. ciguatera poisoning), include the blue-ringed octopus bite (WALKER 1983), some cone shell envenomations {Chapter 13, page 338}, and the occasional sea snake bite (MERCER 1981; SUTHERLAND 1983) {Chapter 17, page 398}. However there has not so far been an unprovoked bite recorded on any fully immersed person by a blue-ringed octopus {but see the note on page 328 of Chapter 13}, nor has such a bite yet been reported when the animal itself was in open water (SUTHERLAND 1983).

The onset of peripheral neurological symptoms such as tingling of lips, paraesthesiae, extreme fatigue and itching due to ciguatera (tropical fish) poisoning {see Chapter 15, page 350} in a diver can suggest DCI (ADAMS 1993). Mild neurological symptoms and signs, again hallmarks of DCI, may also appear after less-serious sea snake or cone shell envenomation. Wet suits (but not stinger suits) offer good protection against sea snake bite (HEATWOLE 1975b).

The Irukandji syndrome {see Chapter 9, page 249} in a diver may present a serious differential diagnostic challenge (WILLIAMSON 1985; BURNETT *et al* 1994). This sting, usually from a tiny cnidarian {Plate 9.2}, seldom leaves a skin mark of any significance {Plate 9.23}, and the onset of symptoms of severe muscle pains, prostration, headache, nausea and vomiting is characteristically delayed some 20–30 minutes following the sting (BARNES 1964; FENNER *et al* 1988).

No marine envenomation requires recompression! A careful history of handling an animal, or of a sighting or 'touch' felt by the diver while in the water, and/or of a recent sting or bite, a careful inspection for bite or sting marks or even adherent tentacle material on skin not protected by a wet suit, and a careful assessment of developing symptoms and signs, will usually enable a correct diagnosis. The Australian 24-hour 'Diver Emergency Service' telephone (DES) — 1-800-088200, or internationally 61-8-83735312 — should be consulted early if in any doubt (WILLIAMSON *et al* 1991). Premature requests for either retrieval services or recompression facilities can be avoided by early expert consultation.

A number of countries are developing 24-hour diving emergency telephone consultation facilities. These countries are collaborating in the establishment of an international network, which aims to provide expert diving medical advice quickly by phone, and to facilitate diver emergency requests for help and advice from most recreational diving locations of the world. Examples are 'DAN (Divers Alert Network) Europe', 'DAN Japan', 'DAN U.S.A.', 'DES Australia' and 'DES New Zealand'.

MANAGEMENT OF MARINE ENVENOMATIONS IN DIVERS

1. RESUSCITATION

As always, resuscitation takes absolute priority over all other immediate measures {see Chapter 5, page 107 and Plate 9.88}. The principles are the same for divers as for all other victims. In a diver in need of resuscitation it will usually be essential, following removal from the water, to remove all diving equipment and release or cut off the top of the patient's wet suit.
The golden rule of resuscitation:
Initial resuscitation is of supreme importance. Those on the spot play the most vital role.

Other than protection of the airway, resuscitation of an unconscious diver should not normally be attempted until the diver is removed from the water; and such removal must be carried out as quickly as possible. The expeditious removal of an unconscious, fully kitted diver from the water onto a boat deck or jetty may be a formidable task for a lone rescuer. Thus the exception to 'no resuscitation attempts in the water' is expired air resuscitation (EAR) for a non-breathing diver on the surface, when immediate removal from the water is impossible [Editors' note: mouth to mask resuscitation is easier to perform when resuscitation is necessary in the water]. Even then, the reliable diagnosis of cardiac arrest (absent carotid pule) in a fully kitted diver in the water is virtually impossible, and apnoea (absent breathing) may or may not be accompanied by cardiac arrest. In any case, chest compression in the water is an impossibility and should never be attempted. Effective assessment and cardio-pulmonary resuscitation (CPR) of any apnoeic diver is possible only out of the water. Added to this, there is an urgent need to place any diver with in-water impaired consciousness into a stable horizontal position (GORMAN 1994).

2. POSTURE AND OXYGEN ADMINISTRATION

It should be remembered that a marine animal injury (actual or perceived) during a dive may produce a panicked ascent to the surface, with the attendant risk of pulmonary barotrauma (pressure injury to the lungs); it should also be remembered that any impairment of conscious state or muscle power in the water may be complicated by the aspiration of water into the lungs (near-drowning). Thus a diver may present with two medical conditions simultaneously: pulmonary barotrauma and any of its consequences, and marine envenomation and any of its consequences. Should any suspicions of pressure injury exist — no matter how faint — first-aiders should place the diver immediately into the horizontal position to reduce the risk of cerebral arterial gas embolism (CAGE), and administer 100% inspired oxygen to the patient (GORMAN 1994).

3. COMPRESSION/IMMOBILISATION BANDAGING

{Figures 5.3A and B, page 110 and 17.2, page 400} Envenomation from sea snake, blue-ringed octopus or cone shell should be followed as soon as possible by immobilisation of the patient and application of compression bandaging over the bite/sting site {see Chapter 5, page 110} (SUTHERLAND 1983, 1994c). This bandage should not obstruct the arterial supply and should not be released until the patient has direct medical supervision {see Plate 17.5}. The application pressure is apparently important (HOWARTH *et al* 1994) {see Chapter 5, page 111}.

4. Pain relief (analgesia)

Pain from any spine or barb injury can be severe, and a modicum of relief can usually be obtained by immersing the envenomated part in hot (not 'boiling') water {see Chapter 5, page 112}.

Do not administer Entonox (50% nitrous oxide with 50% oxygen) to any diver because it may produce a (re-)occurrence of the symptoms of decompression illness (ACOTT & GORMAN 1992; EDMONDS 1994).

The direct application of wrapped ice or cold packs can provide effective analgesia for some cnidarian stings (EXTON *et al* 1989). {But see Chapter 5, page 110 for comments concerning the necessity for avoiding water exposure to dried nematocysts adherent to stung skin.}

Local anaesthesia in selected situations is very effective (FENNER *et al 1989*) but requires skilled medical participation.

Oral or injected analgesics are also effective when used correctly but should never be administered to patients who are not fully conscious and co-operative. Injectable analgesics require skilled medical involvement.

5. More than one diver affected

Some cnidarians occur in swarms; examples include *Physalia* spp., various members of the class Scyphozoa (such as *Linuche* sp.) and *Carukia barnesi*. Several divers simultaneously may sustain such envenomations. (Ciguatera poisoning {see Chapter 15, page 350} may affect several divers simultaneously, but usually not in the water! [ADAMS 1993]). Any Irukandji syndrome requires immediate medical consultation (telephone DES Australia — see above).

Prevention of marine envenomation in divers

- Do not handle marine animals or structures unnecessarily.
- Wear gloves and a full wet suit, or at least a stinger suit {Plate 1.22}, when diving {see the note on page 103 of Chapter 5}. This is particularly important in conditions of surge or current.
- Beware of reaching blindly into underwater crevices.
- Maintain neutral buoyancy underwater at all times, when not intentionally changing depth.
- Inspect any underwater area before lying or sitting on it {Plate 8.11}. This precaution has environmental as well as safety advantages.
- Follow the golden rule of ascent in diving: 'Look up, then reach up, then come up slowly'. When night diving, shine your light upward ahead of your ascent path. In jellyfish conditions, keep a lookout for jellyfish and tentacles when approaching the surface (BURNETT *et al* 1994).
- Follow instructions and advice in the pre-dive briefing concerning the avoiding of marine animal injuries, with emphasis on specific known local hazards.
- Always carry three broad stretch bandages in the diver's first aid kit.
- Treat the underwater environment with care and respect.

These tips are based on the advice of BURNETT *et al* (1994). Apart from envenomation hazards, as mentioned above marine animals may produce a startled reaction in a diver, which can lead to underwater panic and an uncontrolled ascent, with all its attendant dangers (EDMONDS *et al* 1992). A fine example of this is the attraction of a grasping but harmless cuttlefish (as opposed to a Japanese needlefish!) to the shining glass of a divers facemask (MARSH & SLACK-SMITH 1986).

CHAPTER 23

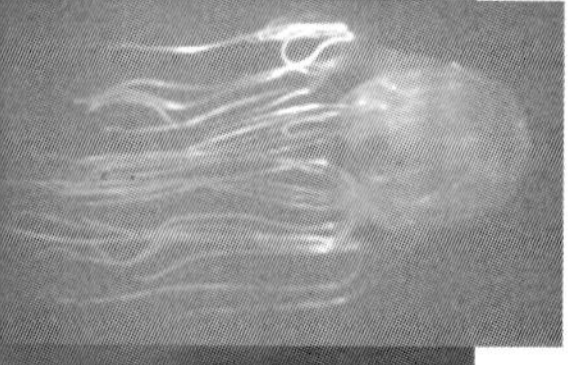

RESEARCH ASPECTS

THE EDITORS

THE INTERNATIONAL CONSORTIUM FOR JELLYFISH STINGS (ICJS)

In 1988, three of the editors (JWB, PJF, JAW), who until then had been working independently, established a collaborative body, initiated and named by JWB (Plate 23.1) 'The International Consortium for Jellyfish Stings' (ICJS) (BURNETT 1990a). A logo has been adopted (Figure 23.1), and the inaugural members include marine biologists, medical clinicians and medical research personnel.

The aims of the ICJS are to foster communication between interested colleagues worldwide in any aspects of marine envenomation, to circulate news of research efforts, and to compile an international database of the epidemiology of marine stings, clinical observations and therapy of such injuries (WILLIAMSON 1993a). To enhance communication internationally, two regular publications were commenced: the *ICJS Newsletter* and the annual 'Sting Report Summary' — and a small international mailing list was established in both Adelaide (Australia) and Baltimore, (Maryland, USA). By December 1994 this contained 160 addressees in 20 countries.*

---

* Since 1992, and at no extra cost to ICJS, the Australasian College of Tropical Medicine (ACTM) in Townsville has graciously circulated copies of each 'ICJS Newsletter' to all its mailing list members.

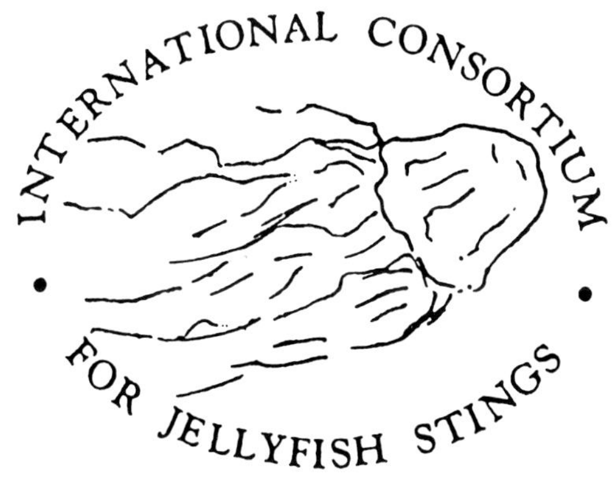

FIGURE 23.1
The logo of the International Consortium for Jellyfish Stings, developed by Professor Joe Burnett, MD, of Baltimore, Maryland, USA.

THE ICJS NEWSLETTER

The newsletter is compiled at least twice yearly in Baltimore, and contains a synoptic survey of relevant publications and research notifications in the international literature over the preceding six months. Requests for copies should be directed to:

Professor J. W. Burnett
Head, Department of Dermatology
University of Maryland Medical School
Baltimore MD 21201
USA
Fax 1-410-3280098
Email for Joseph W. Burnett:
jburnett@umabnet.ab.umd.edu

ICJS STING REPORT SUMMARY

This is compiled annually in Adelaide and Mackay (Australia) from first-hand data collected from sting reports, correspondence, surf life saving and rescue data, and publication material. Only verified data are included, and all sources are referenced. Requests for copies should be directed to:

A. Prof. J. A. Williamson, FANZCA
Hyperbaric Medicine
Royal Adelaide Hospital
Adelaide SA 5000
Australia
Fax 61-8-82324207
Email for John Williamson:
jaws@dove.net.au

or Dr P. J. Fenner, FRCGP
P.O. Box 3080
North Mackay QLD 4740
Australia
Fax 61-79-577824
Email for Peter Fenner:
Peter.Fenner@m130.aone.net.au

The mix of disciplines and nationalities on the ICJS mailing list reflects the international occurrence of marine envenomation problems and the wide range of skills and knowledge necessary to tackle the problems it presents (WILLIAMSON 1992).

THE IMPORTANCE OF FIRST-HAND DOCUMENTATION AND OF CLINICAL PHOTOGRAPHY

As mentioned in Chapter 5 {page 105} the early written recording by first-hand observers of marine envenomation events, and especially early and/or follow-up, high-quality photographic records of animals and stings are of great value in the study and advance of understanding for prevention and treatment of such injuries. Such records should include date, place and time, but without patient identification details; if the material is extracted from a newspaper or periodical, the full name of both the article and the publication in which it appears, date of publication, name of journalist if available, and page and column number, are all required. Some of the material in this book has been obtained by such worthy voluntary efforts.

As has been previously emphasised, any such on-site efforts must never be permitted to interfere with rescue, resuscitation or prevention measures. In addition, unskilled attempts to 'capture' venomous marine animals in order to obtain photographs is to be discouraged; anyone doing this is likely to injure themselves (and someone else) and the animal.

Surf Life Saving Queensland, or any of the editors, would be grateful to receive any such information. Biological (as opposed to medical or first aid) information or queries should be directed to:

Dr Jacquie Rifkin, PhD
Consultant in Marine Biology
105 Riverside Avenue
Karalee. Queensland 4306
Australia
Fax 61-7-33651655
Email for Jacquie Rifkin:
W.Stablum@mailbox.uq.oz.au

All such material and information would be added to the ICJS database and would be published only with the provider's permission and with due acknowledgment.

PUBLIC EDUCATION

This important topic has been discussed in detail elsewhere in Chapter 5, under the heading 'Prevention of marine envenomations' {see page 106}.

From the research aspect, public education remains a challenging frontier (WILLIAMSON 1992b), which is still commonly neglected out of ignorance of its power, of misplaced priorities, or even through cynicism. The tendency of many people in the Western world, particularly, to regard progress in this (or any) field as impossible without a budget of millions of dollars is an example of such ignorance. Surf Life Saving in Australia is a particularly fine example of the success of a sustained and intelligent public education program using a modest budget and the altruism of thousands of ordinary men, women and children (SURF LIFE SAVING AUSTRALIA LTD 1993). However, while the advice offered by most public education is pure 'common sense', it is acknowledged that common sense is far from 'common'! The added incentive of the needs of environmental conservation now demands an organised international research effort in public education (MOORE 1992; PLOTKIN 1992) and the realisation that it is an individual responsibility as well as a corporate one. To this end, a shift in attitude towards responsible citizenry of the world (rather than of just town or state), and the assistance of human behaviour psychologists in the design of programs, are important.

INTERNATIONAL INDEX OF RESEARCHERS

The following scientists are actively engaged in research into the problems of marine envenomations within their respective areas of expertise. They would welcome contact concerning medical aspects (see the annotations 'MD', 'DM', 'FRCGP', or 'FANZCA') or biological aspects ('PhD', 'MSc', or 'MIScT'). The reader is referred to the list of contributors {pages 13–16}, for further professional contact details of many of these people. Here we include their facsimile numbers where available (as of January 1996).

Dr Christopher Acott, FANZCA, DipDHM. South Australia. Fax 61-8-82324207.

Dr Junaid Mahmood Alam, PhD. Karachi, Pakistan. Fax 92-21-5887513 (attn: Prof. Alam)

Mr Phil Alderslade, MSc. Northern Territory, Australia. Fax 61-89-411258.

Professor Paul S. Auerbach, MD. 6855 S Red Road, Suite 400, Coral Gables, Florida 33143-3632, USA. c/o Fax 1-415-7230121.

Professor Joseph W. Burnett, MD. Baltimore, MD, USA. Fax 1-410-3280098.

Associate Professor Vic Callanan, FANZCA, FFIC, ANZCA. North Queensland, Australia. Fax 61-(0)77-726711.

Associate Professor Pratap Chand, MRCP, DM. Sultanate of Oman. Fax 968-513419.

Mr Neville Coleman, Brisbane Australia

Dr Paul F. S. Cornelius, PhD. London, UK. Fax 44-171-9389158.

Associate Professor Bart Currie, FRACP, FACTM. Northern Territory, Australia. Fax 61-89-275187.

Dr David Davies, FANZCA, DipDHM. Perth, Western Australia. Fax 61-(0)9-3704541.

Dr Peter Fenner, MB BS, DRCOG, FRCGP, FACTM. North Queensland, Australia. Fax 61-(0)79-577824.

Professor Dr Thomas Heeger, PhD. Philippines. Fax: 63-32-460351

Professor Ken W. Kizer, MD. Undersecretary of Health, Department of Veteran's Affairs 810 Vermont Avenue N.W. Washington D.C. 20420 USA. Phone (w) 1-202-5357010.

Dr Masahiro Kohama, MD, IE Village Hospital, Okinawa. Japan. Fax 81-980-495851 or 61-8-82324207

Dr Franco Kokelj, MD. Italy. Fax 39-40-910690.

Dr Richard J. Lewis, PhD. Queensland, Australia. Fax 61-7-32033517.

Dr Hermes W. Mianzan, PhD. Argentina. Fax 54-23-861292 or 517442.

Mr Simon Moore, MIScT, FLS. United Kingdom. Fax 44-962-869836.

Dr Fernando C. Ramírez, PhD. Argentina. Fax 54-23-861292 or 517442.

Dr Jacqueline F. Rifkin, PhD, Brisbane, Australia. Fax 61-(0)7-33651655

Dr Ronald V Southcott, DSc, MD, DTM&H, Adelaide, Australia. Fax 61-8-82768873

Associate Professor Struan K Sutherland, MD, DSc. Melbourne, Australia. Fax 61-3-93482048.

Professor Yasuhiro Tomihara, PhD. Okinawa, Japan. Fax 81-98-9466711.

Professor R. Victor, PhD. Sultanate of Oman. Fax 968-513124.

Associate Professor John Williamson, SBStJ, FANZCA, DipDHM. South Australia. Fax 61-8-82324207.

Dr Yuri M. Yakovlev, PhD. Vladivostok, eastern Russia. Fax 7-4232-310900.

Dr Mingliang Zhang, MD. Qingdao, Shandong province. China.

SOME USEFUL ADDRESSES

As well as the contributors to this book {contacts listed on pages 13–16} the following list of colleagues are known to the Editors to be working in the field of marine envenomation or allied fields.

Associate Professor Nor Azila PhD
Department of Biochemistry, Faculty of Medicine
University of Malaya
59100 Kuala Lumpur
Malaysia
Fax 603-7557740

Dr J. T. (Joe) Baker
Office of the Commissioner for the Environment
4th Floor, Health Building
Moore Street
Canberra ACT 2600
Australia

Professor Jean Bouillon
Laboratoire de Zoologie
Université Libre de Bruxelles
50 av. F.D. Roosevelt 1050 Bruxelles
Belgium

Dr Dale Calder
Royal Ontario Museum
100 Queens Park
Toronto, Ontario M5S 2C6
Canada

Dr Ann Cameron
Department of Zoology
University of Queensland
St. Lucia QLD 4067
Australia

Simon Chater
Oceanographic Research Institute
Box 10712
Marine Parade 4056
Republic of South Africa

Timothy Coffer
National Museum of Natural History
10th and Constitution Avenues NW
Mail Stop NHB-163
Washington DC 20560
USA

Ben Cropp
Maritime Museum
Port Douglas QLD 4871
Australia

Dr Malik Fernando
118/2 Cotta Road
Colombo 8
Sri Lanka
Fax (Colombo) 696632

Dr P. Gopalakrishnakone
Department of Anatomy
National University of Singapore
Lower Kent Ridge Road
Singapore 0511

Dr Nicole Gravier-Bonnet
Université de La Réunion
Faculté des Sciences
15 Avenue Rané Cassin — BP 7151
97715 Saint-Denis Messag cedex 9
La Réunion

Professor Charles Griffiths
Zoology Department
University of Cape Town
Rodebosch 7700
Republic of South Africa

Dr Des Gurry
Department of Child Health
Princess Margaret Hospital for Children
GPO Box D184
Perth WA 6001
Australia

Professor W. (Bill) Hamner
Department of Biology
University of California
Los Angeles CA 90024-1606
USA

Dr Bob Hartwick
Senior Lecturer in Marine Biology
James Cook University of North Queensland
Townsville QLD 4810
Australia

P.C. Heemstra
Curator of Marine Fishes
JLB Smith Institute of Ichthyology
Private Bag 1015
Grahamstown 6140
Republic of South Africa

Dr John Hooper
Curator of Lower Invertebrates
Queensland Museum
Brisbane 4101
Australia

Dr Gerhard Jarms
Zoologisches Institut und Zoologisches Museum
Martin Luther King Platz 3
D 20146 Hamburg
Germany

Martin Jones
Marine Wonderworld
Townsville QLD 4810
Australia

Dr Shin Kubota
Seto Marine Biological Laboratory
Kyoto University
Shirahama, Wakayama 649-22
Japan

Dr Lyndon Llewellyn
Australian Institute of Marine Science
P.M.B. No. 3, MC Townsville, Qld 4810
Australia

Professor Jon T. Mader, MD
Department of Marine Medicine
University of Texas Medical Centre
Galveston Texas
USA

Ms Loisette Marsh
Curator, Department of Marine Zoology
Western Australian Museum
Francis Street
Perth WA 6000
Australia

Dr George Matsumoto
School of Biological Sciences
Flinders University of South Australia
GPO Box 2100 Adelaide, SA 5001
Australia

Professor Dr Dietrich Mebs
Zentrum der Rechtsmedizin
Universtät Frankfurt am Main
Kennedyallee 104
60596 Frankfurt Main
Germany

Dr P. del Negro
Laboratorio di Biologia Marina
Strada Costiera, 336-S. Croce
34010 Trieste
Italy

Dr Iekhsan Othman
Department of Biochemistry
Faculty of Medicine
Universiti Malaya
59100 Kuala Lumpur
Malaysia
Fax 603-7557740

Dr Francesc Pagès
Institut de Ciències de Mar (CSIC)
Passeig Nacional s/n
08039 Barcelona
Spain

Dr Robert Pearce
Suite 21 Mount Hospital Medical Centre
146 Mounts Bay Road
Perth WA 6000
Australia

Professor John Pearn
University Department of Child Health
Royal Children's Hospital QLD 4029
Australia

Ms Shirley Slack-Smith
Curator of Molluscs
Western Australian Museum
Francis Street
Perth WA 6000
Australia

Professor David Warrell
Nuffield Department of Clinical Medicine
John Radcliffe Hospital
Headington, Oxford OX3 9DU
United Kingdom

Dr Richard Willan
Museums and Art Galleries of the
Northern Territory
PO Box 4646
Darwin 0801 NT
Australia

CHAPTER 24

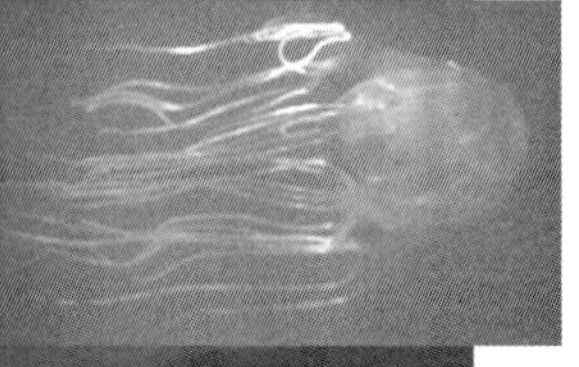

Mystery Syndromes Possibly Associated with Marine Envenomations

Joseph W. Burnett

Anecdotal case reports of various sting-associated syndromes afflicting bathers or people at the seaside have been reported. In many instances definitive causal relationship has not been established, nor have corroborating cases been uncovered. Therefore such syndromes should be classified as 'possibly marine envenomation related', and judgement held in abeyance pending future investigation and corroboration.

These cases have common features, in that they resemble other chronic ill-defined pathological entities affecting predominantly the skin, musculoskeletal or neurological systems. They have occurred all over the world, and their indefinite aetiology is occasionally related to lack of sufficient clinical data or to delay in the patient presenting for medical attention.

Dermatological Syndromes

One patient, who developed granuloma annulare after a *Physalia* sp. sting, was reported (MANDOJANA 1990) {Plate 11.2}. Although other cases of this entity have not been yet reported, some substantiation of its venom aetiology is offered by the existence of a similar injury following contact with a squid (FULGHUM 1986).

Erythema nodosum skin lesions were reported following a jellyfish sting (AUERBACH & HAYS 1987). However, this patient neither had pain nor noticed the jellyfish touch the skin. The patient clearly had deep erythematous nodules in the lower legs, and no conclusion could be made whether this was due to a granulomatous or a septal panniculitis, since a biopsy was not performed. Another patient noted the onset of a lymphomatoid papulosis eruption shortly after being stung by *Linuche* sp. in Yucatan, Gulf of Mexico (BURNETT & BURNETT 1990). A patient with a peripheral eruption diagnosed as vasculitis with small eschars, was also suspected as a delayed result of a *Physalia* sp. sting.

MUSCULOSKELETAL DISEASES

One well-authenticated patient who had a reactive polyarthralgia following a *Physalia* sp. sting has been reported (BRODNER 1988). Another case of vague arthralgias and myalgias following exposure to an unknown jellyfish in New Zealand is known (BURNETT 1992, unpublished observations).

PERIPHERAL NERVOUS SYSTEM AND SPINAL CORD DISORDERS

Ascending paralysis (polyneuritis) (Guillain-Barré syndrome [KOERNER 1993]) and various motor and sensory abnormalities (mononeuritis multiplex [BURNETT *et al* 1994b]) following marine injuries are documented. However, only a few of the case reports known to the editors have a clearcut temporal relationship or indeed closely postdate the sting.

OTHER CENTRAL NERVOUS SYSTEM DISORDERS

Three healthy young males developed various central neurological infarctions after swimming or wading in shallow water in the south-eastern Atlantic coast of the United States (MEYER 1993). No clearcut envenomation was documented, although each individual remembered an unusual encounter with something in the water, prior to the onset of symptoms. One patient, aged 39 years, developed infarction in portions of the cerebellum and brain stem with occlusion of the right posterior inferior cerebellar artery; another aged 15 years developed a basilar artery thrombosis; and the third patient, aged 13 years, was diagnosed (with elements of uncertainty) with non-specific myelitis, which had levels sufficiently high to cause loss of consciousness for a prolonged period of time. In two of the three cases the serum alkaline phosphatase level was double normal values. All were left with permanent neurological disabilities.

OTHER 'SYNDROMES'

Weakness, dysphagia, dysarthria and respiratory distress appeared 48 hours after a *Physalia* sp. sting in Hawaii. A case of transient global amnesia appeared a few weeks after a South Carolina (USA) bather stepped on a jellyfish thought to be a *Physalia* or *Chiropsalmus* species. A difficulty in assigning a causative role of these neurological cases to the sting is their lack of close temporal association. Yet the delay time for neurological symptoms in such envenomations may be longer than suspected. For example, one report of patients stung by *Hymenoptera* (bees, wasps) mentions a three- to ten-day delay before the clinical abnormalities appear (BACHMAN *et al* 1982). The indefinite nature of all the above cases is troublesome, so attention to future patients in order to detect corroborating instances may allow us to assign a more exact causal relationship of these entities with the cnidarian envenomation. Only time and diligent observation and documentation will tell.

CHAPTER 25

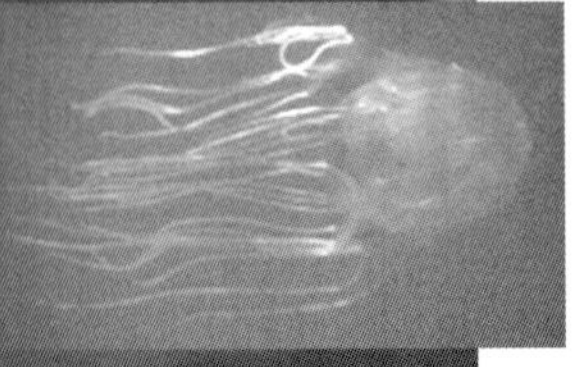

PLATYPUS ENVENOMATION

Editors' note: The inclusion of this subject in a text on marine envenomation could be considered somewhat inappropriate, as the Australian platypus is not a marine animal. Nevertheless, exposure to the risk of platypus envenomation (small though that risk may be) is shared by people who love and explore the Australian coastline where magnificent forests (and the streams that are the platypus's habitat) overlook the ocean. Further, although excellent researches into the biology of and envenomation by this remarkable animal are in progress (TEMPLE-SMITH 1978; GRANT 1989; GRIGG *et al* 1992) {see also Chapter 26}, some clinical and practical aspects of prevention and management of this painful experience remain inadequately documented. First-hand (!) clinical experience (FENNER *et al* 1992) is always valuable. We consider such information of direct relevance to those who may relate to the other material in this text.

THE AUSTRALIAN PLATYPUS *ORNITHORHYNCHUS ANATINUS*

PETER FENNER AND JOHN WILLIAMSON

DISTRIBUTION

The first European to capture a platypus in Australia did so in a lagoon near the Hawkesbury River, New South Wales, in 1797; disbelief in zoological circles followed its early reporting (COLLINS 1798). Platypuses are found in fresh water streams along the eastern seaboard of Australia and survive well in warm streams from Cooktown in northern

Queensland, to icy-cold streams high in the mountains of Tasmania (FLEAY 1980).

APPEARANCE

A platypus is a furry creature with a bill like a duck, webbed feet and a paddle-like tail for swimming, and claws on the webbed feet used for digging burrows (Plate 25.1). A mature male platypus may weigh 2 kg and reach 0.6 m in length, with the tail being the final third. A female reaches two-thirds the size of the male.

As well as having a strange appearance it is a warm-bodied animal (mammal), although the core temperature varies from 29.2 to 34.6°C (usually 32°C) (GRANT & DAWSON 1978; GRANT 1983a; GRIGG *et al* 1992). Unlike other mammals except the echidna, the female of the species lays eggs; she later suckles her young.

The male platypus is the only male mammal in the world, apart from a male echidna, (of which the editors are aware) that has retained venom glands (COLLINS 1798; FLEAY 1980; SUTHERLAND 1983). The male has a spur just beneath each rear webbed foot, connected to a venom sac. Only adult platypus spurs present any threat of envenomation; juvenile spurs (Plate 25.2) are harmless (GRANT 1989). These spurs are used as a defensive mechanism to keep other males away. The spurs are strong and can inject a venom that causes severe pain and morbidity (TEMPLE-SMITH 1978; FENNER *et al* 1992). It is not known if the venom varies in volume or toxicity during the breeding season, which is reported as being July and August in warmer areas, and October in cooler areas (FLEAY 1980).

Platypuses are shy creatures and nocturnal feeders upon worms, insects and small crustaceans on the bottom of their streams. They enter the rivers around dusk, feed actively during the night, and return to their burrows at dawn; it is during these entering and leaving times that they are most commonly sighted (GRANT 1983b; GRIGG *et al* 1992). However, some individual animals may feed in the daytime (see CASE REPORT below). It has recently been shown that the platypus has electroreceptors in the bill (SCHEICH *et al* 1986) which can detect prey via the alternating current emanating from the prey's muscles. It can also locate and differentiate inanimate objects by detection of direct current signals. This was the first time such a phenomenon had been reported in higher vertebrates, although similar receptors have recently been described in the Australian echidna (AUGEE & GOODEN 1993).

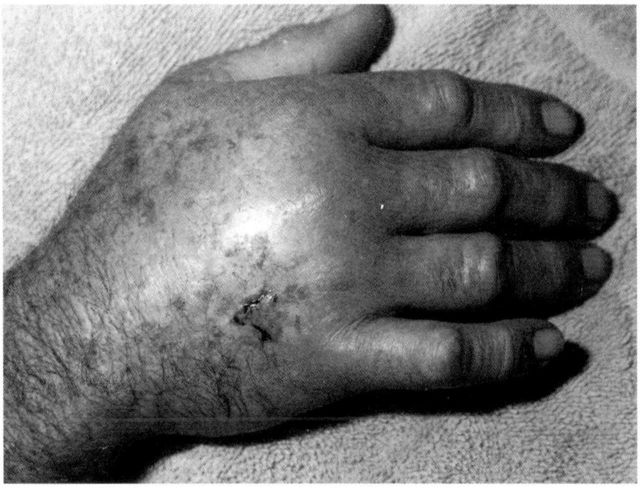

FIGURE 25.1A
The dorsum of the patient's painful, swollen and immobilised right hand, some five hours after platypus envenomation. Notice the absence of bleeding at the spur wound, and of circulatory embarrassment in the hand or fingers. The patient had excruciating local pain. (Copyright ©*Medical Journal of Australia* 1992; 157: 830.)

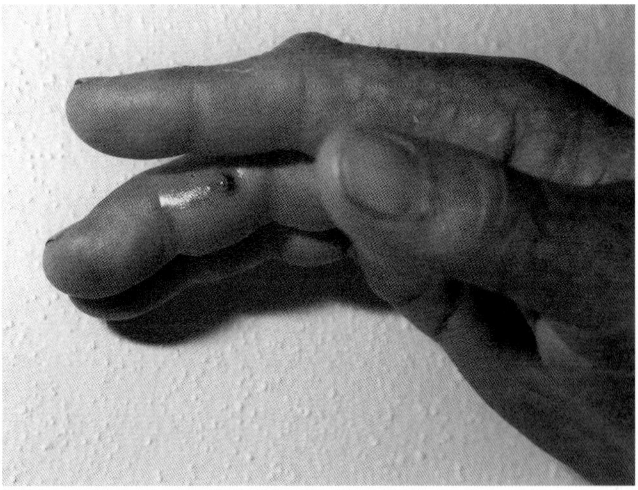

FIGURE 25.1B
The platypus-envenomated right middle finger after five hours. Note the spur puncture wound and the swelling of that finger. (Copyright ©*Medical Journal of Australia* 1992; 157: 830.)

Platypuses will usually dive quickly when approached and may stay underwater for up to 5 minutes. They live in burrows dug with their clawed webs from the muddy sides of their streams. It is not known for sure, but it is believed that the eggs are laid on the mother's abdomen and incubated under the tail (GRANT 1989) before hatching some 10–12 days later. The young are suckled by areolae on the ventral abdomen (GRIFFITHS 1978). They are weaned at about 16 weeks, reach maturity by two and a half years and may live as long as 16 years according to FLEAY (1980), the first person to have reared them in captivity. This feat has been repeated by Dr Noel Walmsley in South Australia.

ENVENOMATION

There has never been a human fatality, or near-fatality recorded, although deaths of hunting dogs have been reported (BURRELL 1928).

Envenomation causes immediate and intractable pain and marked swelling which is aggravated by movement and increased venous pressure (including coughing, sneezing, and any other valsalva manoeuvre). Consequently movement of the envenomed part is carefully avoided by the patient. Swelling further impedes movement.

Elements of reflex sympathetic dystrophy may appear, including exquisite hyperaesthesia of the injured part. Also pain and hyperaesthesia may spread from an envenomed limb into the body and become generalised and distressing. It may last several days before retracting back to the envenomated area in a similar, but slower manner than its spread. Loss of muscle mass has occurred in the envenomated limb (FENNER *et al* 1992.)

THE VENOM

Study of the venom has shown it to have coagulant effects (KELLAWAY & MESSURIER 1935). Little else is known {but see DE PLATER *et al* 1985 in Chapter 26}, although important toxinological studies have been carried out by Dr Peter Temple-Smith in Sydney (TEMPLE-SMITH in GRIFFITHS 1978; FLEAY 1980); Dr Temple-Smith injected himself with platypus venom, under medical supervision, during his studies {see editors' comments, Chapter 1, page 49}.

CASE REPORT (FENNER et al 1992)

A 57-year-old healthy male presented to his doctor's surgery at 10.30am on 30 May 1991 with intractable pain and marked swelling of his right hand four hours after envenomation by a small platypus on the dorsum of his right hand and right middle finger. There was 'immediate, severe pain'. The platypus was difficult to remove from his hand and had to be pulled in the reverse direction to the spurs' entry for detachment. Very little bleeding occurred in either the finger wound or the wound on the back of the hand.

The pain and swelling in the right hand became increasingly severe, reaching to peak severity about an hour after the original envenomation and not letting up thereafter. The victim and his son tried soaking a towel in hot water and applying it directly to the wounds, but this provided no relief. Any movement of the hand greatly aggravated the pain. The area appeared to be hyperaesthetic, as touching it or even the weight of the hot towel when placed on the area caused increased pain. A sling elevating the hand gave little relief. Ice also proved ineffectual, as even the pressure from small chunks of ice on the envenomated skin greatly increased the pain. There was obvious swelling of the right hand (Figure 25.1A) and right third finger (Figure 25.1B), but there was no ischaemia and capillary refill was normal. The pulse, blood pressure, respiratory rate and an ECG were normal. Intravenous (IV) morphine up to 30 mg gave very little pain relief.

Wisely, no attempt was made to slow absorption of the venom from the area by the use of compression/immobilisation techniques {see Chapter 5} because of the severe hyperaesthesia of the envenomated area and absence of systemic signs and symptoms. Tetanus prophylaxis was given before the patient was transferred to hospital, where he arrived at 11.30am, five and a half hours after the original envenomation. On arrival an IV morphine infusion of 1.5 mg per hour was commenced, and a triple-nerve wrist block was performed by an anaesthetist, using 20 ml of plain 0.5% bupivicaine. Appreciable pain relief occurred only after this regional anaesthetic block.

After swabs were taken from both puncture wounds, the patient was commenced on a cephalothin sodium 1 g six hourly IV. The hand was elevated on two pillows. Blood tests at this time included full blood count and ESR, urea and electrolytes, renal function tests, liver function tests and cardiac enzymes with the creatine kinase differentiated into cardiac and skeletal muscle. They all proved normal except for the ESR which was 44 mm/hour (normal 2–10), total protein 56 g/l (normal 60–80) and the serum albumin of 26 g/l (normal 33–50). Hourly observations showed a respiratory rate that remained at 18–20/min, but pulse oximetry breathing air (21% oxygen) gave an oxygen saturation of 87 per cent. Oxygen therapy was commenced, and 2 l/min via nasal prongs increased the oxygen saturation to 97 per cent. Vomiting and nausea began shortly after the morphine infusion was started but was controlled by a single injection of metoclopramide 10 mg IV. The hand remained numb and comfortable, and so the morphine infusion was reduced to 1 mg/hour at about 10pm.

Within two hours of reducing the morphine infusion the patient began to complain that a severe muscle and skin pain was moving 'all over'. The pain 'started moving from the right forearm into the upper arm, the right side of the chest and then it moved all over 'from the top of my head to the tips of my toes'. It was made worse by any pressure on the skin. Touching any area of the body made the pain 'severe'. It was described as 'feeling like a deep painful bruise'. The morphine infusion was increased to 1.5 mg but although this reduced the pain in the hand to a tolerable level, all areas of the body were painful to touch. Oral ibuprofen 400 mg eight-hourly was commenced. As it produced no benefit and as the patient previously suffered from duodenal ulceration and was currently on ranitidine 150 mg twice daily, it was ceased (although it caused no side effects). The patient's temperature was noted to be 37.5°C. He remained mildly pyrexic for the next four days with no evidence of underlying infection.

Two days after the original envenomation the hand and forearm were still 'tender to touch' but were not hyperaesthetic. Ice and cold packs could now be placed on the area. The hand and fingers were still just as swollen, but now some movement of the fingers was possible. Oral dextropropoxyphene napsylate 200 mg four-hourly was commenced. The morphine infusion was slowed to 1 mg/hr and then ceased later that night, although a single intramuscular injection of 10 mg was required in the early hours of the morning after the patient rolled on to his hand.

The next morning the swelling had decreased slightly and there was a little more movement in the fingers. The dorsum of the hand and the fingers were still very painful to touch but the tenderness had retracted down out of the forearm in a reverse fashion to its movement up the forearm, prior to its moving all over the body. The cold packs were very effective whilst applied, but the pain and tenderness returned as the area warmed up. The pain was 'tolerable'.

The patient was discharged from hospital six days later with the arm elevated in a sling. The third finger was still very swollen and had no movement, but the other fingers and the hand were less so and had 25 per cent movement. He was still taking dextropropoxyphene 200 mg, and using cold packs four-hourly for pain relief. The right lower forearm remained sensitive to touch, as well as the whole of the dorsum of the hand and the fingers. The forearm did not appear swollen. Oral prednisone 12.5 mg was commenced twice daily.

Three days later (two weeks after the original envenomation) the patient still complained of pain and tenderness in the right hand. He was unable to put any pressure on the hand and so had trouble cutting up food, dressing himself and signing cheques! The pain woke him at night if he rolled on to the envenomated hand. He considered that he had a 'great improvement' since starting the prednisone and that there was less 'pressure pain' when the hand was used. For the first time since the envenomation he was just able to hold the weight of a light jacket in his fingers. He still complained that he was unable to bear any weight (including a bed-sheet at night) on the back of his hand, as it irritated the hairs on the back of the hand, which caused a strange, painful sensation.

Loss of muscle bulk in his right forearm was noted at this time, and was still visible one month after envenomation (Figure 25.4). The whole arm, despite his being right-handed, was much weaker than the left. It was impossible to test for hand strength due to the pain and swelling. The circumference of the right (dominant) forearm over the compartment muscles was 270 mm compared to 285 mm on the left arm in the same position. Upper arm circumference over the biceps and triceps was 290 mm on the right arm compared to 310 mm on the left. At this time the skin of the dorsum of the hand and over the envenomated finger began to desquamate (peel off). It was a further five months before the hand returned to normal, painless function.

FIRST AID

The usual first aid measures for envenomation, including the application of either heat or cold, appear to be of no benefit in this envenomation. Compression bandaging increases pain, and is not clinically indicated {Chapter 5}.

MEDICAL MANAGEMENT

Hospital admission is essential (FENNER *et al* 1992) for the adequate management of this envenomation, for:

- regional anaesthetic blockade
- high-dose intravenous narcotics
  (It is emphasised that these measures should only be carried out where resuscitation facilities are immediately available).
- tetanus prophylaxis
- short-term systemic steroid therapy, which appears to diminish pain and persistent swelling in the affected part

Anti-inflammatory medication (including non-steroidal anti-inflammatory medications) seem of little benefit.

SECTION

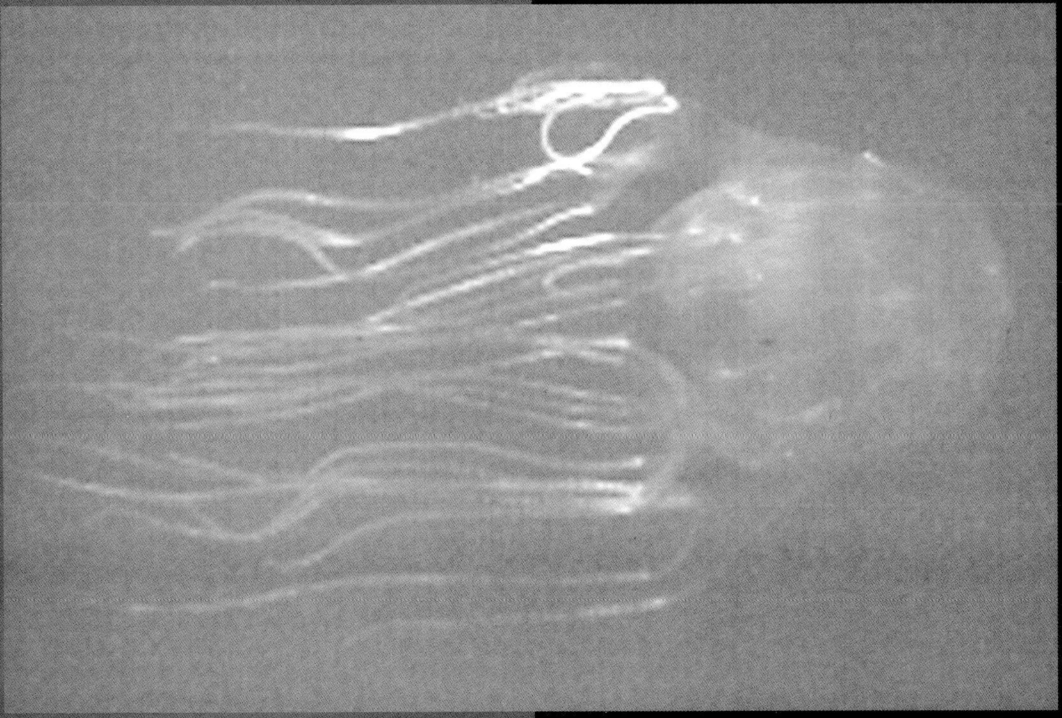

SEVEN

TO FINISH

CHAPTER 26

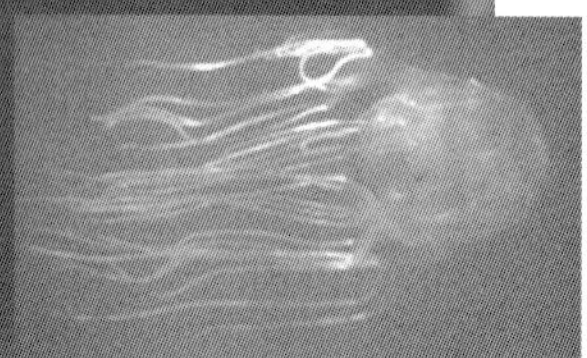

'STOP PRESS'

INTRODUCTION

This chapter attempts to gather together some last minute information of significance which came to the editors' notice too late for inclusion in the body of the text. While the majority of the material is very recent, inevitably some older but 'just-discovered' publications are included. The entries are listed under each subject heading first in chronological order by year, then in alphabetical order for the same year of publication.

CLASSIFICATION OF ANIMALS

FAUTIN DG, LOWENSTEIN JM 1992

SCYPHOMEDUSAE AND THEIR POLYPS ARE THE SAME IMMUNOLOGICALLY: IMPLICATIONS FOR SYSTEMATICS. *COMP BIOCHEM PHYSIOL* 102B:13–14.

This study showed that whole polyp and whole medusa stages of each of the two scyphozoan species *Aurelia aurita* and *Pelagia colorata* were indistinguishable by radioimmunoassay, although they were each distinguishable by radioimmunoassay from all other cnidarians which were tested. These results suggest that at least for some taxa and for some protein analysing techniques, it is unnecessary to require that immunological comparison be carried out using organisms at the same developmental stages.

EPIDEMIOLOGY

FEARON JJ, BOYD AJ, SCHÜLEIN FH. 1992.

VIEWS ON THE BIOMASS AND DISTRIBUTION OF *CHRYSAORA HYSOSCELLA* (LINNÉ, 1776) AND *AEQUOREA* (FORSKÅL, 1775) OFF NAMIBIA, 1982–1989.

Using BONGO nets, it was shown that largest concentrations of these jellyfish occurred within 45 nautical miles of the Namibian coast, with the smaller *Chrysaora* animals found closer to the coast than the larger *C. hysoscella* species. *A. aequorea* and *C. hysoscella* appear in this region where the Benguela Nino upwelling occurs.

BAKER JT. 1995.

PERSONAL COMMUNICATIONS.

1. From 16 November 1994, by reason of the (United Nations) Law of the Sea Convention (LOSC), Australia assumed responsibility for an additional area of marine territory adjoining the coastlines of the Australian continent and all of its islands, and of the Australian Antarctic Territory, known as the 'Marine Jurisdictional Zones of Australia'. These zones are (i) the sea bed for the entire area and (ii) the waters above the sea bed for those areas coloured green (see Plate 26.1); their area is approximately 1.5 times the area of the Australian continent. Australia has the responsibility for the management of these zones and their resources.
2. The Australian Institute of Marine Science (AIMS) in Townsville, north Queensland, has created a fully curated collection of over 7000 marine macro-organisms. This database is accessible for reference internationally, but strictly for scientific research purposes. In addition, AIMS scientists have embarked upon a systematic development of appropriate culture methods for the notoriously fastidious population of marine micro-organisms (some of which are seriously pathogenic to humans — see Chapter 4).

TRITRAKARN T. 1995.

UNPUBLISHED OBSERVATIONS.

From Thailand, Dr Thara Tritrakarn reports data of interest concerning tetrodotoxin (TTX) poisoning.

Freshwater puffer fish in Thailand can be toxic the year round. The flesh contains approximately 300 mouse units of TTX per gram. Sporadic intoxication of Thai villagers can occur in the north-east regions, where the puffer fish are regularly eaten.

Of the 58 cases of TTX poisoning recorded in Thailand, 22 (38%) were fatal; 30 of these (17 [57%] fatal) occurred prior to 1979; since 1979 only 5 out of 28 cases (18%) were fatal (see also Chapter 3, page 73).

GENERAL PHYSIOLOGY

1ST INTERNATIONAL CONGRESS OF ENVENOMATIONS AND THEIR TREATMENTS, INSTITUT PASTEUR, PARIS, 1–4 JUNE 1995.*

FREIRE-MAIA L. *et al* — The pulmonary oedema resulting from scorpion envenomation in Brazil is unlike that resulting from heart failure, hypertension or left ventricular failure. Radiological appearance is of 'butterfly wings' (hilar shadowing). It may be unilateral, and it occurs mainly in envenomed children. In the rat model, the pulmonary oedema often occurs with haemorrhage. Increased vascular permeability appears to be a feature. Clinically bradycardia, then tachycardia/bradycardia, hyperthermia and hyperglycaemia may occur. The envenomation syndrome commences some 30-120 minutes after initial envenomation. Scorpion antivenom does not always control cardiac/respiratory symptoms, which may appear even hours after envenomation, and may be due to small molecules of venom (7000Da) possibly opening Na channels (WHITE J). Australian funnel-web spider envenomation syndrome commences 10-60 minutes following envenomation.

* C. BON and M. GOYFFON (eds). Lyon. 1996.

(Editors' Note: Compare the above descriptions with Irukandji syndrome: see Chapter 9)

HISTORY

PEARN J. 1994.

HUGO FLECKER'S CONTRIBUTIONS TO MEDICINE AND NATURAL HISTORY. *MED J AUST* 161(1):63, 66–7.

This is a long overdue (albeit second) documentation of Dr Hugo Flecker's pioneer and powerful influence upon the science of toxinology in Australia.

IMMUNOLOGY

FAUTIN DG, LOWENSTEIN JM. 1992.

SCYPHOMEDUSAE AND THEIR POLYPS ARE THE SAME IMMUNOLOGICALLY: IMPLICATIONS FOR SYSTEMATICS. *COMP BIOCHEM PHYSIOL* 102B:13-14. (See above under 'Classification of Animals'.)

BURNETT JW, KUMAR S, MALECKI JM, SZMANT AM. 1995.

THE ANTIBODY RESPONSE IN SEABATHER'S ERUPTION. *TOXICON* 33:99–104.

Serological assay of specific IgG antibodies using ELISA showed 36 of 44 patients with sea bather's eruption in Florida, the Bahamas and Aruba had specific such titres against *Linuche unguilata* (thimble jelly) in their serum (even after its storage for 10 years). The specific titres correlated with sting severity or extent of the cutaneous eruption.

RECENT DATA FROM JOSEPH W. BURNETT'S LABORATORY. 1995.

Graduate student David Bloom, in conjunction with Dr Burnett, has demonstrated that an adult healthy male's T-cells which respond *in vitro* to *Chrysaora* (sea nettle) antigen have different Vβ family groups from those which proliferate in response to the herpes simplex antigen. They have also shown that a subfraction of *Chironex* crude nematocyst venom (from mostly mastigophores which had been purified by density gradient centrifugation) further purified by capillary electrophoresis is found in one of four fractions. This subfraction contains the least amount of protein but is the only one capable of causing T-cell proliferation in a previously non-exposed male. This subfraction is also the only one that reacts by ELISA with the Australian Commonwealth Serum Laboratories' commercially available ovine antiserum against *Chironex* and is also the only haemolytically active fraction.

With Heather Houck, MD, and Steven Vanucci, MD, the effect of *Chrysaora* and *Chironex* venoms upon cultured rat liver and kidney cells respectively has been studied. There is significant delay before cellular damage can be demonstrated; this corresponds well with the delay seen in the onset of hepatic and kidney damage in clinical cases.

A case of venous congestion appearing 3 days after envenomation by a jellyfish off the North Carolina shore was reported in 1995 (AL-EBRAHIM *et al* 1995 — see page 447 below). Serological antibodies demonstrated high reactivity to *Chiropsalmus quadrumanus*, which fits in well with this previous report (Editors' Note: See also GUEST 1959, and BENGSTON *et al* 1991).

MANAGEMENT

DAS SK, JOHNSON MB, COHLY HHP. 1995.

CATFISH STINGS IN THE MISSISSIPPI. *SOUTH MED J* 88:809–812.

This is a retrospective review of the occurrence and management of 83 catfish stings in humans. Immersion of the envenomed part in hot (45°-50°C) water provides helpful immediate pain relief. Secondary infections are often penicillin-resistant. Oral cephalosporin therapy is recommended for early infections.

FISHER M. 1995.

TREATMENT OF ACUTE ANAPHYLAXIS. *BMJ* 311:731–733

An authoritative recent paper which postdates other reference to the management of anaphylaxis quoted in this text. It once again underscores that 'Adrenalin is the treatment of choice for clinical anaphylaxis.'

TUN-PE, AYE-AYE-MYINT,KHIN-EI-HAN, THI-HA, TIN-NU-SWE. 1995.

LOCAL COMPRESSION PADS AS A FIRST-AID MEASURE FOR VICTIMS OF BITES BY RUSSELL'S VIPER (*Daboia russelli siamensis*) IN MYANMAR. TRANSACTIONS *ROYAL SOCIETY TROP MED HYGIENE* 89:293–295.

Although this was not a marine envenomation account, it is one of the first reports of the efficacy of compression/immobilisation bandaging {see Figure 17.2, page 400} upon envenomations outside Australia. Central movement of the venom antigen was retarded in 13 of 15 cases.

PROFESSOR DR DIETRICH MEBS, MD. 1996.

PERSONAL COMMUNICATION

Regarding hot water treatment of venomous fish stings, several medical colleagues, mainly in the Mediterranean area, have observed severe scalding effects when this treatment, such as 'keeping the limb in hot water for 30 minutes' is performed in the field. Professor Mebs considers that 'hot water' is an unsatisfactory term as the temperature is difficult to control under such conditions. Also his personal experience does not support the use of hot water for pain relief from *Acanthaster planci* (crown-of-thorns sea star) spine injuries {see Chapter 12, page 317}. (Editors' Note: Further controlled testing is obviously necessary.)

MARINE POISONING

CAMERON J, FLOWERS AE, CAPRA MF. 1991.

ELECTROPHYSIOLOGICAL STUDIES ON CIGUATERA POISONING IN MAN (PART II). *J NEUROLOGICAL SCI* 101(1):93-7.

Slowing of sensory conduction velocity and prolongation of the absolute refractory, relative refractory and supernormal periods were recorded in the sural and common peroneal nerves of 15 patients with acute ciguatera poisoning. These findings indirectly suggest that ciguatoxin causes an abnormally prolonged opening of sodium channels in nerve membranes.

CAMERON J, CAPRA MF. 1993.

THE BASIS OF THE PARADOXICAL DISTURBANCE OF TEMPERATURE PERCEPTION IN CIGUATERA POISONING. *CLIN TOXICOL* 31:571–579.

A study of the gross temperature perception abilities of 5 previously ciguatera-poisoned adults in the Brisbane area showed that such perception remained intact. The so-called 'paradoxical' temperature discomfort so often attributed to ciguatoxin (a lipid-soluble polyether compound of molecular weight 1111.7 Daltons, which causes opening of sodium channel in nerve membrane resulting in runs of spontaneous nerve discharges), appeared to be a misleading description of the sensory symptoms experienced in ciguatoxin poisoning. The specific sensory discomfort is most likely a result of exaggerated nerve depolarisation in small peripheral A-delta myelinated fibres and, in particular, C-polymodal nociceptor fibres.

(Editors Note: See also CAMERON J, FLOWERS AE, CAPRA MF. 1993, in the main Bibliography.)

KATZ AR, TERRELL-PERICA S, SASAKI DM. 1993.

CIGUATERA ON KAUAI: INVESTIGATION OF FACTORS ASSOCIATED WITH SEVERITY OF ILLNESS. *AMER J TROP MED & HYG* 49(4):448–54.

A 1985 outbreak on the island of Kauai in Hawaii, involving 15 persons of various ages, was investigated to determine factors associated with disease severity. The presence of ciguatoxin was detected in leftover portions of the implicated fish by enzyme immunoassay. All cases were medically and epidemiologically investigated and followed prospectively. Ten of the 15 cases demonstrated bradycardia; seven were hospitalized, including two requiring placement in intensive care. Bradycardia was associated with increasing age and body weight ($P < 0.01$ and < 0.05, respectively) as well as the amount of toxic fish consumed ($P < 0.01$). Duration of illness ranged from two to 132 days. Increasing duration of illness was correlated with both increasing age and weight (rs = 0.64 and rs = 0.72, respectively, both $P < 0.01$) and was independent of amount and components of toxic fish consumed. The correlation between increasing age and weight with duration and severity of symptoms may be explained by prior subclinical toxin exposure, and is consistent with the observation that repeated ciguatoxin exposures are associated with more severe illness. The association between amount of toxic fish consumed and bradycardia is consistent with an increased dose of ciguatoxin. These findings support previously unconfirmed observations.

DEFUSCO DJ, O'DOWD P, HOKAMA Y, OTT BR. 1993.

COMA DUE TO CIGUATERA POISONING IN RHODE ISLAND. *AMER J MED* 95(2):240–3.

A patient developed an acute gastrointestinal and neurologic syndrome progressing to coma after ingestion of a fish soup made in Rhode Island. Laboratory studies confirmed the clinical diagnosis of ciguatera poisoning. Physicians should be aware of the diagnostic features of this disease even in non-endemic areas.

HABERMEHL GGM, KREBS HC, RASOANAIVO P, RAMIALHARISOA A. 1994.

SEVERE CIGUATERA POISONING IN MADAGASCAR: A CASE REPORT. *TOXICON* 32:1539–1542.

This report of 98 fatalities among more than 500 patients affected by poisoning from the ingestion of shark flesh, is attributed to ciguatera poisoning on clinical grounds alone. No previous reports are known resulting from shark meat. (Editors' Note: This diagnosis must await biochemical confirmation. See the report by BOISIER *et al* below.)

NOGUCHI T, MATSUI T, MIYAZAWA K, ASAKAWA M, IIJIMA N, SHIDA Y, FUSE M, HOSAKA Y, KIRIGAYA C, WATABE K, USUI S, FUKAGAWA A. 1994.

POISONING BY THE RED ALGA 'OGONORI' (*Gracilaria verrucosa*) ON THE NOJIMA COAST, YOKOHAMA, KANAGAWA PREFECTURE, JAPAN. *TOXICON* 32:1533–1538.

Two females sustained food poisoning (one fatally) from ingestion of the edible red alga 'ogonori'. Indirect evidence suggested that the endogenous production of prostaglandins (mainly PGE_2) from enzyme action upon highly unsaturated fatty acids in the victim's stomach, producing fatal hypotensive shock. The authors comment that PGE_2 seems to act more selectively on females. No toxin was found in the alga itself.

BOISIER P, RANAIVOSON G, RASOLOFONIRINA N, ANDRIAMAHEFAZAFY B, ROUX J, CHANTEAU S, SATAKE M, YASUMOTO T. 1995.

FATAL MASS POISONING IN MADAGASCAR FOLLOWING INGESTION OF A SHARK (*Carcharhinus leucas*): CLINICAL AND EPIDEMIOLOGICAL ASPECTS AND ISOLATION OF TOXINS. *TOXICON* 33(10):1359–1364.

Ingestion of the meat from a single shark from the south-east coast of Madagascar caused hospital admission in 188 persons (morbidity rate 94%) with a mortality of 30%. Neurological symptoms almost exclusively resulted 5 to 10 hours following ingestion; gastrointestinal symptoms were rare. This shark species is commonly eaten in this region without problems. Bacteriological and (contaminant) chemical causes were excluded. Two liposoluble toxins that were isolated from the liver (charcatoxin-A and -B) were chromatographically distinct from ciguatoxin. (Editors' Note: There is no reference in this report to the HABERMEHL *et al* report above. It is unclear whether it is the same poisoning.)

HANNE M, MATSUBAYASHI H, VOGT R, WAKIDA C, HAU S, NAGAI H, HOKAMA Y, SOLORZANO L. 1995.

OUTBREAK OF GASTROINTESTINAL ILLNESS ASSOCIATED WITH CONSUMPTION OF SEAWEED - HAWAII, 1994. *JAMA* 274:1753–1754.

There have been previous reports of a toxic illness resulting from some Pacific seaweed consumption. This is a report of an investigation into an outbreak of acute gastrointestinal illness associated with the consumption of seaweed in Hawaii in September 1994. There were no fatalities. The seaweed was identified as *Gracilaria coronopifolia*; the unidentified toxin was heat (boiling), cold (refrigeration) and marination resistant.

ZLOTNICK BA, HINTZ S, PARK DL, AUERBACH PS. 1995.

CIGUATERA POISONING AFTER INGESTION OF IMPORTED JELLYFISH: DIAGNOSTIC APPLICATION OF SERUM IMMUNOASSAY. *WILDERNESS AND ENVIRONMENTAL MED* 6:288-294.

What is claimed as the first case of ciguatera fish poisoning associated with cnidarian ingestion is reported. The 12-year-old patient in California ate a jellyfish (minus tentacles) which had been carried the previous day from American Samoa, in a jar. Within 3 hours she exhibited ballistic-startle movements of limbs, oculo-gyric type eye movements, mental disorientation, abdominal pain and pruritis. Clinical diagnosis was supported by positive ciguatoxin and 'related polyether toxins (including okadaic acid)' serum detection by a rapid extraction method and solid phase immunobead assay.

MARINE STINGINGS

AMARASEKERA N, JAYAWARDENA A, ARIYARATNAM A, HEWAGE UC, DE SILVA A 1994.

BITE OF A SEA SNAKE (*Hydrophis spiralis*): A CASE REPORT FROM SRI LANKA. J *TROP MED & HYGIENE* 97(4):195-8.

This is claimed to be the first authenticated case of sea-snake envenoming in Sri Lanka (Editors' note: See FERNANDO & GOONERATNE 1983). Unusual clinical features included pain at the bite site, regional lymphadenopathy and absence of muscle pain and tenderness. Polyvalent land-snake antivenom administration was followed by some improvement in neurological signs (sea snake antivenom is unavailable in Sri Lanka; see GAWADE *et al* 1980).

KOKELJ F, TREVISAN G, STINCO G, PISCANC AM. 1994.

SKIN DAMAGE CAUSED BY MUCILAGINOUS AGGREGATES IN THE ADRIATIC SEA. *CONTACT DERMATITIS* 31:257-259.

Brief casual contact in sensitive individuals was sufficient to produce dermatitis. Identification of the source of the aggregates is not described.

MENAHEM S, SHVARTZMAN P. 1994.

RECURRENT DERMATITIS FROM JELLYFISH ENVENOMATION. *CANADIAN FAM PHYS* 40:2116-8.

The authors describe a case of recurrent local dermatitis after jellyfish envenomation and suggest appropriate treatment.

ALDERSLADE P. 1995.

UNPUBLISHED OBSERVATIONS.

The siphonophore suspected to be *Rhizophysa* produced a short-lived but severe sting on a fisheries researcher in Northern Territory (Australia) waters, when the animals became entangled in the trawling long-lines being reeled onboard. (Editors' Note: See also ROBINSON BH. 1995., below.)

AL-EBRAHIM K, TAHIR MZ, RUSTOM M, SHAFEI H. 1995.

JELLYFISH-VENOM-INDUCED DEEP VENOUS THROMBOSIS. A CASE REPORT. *ANGIOLOGY* 46(5):449-51.

The authors describe a case of jellyfish sting that resulted in deep venous thrombosis of the leg and thigh in a healthy 35-year-old man. Successful treatment measures are described.

BURNETT JW. 1995.

UNPUBLISHED OBSERVATIONS.

The sixth and seventh cases of mononeuritis multiplex were reported during 1995. The first followed a sting from a box jellyfish in a 52-year-old woman off the northern shores of Papua New Guinea. The second, a young woman, experienced a left forearm palsy after contact with *Chrysaora quinquecirrha* in Chesapeake Bay, Maryland, USA. There now have been 5 women and 2 men reported affected with this transient defect, which appears to occur near or adjacent to the envenomated area.

Recent capillary electrophoresis studies have shown that hyperimmune serum can bind to isolated protein peaks within either *Chironex* or *Chrysaora* nematocyst venom within a few seconds.

BURNETT JW, BURNETT MG, KAUFFMANN CL. 1995.

ANOTHER SEA PEST. *ARCH DERMATOL* 131:965.

Fragments of tentacles associated with nuisance cutaneous stings were recovered from Boca Raton to Pompano Beach (Florida) waters, during December 26–28, 1994. Microscopic findings were of tentacle fragments containing mastigophores and spirocysts, suggesting an anthozoan as the causative cnidarian in this episode of sea bather's eruption. (See also ROBINSON BH. 1995, below.)

CARLOS DE FRIETAS J, SCHIOZER WA, MALPEZZI ELA. 1995.

A CASE OF ENVENOMING BY PORTUGUESE MAN-O'-WAR FROM THE BRAZILIAN COAST. *TOXICON* 33:859–861.

This sting on the left index finger resulted in skin necrosis, surgical debridement 6 days following the envenomation, and significant loss of finger function (Plate 26.2). The authors of this paper report this case as being, to their knowledge, 'the first clinical case (i.e. of *Physalia physalis* stinging) documented for the Brazilian shores'. (The reader is referred to the information provided courtesy of Dr Hermes MIANZAN, Mar del Plata, Argentina, and Dr Rossana H. PITA VIRGA, University Cat. Santos, Brazil, on pages 194–95.)

CASE REPORT

1995. Sea Anemone Envenomation (unpublished).

A 49-year-old healthy male with no known allergies was gathering mussels from the underside of a pontoon at'Hokimai Bay, Bon Accord Harbour, Kawau Island, New Zealand (36.5°S), at mid-afternoon in November 1994. He made close and repeated contact on his bare chest with numerous small anemones growing on the underside of the pontoon (Plate 26.3A). No immediate stinging was felt, but within minutes he developed a painful rash on his chest, accompanied by increasing localised swelling and burning, and systemic symptoms of giddiness, malaise and "heat". The skin of the chest developed haemorrhagic vesicles in the next 24 hours, with serous discharge. Over the ensuing days he had nausea, inability to sleep or concentrate, and fluctuating giddiness. He noticed significant memory impairment and muscle incoordination which continued for 4 weeks; he was unable to work at his profession during this time. Haematological investigations were all normal. Neurological examination could not exclude a neurotoxin. No specific treatment was found. There was no response to oral antihistamine therapy, but 'cold compresses' and 5 days of topical betamethasone valerate cream were thought to provide 'some relief'. He had recovered completely by the end of the fifth week. The appearance of his chest rash at 11 days following the envenomation is shown in Plate 26.3B. A 54-year-old male companion was envenomed at the same time, developed identical acute localised rash and vesicles, but recovered in several days. Both patients reported reappearance of irritable skin rashes, enhanced by exposure to sunlight, in areas of the body unrelated to contact with the anemones, over the ensuing weeks.

FITZGERALD GJ. 1995.

ANALYSIS OF 24 CASES OF BULLROUT ENVENOMATION. *EMERG MED* 5:199–244

This is a useful hands-on account of southern Queensland experience including description of a successful management regimen using hot water immersion followed by local anaesthetic infiltration.

HALSTEAD R. 1995.

PERSONAL CASE REPORT.

An unusual painful hand sting with a delayed onset of several minutes was sustained by a diver who picked up an attractive sea star, possibly a species of *Nardoa* (family Ophidiasteridae), at 55m depth (Plate 26.4). The suspicion is that this animal may only occur in deeper waters.

ROBINSON BH. 1995.

LIGHT IN THE OCEAN'S MIDWATERS. *SCIENTIFIC AMERICAN* JULY, PP60-64.

Midwaters (100–1000 metres depth) contain many newly described animals, including elongated siphonophores (up to 40 metres). The siphonophores ('scyphonophores') *Batryphsia* and *Colobonema* are discussed; it is stated that these two shed tentacles while fleeing from predators.

DR I OTHMAN. 1996.

PERSONAL REPORT.

A 19kg specimen of *Rhopilema hispidum* produced stings on Dr Othman's arms while he was attempting to bring it into a boat. The application of formaldehyde on the fresh sting fired more nematocysts, and was followed by pain for 24 hours and skin pigmentation in the sting area for more than 7 days. With his colleague Dr Paul Cornelius, he identified the following cnidarians in Malaysian waters: *Phyllorhiza punctata*, *Acromitus flagellatus*, *Versuriga*, *Rhopilema hispidum*, *Chiropsalmus buitendijki*, *Carybdea alata*, *Lobonema smithii*, *Chrysaora quinquecirrha*, *Chrysaora hyoscella*, *Catostylus mosaicus*.

MOLECULAR BIOLOGY AND CHEMISTRY

BURNETT JW.

PARTIALLY UNPUBLISHED DATA. 1975 & 1980.

With the interest in possible cnidarian venom action and blockade involving divalent cations, this laboratory looked at the influence of magnesium (Mg^{++}).

1975 data: $MgCl_2.6H_2O$, 0.01M, does not alter *Chrysaora quinquecirrha* venom lethality at 4°C, nor does it alter *Physalia* venom cardiotoxicity. The Mg^{++} was placed with the venom and simultaneously injected.

1980 data: *Chrysaora quinquecirrha* collagenase activity was hurt by $MgCl_2$ (0.01M), whereas there was no effect on the venoms' kinin activity (5mM $MgCl_2$).

PIATIGORSKY J, HORWITZ J, NORMAN BL. 1993.

J1-CRYSTALLINS OF THE CUBOMEDUSAN JELLYFISH LENS CONSTITUTE A NOVEL FAMILY ENCODED IN AT LEAST THREE INTRONLESS GENES. *J BIOL CHEM* 268(16):11894–901.

The transparent cellular eye lens of the jellyfish *Tripedalia cystophora* contains three major proteins called J1-, J2-, and J3-crystallins. cDNAs were isolated encoding three novel 37-kDa J1-crystallin polypeptides (J1A, J1B, and J1C) sharing 84–98% identity in amino acid sequence among themselves. Each polypeptide is encoded in a separate gene lacking introns. In contrast to the striking similarity of the coding regions, the 5'- and 3'-untranslated sequences of the three J1-crystallin mRNAs are completely different, consistent with an ancient duplication of their genes.

NEMATOCYST FUNCTION

(SEE ALSO RIFKIN UNDER 'RECENT TEXTS', BELOW)

LOTAN A, FISHMAN L, LOYA Y, ZLOTKIN E. 1995.

DELIVERY OF A NEMATOCYST TOXIN. *NATURE* 375:456.

A 16,000Da phospholipase A_2 toxin, with a 60% amino acid homology with a lizard toxin, was isolated from the tentacles of *Rhopilema nomadica*. Nematocyst cutaneous penetration was studied using polyclonal antibodies against the toxin, and emergence of the toxin through the hollow, exteriorised barbs along the emerging tubules of a discharging nematocyst was shown. (Editors' Note: This important paper corroborates the earlier work with *Chironex fleckeri* nematocysts by Endean and Rifkin concerning the opportunity for direct intravascular deposition of nematocyst toxins during cutaneous penetration in human stings {see page 273}.)

TARDENT P. 1995.

THE CNIDARIAN CNIDOCYTE, A HIGH-TECH CELLULAR WEAPONRY. *BIOASSAYS* 17(4):351–363.

The members of the phylum Cnidaria (corals, sea anemones, medusae) are all equipped with stinging cells (cnidocytes, nematocytes), which serve mainly in prey capture and defence. The secretory product of these cells is a most complicated extrusome consisting of a cyst containing a tubule and a liquid matrix. Mechanical stimulation of the cell's cnidocil apparatus by a prey or an offender leads via bioelectrical signal transduction to the explosive discharge of the cnidocyst. In stenoteles of Hydra this process, during which the tubule is everted out of the cyst, takes less than 3 msec. The forces involved are partly due to spring-like tensions stored in the collagenous structural compartment, and partly to an osmotically generated intracapsular pressure, which can amount to 150 bar (1.5×10^7 Pascals). The osmotic machinery depends upon the presence in the cyst's matrix of inorganic cations (either K^+, Mg^{2+} or Ca^{2+}) and rare polyanions (poly-g-L-glutamates), which, so far, have not been reported from evolved eukaryotes. The discharging cyst acts like a self-reloading syringe, injecting poison and other components into the target. Since the cnidocytes are incapable of regenerating their exocytosed cysts, they have to be replaced by new cells derived by differentiation from pluripotential stem cells (interstitial cells).

PLATYPUS

TONKIN MA, NEGRINE J. 1994.

WILD PLATYPUS ATTACK IN THE ANTIPODES. A CASE REPORT. *J HAND SURG* (BRIT VOL) 19(2):162–4.

This is a report of a naturalist stung while trying to study the platypus (*Ornithorhynchus anatinus*) in the wild. An intense local reaction resulted, and the authors advocate that 'warning signs should be erected at air and sea ports warning tourists of the dangers of these venomous Australians'!

RECENTLY PUBLISHED TEXTS AND OTHERS RECENTLY DISCOVERED BY THE EDITORS

A GUIDE TO THE WORLD OF JELLYFISH 1992

TEXT: EILEEN CAMPBELL; DESIGN: KATHY KOPP; ILLUSTRATIONS: ANDREA MCCANN; MULTIPLE PHOTOGRAPHER CONTRIBUTORS. CALIFORNIA. MONTEREY BAY AQUARIUM FOUNDATION. 16 PP.

This lavishly illustrated booklet contains some of the best colour photographs of jellyfish seen by the editors.

HANDBOOK OF CLINICAL TOXICOLOGY OF ANIMAL VENOMS AND POISONS

EDITED BY JÜRG MEIER, JULIAN WHITE. BOCA RATON, FLORIDA. CRC PRESS. 1995. 751PP.

A major new black and white illustrated text, covering terrestrial and marine toxicology worldwide. The snake bite chapters are particularly useful.

KOHAMA M 1995

***THE MARINE STINGER GUIDE* (IN JAPANESE). OKINAWA ISLANDS. 64PP.(FACSIMILE 81.98.936.0110)**

A small handbook with numerous high quality colour photographs of a wide range of marine animals and fresh stings/injuries from the Okinawa environment.

MANDOJANA R 1995

***INTRODUCTION TO AQUATIC DERMATOLOGY*, 31PP, IN: (J. DENNIS, ED.) CLINICAL DERMATOLOGY, 22ND REVISION. PHILADELPHIA. LIPPINCOTT-RAVEN.**

This is a general dermatological overview with some useful illustrations, including an interesting depiction of a Monaco postage stamp commemorating the discovery of anaphylaxis, from the venom of *Physalia physalis* and from sea anemones, which won the Nobel Prize for Medicine for Drs Portier and Richet in 1901 (see PORTIER & RICHET 1902a and 1902b).

RIFKIN JF 1996

***STINGING CELLS FROM CNIDARIANS OF MEDICAL IMPORTANCE*. KARALEE QLD 4306, AUSTRALIA. RIFKIN PUBLISHER. 220 PP; 8 COLOUR PLATES; 60 B AND W PLATES; TABLES AND APPENDICES.**

This book is based upon the author's PhD Thesis with updates and related new information.

Extract: The cnidocytes of Australian cnidarians of medical importance, *Chironex fleckeri* (Class Cubozoa), *Lytocarpus philippinus* (Class Hydrozoa), *Cyanea nozakii* (Class Scyphozoa), a species of *Pachycerianthus* and *Stichodactyla haddoni* (both members of the Class Anthozoa) were studied. The nematocyst complement of each species is described, including two new types of nematocyst. The general histology of the tentacles of each species is examined with emphasis on the epidermal layer and the arrangement of cnidae. All species were examined with respect to their fine structural characteristics. Fibres present in nematocytes of *Chironex fleckeri* are also present in the basal ends of nematocytes of all five species studied, and contained either microtubules or microfilaments. These fibres act as anchors, but may or may not act to expose the apical end of the nematocyte or remove it from the firing line. Like the cnidocytes of *C. fleckeri*, those of *Lytocarpus philippinus* and *Cyanea* each possess a cnidocil apparatus. Cnidocytes of *Pachycerianthus* sp. marginal tentacles and *Stichodactyla* possess the putative sensory structures, ciliary cones, at their apical surfaces.

Studies of isolated mastigophores of all five species have revealed granular electron dense material inside undischarged tubules. This material is released continuously as the tubule everts. Penetrants of *Chironex fleckeri* are closed at their tips. The bulk of the capsular contents are released once the tubule tip is ruptured. Toxic material appears to be deposited directly intravenously in mice; such intravascular deposition is the only rational explanation for the rapid onset of systemic effects in human victims of serious chirodropid stinging.

When subjected to a series of electrical stimuli, the cnidae of each species show different thresholds for cnida discharge. There is also a highly variable response to a group of reagents tested for their ability to discharge cnidae on the tentacles of the five species. The response to each reagent varies even between members of the same class and species. This study suggests that currently there is no universal inactivator for cnida discharge, at least using the experimental procedures employed by the author. Vinegar, advocated for use as a cnida discharge inhibitor for cubozoan stings, is ineffective for use with the cnidae of other cnidarians. These studies of the mode of envenomation, activation and inhibition of cnida discharge and deactivation of venom contained within isolated cnidae may have useful implications for treatment of stings.

WILDERNESS MEDICINE. MANAGEMENT OF WILDERNESS AND ENVIRONMENTAL EMERGENCIES 1995

THIRD EDITION (PAUL S. AUERBACH, ED.). ST LOUIS. MOSBY. 1506PP.

For outdoor medicine this is a definitive, encyclopaedic text, well illustrated in black and white and colour (colour plates grouped together at the back). There are 93 contributors. The marine-related toxinology content is written principally by the editor.

WHITE J. CSL ANTIVENOM HANDBOOK 1995

MELBOURNE. CSL LIMITED. 62PP.

A concise and systematic treatment of first aid and medical management for stings and bites, both terrestrial and marine, and up-to-date information on antivenoms from the Commonwealth Serum Laboratories, one of the world's leading antivenom sources.

TOXINOLOGY

ARILLO A, BURLANDO B, CARLI AM, MARIOTTINI GL 1994

MITOCHONDRIAL ALTERATION CAUSED BY CNIDARIAN TOXINS: A PRELIMINARY STUDY. *J BIOL RES-BOLL SOC IT BIOL SPER* 70:307–313.

Toxin from the tentacles of sea anemone *Anemonia sulcata* altered the function of rat liver mitochondria; the effects were dose- and time-dependent and disappeared when Ca^{++} was absent. The toxin may thus have a phospholipase-like action, as reported for other cnidarian toxins. It may be postulated that the action of this cnidarian toxin upon mitochondrial function may be a particular example of a general mechanism of toxicity affecting biological membranes.

DE PLATER G, MARTIN RL, MILBURN PJ 1995

A PHARMACOLOGICAL AND BIOCHEMICAL INVESTIGATION OF THE VENOM FROM THE PLATYPUS (*Ornithorhynchus anatinus*). *TOXICON* 33:157-169.

This venom exhibited several activities including local sub-plantar oedema and rat uterine relaxation associated with a 4200Da peptide. Three other major proteins (140,000Da, 55,000Da, 16,000Da) were isolated, none of which was found to have any sequence homology with the proteins listed in the SwissProt database.

GARNIER P, GOUDEY-PERRIÈRE F, BRETON P, DEWULF C, PETEK F, PERRIÈRE C 1995

ENZYMATIC PROPERTIES OF THE VENOM OF THE STONEFISH (*Synanceia verrucosa* BLOCH AND SCHNEIDER, 1801) AND PURIFICATION OF A LETHAL, HYPOTENSIVE AND CYTOLYTIC FACTOR. *TOXICON* 33:143–155.

FPLC filtration verrucotoxin was found to be a glycoprotein of 322,000Da comprising 4 subunits (2α — 83,000; 2β — 78,00). Verrucotoxin was lethal to mice, induced a fall in arterial pressure in mice, and haemolysed washed rabbit erythrocytes.

HESSINGER DA 1995

UNPUBLISHED OBSERVATIONS.

A specific anti-calcium cardiotoxin, which is reversibly affected by verapamil and nifedipine has been reported in the venom of *Physalia* species.

JUNAID MAHMOOD ALAM 1995

UNPUBLISHED OBSERVATIONS.

This Karachi researcher, with the assistance of his colleagues, is investigating, with some success, the protective efficacy of certain plant extracts in Pakistan (e.g. *Allium cepa* and *Calotropis procera*) against the crude venom of *Physalia* species from Pakistani waters, both *in vitro* and *in vivo*. He has also recorded the epidemiology and clinical details of mild and severe jellyfish envenomations ('mainly *Physalia*, both single and multiple tentacular species') (including blood chemistry of the severe cases) on Karachi beaches during 1995. In addition he has studied the blood chemistry of 7 non-fatal but clinically significant sea snake envenomations ('mainly *Hydrophis spiralis* and *H. lapemoides*') in the same region. This work will soon be submitted for publication (see also page 302).

KARALLIEDDE L 1995

ANIMAL TOXINS. *BRIT. J. ANAESTH.* 74:319–327

This paper makes some useful observations relating to marine toxins. There is a helpful diagram indicating site of actions of animal toxins.

MUSTAFA MR, WHITE E, HONGO E, OTHMAN I, ORCHARD CH 1995

THE MECHANISM UNDERLYING THE CARDIAC EFFECT OF THE TOXIN FROM THE JELLYFISH *Chironex fleckeri*. *TOXICOL AND APPLIED PHARM* 133:196–202.

The cardiac effects of the toxin of the boxjellyfish *Chironex fleckeri* were investigated by studying the isolated papillary muscles and ventricular myocytes from ferrets and rats. Application of toxins to those preparations resulted in a large increase in intracellular calcium and the adverse symptoms of calcium overload (after contractions, spontaneous contractions, a decrease in developed force and an increase in resting force). Calcium and sodium channel blockers and inhibitors of the sarcoplasmic reticulum Na^+/K^+ ATPase or Na^+/H^+ exchange were ineffective. Removing the sodium from the bathing solution and replacing it with Ni^{++} did block the toxic response. In addition to the calcium effect the toxin also caused an increase in intracellular Na^+ and the appearance of an intracellularly directed, reversible current. Thus the authors hypothesise that the toxin increases Na^+ influx into the cell, which then would increase intracellular Ca^{++} via the Na^+/Ca^{++} exchange mechanism to result in an observed Ca^{++} overload. In this excellent paper the authors corroborate and expand data obtained by Burnett's group on *Physalia* and *Chrysaora* and expand to cover *Chironex* (*Toxicon* 1973;11:341; *Toxicon* 1983;21:562; *Biophysics J* 1983;42:199). An interesting feature of their data is that the calcium channel blocker produced a partial effect which gave way after a period of time. Unfortunately the investigators used nifedipine, an agent which is known to be weaker than verapamil against the *Chironex* jellyfish cardiotoxic effects. Additionally, it should be remembered that Purkinje fibre is more sensitive to the toxin than the cardiac myocyte (*Toxicon* 1973;11:341) (see below).

ROTTINI G, GUSMANI L, PAROVEL E, AVIAN M, PATRIARCA P 1995

PURIFICATION AND PROPERTIES OF A CYTOLYTIC TOXIN IN VENOM OF THE JELLYFISH *Carybdea marsupialis*. *TOXICON* 33:315–326.

A sheep-red-cell-haemolytic toxin from the nematocysts of *Carybdea marsupialis*, captured in the Northern Adriatic Sea, was protein with an apparent molecular weight of 102-107Da, was heat labile, highly unstable in polar media, inactivated by reducing agents and with no phospholipase activity. A pore-forming mechanism of toxin action is suggested.

SUNDARAY JK, VENKATESHVARAN K, ELANCHERAN P. 1995.

TOXICITY OF CERTAIN JELLYFISHES OF BOMBAY COAST. CENTRAL INSTITUTE OF FISHERIES EDUCATION, VERSOVA, BOMBAY 400 061, INDIA.

This paper reports the collection of 5 species of jellyfish between December and May, 1994, in Bombay waters — *Chrysaora quinquecirrha*, *Pelagia noctiluca*, *Aurelia solida*, *Octocanna polynema* and *Eirene madrasensis*. Animal toxicity studies implicated the venom from only *A. solida* and *C. quinquecirrha*. Heart muscle 'constriction' was observed at autopsy in the animals which died.

CONCLUDING HYPOTHESES

BY JOSEPH W BURNETT

After years of investigating the effects of jellyfish venoms on cells and animals, one of the editors (JWB) hypothesises:

1. The dramatic pathogenic action of all jellyfish venoms is similar.
2. The principal action will effect ion transport of Na^+ and/or Ca^{++} as well as producing a pore in active membranes.
3. When the venoms are finally purified a group of active fractions will be detected but one will be dominant. Other agents designed to potentiate, stabilise or activate the principal one will be present.
4. Once envenomation occurs the victim is attacked with agent(s) acting on multiple organs simultaneously. Each action will have a different time schedule.
5. The clinical effect(s) will be determined by dosage and resistance, with the organs most likely to fail being:
   (a) the heart within a few minutes through actions on myocardial cells (conducting and contractile functions — it should be noted that the Purkinje fibres are more sensitive than the papillary muscle cells [Kleinhaus AL, Cranefield PF, Burnett JW. 1973. Toxicon 1:341–349] and/or coronary vascular supply) then,
   (b) the lung through central or peripheral neural failure in minutes, or parenchymal respiratory failure within 1 to 6 hours, then,
   (c) kidney (via tubular damage near the corticomedullary junction in 1 to 3 days, and finally
   (d) the liver after 3 days with mid-zonal cellular necrosis. With the exception of very rare anaphylaxis, the acute post-envenomation disease will be a result of toxicity. In contrast allergy would play a bigger role with recurrent or persistent reactions, especially those on the skin.
6. To survive a major envenomation the untreated patient must progress through the attack on these 4 organs sequentially. Existing antivenom will be effective as long as venom is freely circulating. Verapamil will assist the patient only through the cardiac attack phase and will combat the venom only to the membrane level. It should be emphasised that the effect of verapamil is stereospecific since only the laevo-rotatory (L-) form is effective. The effect of this agent will only be temporary (if important), not curative.
7. Because there is no exact dose-(as indicated by length and width of damaged skin)-to-weight-of-victim relationship affecting sting outcome, other factors modifying human resistance will be uncovered. Some good possibilities include: other diseases, concomitant medications, incident sunlight, gender, menstrual cycle timing, and amount of physical activity (affecting cardiac output and peripheral perfusion, thus venom uptake) at the time of sting. The overall health and activity of the jellyfish are other major prognostic influences.
8. At present protective antibodies have not been detected *in vitro* in humans. However, the highest titre measured can only be diluted about 3000 times (ELISA). These antibodies are present but in very low amounts. Using the Commonwealth Serum Laboratories' *Chironex* antivenom as a model, the ELISA test would have to be at

least 7 times as potent as it is now in order to allow detection in the mouse lethality model system. In nature, humans haven't a chance of building their own immunoprotection by incidental stings. Since active immunisation is preferable to passive, harmless methods of augmenting antibody for bathers frequently exposed to lethal jellyfish will have to be discovered. Augmenting immunological defences is an aim of oncological research; a 'spin-off' from that area of research may assist in toxinology.

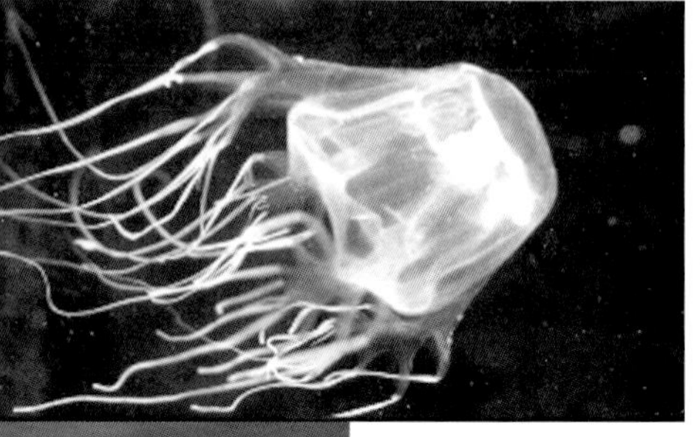

BIBLIOGRAPHY

ABBOTT R T. 1973. Mollusks dangerous to scuba divers. Delaware Med J 45: 161–164

ABORIGINAL COMMUNITIES OF THE NORTHERN TERRITORY OF AUSTRALIA. 1988. Traditional Bush Medicines. An Aboriginal Pharmacopoeia. Victoria, Australia. Greenhouse Publications. 256pp.

ABU-NEMA T, AYYASH K, WAFAII I, AL-HASSAN J, THULESIUS O. 1988. Jellyfish sting resulting in severe hand ischaemia successfully treated with intra-arterial urokinase. Brit. J Accident Surg 19: 294–296.

ACOTT C J. 1983. Sea snake envenomation. SPUMS (South Pacific Underwater Medicine Society, Inc.) J 4: 8–12.

ACOTT C J. 1986. Sea snake envenomation. Med J Aust 144: 448.

ACOTT C, GORMAN D. 1992. Decompression illness and nitrous oxide anaesthesia in a sports diver. Anaesth Intens Care 20: 249–250.

ADAMS A. 1848. In: Narrative of the Voyage of the HMS Samarang, Vol 2, pp 356–357. (Refer also to Clench W.J. and Kondo Y. 1943. The poison cone shell. American J Trop Med 23(1): 105–120).

ADAMS M J. 1993. An outbreak of ciguatera poisoning in a group of scuba divers. J Wilderness Med 4: 304–311.

ADIGA K M. 1984. Brachial spasm as a result of a sting. Med J Aust 140: 180–181.

AGASSIZ L. 1862. *Chiropsalmus*. Contribution Natural Hist USA 4: 174. (380 pp in book)

AGASSIZ L. 1862. *Gonionemus*. Contribution Natural Hist USA 4: 350.

AGASSIZ L. 1865. Illustrated Catalog of the Museum of Comparative Zoology at Harvard College, No 11: North American Acalephae. 234 pp.

AGASSIZ A, MAYER A G. 1902. Medusae. In: Reports on the scientific results of the expedition to the tropical Pacific in charge of Alexander Agassiz, by the US Fish Commission steamer *Albatross* 1899–1900. Mem Mus Comp Zool Harv 26(3): 139–176 (pls 1–3).

ALAM J M, QASIM R. 1991. Toxicology of *Physalia's* (Portuguese man o'war) venom. Pakistan J Pharmaceutical Sci 4(2): 159–168.

ALAM S M, JAMAL Q, IQBAL N, ALAM J M. 1992. Effects of marine toxins on albino rats. J Coll Phys Surg Pakistan 2(3): 73–78.

ALAM J M, ALAM S M, QASIM R. 1993. Pathogenesis of the morphological lesions in various organs in guinea pigs produced by cytotoxic compounds from marine animals. Pakistan J Path 4: 2; 95–99.

ALCALA AC 1983. Recent cases of crab, coneshell and fish intoxication in Southern Negros Island, Philippines. Toxicon Suppl 3: 1–3.

ALENDER C B, FEIGEN G A, TOMITA J T. 1965. Isolation and characterization of sea urchin toxin. Toxicon 3: 9–17.

AL FAHAL (Arabic newspaper). 1984. 'The wasp with the deadly sting', in 'About People' column. (journalist not named). Muscat. 13 October 1984.

AL-HASSAN J M, DYSON M, YOUNG S R, THOMSON M, CRIDDLE R S. 1991. Acceleration of wound healing responses induced by preparations from the epidermal secretions of the Arabian Gulf catfish (*Arius bilineatus* Valenciennes). J Wilderness Med 2: 153–163.

AL-HASSAN J M, THOMSON M, ALI M, FAYAD S, ELKWAHAD A, THULESIUS O, CRIDDLE R S. 1986. Vasoconstrictor components in the Arabian Gulf catfish (*Arius thalassinus*, Ruppell) proteinaceous skin secretion. Toxicon 24: 1009–1014.

ALLEN G R, STEENE R. 1994. Indo-Pacific Coral Reef Field Guide. Singapore. Tropical Reef Research.

ALLISON J. 1991. A Tribute to Oman. Vol 10. Muscat. Apex Publishing. pp 102–107.

ALLSOP J L, MARINI L, LEBRIS H, POLLARD J, WALSH J, HODGKINSON S. 1986. Neurological symptoms and signs of ciguatera: three cases with neurophysiological study and one biopsy. Rev Neurol (Paris) 142: 590–597.

ANDERLINI V. 1984. Jellyfish. Mimeographed information document circulated by Council for Conservation of Environment and Prevention of Pollution. September 1984. pp 1–7.

ANDERSON W. 1776. An account of some poisonous fish in the South Seas. Philos Trans Royal Soc London 66: 544.

ANDREWS G. 1986. The crown of thorns starfish, *Acanthaster planci*, in the Cook Islands. Report submitted to the Great Barrier Reef Marine Park Authority, Townsville. 75 pp.

ANEIROS A, GARCIA I, MARTINEX J R, HARVEY A L, ANDERSON A J, MARSHALL D L, ENGSTROM A, HELLMAN U, KARLSSON E. 1993. A potassium channel toxin from the secretion of the sea anemone *Bunodosoma granulifera*: Isolation, amino acid sequence and biological activity. Biochem et Biophys Acta 1157: 86–92.

ANGELINI G, VENA G A. 1991. Dermatologia acquatica. Milano. Gruppo Lepetit SPA. 117 pp.

ANGKIE J. 1992. 'Killer fish attacks child'. 'Sabah Times' 30 December 1992. pp 1–2.

AOKI T. 1922. Über Medusenstichkrankheit, Japan. Zischr f Dermatol und Urologie 22(10): 835–891 (in Japanese). German summary pp 71–72.

APPLIED SCIENTIFIC RESEARCH CORPORATION OF THAILAND (ASRCT). 1977. The medicinal value of Phakbung-thale. ASRCT Research News, 196 Phahonyothin Road, Bang Khen, Bangkok 9. April 1977. p 109.

ARNESON A C, CUTRESS C E. 1976. Life cycle of *Carybdea alata*. In: (GO Mackie, ed) Coelenterate Ecology and Behaviour. New York. Plenum Press. pp 227–236.

ARNOLD H L Jnr. 1971. Portuguese man-o'-war ('Blue Bottle') stings, treatment with papain. Proc Straub Clinic, Honolulu. 37: 30–33.

AUERBACH P S. 1987. A Medical Guide to Hazardous Marine Life. Jacksonville, Florida. Progressive Printing Co. 47 pp.

AUERBACH P S. 1988. Clinical therapy of marine envenomation and poisoning. In: (A.T.Tu, ed) Handbook of Natural Toxins, Vol 3: Marine Toxins and Venoms. New York. Marcel Dekker. pp 493–565.

AUERBACH PS, Editor. 1995. Wilderness Medicine: Management of Wilderness and Environmental Emergencies. St. Louis, Missouri. Mosby. 1506 pp.

AUERBACH P S, HAYS J P. 1987. Erythema nodosum following a jellyfish sting. J Emerg Med 5: 487–491.

AUERBACH P S, YALKO D M, NOSSOS P S, KIZER K W, McCOSKER J E, GEEHR E C, HADLEY W K. 1987. Bacteriology of the marine environment: implications for clinical therapy. Ann Emerg Med 16: 643–649.

AUGEE M, GOODEN B. 1993. Echidnas of Australia and New Zealand: Australian Natural History Series. Sydney. New South Wales University Press. pp 18 & 41–43.

AUSTIN L, GILLIS R G, YOUATT G. 1965. Stonefish venom: some biochemical and chemical observations. Aust J Exp Biol Med Sci 43: 79–90.

AUSTRALASIAN COLLEGE OF TROPICAL MEDICINE. 1993. Dr Les Ashdown. President's report: Associate Professor Rick Speare. 1 June 1993. pp 12–13.

AUSTRALIAN RESUSCITATION COUNCIL. 1989. Cardiopulmonary Resuscitation. Melbourne. Australian Resuscitation Council. pp 1–30.

AUSTRALIAN SHARK ATTACKS. 1981. Sydney. Percival Publishing Co, a division of General Publishers. 64 pp.

AVIAN M, SPANIER E, GALIL B. 1995. Nematocysts of *Rhopilema nomadica* (Scyfozoa: Rhizostomeae). An immigrant jellyfish in the eastern Mediterranean. *J Morph* 224: 221–231.

AXIAK V. 1983. UNEP: Monitoring of swarming by Scyphomedusae. Publication in connection with Workshop on Jellyfish Blooms in the Mediterranean. Athens, 31 October–4 November 1983. United Nations Environment Programme. Mediterranean action plan: long-term programme for pollution monitoring and research in the Mediterranean (Med Pol-Phase II). 5 September. *Mauretic* 1984 p38.

AXIAK V, CIVILI F S. 1987. Jellyfish blooms in the Mediterranean: causes, mechanisms, impact on man and the environment. A programme review. In: (D. Eaker, T. Wadstrum, eds) Natural Toxins. Oxford. Pergamon Press. pp 1–21.

AZILA N. 1990. Effect of *Catostylus mosaicus* venom on erythrocytes. Toxicon 28: 247.

AZILA N A, OTHMAN I. 1993. Cases of jellyfish envenomation in Malaysia and some cytotoxic studies on their toxins, in (N H Tan, S L Oo, V Thambyrajah, N Azila, eds) Advances in Venom and Toxin Research. Proceedings of the Third Asia Pacific Congress on Animal, Plant and Microbial Toxins, 27 June–1 July, 1993. Kuala Lumpur. Malaysian Society on Toxinology. pp 113–118.

AZNAURIAN M S. 1958. Clinics of stings by the venom medusa at the Far East. Klin Med 36 (No 6): 105–108 (in Russian).

AZNAURIAN M. 1964. The venomous krestovik-medusa. Vladivostok. Dalnevost. KN. Izd. 1–34 (in Russian).

AZUMA H, SEKIZAKI S, SATOH A, NAKAJIMA T, ISHIKAWA M. 1986. Platelet aggregation caused by a partially purified jellyfish toxin from *Carybdea rastoni*. Toxicon 24: 489–499.

BACHAND R. 1983. Fire coral — the stingers. Underwater Naturalist 14: 12–15.

BACHMAN D S, PAULSON G W, MENDELL J R. 1982. Acute inflammatory polyradiculoneuropathy following Hymenoptera stings. JAMA 247: 1443–1445.

BACQ Z M. 1937. L'amphiporine' et la 'némertine' poisons des vers némertiens. Arch Intern Physiol (Liége) 44: 190–204.

BADEN D G. 1988. Public health problems of red tides. In (A T Tu, ed.) Handbook of Natural Toxins, Vol 3: Marine Toxins and Venoms. New York. Marcel Dekker pp 259–277.

BADEN H P. 1987. Injuries from sea urchins. Clin in Dermatol 5(3): 112–117.

BAGNIS R. 1970. Concerning a fatal case of ciguatera poisoning in the Tuamotu Islands. Clin Toxicol 3: 579–583.

BAGNIS R. 1994. Natural versus anthropogenic disturbances to coral reefs: comparison in epidemiological patterns in ciguatera. Memoirs of the Queensland Museum 34(3): 455–460.

BAGNIS R, BENNET J, BARSINAS M, DROLLET J H, JAQUET G, LEGRAND A M, CRUCHET P H, PASCAL H. 1988. Correlation between ciguateric fish and damage to reefs in the Gambier Islands (French Polynesia). In: (Choat J H et al, eds) Proc Sixth Int Coral Reef Symp, Townsville, Vol 2. Townsville. 6th International Coral Reef Symposium Executive Committee. pp 195–200.

BAGNIS R, KUBERSKI T, LAUGIER S. 1979. Clinical observations on 3009 cases of ciguatera (fish poisoning) in the South Pacific. Amer J Trop Med Hyg 28: 1067–1073.

BAIRD J K, WEAR D J. 1987. Cercarial dermatitis: the swimmers itch. In: Clinics in Dermatology. Philadelphia. J B Lippincott. 5(3): 88–91.

BAKER J T, WILLIAMSON J A. 1986. Toxins and beneficial products from reef organisms. Oceanus (American–Australian bicentennial issue. The Great Barrier Reef: Science & Management) 29(2): 109–114.

BANFIELD E J. 1994. Confessions of a Beachcomber. St Lucia. University of Queensland Press. 336 pp. (first published by Fisher, Unwin, London, 1908).

BARME M. 1968. Venomous sea snakes (Hydrophiidae). In: (W Bücherl, E E Buckley, V Deulofeu, eds) Venomous Animals and their Venoms, Vol I: Venomous Vertebrates. New York. Academic Press. pp 285–308.

BARNES J H. 1960. Observations on jellyfish stingings in North Queensland. Med J Aust 2: 993–999.

BARNES J H. 1964. Cause and effect in 'Irukandji' stingings. Med J Aust 1: 897–904.

BARNES J H. 1965. *Chironex fleckeri* and *Chiropsalmus quadrigatus*: morphological distinctions. North Queensland Naturalist 32: 13–22.

BARNES J H. 1966. Studies on three venomous Cubomedusae. In: The Cnidaria and their Evolution: Symposium of the Zoological Society of London, No 16. London. Academic Press. pp 307–332.

BARNES J H. 1967. Extraction of cnidarian venom from living tentacle. In: (F E Russell, P R Saunders, eds) Animal Toxins: A Collection of Papers presented at the First International Symposium on Animal Toxins, Atlantic City, New Jersey, 9–11 April 1966. Oxford. Pergamon Press. pp 115–129.

BARNES R S K, CALOW P, OLIVE P J W, GOLDING D W. 1993. The Invertebrates: A New Synthesis, 2nd edition. London and Edinburgh. Blackwell Scientific Publications. 488 pp.

BARNES J, ENDEAN R. 1964. A dangerous starfish — *Acanthaster planci* (Linné). Med J Aust 1: 592–593.

BARNETT D, DEAN P W, HART R J, LUCAS J S, SALATHE R, HOWDEN M E H. 1988. Determination of contents of steroidal saponins in starfish tissues and study of their biosynthesis. Comp Biochem Phys 90B: 141–145.

BARON E J, FINEGOLD S M. 1990. Bailey and Scott's Diagnostic Microbiology, 8th edition. St Louis. C.V. Mosby Co. 861 pp (313 colour illust.).

BARSS P G. 1982. Injuries caused by garfish in Papua and New Guinea. Brit Med J 284: 77–79.

BARSS P. 1984. Wound necrosis caused by the venom of stingrays: Pathological findings and surgical management. Med J Aust 141: 854–855.

BAXTER E H, GALLICHIO H A. 1974. Cross neutralisation by tiger snake (*Notechis scutatus*) antivenene and sea snake (*Enhydrina schistosa*) antivenene against several sea snake venoms. Toxicon 12: 273–278.

BAXTER E H, MARR A G M. 1969. Sea wasp (*Chironex fleckeri*) venom: lethal, haemolytic and dermonecrotic properties. Toxicon 7: 195–210.

BAXTER E H, MARR A G M. 1974. Sea wasp (Chironex fleckeri) antivenene: neutralising potency against the venom of three other jellyfish species. Toxicon 12: 223–229).

BAXTER E H, MARR A G M. 1975. Sea wasp toxoid: an immunising agent against the venom of the box jellyfish Chironex fleckeri. Toxicon 13: 423–435.

BAXTER E H, MARR A G M, LANE W R. 1968. Immunity to the venom of the sea wasp (Chironex fleckeri). Toxicon 6: 45–50.

BAYER F M, OWRE H B. 1968. The Free-Living Lower Invertebrates. New York. Macmillan. 229 pp.

BEADNELL C E, RIDER T A, WILLIAMSON J A, FENNER P J. 1992. Management of a major box jellyfish (*Chironex fleckeri*) sting: lessons from the first few minutes and hours. Med J Aust 156: 655–658.

BEARUP A J. 1954. Schistosome dermatitis (surfer's itch). Health, Canberra 4(3): 83–86.

BEARUP A J. 1956. Life cycle of *Austrobilharzia terrigalensis* Johnston 1917. Parasitology 46(3,4): 470–479.

BECKMAN E N, LEONARD G L, CASTILLO L E, GENRE C F, PANKEY G A. 1981. Histopathology of marine *vibrio* wound infections. Am J Clin Path 76: 765–772.

BENDT R R, AUERBACH P S. 1991. Foreign body reaction following stingray envenomation. J Wilderness Med 2: 298–303.

BENGSTON K, NICHOLS M M, SCHNADIG V, ELLIS M D. 1991. Sudden death in a child following jellyfish envenomation by *Chiropsalmus quadrumanus*: case report and autopsy findings. JAMA 266: 1404–1406.

BENNETT I. 1987. W J Dakin's Classic Study: Australian Seashores. Revised, expanded deluxe edition. Sydney. Angus & Robertson. 411 pp.

BENSON J. 1956. *Tetraodon* (blowfish) poisoning: a report of two fatalities. J Forensic Sci 14: 119–125.

BERDYSHEV E V, VASKOVSKY V E, PAVLENKO A F, OVODOV Yu S, GORBACH V I. 1991. A possible role of platelet activating factor (PAF) in medusa toxicity (abstract). 10th World Congress on Animal, Plant and Microbial Toxins. Singapore. 248 pp.

BÉRESS L, BÉRESS R. 1971. Reinigung zweier krabbenlähmender toxine aus der seeanemone *Anemonia sulcata*. Kieler Meeresforsch 27(1): 117–127.

BÉRESS L, ZWICK J, KOLKENBROCK H J, KAUL P N, WASSERMANN O. 1983. A method for the isolation of the Caribbean palytoxin (C-PTX) from the coelenterate (zoanthid) *Palythoa caribaeorum*. Toxicon 21: 285–290.

BERGER E W. 1900. Physiology and histology of the Cubomedusae including Dr F S Conant's notes on the physiology. Mem Biol Lab, Johns Hopkins Univ 4: 1–84.

BEST W C, SABLAN R G. 1964. Cymothoidism (sea louse dermatitis). Archives Dermatol 90: 177–179.

BIGELOW H B. 1911. Siphonophorae. In: Reports on the scientific results expedition to the eastern tropical Pacific, in charge of Alexander Agassiz, by the US Fish Commission steamer *Albatross*. Mem Mus Comp Zool Harvard 38(2): 171–402.

BIGELOW H B. 1912. Preliminary account of a new genus and three new species of medusae from the Philippines. Proc US Nat Mus 43: 253–260.

BIGELOW H B. 1913. Medusae and Siphonophorae collected by the US Fisheries steamer *Albatross* in the north west Pacific, 1906. Proc US Nat Mus 44: 1–119.

BIGELOW H B. 1938. Plankton of the Bermuda oceanographic expeditions. VIII: Medusae taken during the years 1929 and 1930. Zoologica NY 23: 99–189.

BIRKELAND C, LUCAS J S. 1990. *Acanthaster planci*: major management problem of coral reefs. Boca Raton, CRC Press. 257 pp.

BLANKENSHIP J E. 1976. Tetrodotoxin: from poison to powerful tool. Perspect Biol Med 19: 509–512.

BLOOMFIELD J. 1959. Know-How in the Surf. Sydney. Angus & Robertson Ltd. 205pp.

BLUMENTHAL K M. 1986. Renaturation of neurotoxin B-IV from the heteronemertine *Cerebratulus lacteus*. Toxicon 24: 63–69.

BLYTHE D G, FLEMING L E, AYYAR D R, DE SYLVA D, BADEN D, SCHRANK K. 1994. Mannitol therapy for acute and chronic ciguatera poisoning. Memoirs of the Queensland Museum 34(3): 465–470.

BOERO F, MINELLI A. 1986. First record of *Carybdea marsupialis* L. (Cnidaria, Cubozoa) from the Adriatic Sea. Boll Mus Civ St Nat Venezia 35: 179–180.

BOKMA H. 1942. Casuïstische mededeeling. Nog eens een beet van een zeeslang. Ned Tijdschr Geneesk 82: 87.

BOSCH D, BOSCH E. 1982. Seashells in Oman. London and New York. Longman Group. pp 123–131.

BOSCHMA H. 1949. The species problem in *Millepora*. Zool Verh Mus Leiden 1: 1–115.

BOTTARD A. 1889. L'appareil à venin des poissons. Compt Rend Acad Sci 108: 534–537.

BOULENGER K L. 1910. On the origin and migration of the stinging cells in craspedote medusae. QL Microsc Sci 55: 763–783.

BOURDY G, CABALION P, AMADE P, LAURENT D. 1992. Traditional remedies used in the Western Pacific for the treatment of ciguatera poisoning. J Ethnopharmacology 36: 163–174.

BOWEN R E, POTTER M E. 1993. Ciguatera: effective control of a growing health problem. J Wilderness Med 4: 237–240.

BOWERMAN M. 1980. Joint research on ciguatera. Australian Fisheries 39(5), and SPUMS J Jul–Sep: 24–26.

BOWMAN P B. 1984. Amitriptyline and ciguatera. Med J Aust 140: 802.

BOYD W. 1984. Sea-wasp antivenom in a toddler. Med J Aust 1: 504.

BRANCH G, BRANCH M. 1981. The Living Shores of Southern Africa. Cape Town. Struik. 272 pp.

BREAULT J L. 1991. Candirú: Amazonian parasitic catfish. J Wilderness Med 2: 304–312.

BREKHMAN I I, MINUT-SOROKHTINA O P. 1951. Pharmacological characteristics of venom of medusa-krestovik. In: Clinical and Experimental Studies of Stings by Toxic Medusa. Vladivostok. Dalnevost. Fil. Akad. Nauk SSR. pp 47–51.

BRINCKMANN-VOSS A. 1974. British Columbia Marine Faunistic Survey. Report on the Hydrozoa. Part 1: Medusae. Tech Report Pacific Biol Station Nanaimo, British Columbia, Canada. 492: 1–21.

BROADBENT G. 1985. Tropical fish poisoning (ciguatera). In: (D Exton, ed) The Marine Stinger Book, 3rd edition. Newstead, Qld. Surf Life Saving Queensland. pp 64–67.

BRODNER R A. 1988. Reactive arthritis following a sting by a Portuguese man-of-war. J Florida Med Assoc 75: 280–281.

BROWN T W. 1973. *Chironex fleckeri*: distribution and movements around Magnetic Island, north Queensland. Free-standing publication as Marine Research Associate, World Life Research Institute, Colton, Calif. USA. 13 August. 26 pp.

BROWNE E T. 1926. Report on the medusae: Zoological results of the Cambridge expedition to the Suez Canal, 1924. Trans Zool Soc Lond 22: 105–115.

BROWNLEE S. 1987. The by-the-wind-sailor: a masterpiece of natural design. Discover 8(8): 46.

BUCK J D, SPOTTE S, GADBAW, J J Jnr. 1984. Bacteriology of the teeth from a great white shark: potential medical implications for shark bite victims. J Clin Microbiol 20: 849–851.

BUCKNALL N C. 1991. Electrical treatment of venomous bites and stings. Toxicon 29: 397–400.

BURNELL D J, APSIMOV J W. 1983. Echinoderm saponins. In: (P J Scheuer, ed.) Marine Natural Products, Vol 5. New York. Academic Press. pp 287–389.

BURNETT H W, BURNETT J W. 1990. Prolonged blurred vision following coelenterate envenomation. Toxicon 28: 731–733.

BURNETT J W. 1987. Dermatological manifestations of marine envenomations (Allergies in Dermatology). Clin Cases in Dermatol 3(1): 11–16.

BURNETT J W. 1990a. Correspondence to the Editors. J Wilderness Med 1: 135–137.

BURNETT J W. 1990b. The use of verapamil to treat box-jellyfish stings. Med J Aust 153: 363.

BURNETT J W. 1991. Clinical manifestations of jellyfish envenomation. Hydrobiologia 216/217: 629–635.

BURNETT J W. 1992a. Human injuries following jellyfish stings. Maryland Med J 41: 509–513.

BURNETT J W. 1992b. Seabather's eruption. Cutis 50: 98.

BURNETT J W. 1992c. Immunological aspects of jellyfish envenomations. In: (P Gopalakrishnakone, C K Tan, eds) Recent Advances in Toxinology Research. Venom and Research Group, Singapore. pp 333–349.

BURNETT J W. 1995. International Consortium for Jellyfish Stings Newsletter No12, January 1995. Baltimore. Department of Dermatology, University of Maryland Medical School. 2.

BURNETT J W, CALTON G J. 1977. Review article: The chemistry and toxicology of some venomous pelagic coelenterates. Toxicon 15: 177–196.

BURNETT J W, CALTON G J. 1983. Response of the box-jellyfish (*Chironex fleckeri*) cardiotoxin to intravenous administration of verapamil. Med J Aust 2: 192–194.

BURNETT J W, CALTON G J. 1984. Box-jellyfish cardiotoxin and IV verapamil. Med J Aust 140: 50.

BURNETT J W, CALTON G J. 1985. Recurrent eruption following a solitary envenomation by the cnidarian *Stomolophus meleagris*. Toxicon 23: 1010–1014.

BURNETT J W, CALTON G J. 1986. Pharmacological effects of various venoms on cutaneous capillary leakage. Toxicon 24: 614–617.

BURNETT J W, CALTON G J. 1987a. Jellyfish envenomation syndromes updated. Annals Emerg Med 16: 1000–1005.

BURNETT J W, CALTON G J. 1987b. Venomous pelagic coelenterates: chemistry, toxicology, immunology and treatment of their stings. Toxicon 25: 581–602.

BURNETT J W, CALTON G J, BURNETT H W. 1986a. Jellyfish envenomation syndromes. J Amer Acad Dermatol 14: 100–106.

BURNETT J W, CALTON G J, FENNER P J, WILLIAMSON J A. 1988a. Serological diagnosis of jellyfish envenomations. Comparative Biochem Physiol 91C: 79–83.

BURNETT J W, CALTON G J, LARSON J B. 1988b. Significant envenomation by *Aurelia aurita*, the moon jellyfish. Toxicon 26: 215–217.

BURNETT J W, CALTON G J, MORGAN R J. 1986b. Venomous sea urchins. Cutis 38: 151.

BURNETT J W, CALTON G J, MORGAN R J. 1987a. Dermatitis due to stinging sponges. Cutis 39: 476.

BURNETT J W, CARGO D G. 1979. Cutaneous irritation induced by crab larvae. J Amer Acad Dermatol 1: 42–43.

BURNETT J W, COBBS C S, KELMAN S N, CALTON G J. 1983a. Studies on the serologic response to jellyfish envenomation. J Amer Acad Dermatol 9: 229–231.

BURNETT J W, FENNER P J, KOKELJ F, WILLIAMSON J A. 1994a. Serious *Physalia* (Portuguese man o'war) stings: implications for scuba divers. J Wilderness Med 5: 71–76.

BURNETT J W, GABLE W D. 1989. A fatal jellyfish envenomation by the Portuguese man-o'-war. Toxicon 27: 823–824.

BURNETT J W, GEAN C J, CALTON G J, WARNICK J E. 1985. The effect of verapamil on the cardiotoxic activity of Portuguese man-o' war (*Physalia physalis*) and sea nettle (*Chrysaora quinquecirrha*) venoms. Toxicon 23: 681–689.

BURNETT J W, HEPPER K P, AURELIAN L. 1986c. Lymphokine activity in coelenterate envenomation. Toxicon 24: 104–107.

BURNETT J W, HEPPER K P, AURELIAN L, CALTON G J, GARDEPE S F. 1987b. Recurrent eruptions following unusual solitary coelenterate envenomations. J Am Acad Dermatol 17: 86–92.

BURNETT J W, ORDONEZ J V, CALTON G J. 1986d. Differential toxicity of *Physalia physalis* (Portuguese man-o'-war) nematocysts separated by flow cytometry. Toxicon 24: 514–518.

BURNETT J W, OTHMAN I B, ENDEAN R, FENNER P J, CALLANAN V I, WILLIAMSON J A. 1990. Verapamil potentiation of *Chironex* (box-jellyfish) antivenom. Toxicon 28: 242–244.

BURNETT J W, PIERCE L H Jr, URIWAN NAWACHINDA, STONE J H. 1968a. Studies on sea nettle stings. Arch Derm 98: 587–589.

BURNETT J W, RUBINSTEIN H, CALTON G J. 1983b. First aid for jellyfish envenomation. Southern Med J 76: 870–872.

BURNETT J W, STONE J H, PIERCE L H, CARGO D G, LAYNE E C, SUTTON J S. 1968b. A physical and chemical study of sea nettle nematocysts and their toxin. J Investigative Dermatol 51: 330–336.

BURNETT J W, WILLIAMSON J, FENNER P. 1994b. Mononeuritis multiplex after coelenterate sting. Med J Aust 161: 320–322.

BURRELL H. 1927. The Platypus. Sydney. Angus & Robertson. 227 pp.

BURY F E. 1922. On the action of medusan venom. Proceedings of the Physician Society of South Ussury Territory. No 1: 3–9 (in Russian).

CADZOW W H. 1960. Puncture wounds of the liver by stingray spines. Med J Aust 1: 936–937.

CALDER D R, PETERS E C. 1975. Nematocysts of *Chiropsalmus quadrumanus* with comments on the systematic status of the Cubomedusae. Helgolander wiss. Meeresunters 27: 364–369.

CAMARASA J G. NOGUE S, ANTICH E, SERRA-BALDRICH S. 1993. Red Sea coral contact dermatitis. Contact Derm 29: 285–286.

CAMERON A M. 1976. Toxicity of coral reef fishes. In (O A Jones, R Endean, eds) Biology and Geology of Coral Reefs, Vol 3: Biology. New York Academic Press. pp 155–176.

CAMERON A M. 1981. Venomous fish hazardous to humans. In: (J Pearn, ed) Animal toxins and man: human poisoning by toxic Australian venomous creatures. Brisbane. Division of Health Education and Information, Queensland. pp 29–45.

CAMERON A. 1988. Stonefishes. In: (J Pearn, J Covacevich, eds) Venoms and Victims. Brisbane. Queensland Museum. pp 43–48.

CAMERON A M, ENDEAN R. 1966. The venom apparatus of the scorpion fish *Notesthes robusta*. Toxicon 4: 111–21.

CAMERON A M, ENDEAN R. 1970. Venom glands in scatophagid fish. Toxicon 8: 171–178.

CAMERON A M, ENDEAN R. 1972a. Refutation of the venomous reputations of three teleost fishes. Toxicon 10: 335–339.

CAMERON A M, ENDEAN R. 1972b. Epidermal secretions and the evolution of venom glands in fishes. Toxicon 11: 401–410.

CAMERON J, FLOWERS A E, CAPRA M F. 1993. Modification of the peripheral nerve disturbance in ciguatera poisoning in rats with lidocaine. Muscle & Nerve 16: 782–786.

CAMPBELL A C. 1989. Seashores and shallow seas of Britain and Europe. London. Hamlyn. 84 pp.

CAMPBELL G D, SMITH E D 1993. The 'problem' of shark attacks upon humans (editorial). J Wilderness Med 4: 5–10.

CAMPBELL S. 1980. A Guide to the Hard Corals of Thai Waters. Hong Kong. Zebra Publishers. 165 pp.

CARGO DG 1971. The sessile stages of a scyphozoan identified as *Rhopilema verrilli*. Tulane Studies in Zool Botany 17(2): 31–34.

CARLGREN O. 1940. A contribution to the knowledge of the structure and distribution of cnidae in the Anthozoa. Lunds Universitets Årsskrift N.F. Avd (2)36(3): 1–62.

CARLISLE D B. 1962. On the venom of the lesser weeverfish *Trachinus vipera*. J Marine Biol Assoc UK 42: 155–162.

CARPENTER C C I. 1990. Other pathogenic vibrios. In: (G L Mandell, R G Douglas, J E Bennett, eds) Principles and Practice of Infectious Diseases, 3rd edition, vol 2. New York. Churchill Livingstone. pp 1646–1649.

CARRÉ D. 1980. Hypothèse sur le mechanisme de l'évagination du filament urticant des cnidocysts. Eur J Cell Biol 20: 265–271.

CARRÉ D, CARRÉ C. 1980. On triggering and control of cnidocyst discharge. Marine Behav Physiol 7: 109–117.

CASTEX M N, LOZA F. 1964. Etiológia de la enfermedad paratrygónoica: estudio anatómico, histológico y funcional del aparato agresor de la raya fluvial americana (gen. *Potamotrygon*). Rev Assoc Med Arg 78(6): 314–324.

CASWELL A. 1992. News in depth: the deathtrap in our northern seas. 'Australian Dr Weekly', 17 January 1992: pp 26–27.

CATALA R. 1964. Carnival Under the Sea. Paris. Sicard.

CHAMBERLAIN D A. 1989. Advanced life support: revised recommendations of the Resuscitation Council (UK). Brit Med J 299: 446–447.

CHAN J C. 1988. Erysipelothrix infections. In: (A Balows, W J Hausler, M Ohashi, A Turano, eds) Laboratory Diagnosis of Infectious Diseases, Vol 1. New York. Springer-Verlag. pp 232–236.

CHANG W J, PIEN F D. 1986. Marine-acquired infections. Postgrad Med 80: 30–41.

CHAPMAN D M. 1985. X-ray microanalysis of selected coelenterate statoliths. J Mar Biol Assoc UK 65: 617–627.

CHIPPAUX J P, GOYFFON M. 1983. Producers of antivenomous sera. Toxicon 21: 739–752.

CHUN K B. 1897. Die Siphonophoren der plankton-expedition. Ergebn. der plankton exp. II: K.b.

CLARK A M. SIMS J K. 1989. Dangerous marine organisms of Hawaii. Sea Grant Advisory Board. University of Hawaii. pp 4–6.

CLARK F E & LANE C E. 1961. Composition of float gases of *Physalia physalis*. Proc. Soc. for Experimental Biol. & Med 107: 673–674.

CLARK S D, COOK C B. 1986. Inhibition of nematocyst discharge during feeding in the colonial hydroid *Halycordyle disticha* (synonym *Pennaria tierella*): the role of previous prey killing. Biol Bull Mar Biol Lab, Woods Hole 171: 405–416.

CLELAND J B, SOUTHCOTT R V. 1965. Injuries to Man from Marine Invertebrates in the Australian Region. Canberra. National Health and Medical Research Council. Special Report Series No 12. 282 pp.

CLELAND J, SOUTHCOTT R V. 1969. Illnesses following the eating of fur seals in Australian waters. Med J Aust 1: 760–763.

CLINICS in DERMATOLOGY. 1987. Symposium Issue on Aquatic Dermatology (R M Mandojana, guest ed.). Philadelphia. J B Lippincott. Vol 5, No 3. 173 pp.

COBBS C S, GAUR P K, RUSSO A J, WARNICK J E, CALTON G J, BURNETT J W. 1983. Immunosorbent chromatography of sea nettle (*Chrysaora quinquecirrha*) venom and characterization of toxins. Toxicon 21: 385–391.

COLEMAN N. 1972. Anonymous anemone. Sea Frontiers 18: 373–375.

COLEMAN N. 1974. Australian Marine Fish in Colour. Sydney. A H & A W Reed. 108 pp.

COLLETTE B B. 1966. A review of the venomous toadfishes, subfamily Thalassophryninae. Copeia. 4: 846–864.

COLLINS D. 1798. An Account of the English Colony in New South Wales. 2 vols. London. Cadell & Davies (1798–1802).

COLLINS S P, COMIS A, MARSHALL M, HARTWICK R F, HOWDEN M E H. 1993. Monoclonal antibodies neutralizing the haemolytic activity of the box jellyfish (*Chironex fleckeri*) tentacle extracts. Comp Biochem Physiol 106B: 67–70.

'COLONIAL TIMES' HOBART. 1831. Coroner's inquest on three victims of toadfish poisoning (journalist unnamed). 29 March, 5 April.

COMIS A, HARTWICK R F, HOWDEN M E H. 1989. Stabilization of lethal and hemolytic activities of box jellyfish (*Chironex fleckeri*) venom. Toxicon 27: 439–447.

CONANT F S. 1897. Notes on the Cubomedusae. Johns Hopkins University Circulars No 132: 8–10.

CONANT F S. 1898. The Cubomedusae. Mem Biol Lab Johns Hopkins Univ 4: 1–61.

COOPER N K. 1985. Two cases of near-fatal box jellyfish stings sustained in littoral waters of the Sultanate of Oman. SPUMS J 5: 16–17.

COOPER N K. 1991. Disarming the box jellyfish in Oman. Medical Corps International 1/91: 26–32.

COOPER P, WAKEFIELD MC. 1974. A sarcoid reaction to injury by sea urchin spines. J Pathol 112: 33.

COPPLESON V M. 1958. Shark Attack. Sydney. Angus & Robertson. 266pp.

CORMIER S M, HESSINGER D A. 1980. Cnidocil apparatus: sensory receptor of *Physalia* nematocytes. J Ultrastruct Res 72: 13–19.

CORRIGAN P, RUSSELL F E, WAINSCHEL J. 1978. Clinical reactions to antivenene. Toxicon 16(Suppl 1.): 457–465.

CORSON E F, PRATT A G. 1943. 'Red moss' dermatitis. Arch Dermatol Syphil (Chicago) 47(4): 574–579.

CRACCHIOLO A, GOLDBERG L. 1977. Local and systemic reactions to puncture injuries by the sea urchin spine and the date palm thorn. Arthritis Rheum 20: 1206–1212.

CREVAUX J, Le JANNE E. 1882. Recit du troisième voyage dans l'Amérique équatoriale. Arch Méd Nav 37: 50–53.

CROFT J A, HOWDEN M E H. 1972. Chemistry of maculotoxin: a potent neurotoxin isolated from *Hapalochlaena maculosa*. Toxicon 10: 645.

CRONE H D. 1976. Chemical modification of the haemolytic activity of extracts from the box jellyfish *Chironex fleckeri* (Cnidaria). Toxicon 14: 97–107.

CROPP B, CROPP L. 1984. The deadliest creature on earth. Scuba Diver 3: 42–48.

CROSS T B. 1976. An unusual stingray injury: the skindiver at risk. Med J Aust 2;25–26: 947–948.

CRUZ L J, GRAY W P, OLIVERA B M. 1978. Purification and properties of a myotoxin from *Conus geographus* Arch Biophys 190: 539–548.

CRUZ L J, JOHNSON D S, OLIVERA B M. 1987. Characterisation of the α-Conotoxin target: evidence for the tissue-specific heterogeneity in calcium channel types. Biochem 26: 820–824.

CURRIE B. 1992. Jellyfish and top end waters. Northern Territory Communicable Dis Bull 1(3): 1–2.

CURRIE B. 1993. Medusae in tropical Australia. Med J Aust 158: 609–615.

CURRIE B. 1994. Review article: clinical implications of research on the box-jellyfish *Chironex fleckeri*. Toxicon 32: 1305–1313.

CURRIE B, ANSTEY N. 1990. Clinical observations of jellyfish envenomation. In: Annual Report 1989–1990, Menzies School of Health Research, PO Box 41096, Casuarina NT 0811, Australia. pp 84–85.

CURRIE B, HO S, ALDERSLADE P. 1993. Box-jellyfish, Coca-Cola and old wine. Med J Aust 158: 868.

CURRIE B, KHANH D M, ALDERSLADE P, WILLIAMSON J, FENNER P. 1991. Jellyfish envenomation in the Northern Territory of Australia. In: Abstracts, 10th World Congress on Animal, Plant and Microbial Toxins, 3–8 November 1991 Singapore. Abstract No 204, p 256.

CURRIE B, KHANH D M, ALDERSLADE P, WILLIAMSON J, FENNER P. 1992. Jellyfish envenomation in the Northern Territory of Australia. Toxicon 30: 501.

CURRIE B J, WOOD Y K. 1995. Identification of *Chironex fleckeri* envenomation by nematocyst recovery from skin. Med J Aust 162: 478–480.

D'ABREU E A. 1931. Effect of a bite from Schneider's water snake, *Hypsyrhina enhydris*. J Bombay Nat Hist Soc 22: 203.

'DAILY TELEGRAPH', Sydney. 1950. Boy and cat die from eating toado (journalist unnamed). 9 December 1950.

DALZELL P. 1994. Management of ciguatera fish poisoning in the South Pacific. Memoirs of the Queensland Museum 34(3): 471–479.

DAS M P, LIM H S, TEH Y E. 1971. Histamine and histamine-like substances in the marine sponge, *Suberites inconstans*. Comp Gen Pharmacol 2(8): 473–475.

DAVIES D H. 1964. About Sharks and Shark Attack. Pietermaritzburg. Shuter & Shooter. 237 pp.

De LAUBENFELS M W. 1932. The marine and freshwater sponges of California. Proc US Nat Mus, Publ No 2927;81(4): 1–140.

De LAUBENFELS M W. 1936. A discussion of the sponge fauna of the Dry Tortugas in particular and the West Indies in general, with material for a revision of the families and orders of the Porifera. Papers from Tortugas Lab (Carnegie Institute, Washington DC) 30: 1–225.

De ORIO G A. 1946. Dermatitis venenata resulting from contact with marine animals (hydroids). Arch Derm Syph 54(6): 637–649.

De QUATREFAGES A. 1854. Mémoire sur l'organisation des Physalies. Ann Sci Nat 2(4): 107–142.

De QUIROS F. 1606. Voyages in the South Sea by the Spanish and Dutch. In: Dalrymple's Collection of Voyages, Vol 1. London. 108 pp.

De VANTIER L M, DEACON G. 1990. Distribution of Acanthaster planci at Lord Howe Island, the southern most Indo-Pacific reef. Coral Reefs 9: 145–148.

DIPPER F, WOODWARD T. 1989. The Living Seas. Dubai. Motivate Publishing. pp. 1–95.

DIROX A, KNUTTGEN H G, TITTEL K. 1988. The Olympic Book of Sports Medicine. International Olympic Committee Publication. Oxford. Blackwell Scientific Publications. Vol 1, pp 457–458.

DOBB G J. 1986. Sea-snake envenomation. Med J Aust 144: 112.

DOYLE A CONAN. Republished 1981. The adventure of the lion's mane. In: The Case Book of Sherlock Holmes. New York. Berkly Books. pp 186–206.

DRURY J K, NOONAN J H, POLLOCK J A, REED W H. 1980. Jellyfish sting with serious complications. Injury 12: 66–68.

DUCHASSAING DE FONBRESSIN P, MICHELOTTI G. 1864. Spongiaires de la Mer Caraïbe. Natuurk Vehr Mij Haarlem 21: 1–124. (see also De Laubenfels 1936).

DUHIG J V, JONES G. 1928. Haemotoxin of the venom of *Synanceia horrida*. Australian J Exp Biol 5(2): 173–179.

DULHUNTY A F, GAGE P W. 1971. Selective effects of an octopus toxin on action potentials. J Physiol 218: 433.

DUNCAN C. 1951. A case of toadfish poisoning. Med J Aust 2: 673–675.

DYBAS C L. 1992. Space-age jewels. National Wildlife Magazine (USA). August–September 1992. pp 18–24.

EASTON R G. 1980. Disarming the box-jellyfish (a letter and a reply). Med J Aust. 1: 335–338.

EDMONDS C. 1969. A non-fatal case of blue-ringed octopus bite. Med J Aust 2: 601.

EDMONDS C. 1978. Dangerous Marine Animals of the Indo-Pacific Region. Newport, NSW. Wedneil Publications. 229 pp.

EDMONDS C. 1989. Dangerous Marine Creatures. Frenchs Forest, NSW 2086. Reed Books. 192 pp.

EDMONDS C. 1990. Marine animal injuries. In: (A A Bove, J C Davis, eds) Diving Medicine, second edition. Philadelphia. W.B.Saunders Co. pp 115–137.

EDMONDS C. 1992. Dangerous marine creatures. In: (C Edmonds, C Lowry, J Pennefather, authors) Diving and Subaquatic Medicine, 3rd edition. Oxford. Butterworth-Heinemann. pp 318–347.

EDMONDS C. 1994. Nitrous oxide induced decompression sickness following shoulder barotrauma. SPUMS J 24: 76–78.

EDMONDS C, LOWRY C, PENNEFATHER J. 1992. Diving and Subaquatic Medicine, third edition. Oxford and Sydney. Butterworth-Heinemann. 565 pp.

EDWARDS C. 1976. A study of erratic distribution: The occurrence of the medusa *Gonionemus* in relation to the distribution of oysters. Adv Mar Biol 14: 251–284.

EDWARDS F P, CAMPBELL A R. 1984. Removal of formaldehyde and xylene fumes from histopathology laboratories: a functional approach to the design of extraction systems. J Clin Pathol 37: 401–408.

ELDILA M T, MUSTAFFA M R, AZILA N, OTHMAN I. 1993. Comparative pharmacological effects of *Carybdea rastoni* and *Chironex fleckeri* on rat aorta. In: (N H Tan, S L Oo, V Thambyrajah, N Azila, eds) Advances in Venom and Toxin Research. Proceedings of the Third Asia Pacific Congress on Animal, Plant and Microbial Toxins. Malaysia, 27 June–1 July 1993. Kuala Lumpur. Malaysian Society on Toxinology. pp 284–288.

ELLIS R. 1975. The Book of Sharks. New York. Grosset & Dunlap.

ENDEAN R. 1961a. The venomous sea urchin *Toxopneustes pileolus*. Med J Aust 1: 320.

ENDEAN R. 1961b. A study of the distribution, habitat, behaviour, venom apparatus and venom of the stonefish. Aust J Mar Freshw Res 12: 177–190.

ENDEAN R. 1981. The box-jellyfish or 'sea-wasp'. In: (J Pearn, ed) Animal Toxins and Man: Human Poisoning by Toxic Australian Venomous Creatures. Brisbane. Queensland Health Department. pp 46–54.

ENDEAN R. 1982. Australia's Great Barrier Reef. St Lucia, Qld. University of Queensland Press. 348 pp.

ENDEAN R. 1987. Separation of two myotoxins from the nematocysts of the box jellyfish (*Chironex fleckeri*). Toxicon 25: 483–492.

ENDEAN R ('Bob'). 1988a. How *Chironex* injects its venom. In: (J Pearn, J Covacevich, eds) Venoms and Victims. Brisbane. Queensland Museum. pp 9–14.

ENDEAN R ('Bob'). 1988b. Venom of *Chironex*, the world's most venomous animal. In: (J Pearn, J Covacevich, eds) Venoms and Victims. Brisbane. Queensland Museum. pp 15–24.

ENDEAN R, DUCHEMIN C, McCOLM D, HOPE FRASER E. 1969. A study of the biological activity of toxic material derived from nematocysts of the cubomedusan *Chironex fleckeri*. Toxicon 6: 179–204.

ENDEAN R, GRIFFITH J K, ROBINS J J, LLEWELLYN L E, MONKS S A. 1993a. Variations in the toxins present in ciguateric narrow-barred Spanish mackerel, *Scomberomorus commersoni*. Toxicon 31: 723–732.

ENDEAN R, GRIFFITH J K, ROBINS J J, MONKS S A. 1993b. Multiple toxins in a specimen of the narrow-barred Spanish mackerel (*Scomberomorus commersoni*). Toxicon 31: 195–204.

ENDEAN R, GYR P, SURRIDGE J. 1977a. The pharmacological actions on guinea-pig ileum of crude venoms from the marine gastropods *Conus striatus* and *Conus magus*. Toxicon 15: 327–337.

ENDEAN R, GYR P, SURRIDGE J. 1979. The effects of crude venoms of *Conus magus* and *Conus striatus* on the contractile response and electrical activity of guinea-pig cardiac musculature. Toxicon 17: 381–395.

ENDEAN R, IZATTS J. 1965. Pharmacological study of the venom of the gastropod *Conus magus*. Toxicon 3: 81–93.

ENDEAN R, MONKS S A, CAMERON A M. 1993c. Toxins from the box-jellyfish *Chironex fleckeri*. Toxicon 31: 397–410.

ENDEAN R, NOBLE M. 1971. Toxic material from the tentacles of the cubomedusan Chironex fleckeri Toxicon 9: 255–264.

ENDEAN R, PARISH G, GYR P. 1974. Pharmacology of the venom of *Conus geographus*. Toxicon 12: 131–138.

ENDEAN R, RIFKIN J. 1975. Isolation of different types of nematocyst from the cubomedusan *Chironex fleckeri*. Toxicon 13: 375–376.

ENDEAN R, RIFKIN J. 1983. Envenomation involving nematocysts of the box jellyfish Chironex fleckeri. Toxicon Suppl 3: 115–118.

ENDEAN R, RIFKIN J F, DADDOW L Y M. 1991. Envenomation by the box-jellyfish *Chironex fleckeri*: how nematocysts discharge. Hydrobiologia 216/217: 641–648.

ENDEAN R, RUDKIN E. 1963. Studies on the venom of some Conidae. Toxicon 1: 49–64.

ENDEAN R, SIZEMORE D J. 1988. The effectiveness of antivenom in countering the actions of box-jellyfish (*Chironex fleckeri*) nematocyst toxins in mice. Toxicon 26: 425–431.

ENDEAN R, SURRIDGE J, GYR P. 1977b. Some effects of crude venom from the cones *Conus striatus* and *Conus magus* on isolated guinea-pig atria. Toxicon 15: 369–374.

ENDO M, NAKAGAWA M, HAMAMOTO Y, ISHIHAMA M. 1986. Pharmacologically active substances from southern Pacific marine invertebrates. Pure Appl Chem 58: 387–394.

ENGLEHARDT U, LASSIG B. 1992. Crown-of-thorns starfish research update 1991/92. Townsville. Great Barrier Reef Marine Park Authority. 19 pp.

ESCHSCHOLTZ F. 1829. System der Acalephen. Berlin. pp vi: 190 (pls 16).

EVANS HM. 1943. Sting-fish and Seafarer. London. Faber & Faber. pp 108–143.

EVERITT B J, JUREVICS H A. 1973. Some pharmacological properties of the crude toxin from the spines of the crown-of-thorns starfish *Acanthaster planci*. Proc Aust Physiol Pharmacol Soc 4: 46–47.

EWERS W H. 1961. A new intermediate host of schistosome trematodes from new South Wales. Nature 190: 283–284.

EXECUTIVE EDITORIAL BOARD, Australian Prescriber. 1989. Medical management of anaphylactoid and anaphylactic reactions. Aust Prescriber 12 (insert into vol. 12): wall chart.

EXTON D. 1987. Echinoderms. In: (J Covacevich, P Davie, J Pearn, Eds) Toxic Plants and Animals: A Guide for Australia. Brisbane. Queensland Museum. pp 123–125.

EXTON D R. 1988. Treatment of *Physalia physalis* envenomation. Med J Aust 149: 54.

EXTON D R, FENNER P J, WILLIAMSON J A H. 1989. Ice packs: an effective first aid treatment for *Physalia* and other painful jellyfish stings. Med J Aust 151: 625–626.

FADDA G. 1988. Mycobacterioses excluding tuberculosis. In: (A Balows, W J Hausler, M Ohashi, A Turano, eds) Laboratory Diagnosis of Infectious Diseases, Vol 1. New York. Springer-Verlag. pp 382–395.

FARBER L, LERKE P. 1963. Studies on the toxicity of *Rhodactis howesii* (matamalu). In: (H L Keegan, W V Macfarlane, eds) Venomous and Poisonous Animals and Noxious Plants of the Pacific Region. Oxford. Pergamon Press. pp 67–74.

FAUST E S. 1924. Coelenterata (zoophyta), pflanzentiere. In Handbuch Exp Pharm. Berlin. pp 1925–1928.

FAUTIN D G, ALLEN G R. 1992. Field Guide to Anemone Fishes and their Host Anemone Sea Anemones. Perth. Western Australian Museum. 157 pp.

FAUTIN DG, MARISCAL RN. 1991. Anthozoa. In: (F W Harrison, J A Westfall, Eds) Lower Invertebrates Vol 2: Placozoa, Porifera, Cnidaria and Ctenophora. New York. Wiley-Liss. pp 267–358.

FENNER P J. 1987. Report of field trip to Philippines by Dr P Fenner and Dr R Hartwick, 1985. Report to Queensland State Centre, Surf Life Saving Australia. 20 pp.

FENNER P. 1990. Medical frontiers in Australian seas. In: (J Pearn, M Cobcroft, eds) Fevers and Frontiers. Royal Children's Hospital, Brisbane. Department of Child Health Publishing Unit, University of Queensland. pp 73–84.

FENNER P J. 1991. Cubozoan jellyfish envenomation syndromes and their medical treatment in northern Australia. Hydrobiologia 216/217: 637–640.

FENNER P. 1993. Marine stingers. In: 63rd Annual Report 1992–93. Newstead, Qld. Surf Life Saving Queensland. pp 39–41.

FENNER PJ. 1995. Stingray envenomation: a suggested new treatment. Med J Aust 163: 655.

FENNER P J, FITZPATRICK P F. 1986. Experiments with the nematocysts of *Cyanea capillata* Med J Aust 145: 174.

FENNER P J, FITZPATRICK P F, HARTWICK R J, SKINNER R. 1985. 'Morbakka', another cubomedusan. Med J Aust 143: 550–555.

FENNER P, RODGERS D, WILLIAMSON J. 1986a. Box jellyfish antivenom and 'Irukandji' stings. Med J Aust 144: 665–666.

FENNER P J, WILLIAMSON J A H. 1987. Experiments with the nematocysts of *Carybdea rastoni*. Med J Aust 147: 258–259.

FENNER P J, WILLIAMSON J A. 1996. World-wide deaths, severe envenomation Med J Aust sub for public.

FENNER P J, WILLIAMSON J A, BLENKIN J A. 1989a. Successful use of *Chironex* antivenom by members of the Queensland Ambulance Transport Brigade. Med J Aust 151: 708–710

FENNER P J, WILLIAMSON J A, BURNETT J W, COLQUHOUN D M, GODFREY S, GUNAWARDANE K, MURTAGH K. 1988. The 'Irukandji syndrome' and acute pulmonary oedema. Med J Aust 149: 150–156.

FENNER P J, WILLIAMSON J A, BURNETT J W, RIFKIN J. 1993a. First aid treatment of jellyfish stings in Australia: response to a newly differentiated species. Med J Aust 158: 498–501.

FENNER P J, WILLIAMSON J A H, CALLANAN V, AUDLEY I. 1986b. Further understanding of, and a new treatment for, 'Irukandji' (*Carukia barnesi*) stings. Med J Aust 145: 569–574.

FENNER P, WILLIAMSON J, CURRIE B. 1993c. Update of marine envenomation in tropical Australia. Presented at 'Tropical Medicine at the Top End'. Mirambeena Tourist Resort, Darwin, Northern Territory, 17–19 June 1993. Townsville, Qld. Australasian College of Tropical Medicine Publication. 88 pp.

FENNER P J, WILLIAMSON J A, MYERS D. 1992. Platypus envenomation: a painful learning experience. Med J Aust 157: 829–832.

FENNER P J, WILLIAMSON J A, SKINNER R A. 1989b. Fatal and non-fatal stingray envenomation. Med J Aust 151: 621–625.

FERNANDO M, GOONERATNE W. 1983. Sea-snake envenoming. Ceylon Med J 28: 131–143.

FIELDS W G, MACKIE G O. 1971. Evolution of the Chondrophora: evidence from behavioural studies on Vellela. J Fish Res Bd Can 28: 1595–1602.

FILLING-KATZ M R. 1984. Mononeuritis multiplex following jellyfish stings. Ann Neurol 15: 213.

FISH C J, COBB M C. 1954. Noxious marine enimals of the western Pacific Ocean. Research Report 36, Fish and Wildlife Service, US Department of the Interior. US Government Printing Office. pp iii and 36: 1–45.

FISHELSON L. 1974. Histology and ultrastructure of the recently found buccal toxic gland in the fish *Meiacanthus nigrolineatus* (Blenniidae). Copeia 2: 386–392.

FISHER A A. 1978. Atlas of Aquatic Dermatology. New York. Grune & Stratton. 113 pp.

FISHER A A. 1984. Toxic vs allergic reactions to jellyfish. Cutis 34: 450–454.

FISHELSON L. 1973. Observations on skin structure and sloughing in the stone fish *Synanceia verrucosa* and related fish species as a functional adaptation to their mode of life. Z. Zellforsch 140: 497–508.

FLEAY D. 1980. Paradoxical Platypus. Brisbane. Jacaranda. p xiii.

FLECKER H. 1936. Cone shell mollusc poisoning with report of a fatal case. Med J Aust 1: 464–466.

FLECKER H. 1945. Injuries by unknown agents to bathers in North Queensland. Med J Aust 1: 98.

FLECKER H. 1952a. Fatal stings to north Queensland bathers. Med J Aust 1: 35–38.

FLECKER H. 1952b. 'Irukandji' sting to north Queensland bathers without production of wheals but with severe general symptoms. Med J Aust 1: 89–91.

FLECKER H. 1956. Injuries from stonefish. Med J Aust 2: 371–373.

FLECKER H, COTTON B C. 1955. Fatal bite from octopus. Med J Aust 2: 329–332.

FLEMING W J, HOWDEN M E H, SALATHE R. 1972. Chemistry of toxins in the crown-of-thorns starfish. Crown-of-thorns Starfish Seminar, University of Queensland, Brisbane. pp 83–95.

FLEMING W J, HOWDEN M E H. 1974. Partial purification and characterization of a lethal protein from *Tripneustes gratilla*. Toxicon 12: 447–448.

FORSTER JR. 1778. Remarks on the Organic Bodies: Observations Made During a Voyage Round the World on Physical Geography, Natural History, and Ethic Philosophy. London. G. Robinson. pp 210 & 642.

FOXALL T L, SHOPTAUGH N H, IKAWA M, SASNER J J. 1979. Secondary intoxication with PSP in cancer (i). In: (D L Taylor, H H Seliger, eds.) Toxic Dinoflagellate Blooms. Amsterdam. Elsevier, North Holland. pp 413–418.

FRANKEL E H. 1992. Seabather's eruption develops following Mexican vacation. Clin Cases Dermatol 4: 6–8.

FREEMAN S E. 1974. Actions of *Chironex fleckeri* toxins on cardiac transmembrane potentials. Toxicon 12: 395–404.

FREEMAN S E. 1976. Electrophysiological properties of maculotoxin. Toxicon 14: 396–399.

FREEMAN S E. 1984. Box-jellyfish cardiotoxin and IV verapamil. Med J Aust 140: 50.

FREEMAN S E, TURNER R J. 1969. A pharmacological study of the toxin of a cnidarian, *Chironex fleckeri* Southcott. Brit J Pharmacol 35: 510–520.

FREEMAN S E, TURNER R J. 1970. Maculotoxin, a potent toxin secreted by *Octopus maculosus* Hoyle. Toxicol Appl Pharmacol 16(3): 681–690.

FREEMAN S E, TURNER R J. 1971. Cardiovascular effects of toxins isolated from the cnidarian *Chironex fleckeri* Southcott. Brit J Pharmacol 41: 154–166.

FREEMAN S E, TURNER R J. 1972a. Cardiovascular effects of cnidarian toxins: a comparison of toxins extracted from *Chiropsalmus quadrigatus* and *Chironex fleckeri*. Toxicon 10: 31–37.

FREEMAN S E, TURNER R J. 1972b. A myotoxin secreted by some piscivorous *Conus* species. Brit J Pharmacol 46: 329–343.

FREEMAN S E, TURNER R J, SILVA S R. 1974. The venom and venom apparatus of the marine gastropod *Conus striatus* Linné. Toxicon 12: 587–592.

FREIRE-MAIA L, CAMPOS J A. 1987. Response to the letter to the editor by Gueron and Ovsyshcher on the treatment of the cardiovascular manifestations of scorpion envenomation. Toxicon 25: 125–130.

FREUDENTHAL A R. 1991. Seabather's eruption: range extended northward and a causative organism identified Rev Internationale D'Oceanographique Medicale. 101–104 (in French). Nassau County Department of Health, Office of Marine Ecology, Bureau of Water Pollution Control, Mineola, NY 11501, USA. pp 137–147.

FREUDENTHAL A R, JOSEPH P R. 1993. Seabather's eruption. New Eng J Med 329: 542–544.

FREYVOGEL T A. 1972. Poisonous and venomous animals of East Africa. Acta Trop (Basel) 29: 401–451.

FRIESS S L. 1963. Some pharmacological activities of the sea cucumber neurotoxin. Amer Inst Biol Sci Bull 13(2): 41.

FUJIWARA T. 1935. On the poisonous pedicellaria of *Toxopneustes pileolus* (Lamarck). Annot Zool Japan 15: 62–69.

FUJIWARA M, MURAMATSU I, HIDAKA H, IKUSHIMA L, ASHIDA K. 1979. Effects of Goniopora toxin, a polypeptide isolated from coral, on electromechanical properties of rabbit myocardium. J Pharmacol Exp Ther 210: 153–157.

FUKUDA T. 1951. Marked increase in cases of puffer poisoning (in Japanese). Clinics and Studies 29(2): 99–106.

FULDE G W O, SMITH F. 1984. Sea snake envenomation at Bondi. Med J Aust 141: 44–45.

FULGHUM D D. 1986. Octopus bite resulting in granuloma annulare. Southern Med J 79(11): 1434–1436.

FUSETANI N, HASHIMOTO K, TAKAHASHI K. 1982. Occurrence of potent neurotoxins in the horseshoe crab *Carcinoscorpius rotundicauda*. Toxicon 20: 662–664.

GALIL B, SPANIER E, FERGUSSON W W. 1990. The Scyphomedusae of the Mediterranean coast of Israel including two Lessepsian migrants new to the Mediterranean. Zool Meded Leiden 64: 95–106.

GALLAGHER M. 1990. Snakes of the Arabian Gulf and Oman. M. Gallagher publication, P O Box 668, Muscat, Sultanate of Oman. pp1–16.

GARCIA P J, SCHEIN R M H, BURNETT J W. 1994. Fulminant hepatic failure from a sea anemone sting. Annals Int Med 120(8): 665–666.

GASPERETTI J. 1988. Snakes of Arabia. In: (D W Buttiker, F Krupp eds) Fauna of Saudi Arabia. Saudi Arabia. 9: 169–450.

GAUR P K, ANTHONY R L, CALTON G J, BURNETT J W. 1982. Isolation of hybridomas secreting antibodies against *Physalia physalis* (Portuguese Man-o'-war) nematocyst venom. Toxicon 20: 419–425.

GAUR P K, ANTHONY R L, CODY T S, CALTON G J, BURNETT J W. 1981. Production of a monoclonal antibody against the sea nettle venom lethal factor. Proc Soc Exper Biol Med 167: 374–377.

GAWADE S P, BHIDE M B. 1977. Enzymatic activities of the venom of *Enhydrina schistosa* (common sea snake) from the western coast of India. Bull Haffkine Inst 5: 48–51.

GAWADE S P, BUDUK D P, GAITONDE B B. 1980. Preparation of monovalent Common Indian sea snake (*Enhydrina schistosa*) antivenin. Indian J Med Res 72: 747–752.

GENENAH A A, SHIMIZU Y. 1981. Specific toxicity of paralytic shellfish poisons. J Agric Food Chem 29: 1289–1291.

GEORGE J D, GEORGE J J. 1979. An Illustrated Encyclopaedia of Invertebrates in the Sea. Sydney. Rigby. (prepared by Lionel Leuenthal Ltd, London, 1979). 288 pp.

GHOSH T K, GOMES A, NAG CHAUDHURI A K. 1990a. Pharmacological actions of tentacle extract of the jellyfish *Acromitus rabanchatu* occurring in the Bay of Bengal. Indian J Expl Biol 28: 39–42.

GHOSH T K, GOMES A, NAG CHAUDHURI A K. 1990b. Glycaemic effects of tentacle extract of the jellyfish *Acromitus rabanchatu* (Annandale). Indian J Expl Biol 28: 441–443.

GHOSH T K, GOMES A, NAG CHAUDHURI A K. 1993. Isolation of a toxin from jellyfish *Acromitus rabanchatu* and its effects on skeletal muscle. Toxicon 31: 873–880.

GILLESPIE N. 1987. Ciguatera poisoning. In: (J Covacevich, P Davie, J Pearn, eds) Toxic Plants and Animals: A Guide for Australia. Brisbane. Queensland Museum. pp 160–169.

GILLESPIE N C, LEWIS R J, PEARN J H, BOURKE A T C, HOLMES M J, BOURKE J B, SHIELDS W J. 1986. Ciguatera in Australia: occurrence, clinical features, pathophysiology and management. Med J Aust 145: 584–590.

GLADFELTER W B. 1970. Swimming in Medusae. PhD thesis, Stanford University, Stanford, Calif.

GLADFELTER W B. 1972. Structure and function of the locomotory system of the scyphomedusa *Cyanea capillata*. Mar. Biol. 14: 150–160.

GLASSER D B, BURNETT J W, KATHURIA S S, RODRIGUES M M. 1993. A guinea-pig model of corneal jellyfish envenomations. Toxicon 31: 808–812.

GLASSER D B, NOELL M J, BURNETT J W, KATHURIA S S, RODRIGUES M M. 1992. Ocular jellyfish stings. Ophthalmology 99: 1414–1418.

GLAZIOU P, CHINAIN M, LEGRAND A-M. 1995. Clinical toxicology of ciquatera poisoning. In: (J Meier, J White, eds) Handbook of Clinical Toxicology of Animal Venoms and Poisons. Boca Ratou, Florida. CRC Press. pp 59-74.

GODKNECHT A, TARDENT P. 1988. Discharge and mode of action of the tentacular nematocysts of *Anemonia sulcata* (Anthozoa: Cnidaria). Marine Biol 100: 83–92.

GOETTE A. 1886. Verzeichniss der Medusen welche von Dr Sander auf SMS 'Prinz Adalbert' gesammelt wurden. BS Preuss Akad Wiss Bd 7: 831–837.

GOLLAN J A. 1968. The dangerous Portuguese man-o'-war. Med J Aust 1: 973.

GONZAGA R A F. 1985a. Venomous fish stings on the European seashore. Postgrad Med 77: 146–147.

GONZAGA R A F. 1985b. Mordeduras e Picadas por Animas da Fauna Portuguesa. Porto. Prémio Bial de Medicina Clínica 1984. 198 pp.

GONZAGA R A F. 1985c. Abortion after a weever fish sting. Brit Med J 290: 518.

GOONEWARDENE R. 1985. Marine Hazards: Common Venomous Fish in Oman. medical newsletter, Ministry of Health, Sultanate of Oman 85/1(14): 11–14.

GOONEWARDENE R. 1986. Marine Hazards, Part 2 'Venomous coelenterates'. Medical newsletter. Ministry of Health, Sultanate of Oman. 86/2(1): 23–28.

GOPALAKRISHNAKONE P, GWEE M C E. 1993. The structure of the venom gland of stonefish *Synanceia horrida*. Toxicon 31: 979–988.

GORMAN D. 1991. A new classification for the decompression illnesses. SPUMS J 21: 74–75.

GORMAN D. 1994. Rescue and resuscitation of divers. In: (D Gorman, C Acott, J Williamson, eds) Diving and Hyperbaric Medicine. Adelaide. Hyperbaric Medicine Unit. 22.1–22.4.

GOSLINE W A, BROCK V E. 1960. Handbook of Hawiian Fishes. Honolulu. University of Hawaii Press. 372 pp.

GOY J. 1983. Les concentrations de meduses en Mer Ligure. Workshop on jellyfish blooms in the Mediterranean. Athens, 31 October–4 November 1983. United Nations Environment Programme. pp 109–120.

GOY J. 1990. Preliminary observations on the zooplankton of Shark Bay, Western Australia, with emphasis on medusae and description of a new rhizostome jellyfish (Cnidaria, Scyphozoa) In: (P F Berry, S D Bradshaw, B R Wilson, eds) Research in Shark Bay. Report of the France-Australe Bicentenary Expedition Committee. Perth. Western Australian Museum. 325 pp.

GRAINGER C R. 1985. Sting ray injuries. Trans Roy Soc Trop Med Hyg 79: 443–444.

GRANT E M. 1975. Guide to Fishes, 3rd edition. Brisbane. Department of Harbours and Marine. 768 pp.

GRANT E M. 1978. Guide to Fishes, 4th edition. Brisbane. Department of Harbours and Marine. 640 pp.

GRANT T R. 1983a. Body temperature of free-ranging platupuses, *Ornithorhynchus anatinus*, with observations on their use of burrows. Australian J Zoology 31: 117–122.

GRANT T R. 1983b. The behavioural ecology of monotremes. In: (J F Eisenberg, D G Kleiman, eds) Advances in the Study of Mammalian Behaviour. American Society of Mammalogists special publication No7.

GRANT T R. 1989. The Platypus: A Unique Mammal. Sydney. New South Wales University Press. 3rd edition (1995) 96 pp.

GRANT T R, DAWSON T J. 1978. Temperature regulation in the platypus, Ornithorhynchus anatinus, maintenance of body temperature in air and water. Physiological Zoology 51: 1–6.

GRAUER F H, ARNOLD H L. 1961. Seaweed dermatitis. First report of a dermatitis-producing marine alga. Arch Dermatol 84: 720–732.

GRAY W R, LUQUE A, OLIVERA B M, CRUZ L J. 1977. Muscle-relaxing peptides from *Conus* venoms. Proc Fed Amer Soc Exp Biol 36: 684.

GRAY W R, OLIVERA B M, BARRETT J, CRUZ L J. 1981. Peptide toxins from *Conus geographus* venom. J Biol Chem 256: 4734–4740.

GREIG D M. 1930. Thanatophoric fish. Edinburgh Med J 37: 638–640.

GRIFFITH M. 1978. The Biology of Monotremes. New York. Academic Press. 367pp.

GRIGG G, BEARD L, GRANT T, AUGEE M. 1992. Body temperature and diurnal activity patterns in the platypus (*Ornithorhynchus anatinus*) during winter. Aust J Zool 40: 135–142.

GUDGER E W. 1929. Nicholas Pike and his unpublished paintings of the fishes of Mauritius, western Indian Ocean, with an index to fishes. Bull Amer Mus Natur Hist 58(9): 489–530.

GUERON M, MARGULIS G, ILIA R, SOFER S. 1993. The management of scorpion envenomation. Toxicon 31: 1071–1076.

GUESS H A, SAVETEER P L, MORRIS C R. 1982. Hemolysis and acute renal failure following a Portuguese man-o'-war sting. Pediatrics 70: 979–81.

GUEST W C. 1959. The occurrence of the jellyfish *Chiropsalmus quadrumanus* in Matagorda bay, Texas. Bull Mar Sci Gulf & Carib 9: 79–83.

GULLIKSEN B. 1971. A new record of *Gonionemus vertens* Agassiz (Limnomedusae) in Norway. Kongelige Norske Videnskabers Selskabm Skrifter 12: 1–4.

HAACKE W. 1887. Die Scyphomedusen des St Vincent Golfes. Jena Zeitschr Naturwiss 20: 588–638.

HACH-WUNDERLE V, MEBS D, FREDERKING K, BREDDEN H K. 1987. Vergiftung durch Feuerqualle. Deutch Med Wochenschr 112: 1865–1868.

HAECKEL E. 1880. System der Acraspeden. Zweite Halfte des Systems der Medusen. Denkschr Med-Naturwiss Ges Jena 1: 361–672.

HALE H M. 1929. The Crustaceans of South Australia, Part 2. Adelaide. Government Printer. pp 201–380.

HALLEGRAEFF G M. 1993. A review of harmful algal blooms and their apparent global increase. Phycologia 32: 79–99.

HALSTEAD B W. 1965, 1967, 1970. Poisonous and Venomous Marine Animals of the World, Vol I: Invertebrates (1965) Vol 2 (1967) Vol 3 (1970). Washington, DC. US Government Printing Office.

HALSTEAD B. 1971. Sea urchin injuries. In: (W Burcherl, E Buckley, eds) Venomous Animals and their Poisons, Vol 3.

HALSTEAD B W. 1978. Coelenterates. In: Poisonous and Venomous Marine Animals of the World (first revised edition). Princeton, NJ. Darwin Press. pp 87–140.

HALSTEAD B W. 1982. Paralytic Shellfish Poisoning Guide. Geneva. WHO Publication.

HALSTEAD B W. 1987. Coelenterate (cnidarian) stings and wounds. Clin in Dermatol: Aquatic Dermatology 5(3): 8–13.

HALSTEAD B W. 1988. Poisonous and Venomous Marine Animals of the World (2nd revised edition). Princeton, NJ. Darwin Press. 1168 pp; plates 288 pp.

HALSTEAD B W, AUERBACH P S, CAMPBELL D. 1990. A Colour Atlas of Dangerous Marine Animals. London. Wolfe Medical Publications. 192 pp.

HALSTEAD B W, BUNKER N C. 1953. Stingray attacks and their treatment. Amer J Trop Med Hyg 2(1): 115–128.

HALSTEAD B W, DANIELSON D D, BALDWIN W J, ENGEN P C. 1972. Morphology of the venom apparatus of the leatherback fish *Scomberoides sanctipetri* (Cuvier). Toxicon 10: 249–258.

HALSTEAD B W, ENGEN P C, DANIELSON D D. 1971. Morphology of the venom organs of the rabbitfishes (family Teuthidae), p 121–157, 29 figs. In: (A de Vries and E Kochva eds) Toxins of Animal and Plant Origin, Vol 1. New York. Gordon & Breach. pp 121–157.

HALSTEAD B W, KUNINOBU L, HEBARD H. 1953. Catfish stings and the venom apparatus of the Mexican catfish *Galeichthys felis* Linnaeus. Trans Amer Microscop Soc 4: 297–314.

HALSTEAD B W, MITCHELL L R. 1963. A review of the venomous fishes of the Pacific area. In: (H L Keegan and W V MacFarlane eds) Venomous and Poisonous Animals and Noxious Plants of the Pacific Region. New York. Pergamon Press. pp 173–202.

HALSTEAD B W, MODGLIN R F. 1958. Weeverfish stings and the venom apparatus of weevers. Z. Tropenmed Parasitol. 9: 129–146.

HALSTEAD B W, VINCI J M. 1988. Biology of poisonous and venomous marine animals. In: (A T Tu, ed) Handbook of Natural Toxins, Vol 3: Marine Toxins and Venoms. New York. Marcel Dekker Inc. pp 1–30.

HAMNER W H, DOUBILET D. 1994. Australia's box jellyfish. National Geographic Magazine 186(2): 116–130.

HAMNER W M, HAURIE I R. 1981. Long-distance horizontal migrations of zooplankton (Scyphomedusae: *Mastigias* Limnol Oceanogr 31: 414–423)

HANSEN P A, HALSTEAD B W. 1971. The venomous sea anemone *Actinodendron plumosum* Haddon of South Vietnam. Micronesica 7: 65–70

HARRIS J. 1981. Tetrodotoxin. In: (J Pearn, ed) Animal Toxins and Man. Brisbane. Division of Health, Education and Information, Queensland. pp 24–28.

HARRIS J, PEARN J. 1987. Bullrout stings. In: (J Covacevich, P Davie, J Pearn eds) Toxic Plants and Animals: A Guide for Australia. Brisbane. Queensland Museum. pp 155–159.

HARRIS L F, STRIPLIN W H, BURNSIDE R C. 1991. Aquatic hazard *Mycobacterium marinum* infection. Ala Med 61: 8–10.

HARTMAN W D. 1967. Revision of *Neofibularia* (Porifera, Desmospongiae), a genus of toxic sponges from the West Indies and Australia. Peabody Museum of Natural History, Yale University. pp 1–41.

HARTWICK B. 1987. The box-jellyfish. In (J Covacevich, P Davie, J Pearn, eds) Toxic Plants and Animals: A Guide for Australia. Brisbane. Queensland Museum. pp 98–105.

HARTWICK R F. 1991a. Observations on the anatomy, behaviour, reproduction and life-cycle of the cubozoan *Carybdea sivickisi*. Hydrobiologia 216/217: 171–179.

HARTWICK R F. 1991b. Distributional ecology and behaviour of the early life stages of the box-jellyfish *Chironex fleckeri*. Hydrobiologia 216/217: 181–188.

HARTWICK R, CALLANAN V, WILLIAMSON J. 1980a. Disarming the box-jellyfish: nematocyst inhibition in *Chironex fleckeri*. Med J Aust 1: 15–20.

HARTWICK R F, CALLANAN V I, WILLIAMSON J A. 1980b. Disarming the box-jellyfish (letter in reply). Med J Aust 1: 336–337.

HASHIMOTO Y. 1979. Marine Toxins and their Bioactive Marine Metabolites. Tokyo. Japan Scientific Societies Press. 369 pp.

HASHIMOTO Y, OKAICHI T. 1960. Some chemical properties of nereistoxin. Ann New York Academy Sci 90: 667–673.

HAZEN T C, FLIERMANS C B, HIRSCH R P, ESCH G W. 1978. Prevalence and distribution of *Aeromonas hydrophila* in the United States. Appl Environ Microbiol 36: 731–738.

HEATWOLE H. 1975a. Voluntary submergence times of marine snakes. Marine Biol 32: 205–213.

HEATWOLE H. 1975b. Attacks by sea snakes on divers. In (W A Dunson, ed.) The Biology of Sea Snakes. Baltimore. University Park Press: 530pp, p32.

HEATWOLE H. 1978. Sea snake attack: myth or menace. Skindiving in Australia and New Zealand 8: 41–45.

HEATWOLE H, SEYMOUR R. 1975. Pulmonary and cutaneous oxygen uptake in sea snakes and a file snake. Comp Biochem Physiol 51: 399–405.

HEDGPETH J W. 1954. Scyphozoa. In: The Gulf of Mexico, its Origin, Waters, and Marine Life. Fishery Bull Fish Wildlife Service US 55(89): 277–278.

HEEGER T, MÖLLER H. 1987. Ultrastructural observations on prey capture and digestion in the scyphomedusa *Aurelia aurita*. Marine Biology 96: 391–400.

HEEGER T, MÖLLER H, MROWIETZ U. 1992. Protection of human skin against jellyfish (*Cyanea capillata*) stings. Marine Biol 113: 669–678.

HEISKANEN L P, JUREVICS H A, EVERITT B J. 1973. The inflammatory effects of the crude toxin of the crown-of-thorns starfish *Acanthaster planci*. Proc Aust Physiol Pharmacol Soc 4: 57.

HENDERSON D. 1980. Disarming the box-jellyfish (a letter with a reply). Med J Aust 1: 383–386.

HENDERSON D, EASTON R G. 1980. Stingose, a new and effective treatment for bites and stings. Med J Aust 2: 146–150.

HERCUS J P. 1944. An unusual eye condition (a letter). Med J Aust 1: 98–99.

HERMITTE L C D. 1946. Venomous marine molluscs of the genus Conus. Trans Royal Soc Trop Med Hyg 39: 485–512.

HERRE A W. 1952. A review of the scorpaenoid fishes of the Philippines and adjacent seas. Philippine J Sci 80(4): 381–482.

HESSINGER D A. 1988. Nematocyst venoms and toxins. In: (D A Hessinger, H M Lenhoff eds) The Biology of Nematocysts. New York. Academic Press. pp 333–369.

HICKSON S J. 1906. Coelenterata and ctenophora. In: The Cambridge Natural History, Vol 1. London. Macmillan. pp 245–424.

HIDAKA M, AFUSO K. 1994. Effects of calcium on the mechanical properties of the capsule wall of isolated nematocysts from Calliactis polypus. Comp Biochem Physiol 107A: 31–36.

HIGA H, UEZATO H, ARAKI Y. 1990. A case of sea snake bite in Okinawa, Japan. The Snake 22: 100–105.

HILLE B. 1975. The receptor for tetrodotoxin and saxitoxin: a structural hypothesis. Biophys J 15: 615–619.

HINDE B. 1885. In: J C Cox, 1885, Notes and exhibits (poisonous effects of the bite inflicted by *Conus geographicus* [sic] Linn.). Proc Linn Soc NSW 9: 944–946.

HIRASE S, et al (additional authors unknown). 1927. Illustrated Japanese Zoological Dictionary. Tokyo. Hokuryukwan & Co. 2235 pp.

HIRATA Y, UEMURA D, OHIZUMI Y. 1988. Chemistry and pharmacology of palytoxin. In: (A T Tu, ed) Handbook of Natural Toxins, Vol 3: Marine Toxins and Venoms. New York. Marcel Dekker. pp 241–258.

HOLMES M J, LEWIS R J. 1992. Multiple gambiertoxins (ciguatoxin precursors) from an Australian strain of *Gambierdiscus toxicus* in culture. In: (P Gopalakrishnakone, C K Tan, eds) Recent Advances in Toxinology Research. Singapore. Venom and Toxin Group. pp 520–529.

HOLMES M J, LEWIS R J, POLI M A, GILLESPIE N C. 1991. Strain dependent production of ciguatoxin precursors (gambiertoxins) by *Gambierdiscus toxicus* (Dinophyceae) in culture. Toxicon 29: 761–775.

HOLMES M J, LEWIS R J, SELLIN M, STREET R. 1994. The origin of ciguatera in Platypus Bay, Australia. Memoirs of the Queensland Museum 34(3): 505–512.

HOLSTEIN T, TARDENT P. 1984. An ultra high-speed analysis of exocytosis: nematocyst discharge. Science 223: 830–833

HOMMA M, OKONOGI T, MISHIMA S. 1964. Studies on sea snake venom, 1: Biological toxicities of venoms possessed by the three species of sea snakes captured in coastal waters of Amami Oshima. Gunma J Med Sci 13(4): 283–296.

HOOPER J N A, CAPON R J, HODDER R A. 1991. A new species of toxic marine sponge (Porifera: Desmospongiae: Poecilosclerida) from Northwest Australia. The Beagle. Records of the Northern Territory Museum of Arts and Sciences 8: 27–36.

HOOPER J N A, LÉVI C. 1993. Poecilosclerida (Porifera: Demospongia from the New Caledonia Lagoon. Invertebr Taxon 7: 1221–1302.

HOOPER J N A, WIEDENMAYER F. 1994. Porifera. In: (A. Wells, ed) Zoological Catalogue of Aust. Melbourne. CSIRO Australia. 12. 625 pp.

HOPKINS D G. 1964. Venomous effects and treatment of octopus bite. Med J Aust 1: 81–82.

HORNE T W. 1988. Box-jellyfish envenomation. Med J Aust 148: 540.

HORRIDGE G A. 1969. Statocysts of medusae and evolution of stereocilia. Tissue Cell 1: 341–353.

HORST R. 1907. On a new Cubomedusa from the Java Sea: *Chiropsalmus buitendijki*. Notes Leyden Museum 29(2): 101–106.

HOVERD W A. 1985. Occurrence of the order Cubomedusae (Cnidaria: Scyphozoa) in New Zealand: collection and laboratory observations of *Carybdea sivickisi*. New Zealand J Zool 12: 107–110.

HOWARTH D M, SOUTHEE A E, WHYTE I M. 1994. Lymphatic flow rates and first-aid in simulated peripheral snake or spider envenomation. Med J Aust 161: 695–700.

HOWDEN M E H, WILLIAMS P A. 1974. Occurrence of amines in the posterior salivary glands of the Octopus *Hapalochlaena maculosa* (Cephalopoda). Toxicon 12: 317–320.

HUANG YAO-WEN. 1998. Cannonball jellyfish (*Stomolophus meleagris*) as a food resource. J Food Science 53: 341–343.

HULET W H, BELLEME J L, MUSIL G, LANE C E. 1974. Ultrastructure of *Physalia* nematocysts. In: (H J Humm, C E Lane eds) Bioactive Compounds from the Sea. New York. Marcel Dekker. pp 99–114.

HUMPAGE A R, ROSITANO J, BAKER P D, NICHOLSON B C, STEFFENSEN D A. 1993. Paralytic shellfish poisons from freshwater blue-green algae. Med J Aust 159: 423.

HUTCHINGS P. 1989. Voyage to Elizabeth and Middleton reefs. Aust Nat Hist 22: 485–488.

HUXLEY T H 1859. The Oceanic Hydrozoa. London, Royal Society. Huxley said of Quatrefages work that though the single specimens observed by Quatrefages collapsed, 'One of my Physaliae, which had already served for many investigations, suddenly began to collapse...The bladder collapsed completely and floated on the surface of the vessel, without giving other sign of life . Later, this animal was distended anew, though smaller than before.' However, current workers (Mackie *et al* 1987) believe that only very small specimens are capable of diving.

HWANG D F, LIN L C, JENG S S. 1993. Paralytic toxins isolated from gastropod mollusk *Nassariidae*. In: (N H Tan, S L Oo, V Thambyrajah, N Azila, eds) Advances in Venom and Toxin Research. Proceedings of the Third Asia Pacific Congress on Animal, Plant and Microbial Toxins Malaysia, 27 June–1 July 1993. Kuala Lumpur. Malaysian Society on Toxinology. pp 216–219.

HWANG D F, NOGUCHI NAGASHIMA Y, LIAO I C, HASHIMOTO K. 1987 Occurrence of paralytic shellfish poison in the purple clam *Soletellina diphos* (bivalve). Nippon Suisan Gakkaishi 53: 623–626.

HYMAN L H. 1940. The Invertebrates: Protozoa through Ctenophora, Vol 1. New York. McGraw-Hill. pp 365–661.

ICHINOTSUBO D, ASAHINA A Y, TITUS E, CHUN S, HONG T L W P, SHIRAI J L, HOKAMA Y. 1994. Survey for ciguatera fish poisoning in west Hawaii. Memoirs of the Queensland Museum 34(3): 513–522.

INTERNATIONAL CONSORTIUM FOR JELLYFISH STINGS. 1993. Some Australian and international marine envenomation reports: progress summary to October 31 1993 (P Fenner, J Williamson, J Burnett eds). Adelaide. Hyperbaric Medicine Unit, Royal Adelaide Hospital. 25 pp.

ICJS. 1994. International Consortium for Jellyfish Stings, Jellyfish Newsletter No 10, January 1994. (J W Burnett, ed.). Bulletin on Marine Science, item 2: 3.

INGRAM D M, SHEINER H J, GINSBERG A M. 1992. Mondor's disease of the breast resulting from jellyfish sting. Med J Aust 157: 836–837.

INTERNATIONAL SYMPOSIUM ON DIVING IN HAZARDOUS WATER. 1991. Selected papers from the Symposium of December 1988, Bethesda, Maryland. Undersea Biomed Res 18(3): 175–266.

ITAKURA Y, KOMORI T. 1988. Biologically active glycosides from Asteroidea, X. Steroid oligoglycosides from the starfish *Acanthaster planci* L., 3. Structures of four new oligoglycoside sulfates. Liebigs Ann Chem 3: 499.

IVANOV A V, STRELKOV A A. 1949. Commercially Valuable Invertebrates of Far East Seas: A Description of the Structure and an Anatomical Atlas. Vladivostok. Primizdat 103 pp (in Russian).

IZUMI A K, MOORE R E. 1987. Seaweed (Lyngbya majuscula) dermatitis. In: Clinics in Dermatology. Philadelphia. J B Lippincott. Vol 5, No 3, pp 92–100.

JAKOWSKA S, NIGRELLI R F. 1960. Antimicrobial substances from sponges. Ann New York Acad Sci 90(3): 913–916.

JANDA J M, DUFFEY P S. 1988. Mesophilic aeromonads in human disease: current taxonomy, laboratory identification, and infectious disease spectrum. Rev Infectious Diseases 10(5): 980–997.

JARMS G. 1991. Taxonomic characters from the polyp tubes of coronate medusae (Scyphozoa, Coronatae). Hydrobiologia 216/217: 463–470. Or (R B Williams, P F S Cornelius, R G Hughes & E A Robson, eds) Coelenterate Biology: Recent Research on Cnidaria and Ctenophora. Dordrecht, Netherlands. Kluwer Academic Publishers. 742 pp.

JARVIS M W, CRONE H D, FREEMAN S E. TURNER R J. 1975. Chromatographic properties of maculotoxin, a toxin secreted by *Octopus* (*Hapalochlaena*) *maculosus*. Toxicon 13: 177–181.

JELINEK G A, HAMILTON T, HIRSCH R L. 1991. Admissions for suspected snake bite to the Perth adult teaching hospitals, 1979 to 1988. Med J Aust 155: 761–764.

JOHNSON C R, STABLUM W J. 1971. Observations on the feeding behavior of *Conus geographus* (Gasstropoda: Toxoglossa). Pacific Science 1: 109–111.

JOHNSON R A. 1992. Bacterial skin and soft tissue infections. Current Opinions in Infectious Diseases 5: 695–702.

JONAS-DAVIES J, LISTON L. 1985. The occurrence of PSP toxins in intertidal organisms. In: DM Anderson, AW White, DG Baden, eds) Elsevier, New York, pp 467–472.

JONES C, DERBYSHIRE K. 1987. Prawn trawl by catch. In: (J Covacevich, P. Davie, J. Pearn, eds.) Toxic Plants and Animals: A Guide for Australia. Brisbane. Queensland Museum. 137–145.

JONES M M. 1991. Toxic algae. In: Marine Organisms Transported in Ballast Water: A Review of the Australian Scientific Position. Bureau of Rural Resources, Bulletin No 11. Canberra. Australian Government Publishing Service. pp 25–31.

JONES C, DERBYSHIRE K. 1987. Prawn trawl bycatch. In: (J. Covacevcek, P. Davie, J. Pearn, eds) Toxic Plants and Animals: A Guide for Australia. Brisbane. Queensland Museum. 137–145.

KAKU N, MEIER J. 1995. Clinical toxicology of Fugu poisoning. In: (J Meier, J White, eds.) Clinical Toxicology of Animal Venoms and Poisons. Boca Raton, Florida. CRC Press. pp 75–83.

KALY U L, JONES G P. 1994. Test of the effect of disturbance on ciguatera in Tuvalu. Memoirs of the Queensland Museum 34(3): 523–532.

KAO C Y. 1966. Tetrodotoxin, saxitoxin and their significance in the study of the excitation phenomena. Pharmacol Review 18: 997–1049.

KAO C Y. 1983. New perspectives on the interactions of tetrodotoxin and saxitoxin with excitable membranes. Toxicon (Suppl.)3: 211–219.

KARASUDANI I, IGARI J, ANIYA Y. 1991. Smooth muscle contractile action of venom from the spines of the crown-of-thorns starfish, *Acanthaster planci*. Jap J Pharm 55(Suppl. 1): 322.

KAUFMAN B, WRIGHT D C, BALLOU W R, MONHEIT D. 1991. Protection against tetrodotoxin and saxitoxin intoxication by a cross-protective rabbit anti-tetrodotoxin antiserum. Toxicon 29: 581–587.

KAUL P N. 1981. Compounds from the sea with actions on the cardiovascular and central nervous systems. Fed Proc 40: 10–14.

KAWABATA T, HALSTEAD B W, JUDEFIND T F. 1957. A report of a series of recent outbreaks of unusual cephalopod and fish intoxications in Japan. Amer J Trop Med Hyg 6(5): 935–939.

KEEN T E B. 1970. Recent investigations on sea-wasp stingings in Australia. Med J Aust I: 266–270.

KEEN T E B. 1971. Comparison of tentacle extracts from *Chiropsalmus quadrigatus* and *Chironex fleckeri*. Toxicon 9: 249–254.

KEEN T E B, CRONE H D. 1969. Dermatonecrotic properties of extracts from the tentacles of the cnidarian *Chironex fleckeri*. Toxicon 7: 173–180.

KELLAWAY C H, Le MESSURIER D H. 1935. The venom of the platypus (*Ornithorhynchus anatinus*). Aust J Exp Biol Med Sci 13: 205–221.

KEM W R. 1971. A study of the occurrence of arabaseine in Paranemertes and other nemertenes. Toxicon 9: 23–32.

KEM W R. 1988. Worm toxins. In: (A T Tu, ed) Handbook of Natural Toxins; Vol 3: Marine Toxins and Venoms. New York. Marcel Dekker. pp 353–378.

KIMURA S, MIURA S, PARK Y-H. 1983. Collagen as the major edible component of jellyfish (*Stomolophus nomurai*). J Food Sci 48: 1758–1760.

KING PF. 1946. Erysipeloid: survey of 115 cases. Lancet ii: 196–197.

KING GK. 1991. Acute analgesia and cosmetic benefits of box-jellyfish antivenom. Med J Aust 154: 365–366.

KINGSTON C W, SOUTHCOTT R V. 1960. Skin histopathology in fatal jellyfish stinging. Trans Royal Soc Trop Med and Hyg. 54: 373–384.

KINJO N, NOGUCHI K, ANIYA Y, HIRAYAMA K, SAKANASHI M. 1991. Cardiovascular effects of crude venom of the crown-of-thorns starfish (*Acanthaster planci*) in dogs and rats. Jap J Pharm 55(Suppl. 1): 340.

KINSEY B. 1986. Barnes on Box Jellyfish. Townsville. James Cook University of North Queensland, Sir George Fisher Centre for Tropical Marine Studies. 90 pp of tape transcripts.

KINSEY B. 1988. More Barnes on Box Jellyfish. Townsville. James Cook University of North Queensland, Sir George Fisher Centre for Tropical Marine Studies. 115 pp.

KISHINOUYE 1899, in MAYER A G. 1910. Medusae of the World, Vol 3: The Scyphomedusae. Washington DC. Carnegie Institution. No 109. 3: 706. Also: Edible medusae. In: (J W Spengel, ed.) Zoologische Jahrbücher Jena 1988: 205–211.

KISHINOUYE K. 1910. Some medusae of Japanese waters. J Coll Sci Tokyo 27(9): 1–35.

KIZER K W. 1984a. Marine envenomations. J Toxicol Clin Toxicol 121: 527–555.

KIZER K W. 1984b. Case report: decompression sickness or Portuguese man-o'-war envenomation. Wilderness Med 1: 7–8.

KIZER K W. 1990. When a stingray strikes: treating common marine envenomations. Physician and Sports Med 18: 93–109.

KIZER K W, McKINNEY H E, AUERBACH P S. 1985. Scorpaenidae envenomation: a five year poison center experience. JAMA 253: 807–810.

KIZER K W, PIEL M. 1982. Arterial blood gas exchange with bluebottle envenomation. Hawaii Med J 41: 193–194.

KLONTZ K C, MULLEN R C, CORBYONS T M, BARNARD W P. 1993. *Vibrio* wound infections in humans following shark attack. J Wilderness Med 4: 68–72.

KLUG M, WEBER J, TARDENT P. 1988. Direct observation of haemolytic activity associated with single nematocysts. In: (D A Hessinger, H M Lenhoff eds) The Biology of Nematocysts. New York. Academic Press. pp 333–369.

KNIGHT J. 1987. Scromboid poisoning. SPUMS J 17: 116–118.

KNIGHT J. 1989. Obituary: Andonis Neofitou, better known as Anthony Newly. SPUMS J 19: 197–198.

KOBAYASHI J, NAKAMURA H, HIRATA Y, OHIZUMI Y. 1982. Isolation of Eburnetoxin, a vasoactive substance from the *Conus eburneus* venom. Life Sci 31: 1085–1091.

KOBAYASHI J, OHIZUMI Y, NAKAMURA H, HIRATA Y. 1981. Pharmacological study on the venom of the marine snail *Conus textile*. Toxicon 19: 757–762.

KOERNER MK. 1993. Guillain-Barré syndrome after sea urchin envenomation. J Wilderness Med 4: 463–464.

KOHN AJ. 1958. Cone shell stings: Recent cases of human injury due to venomous marine snails of the genus *Conus*. Hawaiian Med J 17(6): 528–532.

KOHN AJ. 1963. Venomous marine snails of the genus Conus. In: (H L Keegan, W B Macfarlane, Eds) Venomous and Poisonous Animals and Noxious Plants of the Pacific Region. Oxford. Pergamon Press. pp 83–.

KOHN A J, SAUNDERS P R, WIENER S. 1960. Preliminary studies on the venom of the marine snail *Conus*. Annals NY Acad Science 90: 706–725.

KOKELJ F. 1992. Contact dermatitis due to the Adriatic Sea jellyfish. Presented to 2nd Congress of the European Academy of Dermatology and Venereology, Athens, October 1991.

KOKELJ F, AVIAN M, SPANIER E, GALIL B. 1995a. Dermatotoxicity of 2 nematocyst preparations of the jellyfish *Rhopilema nomadica*. Contact Dermatitis 32: 244.

KOKELJ F, BURNETT J W. 1990. Treatment of a pigmented lesion induced by a *Pelagia noctiluca* sting. Cutis 46: 62–64.

KOKELJ F, DEL NEGRO P, MONTENERI G. 1992. Jellyfish dermatitis due to *Carybdea marsupialis*. Contact Dermatitis 27: 195.

KOKELJ F, MIANZAN H, AVIAN M, BURNETT J W. 1993. Dermatitis due to *Olindias sambaquiensis*: a case report. Cutis 51: 339–342.

KOKELJ F, STINCO G, AVIAN M. 1993. Dermatossicitá di una preparazione di nematocisti di *Carybdea marsupialis*. Ann Ital di Dermatol Clini e Sperimentale 47: 265–267.

KOKELJ F, STINCO G, AVIAN M, MIANZAN H, BURNETT J. 1995b. Cell-mediated sensitization to jellyfish antigens confirmed by positive patch test to *Olindias sambaquiensis* preparations. J Amacad Dermatol 33: 307–309 (in press) p35.

KOKELJ F, TUBARO A, DEL NEGRO P. 1989. Effetto dermatossico da *Chrisaora hysoscella*: presentazione di un caso. Giorn It Dermatol 124: 297–298.

KOMAI T. 1935. On *Stephanoscyphus* and *Nausithoë*. Mem Coll Sci Kyoto Imp Univ B 10(5): 289–339.

KRAA E, CAMPBELL B, LEVY M, STEWART G. 1994. Delayed diagnosis and reporting during an outbreak of ciguatera poisoning. Med J Aust 161: 510.

KRAMP P L. 1955. The Medusae of the tropical west coast of Africa. Atlantide-report 3. Copenhagen. University of Copenhagen and British Museum (Natural History). pp 288–292.

KRAMP P L. 1956. Medusae of the Iranian Gulf. Videnskabelige meddelelser fra Dansk Naturhistorisk Forening i København 118: 235–242.

KRAMP P L. 1959. The Hydromedusae of the Atlantic Ocean and adjacent waters. Copenhagen: Andr. Fred Host & Son. Dana Report No 46: 1–278.

KRAMP P L. 1961. Synopsis of the medusae of the world. J Mar Biol Assoc UK 40: 1–469.

KRAMP P L. 1968. The hydromedusae of the Pacific and Indian Oceans. Copenhagen. Andr. Fred Host & Son. Dana Report No 72: 1–200.

KRAMP P L. 1970. Marine biological investigation in the Bahamas: some Medusae from the Bahamas. Sarsia 44: 59–68.

KRASINSKA S. 1914. Beitrage zur Histologie der Medusen. Z Wiss Zool Liepzig 109: 256–348.

KREBS H C. 1986. Recent development in the field of marine natural products with emphasis on biologically active compounds. Prog Chem Organic Natural Prod 49: 151–363.

KREGER A S. 1991. Detection of a cytolytic toxin in the venom of the stonefish (*Synanceia trachynis*). Toxicon 29: 733–743.

KRISTENSON A. 1949. Some observations concerning the pathophysiological effects on the human skin caused by the stinging jellyfish (*Cyanea capillata*). Acta Physiol Scand 18(2–3): 151–156.

KRUMBACH T. 1925. 'Scyphozoa' In: (W Kukenthal, T Krumbach, eds) Handbuch der Zoologie: Eine Naturgeschichte der Stämme des Tierreiches 1: 522–686.

KUDOH Y. 1988. Vibriosis other than cholera. In: (A Balows, W J Hausler, M Ohashi, A Turano eds) Laboratory Diagnosis of Infectious Diseases, Vol 1. New York. Springer-Verlag. pp 533–539.

KUMAGAI M, YANAGI T, MURATA M et al. 1986. Okadaic acid as the causative toxin of diarrhetic shellfish poisoning in Europe. Agric Biol Chem 50: 2853–2857.

KURIGAN J. 1995. Snail toxins yield fast-paced advances in drug research. The journal of NIH research 7:30–32.

KUSUNOKI SHIN. 1967 (approx.). The clinical study of poisonous jellyfish stingings in the Hokkaido (Lake of Saroma). Publication of Engaru Welfare General Hospital. (Summary in English.) p 29

LAKE J S. 1971. Freshwater Fishes and Rivers of Australia. Melbourne. Nelson. 61 pp.

LAMBERT H P. 1973. Paralytic shellfish poisoning. In: (P B Beeson, W McDermott, J B Wyngaarden, eds) Cecil Textbook of Medicine, 15th edition. Philadelphia. W B Saunders & Co. pp 68–71.

LANE C E. 1960a. The Portuguese man-o'-war. Scientific American 202: 158–168.

LANE C E. 1960b. The toxin of *Physalia* nematocysts. Ann New York Acad Science 90(3): 743–750.

LANE C E, DODGE E. 1958. The toxicity of *Physalia* nematocysts. Biol Bull 115: 219–226.

LANE F W. 1960. Kingdom of the Octopus: The Life History of the Cephalopods. New York. Sheridan House. 300 pp and 46 plates.

LANGE W R. 1993. Severity rating scales for ciguatera fish poisoning. Toxicon 31: 777–781.

LARSEN J B, BILLINGS J D. 1986. Cardiac effects of *Stomolophus meleagris* (cabbage head jellyfish) toxin. 2nd American Symposium on Animal, Plant and Microbial Toxins, Tempe, Arizona, May 1986.

LARSEN N P. 1925. Fish poisoning. Queen's Hospital Bull 2: 1–3.

LARSON R J. 1976a. Cubomedusae: feeding-functional morphology, behavior and phylogenetic position. In: (G O Mackie, ed) Coelenterate Ecology and Behaviour. New York. Plenum. pp 237–245.

LARSON R J. 1976b. Marine flora and fauna of the northeastern United States. Cnidaria: Scyphozoa. NOAA Tech. Rpt NMFS Circ. No 397.

LASSIG B R, KELLEHER G. 1991. Crown-of-thorns starfish on the Great Barrier Reef. In: Environmental Research in Australia, III: Case Studies. Canberra. Australian Science and Technology Council. pp 39–62.

LAST P R, STEVENS J D. 1994. Sharks and Rays of Australia. Melbourne. Commonwealth Scientific and Industrial Research Organisation (CSIRO). 513 pp.

LEHMANN D F, HARDY J C. 1993. Stonefish envenomation. New Eng J Med 329: 510–511.

LENDENFELD R von. 1882. Über Coelenteraten der Südsee. Z Wiss Zool 37(4): 465–552.

LENDENFELD R von. 1887. Descriptive Catalogue of the Medusae of the Australian Seas. Sydney. Australian Museum. pp 1–49.

LENNETTE E H, BALOWS A, HAUSLER, W J Jr, SHADOMY H J. 1985. Manual of Clinical Microbiology, 4th edition. Washington, DC. American Society for Microbiology.

LESH-LAURIE G E, SUCHY P E. 1991. Cnidaria: Scyphozoa and Cubozoa. In: (F S Harrison, J A Westfall, eds) Microscopic Anatomy of Invertebrates, Vol 2: Placozoa, Porifera, Cnidaria and Ctenophora. New York. Wiley-Liss pp 185–266.

LESSIOS H A. 1983. Sea urchins of Panama, the lesions they cause and their treatment. Rev Med Panama 8: 56–71.

LESSON R P. 1830. 'Voyage Autour du Monde Exécuté par Ordre du Roi, sur la Corvette de Sa Majesté, *La Coquille*, pendant des Années 1822, 1823, 1824 et 1825, Zoologie.' (Pt 2, Div 2, Zoophytes). 135 pp. Also a companion Atlas (1826). Zoophytes: 16 plates.

LEVINE W C, GRIFFIN P M and The Gulf Coast Working Group. 1993. Vibrio infections on the Gulf Coast: results of first year of regional surveillance. J Infectious Diseases 167: 479–483.

LEVY S, MASRY D, HALSTEAD B W. 1970. Report of stingings by the sea anemone *Triactis producta* Klunzinger from Red Sea. Clin Toxicol 3(4): 637–643.

LEWIS R J. 1987. Ciguatera in south-eastern Queensland. In: (J Covacevich, P Davey, J Pearn eds) Toxic Plants and Animals in Australia. Brisbane. Queensland Museum. pp 180–187.

LEWIS R J. 1992a. Socioeconomic impacts and management of ciguatera in the Pacific. Bulletin de la Société de Pathologie Exotique 85: 427–434.

LEWIS R J. 1992b. Ciguatoxins are potent ichthyotoxins. Toxicon 30: 207–211.

LEWIS R J. 1994a. Immunological, biochemical and chemical features of ciguatoxins: implications for the detection of ciguateric fish. Memoirs of the Queensland Museum 34(3): 541–548.

LEWIS R J. 1994b. Impact of a validated, cost effective screen for ciguateric fish. Memoirs of the Queensland Museum 34(3): 549–553.

LEWIS R J, BRERETON I M. 1994. Inverse-detected NMR of ciguatoxin: quaternary carbon locations confirmed in CTX-1. Memoirs of the Queensland Museum 34(3): 555–559.

LEWIS R J, CHALOUPKA M Y, GILLESPIE N C, HOLMES M J. 1988. An analysis of the human response to ciguatera in Australia. In: (J H Choat et al Eds) Proc 6th Int Coral Reef Symp, Townsville; Vol 3. 6th International Coral Reef Symposium Executive Committee. pp 67–72.

LEWIS R J, HOLMES M J. 1993. Origin and transfer of toxins involved in ciguatera. Comp Biochem Physiol 106C(3): 615–628.

LEWIS R J, SELLIN M. 1992. Multiple ciguatoxins in the flesh of fishes. Toxicon 30: 915–919.

LEWIS R J, SELLIN M. 1993. Recovery of ciguatoxin from fish flesh. Toxicon 31: 1333–1336.

LEWIS R J, SELLIN M, POLI M A, NORTON R S, MacLEOD J K, SHEIL M M. 1991. Purification and characterization of ciguatoxins from moray eel (*Lycodontis javanicus*, Muraenidae). Toxicon 29: 1115–1127.

LEWIS R J, WONG HOY A W, McGIFFIN D C. 1992. Action of ciguatoxin on human atrial trabeculae. Toxicon 30: 907–914.

LIGGINS J B. 1939. An unusual bathing fatality. New Zealand Med J 203: 27–29.

LIGHT S F. 1914a. Another dangerous jellyfish in Philippine waters. Philippine J Sci B9(3): 291–295.

LIGHT S F. 1914b. Some Philippine Scyphomedusae, including two new genera, five new species, and one new variety. Philippine J Sci D9(3): 195–231.

LIGHT S F. 1921. Further notes on Philippine scyphomedusae jellyfishes. Philippine J Sci 18(1): 25–45.

LIMPUS C J. 1978. Toxicology of the venom of sub-tropical Queensland Hydrophiidae. Toxicon (Suppl 1): 39–70.

LIMPUS C. 1987. Sea snakes. In: (J Covacevich, P Davie, J Pearn, eds) Toxic Plants & Animals: A Guide for Australia. Brisbane. Queensland Museum. pp 195–203.

LINARES DEL RIO F, MONICHE GARCIA PUMARINO M, HERRUEZO PEREZ A. 1989. Therapeutic application of anaesthetic blocks in weever-fish stings. Rev Esp Anestesiol Reanim 36: 57–59.

LINAWEAVER P G. 1967. Toxic marine life. Military Med 131: 437–442.

LINAWEAVER P G, BOVE A A. 1990. Physical examination of divers. In: (A A Bove, J C Davis, eds) Diving Medicine, second edition. Philadelphia. W B Saunders Co. pp 302–309.

LINNÉ C von. 1758. Systema Naturae, edition 10. reformata (2 vols). Holmiae 1758–1759.

LINSTOW O V. 1894. Mollusca. In: Die Gifttiere und Ihre Wirking auf den Menschen. Berlin. August Hirschwald. pp 130–145.

LLEWELLYN L E, ENDEAN R. 1988. Toxic coral reef crabs from Australian waters. Toxicon 26: 1085–1088.

LLEWELLYN L E, ENDEAN R. 1991. Paralytic shellfish toxins in the xanthid crab *Atergatis floridus* collected from Australian coral reefs. J Wilderness Med 2: 118–126.

LONG K O, BURNETT J W. 1989. Isolation, characterization, and comparison of hemolytic peptides in nematocyst venoms of two species of jellyfish (*Chrysaora quinquecirrha* and *Cyanea capillata*). Comparative Biochem Physiol 94B: 641–646.

LONG-ROWE K O, BURNETT J W. 1994a. Characteristics of hyaluronidase and hemolytic activity in fishing tentacle nematocyst venom of *Chrysaora quinquecirrha*. Toxicon 32: 165–174.

LONG-ROWE KO, BURNETT J W. 1994b. Sea nettle (*Chrysaora quinquecirrha*) lethal factor: purification by recycling on m-aminophenyl boronic acid acrylic beads. Toxicon 32: 467–478.

LOTAN A, FISHMAN L, LOYA Y, ZLOTKIN E. 1995. Delivery of a nematocyst toxin. Nature. 375:456

LUCAS J S, HART R J, HOWDEN M E, SALATHE R. 1979. Saponins in eggs and larvae of *Acanthaster planci* (L.) (Asteroidea) as chemical defenses against planktivorous fish. J Exp Mar Biol Ecol 40: 155–165.

LUMLEY J, WILLIAMSON J A, FENNER P J, BURNETT J W, COLQUHOUN D M. 1988. Fatal envenomation by *Chironex fleckeri*, the north Australian box jellyfish: the continuing search for lethal mechanisms. Med J Aust 148: 527–534.

LYN P C. 1985. Puffer fish poisoning: four case reports. Med J Malaysia 40: 31–34.

McCAFFREY E J. 1983. Biologically active material from a marine sponge. Toxicon (Suppl.3): 277–279.

McCLESKEY E W, FOX A P, FELDMAN D H, CRUZ L J, OLIVERA B M, TSIEN R W. 1987. ω-Conotoxin: direct and persistent blockade of specific types of calcium channels in neurons but not muscle. Proc Nat Acad Sci USA 84: 4327–4331.

McFARLANE A C. 1989. The prevention and management of the psychiatric morbidity of natural disasters: an Australian experience. Stress Med 5: 29–36.

McFARLANE A C. 1991. Post-traumatic stress disorder. Internat Rev Psych 3: 203–213.

McFARREN E F, SCHAEFER M L, CAMPBELL J E, LEWIS K H, JENSEN E T, SCHANTZ E J. 1957. (Title unknown). In: Proc Nat Shellfisheries Assoc 47: 114, as reported in MURTHA E F. 1960. Pharmacology study of poisons from shellfish and pufferfish. Annals New York Acad Sci 90(3): 820–836.

MacGOWAN A J. 1887. Poisonous fish in China. Bull US Fish Comm 6: 130–131.

MACHADO O. 1943. Catálogo sistemático dos animals urticantes e peconhentos do Brasil. Bol Inst Vital Brazil 25: 41–65.

McKAY R J. 1987. A guide to common venomous fishes. In: (J Covacevich, P Davie, J Pearn eds) Toxic Plants and Animals: A Guide for Australia. Brisbane. Queensland Museum. pp 147–153.

MACKERRAS I M. 1958. Identical skin eruption in five men after massive exposure to jellyfish (*Aurelia*) (letter). Med J Aust 1: 56.

MACKIE G O, PUGH G O, PURCELL J E. 1987. Siphonophore biology. Advances in Marine Biology 24: 97–262

McKINSTRY D M. 1993. Catfish stings in the United States: case report and review. J Wilderness Med 4: 293–303.

McNEILL F A. 1948. Deserters from the sea. Aust Mus Magazine 9(8): 258–262.

McNEILL F A, POPE E C. 1943. A venomous medusa from Australian waters. Aust J Science 5(6): 188–191.

McNEIL I Y. 1985. Hypersensitivity reaction to mannitol. Drug Intell Clin Pharm 19: 552–553.

MAGEROY N, BAERHEIM A. 1991. Ratfish (*Chimera monstrosa*) sting. Tidsskr Nor Laegeforen 111: 2102–2103.

MAGUIRE E J. 1968. *Chironex fleckeri* ('sea wasp') sting. Med J Aust 2: 1137–1138.

MAHNIR V M, SHUMLOV U N, KOVALEVSKAYA A M, ROMANENKO L A, GREBELNY S D. 1993. Evidence for several types of biologically active substances in the north Pacific sea anemones. Comp Biochem Physiol 106C: 661–665.

MAIER L, RATHMAYER W. 1982. Lokalisierung von anemonentoxin in den tentakeln der wachsrose *Anemonia sulcata* (Coelenterata) mit hilfe spezifischer antikörper. Verh Dtsch Zool Ges 1982: 281.

MALASIT P, WARRELL D A, CHANTHAVANICAH P, et al. 1986. Prediction, prevention, and mechanism of early (anaphylactic) antivenom reactions in victims of snakebite. Brit Med J 292: 17–20.

MANDOJANA R M. 1987. Aquatic Dermatology. In: Clinics in Dermatology. Philadelphia. J B Lippincott. Vol 5, No 3, pp i–xvi and 1–173.

MANDOJANA R M. 1990. Granuloma annulare following a blue bottle jellyfish (*Physalia utriculus*) sting. J Wilderness Med 1: 220–224.

MANDOJANA R M, SIMS J K. 1987. Miscellaneous dermatoses associated with the aquatic environment. In: Clinics in Dermatology. Philadelphia. J B Lippincott. Vol 5, No 3, pp 135–145.

MANN J W, WERNTZ J R. 1991. Catfish stings to the hand. J Hand Surg Am 16: 318–321.

MANOWITZ N R, ROSENTHAL R R. (1979). Cutaneous-systemic reactions to toxins and venoms of common marine organisms. Cutis 23: 450–454.

MÅNSSON T, RANDLE H W, MANDOJANA R M, CALTON G J, BURNETT J W. 1985. Recurrent cutaneous jellyfish eruptions without envenomation. Acta Derm Venerol 65: 72–75.

MANSUETTI R. 1963. Symbiotic behaviour between small fishes and jellyfishes, with new data on that between the stromateid, *Peprilus alepidotus*, and the scyphomedusa, *Chrysaora quinquecirrha*. Copeia: pp 38, 40–80.

MARETIĆ Z. 1975. Zivitinje otrovnice i otrovne Zivitinje Jadranskog mora. Jugosl Akadem Zagreb Univ. 11–22.

MARETIĆ Z. 1984. The bloom of jellyfish *Pelagia noctiluca* along the coast of Pula and Istria in 1977–1983 with special reference to epidemiology, clinics and treatment. Workshop on jellyfish blooms in the Mediterranean, Athens, 31 October—4 November 1983. United Nations Environment Programme. pp 83–87.

MARETIĆ Z. 1988. Fish venoms. In: (A T Tu, ed) Handbook of Natural Toxins, Vol 3, Marine Toxins and Venoms. New York. Marcel Dekker. pp 445–478.

MARETIĆ Z., MATIC-PLANTANIDA D, LADAVAC J. 1987. The bloom of the jellyfish *Pelagia noctiluca* in the Mediterranean and Adriatic and its impact on human health. Proceedings of the 2nd Workshop on Jellyfish in the Mediterranean Sea, Trieste, 2–5 September 1987. Mediterranean Action Plan of the United Nations Environmental Programme, Athens, 1991. pp 260–267.

MARETIĆ Z., RUSSELL F E. 1983. Stings by the sea anemone *Anemonia sulcata* in the Adriatic Sea. J Trop Med Hygiene 32: 891–896.

MARETIĆ Z., RUSSELL F E, LADAVAC J. 1980. Epidemic of stings by the jellyfish *Pelagia noctiluca* in the Adriatic. In: (D Eaker, T Wadstrum, eds) Natural Toxins. Oxford. Pergamon Press. pp 77–82.

MARINKELLE C J. 1966. Accidents by venomous animals in Colombia. Ind Med Surg 35: 988–994.

MARMAR C R, FOY D, KAGAN B, PYNOOS R S. 1993. An integrated approach for treating post-traumatic stress. In: (J M Oldham, M B Riba, A Tasman, eds) Review of Psychiatry. Washington DC. American Psychiatric Press. pp 239–272.

MARR J J. 1967. Portuguese man-o'-war envenomization: a personal experience. JAMA 199: 115–116.

MARSDEN A T H, REID H A. 1961. Pathology of sea snake poisoning, Brit Med J 1: 1290–1293.

MARSH L, SLACK-SMITH S, GURRY D. 1986. Sea Stingers. Perth. Western Australian Museum. 133 pp.

MARTIN C, AUDLEY I. 1990. Cardiac failure following Irukandji envenomation. Med J Aust 153: 164–166.

MARTIN C J, TIDSWELL F. 1894. Observations on the femoral gland of *Ornithorhynchus* and its secretions; together with an experimental enquiry concerning its supposed toxic action. Proc Linn Soc NSW 9(2): 471–500.

MARTIN E J. 1960. Observations on the toxic sea anemone, *Rhodactis howesii* (Coelenterata). Pacific Sci 14(4): 403–407.

MARTIN EJ. 1963. Toxicity of dialyzed extracts of some California anemones (Coelenterata). Pacific Sci 17: 302–304.

MATSUOKA N, ASANO H, HAGEWARA Y. 1992. A preliminary study of the biochemical systematics of Hydromedusae: electrophoretically detectable enzymes and their isozyme patterns in *Gonionemus oshoro*. Sci Reports Hirosaki Univ 39(2): 103–110 (in Japanese).

MATUSOW R J. 1980. Oral inflammatory response to a sting from a Portuguese man-o'-war. J Am Dent Assoc 100: 73–75.

MAYER A G. 1910. Medusae of the World, Vol 3. The Scyphomedusae. Washington, DC. Carnegie Institution. No 109. i–iv: 499–735.

MEBANE G Y. 1994. Debunking the sea lice myth: the latest information on sea bather's eruption. 'Alert Diver' January/February 1994: 34.

MEBS D. 1991. A myotoxic phospholipase A_2 from the crown of thorns starfish *Acanthaster planci*. Toxicon 29: 289.

MEBS D. 1994. Clinical toxicology of sea urchin and starfish injuries. In: (J Meier, J White, eds) Clinical Toxicology of Animal Venoms and Poisons. Boca Raton, Florida. CRC Press. pp 129–133.

MEBS D, GEBAUER E. 1980. Isolation of proteinase inhibitory, toxic and hemolytic polypeptide from a sea anemone, *Stoichactis* sp. Toxicon 18: 97–106.

MEIER J. 1995. Commercially available antivenoms and antivenom therapy ('hyperimmune sera', 'antivenins' and 'antisera'). In (J Meier, J White, eds) Clinical Toxicology of Animal Venoms and Poisons. Boca Raton, Florida. CRC Press pp 689–721.

MELTON R J, RANDALL J E, FUSETANI N, et al. 1984. Fatal sardine poisoning. Hawaii Med J 43: 114, 116, 118–120, 124.

MEMOIRS of the QUEENSLAND MUSEUM, PART 3. 1994. Proceedings of the International Workshop on Ciguatera Management, Bribie Island, 12–16 April 1993 (R J Lewis Chairman of Workshop). Brisbane. Queensland Museum. 34: pp 639.

MENON M G K. 1930. The scyphomedusae of Madras and the neighbouring coast. Bull Madras Govt Mus 3: 1–3.

MENON M G K. 1936. Scyphomedusae of Krusadai Island. Bull Madras Govt Mus, NS Nat Hist Sect (I): 2;1–9.

MERCER H P, McGILL J J, IBRAHAM R A. 1981. Envenomation by sea snakes in Queensland. Med J Aust 1: 130–132.

MEYER K F. 1953. Medical progress: food poisoning. New Eng J Med 249: 765–773, 804–812, 843–852.

MEYER K F, SOMMER H, SCHOENHOLZ P. 1928. Mussel poisoning. J Preventive Med 2: 365–394.

MEYER PK. 1993. Seastroke: a new entity. Southern Med J 86: 777–779.

MIANZAN H. 1989. Distribución de *Olindias sambaquiensis* Müller 1861 (Hydrozoa, Limnomedusae) en el Atlántico Sudoccidental. Iheringia ser Zool 69: 155–157.

MIANZAN H, OLAGÜE G, MONTERO R. 1988. Scyphomedusae de las aguas Uruguayas. Spheniscus 6: 1–9.

MIANZAN H, RAMIREZ F. 1996. Ecological studies on the stinging hydromedusae *Olindias sambaquiensis* Müller 1861 (Limnomedusae, Olindiidae) at Blanca Bay, Argentina. Distribution, abundance, gonadal cycle, population structure Mar del Plata. INDEP, Instituto Nacional de Investigación y Desarrollo, Pesquero (in press).

MIANZAN H, ZAMPONI M. 1988. Estudio bioecológico de *Olindias sambaquiensis* Müller 1861 (Limnomedusae, Olindiidae) en el área de Monto Hermoso. II: factores meteorológicos que influyen en su aparición. Iheringia ser Misc 02: 63–68.

MICHEL H B, BEHBEHANI M, HERRING D, ARAR M, SHOUSHANI M, BRAKONIECKI T. 1986. Zooplankton diversity, distribution and abundance in Kuwait waters. Kuwait Bull Marine Sci 8: 37–105.

MIGAS E A, POPOVA G A. 1988. Instructions on rendering medical assistance to persons stung by a venomousous krestovik-medusa from the Sea of Japan. Vladivostok. pp 1–10 (in Russian).

MIKULITCH L V, NAUMOV D V. 1974. Venomous medusa 'krestovik', its taxonomic position, morphology, life history, biology and distribution. In: Studies on a Venomous Medusa-Krestovik. Vladivostok. pp 27–33 (in Russian).

MILLINGTON J T, WILHELM P. 1993. Marine microbiology of Rocas Alijos. J Wilderness Med 4: 384–390.

MILLS C E. 1983. Vertical migration and diel activity patterns of hydromedusae: studies in a large tank. J Plankton Res 5: 619–635.

MILLS C E. 1984. Diversity of swimming behaviors in hydromedusae as related to feeding and utilization of space. Marine Biology 64: 185–189.

MINALE L, CIMINO G, De STAFANO S, SODANO G. 1976. Natural products from Porifera. Fortschritte der Chem Org Naturst 33: 1–72.

MINE-GOLOMB D, EIMERL S, SCHRAMM M. 1989. Cysteine sulfinic acid-induced release of D-[^{3}H] aspartate and [^{14}C] GABA in hippocampus slices: the role of sodium channels and cAMP. Brain Res 490: 205–211.

MING C L, GOPALAKRISHNAKONE P, TAN L, MAN Y C, HOW J, SIANG N C. 1990. Venomous fishes. In: (P Gopalakrishnakone, ed) A Colour Guide to Dangerous Animals. Singapore University Press. pp 53–66.

MINGLIANG ZHANG. 1993. A study on nematocyst dermatitis caused by jellyfish *Stomolophus nomurai* stings in North China Sea. Acta Acad Med Qingdao 29(4) 253–268.

MINGLIANG ZHANG. 1988. Study on jellyfish *Stomolophus nomurai* stings on Beidehe. Nat Med J China 9: 499.

MINGLIANG ZHANG. 1992. The Marine Stinger. Qingdao. Oceanology University Press. p50 68pp.

MINGLIANG ZHANG, QIN SHEDE. 1990. Nematocyst dermatitis in Huanghai and Bohai (North China Sea). Proc 2nd China–Japan Meeting of Dermatology. Shanghai. 50 pp.

MINGLIANG ZHANG, QIN SHEDE. 1991. The marine stinger in South China Sea. Chinese J Marine Drugs 4: 36.

MINGLIANG ZHANG, QIN SHEDE. 1992. Nematocyst dermatitis in China Sea. Institute of Dermatol, Qingdao Medical College. 51 pp.

MINTON S A. 1975. Venomous marine invertebrates. In: (W T Hubbert, W F McCulloch, P R Schurrenberger, eds) Diseases Transmitted from Animals to Man, 6th edition. Springfield, Illinois. C C Thomas. p1031.

MITCHELL J H. 1962. Eye injuries due to jellyfish (*Cyanea annaskala*). Med J Aust 2: 303–305.

MIURA S, BURNETT J W, AURELIAN L. 1993. Immunity to jellyfish venoms: suppression of venom-induced immune responses in ultraviolet B-irradiated mice. Toxicon 31: 1415–1422.

MIURA S, BURNETT J W, AURELIAN L. 1996. Ultraviolet light and immunity to coelenterate venom. Cutis. 57: 201–204.

MIYAZAWA K, JEON J K, MARUYAMA J, NOGUCHI T, ITO K, HASHIMOTO K. 1986. Occurrence of tetrodotoxin in the flatworm *Planocera multitentaculata*. Toxicon 24: 645–650.

MOATS W E. 1992. Case report: fire coral envenomation. J Wilderness Med 3: 284–287.

MOORE C T. 1992. Reflections on Rio: the Earth Summit and the Wilderness Medical Society. J Wilderness Med 3: 331–333.

MOORE H F. 1908. The commercial sponges and the sponge fisheries. US Fish Wildlife Service, Fish Bull 28(1): 403–511.

MOORE R E, SCHEUER P J. 1971. Palytoxin: a new marine toxin from a coelenterate. Science 172: 495–498.

MOORE S J. 1988. A new species of cubomedusan (cubozoa: cnidaria) from northern Australia. The Beagle, Records of the Northern Territory Museum of Arts and Sciences. 5: 1–4.

MORAN P J. 1986. The *Acanthaster* phenomenon. Oceanogr Mar Biol Ann Rev 24: 379–480.

MORAN P J. 1988. Crown-of-thorns starfish questions and answers. Australian Institute of Marine Science, Townsville. 35pp.

MORAN P J, BRADBURY R H, REICHELT R E. 1988. Distribution of recent outbreaks of the crown-of-thorns starfish (*Acanthaster planci*) along the Great Barrier Reef: 1985–1986. Coral Reefs 7: 125–137.

MORAN P J, DAVIES S, BAKER V J. 1991. Acanthaster planci: an annotated bibliography. The Crown-of-thorns Study. Australian Institute of Marine Science. 95 pp.

MORAN P J, WILLIAMSON J. 1986. Toxic reactions to injuries caused by the spines of the crown of thorns starfish (*Acanthaster planci*). SPUMS J 16: 91–95.

MORDECAI E R. 1860. Wound inflicted by a stingray. New Orleans Med News Hospital Gaz 7: 679–680.

MORTENSEN T. 1935–1943. A Monograph of the Echinoidea, Vols II (647 pp), III (446 pp). Carlsberg-Fund. Copenhagen. C A Reitzel.

MOSCHELLA S L. 1951. Further clinical observations on seabather's eruption. Arch Dermatol 64: 55–56.

MOUNTFORD C P. 1956. Records of the American-Australian Scientific Expedition to Arnhem Land, 1: Art, myth and symbolism. Melbourne. Melbourne University Press. 252.

MUHVICH K H, SENGOTTUVELU S, MANSON P N, MYERS R A M, BURNETT J W. 1991. Pathophysiology of sea nettle (*Chrysaora quinquecirrha*) envenomation in a rat model and the effects of hyperbaric oxygen and verapamil treatment. Toxicon 29: 857–866.

MÜLLER F. 1859. Zwei neue quallen von Santa Catharina (Brasilien). Abhandl Naturf Ges Halle 5: 1–12.

MÜLLER F. 1861. Polypen und Quallen von Santa Catharina. *Olindias sambaquiensis* n. sp. Diesem Archiv BD. IS. 314: 6p.

MULLINS J F, WILSON C J, BEST W C. 1957. Cryotherapy in the treatment of stingray wounds. Southern Med J 50: 533–535.

MURATA M, LEGRAND A M, ISHIBASHI Y, FUKUI M, YASUMOTO T. 1990. Structures and configurations of ciguatoxin from moray eel *Gymnothorax javanicus* and its likely precursor from the dinoflagellate *Gambierdiscus toxicus*. J Am Chem Soc 112: 4380–4386.

MURPHEY D K, SEPTIMUS E J, WAAGNER D C. 1992. Catfish-related injury and infection: report of two cases and review of the literature. Clin Infect Dis 14: 689–693.

MUSTAFA M R, WHITE E, HONGO K, OTHMAN I, ORCHARD C H. 1995. The mechanism underlying the cardiotoxic effect of the toxin from the jellyfish, *Chironex fleckeri*. Toxicology Applied Pharmacol 133: 196–206.

NAGAO Z. 1973. The life history of Eperetmus typus Bigelow and the systematics of the family Olindiadidae (Limnomedusae). Publ Seto Marine Biol Lab 20: 89–102.

NAGUIB A M F, BANSAL J, CALTON G J, BURNETT J W. 1988. Purification of *Chironex fleckeri* venom components using Chironex immunoaffinity chromatography. Toxicon 26: 387–394.

NAIR K K. 1951. Medusae of the Trivandrum coast. I Systematics. Bull Research Inst Univ Travencore. Sect C, Nat Sci 2(1): 71.

NAKAMURA H, SATO S, KOBAYASHI J, OHIZUMI Y, HIRATA Y. 1983. Structures and actions of the peptide toxins, Geographutoxin I and II isolated from the cone shell *Conus geograpichus*. In: (E Munekata, ed) Peptide Chemistry. Osaka. Research Foundation. pp 197–202.

NARAHASHI T, MOORE J W, POSTON R N. 1967. Tetrodotoxin derivatives: chemical structure and blockage of nerve membrane conductance. Science 156: 976–979.

NAUMOV D V. 1955. The systematic position of a poisonous krestovik-medusa and a revision of the genus *Gonionemus* A. Agassiz. Transactions Zoolog Inst USSR Acad Sci 21: 102–109.

NAUMOV D V. 1960. Sea, brackish and freshwater hydroids and hydromedusae of the USSR. Guide to the fauna of the USSR. Moscow and Leningrad. Nauka Press. 70: 1–585 (in Russian).

NAVY MEDICINE. submitted for publication 1 Nov. 1965. Poisonous Snakes of the World. A manual for use by US amphibious forces, Navmed P-5099. United States Government Printing Office.

NELSON J S. 1984. Fishes of the world. New York. John Wiley & Son. 523 pp.

NICHOLLS D. 1992. Sea anemone sting while SCUBA diving. New Zealand Med J 105: 245–246.

NITTA S. 1934. In Yakugaku-Zasshi, quoted by Hashimoto and Okaichi. J Pharmacol Soc Japan 54: 648.

NOVAK V, SKET D, CANKAR G, LEBEZ D. 1973. Partial purification of a toxin from tentacles of the sea *Anemone sulcata*. Toxicon 11: 411–417.

ODOM C B, FISCHELMANN E A. 1972. Crown-of-thorns starfish wounds: some observations on injury sites. Hawaii Med J 31: 99–100.

OH T E. 1990. Sedation in intensive care. In: (T E Oh, ed) Intensive Care Manual, 3rd edition. Sydney. Butterworths. pp 313–317.

OHTAKI N, OKA K, SUGIMOTO A, AKIZAWA T, YASUHARA T, AZUMA H. 1990. Cutaneous reactions caused by experimental exposure to jellyfish, *Carybdea rastoni*. J Dermatol 17: 108–114.

OHTAKI N, SATOH A, NAKAZIMA T. 1986. Delayed flare-up reactions caused by jellyfish. Dermatologica 172: 98–103.

OLAGÜE G, FAILCA G, MIANZAN H, BALDAÑA J. 1990. Relación entre la distribución de vientos y la aparición de Scyphomedusan en el Puerto de Punta del Este (R O Uruguay). Geofisica 33: 69–73.

OLD H H. 1908. A report of several cases with unusual symptoms caused by contact with some unknown variety of jellyfish (Scyphozoa). Philippine J Sci B3(4): 329–333.

OLIVERA B M, GRAY W R, CRUZ L J. 1988. Marine snail venoms, in (A T Tu, ed) Handbook of Natural Toxins: Vol 3: Marine Toxins and Venoms. New York. Marcel Dekker. pp 327–349.

OLSEN C E, HEARD M J, CALTON G J, BURNETT J W. 1985. Interrelationships between toxins: studies on cross-reactivity between bacterial or animal toxins and monoclonal antibodies to two jellyfish venoms. Toxicon 23: 307–316.

OLSON C E, POCKL E E, CALTON C J, BURNETT J W. 1984. Immunochromatographic purification of a nematocyst toxin from the cnidarian *Chironex fleckeri* (sea wasp). Toxicon 22: 733–742.

O'NEAL R L, HALSTEAD B W, HOWARD L D. 1964. Injury to human tissues from sea urchin spines. Califor Med 101: 199–202.

OSMENT L S. 1976. Update: seabather's eruption and swimmer's itch. Cutis 18: 545–547.

OTHMAN I, AINI Y, YUSOFF A W, BURNETT J W, AZILA N. 1994. Biochemical studies of *Chironex fleckeri* and the treatment of *Chironex* envenomation with traditional preparations. Toxicon 32: 531–532.

OTHMAN I, BURNETT J W. 1990. Techniques applicable for the purifying of *Chironex fleckeri* (box-jellyfish) venom. Toxicon 28: 821–835.

OTHMAN I, BURNETT J W. 1994. Cases of jellyfish envenomation in Malaysia and some cytotoxic studies of their toxins. Toxicon 32: 532.

OTSURU M, SEKIKAWA H, HIROII Y, SUZUKI T, SATO Y, SHIRAKI T, NAGASHIMA Y. 1974. Observations on the sting occurring among swimmers in the rocky seashore. Jap J Zool 24: 225–235 (in Japanese). Aquat Sci Fish Abst 1974: 4Q10805M (in English).

OWRE H B. 1956. Coelenterate nematocysts as a cause of certain kinds of sea stings. Bull Marine Sci Gulf and Caribbean 6(4): 309–314.

PACY H. 1957. Identical skin-eruption in five men after exposure to jellyfish (*Aurelia*). Med J Aust 2: 580–582.

PACY H. 1962. Stingray and catfish injuries in New South Wales. Med J Aust 1: 119–120.

PAGÈS F, GILI J M, BOUILLON J. 1992. Planktonic cnidarians of the Benguela current. Scientia marina 56 (supl 1);1–444.

PAIJMANS K. 1976. New Guinea Vegetation. Amsterdam. Elsevier Scientific Publishing Co.

PALAFOX N A, JAIN L G, PINANO A Z, GULICK T M, WILLIAMS R K, SCHATZ I J. 1988. Successful treatment of ciguatera fish poisoning with intravenous mannitol. JAMA 259: 2740–2742.

PANTIN C F A. 1942a Excitation of nematocysts. Nature, London. 149: 109.

PANTIN C F A. 1942b. The excitation of nematocysts. J Exp Biol 19: 294–310.

PARADICE W E J. 1924. Injuries and lesions caused by the bites of animals and insects. Med J Aust 2: 650–652.

PARK D L. 1994. Reef management and seafood monitoring programs for ciguatera. Memoirs of the Queensland Museum 34(3): 587–594.

PARKER P. 1980. Family scorpaenidae, Scorpionfishes. In: (R H McDowall ed) Freshwater Fishes of Southeastern Australia. Sydney. Reed. 208 pp.

PARRA A. 1787. Descripción de diferentes piezas de historia natural las más del ramo maritimo, representadas en setenta y cinco láminas. Havana.

PARRISH D O. 1993. Seabather's eruption or diver's dermatitis. JAMA 270: 2300–2301.

PARRISH H M. 1959. Deaths from bites and stings of venomous animals and insects in the United States. Arch Int Med 104: 198–207.

PASSANO L M. 1982. Scyphozoa and Cubozoa. In: (G A B Shelton ed) Electrical Conduction and Behaviour in 'Simple' Invertebrates. Oxford. Clarendon Press. pp 149–202.

PATTERSON L, SWALLOW S. 1991. Sea snake envenomation. Med J Aust 155: 850.

PEARN J H. 1994. Ciguatera: dilemmas in clinical recognition, presentation and management. Memoirs of the Queensland Museum 34(3): 601–604.

PEARN J H, LEWIS R J, RUFF T, TAIT M, QUINN J, MURTHA W, KING G, MALLETT A, GILLESPIE N C. 1989. Ciguatera and mannitol: experience with a new treatment regimen. Med J Aust 151: 77–80.

PEARSE J S, PEARSE V B. 1978. Vision in cubozoan jellyfishes. Science 199: 458.

PEARSE V, PEARSE J, BUCHSBAUM M, BUCHSBAUM R. 1987. Living Invertebrates. Palo Alto and Pacific Grove. Blackwell Scientific Publications and the Boxwood Press. pp. 1–848.

PEARSON R B. 1967. The venom apparatus of selected Queensland stingrays. Thesis, University of Queensland, Brisbane. 139 pp.

PEEL N, KANDLER R. 1990. Localised neuropathy following jellyfish sting. Postgrad Med J 66: 953–954.

PERL T M, BEDARD L, KOSATSKY T, HOCKIN J C, TODD E C D, REMIS R S. 1990. An outbreak of toxic encephalopathy caused by eating mussels contaminated with domoic acid. N Eng J Med 322: 1775–1780.

PÉRON F, LESUEUR C A. 1810. Historie générale et particulière de tous les animaux qui composent la famille des Méduses. Annu Mus Hist Nat 14: 312–366.

PÉRON F, and PÉRON F, LESUEUR CA. 1807. 'Voyage de découvertes aux terres Australes exécuté sur les corvettes le Géographe, le Naturaliste, et la goelette le Casuarina, pendant les années 1800–1804.' Paris. 15 and 496 pp, with an atlas of 41 plates. 'Mollusques et Zoophytes', plates 29–31 by Lesueur.

PERRIERE C, GOUDEY-PERRIERE F. 1989. Origin and function of supporting cells in the venom glands of the lesser weeverfish (*Trachinus vipera*). Toxicon 27: 287–295.

PETRAUSKAS L E. 1955. A case of cone shell poisoning by 'bite' in Manus Island. Papua New Guinea Med J 1: 67–68.

PHILLIPS P J, BURKE W D. 1970. The occurrence of sea wasps (cubomedusae) in Mississippi Sound and the northern Gulf of Mexico. Bull Marine Sci 20: 853–859.

PHOON W O, ALFRED E R. 1965. A study of stonefish (Synanceia) stings in Singapore with a review of the venomous fishes of Malaysia. Singapore Med J 6: 158–163.

PIATIGORSKY J, HORWITZ T, KUWABARA T, CUTRESS C E. 1989. The cellular eye lens and crystallins of cubomedusan jellyfish. J Comp Physiol (A) 164: 577–587.

PICARD J. 1951. Notes sur les hydromeduses Mediterranéennes de la famille des Olindiadidae. Arch Zool Exp Gen 88: 39–48.

PICCINETTI MANFRIN G, PICCINETTI C. 1986. Distribuzione di *Pelagia noctiluca* in Adriatico negli anni 1983–84. Proc of workshop on jellyfish in the Mediterranean Sea, Trieste 1984. Nove Thalassia 8(2)(Suppl 2): 99–102.

PICKEN L E R, SKAER R J. 1966. A review of researches on nematocysts. Symp Zool Soc London 16: 19–50.

PIEN F D, ANG K S, NAKASHIMA N T, EVANS D G, GROTE J A, HEFLEY M L, KUBOTA E A. 1983. Bacterial flora of marine penetrating injuries. Diagn Microbiol Infect Dis 1: 229–232.

PIERCE R. 1986. Red tide (*Ptychodiscus brevis*) aerosol toxins: a review of current knowledge. Toxicon 24: 955–966.

PIOLATTO A, TARTAGLINO B. 1986. Punture di Celenterate. Federazione Medica XXXIX. 4: 617–618.

PLOTKIN M J. 1992. Thoughts on saving the world's rain forests. J Wilderness Med 3: 348–350.

PLUM F, POSNER J F. 1986. Disorders of consciousness and higher brain function, Part B: Disorders of mood and behaviour. In: (T E Andreoli, C C J Carpenter, F Plum, L H Smith Jr, eds) Cecil Essentials of Medicine. Philadelphia. W B Saunders Company. 1986: 664–671.

PONGPRAYOON U, BOHLIN L, WASUWAT S. 1991. Neutralization of toxic effects of different crude jellyfish venoms by an extract of *Ipomoea pes-caprae* (L) R Br J Ethnopharmacol 35: 65–69.

POPE E C. 1947. Some sea animals that sting and bite. Aust Mus Mag 9(5): 164–168.

POPE E. 1953. Marine stingers. Aust Museum Mag 11(4): 111–115.

POPE E C. 1963. Some noxious marine invertebrates from Australian seas. In: International Convention on Life Saving Techniques, Sydney, 11–20 March 1960. Republished as Proc 1st Int Convention on Life Saving Techniques, Part 3: Scientific Section. Suppl of Bull Post Grad Committee in Med, Univ Sydney 18(12): 91–102.

POPE E C. 1964. A stinging by a crown of thorns starfish. Aust Nat Hist 14: 350.

POPE E C. 1968. Venomous starfish in Sydney Harbour. Aust Nat Hist 16: 26.

PORTIER P, RICHET C. 1902a. De l'action anaphylactique de certains venins. Compt Rend Soc Biol 54: 170–172.

PORTIER P, RICHET C. 1902b. Nouveaux faits d'anaphylaxie, ou sensibilisation aux venins par doses réitérées. Compt Rend Soc Biol 54: 548–551.

POWELL AWB. 1946. The Shellfish of New Zealand. Auckland. Whitcomb and Tombs. 186pp.

PRATAP CHAND R, SELLIAH K. 1984. Reversible parasympathetic dysautonomia following stinging attributed to the box jellyfish (*Chironex fleckeri*). Aust NZ J Med 14: 673–675.

PRITCHARD M. 1991. Lice attacks spark fear for swimmers. West Australian, 13 May 1991. p 9.

PROKSCH P. 1994. Defensive roles for secondary metabolites from marine sponges and sponge-feeding nudibranchs. Toxicon 32: 639–655.

PUGH W S. 1913. Report of a case of poisoning by sea urchin. U.S. Naval Med Bull 7(2): 245–255.

PUGILEVSKY S V, MICHALEV P V. 1969. Poisoning by the medusa *Gonionemus vertens* in the Sea of Japan. Toxicon 7: 145–149.

QIN SHEDE (SHIDE), MINGLIANG ZHANG. 1992. Nematocyst Dermatitis in China Sea. Qingdao. Qingdao medical College, Institute of Dermatology. 51 pp.

QUOY J R C, GAIMARD J P. 1824. Voyage autour de monde...executé sur l'Uranie et la Physicienne... 1817–1820. Zoologie IV: 712.

QUOY J R C, GAIMARD J P. 1833. Voyage de découvertes de l'Astrolabe exécuté par ordre du roi pendant les années 1826–1829 sous le commandement de M.J. Dumont D'Urville. Zoologie. (4 vols) Paris. Vol 4: 390 pp.

RAJAGOPAL S, NAIR K V K, AZARIAH J. 1989. Some observations on the problem of jelly fish ingress in a power station cooling system at Kalpakkam, east coast of India. Mahasagar (Quarterly Journal in Oceanography, National Institute of Oceanography, Goa, India) 22(4): 151–158.

RANDALL J E, MILLINGTON J T. 1990. Triggerfish bite: a little-known marine hazard. J Wilderness Med 1: 79–85.

RANSON G. 1945. Les scyphoméduses de la collection du Muséum National d'Histoire Naturelle Paris, II: Catalogue raisonné; origine des récoltés. Bull Mus Nat Hist Paris; ser 2 Tome 16: 314.

RAO H S. 1931. Notes on the Scyphomedusae in the Indian Museum. Rec Indian Museum 33: 25–62.

RATHJEN W F, HALSTEAD B W. 1969. Report on two fatalities due to stingrays. Toxicon 6: 301–302.

READ B E. 1939. Chinese materia medica. Fish drugs. Peiping Nat Hist Bull. 136 pp.

REED K M. BRONSTEIN B R, BADEN H P. 1984. Delayed and persistent cutaneous reactions to coelenterates. J Am Acad Dermatol 10: 462–466.

REID H A. 1956a. Sea snake bite research. Trans Roy Soc Trop Med Hyg 50: 517–542.

REID H A. 1956b. Sea snake bites. Brit Med J 2: 73–78.

REID H A. 1957. Antivenene reaction following accidental sea snake bite. Brit Med J 2: 26–29.

REID H A. 1959. Sea snake bite and poisoning. The Practitioner 183: 530–534.

REID H A. 1961a. Myoglobinuria and sea snake poisoning. Brit Med J 1: 1284–1289.

REID H A. 1961b. Diagnosis, prognosis and treatment of sea snake bite. Lancet 2: 399–402.

REID H A. 1962. Sea snake antivenene: successful trial. Brit Med J 2: 576–579.

REID H A. 1975a. Epidemiology and clinical aspects of sea snake bites. In: (W A Dunson, ed.) The Biology of Sea Snakes. Baltimore. University Park Press. pp 417–462.

REID H A. 1975b. Antivenom in sea-snake bite poisoning. Lancet 1: 622–623.

REID H A, LIM K J. 1957. Sea-snake bite: a survey of fishing villages in northwest Malaya. Brit Med J: 1266–1272.

REISMAN R E. 1994. Review article: Insect stings. New Eng J Med 331: 523–527.

RICCI A. 1982. Sala absidata - Il pavimento. In: (A Carandini, A Ricci, M de Vos, eds) Filosofiana la Villa di Piazza Armerina, imagine di un aristocratico romano al tempo di Costantino. Palermo. S F Flaccovio. (Fig 157, page 260).

RICE N E, POWELL W A. 1972. Observations on three species of jellyfishes from Chesapeake Bay with special reference to their toxins. II. *Cyanea capillata*. Biol Bull 143: 617–622.

RICE R D, HALSTEAD B W. 1968. Report of fatal cone shell sting by *Conus geographus* Linnaeus. Toxicon 5: 223–224.

RICHARDSON B W. 1893. Fish poisoning and the disease 'siguatera'. Asclepiad, London 10: 38–42.

RICHARDSON J. 1861. On the poisonous effect of the small portion of the liver of a *Diodon* inhabiting the seas of southern Africa. J Linn Soc London 5: 213–216.

RICHET C. 1903. Des poisons contenus dans les tentacules des actinies (congestine et thalassine). Compt Rend Soc Biol 55: 246–248.

RIFKIN J F. 1982. Use of electrical stimulation for discharging cnidom components of a species of *Cerianthus* (Anthozoa: Cnidaria). Marine Biology 69: 31–36.

RIFKIN J F. 1988a. Studies of the structure, arrangement and mode of operation of cnidae from cnidarians belonging to the classes Cubozoa, Hydrozoa, Scyphozoa and Anthozoa. PhD thesis. Department of Zoology, University of Queensland, Brisbane.

RIFKIN J F. 1988b. Animal toxins and humans. Coronary Care Pre-Hospital Life Support National Seminar proceedings. 29, 30 April, 1 May. (unnumbered pages).

RIFKIN J F. 1991. A study of the spirocytes from the Ceriantharia and Actiniaria (Cnidaria: Anthozoa) Cell Tissue Res 266: 365–373.

RIFKIN J. 1993. The amazing nematocyst. SPUMS J 23: 107–111.

RIFKIN J F. 1996. Revision of the Australian cubomedusae. Memoirs of the Queensland Museum. (In press).

RIFKIN J, ENDEAN R. 1983. The structure and function of the nematocysts of *Chironex fleckeri* Southcott 1956. Cell Tissue Research 233: 563–577.

RIFKIN J F, ENDEAN R. 1988. Arrangement of accessory cells and nematocysts bearing mastigophores in the tentacles of the Cubozoan *Chironex fleckeri*. J Morphology 195: 103–115.

RIFKIN J F, FENNER P J, WILLIAMSON J A H. 1993a. First aid treatment of the sting from the hydroid *Lytocarpus philippinus*: the structure of, and in vitro discharge experiments with its nematocysts. J Wilderness Med 4: 252–260.

RIFKIN J, WILLIAMSON J, FENNER P, BURNETT J. 1993b. Disarming the box jellyfish (letter). Med J Aust 158: 647–648.

RIVERA VR, POLI MA, BIGHAMI GS. 1995. Prophylaxis and treatment with monoclonal antibody of tetrodotoxin poisoning in mice. Toxicon 33: 1231–1237.

ROSCO M D. 1977. Cutaneous manifestations of marine animal injuries including diagnosis and treatment. Cutis 19: 507–510.

ROTTINI G, GUSMAN L, PAROVEL E, AVIAN M, PATRIARCA P. 1995. Purification and properties of a cytolytictoxin in venom of the jellyfish *Carybdea marsupialis*. Toxicon 33: 315–326

ROTTINI SANDRINI I, STRAVISI F, PIERI G. 1980. A recent shift in the wind distribution and the appearance of unusual planktonic organisms in the Gulf of Trieste. Bull Soc Adr Sci LXIV: 77–84.

RUFF T A, LEWIS R J. 1994. Clinical aspects of ciguatera: an overview. Memoirs of the Queensland Museum 34(3): 609–619.

RUMPHIUS G E. 1705. 'D'Amboinsche Rariteitkammer', Translated by W S S Van Benthem Jutting. Rumphius Memorial Volume (H C D De Witt, ed) 1959: 185.

RUSSELL F E. 1965. Marine toxins and venomous and poisonous marine animals. In: (F S Russell, ed) Advances in Marine Biology, Vol III. London. Academic Press. pp 255–384.

RUSSELL F E. 1966. *Physalia* stings: a report of two cases. Toxicon 4: 65–67.

RUSSELL F E. 1971. Poisonous Marine Animals. New York. Neptune City. TFH Publications.

RUSSELL F E. 1972 The Stingray. Natural History, Venom Apparatus, Chemistry and Toxicology, and Clinical Problems. Mote Marine Laboratories, Sarasota, Florida.

RUSSELL F E. 1984. Marine toxins and venomous and poisonous marine plants and animals (Invertebrates). In: (J H S Blaxter, F S Russell, M. Yonge eds) Advances in Marine Biology 21: 60–194.

RUSSELL F E. 1987. Another warning about electric shock for snakebite (Letter). Postgrad Med 82: 32.

RUSSELL F E, EGAN N B. 1991. Ciguateric fishes, ciguatoxin (CTX) and ciguatera poisoning. Toxin reviews, J Toxicol 10(1): 37–62.

RUSSELL F E, EMERY J A. 1960. Venom of the weevers *Trachinus draco* and *Trachinus vipera*. Ann NY Acad Sci 90(3): 805–819.

RUSSELL F E, FAIRCHILD M D, MICHAELSON J. 1958a. Some properties of the venom of the stingray. Med Arts Sci 12: 78–86

RUSSELL F E, PANOS T C, KAHN X W, WARNER A M, COLKET T C. 1958b. Studies on the mechanism of death from stingray venom: a report of two fatal cases. Amer J Med Sci 235: 566–584.

RUSSELL F E, Van HARREVELD A. 1954. Cardiovascular effects of the venom of the round stingray, *Urobatis halleri*. Arch Int Physiol 62: 232–233.

RUSSELL F S. 1953. The Medusae of the British Isles. Cambridge University Press. 530 pp.

RUSSELL F S. 1964. On scyphomedusae of the genus *Pelagia*. J Mar Biol Assn of UK 44: 133–136.

RUSSELL F S. 1970. The Medusae of the British Isles, Vol II: Pelagic Scyphozoa, with supplement to Vol I on Hydromedusae. Cambridge University Press. 284 pp.

RUSSO A J, CALTON G J, BURNETT J W. 1983. The relationship of the possible allergic response to jellyfish envenomation and serum antibody titres. Toxicon 21: 475–480.

RYAN K C, CARAVATI E M. 1994. Life-threatening anaphylaxis following envenomation by two different species of Crotalidae. J Wilderness Med 5: 263–268.

SAGI A, ROSENBERG L, BEN-MEIR P, HAUBEN D J. 1987. Fire coral *Millepora dichotoma* as a cause of burns: a case report. Burns 13: 325–326.

SAITANU K, LAOBHRIPATR S, LIMPAKARNJANARAT K, SANGWANLOY O, SUDHASANEYA S, ANUCHATVORAKUL B, LEELASITORN S. 1991. Toxicity of the freshwater puffer fish *Tetraodon fangi* and *T. palembangensis* from Thailand. Toxicon 29: 895–897.

SAKAI M. 1970. Nereistoxin and its derivitives: their ganglionic blocking and insecticidal activity. In: (R D O'Brien, I Yamamoto, eds) Biochemical Toxicology of Insecticides. New York. Academic Press. 33–40.

SALDANHA L. 1980. Fauna Submarina Atlântica. Lisboa, Portugal. Publicações Europa América. 202 pp.

SALLEO A, LaSPADA G, FALZEA G, DENARO N G. 1984. Discharging effects of anions and the inhibitory effect of divalent cations on isolated nematocysts of *Pelagia noctiluca*. Molecular Physiol 5: 25–26.

SALLEO A. LaSPADA G, BARBERA R. 1994. Gadolinium is a powerful blocker of the activation of nematocysts of *Pelagia noctiluca*. J Exp Biol. 187: 201–206.

SAMS W M. 1949. Seabather's eruption. Arch Dermatol 60: 227–237.

SATO S, NAKAMURA H, OHIZUMI Y, KOBAYASHI J, HIRATA Y. 1983. The amino acid sequences of homologous hydroxyproline-containing myotoxins from the marine snail *Conus geographus* venom. FEBS Letters 155(2): 277–280.

SATOH A. 1985. Studies on the toxins of *Carybdea rastonii*. Ochanomizu Med J 33: 131– (in Japanese).

SATOH A, YASUHARA T, NAKAJIMA T. 1983. Purification and characterization of nematocyst toxin from *Carybdea rastonii*. J Biochem, Tokyo 55: 1019– (in Japanese).

SATOH A, YASUHARA T, NAKAJIMA T. 1984. Purification of Carybdea rastonii toxin. In: Proceedings of the 10th Annual Meeting of the Japanese Pharmaceutical Society. (Abstract). Vol p 1019 (in Japanese).

SATTERLIE R A. 1979. Central control of swimming in the cubomedusan jellyfish *Carybdea rastonii*. J Comp Physiol 133: 357–367.

SATTERLIE R A, SPENCER A N. 1979. Swimming control in a cubomedusan jellyfish. Nature 281: 141–142.

SAUNDERS P R. 1959. Venom of the stonefish *Synanceia verrucosa*. Science 129(3344): 272–274.

SAUNDERS P R, TAYLOR P B. 1959. Venom of the lionfish *Pterois volitans*. Amer J Physiol 197: 437–440.

SAUVIAT M-P, GARNIER P, GOODEY-PERRIERE F, PERRIERE C. 1995. Does crude venom of the stonefish (*Synanceia vettucosa* activate β-adrenoceptors in the frog heart muscle? Toxicon 33: 1207–1213.

SAVAGE I V E, HOWDEN M E H. 1977. Hapalotoxin: a second lethal toxin from the octopus *Hapalochlaena maculosa*. Toxicon 15: 463–466.

SAVILLE-KENT W. 1893. The Great Barrier Reef of Australia: Its Products and Potentialities. London. W H Allen & Co. 387 pp. (Modern version CURRY J. 1970 Melbourne. O'Neil. 387 pp.

SCARPA C, KOKELJ F, DEL NEGRO P, TUBARO A. 1987. Valutazione dell'effetto irritante sulla cute umana di una preparazione di nematocisti di *Pelagia noctiluca*. Ann It Derm Clin Sper 41: 337–341.

SCHANTZ E J. 1971. Paralytic shellfish poisoning and saxitoxin. In: (L L Simpson, ed) Neuropoisons: Their Pathophysiological Actions, Vol 1: Poisons of Animal Origin. New York. Plenum Press. pp 159–168.

SCHANTZ E J, MOLD J D, HOWARD W L, BOWDEN J P, STANGER D W, LYNCH J M, WINTERSTEINER O P, DUTCHER J D, WALTERS D R, REIGEL B. 1961. Paralytic shellfish poison, VIII: Some chemical and physical properties of purified clam and mussel poisons. Can J Chem 39: 2117–2123.

SCHEICH H, LANGNER G, TIDEMANN C, COLES R B, GUPPY A. 1986. Electroreception and electrolocation in platypus. Nature 379: 401–402.

SCHEUER P J. 1964. The chemistry of toxins isolated from some marine organisms. Fortsch Chem Organ Naturst 22: 265–278.

SCHEUER P J. 1994. Ciguatera research — an historical perspective (abstract). Memoirs of the Queensland Museum 34(3): 586.

SCHMIDT H. 1974. Evolution in the Anthozoa. In: (AM CAmeron, B M Campbell, A B Cribb, R Endean, J S Jell, O E Jones, P Mather, F H Talbot eds) Proc Second International Symposium on Coral Reefs. Brisbane. 533–563.

SCHNADIG V J, BURKE S S, BENGSTON K, DAVIE K D, NICHOLS M M. 1991. Cytologic identification of jellyfish nematocysts in skin scraping. 39th Annual Scientific Meeting of the American Socety of Cytology, Los Angeles, Calif. 5–10 November 1991. Acta Cytol 35(5): 1991.

SCHUETT W, RAPOPORT H. 1962. Saxitoxin, the paralytic shellfish poison. Degradation to a pyrrolopyrimidine. J Amer Chem Soc 84: 2266.

SCOGGIN C H. 1975. Catfish stings. JAMA 231: 176–177

SEARLE A G. 1957. An introduction to Malayan jellyfish. Malayan Nature J 11(3): 70–93.

SHANNON L V, CHAPMAN P. 1983. Incidence of *Physalia* on beaches in the south western Cape Province during January 1983. Sth Afr J Sci 79: 454–458.

SHARMA G M, VIG B. 1972. Studies on the antimicrobial substances of sponges, VI: Structures of two antibacterial substances isolated from the marine sponge *Disidea herbacea*. Tetrahedron Letters. 17: 1715–1718.

SHEPPARD C, PRICE A, ROBERTS C. 1992. Marine Ecology of the Arabian Region. Patterns and Processes in Extreme Tropical Environments. London. Academic Press, Harcourt Brace Jovanovich Publishers. 359 pp.

SHEUMACK D D, HOWDEN M E H, SPENCE I, QUINN R J. 1978. Maculotoxin: a neurotoxin from the venom glands of the octopus *Hapalochlaena maculosa* identified as tetrodotoxin. Science 199: 188–189.

SHIOMI K, ITOH K, YAMANAKA H, KIKUCHI T. 1985. Biological activity of crude venom from the crown-of-thorns starfish, *Acanthaster planci*. Nippon Suisan Gakkaishi 51: 1151–1154.

SHIOMI K, KATSUYUKI N, YAMANAKA H, KIKUCHI T. 1989. Inhibitory effects of anti-inflammatory agents on cutaneous capillary leakage induced by six marine venoms. Nippon Suisan Gakkaishi 55: 131–134.

SHIOMI K, SAITSU N, KIKUCHI T. 1991. Hepatotoxic effects induced in rats by the crown-of-thorns starfish Acanthaster planci lethal factor. Nippon Suisan Gakkaishi 57: 341–344.

SHIOMI K, TAKAMIYA M, YAMANAKA H, KIKUCHI T, SUZUKI Y. 1988a. Toxins in the skin secretion of the oriental catfish (*Plotosus lineatus*): immunological properties and immunocytochemical identification of producing cells. Toxicon 26: 353–361

SHIOMI K, YAMAMOTO S, YAMANAKA H, KIKUCHI T. 1988b. Purification and characterization of a lethal factor in venom from the crown-of-thorns starfish (*Acanthaster planci*). Toxicon 26: 1077–1083.

SHIOMI K, YAMAMOTO S, YAMANAKA H, KIKUCHI T, KONNO K. 1990. Liver damage by the crown-of-thorns starfish *Acanthaster planci* lethal factor. Toxicon 28: 469–476.

SHOKITA S. 1986. Dangerous Animals and Plants of Okinawa. 1st edn. Furujima 47–1, Naha-shi, Okinawa. Okinawa-Shuppan Publishing Co. Ltd. pp 1–23.

SIMPSON C, GREY K A. 1989. Surveys of crown-of-thorns starfish and coral communities in the Dampier Archipelago, Western Australia. Environmental Protection Authority Series 25, Environmental Protection Authority, Perth. 24 pp.

SIMS J K, ENOMOTO P I, FRANKEL R I, LIVINGSTON M F. 1983. Marine bacteria complicating seawater near drowning and marine wounds: a hypothesis. Ann Emerg Med 12: 212–216.

SIMS J K, OSTMAN D C. 1986. Pufferfish poisoning: emergency diagnosis and management of mild human tetrodotoxication. Annals Emerg Med 15: 1094–1098.

SINGLA CL. 1975. Statocysts of hydromedusae. Cell Tissue Res 158: 391–407.

SKEIE E. 1962. Toxin of the weeverfish (*Trachinus draco*). Experimental studies on animals. Acta Pharmacol Toxical 19: 107–120.

SKEIE E. 1966. Weeverfish stings: frequency, occurrence, clinical course, treatment and studies on the venom apparatus of the weeverfish, the nature of the toxin, and immunological aspects. Dan Med Bull 13: 119–121.

SMART D. 1995. Clinical toxicology of shellfish poisoning. In: (J Meier, J White, eds) Clinical Toxicology of Animal Venoms and Poisons. Boca Raton, Florida. CRC Press. pp 33 57.

SMITH G. 1983. Erysipelothrix and Listeria. In: (G Wilson, A Miles, M T Parker, eds) Topley and Wilson's Principles of Bacteriology, Virology and Immunity, 7th edition, Vol 2, London. Edward Arnold. pp 50–59.

SMITH J L B. 1951. A case of poisoning by the stonefish, *Synanceia verrucosa*. Copeia 3: 207–210.

SMITH J L B. 1957. Two rapid fatalities from stonefish stabs. Copeia 3: 249.

SMITH M. 1926. Monograph of the sea snakes (Hydrophiidae). London. British Museum. 130 pp.

SMITH M M. 1977. Sea and Shore Dangers: Their Recognition, Avoidance and Treatment. Grahamstown, South Africa. J L B Smith Institute of Ichthyology, Rhodes University. pp 21–31.

SMITH DS, CAYER M L, RUSSELL FE. 1974. Membrane-limited microtubular aggregates in the venom secreting cells of a stingray. Toxicon 12: 331–335.

SOUTHCOTT R V. 1956. Studies on Australian cubomedusae, including a new genus and species apparently harmful to man. Aust J Mar Freshw Res 7: 254–280.

SOUTHCOTT R V. 1959. Tropical jellyfish and other marine stinging. Military Med 124(8): 569–579.

SOUTHCOTT R V. 1960. Venomous jellyfish. Good health for South Australia. 113: 18–23.

SOUTHCOTT R V. 1963. Identification of jelly-fishes. The hazards of dangerous marine creatures. Committee of the 1960 International Convention on Life Saving Techniques. pp 1–131. Republished as Proc 1st Int Convention of Life Saving Techniques, Part 3, Scientific Section: Suppl. to Bull Post Grad Comm in Medicine, Univ of Sydney 18(12): 1–131.

SOUTHCOTT R V. 1967. Revision of some Carybdeidae (Scyphozoa: Cubomedusae), including description of the jellyfish responsible for the 'Irukandji syndrome'. Aust J Zool 15: 651–671.

SOUTHCOTT R V. 1970. Notes on stings of some venomous Australian fishes. Med J Aust 2: 722–725.

SOUTHCOTT R V. 1975. Australian Venomous and Poisonous Fishes. Adelaide. R V Southcott. 20 pp.

SOUTHCOTT R V. 1979. Portuguese man-o'-war or bluebottle. Adelaide. South Australian Museum. pp 1–8.

SOUTHCOTT R V. 1982a. Jellyfishes (Classes Scyphozoa and Hydrozoa). In: (S A Shepherd, I M Thomas, eds) Marine Invertebrates of Southern Australia, Part I. Adelaide. Government Printer. pp 115–159.

SOUTHCOTT R V. 1982b. Vitamin A content of the liver of the Australian sea lion *Neophoca cinerea* (Peron) and its toxicological significance. Trans Royal Soc South Australia, Vol 106, Pt 3, pp 85–91.

SOUTHCOTT R V. 1985. The 'morbakka'. Med J Aust 143: 324.

SOUTHCOTT R V. 1987a. Sponges. In: (J Covacevich, P Davie, J Pearn, eds) Toxic Plants and Animals: A Guide for Australia. Brisbane. Queensland Museum. pp 73–76.

SOUTHCOTT R V. 1987b. Jellyfish of medical importance. In: (J Covacevich, P Davie, J Pearn, eds) Toxic Plants and Animals: A Guide for Australia. Brisbane. Queensland Museum. pp 93–98.

SOUTHCOTT R V. 1987c. Anemones, corals and their relatives. In: (J Covacevich, P Davie, J Pearn, eds) Toxic Plants and Animals: A Guide for Australia. Brisbane. Queensland Museum. pp 107–111.

SOUTHCOTT R V, COULTER J R. 1971. The effects of the southern Australian marine stinging sponges, *Neofibularia mordens* and *Lissodendoryx* sp. Med J Aust 2: 895–901.

SOUTHCOTT R V, KINGSTON C W. 1959. Lethal jellyfish stings: a study in 'sea wasps'. Med J Aust 1: 443–444.

SOUTHWARD A J. 1955. Observations on the ciliary currents of the jelly-fish *Aurelia aurita* L. J Mar Biol Assoc UK 34: 201–216.

SPELMAN F J, BOWE E A, WATSON C B. 1982. Acute renal failure as a result of *Physalia physalis* sting. South Med J 75: 1425–26.

SPENCE I, GILLENSEN D, GREGSON R P, QUINN R J. 1977. Characterization of the neurotoxic constituents of *Conus geographus* (L) venom. Life Sci 21: 1759–1770.

SPENCER A N. 1981. The parameters and properties of a group of electrically coupled neurones in the central nervous system of a hydrozoan jellyfish. J Exp Biol 93: 33–50.

STANISIC J. 1987. Cone Shells. In: (J Covacevich, P Davie, J Pearn, eds) Toxic Plants and Animals: A Guide for Australia. Brisbane. Queensland Museum. pp 113–117.

STARK K, MOSS K. 1986 approx (undated). Stinger Resistant Enclosures. Townsville. Civil & Systems Engineering, James Cook University of North Queensland. 4 pp.

STEIN M R, MARRACCINI J V, ROTHSCHILD N E, BURNETT J W. 1989. Fatal Portuguese man-o'-war (*Physalia physalis*) envenomation. Ann Emerg Med 18: 312–315.

STEPHENSON T A, STEPHENSON A, TANDY G, SPENDER M. 1931. The structure and ecology of Low Isles and other reefs. Sci Reports, Great Barrier Reef Exped 3(2): 17–112.

STEPHENSON W. 1962. A peculiar fish from Moreton Bay. Qld Naturalist 16(5,6): 94.

STEVENS J D (ed). 1987. Sharks. Mc Mahons Point, Sydney. Weldon Owen. 240 pp.

STEWART R D. 1988. Nitrous oxide. In: (P M Paris, R D Stewart eds) *Pain Management in Emergency Medicine*. Connecticut. Appleton & Lange. pp 221–238.

STIASNY G. 1919. Die Scyphomedusen-Sammlung des Naturhistorischen Reichsmuseums in Leiden. I. Die Carybdeiden (Cubomedusen). Zoöl Meded Deel 5: 33–58.

STIASNY G. 1922. Die Scyphomedusen-Sammlung von Dr. Th. Mortensen nebst anderen Medusen aus dem zoologischen Museum der Universität in Kopenhagen. Vidensk. Meddr. dansk naturh. Foren. 77: 33–558.

STIASNY G. 1924. Ueber einige Scyphomedusen von Sydney (Port Jackson). Zool Meded Deel 8: 55–72.

STIASNY G. 1926a. Über einige Scyphomedusen von Puerto Galera, Mindoro (Philippinen) Zool. Meded Leiden 9: 239–248.

STIASNY G. 1926b. Alte und neue Scyphomedusen von Australien. Zool Meded 9: 249–257.

STIASNY G. 1931. Die Rhizostomeen-Sammlung des British Museum (Natural History) in London. Zoöl Meded Deel 14: 139.

STIASNY G. 1934. Scyphomedusae. Discovery Reports 8: 329–396.

STIASNY G. 1937. Scyphomedusae. The John Murray Expedition Scientific Reports 4(7): 203–242 (& pl. 1).

STIASNY G. 1940 Die Scyphomedusen. Dana Rep 18: 5.

STONIK V A, ELYAKOV G B. 1988. Structure and biologic activities of sponge and sea cucumber toxins. In: (A T Tu, ed) Marine Toxins and Venoms. New York. Marcel Dekker. pp 107–120.

STORER T I, USINGER R L. 1957. General Zoology, 3rd Edition. New York. McGraw Hill. 664 pp.

STRAUS J S. 1956. Seabather's eruption. Arch Dermatol 74: 293–295.

STRONG R P. 1942. Stitt's Diagnosis, Prevention and Treatment of Tropical Diseases, 6th edition, March 1943 (2 vol). Philadelphia. Blakiston. 1550 pp.

STRUTTON G, LUMLEY J. 1988. Cutaneous light microscopic and ultrastructural changes in a fatal case of jellyfish envenomation. J Cutan Pathol 15: 249–255.

STUDEBAKER J P. 1972. Development of the cubomedusa, *Carybdea marsupialis*. MSc thesis. Dept Mar Sci, Univ Puerto Rico, Mayaguez. 59 pp.

STYS P K, SONTHEIMER H, RANSOM B R, WAXMAN S G. 1993. Nonactivating, tetrodotoxin-sensitive Na+ conductance in rat optic nerve axons. Proc National Acad Sci USA 90: 6976–6980.

SUENAGA K, KOTOKU S. 1980. Detection of tetrodotoxin in autopsy material by gas chromatography. Arch Toxicol 44: 291–297.

SUN K, WAT J, SO P. 1994. Puffer fish poisoning. Anaesth Intens Care 22: 307–308.

SUN L, GOPALAKRISHNAKONE P, KHOO KS. 1993. Cytotoxic properties of a novel marine toxin — MTX1 on human blood cells. In: (N H Tan, S L Oo, V Thambyrajah, N Azila, eds) Advances in Venom and Toxin Research. Proc 3rd Asia Pacific Congress on Animal, Plant and Microbial Toxins Malaysia, 27 June–1 July, 1993. Kuala Lumpur. Malaysian Society on Toxinology. pp 228–233.

SURF LIFE SAVING AUSTRALIA LTD. 1993. Annual Report 1993. Brighton-Le-Sands NSW 2216. 54 pp.

SUTHERLAND S K. 1977. Serum reactions: an analysis of commercial antivenoms and the possible role of anticomplimentary activity in the de-novo reactions to antivenoms and antitoxins. Med J Aust 1: 613–615.

SUTHERLAND S K. 1983. Australian Animal Toxins: The Creatures, Their Toxins and the Care of the Poisoned Patient. Melbourne. Oxford University Press. 540 pp.

SUTHERLAND S K. 1990. The use of verapamil to treat box jellyfish stings. Med J Aust 153: 363.

SUTHERLAND S K. 1992. Antivenom use in Australia: premedication, adverse reactions and the use of venom detection kits. Med J Aust 157: 734–739.

SUTHERLAND S K. 1994a. Cerebrovascular accident after taipan bite (letter in reply). Med J Aust 160: 94.

SUTHERLAND S K. 1994b. Antivenom research in Australia. Med J Aust 161: 48–50.

SUTHERLAND S K. 1994c. The pressure immobilisation technique. Med J Aust 161: 700–701.

SUTHERLAND S K, BROAD A J, LANE W R. 1970. Octopus neurotoxins: low molecular weight, non-immunogenic toxins present in the saliva of the blue-ringed octopus. Toxicon 8: 249–250.

SUTHERLAND S K, COULTER A R, HARRIS R D. 1979. The rationalisation of first-aid measures for elapid snakebite. Lancet 1: 183–186.

SUTHERLAND S K, DUNCAN A W. 1980. New first-aid measures for envenomation: with special reference to bites by the Sydney funnel-web spider (*Atrax robustus*). Med J Aust 1: 378–379.

SUTHERLAND S K, KING K. 1991. Management of snake-bite injuries. Monograph series No 1. Royal Flying Doctor Service of Australia (Federal Council), 43 Bridge Street, Hurstville, NSW 2220. 12 pp.

SUTHERLAND S K, LANE W R. 1969. Toxins and mode of envenomation of the common ringed or blue-banded octopus. Med J Aust 1: 893–898.

SUTHERLAND S K, LOVERING K E. 1979. Antivenoms: Use and adverse reactions over a 12-month period in Australia and Papua New Guinea. Med J Aust 2: 671–674.

SWAROOP S, GRAB B. 1954. Snakebite mortality in the world. Bulletin World Health Organ 10: 35–76.

TAHARA Y. 1910. Über das tetrodongift. Biochem Z. 10: 255–275.

TAIRA E, TANAHARA N, FUNATSU M. 1975. Studies on the toxin in the spines of the starfish *Acanthaster planci*, I. Isolation and some properties of the toxin found in the spines. Science Bulletin of the College of Agriculture, University of the Ryukyus, Okinawa 22: 203–213.

TAKAHASHI T, SERISAWA S, OKA S, SAKITA T, HOSAKA H. 1955. Experimental cirrhosis of the liver induced by shellfish poison, with special reference to the pathogenesis of fibrosis (in Japanese). Sogo Igaku 12: 81–86.

TAMBS-LYCHE H. 1964. *Gonionemus vertens* L. Agassiz (Limnomedusae): a zoogeographical puzzle. Sarsia 15: 1–8.

TAMKUN M M, HESSINGER D A. 1981. Isolation and partial characterisation of a hemolytic and toxic protein from the nematocyst venom of the Portuguese man-o'-war. Biochem Biophys Acta 667: 87–98.

TANGE Y. 1955. Beitrage zur Kenntnis der Morphologie DES Giftapparates bei den japanischen Fischen. XIII. Uber den Giftapparat bei *Liobagris reinii* (Hilgendorf). Yokahama Med Bull. 6: 255–265.

TANI T. 1945. *Toxicological Studies on Japanese Puffer* (in Japanese). Tokyo. Teikoku Tosho. 103 pp.

TAYLOR JD, KANTOR YI, SYSOEV AV. 1993. Foregut anatomy, feeding mechanisms, relationships and classification of the Conoidea (=Toxoglossa) (Gastropoda). Bull Nat Hist Mus Lond. (Zool) 59 (2):125–170.

TEMPLE-SMITH P. 1978. In: GRIFFITH M. 1978. The Biology of the Monotremes. New York. Academic Press.

TENENBEIN M. 1991. Multiple doses of activated charcoal: time for reappraisal? Ann Emerg Med 20: 529–531.

THEAKSTON R D G, WARRELL D A. 1991. Antivenoms: a list of hyperimmune sera currently available for the treatment of envenoming by bites and stings. Toxicon 29: 1419–1470.

THIEL M E. 1928. Die Scyphomedusen des Zoologiischen Staatsinstituts und Zoologischen Museums im Hamburg. I: Cubomedusae, Stauromedusae und Coronatae. Mitt Zool St Inst Hamb 43: 1–34.

THIEL M E. 1936. Scyphomedusae. In: (H G Bronns, ed) Klassen und Ordnungen des Tierreichs, 2Bd 2Abt 2Buch 2Lfg, Cubomedusae. pp 173–308.

THOMAS M B, EDWARDS N C. 1991. Cnidaria: Hydrozoa. In: (F W Harrison, J A Westfall, eds) Microscopic Anatomy of Invertebrates. Vol 2: Placozoa, Porifera, Cnidaria and Ctenophora. New York. Wiley-Liss. pp 91–183.

THOMPSON T E, BENNETT I. 1970. Observations on Australian Glaucidae (Mollusca: Opisthobranchia) Zoological Linnean Soc 49: 187–197.

THORINGTON G U, HESSINGER D A. 1988. Control of cnida discharge, I: Evidence for two classes of chemoreceptor. Biol Bull 174: 163–171.

TIBBALLS J. 1988. Severe tetrodotoxic fish poisoning. Anaesth Intens Care 16: 215–217.

TIBBALLS J. 1994. Premedication for snake antivenom. Med J Aust 160: 4–7.

TODD E C, KIER A, EBELING A W. 1966. *Gonionemus vertens* L. Agassiz (Hydrozoa; Limnomedusae) in Southern California. Bull Sthn Calif Acad Sci 65: 205–210.

TOGIAS A G, BURNETT J W, KAGEY-SOBOTKA A, LICHTENSTEIN L M. 1985. Anaphylaxis after contact with a jellyfish. J Allergy Clin Immunol 75: 672–675.

TOMCHIK R S, RUSSELL M T. 1993. Seabather's eruption 'sea lice'. J Florida Med Assoc 80: 813–816.

TOMCHIK R S, RUSSELL M T, SZMANT A M, BLACK N A. 1993. Clinical perspectives on seabather's eruption, also known as sea lice. JAMA 269: 1669–1672.

TONGE J I, BATTEY Y, FORBES J J. 1967. Ciguatera poisoning: a report of two outbreaks and a possible fatal case in Queensland. Med J Aust 2: 1088–1090.

TOOM P M, CHAN D S. 1972. Preliminary studies of nematocysts from the jellyfish *Stomolophus meleagris*. Toxicon 10: 605–610.

TOOM P M, PHILLIPS T D. 1975. Effect of purified components of jellyfish toxin (*Stomolophus meleagris*) on active sodium transport. Toxicon 13: 261–271.

TOOM P M, PHILLIPS T D, KOCH R B. 1975. Effects of purified components of jellyfish toxin (*Stomolophus meleagris*) on adenosine triphosphatase activities. Biochem Pharmac 25: 551–555.

TORDA T A, SINCLAIR E, ULYATT D B. 1973. Puffer fish (tetrodotoxin) poisoning: clinical record and suggested management. Med J Aust 1: 599–602.

TOTTON A K, MACKIE G O. 1960. Studies on *Physalia physalis*. Discovery Reports 30: 301–408.

TOTTON A K. 1965. A synopsis of the Siphonophora. London. British Museum (Natural History). 230 pp.

TREE-TRAKAN T, SATITKARNMANEE T, DHORRANINTRA B, CHAREANTHAITAWEE P. 1979. Long acting local anaesthetic properties of tetrodotoxin for spinal and epidural anesthesia in dogs. Mahidol University Annual Research Abstracts 1980 (1 January–31 December 1979). Mahidol University Development and Planning Office, Bangkok 10700, Thailand. 132 pp.

TRUITT R V. 1939. Our water resources and their conservation. Contract No 27. Chesapeake Biology Laboratory, Maryland Conservation Department. 17 pp.

TSUNENARI S, UCHIMURA Y, KANDA M. 1980. Puffer poisoning in Japan: a case report. J Forensic Sci 25: 240–245.

TU A T, FULDE G. 1987. Sea snake bites. In: Aquatic Dermatology. Clin in Dermatol 5(3): 118–126.

TURNER B, SULLIVAN P, PENNEFATHER J. 1980. Disarming the bluebottle: treatment of *Physalia* envenomation. Med J Aust 2: 394–395.

TURNER R J, FREEMAN S E. 1969. Effects of *Chironex fleckeri* toxin on the isolated perfused guinea pig heart. Toxicon 7: 277–286.

TURNER R J, FREEMAN S E. 1974. Factors affecting the muscle depolarisation due to *Conus achatinus* toxin. Toxicon 12: 49–55.

UCHIDA T. 1929a. Studies on Japanese hydromedusae, 3: Olindiadae. Ann Zool Japan 12: 351–371.

UCHIDA T. 1929b. Studies on the Stauromedusae and Cubomedusae, with special reference to their metamorphosis. Jap J Zool 2: 103–109.

UCHIDA T. 1938. Medusae from Mutsu Bay (revised report). Sci Rep Tohoku Imp Univ 4th Ser 13;1.

UCHIDA T. 1970. Revision of Japanese Cubomedusae. Publications Seto Marine Biol Lab 17: 289–297.

ULMER N. 1865. Tod durch den Stich eins Trachinus draco. Allg Mil Arztl Ztg 6: 329–332.

UNDERHILL D. 1987. Australia's Dangerous Creatures. Sydney. Readers Digest. 368 pp.

UVNÄS B. 1960. Mechanism of action of a histamine-liberating principle in jellyfish (*Cyanea capillata*). Ann NY Acad Sci 90(3): 751-759.

VALENTINE G H. 1982. Hypersensitivity after a sea urchin sting (letter) British Med J (Clin Res) 285: 1656.

VANHÖFFEN E. 1902. Die Acraspeden Medusen der Deutschen Tiefsee-Expedition 1898–1899, I: Trachymedusen. Wiss. Ergebn. 'Valdivia', Bd 3, pp 1–52, 55–86, pls 1–12.

VANNUCCI M. 1951. Hidrozoa e Scyphozoa existentes no Instituto Paulista de Oceanographia. I. Bol Inst Paulista de Oceanogr 2(1): 67–98.

VANNUCCI M. 1954. Hidrozoa e Scyphozoa existentes no Insituto Oceanografico. Bol Inst Paul Oceanogr. Tomo 5: 120.

VANUCCI M, TUNDISI J. 1962. Las medusas existentes en los museos de La Plata y Buenos Aires. Com Mus Arg Cs Nat B Rivadavia CS Zool 3(8): 203–215.

VERNOUX J P. 1994. The mouse ciguatoxin bioassay: directions for use to control fish for consumption. Memoirs of the Queensland Museum 34(3): 625–629.

VERNOUX J P, LEJEUNE J. 1994. Ciguatera in the French West Indies. Memoirs of the Queensland Museum 34(3): 631–638.

VERON J E N. 1986. Corals of Australia and the Indo-Pacific. Sydney. Angus & Robertson. 644 pp.

VERRILL AE. 1907. The Bermuda Islands, V: Characteristic life of the Bermuda coral reefs. Trans Connecticut Acad Arts Sci 12: 340.

VIMPANI G, DOUDLE M, HARRIS R. 1988. Child accident-mortality in the Northern Territory, 1978–1985. Med J Aust 148: 392–395.

VLACHOS P, KONTOS P. 1987. Epidemiology and therapeutic methods of 'jellyfish' poisoning in Greece, Athens. Jellyfish Blooms in the Mediterranean. Proc 2nd Workshop on Jellyfish in the Mediterranean Sea, Trieste, 2–5 September 1987. United Nations Environment Programme; Mediterranean Action Plan, Tech Rep Series No 47. pp 302–308.

VUCETIC T. 1985. Changes in the distribution and size of the population of *Pelagia noctiluca* in the Mediterranean. Athens. United Nations Environment Programme editions for the Mediterranean Action Plan. Long-term Programme for Pollution Monitoring and Research in the Mediterranean Sea. 1: 37.

WACHSMAN M, AURELIAN L, BURNETT J W. 1991. Human immunosuppression induced by sea nettle (*Chrysaora quinquecirrha*) venom. Toxicon 29: 386–390.

WADE H W. 1927. Post-mortem findings in acute jellyfish poisoning with sudden death in status lymphaticus. Monthly Bull Philipp Health Service 7(9): 479.

WAISMAN M. 1989. Seabather's eruption. In: (D Joseph Demis, chief ed.) Clinical Dermatology, Vol 4 (16th revision). Philadelphia. J B Lippincott Company. 18–22: 1–3.

WALKER D G. 1983. Survival after severe envenomation by the blue ringed octopus (*Hapalochlaena maculosa*). Med J Aust 2: 663–665.

WALKER M J A. 1977. Pharmacological and biochemical properties of a toxin containing material from the jellyfish, *Cyanea capillata*. Toxicon 15: 3–14.

WALKER M J A. 1988. Coelenterate and Echinoderm toxins: mechanisms and actions. In: (A T Tu, ed) Marine Toxins and Venoms, Vol 3. New York. Marcel Dekker. pp 279–325.

WALLS J G. 1980. Cone Shells: A Synopsis of the Living Conidae. Hong Kong. TFH Publications. 1011 pp.

WARRELL D A, FENNER P J. 1993. Venomous bites and stings. British Med Bull 49: 423–439.

WASUWAT S. 1970. Extract of *Ipomoea pes-caprae* (Convolvulaceae) antagonistic to histamine and jellyfish poison. Nature 225 (5234, 21 February): 758.

WATABE S, SATO Y, NAKAYA M, HASHIMOTO K, ENOMOTO A, KAMINOGAWA S, YAMAUCHI K. 1989. Monoclonal antibody raised against tetrodonic acid, a derivative of tetrodotoxin. Toxicon 27: 265–268.

WATERBURY J B. 1992. Viruses of marine bacteria. Oceanus (Fall 1992): 107–108.

WATKINS A B K. 1969. Bullrout stings. Med J Aust 1: 212.

WATSON G M, HESSINGER D A. 1989. Cnidocyte mechanoreceptors are tuned to the movements of swimming prey by chemoreceptors. Science Reports 243: 1589–1591.

WATSON G M, HUSON R R. 1994. Frequency and amplitude tuning of nematocyst discharge by proline. J Exp Zool 268: 177–185.

WEBER J N. 1969. Disaster at Green Island: other outer Pacific Islands may share its fate. Earth and Mineral Sciences, Pennsylvania State Univ 38: 37–41.

WEILL R. 1934. Contribution à l'étude des cnidaires et de leurs nématocystes: (2 vols). Trav Sta Zool Wimereux I 10: 1–347; II 11: 348–701.

WEILL R. 1938. Une meduse tropicale Indopacifique *Gonionemus suvaensis* Ag et Mayer dans l'ocean Atlantique (les Bermudes). Bull Soc Zool 63: 33–41.

WEINBERG S R. 1988. Reactive arthritis following a sting by a Portuguese man-o'-war. J Florida Med Assoc 75: 280–281.

WELSH J H. 1956. On the nature and action of coelenterate toxins. Deep Sea Res Suppl 3: 287–297.

WERNER B. 1973. Spermatozeugmen und Paarungsverhalten bei Tripedalia cystophora (Cubomedusae). Mar Biol 18: 212–217.

WERNER B. 1975. Bau und Lebensgeschichte des Polypen von *Tripedalia cystophora* (Cubozoa, class. nov., Carybdeidae) und seine Bedeutung fur die Evolution der Cnidaria. Helgolander wiss. Meeresunters 27: 461–504.

WERNER B, CUTRESS C E, STUDEBAKER J P. 1971. Life cycle of *Tripedalia cystophora* Conant (Cubomedusae). Nature, London 232: 582–583.

WESTFALL J A. 1970. The nematocyte complex in a hydromedusan *Gonionemus vertens*. Z Zellforsch 110: 457–470.

WHITE J. 1991. Snakebite: an Australian perspective. J Wilderness Med 2: 219–244.

WHITE J. 1995. Clinical toxicology of sea snakes. In: (J Meier, J White, eds) Clinical Toxicology of Animal Venoms. Florida. CRC Press. pp 159–170.

WHO OFFSET PUBLICATION. 1981. Report of working group on public health aspects of marine food fish poisoning. Suva, Fiji. 23–25 February 1981.

WHYTE J M, ENDEAN R. 1962. Pharmacological investigation of the venom of the marine snails *Conus textile* and *Conus geographus*. Toxicon 1: 25–31.

WIENER S. 1958. Stone-fish sting and its treatment. Med J Aust 2: 219–222.

WIENER S. 1959a. Observations on the venom of the stonefish (*Synanceja trachynis*). Med J Aust 1: 620–627.

WIENER S. 1959b. The production and assay of stonefish antivenene. Med J Aust 2: 715–719.

WIENER S. 1965. A case of stone-fish sting treated with antivenene. Med J Aust 1: 191.

WILKS J. 1993. Scuba safety in Queensland. SPUMS J 23: 139–141.

WILLIAMSON D E. 1968. The coral predator. Skin Diver 17: 26.

WILLIAMSON J A. 1981. Some Australian Marine Stings, Envenomations and Poisonings, 2nd edition. Brisbane. Surf Life Saving Association of Australia, Queensland State Centre. 52 pp.

WILLIAMSON J. 1983. Children and box-jellyfish in north Australia. In: (J Pearn, ed) Accidents to children: Their Incidence, Causes and Effects. Melbourne. Child Accident Prevention Foundation of Australia. College of Surgeons. pp 49, 93–102.

WILLIAMSON J A. 1985a. The Marine Stinger Book, 3rd edition (D Exton, ed) Brisbane. 4006. Queensland State Centre, Surf Life Saving Australia. 85 pp.

WILLIAMSON J A. 1985b. 'Irukandji' syndrome or decompression sickness or cerebral arterial gas embolism? A differential diagnostic trap for practitioners of diving medicine in north Queensland. SPUMS J 15: 38–39.

WILLIAMSON J A H. 1987. The blue-ringed octopus bite and envenomation syndrome. In: (R M Mandojana, guest ed) Aquatic Dermatology. Clin in Dermatol 5(3): 127–133.

WILLIAMSON J. 1988a. Jellyfish envenomation: what diving medical physicians should know. SPUMS J 18: 118–121.

WILLIAMSON J. 1988b. Multi-tentacled box jellyfish. In: (J Pearn & J Covacevich, eds) Venoms and victims. Brisbane. Queensland Museum. pp 1–8.

WILLIAMSON J. 1990. Ciguatera and mannitol: a successful treatment. Med J Aust 153: 306–307.

WILLIAMSON J. 1992a. Venomous and poisonous marine animals. In: (J B Wyngaarden, L H Smith Jr, J C Bennett, eds) Cecil Textbook of Medicine, 19th edition. Philadelphia and London. W B Saunders Company. pp 2030–2032.

WILLIAMSON J. 1992b. Current challenges in marine envenomation: an overview. J Wilderness Med 3: 422–431.
WILLIAMSON J. 1993a. The work of the International Consortium for Jellyfish Stings. SPUMS J 23: 111–113.
WILLIAMSON J A. 1993b. The diver, the Great Barrier Reef and our planet. SPUMS J 23: 57–59.
WILLIAMSON J A. 1995. Diving accidents and first aid. In: (J Pearn, P Bowler, eds) The Science of First Aid. Canberra. St John Ambulance Australia. (in press).
WILLIAMSON J, ACOTT C, WEBB R, CAPPS R, GILLIGAN F, GORMAN D. 1991. Australia's 'Diver Emergency Service' (DES) 008-088-200: analysis of recorded usage over a 35-month period during 1987–1990. SPUMS J 21: 17–21.
WILLIAMSON J A, BURNETT J W, FENNER P J, HACH-WUNDERLE V, HOE L Y, ADIGA K M. 1988. Acute regional vascular insufficiency after jellyfish envenomation. Med J Aust 149: 698–701.
WILLIAMSON J A, CALLANAN V I, HARTWICK R F. 1980. Serious envenomation by the northern Australian box-jellyfish (*Chironex fleckeri*). Med J Aust 1: 13–15.
WILLIAMSON J A, CALLANAN V I, UNWIN M L, HARTWICK R F. 1984a. Box-jellyfish venom and humans. Med J Aust 140: 444–445.
WILLIAMSON J, FENNER P, ACOTT C. 1987. Medical aspects of marine envenomation and poisoning. In (J Covacevich, P Davie, J Pearn eds) Toxic Plants and Animals: A Guide for Australia. Brisbane. Queensland Museum. pp 214–240.
WILLIAMSON J A, FENNER P J, BURNETT J W, ALAM J M, YAKOVLEV Y M, VASKOVSKY V E. 1993. Jellyfish venoms and the diver: lessons for diving medicine (Abstract). Program and Abstracts, Undersea and Hyperbaric Medical Society Annual Scientific Meeting, 7–10 July 1993. Undersea Hyperbaric Med 20(Suppl.): 78.
WILLIAMSON J A, Le RAY L E, WOLFHART M, FENNER P J. 1984b. Acute management of serious envenomation by box-jellyfish (*Chironex fleckeri*). Med J Aust 141: 851–853.
WILSON B R, GILLETT K. 1980. Australian Shells. Sydney. A H & A W Reed.
WONG D E, MEINKING T L, ROSEN L B, TAPLIN D, HOGAN D J, BURNETT J W. 1994. Seabather's eruption: clinical, histological, and immunological features. J Am Acad Derm. 30: 399–406.
WOOD J G. 1874. Out of Doors: A Selection of Original Articles on Practical Natural History. London. Longmans, Green & Co.
WOODLAND D J, ALLAN G. R. 1977. *Siganus trispilos*, a new species of Siganidae from the eastern Indian Ocean. Copeia 4: 617–620.
WRIGHT-SMITH R J. 1945. A case of fatal stabbing by stingray. Med J Aust 2: 466–467.
YAFFEE H S. 1970. Irritation from red sponge. N Eng J Med 282: 51.
YAKOVLEV Yu M, VASKOVSKY V E. 1993. A toxic medusa-krestovik (*Gonionemus vertens*). Biol Morya 5–6: 3–16 (in Russian).
YAMAGUCHI M, HARTWICK R. 1980. Early life history of the sea wasp, *Chironex fleckeri* (Class Cubozoa). In: (P Tardent, R Tardent, eds) Development and Cellular Biology of Coelenterates. Amsterdam. Elsevier/North Holland Biomedical Press. pp 11–16.
YAMANOUCHI T. 1943. On the poison contained in *Holothuria vagabunda*. (in Japanese) Folia Pharmacol Japon 38(2): 115.
YAMANOUCHI T. 1955. On the poisonous substance contained in holothurins. Publ Seto Marine Biol Lab 4(2–3): 183–203.
YAMASU T, YOSHIDA M. 1976. Fine structure of complex ocelli of a cubomedusan, *Tamoya bursaria* Haeckel. Cell Tissue Res 170: 325–339.
YANAGITA T M. 1960a. Physiological mechanism of nematocyst responses in sea anemone, III: Excitation and anaesthetisation of the nettling response system. Comp Biochem Physiol 1: 123–139.
YANAGITA T M. 1960b. Physiological mechanism of nematocyst responses in sea-anemone-IV: Effects of surface-active agents on the cnidae *in situ* and in isolation. Comp Biochem Physiol 1: 123–139.
YANAGITA T M. 1960c. The physiological mechanism of nematocyst responses in sea anemone V: The effects of lipoid solvents on the cnidae in situ and in isolation. Annot Zool Japan 33: 203–210.
YANAGITA T M, WADA T. 1954. Effects of trypsin and thioglycollate upon the nematocysts of the sea-anemone. Nature, London. 173, plate 171.
YANAGITA T M, WADA T. 1959. Physiological mechanism of nematocyst responses in sea-anemone, II: A note on the microscopical structure of acontium with special reference to the situation of cnidae within its surface. Cytologia (Tokyo) 24: 81–97.
YARA A, NAGUCHI K, NAKASONE J, KINJO N, HIRAYAMA K, SAKANASHI M. Cardiovascular effects of *Acanthaster planci* venom in the rat: possible involvement of PAF in its hypotensive effect. Toxicon 1992; 30: 1281–1289.
YASHIRO H. 1939. Fatal bite of *Conus geographus*. Venus IX (3–4) October 1939 (in Japanese).
YASUMOTO T, MURATA M, OSHIMA Y, MATSUMOTO GK, CLARDY J. 1984. Diarrhetic shellfish poisoning. In: (E P Ragelis, ed) Seafood Toxins. Washington DC. American Chemical Society. 207.
YASUMOTO T, OSHIMA Y, TAJIRI M, KOTAKI Y. 1983. Paralytic shellfish toxins in previously unrecorded species of coral reef crabs. Bull Jap Soc Sci Fish 49(4): 633–636.
YENTSCH C M. 1981. Flow cytometric analysis of cellular saxitoxin in the dinoflagellate *Gonyaulax tamarensis* var. *excavata*. Toxicon 19: 611–621.
ZAMPONI M, FACAL H. 1987. Estudio bioecológico de *Olindias sambaquiensis* Müller 1861 (Limnomedusae, Olindiidae) en el área de Monte Hermoso. I: Ciclo de vida. Neotrópica 33: 119–126.
ZANN L, EAGER E. 1987. The crown-of-thorns starfish. Aust Sci Mag 1987;3: 14–55.
ZANN L P, MORAN P J. 1988. A coordinated research program on the *Acanthaster* phenomenon in Australia. Proc 6th Int Coral Reef Symp 2: 177–182. p 50.

ZELICKMAN E A. 1976. Fauna and ecological notes on Coelenterata of Bousset Lagoon (Southern Sakhalin, Aniva Bay, Sea of Okhotsk). Tr Inst Okeanol Akad Nauk SSSR 105: 198–209 (in Russian).

ZERVOS S. 1903. La maladie des pêcheurs d'éponges. Semaine Méd (24th June, 1903) 25: 208–209.

ZHENG ZHONG *et al* 1989. Marine Planktonology. Beijing: China Ocean Press. Berlin: Springer-Verlag. 454 pp.

ZIMMERMAN S E, HEATWOLE H. 1987. Olive sea snake venom. In: (J Covacevich, P Davie, J Pearn, eds) Toxic Plants and Animals: A Guide for Australia. Brisbane. Queensland Museum. pp 205–213.

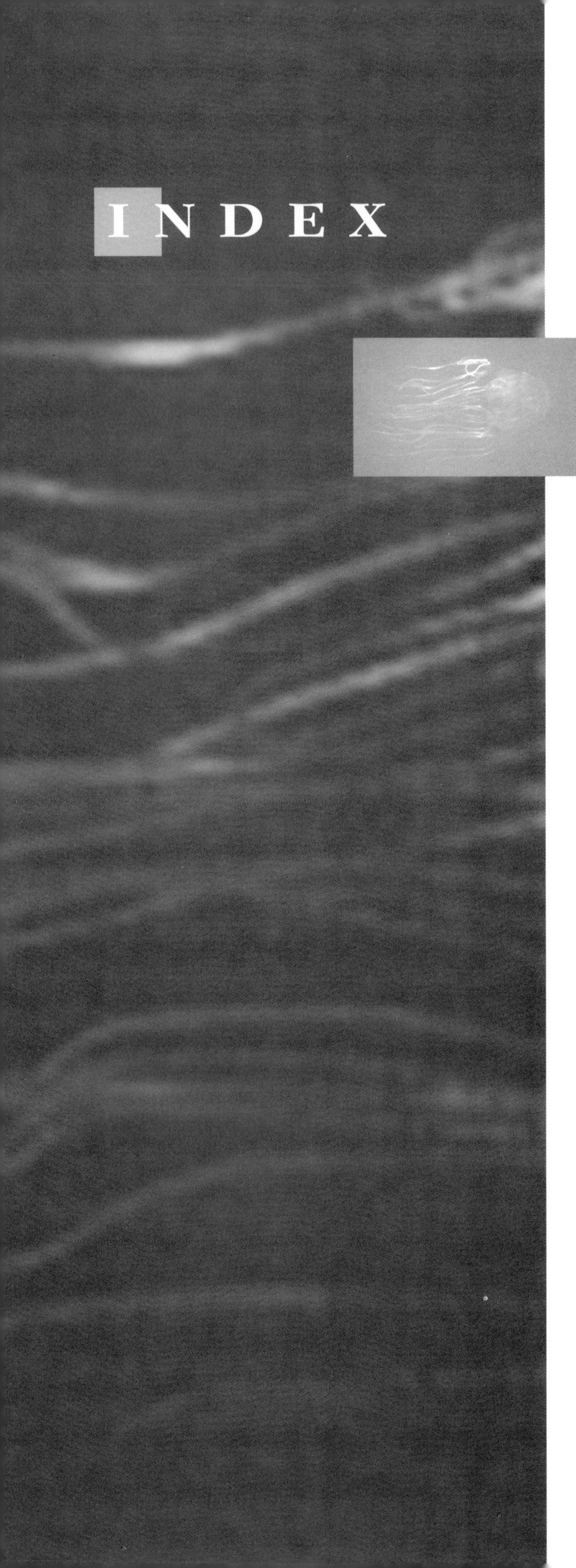

INDEX

Bold numbers indicate colour plates located at the back of the book.